LASER INTERACTION AND RELATED PLASMA PHENOMENA

12th International Conference

LASER INTERACTION AND RELATED PLASMA PHENOMENA

12th International Conference

Osaka, Japan   April 1995

EDITORS
Sadao Nakai
Osaka University

George H. Miley
University of Illinois

American Institute of Physics

AIP CONFERENCE
PROCEEDINGS 369
PART ONE

Woodbury, New York

Authorization to photocopy items for internal or personal use, beyond the free copying permitted under the 1978 U.S. Copyright Law (see statement below), is granted by the American Institute of Physics for users registered with the Copyright Clearance Center (CCC) Transactional Reporting Service, provided that the base fee of $6.00 per copy is paid directly to CCC, 222 Rosewood Drive, Danvers, MA 01923. For those organizations that have been granted a photocopy license by CCC, a separate system of payment has been arranged. The fee code for users of the Transactional Reporting Service is: 1-56396-445-7/ 96 /$6.00.

© 1996 American Institute of Physics

Individual readers of this volume and nonprofit libraries, acting for them, are permitted to make fair use of the material in it, such as copying an article for use in teaching or research. Permission is granted to quote from this volume in scientific work with the customary acknowledgment of the source. To reprint a figure, table, or other excerpt requires the consent of one of the original authors and notification to AIP. Republication or systematic or multiple reproduction of any material in this volume is permitted only under license from AIP. Address inquiries to Office of Rights and Permissions, 500 Sunnyside Boulevard, Woodbury, NY 11797-2999; phone 516-576-2268; fax: 516-576-2499; e-mail: rights@aip.org.

L.C. Catalog Card No. 96-85009
ISBN 1-56396-445-7 (set)
ISBN 1-56396-623-9 (Part One)
ISBN 1-56396-624-7 (Part Two)
DOE CONF- 950476

Printed in the United States of America

CONTENTS

PART ONE

Preface xxi
   Introduction xxi
   Conference Staff xxii
   Edward Teller Medal xxiv
   Technical Sessions xxv
   Future Conferences and Teller Award Nominations xxvi
   Acknowledgments xxvii
   Congratularory Address: Mr. Y. Kitao xxix
Welcome Address: Dr. J. Kanamori xxxi
Opening Address: Dr. T. Miyajima xxxii
Keynote Address: Dr. Edward Teller xxxiv
1995 Teller Award Recipients xxxvi

ICF MEMORIAL LECTURE

Past, Present, and Future of Laser Fusion Research 3
   C. Yamanaka

1. ICF PROGRAMS AND ENERGY DRIVERS

Progress in the US ICF Program 21
   M. M. Sluyter
Progress of Laser Fusion at ILE and ICF: Research in Japan 30
   S. Nakai
Overview of ICF Program at Centre D'Etudes de Limeil-Valenton 40
   CEL-V Laser Team
Path to Ignition: US Indirect Target Physics 53
   M. Cray and E. M. Campbell
Inertial Confinement Fusion Program at CAEP 61
   H. Peng
Experiments on the OMEGA Laser to Validate High-Gain, Direct-Drive
Performance on the National Ignition Facility 71
   R. L. McCrory, J. M. Soures, C. P. Verdon, T. R. Boehly, D. K. Bradley,
   R. S. Craxton, J. A. Delettrez, R. Epstein, P. A. Jaanimagi, S. D. Jacobs,
   R. L. Keck, J. H. Kelly, T. J. Kessler, H. Kim, J. P. Knauer, R. L. Kremens,
   S. A. Kumpan, S. A. Letzring, F. J. Marshall, P. W. McKenty, S. F. B. Morse,
   A. Okishev, W. Seka, R. W. Short, M. D. Skeldon, S. Skupsky, M. Tracy,
   and B. Yaakobi
Status of HIB Fusion in Europe 80
   D. H. H. Hoffmann and J. Jacoby

2. CRITICAL ELEMENTS FOR IGNITION

2.1. TARGET EXPERIMENT

High Convergence, Indirect Drive Inertial Confinement Fusion Experiments at NOVA 89
    R. A. Lerche, M. D. Cable, S. P. Hatchett, J. A. Caird, J. D. Kilkenny,
    H. N. Kornblum, S. M. Lane, C. Laumann, T. J. Murphy, J. Murray,
    M. B. Nelson, D. W. Phillion, H. Powell, and D. Ress

Symmetry Experiments in Gas Filled Hohlraums at NOVA 95
    N. D. Delamater, T. J. Murphy, A. A. Hauer, R. L. Kauffman,
    A. L. Richard, E. L. Lindman, G. R. Magelssen, B. H. Wilde,
    L. V. Powers, S. M. Pollaine, L. J. Suter, R. Chrien, D. B. Harris,
    M. B. Nelson, M. D. Cable, J. B. Moore, K. Gifford, and R. J. Wallace

Implosion Experiments with Uniformity-Improved GEKKO XII: Overview .. 101
    Y. Kato, H. Azechi, H. Takabe, N. Miyanaga, T. Kanabe, T. Norimatsu,
    H. Nishimura, H. Shiraga, M. Nakai, R. Kodama, K. A. Tanaka, M. Takagi,
    M. Nakatsuka, K. Nishihara, K. Mima, T. Yamanaka, and S. Nakai

Improvement of the Imploded Core Performance with Uniform GEKKO-XII Green Laser System 108
    H. Shiraga, M. Heya, M. Nakasuji, H. Azechi, N. Izumi, T. Yamagajo,
    M. Saito, T. Urano, Y. Kitagawa, S. Miyamoto, Y. Kato, H. Takabe, K. Mima,
    K. Nishihara, and S. Nakai

Acceleration of Planar Foils by the Indirect-Direct Drive Scheme 113
    J. J. Honrubia, J. M. Martinez-Val, J. L. Bocher, and G. Faucheux

A Nonsupported Pellet for Laser Fusion Scheme 119
    Y. Sakagami, H. Yoshida, T. Mizutani, S. Miyagawa, M. Sekimura,
    and K. Yasufuku

Dynamics of Gas-Filled Hohlraums 125
    T. J. Orzechowski, R. L. Kauffman, R. K. Kirkwood, H. N. Kornblum,
    W. K. Levedahl, D. S. Montgomery, L. V. Powers, T. D. Shepard, G. F. Stone,
    L. J. Suter, R. J. Wallace, J. M. Foster, and P. Rosen

Measurements of Absorption Distribution by Second Harmonic and X-Ray Images 131
    K. A. Tanaka, T. Ohnishi, A. Ozaki, T. Matsushita, S. Nakaji, H. Hashimoto,
    R. Kodama, Y. Oshikane, M. Nakai, M. Tsukamoto, S. Matsuoka,
    K. Tsubakimoto, M. Murakami, K. Shimada, H. Azechi, H. Nishimura,
    N. Miyanaga, H. Shiraga, Y. Kato, K. Mima, and S. Nakai

$\rho\Delta R$ Measurement of Imploded Cryogenic Foam Target by DD-Protons 136
    Y. Kitagawa, K. A. Tanaka, M. Nakai, T. Yamanaka, K. Nishihara,
    H. Azechi, N. Miyanaga, T. Norimatsu, T. Kanabe, C. Chen, A. Richard,
    M. Sato, H. Furukawa, and S. Nakai

Hot Spark Structure in Laser-Imploded Core Plasmas Observed with 10-ps-Resolved X-Ray Imaging 143
    M. Heya, H. Shiraga, N. Miyanaga, K. Nishihara, K. Tanaka, Y. Kato,
    T. Yamanaka, and S. Nakai

Preliminary Analysis of Core Capsule X-Ray Spectroscopy and Image Results for Medium-to-High Growth Factor Implosions 149
    G. Pollak, N. Delamater, N. Landen, B. Hammel, and C. Keane

2D and 3D Ablation Front Hydrodynamic Instability Experiments on Nova 155
    B. A. Remington, M. M. Marinak, S. V. Weber, K. S. Budil, O. L. Landen, S. W. Haan, J. D. Kilkenny, and R. J. Wallace

Experimental Analyses of Rayleigh-Taylor Growth in Cylindrical Implosions 160
    J. B. Beck, W. H. Hsing, N. M. Hoffman, and C. K. Choi

Multimode Hydrodynamic Stability Calculations for National Ignition Facility Capsules 166
    N. M. Hoffman, D. C. Wilson, W. S. Varnum, W. J. Krauser, and B. H. Wilde

2.2. TARGET PHYSICS AND DESIGN

Volume Ignition for Inertial Confinement Fusion Tested by Different Stopping Power Models 172
    H. Hora, S. Eliezer, J. J. Honrubia, R. Höpfl, J. M. Martinez-Val, M. Peria, and G. Velarde

Effects of Nonlocal Heat Transport on Laser Implosion 179
    K. Mima, M. Honda, S. Miyamoto, and S. Kato

Integrated Ignition Calculations for Indirectly Driven Targets 186
    W. J. Krauser, B. H. Wilde, D. C. Wilson, P. Bradley, and F. Swenson

The Computational Optimization of Indirect-Driven ICF Targets 194
    V. A. Lykov, E. N. Avrorin, N. G. Karlykhanov, V. E. Chernyakov, M. Y. Kozmanov, V. A. Murashkina, and Y. Z. Kandiev

Physics Issues Related to the Confinement of ICF Experiments in the U.S. National Ignition Facility 200
    M. Tobin, A. Anderson, J. Latkowski, M. Singh, C. Marshall, and T. Bernat

Large-Scale Laser Target Design. Alternative Approaches 207
    L. P. Feoktistov, I. G. Lebo, V. B. Rozanov, and V. F. Tishkin

Fast Ignitor Concept. Numerical Simulation 213
    A. Pukhov and J. Meyer-ter-Vehn

Thermal Smoothing and Hydrodynamical Compensation of the Non-uniformities of Laser Energy Deposition in a Direct-Driven Target 219
    S. Gus'kov, V. Rozanov, I. Lebo, G. Vergunova, V. Tishkin, N. Zmitrenko, T. Kozubskaya, I. Popov, and V. Nikishin

Kinetic Effects on the Electron Thermal Transport in Ignition Target Design 225
    M. Honda, A. Nishiguchi, K. Mima, H. Takabe, H. Azechi, and S. Nakai

Modelling of Initial Imprinting Caused by Laser-Intensity Nonuniformities in Ablative Plasmas 231
    Y. Shimuta and K. Nishihara

Ablation Effects in Weakly Nonlinear Stage of the Ablative
Rayleigh-Taylor Instability 237
    S. Hasegawa and K. Nishihara

A New Instability of a Contact Surface Driven by a Nonuniform
Shock Wave... 243
    R. Ishizaki, K. Nishihara, H. Sakagami, and M. Murakami

Two Dimensional Simulation of Turbulent Mixing in Stagnation
Dynamics of Implosion 249
    A. Sunahara, H. Takeuchi, and H. Takabe

The Design and Characterization of Toroidal-Shaped NOVA Hohlraums
that Simulate National Ignition Facility Plasma Conditions for Plasma
Instability Experiments.. 255
    B. H. Wilde, J. C. Fernandez, W. W. Hsing, J. A. Cobble, N. D. Delamater,
    B. H. Failor, W. J. Krauser, and E. L. Lindman

Modelling of Drive-Symmetry Experiments in Gas-Filled Hohlraums
at NOVA 263
    E. L. Lindman, N. D. Delamater, A. A. Hauer, R. L. Kauffman,
    G. R. Magelssen, T. J. Murphy, S. M. Pollaine, L. V. Powers, L. J. Suter,
    and B. H. Wilde

Relevance of the U.S. Natinal Ignition Facility for Driver and Target
Options to Next-Step Inertial Fusion Test Facilities 269
    B. G. Logan

A New Drive for ICF: Externally Guided Implosions 277
    J. M. Martínez-Val, M. Piera, and P. M. Velarde

Implosion and Ignition of Stagnation-Free Targets......................... 283
    J. M. Martínez-Val, G. Velarde, S. Eliezer, H. Hora, J. J. Honrubia,
    E. Mínguez, M. Perlado, M. Piera, and P. Velarde

Effect of Fusion Reaction Products Heating on the Volume Ignition
of DT and D^3He Fuel Pellets 289
    R. Khoda-Bakhsh

Implosion Dynamics of a Hot Core... 297
    M. Murakami, M. Shimoide, and K. Nishihara

Kinetic Model for DT Ignition and Burn in ICF Targets 302
    S. I. Anisimov, A. M. Oparin, and J. Meyer-ter-Vehn

Emission of Highly Energetic Neutrons from Laser-Imploded D-T Pellets
and its Applicability to Pellet Diagnosis 309
    Y. Tabaru, Y. Nakao, H. Nakashima, and K. Kudo

Neutron Heating in Ignition and Burn Phases of Laser-Imploded
D-T Pellets... 315
    Y. Nakao, T. Johzaki, H. Nakashima, A. Oda, and K. Kudo

A Moderate-Gain ICF Model from LTE Ignition to Non-LTE Burn 321
    X. T. He, Y. S. Li, and M. Yu

Practical Large-Size-Pellet ICF 327
    S. Kawata and K. Kurawaki

Development of High Yield Low Level Tritium Targets for Inertial
Fusion Reactor Systems 333
    N. A. Tahir and D. H. H. Hoffmann

DT-Fuel Concentration in an ICF Pellet................................. 339
    S. Kawata

3. LASER-MATTER INTERACTION PHYSICS

3.1. INTERACTION PHYSICS

Options for Laser Compression of Matter to Study Dense-Plasma Phases at Low Entropy, Including Metallization of Hydrogen.................... 347
    J. Meyer-ter-Vehn, A. Oparin, and T. Aoki

Simulation of Astrophysical Plasma Dynamics in the Laser Experiments..... 357
    Y. P. Zakharov, V. M. Antonov, A. V. Melekhov, S. A. Nikitin,
    A. G. Ponomarenko, V. G. Posukh, V. O. Stoyanovsky,
    and I. F. Shaikhislamov

Anomalous Spectral Signatures of High-Intensity Stimulated Raman Backscattering 363
    M. M. Škorić, M. S. Jovanović, and M. R. Rajković

Investigation of the Dynamic Fracture Process at Ultrahigh Strain Rate caused by Laser-Induced Shock Waves in Solid Targets 369
    V. I. Vovchenko, I. K. Krasyuk, P. P. Pashinin, and A. Y. Semenov

Extremely Cold Plasma Production by Optical-Field-Induced Ionization of a Preformed Plasma.. 374
    Y. Nagata, K. Midorikawa, M. Obara, and K. Toyoda

SDOSS: A Spatially Discriminating, Optical Streaked Spectrograph......... 380
    J. A. Cobble, S. C. Evans, J. C. Fernández, J. A. Oertel, R. G. Watt,
    and B. H. Wilde

Two-Temperature Model of Atoms in Dense Plasmas 386
    Y. Furutani and A. Fukuyama

Photoionization in Laser-Produced Hot Dense Plasmas.................... 392
    H. Furukawa

Pressure Ionization of Dense Plasmas in Spherical Ion-Cell Model with Spin-Orbit Interactions 398
    K. Ishikawa, T. Blenski, H. Takahashi, T. Iguchi, and M. Nakazawa

Time Dependent Ionization of Carbon Atom and Ions in Hot Dense Plasmas 404
    T. Kato, T. Fujimoto, T. Kawachi, J. Dubau, and U. Safronova

A Relativistic WKB Method for Bound States of an Electron in Screened Coulomb Potentials 410
    K. Kosaka

Self-Consistent Continuum Lowering Model.............................. 416
    K. J. LaGattuta

Improved Screened Hydrogenic Model................................... 422
    T. Nishikawa

Excitation of Intense Shock Waves by Soft X-Radiation from a Z-Pinch Plasma 428
    V. E. Fortov, K. Dyabilin, M. Lebedev, O. Y. Vorobiev, E. Grabovskij,
    V. Smirnov, and B. Goel

Modelling of the Shock-Induced Luminescence of Free Metal Surface 434
 A. Y. Semenov
Collisional High Temperature or Solid State Plasma with Superconductivity Based on Laser-Plasma Interaction 443
 H. Hora
N-Body Lyapunov Expansion Rates in One Component Strongly Coupled Plasmas 447
 Y. Ueshima, K. Nishihara, D. M. Barnett, T. Tajima, and H. Furukawa
Analysis of Stimulated Raman Scattering Light Spectrum and Plasma Temperature and Density for Cavity Targets. 453
 J. T. Zhang
Stimulated Raman Scattering From Two Overlapped 527 nm Laser Beams 459
 M. Tsukamoto, K. A. Tanaka, K. Mima, R. Kodama, M. Kado,
 S. Nakaji, A. Nishiguchi, M. Nakai, T. Norimatsu, K. Nishihara,
 T. Yamanaka, and S. Nakai
The Investigation of Stimulated Raman Scattering in Laser Produced Plasma for Heating Wavelength of 0.53 μm 465
 E. A. Bolkhovitinov, V. Y. Bychenkov, M. O. Koshevoi, M. V. Osipov,
 A. A. Rupasov, A. S. Shikanov, V. T. Tikhonchuk, A. V. Kilpio,
 N. G. Kiselev, D. G. Kochiev, P. P. Pashinin, E. V. Shashkov,
 and Y. A. Suchkov
Anomalous High Energy Electron Emission from Laser Plasma 471
 V. V. Ivanov, A. K. Knyazev, A. V. Koutsenko, A. A. Matzveiko,
 Y. A. Mikhailov, V. P. Osetrov, A. I. Popov, G. V. Sklizkov,
 and A. N. Starodub
Nonlinear Interaction of High Power Microwave with an Inhomogeneous Plasma 477
 X. Xu, H. Itoh, N. Yugami, B. Cros, and Y. Nishida
Interaction of 1.05 μm and 0.53 μm lasers with Gold Disks 483
 S. Liu, Y. Ding, Z. Zheng, and D. Tang
Instabilities of Nuclear Flames in Thermonuclear Supernovae. 489
 K. Nomoto, K. Iwamoto, T. Shigeyama, and H. Takabe
Numerical Simulation of JET-Like Structures in Laser Plasma 499
 A. Y. Semenov and S. F. Goncharov
PIC Code Simulation and Theoretical Analysis on Generation of Weibel-Type Instabilities. 505
 K. Satou and T. Okada

3.2. DIAGNOSTICS

Laser Diagnostics of Tokamak Plasmas 511
 T. Fukuda and T. Matoba
Properties of High-Z Laser-Produced Plasma Determined by Means of Ion Diagnostics. 521
 J. Wołowski, P. Parys, E. Woryna, L. Láska, K. Mašek, K. Rohlena,
 W. Mróz, and J. Farny

ICF Burn-History Measurements Using 17-MeV Fusion Gamma Rays 527
    R. A. Lerche, M. D. Cable, and P. G. Dendooven
Target Laser Plasmas Scale Effect in Electron Temperature Measurements by Electric Potential Probe Technique 533
    I. A. Bufetov, G. A. Bufetova, V. B. Fedorov, and S. B. Kravtsov
X-Ray Emission Computed Tomography with Attenuation Correction for ICF Research 538
    Y.-W. Chen
Two Dimensionally Space-Resolved Electron Temperature Measurement of Fusion Plasma by X-Ray Monochromatic Imaging Method 544
    K. Fujita, H. Nishimura, I. Uschmann, E. Förster, H. Takabe, Y. Kato, and S. Nakai
Development of Multi Channel Neutron Spectrometer at GEKKO XII Laser Fusion Facility 550
    N. Izumi, T. Yamagajo, T. Nakano, T. Kasai, H. Azechi, Y. Kato, S. Nakai, and T. Iida

4. HIGH INTENSITIES, SHORT PULSE INTERACTIONS

Generation of Coherent XUV Radiation with Sub-Picosecond KrF Lasers 559
    A. A. Offenberger, S. G. Preston, M. Zepf, C. G. Smith, W. J. Blyth, M. H. Key, J. S. Wark, A. Djaoui, and D. Neely
Scenarios of Plasma Formation with Intense fs Laser Pulses 565
    P. Mulser, A. Al-Khateeb, D. Bauer, A. Saemann, and R. Schneider
Vlasov Simulation of Superintense Laser-Solid Interaction 575
    H. Ruhl
Radiative Pumping Generated by a Hot Electron $K\alpha$ X-Ray Source in Femtosecond Laser-Produced Plasma Experiments 584
    A. Rousse, J. P. Geindre, P. Audebert, F. Fallies, J. C. Gauthier, A. Dos Santos, G. Grillon, A. Mysyrowicz, and A. Antonetti
Ultra-Intense, Short Pulse Laser-Plasma Interactions with Applications to the Fast Ignitor 590
    S. C. Wilks, W. L. Kruer, P. E. Young, J. Hammer, and M. Tabak
Large Amplitude Wakefield Excitation and Particle Acceleration in High Density Plasma for Plasma Based Accelerator 597
    Y. Nishida
Self-Modulation of Short Intense Laser Pulse: Opportunities for Wake-Field Accelerator 603
    N. E. Andreev, L. M. Gorbunov, and V. I. Kirsanov
Plasma-Core Induced Self-Guiding of an Ionizing Ultra-Short Pulse in Gases 609
    A. V. Kim, D. Anderson, M. Lisak, V. A. Mironov, A. M. Sergeev, and L. Stenflo
Anomalous Attenuation of Laser Light Due to Ion Acoustic Decay Instability (IADI) in Large Scale Plasma Relevant to Laser Fusion 615
    K. Mizuno, P. E. Young, and K. Estabrook

Evolution of fs High-Power Laser-Produced Plasma **621**
  K. Shimizu, M. Kuwabara, K. Go, H. Takakuwa, H. Kitazawa,
  and S. Karashima

**Theoretical Analysis of the Multi-Species Ion Plasma Interacting
with an Ultra Intense Laser** **627**
  S. Miyamoto, S. Kato, and K. Mima

**Higher Harmonics Generation in Dense Plasmas with an Intense
Ultra-Short Pulse Laser** **633**
  S. Kato, A. Nishiguchi, S. Miyamoto, and K. Mima

**Super-Strong Laser Field Generation and Their Interaction with Solid
Target in Vacuum** **639**
  A. A. Andreev, V. I. Bayanov, A. B. Vankov, A. A. Kozlov, V. A. Komarov,
  I. V. Kurnin, N. A. Solovyev, S. A. Chizhov, and V. E. Yashin

**Simulation Study for Interaction Between Ultra Power Laser and Dense
Plasma Slab** **646**
  H. Sakagami and K. Mima

**Study of Supra-Thermal Electrons and K-α X-Rays from High Intensity
500 fs Laser-Produced Plasmas** **652**
  J. Dunn, B. K. F. Young, A. K. Hankla, A. D. Conder, W. E. White,
  and R. E. Stewart

Radiation Properties from Ultra Short Pulse Laser Produced Plasma **660**
  N. Hasegawa, H. Nakagawa, H. Yoneda, K. Ueda, and H. Takuma

**Dependence of X-Ray Yield from Aluminum Plasma Produced
by a Pair of Femtosecond Ti: Sapphire Laser Pulses on Pulse Time
Separation** **666**
  H. Nakano, T. Nishikawa, H. Ahn, and N. Uesugi

**Applications of a 30-fs Multiterawatt Laser (A): Generation
and Time-Gated Imaging of Laser-Produced X-Rays for Medical
Applications** **672**
  C. P. J. Barty, C. L. Gordon III, B. E. Lemoff, G. Y. Yin, and P. M. Bell

**Electron Acceleration by Transverse Electromagnetic Wave Supplemented
with Crossed Static Magnetic Field** **678**
  N. Yugami, S. Sanjou, and Y. Nishida

**Large Amplitude Wakefield in an Ion Wave Regime Excited by Short
Microwave Pulse** **684**
  N. Yugami, Y. Nishida, B. Cros, and G. Matthieussent

**Thomson Scattering Measurement of the Beat-Wave Excited Relativistic
Plasma Waves** **690**
  Y. Kitagawa, S. Watanabe, K. Kawase, K. Sawai, and S. Nakai

**Electron Acceleration by Longitudinal Electric Field Generated
by Colinearly Overlapped Two Laser Beams** **695**
  S. Takeuchi

**Properties of Spectra of the Reflected and Transmitted Radiation
During Propagation of Relativistically Strong Laser Pulses in Underdense
Plasmas** **701**
  S. V. Bulanov, T. Z. Esirkepov, and N. M. Naumova

Stimulated Raman Scattering and Short Laser Pulse Evolution 707
    N. E. Andreev and L. M. Gorbunov
Stimulated Scattering of Radiation at Interaction of Short Laser Pulse
with Dense Plasma.. 713
    A. A. Andreev and A. N. Sutyagin
Problem of Plasma Wakefields with Low Phase Velocities and Relativistic
Self Focusing of Short Laser Pulses 719
    L. N. Tsintsadze
The Inertia of Tunneling Ionization and High-Order Harmonic Shifting
in the Nonlinear Single-Atom Response 728
    A. V. Kim, E. V. Vanin, A. M. Sergeev, D. Farina, M. Lontano,
    and M. C. Downer

PART TWO

5. LASERS AND OPTICAL TECHNOLOGIES

5.1. X-RAY LASERS

The X-Ray Laser as a Tool for Imaging Plasmas 737
    S. B. Libby, L. B. DaSilva, T. W. Barbee, Jr., R. Cauble, P. Celliers,
    R. A. London, D. L. Matthews, S. Mrowka, J. C. Moreno, D. Ress,
    J. E. Trebes, A. S. Wan, and F. Weber
Efficient Production of the Nickel-Like Soft X-Ray Lasers 743
    H. Daido, Y. Kato, K. Murai, S. Ninomiya, R. Kodama, H. Takabe,
    and F. Koike
Emission from Line-Focused Laser-Heated Multi-Groove Target 749
    R. Li, Z. Xu, Z. Zhang, P. Fan, S. Han, P. Lu, L. Zhang, X. Wang,
    and Y. Liu
Compact Soft X-Ray Laser Pumped by a Pulse-Train Laser 755
    T. Hara, K. Ando, M. Aoyama, and Y. Aoyagi
Population Inversion and Gain in Recombination Pumped He-Like Soft
X-Ray Lasers 760
    T. Ozaki, S. Orimo, and H. Kuroda
Nuclear Two-Quanta Gamma-Ray Lasing by Strong-Field Ignition 766
    L. A. Rivlin
X-Ray Laser Experiments by Using a Gas Puff Target with the
ASTERIX IV Facility 772
    H. Fiedorowicz, A. Bartnik, E. Fill, Y. Li, G. Pretzler, and M. Szczurek
Impact of Three-Dimensional Nonuniformity on the Germanium
X-Ray Laser Output 778
    A. S. Wan, R. W. Mayle, Y. Kato, and A. L. Osterheld
Applications of a 30-fs Multiterawatt Laser (B): Field-Ionization-Driven
Electron-Pumped XUV Lasers.. 784
    C. P. J. Barty, C. L. Gordon III, B. E. Lemoff, G. Y. Yin, and S. E. Harris

Deexcitation and Recombination of Excited Ions Involving Doubly
Excited States in Dense Recombining Plasma: Lithiumlike Ion X-Ray
Laser 790
    T. Fujimoto and T. Kawachi

Soft X-Ray Amplification in Laser-Produced Recombination Plasma 796
    H. Takakuwa, Y. Kobayashi, O. Nozawa, K. Shimizu, K. Go,
    H. Kitazawa, and S. Karashima

2D Simulation of Recombination Al Balmer-α Lasers 802
    A. Sasaki, H. Yoneda, K. Ueda, and H. Takuma

Heat Resistance of Mo-Based and W-Based Multilayer Soft X-Ray
Mirrors 808
    H. Takenaka, T. Kawamura, and Y. Ishii

5.2. FREE ELECTRON LASERS

Development of High-Brightness Photon Sources Using Enhanced
Compton Scattering in the Supercavity 814
    M. Fujita, T. Asakuma, A. Moon, T. Minamiguchi, M. Asakawa, J. Chen,
    K. Imasaki, C. Yamanaka, P. K. Roy, S. Nakai, N. Ohigashi,
    and Y. Tsunawaki

Compact High Brightness Radiation Sources 820
    K. Imasaki, M. Fujita, J. Chen, M. Asakawa, S. Nakai, and C. Yamanaka

RF-Linac Driven VUV Free Electron Laser Based on SASE 830
    J. Chen, M. Fujita, M. Asakawa, K. Imasaki, C. Yamanaka, K. Mima,
    and S. Nakai

5.3. LD AND LD PUMPED LASERS

Feasibility Experiments on the Laser-Diode Pumped Solid-State Laser
for ICF Driver 836
    M. Ohmi, H. Kiriyama, N. Srinivasan, T. Kimura, M. Yamanaka,
    M. Nakatsuka, Y. Izawa, T. Yamanaka, S. Nakai, and C. Yamanaka

New Optical Gain Mechanism for Wide-Gap Semiconductor Laser Diodes ... 843
    T. Uenoyama

Mode Control of Semiconductor Laser with Diffraction and Dispersion
Feedback 849
    G. Xu, R. Tsuji, K. Fujii, S. Nakayama, S. Amano, H. Kiyono, Y. Uchiyama,
    Y. Tokita, Y. Hanasawa, S. B. Mirov, M. J. McCutcheon, and J. R. Whinnery

5.4. GAS LASERS

Two-Dimensional Beam Smoothing Techn;ique for KrF Laser Systems 854
    I. Matsushima, H. Yashiro, T. Tomie, I. Okuda, Y. Matsumoto, E. Miura,
    E. Takahashi, and Y. Owadano

High Power Wide Aperture UV and IR Gas Lasers...................... 860
    V. F. Tarasenko, V. S. Skakum, and A. V. Fedenev

He-Pumped Iodine Laser for Plasma and High Intensity Interactions 866
G. A. Kirillov, G. G. Kochemasov, S. M. Kulikov, S. N. Pevny,
and S. A. Sukharev

Pulse Broadening and Recycle of Copper Vapor in Copper-Halides Lasing System... 872
K. Oouchi, H. Kato, K. Fujii, H. Kitatani, M. Suzuki, T. Takahashi,
and H. Saito

Enhancement of Population Inversion by Effective Use of Potential Energy in Self-Terminating Laser.. 878
H. Saito, Y. Masumura, M. Ishibashi, H. Kato, K. Oouchi, R. Tsuji,
K. Fujii, M. Suzuki, H. Kitatani, T. Takahashi, and M. Amano

5.5. NUCLEAR PUMPED LASERS

Prospects for Fusion Neutron NPLs 883
M. Petra, G. H. Miley, E. Batyrbekov, D. L. Jassby, and D. McArthur

Concept of Nuclear Reactor Pumped Laser for ICF 894
P. P. Dyachenko

Application of Nuclear Pumped Laser to an Optical Self-Powered Neutron Detector 900
N. Yamanaka, H. Takahashi, T. Iguchi, M. Nakazawa, T. Kakuta,
H. Yamagishi, and M. Katagiri

Ignition Experiment Design Based on γ-Pumping Gas Lasers 906
E. K. Bonyushkin, R. I. Il'kaev, A. P. Morovov, A. I. Pavlovskii,
B. V. Lazhintsev, N. Basov, S. Y. Gus'kov, V. B. Rosanov, and N. V. Zmitrenko

Optimization of Nuclear-Pumped Laser Active Media by Electron Beam 915
V. F. Tarasenko and A. V. Fedenev

Perspectives of Using (Ar-Xe) Direct Nuclear Pumped Laser at High Temperatures 921
A. A. Mavlyutov and A. I. Mis'kevich

Experimental Study of Nuclear-Induced Plasma Kinetics 927
E. D. Poletaev and Y. A. Dyuzhov

Numerical Simulation of the Power Characteristics of Twin-Core Pulse Reactor-Pumped Laser System 933
A. V. Gulevich, A. P. Barzilov, P. P. Dyachenko, A. V. Zrodnikov,
O. F. Kukharchuk, B. V. Kachanov, S. G. Kolyada, and E. A. Pashin

5.6. SHORT PULSE LASERS

Next Generation Ultrashort Pulse Lasers: Terawatts to Petawatts........... 939
C. P. J. Barty, C. L. Gordon III, G. Korn, B. E. Lemoff, R. Raksi,
C. Rose-Petruck, J. Squier, K. R. Wilson, V. V. Yakovlev, and K. Yamakawa

Terrawatt KrF/Ti:Sapphire Hybrid Laser System and its Application 945
K. Kondo, Y. Kobayashi, A. Sagisaka, Y. Nabekawa, and S. Watanabe

Subpicosecond Nd:Glass CPA Laser System for Kilovolt X-Ray Generation .. 951
    T. Zhang, L. B. Sharma, H. Daido, H. Uematsu, S. Ninomiya, K. Murai,
    Y. Kato, M. Nakatsuka, Y. Izawa, S. Nakai, T. Kitada, and C. Yamanaka

Efficient Direct Amplification of Powerful Picosecond Pulses in Nd-Glass Laser 957
    V. V. Ivanov, Y. A. Mikhailov, V. P. Osetrov, A. I. Popov, and G. V. Sklizkov

5.7. OPTICAL TECHNOLOGIES

Improvement of Irradiation Uniformity of New Gekko XII: Partially Coherent Light and Power Balance 963
    M. Nakatsuka, N. Miyanaga, T. Jitsuno, T. Kanabe, K. Tsubakimoto,
    S. Matsuoka, and S. Nakai

Phase Conjugation in Short Pulse Megajoule Class Lasers 969
    D. Eimerl, V. M. Chernyak, M. I. Pergament, R. V. Smirnov, and V. I. Sokolov

Suppression of Interference Speckles Due to Random Phase Plate by Liquid Crystal Polarization Rotator Array 975
    K. Tsubakimoto, C. Yamanaka, N. Miyanaga, M. Nakatsuka, T. Jitsuno,
    and S. Nakai

Two-Dimensional Multi-Lens Array with Circular Aperture Spherical Lens for Flat-Top Irradiation of ICF Target 981
    N. Nishi, T. Jitsuno, M. Nakatsuka, and S. Nakai

Design of Multi-Beam Laser Irradiation System and Uniformity Improvement 989
    M. Murakami

New Nd:Doped SiO_2 Material for High Average Power Laser 992
    Y. Fujimoto, M. Nakatsuka, K. Tsubakimoto, T. Kanabe, and S. Nakai

New Nonlinear Optical Crystal $CsLiB_6O_{10}$ for Laser Fusion 998
    Y. Mori, S. Nakajima, A. Taguchi, A. Miyamoto, M. Inagaki, T. Sasaki,
    H. Yoshida, and S. Nakai

High Performance of Phase Conjugated Stimulated Brillouin Scattering Mirror Based on High Purity Liquid Heavy Fluorocarbons 1004
    H. Yoshida, V. Kmetik, H. Fujita, K. Yoshida, T. Yamanaka, and S. Nakai

Radiation-Hardening of Final Optics for an ICF Reactor 1009
    S. G. DelMedico, O. M. Barnouin, M. Petra, and G. H. Miley

Efficient KTP Optical Parametric Oscillator Amplifier and its Applications .. 1022
    N. Srinivasan, T. Kimura, H. Kiriyama, M. Ohmi, M. Yamanaka, Y. Izawa,
    S. Nakai, and C. Yamanaka

6. PARTICLE BEAMS

6.1. LIGHT ION BEAM FUSIONS

Experimental Studies of Laser-Created Plasma as a Source of Highly Charged Ions 1029
    W. Mróz, P. Parys, J. Wołowski, E. Woryna, L. Láska, K. Mašek,
    K. Rohlena, J. Collier, H. Haseroth, K. L. Langbein, T. R. Sherwood,
    O. B. Shamaev, B. Y. Sharkov, and A. V. Shumshurov

Ablation Pressure and Temperature Diagnostics in TW Proton Beam Target Experiments on KALIF 1035
    K. Baumung, H. Bluhm, B. Goel, P. Hoppé, H. U. Karow, G. Meisel, D. Rusch,
    O. Stoltz, J. Singer, G. J. Kanel, A. V. Utkin, and S. V. Razorenov

Light Ion Hohlraum Target Experiments on PBFA II and Nova 1042
    R. J. Leeper, J. E. Bailey, T. L. Barber, A. L. Carlson, G. A. Chandler,
    D. L. Cook, M. S. Derzon, R. J. Dukart, D. E. Hebron, D. J. Johnson,
    M. K. Matzen, T. A. Mehlhorn, A. R. Moats, T. J. Nash, D. D. Noack,
    R. W. Olsen, R. E. Olson, J. L. Porter, J. P. Quintenz, C. L. Ruiz, M. A. Stark,
    J. A. Torres, and D. F. Wenger

Two-Stage Extraction Ion Diode Experiments on Reiden-SHVS for Light Ion Fusion 1048
    S. Miyamoto, K. Yasuike, T. Ochi, T. Yamashita, H. Urai, K. Imasaki,
    K. Kasuya, T. Aoki, K. Horioka, T. Fujii, M. Murakami, K. Nishihara,
    C. Yamanaka, and S. Nakai

Characteristics of Laser Produced Plasmas and Lasers for Pulsed Ion Sources 1054
    K. Kasuya, T. Suzuki, Y. Itoh, T. Kamiya, M. Watanabe, Y. Kawakita,
    K. Shioda, and H. Kanazawa

Intense Shock Waves in Hot Dense Matter Generated by High-Power Light Ion Beams 1060
    V. Fortov, G. Kanel, A. Utkin, O. Vorobiev, G. Kessler, H. Karow,
    K. Baumung, B. Goel, and V. Light

Effects of Virtual Anode Formation on the Beam Optics of Grid-Controlled Vacuum Arc Ion Source 1066
    K. Horioka, J. Hasegawa, M. Nakajima, and H. Iwasaki

Multi-Dimensional Diagnostics for Intense Ion Beams 1072
    K. Yasuike, T. Yamashita, T. Ochi, M. Urai, S. Miyamoto, K. Imasaki,
    C. Yamanaka, and S. Nakai

Light-Ion Beam Propagation in Plasmas with External Magnetic Field 1078
    T. Okada

A Proposal of Proton-Beam-Fusion Power Plant with a New Method of Beam Transport 1084
    K. Niu

Pulsed Power Generators Using an Inductive Energy Storage System 1090
    H. Akiyama, T. Sueda, U. Katschinski, S. Katsuki, and S. Maeda

Measurement of Energy Spectra of Ion Beams Produced in Gas-Puff
Z-Pinch Plasma Device 1096
    X. M. Guo and C. M. Luo
Inductive Particle Acceleration with a High $\partial B/\partial t$ 1103
    S. Kawata and S. Nishiyama
Study on Transport Control of High-Current Electron Beam............... 1109
    S. Nishiyama, S. Kawata, K. Naito, and S. Kato

6.2. HEAVY ION BEAM FUSIONS

Stopping Power of an Ion-Cluster with Two-Ion Correlation Effects 1115
    J. D'Avanzo, M. Lontano, and F. Raimondi
RFQ-Linac System for Intense Heavy Ion Beams and Beam-Target
Interaction Experiments 1121
    K. Sasa, Y. Oguri, M. Okamura, M. Okada, T. Ito, and T. Hattori
Application of the Two Stable Phases Phenomenon for Improvement
of the Longitudinal Stability in RFQ-Like Funneling System............... 1127
    V. Kapin, M. Inoue, Y. Iwashita, and A. Noda
Experiments on Interaction Between Heavy-Ions and Plasma at HIMAC 1133
    M. Ogawa, M. Nakajima, T. Hosokai, O. Iwase, T. Nakamura, T. Endou,
    K. Fujii, T. Aoki, K. Horioka, and K. Murakami
Interaction of Heavy-Ion Beams with Target Plasma...................... 1138
    R. Takahashi, T. Hotta, and T. Okada

7. APPLICATIONS OF LASER AND PLASMA

7.1. PLASMA SOURCE AND APPLICATION

Laser Ion Sources for Particle Accelerators.............................. 1145
    T. R. Sherwood
Railgun System Using a Laser-Induced Plasma Armature 1153
    M. Onozuka, Y. Oda, and K. Azuma
Precise Determination of Electric Dipole Moment in Atomic Transitions..... 1159
    T. Yoshida, A. Kuwako, I. Yoguchi, and T. Watanabe
Spatial and Temporal Behaviors of Laser Beam Propagating in Atomic
Vapor 1165
    K. Nomaru, Y.-W. Chen, Y. Izawa, S. Nakai, and C. Yamanaka
Some Design Aspects of Laser Fusion Rocket (II) 1171
    H. Nakashima, H. Shoyama, Y. Nagamine, T. Fusuki, Y. Kanda,
    N. Yoshimi, and Y. Nakao
Characteristics of a Plasma Production and a Laser-Induced Discharge
by a CO_2 Laser on DC Electric Field................................... 1177
    S. Ihara, T. Maiguma, S. Satoh, M. Ishimine, and C. Yamabe

Basic Study on Discharge Induction by Laser-Produced Plasmas in Atmospheric Air........ 1183
    T. Tsuji, C. Honda, M. Uchiumi, T. Tanaka, K. Muraoka, T. Takuma,
    F. Kinoshita, O. Katahira, and M. Akazaki

Experimental Study on Artifically Triggered Lightning using High Power Lasers........ 1189
    S. Uchida, Y. Shimada, H. Yasuda, C. Yamanaka, H. Fujita, Y. Izawa,
    T. Yamanaka, D. Wang, Z. Kawasaki, K. Matsu-ura, Y. Ishikubo, and M. Adachi

Gas Dynamics of UV-Laser Produced Vapor Plumes........ 1197
    S. I. Anisimov, B. S. Luk'yanchuk, and A. Luches

Temporal Process of Plasma Discharge by an Electron Beam........ 1205
    M. Sugawa, R. Sugaya, S. Isobe, A. Kumar, and H. Honda

Laser Spectroscopy of ECR Plasma Source using Visible Laser Diode........ 1211
    S. Satori, K. Nishiyama, H. Kuminaka, and K. Kuriki

Relativistic Electron Beam Acceleration by Nonlinear Landau Damping of Electromagnetic Waves........ 1216
    R. Sugaya and M. Sugawa

Strong Langmuir Turbulence Driven by an Intense Relativistic Electron Beam........ 1222
    M. Masuzaki, R. Ando, M. Yoshikawa, H. Koguchi, Y. Hamada,
    K. Miyajima, A. Ikeda, S. Takahata, H. Yoshida, and K. Kamada

From Microheterogeneous Targets to Microheterogeneous Plasma—A Way to Long-Evolution Plasma in Laser Shot........ 1228
    N. G. Borisenko and Y. A. Merkuliev

Photon Density and The Correspondence Principle of Electromagnetic Interaction........ 1234
    B. W. Boreham, H. Hora, and P. R. Bolton

Possible In-Lattice Confinement Fusion (LCF)........ 1244
    Y. Kawarasaki

7.2. LASER PROCESSING

Molecular Mechanism of Porphyrin-Sensitized Laser Ablation of Polymeric Materials........ 1250
    H. Fukumura, N. Mibuka, H. Fukumoto, and H. Masuhara

Molecular Dissociation Dynamics in Intense IR Laser Field........ 1256
    B. A. Grishanin and V. N. Zadkov

Pulsed Laser Deposition of ND:YAG Crystalline Thin Films........ 1262
    M. Ezaki, H. Kumagai, K. Kobayashi, K. Toyoda, and M. Obara

Laser Applications for Micromachining........ 1268
    M. Esashi and K. Minami

Deep Melting of Metals by Specially Profiled Laser Pulses........ 1274
    Y. V. Afanasiev, I. N. Zavestovskaya, A. P. Kanavin, and S. V. Kayukov

Deposition of Diamond-Like Carbon Films by Laser Ablation........ 1280
    M. Hanabusa and K. Tsujihara

Dynamics of Laser-Ablated Particles in Newly Developed Eclipse PLD........ 1286
    H. Nakatsuka, H. Ishibashi, and T. Kobayashi

Wavelength Dependence of Boron Nitride Ablation by TEA CO_2 Lasers 1291
    T. Sumiyoshi, H. Tomita, A. Takahashi, M. Obara, and K. Ishii
Structural Change of Metal Targets and Particles by Excimer Laser Ablation 1297
    Y. Nishikawa, Y. Yoshida, and S. Mizuguchi
Laser-Assisted Cluster Source 1303
    K. Koyama, N. Saito, Y. Iwata, and M. Tanimoto
UV Laser Ablative Figuring of Optical Plastics for Precise Optics 1309
    T. Jitsuno, K. Tokumura, N. Nishi, Y. Takigawa, N. Nakashima, M. Nakatsuka, and S. Nakai
Preparation of High Quality Barium Titanate Thin Films by KrF Laser Ablation 1314
    H. Tomita, Y. Ninomiya, A. Ito, and M. Obara

8. EDWARD TELLER AWARD LECTURES[1]

Teller Award Acceptance Speech 1323
    Professor Robert L. McCrory
Teller Award Acceptance Speech 1332
    Gennady A. Kirillov
Patience and Optimism 1334
    George H. Miley

Technical Sessions 1353
List of Attendees 1363
Author Index A-1

---

[1]The lecture by Dr. E. M. Campbell was not presented due to his unavoidable illness.

PREFACE

The 12th International Conference on Laser Interaction and Related Plasma Phenomena was held April 24–28, 1995 at the Life Science Center, Osaka, Japan. The International Atomic Energy Agency (IAEA) and Osaka University were the co-organizers of the meeting, while the Institute of Laser Engineering (ILE) organized and ran the meeting. This 12th Conference was a landmark in some ways. While the participation has always been very international in character, this was the first time since the first workshop at Rensellaer Polytechnic Institute in June 1969, that the meeting was held outside of the United States. In addition, the name of the meeting was changed from International *Workshop to International Conference*—a title which seemed more descriptive of the character of the meeting. These new trends might be viewed as a natural reflection of the growing status of lasers and high energy density physics. The size of the international community continues to expand, new areas of research and technology are emerging, and the field is growing in depth and breadth—ranging from basic science to increasingly sophisticated industrial applications.

The printed proceedings from the previous eleven meetings in this series have continued to provide the scientific community with a vital documentation of the newest and most exciting contemporary results and research directions. Following the custom of the prior meetings, conference participants were encouraged to bring new key results to the meeting as a basis for interactive, brainstorming discussions interspersed throughout the formal presentations. The Conference Organizers and Advisory Board members are pleased to add another important chapter to the history of the field of Laser Interaction and Related Plasma Phenomena.

INTRODUCTION

As was true for previous meetings in the series, and as was anticipated, given the recent growth of the field, the Conference retained a broad international participation. Approximately 310 experts from 17 countries (Australia, Canada, China, France, India, Iran, Israel, Italy, Japan, Kazakhstan, Korea, Poland, Russia, Spain, the United Kingdom, the United States, and Yugoslavia) attended the meeting, presenting 287 papers. The topics covered by the Conference included:

1. Inertial confinement fusion programs.
2. Critical elements for ignition.
3. Energy drivers.
4. Laser-matter interactions.
5. High-intensity physics.
6. Laser applications, and other new fields.

Conference staff

Conference Chairmen
S. Nakai   G.H. Miley

International Advisory Board
G.H. Miley (Chair)   H. Hora (Honorary Chair, Emeritus)
D.L. Banner   N.G. Basov   E.M. Campbell   R. Dautray   S. Eliezer
M.H. Key   R.L. McCrory   A. Guenther   S. Nakai   B.H. Ripin
P. Mulser   G. Velarde   C. Yamanaka   S. Singer

Organizing Committee
C. Yamanaka (Chair)
S. Harada   T. Hiruma   K. Husimi   A. Iiyoshi   M. Ishiou
S. Kamiyama   J. Kanamori   M. Iwamoto   C. Kojima   J. Kondo
H. Kashiwagi   T. Miyajima   S. Mori   S. Minagawa   S. Namba
J. Nishidai   S. Nakai   J. Nishizawa   T. Nitta   K. Nishikawa
T. Sekiguchi   H. Shimayama   K. Shimoda   S. Susei   H. Takuma
S. Tomiyama   K. Yamamoto   N. Yumoto

National Steering Committee
S. Nakai (Chair)
H. Akiyama   K. Fujima   T. Fujimoto   J. Fujita   T. Gonda
M. Hakoda   M. Hanabusa   T. Hattori   K. Imasaki   K. Kasuya
T. Katayama   T. Kawai   Z. Kawasaki   U. Kubo   M. Maeda
H. Masuhara   M. Masuzaki   T. Matoba   S. Miyake   Y. Miyake
T. Mochizuki   N. Moribe   K. Muraoka   H. Nakajima   Y. Nishida
K. Nomoto   M. Obara   A. Ogata   Y. Owadano   S. Ohtani
Y. Sakagami   T. Sasaki   T. Suzuki   M. Takao   K. Takayama
K. Toyoda   K. Ueda   S. Watanabe   T. Yabe   C. Yamabe
M. Yamaguchi   Y. Yasojima   K. Yatsui   O. Yoneyama   T. Yoshida

Program Committee
Y. Kato (Chair)   K. Mima (Chair)
1. Inertial Confinement Fusion Programs
   Y. Owadano (Organizer)   T. Yamanaka (Organizer)
   J. Fujita   T. Matoba   S. Miyamoto   K. Yatsui
2. Critical Elements for ICF
   K. Nomoto (Organizer)   H. Takabe (Organizer)
   H. Azechi   H. Nakashima
3. Energy Drivers
   K, Kasuya (Organizer)   K. Ueda (Organizer)   M. Nakatsuka (Organizer)
   H. Akiyama   T. Hattori   T. Katayama   M. Maeda   M. Masuzaki
   N. Moribe   T. Sasaki   T. Suzuki   M. Yamanaka
4. Laser-matter Interactions
   K. Mima (Organizer)   Y. Sakagami (Organizer)
   K. Fujima   T. Fujimoto   Y. Miyake   N. Miyanaga   K. Muraoka
   S. Ohtani   K. Takayama   T. Yabe
5. High Intensity Physics
   Y. Kato (Organizer)   S. Watanabe (Organizer)
   H. Daido   Y. Nishida   A. Ogata
6. Laser Applications
   Y. Izawa (Organizer)   K. Toyoda (Organizer)
   T. Gonda   M. Hanabusa   M. Hakoda   K. Imasaki   T. Kawai
   Z. Kawasaki   U. Kubo   H. Masuhara   S. Miyake   T. Mochizuki
   N. Nakashima   M. Obara   M. Takao   C. Yamabe   M. Yamaguchi
   Y. Yasojima   O. Yoneyama   T. Yoshida

Local Steering Committee

H. Azechi	H. Daido	H. Fujita	M. Fujita	H. Furukawa
Y. Izawa	T. Jitsuno	T. Kanabe	Y. Kato	Y. Kitagawa
R. Kodama	S. Kuruma	K. Mima	S. Miyamoto	N. Miyanaga
M. Murakami	M. Nakai	N. Nakashima	M. Nakatsuka	K. Nishihara
H. Nishimura	T. Norimatsu	S. Sakabe	H. Shiraga	H. Takabe
K. A. Tanaka	S. Uchida	M. Yamanaka	T. Yamanaka	

Conference secretariat

K. Nishihara
C. M. Elliot
T. Norimatsu (Publication)

Edward Teller Medal

The Edward Teller Medal is presented to scientists in the laser/inertial confinement fusion (ICF) community "in recognition of pioneering research and leadership in the use of lasers and particle beams to produce unique high-energy density matter for scientific research and for controlled thermonuclear fusion."

This was the third Conference where Edward Teller Medals were awarded. The tradition began with the 10th Workshop in 1991, followed by another round of awards announced at the 11th Workshop in 1993, both held in Monterey, California, USA. The 1995 recipients were Dr. R. L. McCrory (Laboratory for Laser Energetics, University of Rochester), Dr. E. M. Campbell (Lawrence Livermore National Laboratory), Dr. G. A. Kirillov (Russian Federal Nuclear Center-institute of Experimental Physics [VNIIEF], Arzamas-16), and Dr. G. H. Miley (Fusion Studies Laboratory, University of Illinois). The citations for each of these outstanding scientists are included in this proceedings.

Dr. Edward Teller attended the Conference and Award ceremony, where he personally presented the Medals to each recipient. He then captivated the audience with a lively, stimulating lecture of keen insight and enduring quality. Among his comments and predictions was a discussion of the potential importance to the scientific community of obtaining insight into the entirely new state of matter afforded by the high densities and temperatures possible in the next series of laser-target experiments planned around the world. This exciting capability to probe ultra-high density matter should stimulate and further focus the interest of the scientific community in this area.

Technical Sessions

The Conference program, in the form of a chronological schedule of all presentations and events, is included in this proceedings to provide a flavor of the meeting. To accommodate the large number of papers while maintaining the opportunity for active interaction among the participants, poster sessions were introduced and some parallel oral sessions were scheduled for the first time in this series.

A warm greeting was given to participants at the opening session by Mr. Z. Kitao, representing the Ministry of Education, Science and Culture (Monbusho). to commemorate the Conference. Dr. J. Kanamori, President of Osaka University, provided a cordial welcome to the University, where ICF research is strongly promoted through the activities at ILE. Finally, Dr. T. Miyajima, Chairman of the Fusion Council of Japan and the former Japanese representative to the international Fusion Research Council (IFRC) of the lAEA. expressed the strong support of ICF research and this Conference by the Fusion Council of Japan and the IAEA.

Two memorial lectures were presented following the opening ceremony. The first was presented by one of the original Teller medalists, Dr. C. Yamanaka, whose talk was titled "Past, present, and future of laser fusion research." The second lecture, entitled "Progress in the U.S. ICF program," was presented by Dr. M.M. Sluyter of the U.S. Department of Energy. The juxtaposition of the two lectures vividly illustrated the steady progress of ICF research, as well as the field's continuing contributions to modern physics. The feasibility of a near-term ignition and burn demonstration was stressed and is scheduled as a key goal of the next ICF projects in the United States, France, and Japan.

Future Conferences and Teller Award Nominations

The International Advisory Board met on Tuesday, 25 April 1995. In attendance were G. H. Miley, Chair, H. Hora, Teller Award Chair, S. Eliezer, R. L. McCrory, S. Nakai, P. Mulser, G. Velarde, and C. Yamanaka. M. Cray attended as E. M. Campbell's designate, M. DeCroisette represented R. Dautray, and G. A. Krokhin substituted for N. G. Basov. Plans for future Conferences and the Teller Medal nomination procedures were discussed.

The next Conference, the 13th, will be held 13 - 18 April 1997 in Monterey, California, USA. The exact date will be announced in the fall of 1995. It will be hosted by the Naval Postgraduate School (NPS) and the Lawrence Livermore National Laboratory (LLNL); Dr. E. M. Campbell of LLNL and Dr. W. Colson of NPS will serve as the local organizers. Dr. G. Miley will serve as the Conference Chair, and C. M. Elliott will be the Conference Secretary. Plans call for the 14th Conference to be held in Europe.

The International Advisor Board has strongly requested nominations for recipients of the 1996 Teller Medal from the community. All nominees will be reviewed and voted on by the International Advisory Board members. Dr. John Nuckolls, Teller medalist was elected to serve as Chair for the 1996 selection committee. Additional details for submitting a Teller Medal nomination are available from the Conference Director, Professor G. H. Miley, Fusion Studies Laboratory, 100 NEL, 103 South Goodwill Avenue, Urban, IL 61801-2984 USA; 1-217-333-3772 (telephone), 1-217-333-2906 (fax), *g-miley@uiuc.edu*.

Acknowledgments

The success of the 12th Conference was due to the strong and warm support of many organizations and to the enthusiasm and effort of many people. The IAEA played an essential role for the success of the Conference, together with the full support of Osaka University as the Co-Organizer. The more than 300 Conference attendees and the many distinguished guests enriched the meeting, not only with their excellent presentations, but also with their wholehearted participation in the many stimulating interactions that led to a truly collegial atmosphere. The dedication of the Conference staff, the International Advisory Board, the Organizing Committee, the National Steering Committee, the Program Committee, and the Local Organizing Committee are especially acknowledged in recognition of their essential contributions to the success of the meeting. The good working relationship and collaborative exchange of information between Celia Elliott, the continuing Conference Secretary, and Katsunobu Nishihara, the present Conference Secretary, was one of the many excellent examples of Japan-US cooperation. The financial support from many foundations enabled the meeting to take its place as the premier forum for presentation of scientific progress at it highest levels.

This meeting was supported by
    The 50th Anniversary Fund of Osaka University,
    Commemorative Association for the Japan World Exposition (1970),
    Inertial Confinement Fusion Forum,
    International Science Foundation,
    Inoue Foundation for Science,
    Kansai Research Foundation for Technology Promotion,
    The Iwatani Naoji Foundation,
    Research Foundation for the Electrotechnology of Chubu,
    American Nuclear Society,
    Atomic Energy Society of Japan,
    The Institute of Electrical Engineers of Japan,
    The Japan Society of Applied Physics,
    The Japan Society of Plasma Science and Nuclear Fusion Research, and
    The Laser Society of Japan.

In closing, we wish to again acknowledge the support of the participants, whose contributions are essential not only for the success of this Conference, but also for the progress of science at large.

Sadao Nakai, Conference Chair
INSTITUTE of LASER ENGINEERING
Osaka University
Osaka, Japan

George H. Miley, Conference Director
FUSION STUDIES LABORATORY
University of Illinois
Urbana, Illinois, USA

Congratulatory address: Mr. Y. Kitao
International Project Director
Science Institutes Division
Ministry of Education, Science and Culture

Before I begin my speech, I would like to express my sincere condolences to people who have passed away by the big earthquake occurred in Kobe. I also express my deep sympathy to sufferers.

Ladies and gentleman, on behalf of the Ministry of Education, Science and Culture, it is my great pleasure to express words of congratulations on the occasion of "The 12th International Conference, Laser Interaction and Related Plasma Phenomena which will be held for 5 days. I would like to exchange our warm greetings to all of you both from overseas and from Japan who are attending this conference.

Today, people are actively working on the nuclear fusion research all over the world to assay energy resources in the future. In Japan, we intensively promote researches at universities utilizing various kind of apparatus as project of major basic researches to solve crucial problems and realize nuclear fusion reactor.

In the light of the present situation, it is certainly significant both scientifically and socially to hold this conference here in collaboration with IAEA and Osaka University and to exchange new results on inertial confinement fusion, plasma physics, and high intensity laser.

The main organizer of this conference, Institute of Laser Engineering at Osaka University, was established in 1976 as the center of inertial confinement fusion research in Japan. Since then, in close contact with relevant research institute in other countries, they have obtained a number of excellent results through implosion experiment using the meeting will make important contribution to enhance research on inertial confinement fusion. I also hope that researchers can understand the aim of the conference develop further cooperation.

Finally, I would like to express my deep respect to Osaka University, IAEA, and the organizing committee for organizing this important meeting and truly wish this meeting will be a great success. I also hope that participants from abroad will enjoy this lovely spring time and get acquainted with the culture and society of Japan.

Thank you.

Welcome address: Dr. J. Kanamori,
The President of Osaka University

Distinguished guests, ladies and gentlemen.

On behalf of Osaka university, I would like to extend our heartfelt welcome to all of you today at the opening of the "12th International conference Laser Interaction and Related Plasma Phenomena".

Our Osaka University has the honor to support this conference financially on the basis of the special found commemorating the 50th anniversary of the university. Incidentally this 50th anniversary of the Osaka University was 1981.

The Institute of Laser Engineering of Osaka University was established in 1972. Since then, the people of this institute have widely explored research field of laser and its application, not only experimental and theoretical work in laser fusion research. We are proud of the fact that these contributions and achievements made by the institute have been highly appreciated world-wide.

We, Osaka University, will continue to provide positive support for the further development of research of laser energy science and laser fusion.

I am very pleased to learn from the program of this conference that the laser technology have been used to explore various new fields of physics recently. I sincerely hope that plasma research will make progress to find a turning point of the history at this conference.

Finally, I would like to express my deep appreciation to all the people who made efforts for holding this conference. Also, I hope that you would visit not only the Institute of Laser Engineering, but also other institutes of our university on this opportunity, and get acquainted well with our University.

Wishing you a successful meeting and a pleasant stay in Osaka, I would like to conclude my welcome address.

Thank you

Opening address: Dr. T. Miyajima
Nuclear Fusion Council of Japan

Distinguished guests, ladies and gentlemen, it is a great pleasure for me to welcome all the participants to this conference on behalf of the Nuclear Fusion Council of Japan.

Assuming that half a century may be required to complete fusion commercial reactors, today may not be a good time to discuss the merit and demerit of the fusion energy compared to other energy sources such as fossil ones under the conditions currently available. Many things could change during this long period. Though people believe that the sun will rise tomorrow, this may not be 100 % certain. It's quite likely that some drastic social phase change might happen. If no fossil energy is allowed to use from the safety reasons in the future, it is nonsense to compare the cost of fusion with that of fossil energy. Concerning this unpredictability we should recall many natural or social happenings occurred in this century and were not well predicted. These were, for instance, DNA, unusual climate, earthquake, crisises, or typhoon and fate of east-west confrontation, and world-wide starvation.

What are the reasons for this unpredictability? Though any phenomena should follow a natural law, we are ignorant about many things. We know little about the earth, society, people, our body and mind. If we would like to know more precisely, enormous amount of measurements and studies are necessary, resulting in tremendous amount of time and money, mental and physical effort, not affordable for the human beings. Thus we shall have to give up the precise prediction of our future and have to choose simple and secure way. In this way we can afford to increase predictability and can avoid rapid change of population, fossil energy consumption in order to diminish occurrence of unexpected happenings. With these efforts we should be able to cooperate with unexpected changes. This is rarely realised since this in a way is against the free competition. We have to be more patient and to understand one another more closely.

The Institute of Laser Engineering of Osaka University has been the only center of ICF in Japan, and has shown remarkable achievements by conducting national and international collaborations with other scientific and industrial institutes. Since a next step of ICF in Japan may be a quite big one, I feel that it is now to establish our national policy for ICF. I am sure that an international collaboration such as the one in INTOR of MCF should be considered. I believe that the conference will be very productive with great success.

Thank you.

Keynote address: Dr. Edward Teller

Dr. Edward Teller
Lawrence Livermore National Laboratory
Livermore, California 94550

Ladies and gentlemen, I will try to be very brief. I have to announce a meeting on collision with asteroids that will take place in a little more than three weeks in Livermore, the Planetary Defense Workshop.

There have been meetings of that kind before; one of the earliest and the best in our sister laboratory in Los Alamos, and another very important one in Russia at Chelyabinsk.

The fact is, we all know about the collision of a big object with Jupiter. We know about a big collision 65 million years ago, at the end of the Mesozoic Age and the beginning of the new age, the Cenozoic.

Here is something that does not occur as often as the hurricane. But when it occurs it's worse, so the danger is comparable. And today we know that we can predict it and there are all kinds of methods, including possibly nuclear methods, to prevent it.

We scientists must understand what is at stake, understand what we can do and use this for a very important double purpose. Number one; prevent a possibly very big catastrophe. Number two; use this as a practice to get together, to understand each other, to get things done together. So, we welcome you all to come and stay with us for the conference in Livermore from, I think, the 22nd to 26th of May.

Now having said this, I want to get back to the great thing that we are celebrating here. I want to first remind you again of this remarkable accomplishment of Campbell in developing the area of laser driven fusion. Lasers have the peculiar ability to concentrate energies that start from a conventional source all the way down to volumes small enough to drive nuclear fusion. Campbell is honored for his very important work in the development and use of lasers that have brought us now to the threshold of ignition experiments.

Now, we come to the work of McCrory. McCrory leads the University of Rochester effort in inertial fusion. And he has made important contributions to understanding the instabilities that are a very hard problem, maybe the main problem in inertial fusion.

Did I mention the work in Russia of Kirillov and others doing experiments with pulsed iodine lasers at a surprisingly low cost? I think many important pieces of knowledge will come from this avenue of work.

And finally, I have to thank Miley not only for his many contributions to the field, but for getting us all together and organize the things.

I have no doubt that ignition will work. I don't know yet who, I don't know yet how, I don't know whether it will be inertial confinement or whether it will be a magnetic ignition. But this development has gone steadily. I have talked with some people who say it will take a long time, while others have promised me and I believe them, that ignition will work, not only some time in the distant future, but before I am hundred years old, and I want to be there.

1995 Teller Award Recipients

Edward Michael Campbell

Dr. Edward Michael Campbell is the Associate Director for Lasers at the Lawrence Livermore National Laboratory (LLNL), Livermore, California. He was born in 1950 in Philadelphia, Pennsylvania, and received his Ph.D. in 1977 from Princeton University. During his seventeen-year career in inertial confinement fusion (ICF), he has made a broad range of contributions, both as a scientist and as a science administrator. Dr. Campbell's earliest work on ICF involved the development of radiation chemical tracers. This work played a key role in verifying the first high-density (100 times) laser-driven implosions of ICF targets. He carried out the first experiments on hohlraums, demonstrating improved coupling with frequency doubled and tripled light emission from the Argus laser, which led to the development of the Novette and the Nova lasers at LLNL.

In 1983, Dr. Campbell was co-inventor of the imploding foil target, which has been used for significant x-ray laser action. He has held a Guggenheim Fellowship and was awarded the 1990 *Award for Excellence in Plasma Physics Research* by the Plasma Physics Division of the American Physical Society.

Under Dr. Campbell's leadership, experiments on the Nova laser in the 1980s demonstrated the drive, symmetry, and implosion performance experimental basis for the proposed National Ignition Facility. Campbell is a co-inventor of the "Fast Ignitor" approach to ignition and made substantial contributions to understanding the physics basis of this technique.

Robert L. McCrory

Professor Robert L. McCrory is the Director of the Laboratory for Laser Energetics (LLE) at the University of Rochester. He was born in Lawton, Oklahoma, in 1946, and received both the B.Sc. and Ph. D. degrees from the Massachusetts Institute of Technology, where he was an Alfred P. Sloan National Scholar and an Atomic Energy Commission Special Fellow. Professor McCrory began his research in inertial fusion as a staff member at the Los Alamos National Laboratory in 1972. He has made numerous contributions to inertial fusion, beginning with his work on the wavelength dependence of the hydrodynamic efficiency of laser-driven targets and hydrodynamic stability theory. McCrory was elected a Fellow of the American Physical Society in 1985, for his many contributions to fundamental understanding of hydrodynamic instability and thermal transport in laser-driven plasmas.

He has led the University of Rochester effort to become the leading research laboratory to investigate direct-drive laser fusion. LLE demonstrated high-density implosions of directly driven cryogenically cooled capsules to densities greater than 200 times the Liquid density of deuterium tritium in 1988. Under McCrory's leadership, LLE is finishing construction of a 60-beam, 30-kJ ultraviolet (351 nm) laser, the OMEGA Upgrade, to experimentally verify the higher efficiency of direct-drive laser fusion over indirect drive. Upon completion, this will be one of the most powerful lasers in the world and will allow the extension of laser fusion into new regimes.

Professor McCrory has been a leader in many areas of U.S. policy, including service on several National Academy of Science committees on military space policy and plasma science. He serves as a consultant to several government and private sector organizations.

Gennady A. Kirillov

Professor Gennady A. Kirillov is Deputy Director of the Federal Russian Nuclear Research Institute for Experimental Physics at *Arzamas-16,* Russia, and is Head of the Optical Physics Department. He was born in 1933 in Nizhni Novgorod and received the Dr. Sc. in physics and mathematics in 1972. In the last twenty years, he has developed several laser systems, and, expanding the earlier work of K. Hohla and S .B. Kormer, has developed the photochemical iodine lasers Iskra-4 and Iskra-5 at Arzamas-16. These lasers produce 120-TW pulses of 250-ps duration, representing the most energetic lasers in this range. Laser fusion target experiments with them have produced a record 10^{10} neutrons per pulse, corresponding to a 7-keV ion temperature from D-D capsules using indirect drive.

Professor Kirillov is not only an internationally recognized leader in this field and the director of one of the large laser fusion experimental teams, but he has also made very significant contributions to the physics and operation of photodissociation iodine, gas dynamics, and chemical lasers, including chemical-explosion-driven lasers. This represents the construction of the largest laser facility in the world, designed to study the properties of high-temperature dense plasmas and to investigate laser-fusion energy problems.

George H. Miley

George H. Miley, Professor of Nuclear Engineering and of Electrical and Computer Engineering, is Director of the Fusion Studies Laboratory at the University of Illinois, Urbana-Champaign. He is internationally-known as a pioneer in nuclear-pumped lasers and advanced fuel fusion. Born in Shreveport, Louisiana, he received a B.Sc. degree from Carnegie Mellon University, and M.S. and Ph.D. degrees from the University of Michigan.

In 1963, he proposed one of the earliest concepts for a nuclear-pumped laser. Shortly thereafter, he was the first to use high-energy electron beams from a pulsed-power diode to directly pump a laser. Later he and his students developed a unique $Ne-N_2$ nuclear-pumped laser, the forerunner of a series of so-called "impurity" nuclear-pumped lasers, including the important carbon laser. Then, in collaboration with scientists at Sandia National Laboratories, he and his students developed the first visible nuclear-pumped laser, using He-Hg, initiating a sustained worldwide search for low-threshhold nuclear-pumped lasers. Most recently, his group reported the discovery of a unique $He-Ne-H_2$ visible laser with a record-low threshold requirement. His group holds the record for having developed the greatest variety of nuclear-pumped lasers.

At the May 1992, international meeting on nuclear-pumped lasers in Obninsk, Russia, the Russian scientists disclosed a large nuclear-pumped laser program in the former Soviet Union. Several of their key laser studies followed closely Prof. Miley's pioneering work, leading to their recognition of him as one of the founding "fathers" of the field.

Professor Miley has had a long-standing involvement in advanced concepts for fusion. He is the author of the seminal book, Fusion Energy Conversion, (American Nuclear Society, Hinsdale, IL [1976]). While initially dismissed as too visionary, this book has had a deep influence on the fusion community. It generated a recognition of the importance of developing advanced fuel fusion, e.g. use of D-^{3}He and p-^{11}B fuels, as the ultimate goal for fusion research, in order to reduce radioactivity and increase the energy conversion efficiency of future fusion power plants. He proposed the pioneering concept of D-T spark-ignited advanced fuel laser fusion targets. His continuing search for alternate approaches to fusion has led to his recent development of an inertial electrostatic confinement device as a unique portable low-level neutron source.

He has held a Guggenheim Fellowship and is a Fellow of the American Nuclear Society, the American Physical Society, and the Institute of Electrical and Electronics Engineers. He has received the Exceptional Service and the Outstanding Achievement Awards of the American Nuclear Society's Fusion Energy Division. He also recently received a Senior Fellow Award from the Japan Society for the Promotion of Science. He is the editor of three professional journals, LASER and PARTICLE BEAMS, FUSION TECHNOLOGY, and JOURNAL of PLASMA PHYSICS.

ICF Memorial Lecture

Past, Present and Future of Laser Fusion Research

C. Yamanaka

Institute for Laser Technology, Yamadaoka Suita, Osaka 565
Himeji Institute of Technology, Shosha Himeji 671
Japan

Abstract. The concept of laser fusion was devised very shortly after the invention of laser. In 1972, the Institute of Laser Engineering, Osaka University was established by the author in accordance with the Edward Teller's special lecture on "New Internal Combustion Engine" for IQEC at Montreal which predicted the implosion fusion. In 1975 we invented the so called indirect drive fusion concept "Cannonball Target" which became later to be recognize as a same concept of "Hohlraum Target" from Livermore. As well known, ICF research in the US had been veiled for a long time due to the defense classification. While researchers from Japan, Germany and elsewhere have concentrated the efforts to investigate the inertial fusion energy which seems to be very interesting for a future civil energy. They were publishing their own works not only on the direct implosion scheme but also the indirect implosion experiment. These advanced results often frustrated the US researchers who were not allowed to talk about the details of their works. In 1988, international members of the ICF research society including the US scientists gathered together at ECLIM to discuss the necessity of freedom in the ICF research and concluded to make a statement "Madrid Manifest" which requested the declassification of the ICF research internationally. After 6 years of halt, the US DOE decided to declassify portions of the program as a part of secretary Hazel O'Leary's openness initiative. The first revealed presentation from the US was done at Seville 1994, which however were well known already. Classification impeded the progress by restricting the flow of information and did not allow the ICF work to compete by the open scientific security.

The implosion experiments by GEKKO XII Osaka demonstrated a high temperature compression of DT fuel up to 10keV, neutron yield 10^{13} and a high density compression of CDT hollow shell pellet to reach $1000 g/cm^3$ respectively. These results gave us a strong confidence to reach the ignition and burn in near future.

The international collaboration is now highly expected.

1. INTRODUCTION

It is my great pleasure to present a memorial lecture at the beginning of the 12th Conference on Laser Interaction and Related Plasma Phenomena which was initiated by Helmut Schwarz and Heinrich Hora in 1969 at Rensselaer Polytechnic Institute. During these 26 years, the laser fusion research has produced a lot of interesting results. In 1973, at the 3rd meeting, E. Teller delivered a plenary talk on "Futurology of High Intensity Lasers". He said not only because of the required laser intensities, but also because of the compression must go on in quite a symmetrical fashion, one can be optimistic or pessimistic. I think that when you have a long way to go we should pick flowers on the way. Lasers could tell us a lot about the equation of state of all kinds of matter. Sometimes to take a detour, to be deflected, is the most efficient way.

And after 18 years at the 10th meeting, in 1991 he said the 1000 times compression of normal density done by Yamanaka means the race of fusion research has turned around the 3rd corner. In Figure 1 the progress of the compressed target experiment is shown.

Now I would like to recall these 25 years progress. First of all I should say about the comparison of MFE and IFE. The former has a strong international collaboration to maintain the research activity. The latter is not. And also the two ways are quite different, one is concerning very thin plasma and another is extremely dense plasma related a lot about astronomy. The advantages for IFE are as shown in Table 1. These features are very important in the final stage of application to the civil energy.

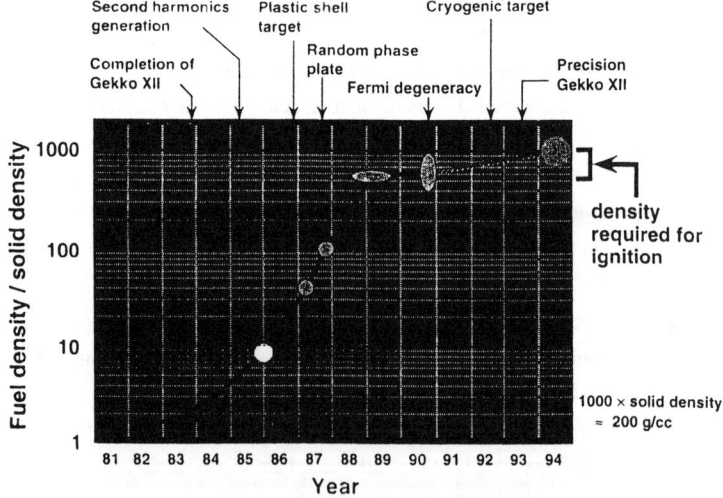

FIGURE 1. Progress of Compressed Fuel Density

TABLE 1. Comparison of Fusion Research MFE and IFE

Advantages for IFE
(1) A driver and reactor chambers keep a large stand off distance. (2) The breeding blanket is inside of structural walls. (3) Materials to produce nuclides are avoided. (4) The tritium inventory is minimized to isolate the pellet.

2. PROGRESS IN THE LAST 25 YEARS

Recent inertial fusion experiments on the direct driven fusion at Osaka have successfully got the high fusion neutron yield 10^{13} and the high density compression of 1000 times normal density. The electron degeneracy of core plasma is also observed. The U.S. Halite / Centurion program informed us of indirect driven fusion which will be attainable the high gain for less than the 10MJ of driver. However, the data base is not yet clear to determine the details for high gain. This question can only be solved by a large laser facility. The U.S. policy on indirect driven fusion program has come to provide the National Ignition Facility with the declassificaton of the experimental data. Experimental and theoretical progress in ICF in the international community has suggested that the time has come to eliminate unnecessary restrictions on information relevant to the energy applications of ICF. Now ICF is in the second stage of the development. The ignition and breakeven are in a scope of the program. The international collaboration will be initiated.

Since 1972, our continuous efforts have been performed to organize the international ICF community concerning the IAEA activity and other authorities. In 1994, the US Secretary of Energy Hazel O'Leary decided the declassification of the ICF information except the weaponary concern. This was welcomed by all the persons. A fundamental change is expected. Now, to show the importance of the international collaboration, the world progress of inertial fusion is briefly reviewed setting particular remarks on the Japanese efforts. The main steps in the history of ICF are shown in Table 2.

At the Levedev Institute in USSR, subsequent pioneering research on ICF had been performed. It led to the disclosure of laser fusion concepts at the International Quantum Electronics Conference in 1963, followed by a presentation in 1968 of the first detection of fusion neutrons from laser irradiated lithium hydride targets. These progress was due to the development of the lasers, Kalmar and Delfin.

At the Lawrence Livermore National Laboratory the pioneering theoretical and experimental works have been performed in the last three decades. Especially a

TABLE 2. Main Steps in the History of ICF

Year	Institution	Event
1963	Lebedev Physics Institute (USSR)	Proposal to use lasers for controlled fusion
1968	Lebedev Physics Institute (USSR)	Registration of thermonuclear neutrons in laser-produced plasma
1970	CEA Lemeil (France)	Definite neutron yield observed
1972	Livermore Natl. Lab. (USA) Los Alamos Natl. Lab. (USA) ILE (Japan)	Starting date for the financing for a national ICF program in the USA ICF program in Japan
1974	Lebedev Physics Institute (USSR) (FIAN) Institute of Applied Mathematics	Concept of low entropy compression of shell targets
1975	ILE (Japan)	Indirect drive Cannonball target concept
1977	Livermore Natl. Lab. (USA)	Launching of 10kJ Nd-laser "Shiva"
1978	Los Alamos Natl. Lab. (USA) ILE (Japan)	Launching of 10kJ CO_2-laser "Helios" Launching of 2kJ Nd-laser "GEKKO IV"
1979	Livermore Natl. Lab. (USA)	Density of compressed fuel reached 20g/cm^3
1983	Livermore Natl. Lab. (USA)	Launching of a 20kJ "Novette" Nd-laser
1983	ILE (Japan)	Launching of 30kJ Nd-laser "GEKKO XII"
1985	ILE (Japan)	Neutron yield 10^{13} LHART target
1985-1989	Livermore Natl. Lab. (USA)	Launching of 130kJ Nd-laser "Nova" (fuel density, 30g/cm^3; neutron yield, 3×10^{13})
1987-1991	ILE (Japan)	~1000g/cm^3 matter density reached with Gekko XII laser facility
1994	Livermore Natl. Lab. (USA)	Approval of DOE NIF and declassification

disclosure in 1972 of the implosion physics at the International Quantum Electronics Conference, Montreal by Edward Teller showed that the laser fusion targets could be ignited with much less energy than predicted there-to-fore if the fuel was compressed up to 1000 times of the normal density. And also the understanding of hydrodynamics and instability phenomena associated with the strong compression of ICF targets were prevailed. Series of the Nd glass lasers, Augus, Shiva and Nova were developed to perform the laser fusion experiments. The major glass suppliers for the large Nd glass lasers are in Japan and Germany. LLNL and also LANL have been performed a lot of interesting works in the ICF research which provided an essential guidance for the ICF program in the world. However the US classification policy of inertial fusion especially on the indirect driven fusion was a crucial problem. It hurt the morale of the US scientists who were unable to take credit for their creative work and often must endure the vexation of seeing nearly identical work published in the open literature by workers in Japan, Europe or the Soviet Union. Classification impeded the progress by restricting the flow of information, and did not allow all ICF work to benefit by the open scientific scrutiny. According to the patient efforts of international movement of several countries, the world situation for cooperation is changing to promote.

In France, the CEA laboratory at Lemeil built a Nd glass laser, Phebus, 20kJ which is now open to the public use. Smaller Nd glass laser facilities exist at the

University of Rochester, Ecole Polytechnique Palaiseau, the Shanghai Institute for Optics and Fine Mechanics and at several other places around the world. The significant progress has been made over the last five years. Since there is in principle no obstacle to prevent the goal in physics, more intense international efforts should be provided for exploring abundant and affordable energy.

The scientific works at Osaka is internationally recognized. In particular our contributions to ICF theory and experiment culminated in 1987 with remarkable achievement of the record of the compression density approaching $1kg/cm^3$.

In 1972, the Institute of Laser Engineering, Osaka University was established in accordance with the Edward Teller's special lecture on "New Internal Combustion Engine" at Montreal. And also we had timely the first Japan-US Scientific Seminar at Kyoto by the Japan Society for Promotion of Science which was an origin of the international collaboration on the inertial fusion research where 30 scientists from the U.S., Germany, Britain, Soviet Union and Japan gathered together which are shown in Figure 2. Our research on the laser plasma initiated in 1963 using ruby lasers and Nd glass lasers. The first issue of research was the laser-plasma coupling. The absorption mechanisms were thoroughly investigated a result of which was to propose the anomalous absorption caused by the plasma parametric instability. Nonlinear plasma instability due to the laser irradiation became a worldwide popular subject. We also investigated the self phase modulation of laser light by plasmas.

FIGURE2. The 1st Japan US Seminar'72 on Laser Interaction with Matter

In 1975, we invented the so called indirect driven fusion concept "Cannonball Target" at our Daisen Summer Seminar which became later the Institute very popular in the world. The various targets of ILE are shown in Figure 3. At the age of oil crisis, the importance of new energy sources was well understood throughout the country. In a fair wind to fusion research we set LEKKO CO_2 lasers to the Los Alamos group and competed with the Livermore program by GEKKO glass lasers and compared the ideas to the Sandia team by REDEN beam machines. The development of glass lasers is shown in Figure 4.

In 1983 the world largest of glass laser GEKKO XII was completed by the cooperation of the NEC. As for the direct driven ICF, it is potentially more efficient but has significantly more stringent requirements on driver beam uniformity and the control of hydrodynamic instabilities. We had significant progress in this field using a novel type of uniform shell target and a random phasing smooth laser beam. The beam smoothing technologies are given in Table 3.

Target	η_{abs}	η_c
Ablative target	50%	1%
	Higher efficiency by blue	
	High uniformity requirement	
LHART	70%	6%
	Higher efficiency by blue	
	Stagnation free	
Hohlraum	50%	5%
	High efficiency	
	Uniformity expected	
Cannon ball	70%	6%
	High efficiency	
	Uniformity improved	
Double shell in cannon	70%	1%
	High efficiency	
	Uniformity greatly improved	

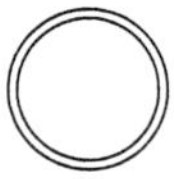

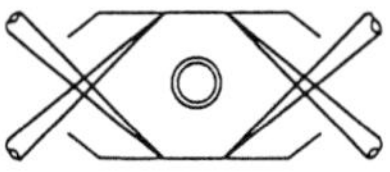

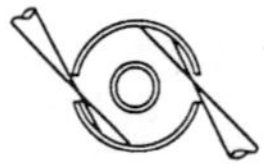

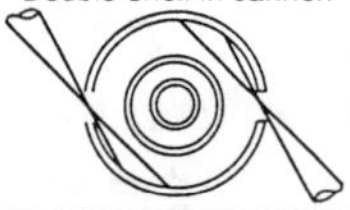

Several types of laser fusion target. Laser absorption η_{abs} and core coupling efficiency η_c are given.

FIGURE 3. Several Types of Laser Fusion Target at ILE Osaka

In 1985, the new idea of LHART (Large High Aspect Ratio Target) was devised by using an implosion simulation code of pusher-fuel mixing free. It could record a super shot of DT fusion neutron yield 10^{13} which was hurriedly after traced by the LLNL group.

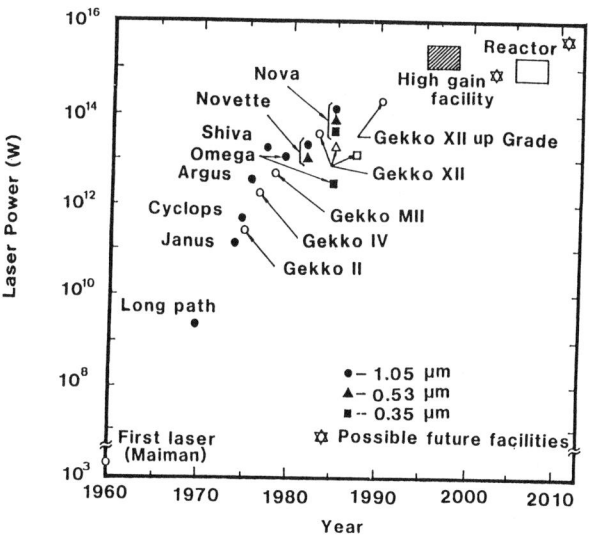

FIGURE 4. GEKKO Laser Systems

In 1987, The green light random phasing 12 beam of GEKKO XII glass laser irradiated a plastic shell target of nearly perfect sphericity to attain the 1000 times normal density. The D-T fuel density reached 200gr/cm^3 in absolute. The plasma is some what Fermi degenerated. These details were reported at the IAEA conference in Nice at 1988. Ablative pressure generation and hydrodynamic behavior of compressed fuel were experimentally and theoretically investigated. The implosion performance was optimized by using an appropriate aspect ratio of the target and a suitable laser pulse. The uniformity of laser irradiation as well as the pellet structure were essentially important to avoid the growth of instability.

3. FUTURE PROSPECTS

Until now the laser fusion research has been developed as shown in Table 4. As for the high temperature demonstration, the LHART experiment produced the high Neutron yield by using a stagnation free compression scheme. The thin shell target was imploded without mixing of thin pusher and fuel which could be applied to the volume ignition. Concerning the high density demonstration of low isentropic compression, we used a deutrated plastic shell target which was prepared under no gravity condition to form an excellent spherosity of the pellet. And also the uniform irradiation of the laser beams was essentially important. In our case, non uniformity

TABLE 3. Development of Laser Beam Smoothing Technique

				near future
ILE	Laser+RPP (1983) beam segmentation	ASE+RPP (1989) broadband spatial incoherency	ASE+RPP with angular dispersion (1991)	envelope control polarization control multi aspherical lens array
NRL	ISI (1983) by echelon	echelon-free ISI (1987)	ASE with complete image relay	envelope control
LLE		1D E-O SSD +RPP (1989)	2D E-O SSD +RPP spectral angular dispersion	60 beams zero correlation mask
LLNL		Noisy SSD +RPP		envelope control Kinoform phase plate
Limeil		Optical fiber smoothing temporal and spatial incoherency by optical fiber		

key issue	beam segmentation speckle pattern	→ spatial and temporal incoherency →	beamlet profile control multi beam irradiation

was kept below 3%. The beam smoothing technique such as the random phase plate has been fully investigated at ILE as shown in Table 3. The key issue for the high density is the uniform ablation. The famous high compression of 1000 times of normal density could not yet reach the ignition. The neutron yield was 100 times smaller than the expected value. The reason for this was due to the fuel mixing in the final compression stage to reduce the temperature of fuel. The method to overcome this problem is to improve the uniformity of ablation.

In Figure 5, a scenario of the ICF research to the scientific feasibility is indicated. To reduce the mixing, the laser irradiation modes shall be improved by using the larger numbers of beams, the optimized arrangement of beams, the beam smoothing as shown before, the envelop profile control for each beam and the power balance of the beams. As shown in Figure 6, we can keep the absorption nonuniformity better than 1% when the beam number is 60, the power imbalance is less than 1% and also the partial coherent laser with the precision RPP is adopted. In these conditions one will be able to expect the spark ignition judging from the data of simulation using the α heating to smoothing the turbulence.

A new method to get the ignition is to use a petta watt ignition laser such as 10^{20}W/cm^2 of a few kJ which may deliver the energetic electrons into the compressed core to heat the fuel. This new way is now an intense target of the ICF research. In this case the main compression laser is enough to have a few 100kJ level and the ignitor laser of petta watt will be a few kJ. As for the fusion demonstration, we shall consider the scale of targets from the present level to future. In Figure 7, we show the targets for the 10kJ laser irradiation, such as LHART, CD-

TABLE 4. Progress of Laser Fusion with GEKKO XII

FY	Achievement	Driver	Pellet & Diagnostics	Theory & Simulation
	Interaction and Ablation			
1983	$Y_n=10^{10}$	Completion of GEKKO XII	Glass micro-balloon (GMB)	Hot electron
	High Temperature Demonstration (Stagnation free compression)			
1984	h_{abs}: Absorption P_{abl}: Ablation	SHG ($\omega \to 2\omega$)	LHART	
1985	$Y_n=10^{12}$			Shock multiplexing
1986	$Y_n=10^{13}$ $T_i=10\,keV$		Fusion yield calibration	
	High Density Demonstration (Low isentropic compression)			
1987		THG ($\omega \to 3\omega$)	CD shell	
1988	$\rho=1000\times\rho_s$ (main fuel)	Random phase plate	CDT shell	Nonlocal transport
1989			2nd. reaction p knock-on Si activation	
1990	Fermi degeneracy High density compression			2D, 3D codes
	Ignition and Burn Demonstration			
1991	Thermal smoothing	Partially coherent light	Cryogenic target	
1992	RT instability R-M instability			α heating
1993	Impression stability	GEKKO XII	Ultra-fast frame	
1994		Up grade	Cannonball	

shell and Foam Cryogenic ones. The next stage is the 100kJ driver experiment for the ignition demonstration. The final stage will be a few MJ driver for the high gain. The 1D simulation r-t diagrams are given in the figures. From these data we can imagine the way to the ignition. As for the pellet gain scaling, Figure 8 shows the necessary laser energy for the different implosion velocities by the ILESTA-1D code. For the ignition demonstration a few 100kJ laser energy is necessary and for the high gain at least a few MJ is expected. According to the ILESTA-2D fuel simulation for 300kJ ignition experiment of Figure 9, the pellet gain 1 can be attained by the laser absorption uniformity better than 1%. These experimental condition will be attained as shown in Figure 6.

As well known the growth rate of Rayleigh Taylor instability in the acceleration phase of implosion can be suppressed by the ablation flow, however in the deceleration phase the only way to prevent the fuel mixing is to keep the uniform ablation condition slowing down the growth of instability. When the ignition starts, the α particle heating in the compressed fuel will produce the fire polishing to kill the

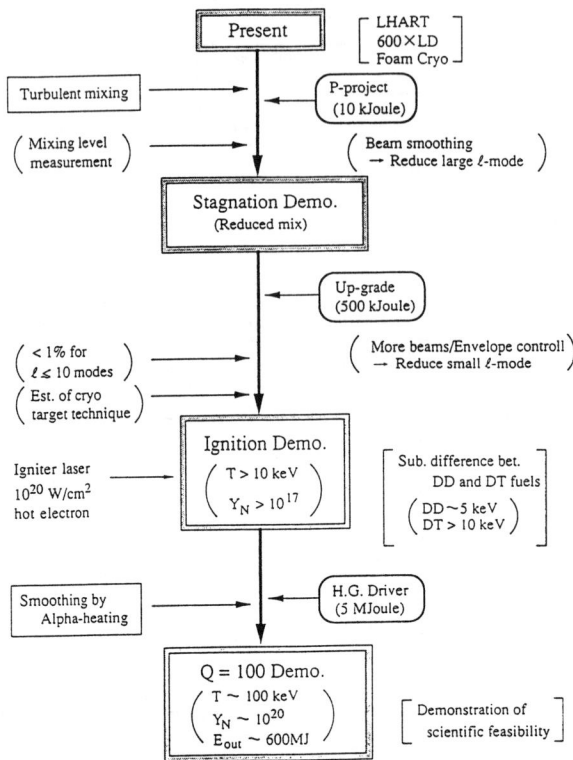

FIGURE 5. ICF Research Scenario

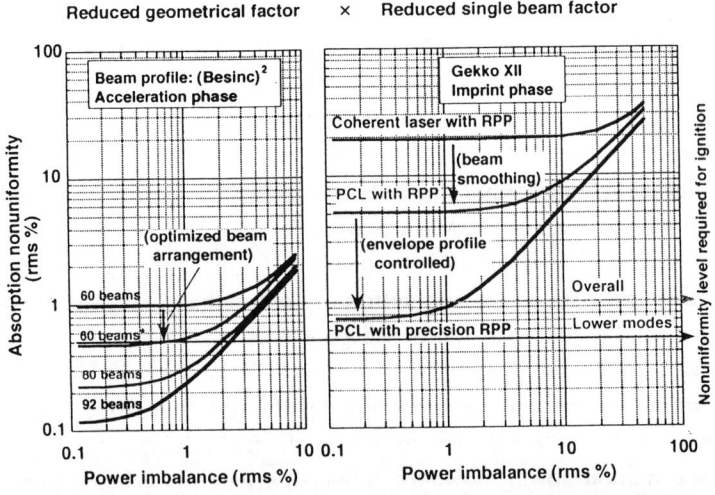

FIGURE 6. The Laser Beam Smoothing for Uniform Irradiation

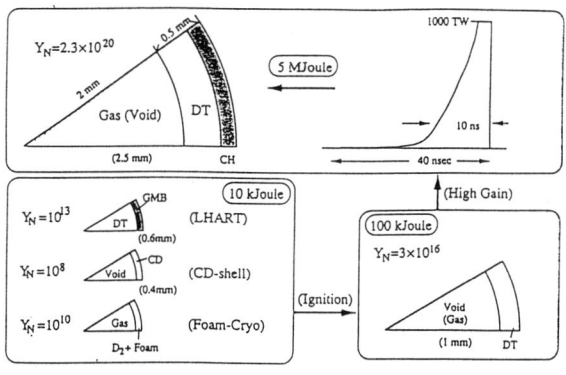

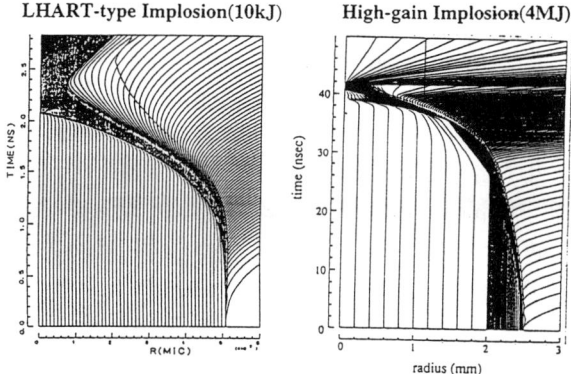

FIGURE 7. Scale of ICF Targets from Present to Future

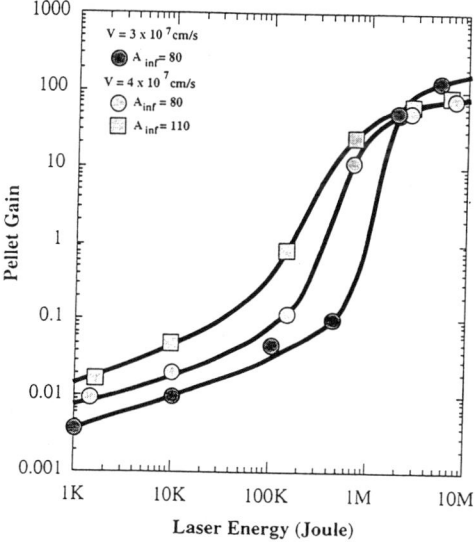

FIGURE 8. Pellet Gain Scalings ILESA-1D Code

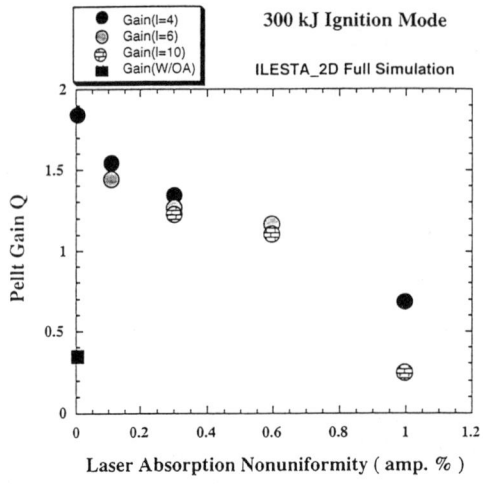

FIGURE 9. 300 kJ Driver Ignition Mode ILESA-2D Fuel Simulation

instability. Figure 10 indicates the fire polishing by α particles.

To observe the implosion process precisely the new diagnostics of high resolution framing picture MIXS has been developed at ILE. The obliquely arranged pinholes on the slit of streak camera in soft x ray range can produce the sectors of the images of a compressed core which are computer processed to make a series of core images. The spatial resolution is 15μm, the temporal resolution is 11.7 psec and the frame interval is 8.7 psec. The MIXS can give us the instantaneous behaviors of the core to analyze the implosion stability. In Figure 11, X ray emission images of the hot spark of the cooled CH shell filled D_2 50 atm + Ar 0.1 atm are given irradiated by GEKKO XII, 5.66kJ, green light. The development of novel diagnostics is essentially important in future researches.

4. CONCLUSION

The inertial fusion research is now preparing for the first ignition and burn of fusion target which will be in about 10 years from now with a MJ glass laser by the indirect implosion and or a 300 kJ laser by the direct implosion. The Peta watt ignition laser seems very interesting for ignition. Key problems are still symmetry and stability of the implosion which cause the fuel mixing to reduce the temperature of the fuel. The significant progress made in smoothing laser beams as well as the development of high resolution diagnostics gives us a great tool to proceed for ignition and burn. Successful numerical simulation predicts the way to the goal.

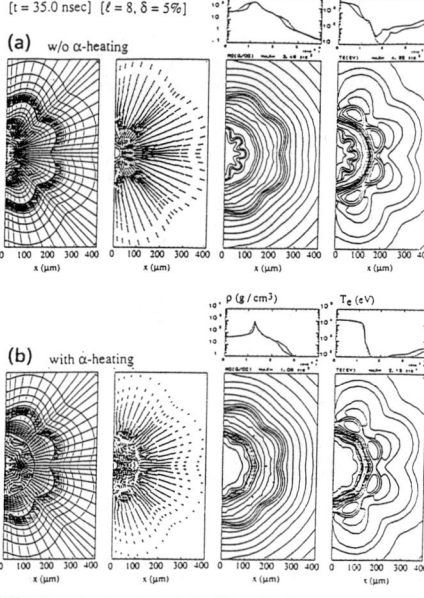

FIGURE 10. Alpha Particle Fire Polishing ILESA-2D Code

The volume ignition scheme is also reconsidered to eliminate the fuel mixing by pusher. The energy driver for ignition is considered to be lasers and these for the energy production might be heavy ion accelerators in the longer time scale.

International collaboration has just started to promote the mutual benefit on ICF research. The declassification steps are hoped to speed up the ICF research. The scaling of high gain pellet from present targets to reactor ones is shown in Figure 12.

Applications of inertial confinement fusion include not only civil energy production but also physics at the laser-atom interaction, nuclear matter under extreme conditions, cosmology, special isotope separation, food preservation, hydrogen production and advanced space propulsion. The pursuit of ICF will contribute substantially to overall scientific strength in several areas.

In the international collaboration, the essential advancement in research and the technology development for fusion shall be carried out in the following items,

(1) High-average-power fusion drivers, lasers as well as heavy ion beams
(2) ICF target for power-plant and fueling technology including cryogenic methods
(3) Material for a reactor chamber and the technology of energy conversion

The development of the ICF reactor is essentially important. The material research including target, tritium and structural materials need an intense technology

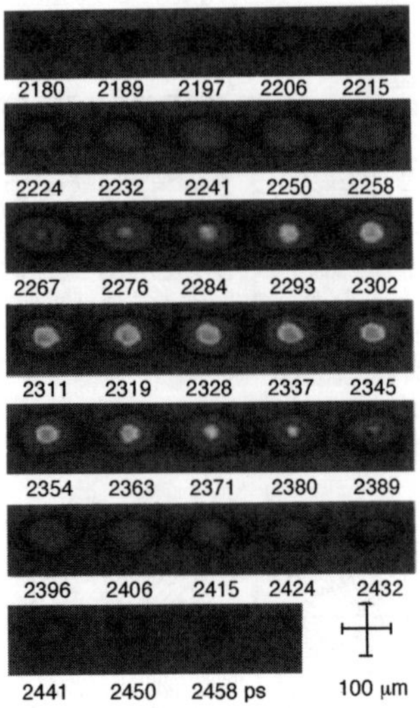

Target: cooled CH shell.diameter=486µm, shell thickness=7.23µm, filled with DD=50atm+Ar=0.1atm
Laser: GekkoXII.wave length=0.527µm, pulse duration=1.34nsec, energy=5.66kJ, energy imbalance=+9.8/-50% with RPP,
MIXS: temporal resolution=11.7psec, flame interval=8.7psec, spatial resolution=15µm, observed spectral range:hv=2.5~4.9keV

FIGURE 11. X-ray Framing Camera for Hot Spark — MIX

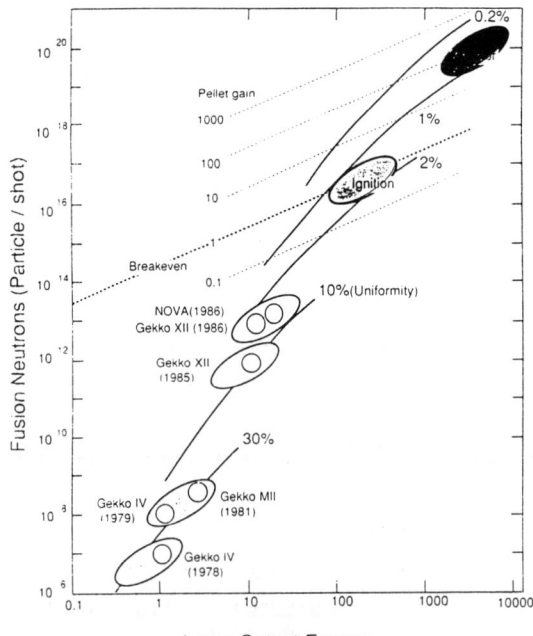

FIGURE 12. High Gain Pellet Design from Present to Future

development to realize a fusion power plant by 2025. The international center for integrating a demonstration power plant of ICF shall be considered. No other alternate energy source holds a bright promise than fusion and none has ever presented such formidable scientific and engineering challenges.

REFERENCES

1. C. Yamanaka : "Present status and future prospects of laser fusion research at Osaka" and "Progress in inertial fusion research" 1st International Symposium on Evaluation of Current Trends in Fusion Research, Washington D. C. 14-18, November (1994).
2. C. Yamanaka, *Introduction to Laser Fusion*, harwood academic publisher (1991).

1. ICF Programs and Energy Drivers

Progress in the US ICF Program

Marshall M. Sluyter

Office of Research and Inertial Fusion (DP-11)
U.S. Department of Energy,
19901 Germantown Road, Germantown, MD 20874

Abstract. The ICF Program has made exciting progress in the past year towards its goal of the achievement of fusion ignition and gain in the laboratory. A series of experiments on the Nova laser facility has resolved the major technical issues involved in the design of an ignition target. A baseline target has been designed that ignites (calculationally) with a nominal drive of 1.35 MJ (at 351 nm). In parallel, a detailed conceptual design for the National Ignition Facility (NIF - a 1.8 MJ glass laser) has been completed and a successful laser beam line prototype has validated its architecture. As a result, the Department of Energy has requested funding for the preliminary design for the NIF from the U.S. Congress. With these developments, the attainment of the long-sought goal is in sight. In addition, two new laser facilities (OMEGA Upgrade and Nike) have recently been completed, and ion-beam fusion driver development is encouraging. Their availability expands the capability of the program to perform advanced ICF and plasma experiments.

INTRODUCTION

After nearly three decades of intense effort, accompanied by substantial technical progress, it is now highly probable that the long-sought goal of inertially confined fusion ignition and gain in the laboratory can be achieved. Due to progress from a national Inertial Confinement Fusion (ICF) Program, all of the elements necessary to achieve this goal are now either in-hand or proposed based on prototype results.

As a result of an intensive experimental target physics campaign conducted on the Nova laser, many scientific issues associated with indirect drive laser fusion targets have been resolved. Integrated computer models that have been benchmarked against experiments have been used to design several targets that achieve ignition with a nominal laser energy of 1.3 MJ. In parallel with these efforts, a detailed conceptual design of the National Ignition Facility (NIF - a 1.8 MJ glass laser) has been successfully completed.(1) This design provides a substantial margin of confidence above the nominal design point at a reasonable cost. The key technical elements of the laser design have been successfully verified on a full-scale scientific laser beam line prototype, Beamlet, resulting in high confidence that the laser can meet design requirements. As a consequence of these achievements, the Department of Energy has recommended the start of Title I (preliminary engineering design) for the project, has established a NIF Project

Office to manage construction and has requested $61M for FY96 from the U.S. Congress to begin pre-construction activities. When the NIF commences target operations in 2004, the first laboratory demonstration of ignition should be obtained.

The U.S. ICF program is also pursuing a parallel path to ignition and high gain through direct drive experiments on the OMEGA Upgrade and Nike lasers. Both of these lasers have recently become operational and will provide significantly enhanced capability for the study of issues related to direct drive laser fusion.

Finally, the program supports research on indirect drive utilizing light ion beam drivers. Light ions hold great promise for the attainment of high yield fusion and progress has also occurred along this line of research.

INDIRECT DRIVE TARGET PHYSICS

The baseline design for an indirect drive laser-driven ignition target is shown in Figure 1. The design consists of a 1.11 mm radius capsule with a 160 μm thick Br-doped CH ablator layer, an 80 μm thick cryogenic fuel layer of solid DT and a DT gas fill of 0.3 mg/cc. The capsule is centered in a gas-filled, gold hohlraum of dimensions 5.5 mm diameter and 9.5 mm length. The hohlraum is sealed at the laser entrance holes with 1 μm CH windows and filled with a mixture of helium and hydrogen to a pressure of 5 atmospheres. Two cones of laser beams, each cone having separate timing, impinge on each side of the hohlraum in order to improve time dependent symmetry. This design is based on a nominal incident laser energy of about 1.3 MJ. The model indicates that about 150 kJ are absorbed by the capsule and, with a gain of order 10, 12-15 MJ are produced. The calculations which verify ignition and gain are performed using an integrated modeling package which includes the following physics: 3D laser ray tracing of energy into the hohlraum; radiation transport from the hohlraum walls to the capsule; Lagrangian hydrodynamics of the hohlraum wall and the capsule; non-LTE physics of the hohlraum wall; and, capsule ignition and burn. Independent

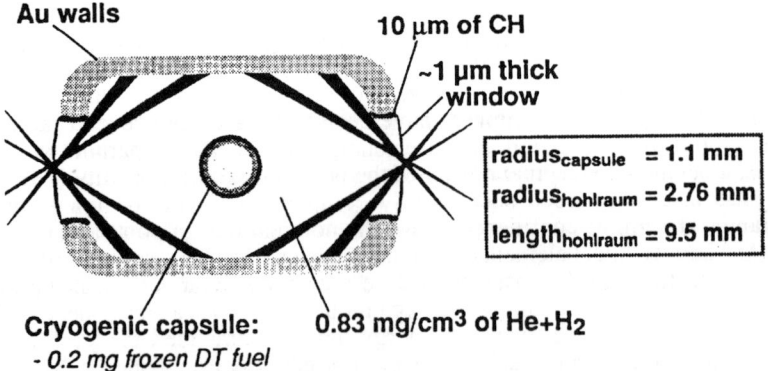

FIGURE 1. Baseline Indirect Drive Ignition Target

calculations done at Lawrence Livermore National Laboratory (LLNL) and Los Alamos National Laboratory (LANL) indicate that this design and several other variants will ignite.(2)

Early hohlraum experiments determined that the inner wall moved inward during the laser pulse. This results in a time-dependent change of the symmetry of the illumination (caused by the varying location of the laser beam absorption). In order to prevent this asymmetry a low-Z liner was placed inside the hohlraum. The liners suppressed the wall motion, but resulted in a plasma which collapsed and stagnated on the capsule, resulting in implosion asymmetry. The solution was a gas fill which suppresses the wall motion while eliminating the hydrodynamic coupling. The choice of gas is dictated by the need to minimize non-linear laser-plasma interactions.

The ignition target baseline design assumes an incident energy (at 351 nm) of 1.3 MJ at a peak power of 410 TW. Ignition occurs with as little as 0.9 MJ, providing a significant confidence factor given the nominal NIF design requirements of 1.8 MJ and 500 TW peak power.

The integrated codes which have been used to evaluate the performance of target designs have been benchmarked against data acquired during an extensive experimental campaign which has been conducted on the Nova laser facility.(3) This campaign, known as the Nova Technical Contract (NTC), arose out of the review of the ICF program that was conducted by the National Academy of Sciences in 1989-90. The campaign was divided into two broad areas: Hohlraum and Laser Physics, which explored laser-hohlraum coupling, generation and transport of x-rays and laser-plasma coupling instabilities; and, hydrodynamically equivalent physics, which explored capsule implosion physics in regimes hydrodynamically equivalent (in terms of hydrodynamic instabilities and convergence) to ignition capsules. These experiments demonstrated adequate control of the symmetry of illumination, acceptably low levels of non-linear laser-plasma instabilities (filamentation, stimulated Brillouin scattering [SBS] and stimulated Raman scattering [SRS]), control of the hohlraum wall motion, and adequate control of hydrodynamic instabilities.

The remaining issues that need to be addressed basically consist of refinement of the understanding within the same general areas that were studied in the NTC. Therefore a new experimental and theoretical effort will be conducted during the next several years to increase our confidence in preparation for NIF operations, to improve performance of ignition targets and to evaluate the sensitivity of target designs to variations in experimental conditions. This work will be concentrated in three general areas: (1) time-dependent symmetry in gas-filled hohlraums; (2) laser-plasma instabilities; and, (3) evaluation of hydrodynamic instabilities and mix at medium to high growth factors and high convergences. A brief description of the status of present understanding, experimental techniques and plans for future work in each of these areas follows.

Currently, with integrated modeling only gas-filled hohlraums have successfully reached ignition. Therefore, gas filling is the baseline technique for suppression of wall motion. Nova experiments have verified that the gas fill effectively suppresses wall blowoff, with higher density gas resulting in a longer suppression of the motion. The effects of the gas on the symmetry of implosions have begun to be studied using x-ray imaging along several lines of sight. These experiments have shown that the symmetry is a sensitive function of the laser pointing and that agreement is good between experimental results and modeling for vacuum hohlraums. However, the same type of data with gas fills indicate an inward shift of the pointing scan by about 150 µm from the previous vacuum data as shown in Figure 2. While adequate symmetry can still be obtained experimentally, the empirically determined pointing location does not match the calculations. Further experiments are underway using these techniques in order to improve the ability to model symmetry.

Preliminary measurements of the backscattered SBS and SRS light from gas-filled hohlraums have indicated that these effects are quite low. However, more refined measurements are needed to enhance our confidence that the levels of scattered light are both low enough for successful operation of the NIF and to improve theoretical understanding.

Progress has been made analyzing the hydrodynamic instability growth in spherical implosion experiments conducted on Nova.(4) Initial experiments carried out using capsules with plain CH ablators and relatively low Rayleigh-Taylor instability growth were in good agreement with modeling. More recent experiments have used CH ablators doped with Ge at the 1-2% level. The doping serves as a preheat shield and results in peak RT growth factors at the pusher-fuel interface of about 150. In this regime saturation effects are very important and these targets are thus in the NIF relevant regime. The effects of instability growth are examined in

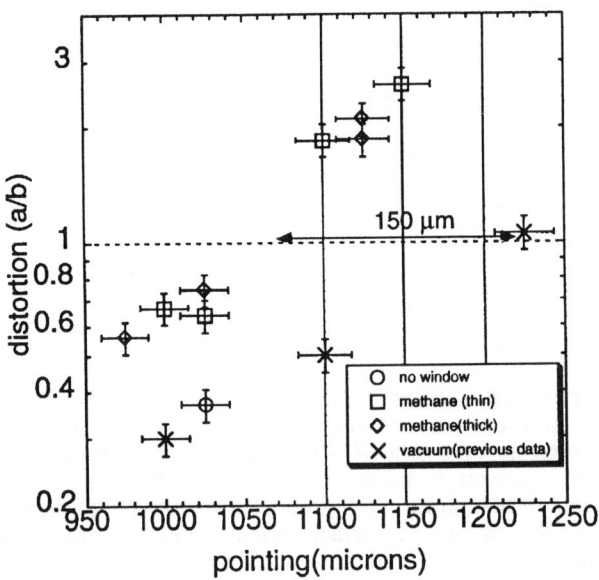

FIGURE 2. Effect of pointing on symmetry in vacuum and gas-filled hohlraums

Nova experiments by measuring the capsule performance as function of the RMS roughness of externally imposed outer surface perturbations. These outer surface perturbations are put in place using a novel laser ablation technique, which allows both the amplitude and mode structure of the perturbation to be controlled. Smooth unroughened capsules have a RMS surface roughness from 0.02-0.04-μm. Experiments to date have used multimode outer surface perturbations with RMS amplitudes as large as 1-μm. The variation in neutron yield with outer surface roughness has been measured and is in reasonable agreement with calculation for surface roughnesses > 0.1-μm. For smoother targets measured yields are lower than predicted by one-dimensional calculations. However, asymmetry and inner surface roughness effects are important in these experiments, and it is expected that improved agreement with simulation will be observed when these effects are included. As a final point, the yield has been observed to increase with the level of Ge doping in the ablator. This represents the first demonstration of the positive effects of preheat shielding for indirectly driven capsules.

National Ignition Facility (NIF)

The National Ignition Facility (NIF) is the next major ICF facility and is designed to achieve fusion ignition in the laboratory.(1) It will consist of a 1.8 MJ glass laser (frequency tripled to 351 nm) comprised of 192 beamlets that will deliver a temporally shaped pulse of 20 ns maximum duration and a peak power of 500 TW. The beamlets are arranged in groups of four symmetric cones that will illuminate a hohlraum in pairs through opposing entrance windows on opposite sides of the hohlraum. Each of the pair of cones on a side can be independently pointed and timed to improve the drive symmetry. The facility includes a target chamber and target handling equipment that will provide the cryogenic targets needed for ignition and handle a fusion yield of up to 45 MJ.

In order to reduce space requirements and cost, the NIF is designed with a new glass laser architecture that utilizes multi-pass amplification. This architecture required several innovations (such as large area Pockels cells) and involved some technical risk. Therefore, a scientific prototype called Beamlet was constructed and tested. This prototype was designed with similar architecture to that proposed for the NIF and was designed to operate at near NIF-scale beam sizes and fluences. The Beamlet has met or exceeded all of its design goals for energy (at 351 nm), power and pulse shaping and fluence. It has therefore validated the scientific soundness of the concept. Simultaneously, a comprehensive conceptual design was performed for the whole facility, including cost estimates. Based on these successes, the U.S. Department of Energy (DOE), on October 1994, approved what is called Key Decision 1. This decision is the second in a series that are required for major DOE construction projects. The effect of this decision is to permit the (estimated $1.07B) project to proceed to Title I design (preliminary engineering efforts), contingent upon congressional funding. The DOE has established a project office to guide the NIF to completion and has requested $61M from the U.S. Congress for preliminary engineering work in FY96. The current schedule calls for the laser to commence operations in 2002, with the first ignition experiments scheduled for 2004-2006, following successful attainment of full laser power.

DIRECT DRIVE

The program also has a vigorous effort in the pursuit of direct drive fusion. The direct drive approach has the advantages of simplicity and potential lower drive energy requirements, but requires significantly smoother drive beams in order to suppress the hydrodynamic instabilities.

OMEGA Upgrade

In the past year, two major new facilities to examine the physics of direct drive with lasers have been completed. At the University of Rochester/Laboratory for Laser Energetics (UR/LLE), the OMEGA Upgrade laser has been completed and is nearing operational status.(6) OMEGA Upgrade is a glass laser consisting of 60 temporally shaped beams that will uniformly illuminate a spherical capsule with up to 30 kJ of 351 nm light. Several laser beam smoothing techniques, (such as Smoothing by Spectral Dispersion (SSD)) that have demonstrated the capability of producing nonuniformity levels of a few percent, have been incorporated into the laser design. The entire laser is now complete and all beams have been directed into the target chamber. The target alignment and focusing system is also complete and a complete array of target diagnostics have been installed. Following intensive laser testing and activation, direct drive hydrodynamics experiments will begin. The OMEGA Upgrade can also be configured to deliver up to 20 kJ utilizing 40 of its 60 beams for indirect drive experiments and such use is planned for the near future as well.

Nike Laser

In addition, the U.S. program is investigating the applicability of KrF lasers for direct drive implosions.(7) An alternate beam smoothing technique applicable to KrF lasers called Induced Spatial Incoherence has been developed at the Naval Research Laboratory(NRL). This technique achieved non-uniformity levels on the order of 1% or less. A full scale laser facility based on this technique, Nike, has recently been completed at NRL. The system consists of 56 beams which are divided into 44 beams for direct single-sided illumination of planar targets and 12 beams for production of an x-ray backlighter. The system uses several discharge and two electron beam amplifiers in an angularly multiplexed geometry to produce greater than 3 kJ of laser light in a four ns pulse at the 248 nm wavelength. The entire system has been operated and has exceeded its design specifications for energy, has achieved better that 1.2% nonuniformity in a high power beam and has demonstrated significant reliability by operating at full power for 5 shots separated by 45 min. intervals. All 44 beams are now being overlapped in the target chamber where their combined level of nonuniformity should be as low as 0.2%. These levels should enable exciting new Rayleigh-Taylor instability experiments and studies of high gain direct drive experiments.(8)

LIGHT ION STATUS

Experiments on Particle Beam Fusion Acceleration (PBFA) II at Sandia National Laboratories have focused lithium ions to power densities equivalent to 1.4 TW/cm^2 averaged over the surface of a 6 mm diameter sphere. (9) These experiments use 9 MeV, 7-9 TW, lithium ion beams to achieve specific power depositions of 800-1400 TW/gm. These beams have been used to drive hohlraum targets, either conical or cylindrical, with a 6 mm mid-plane diameter and 6 mm height. The hohlraum consists of a low-density-foam energy-deposition region surrounded by a thin gold wall. This design improves the uniformity of capsule irradiation. To date, target temperatures of 58 eV have been measured. It was demonstrated that the foam became optically thin to the x-rays and that the foam tamped expansion of the hohlraum wall. By diagnosing the incident ion beam uniformity and then measuring the radiation uniformity in the hohlraum, it was determined that the radiation uniformity was significantly better than the incident beam. This confirms that foam-filled hohlraums provide for radiation smoothing when the foam becomes optically thin.

Improvements of the lithium ion intensity is limited by a parasitic load that restricts ion power and by the total beam divergence.(10) The parasitic load is defined as the difference in the ion currents measured by gas cell B-dot monitors and by filter Faraday cups and by on-axis diagnostics. Experiments have identified the parasitic load is carried by ion contaminants, higher-Z (e.g., C,N,O,F) or molecular ions with energies >2 MeV, and by protons. Vacuum surface cleaning experiments are in progress that attempt to eliminate the parasitic load. Cleaning techniques employ plasma discharge cleaning, anode heating, additional vacuum pumping and cryogenic cathodes. Preliminary cleaning experiments on the Sabre diode have demonstrated a 2-3 improvement in the beam. Some of these techniques are being adapted to PBFA II.

Visible spectroscopy is used to measure ~10 MV/cm peak electric from the Stark-shift of neutral lithium lines, an order of magnitude higher than any previous laboratory Stark-shift measurement.(11) Fields measured near the anode surface are 7-10 MV/cm. This shows the LiF ion source to be a field-threshold emitter, rather than a space-charge limited source. Azimuthal electric field non-uniformities of 10-20% have also been measured. Possible mechanisms contributing to the non-uniformities are electromagnetic fluctuations induced by instabilities, non-uniformities in the ion emission, and nonuniform injection of electrons from the magnetically insulated transmission line or cathode tip plasmas.

High-yield target designs utilizing light ion beams require multiple beams from extraction diodes to be transported over distances of several meters onto a high-yield target. Several areas of research and technology development must be investigated before high-yield designs become possible. These include ion beam transport and focusing, generation of extracted ion beams, active ion source development such as lithium Exploding Metal Foil Active Anode Plasma Source (EMFAAPS) diodes, and 2-stage diode development. Self-pinched transport and focusing using an achromatic lens system is the baseline transport method. Generation of extracted ion beams has begun on the Sabre device and will be developed further with an extraction diode option developed for PBFA II. A lithium EMFAAPS ion source has been designed for Sabre. Initial 2-stage diode experiments have been performed on the Alias device and on PBFA II.

CONCLUSIONS

The U.S. ICF Program has coupled top notch national resources to make exciting progress in the past year towards its goal of achieving fusion ignition and gain in the laboratory. The Nova Technical Contract, a series of experiments on the

Nova laser facility, at Livermore National Laboratory, has resolved successfully the major technical issues involved in the design of an ignition target. Several designs have predicted ignition performance when illuminated by the NIF laser. NIF's conceptual design has been completed and approved and the Department of Energy has recommend start of Title I (preliminary engineering design) in FY 96, and anticipating congressional approval, a NIF project office has been established. The key technical elements of the NIF design have been successfully verified on a full-scale scientific laser beam line prototype, Beamlet, resulting in high confidence that the NIF can meet its specified goals from both a technical and budget perspective.

Significant progress has been achieved in two major facility projects with excellent performance characteristics. The OMEGA Upgrade laser, at the University of Rochester, consisting of 60 temporally shaped beams that will uniformly illuminate a spherical capsule with up to 30 kJ of 351 nm light, has been completed and is nearing operational status. OMEGA will play a key role in studying direct drive concepts. Nike, at the Naval Research Laboratory, a 56 beam KrF laser at 248 nm wavelength, is operational, has exceeded its design specifications for energy, and has achieved better that 1.2% nonuniformity in a high power beam. Nike's reliability is evidenced by operating at full power for 5 shots separated by 45 minute intervals. Nike will provide valuable data from flat foil experiments which can be well instrumented. This data may be important in improving our codes.

The light ion program, at Sandia National Laboratory, has successfully focused lithium ions to power densities equivalent to 1.4 TW/cm^2 . These experiments use 9 MeV, 7-9 TW, lithium ion beams to achieve specific power depositions of 800-1400 TW/gm. Foam-filled hohlraums have demonstrated significantly radiation smoothing and target temperatures of 58 eV have been obtained. Efforts to reduce beam divergence and cleaning techniques to reduce parasitic loads hold promise to further improve ion beam performance.

The national cryogenic target development, which includes General Atomics, Los Alamos, Livermore, and Rochester is an excellent example of the national teaming to solve a key problem. This national team is investigating beta layering to reduce surface roughness as well as techniques to monitor internal capsule roughness.

With recent declassification and the development of top notch facilities the U.S. ICF program looks forward to enhanced interactions with the entire ICF community providing valuable insights and data as ignition in the laboratory is achieved.

ACKNOWLEDGMENTS

It gives me pleasure to acknowledge the fine contributions of many participants in the US ICF Program. I especially would like to note the teams at Los Alamos, Lawrence Livermore, and Sandia National Laboratories as well as the University of Rochester, Naval Research Laboratory, and General Atomics. I also thank Tom Haill, Chris Keane, Ted Saito, and George York.

References

1. *National Ignition Facility Conceptual Design Report,* UCRL-PROP-117093, NIF-LLNL-94-113, L-169731, Livermore, CA , May 1994.

2. S.W. Haan et al., to be published in *Physics of Plasmas.*

3. E.M. Campbell, *Laser and Particle Beams* , **9**, 209 (1991).

4. Dittrich, T.R., Hammel, B.A., Keane, C.J., McEachern, R., Turner, R.E., Haan, S.W. and Suter, L.J. , *Phys. Rev. Lett.* **73**, 2324 (1994).

5. Keane, C.J. et. al. "Diagnosis of Pusher-Fuel Mix in Spherical Implosions Using X-ray Spectroscopy" submitted to *Rev. Sci. Inst..*, (1994)

6. Soures, J. "The OMEGA Upgrade Laser Facility for Direct Drive Experiments" *J. Fus. Energy* **10**, 295 (1991) and Boehly, T. et. al "The Upgrade to the OMEGA Laser System" *Rev. Sci. Inst.* **66**, 508 (1995)

7. Dahlburg, J. editor, *Nike KrF Laser Annual Report* , NRL-6730-94-264, Washington, DC June 94

8. Bodner, S.B. et. al. "Uniform Target Illumination & High Gain Direct Drive Target Performance using KrF Lasers" 15th AIEA Conference on Plasma Physics and Controlled Nuclear Fusion Research, Seville Spain, *Nuclear Fusion,* 1994.

9. Filuk, A. B., et al., "Progress in Ion Beam Power Coupling and Ion-Driven Hohlraums on PBFA II," *Proceedings IAEA Technical Committee Meeting on Drivers for Inertial Confinement Fusion*, Paris, France, November 1994.

10. Mehlhorn, T. A., et al., "Progress in Lithium Beam Power, Divergence, and Intensity at Sandia National Laboratories," *Proceedings of the Tenth International Conference on High Power Particle Beams,* San Diego, CA, June 1994.

11. Bailey, J. E., "Measurements of Acceleration Gap Dynamics in a 20-TW Applied-Magnetic-Field Ion Diode," *Phys. Rev. Lett.,* 74 (10) 1771, March 6, 1995.

Progress of Laser Fusion at ILE and ICF Research in Japan

S. Nakai

Institute of Laser Engineering, Osaka University
2-6 Yamada-oka, Suita, Osaka, 565 Japan

Abstract. The progresses of laser fusion experiment by GEKKO XII and technology developments for inertial fusion energy in Japan are reviewed. New strategy of IFE addressing to reach the goal of IFE, power plant, is described.

1. INTRODUCTION

The progress of laser fusion research by GEKKO XII is summarized in Table 1. Since the completion of GEKKO XII and its first shot in December 1983, the experimental campaign on the high temperature demonstration and the high density demonstration have achieved the respective goal plasma parameters and revealed the physics of implosion. The achieved plasma parameters such as temperature [1] and density [2] have reached respectively to the conditions which are required for the fusion ignition. The high density compression to the density more than $600 g/cm^3$ was really the breakthrough which gave us the confidence of laser fusion as an feasible approach toward fusion energy development.

The research program at Institute of Laser Engineering have turned to new phase, that is Ignition and Burn Demonstration where the hot spark formation surrounded by cold main fuel is essential.

Examining and reviewing the recent achievements on the implosion physics which is mainly concerned with stability issues, and on the high power laser technology concerned with uniformity improvement, ILE has started an intensive research and development (R&D) project addressing a number of physics and technical issues related to ignition and burn. They are ① development of new concept on simulation code and numerical experiment of the overall implosion process, with the reinforcement of code and numerical modeling by improved and sophisticated experiments with New GEKKO XII, and if possible, with super uniform implosion facility, ② ignition and burn physics investigation to demonstrate high gain, and ③ conceptual design study of laser fusion reactor and technology development for key issues such as driver, fuel pellet fabrication and reaction chamber.

One beam proto-model for KONGHO laser which can deliver 8 kJ/ω 3ns out put is under construction and will be completed in 1995. This provides us the basic technologies for the design and construction of a few hundreds kJ system for ignition and burn physics investigation and MJ system for high gain demonstration. The replacement of power amplifiers with those of LD pumped high repetition specification will lead to the driver for Engineering Test Facility.

TABLE 1. Progress of implosion by GXII and laser fusion research at ILE, Osaka University.

FY	Achievement	Driver	Pellet & Diagnostics	Theory & Simulation
	Interaction and Ablation			
1983	$Y_n=10^{10}$	Completion of GEKKO XII	Glass micro-balloon (GMB)	Hot electron
	High Temperature Demonstration (Stagnation free comp.)			
1984	η_{abs} & P_{abl}	SHG	LHART	
1985	$Y_n=10^{12}$			Shock multiplexing
1986	$Y_n=10^{13}$ $T_i=10$keV		Fusion yield calibration	
	High Density Demonstration (Low isentropic comp.)			
1987		THG	CD shell	
1988		Random phase plate	CDT shell	Nonlocal transport
1989	$\rho=600\times\rho_s$ (main fuel)		Sec. reaction p knock-on Si activation	
1990	Fermi degeneracy			2D, 3D codes
	Japan-US workshop for high density comression			
	Ignition and Burn Demonstration			
1991	Thermal smoothing	Partially coherent light	Cryogenic target	
1992	RT instability R-M instability		80ps x-ray frame camera	α heating
1993		new GEKKO XII		turburent mixing
1994	stabilized implosion	improvement of irradiation uniformity	10ps x-ray frame camera	ignition & high gain scaling
1995	initial imprinting	one beam proto-model with 350mm disk amplifier.		

2. RECENT PROGRESSES OF LASER IMPLOSION

Since 1991 when the main direction of research program was turned aiming at ignition and burn investigation, various new research and developments have been proceeded in pellet implosion and plane target experiments, new diagnostics, cryogenic target with better uniformity, driver technologies, and theoretical modeling and numerical simulation. The results and achievement have been reported at this Conference and presented in this proceeding.

2.1 Pellet implosion and plane target experiments

The physical processes which are related to the implosion stability are schematically shown in Fig. 1. The elementary processes such as initial imprint, shock induced instability, thermal smoothing, Richtmeier-Meshkov (R-M) instability and Rayleigh-Taylor instability have been experimentally investigated by using plane target. The experimental results are compared with numerical modeling and computer simulation to obtain the quantitative and generalized data base which can be installed in simulation code to increase the reliability and credibility of the simulation.

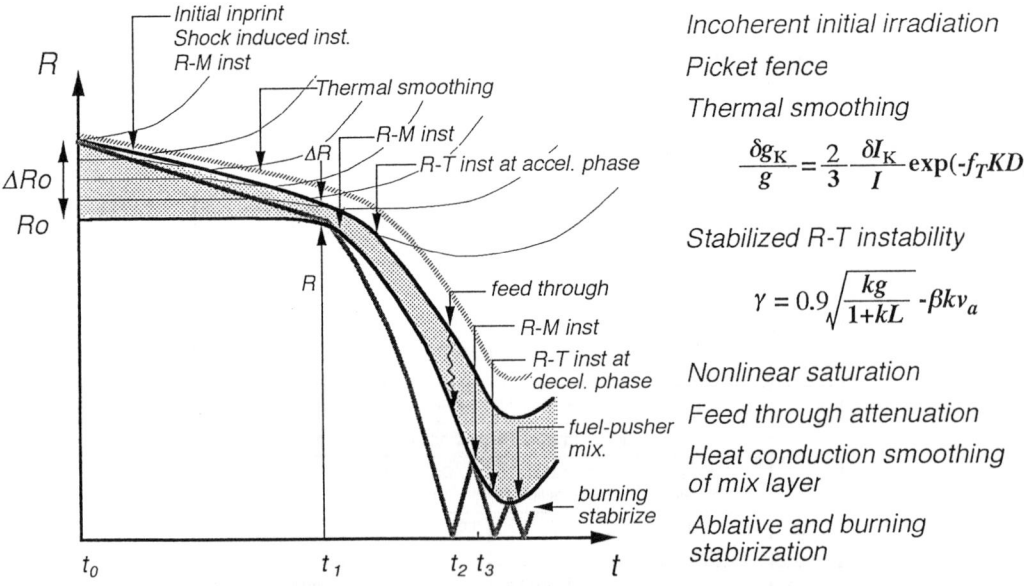

FIGURE 1. Implosion and physical processes related to the stability

As for the stability of spherical implosion, we have performed advanced experiments with the following improvements in (1) driver: irradiation uniformity improvement and pulse shape control, (2) fuel pellet: cooled plastic shell with variable filling gas pressure, which can control the convergence keeping the quality of the pellet good enough for the stability investigation, and (3) new diagnostics: Multi-Imaging X-ray Streak Camera (MIXS) to observe the dynamic behavior of the compressed core with time and space resolution of 11 ps and 15µm respectively, temperature mapping of compressed core by energy resolved x-ray imaging with bent crystal microscope, and advanced Kirkpatrick-Baez X-ray microscope having a 3µm resolution to see the micro-structure of compressed core.

Through these experiments and quantitative comparisons with the simulation results, it can be concluded that the pellet implosions have been improved closer to 1D hydrodynamics with better uniformity, and that neutron yield reduction with increased converging ratio of direct drive configuration seems to be comparable with indirect drive configuration [3], [4]. The important achievement is the demonstration of stability control by changing the pulse shape of driver as shown in Fig. 2.

2.2 Driver development

GEKKO XII has been renewed in 1993~94 mainly being aimed at improving the irradiation uniformity on the target [5]. The new functions which were added to GEKKO XII are ① beam smoothing, ② power balance control of twelve beams, ③ precision waveform monitors on all the twelve beams, and ④ diagnostics of intensity distribution on the target.

For the quantitative diagnostics of the improvement of implosion characteristics, a new laser chain, which generate synchronized high intensity, short pulse laser pulse, has been designed and the construction are scheduled to complete in 1995 fiscal year. The design of this new chain is identical to the one beam of KONGOH laser, and is considered to be proto-model of the full system.

2.3 Numerical simulation and gain scaling of laser fusion implosion

Figure 3 shows the pellet gain scaling of direct drive scheme predicted by 1D simulation. The ignition energy is expected as low as order of 100 kJ for an implosion mode of high velocity, and high inflight aspect ratio. However on actually required energy increases due to various 2D effects. Figure 4 shows the gain reduction normalized by the value of the 1D simulations as a function of the laser irradiation nonuniformity at high gain region (a) and at marginal ignition region (b), which were evaluated for low l mode non-uniformity.

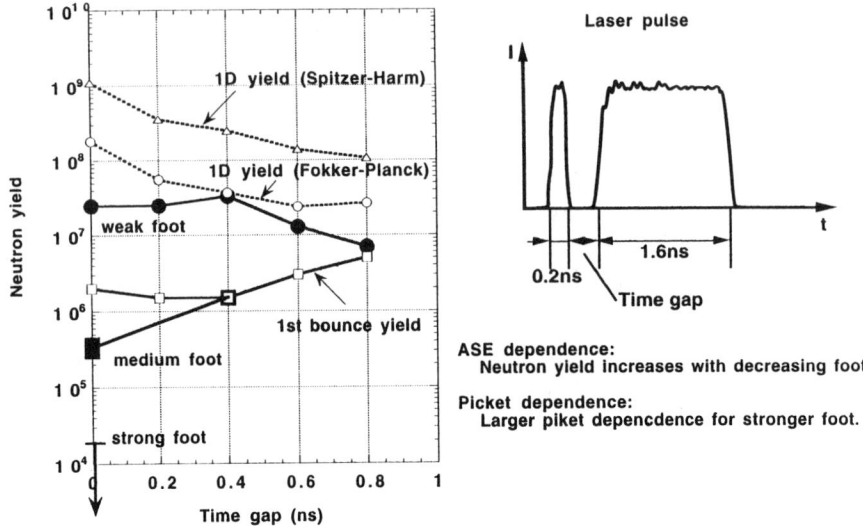

FIGURE 2. Neutron yield vs. foot-pulse-intensity and picket-time-gap

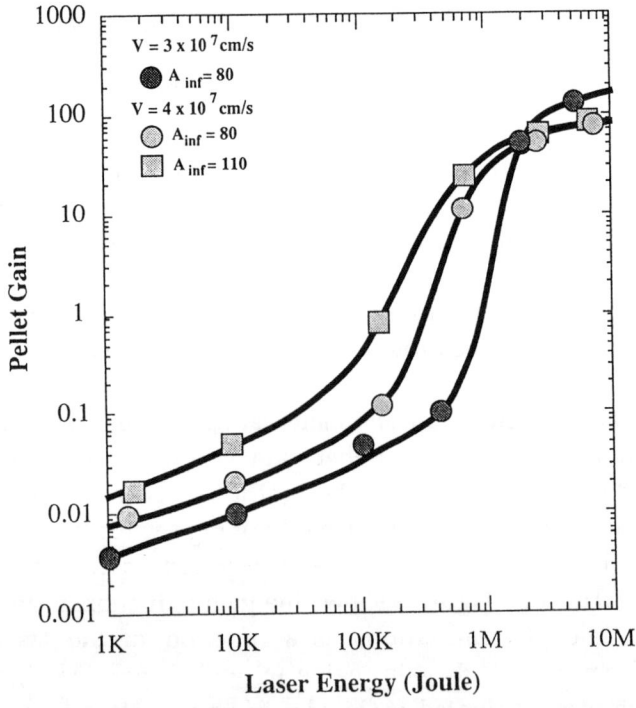

FIGURE 3 Pellet gain scaling on laser energy with different implosion mode

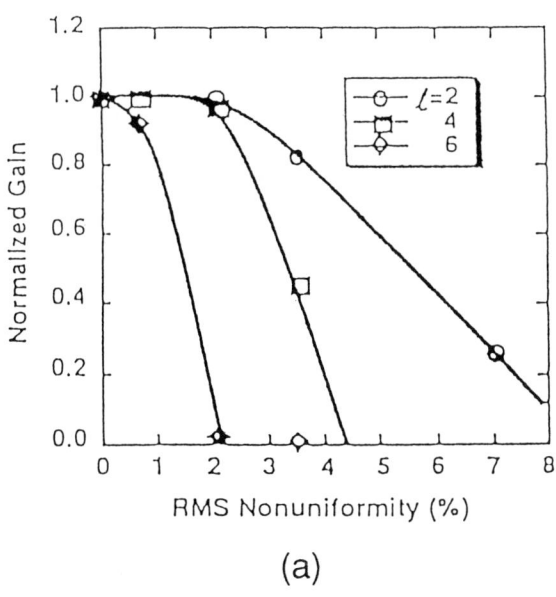

(a)

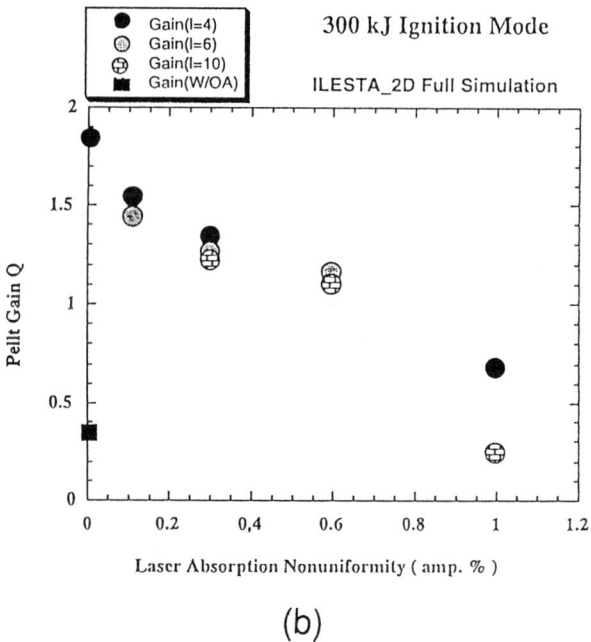

(b)

FIGURE 4. Pellet gain sensitivity with the irradiation nonuniformity.

In order to evaluate the effect of short-wavelength instability, we have developed a turbulent mixing model, the results of which have been compared to the experimental results, and good agreement have been obtained.

For full implosion simulation which contains low, medium and high l mode instability and turbulent mixing at the plasma condition where radiation and nonlocal energy transports are effective, we have to investigate new physical models which have to be examined by experimental results [6].

3. STRATEGY OF IEF TOWARD FUTURE

The strategy of IFE has been re-examined including the recent progress of implosion physics and technology development. It consists of three main programs aiming to the main goals of
(1) basic physics investigation,
(2) high gain demonstration, and
(3) development of reactor technology .
The flow chart of these programs to reach the final goal "Commerical Power Plant" is shown in Fig. 5.

3.1 basic physics investigation

This program consists of two main parts of (1) modeling and numerical code development and (2) basic fusion plasma and hydrodynamic experiments. New GEKKO XII which was improved in the irradiation uniformity and pulse shape controllability is the main facility for the advanced implosion experiment in the next few year. New facility which have the capability of ultra-uniform irradiation is important to facilitate basic physics experiment and also to develop new concept of laser driver technology for future program.

3.2 high gain demonstration

The progress of LD (Laser Diode) pumped solid state laser is very rapid and goes beyond our expectation. It gives us feasible prospect toward reactor driver in the technical and economical aspects.

The main point of the modification of strategy is to introduce the LD pumped SS laser technology into the system even for the near term facility. By this modification, ignition and high gain demonstration facility can have the technical continuation toward the reactor driver in the future.

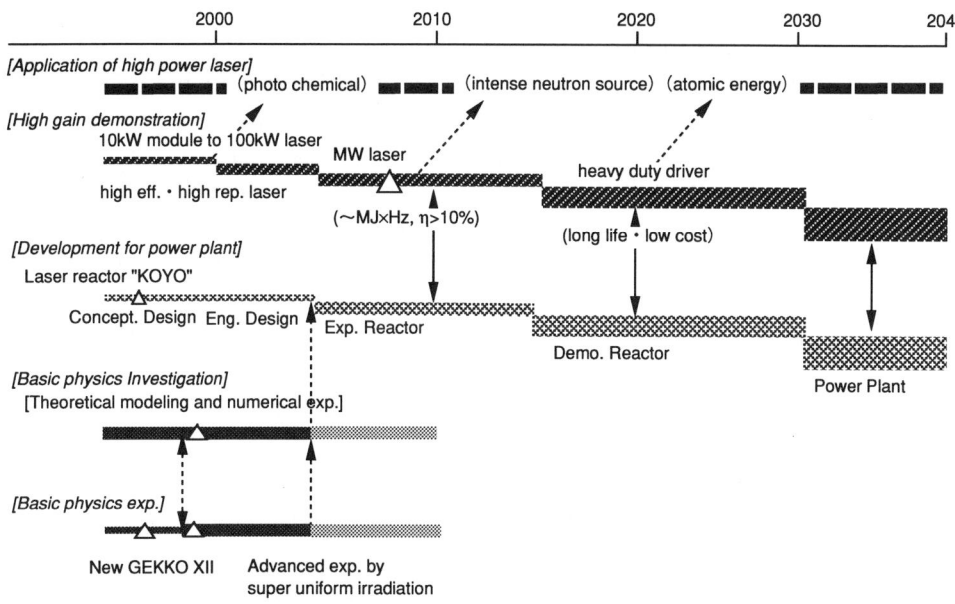

FIGURE 5. Development strategy for inertial fusion by laser

The main target of this channel is the development of driver technologies. We have already succeeded in developing high power LD technology and achieved a stacked LD of 1 cm^2 which can deliver 3kW/cm^2 with a efficiency of 50%. A small scale MOPA system has been constructed and tested to cralify the technical issues of LD pumped SS laser in the repetitive operation with high overall efficiency.

The development of 10 kJ module is the basic step to proceed. The construction of a few hundreds kJ system and MJ system will follow depending on the development program in implosion physics issues and engineering development issues toward the final goal, that is the demonstration of high gain with the driver and reactor chamber which have the technical continuation toward demonstration power plant.

3.3 Development of reactor technology

Inertial fusion reactor differs essentially from magnetic fusion reactor in its system integration and also in its basic technologies which consist the power plant system. IFE power plant consist of (1) driver system, (2) reactor chamber, (3) fuel pellet system, and (4) remainder. The most important feature of IFE system is the separability of the subsystems and frexibility in optimization of performance of the total system. It should also be noted that the subsystems can be developed separately and in paralell each other and then can be combined to construct total power plant system.

The conceptual design study on laser fusion power plant "KOYO" was started in 1991 and was completed in 1993 at its first phase. The interim report was publish in 1994 [7]. For this design study, a Laser Fusion Reactor Design Committee has been organized, which is a joint committee of members from universities, national research institutes and industries. Through the activity of the committee for the design studies on "KOYO" , a collaborative research network on IFE reactor technology has been formed. It could be formalized as followed

Driver	DPSSL
	Excimer gas laser
	Ion beam
	Interface between driver and chamber
Fuel pellet	Design
	Fabrication and Handling
	(Injection and tracking)
	Tritium handling
Reactor Chamber	Structural design
	Liquid first wall and blanket
	Heat and Nuclear design
Remainder	System
	Power generation
	Miseranious

52 members from 12 universities, 4 national laboratories and 8 industrial companies have joined to above working. New network organization is now under consideration toward detail design studies.

4. INTERNATIONAL COLLABORATION

International collaborative work to review the present state and future prospect of IFE was started in 1991 under the auspice of IAEA. Three years working with the collaboration of 81 specialist from all over the world completed the review in 1994 and the report has been publish in April 1995 from IAEA [8]. This is the first products of world wide collaboration in IFE which has been completed with the individual dedication and with the desire to develop fusion energy for the future human beings on the earth.

It has been recognized through the collaborative review work that IFE research and development has reached the stage where international joint project could be organized for the efficient and rapid progress toward IFE realization .

Two major approaches to fusion energy, Magnetic and Inertial fusion, should be proceeded in parallel keeping the open discussions and collaborative interactions for our common goal of developing new energy which is safe and environmentally clean, abundant, and economically reasonable. The role of IAEA is becoming more and more important.

REFERENCES

1. Yamanaka, C., Nakai,S. : Nature 319, (1986) 757.
2. Nakai, S. :Bull. Am. Phys. Soc. 34, (1989) 2040
   Nakai, S., et al., : 13th Int. Conf. on Plasma Phys. Controlled Nucl. Fusion Res., Washington DC, USA, Oct. (1990), paper No IAEA-CN-53/B-1-3.
   Azechi, H., et al., : Laser and Particle Beams 9, (1991) 193.
3. Kato, Y., et al., : Implosion Experiments at Uniformity-Improved GEKKO XII : Overview, 12th Int. Conf. Laser Interaction and Related Plasma Phenomena, Osaka, Japan, April 24-28, 1995, Paper No. FrIa-4.
4. Shiraga, H., : Improvement of the imploded core performance with uniform GEKKO XII green laser system ibid FrIa-5.
5. Nakatsuka, M., et al., : Improvement of irradiation uniformity of GEKKO XII -power balance-, ibid TuI-1.
6. Takabe., H., : Design and stability analysis of ignition and high gain targets, ibid TuII-2.
7. Nakai, S., Mima, K., Kitagawa, Y., Sakabe, S., "Conceptual Design of Laser Fusion Power Plant 「KOYO」", Report of Committee on Laser Fusion Reactor Technology, Institute of Laser Engineering, December (1994).
8. Hogan, W. J., Coutant, J., Nakai, S., Rozanov, V.B., Velarde, G., : Energy from Inertial Fusion IAEA Vienna, (1995).

Overview of ICF Program at Centre D'Etudes de Limeil-Valenton

CEL-V LASER TEAM

Centre d'Etudes de Limeil-Valenton
94195 Villeneuve Saint Georges Cedex

Abstract : The major objectives of the CEA-DAM laser program is to determine the various requirements to achieve thermonuclear fusion in laboratory. We report here recent results obtained at Centre d'Etudes de Limeil-Valenton on high density X-Ray implosions, radiative transfer processes, hydrodynamic instabilities and laser-plasma interaction involved in cavity physics.
Ignition and a moderate gain appears to be achievable with a laser energy of about 1.5 - 2 MJ delivered at $\lambda = 0,35$ µm with a shaped pulse (duration~ 16 ns). The construction of such a laser is realizable and a conceptual design is under preparation.

INTRODUCTION

A major objective of the laser program in which the CEA/DAM has engaged since 1962 is to ignite and burn a small sphere of deuterium-tritium (DT) [1].

The indirect drive has been chosen as the most promising way to reach the level of irradiation uniformity requested for high gain implosions : in this scheme, the laser radiation enters a casing containing the DT sphere ; a series of absorption-reemission processes give rise to a nearly Planckian X-ray radiation which implodes the capsule.

The determination of the laser energy needed to get ignition and combustion requires a deep knowledge of hohlraum and implosion physics, to the study of which the experimental programs of PHEBUS and OCTAL are mainly devoted.

INDIRECTLY DRIVEN IMPLOSIONS

X-ray driven implosion are currently performed using the two beams laser facility Phebus. In spite of the small number of laser beams, representative implosions are obtained using the target geometry given on figure 1. The laser light is focused on two shields apart the capsule. These shields are gold disks or cones and they act as X-ray converted. A fraction of laser energy is also refracted toward the cavity walls and contributes to the hohlraum heating. We can notice that this target presents some similarity with advanced targets for heavy ions ICF [2].

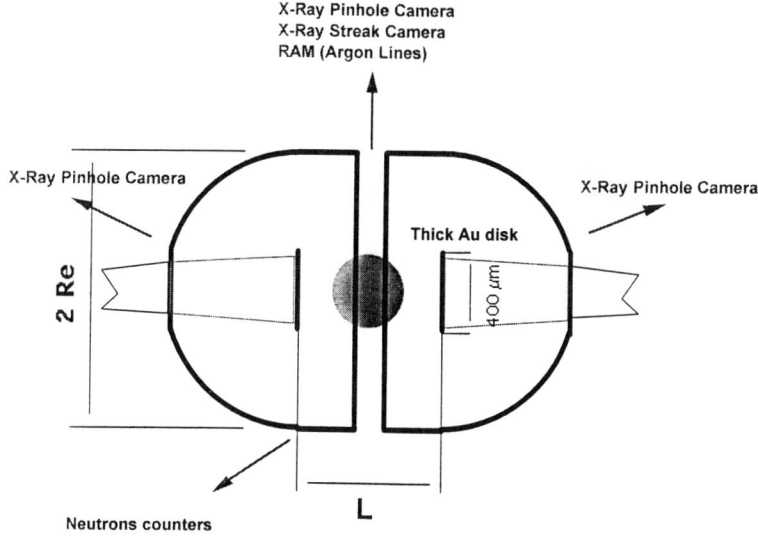

FIGURE 1 : target geometry and diagnostics for indirect drive experiments. For most shots L/2Re = 0.6. Conical X-ray converters are also used.

As the experiment is axially symmetric, the results can be compared to 2D Lagrangian numerical simulations. The objectives of experiments are to examine the energy transfer efficiency and measure the radiation temperature around the capsule. Another point is to control the capsule irradiation symmetry and to check the influence of hydrodynamical instabilities generated either by irradiation non-uniformities or capsule roughness. These capsules are DT filled - CH coated glass microballoons or D_2 + Ar in plastic shells.

Specific experiments were design to infer the temperature distribution along the hohlraum wall. We have used a cavity without microballoon and with an observation slit between the shields. One side of the slit was covered by a thin gold foil. The slit was imaged by a soft X-ray (hν ~ 200 eV) streak camera. From the observation through the uncovered part of opposite wall, it is possible to get a measurement of temperature uniformity. In the example given figure 2 (disk converter and 1.3 ns laser pulse), in early times the temperature is, as expected, higher near the heated disks, but about 300 ps after the beginning of the laser pulse, it is almost uniform along the slit.

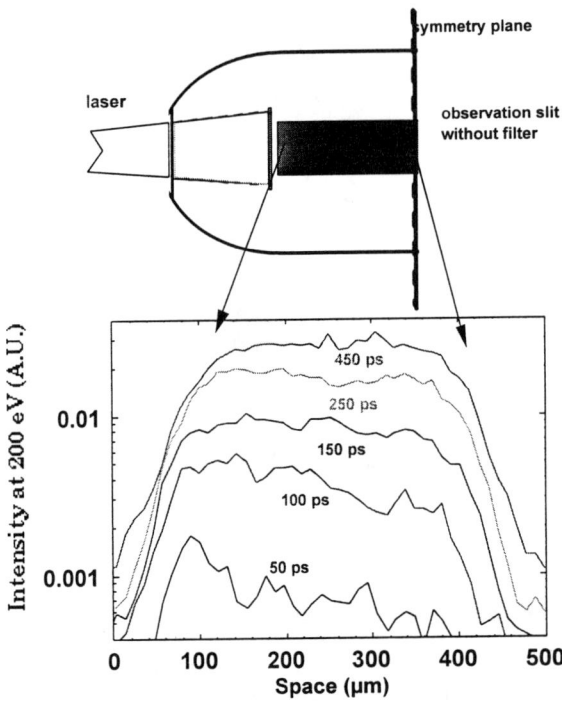

FIGURE 2 : time and space resolved soft X-ray from the inner wall of the cavity, observation is done through a slit along the target.

The radiation temperature of the cavity wall is deduced from the transit time of the Marshak wave through the gold foil [3]. Temperatures of the order of 180 eV are obtained in 1 mm 1D cavities (fig. 3).

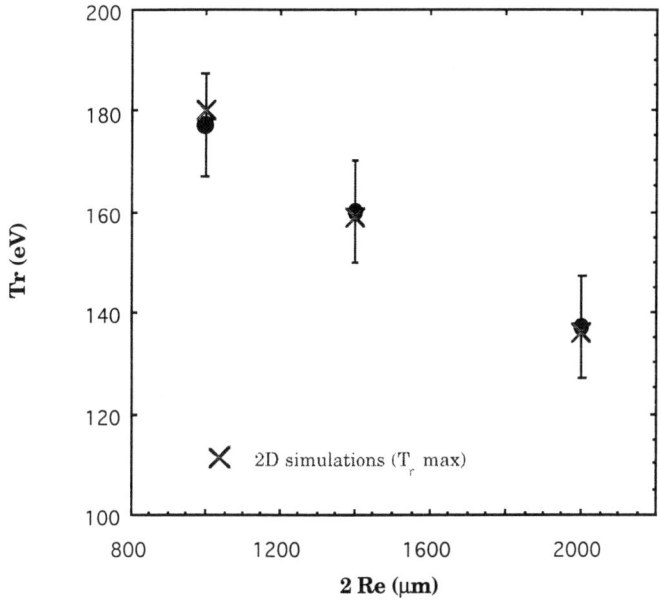

FIGURE 3 : comparison of experimental and calculated cavity temperatures

For implosion experiments, measurements include implosion time, neutron yield and X-ray imaging of compressed fuel. These experimental results are compared to 2D numerical simulations (FCI-2 code). FCI-2 is a Lagrangian code with Monte-Carlo radiation transfer and non-LTE atomic physics [4]. A flux limiter f = 0.03 is used for electron conduction.

The agreement between experimental and calculated implosion time insure that the radiative transfer in the cavity is correctly modeled. The X-ray emission of the compressed capsule is post-processed and compared to the time integrated or gated pinhole images to get further information on capsule irradiation symmetry. Calculations show that the dominant modes are l = 2 and 4, and that their coefficient γ are time dependent. Fig. 4 give the temporal evolution of γ_2 and γ_4 for 2 geometries (2 Re = 1000 µm, 2 Rm = 300 µm ; 2Re = 1400 µm, 2 Rm = 350 µm with disk converters in both cases, where Rm stands for the capsule radius). As expected the best symmetry is for the lowest Rm/Re ratio ; by the end of laser pulse the uniformity is of the order of 5 %.

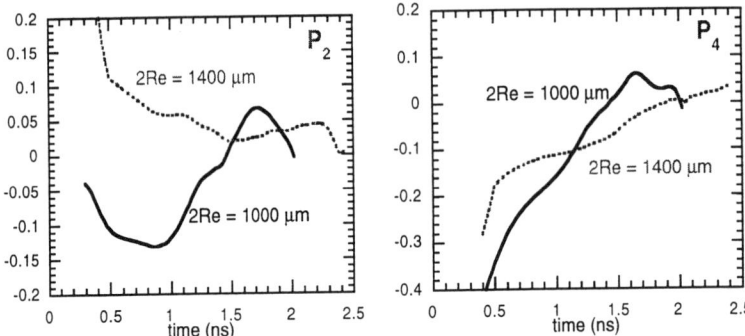

FIGURE 4 :Implosion experiments with PHEBUS ; temporal evolution of P_2 and P_4 modes for two hohlraum dimensions.

Neutron yield of DT-filled glass microballoons is sensitive to the symmetry of the implosion. Keeping the drive conditions unchanged we have varied the implosion convergence ratio (CR = initial DT radius / final radius) by using different fill pressure.

Measured yield is compared to 1 and 2D simulations for 2 pressures (5 and 30 bars) on figure 5. In both cases 1D calculations overestimate the neutron emission. In 2D simulations the calculated yield is reduced due the fuel-pusher distortions induced mainly by P_4. A good agreement is then obtained for the high pressure, low convergence experiments, but the 2D calculation is still 2 orders of magnitude too high for low fill pressure cases. The discrepancy may be attributed to the effect of short wavelength non-uniformities (interface roughness) not considered in the simulations.

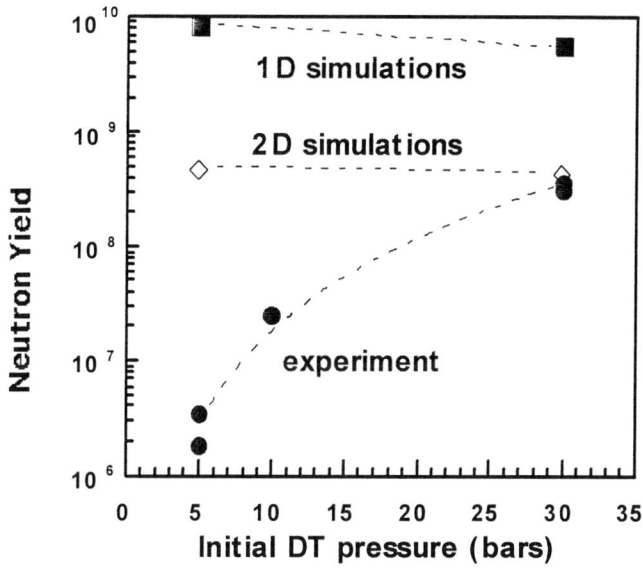

FIGURE 5 : Effect of initial DT pressure on neutron yield

HYDRODYNAMIC INSTABILITIES

A most important issue in high gain implosion experiments is the development of hydrodynamic instabilities, and particularly Rayleigh-Taylor instabilities which appear at the ablation front of accelerates targets [5]. Experiments have been performed at CEL-V in collaboration with the Imperial College (U.K.), to verify if the stabilization by ablation process occurs for small modulation wavelengths (5-30 µm) [6]

1 - Ablation front instability experiments in x-ray drive

The rear side x-ray emission of a thin gold converter is used to accelerate CH or CHBr targets with thickness between 6 and 30 µm. The x-ray irradiance is of the order of 10^{14} W cm^{-2}. The target acceleration is measured with a good accuracy by means of a side-on x-ray backlighting system at 200 eV with a spatial resolution of 2 µm. Acceleration as high as $1.6 \; 10^{16}$ cm s^{-2} is measured for 20 µm thick CHBr targets.

The growth rates are measured by using a front x-ray backlighting system at 1.5 keV with a 10 µm spatial resolution. A typical result is presented in figure 6. The thickness of the CH target is 28 µm. The wavelength of the sinusoidal modulations is 50 µm and the initial amplitude is 1.2 µm. The growth rate deduced from the temporal evolution is 1.2 ns^{-1}. This value is slightly lower than the expected value either from the Takabe model[6] or from the formula 0.6 $(Akg)^{1/2}$.

First results have been obtained for a small modulation wavelength (12 µm) with another backlighting system and a better resolution. The growth rate is of the order of 2 ns^{-1}. So, the smoothing expected from the Takabe model for the small wavelenghts seems to be not verified.

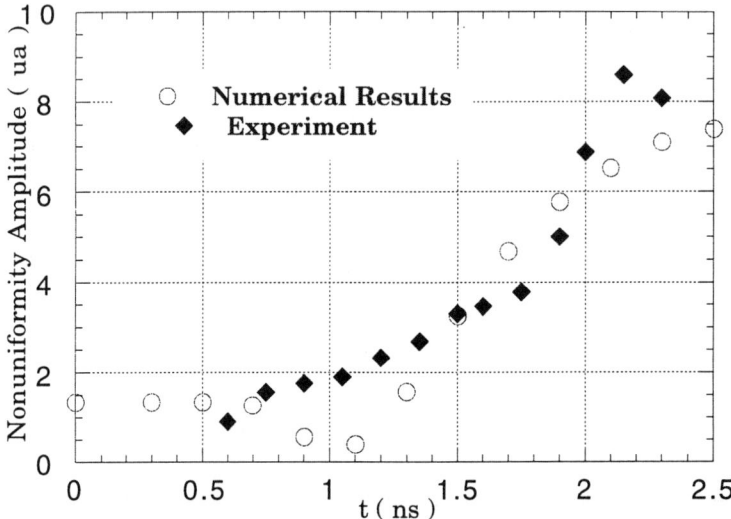

FIGURE 6 : Ablation front instability. Temporal evolution of the amplitude of the modulation. CH target thickness : 28 µm. Modulation wavelength : 50 µm. Initial amplitude : 1.2 µm

2 - Numerical simulations

The ablation front instability is simulated by means of the purely Lagrangian two dimensional FCI2 code. The growth rate is deduced either from the temporal evolution of mass variations through the target or from the temporal evolution of the position of the isodensity $\rho_{max}/2$. A result is presented in figure 6 for x-ray driven targets with the initial conditions of the experiment. First, the modulation amplitude decreases during the shock transit time. Then, the amplitude increases during the acceleration. The increase is slower after the end of the x-ray pulse. Such details are not precisely seen experimentally, but the general accordance is quite good.

Several sets of numerical simulations have been performed in direct drive. A typical result is shown in figure 7. The isodensity contours are plotted for a 10 μm modulation wavelength and a 0.1 μm initial amplitude. The target thickness is 25 μm, the laser irradiance is $8\ 10^{14}$ W cm^{-2}. The beginning of the bubble-and-spike structure can be seen.

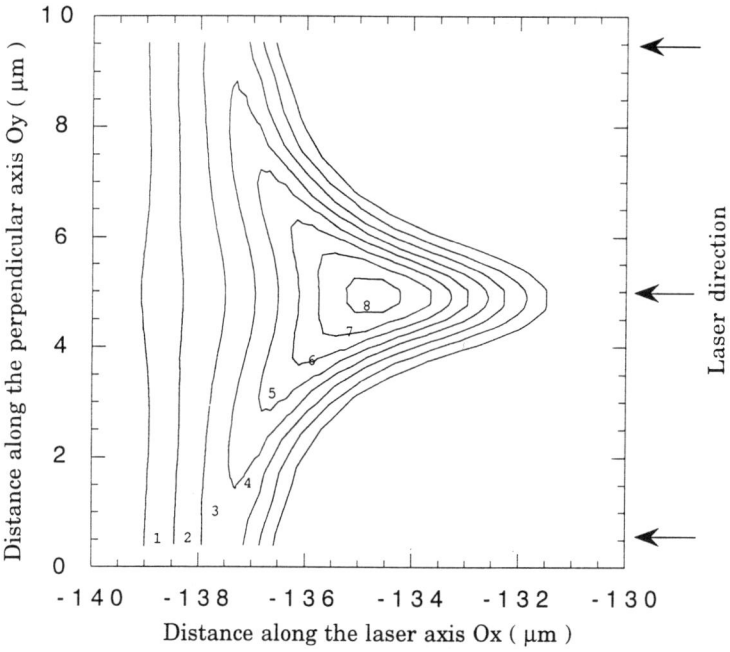

FIGURE 7 : Numerical simulation of the ablation front instability in direct drive. Isodensity contours at 1.5 ns from 4 to $8.66\ 10^{23}$ cm^{-3} by step of $0.66\ 10^{24}$ cm^{-3} for 25 μm thick CH targets and $8\ 10^{14}$ W cm^{-2} irradiance. Modulation wavelength : 10 μm. Initial amplitude : 0.1 μm.

The growth rate is deduced from the mass modulations through the target First, a fit of the numerical results to the Takabe model is obtained for 12 μm thick targets and $3\ 10^{16}$ cm s^{-2} acceleration. The growth rate is reduced to $(A k g)^{1/2} - \beta k V_a$ with $\beta = 3 - 4$ as generally deduced elsewhere [6].

For 25 μm thick targets and $1.5\ 10^{16}$ cm s^{-2} acceleration, the same fit using the same value of β is impossible, as shown in figure 8. The values deduced from the Takabe model with $\beta = 4$ are largely lower than the classical values. The model by Munro [7] taking into account a length of the unstable region reaching 8 μm leads to intermediate values. The results of the numerical simulations give a growth rate which is 0.6 times the classical value. So, we do not observe a smoothing of the ablation front instability for the small wavelengths. Precise experimental results in this range are necessary.

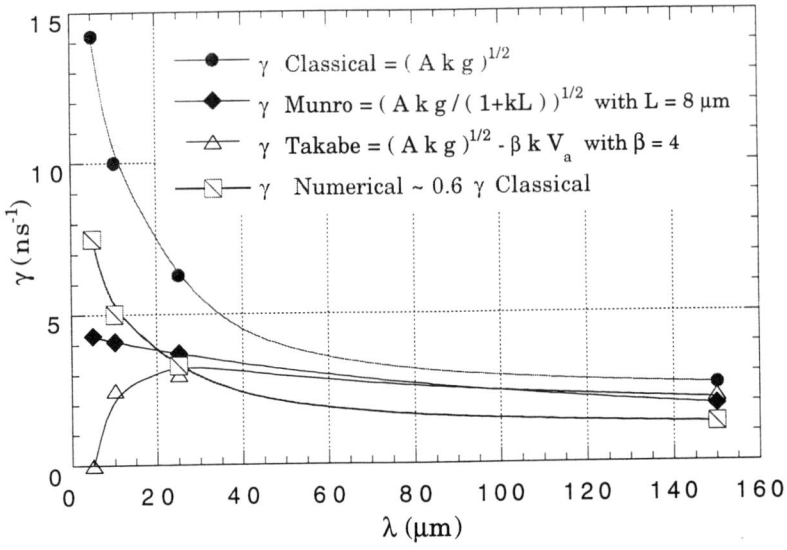

FIGURE 8 : Growth rate of the ablation front instability versus modulation wavelength for 25 µm thick CH targets and 8×10^{14} W cm^{-2} irradiance.

STIMULATED RAMAN SCATTERING

Parametric instabilities such as Brillouin and Raman scattering can appear during laser interaction with the low density plasma which fills up the cavity. Their development is expected to depend on the uniformity (hot spots above threshold) and spectral width of the laser radiation.

Backward stimulated Raman scattering (SRS) experiments have been performed using the optical fiber smoothing implemented on the high-power Phebus laser facility [8]. The interaction took place in low-Z plasmas presenting either exponential density profiles (solid targets) or Gaussian-type profiles (subquarter-critical plasmas). The experiments carried out at a laser wavelength of 0.53 µm show that high Raman reflectivity, up to 10%, is produced whatever the density profile may be. They indicate on the one hand that the Raman instability is not sensitive to the optical-fiber smoothing in large subquarter-critical plasmas. On the other hand, smoothing becomes effective in exponential density profiles. Figure 9 shows that the SRS reflectivity sharply decreases from 10^{-2} to 10^{-6} as the average intensity of the smoothed beam is reduced from 1.5×10^{15} W/cm^2 to

4.10^{14} W/cm², whereas it remains at a few percent in this intensity range and in the standard configuration.

The measurements could be interpreted as follows [9] : first, the spectral bandwidth, $\Delta\omega/\omega = 3.10^{-3}$ at 0.53 µm, is too narrow to affect SRS in our experimental conditions, i.e. the temporal incoherence of the laser pump is too large to alter the backward-SRS growth-rate ; this is in agreement with theoretical predictions. Second, the 2D spatial smoothing of the laser beam significantly improves the intensity distribution in the focal spot : the modification of SRS behavior in plasmas characterized by an exponential profile may be due to the reduction of the hot spots, of induced filamentation and/or of localized microstructures inside the plasma. However, the improvement of the intensity distribution is not sufficient to reduce SRS in more homogeneous Gaussian density profiles, because of the low threshold in these plasmas.

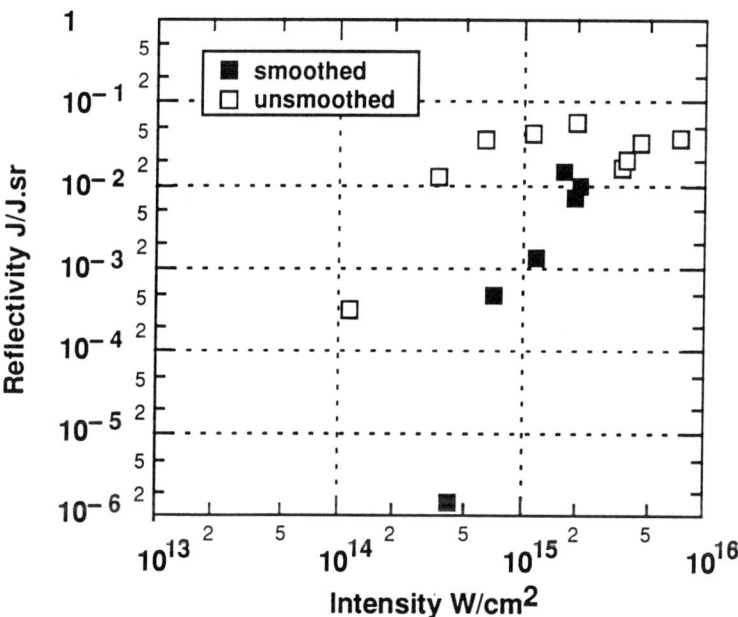

FIGURE 9 : Comparison of the Raman efficiencies obtained at 0.53 µm laser wavelength with and without smoothing, as a function of the laser intensity in CH solid target plasmas (exponential profiles).

IGNITION AND GAIN - LMJ PROJECT

The laser energy required to reach DT ignition and obtain significant gain from X-ray driven targets has been evaluated from numerical simulations with the Lagrangian FCI1 code, and also an analytical model which takes into account four types of processes :

- hohlraum physics,

- Capsule implosion, described in spherical geometry by means of a steady-state ablation model ;

- Hot spot formation

- Burn propagation, and release of thermonuclear energy.

In order that such a description leads to realistic target configurations, several phenomenological criteria have been included, which limit the extend of parameters such as the fuel preheat and the in-flight aspect ratio of the capsule.

Ignition and a moderate gain appears to be achievable with a laser energy of about 1,5 - 2 MJ delivered at $\lambda = 0,35 \mu m$ with a shaped pulse (duration ~ 16 ns). [10] The basic scheme of this Nd glass laser (LMJ) is composed of 240 square beamlets frequency converted at the third harmonic. The high energy level which is wanted cannot be obtained using usual Master Oscillator Power amplifier schemes (MOPA). In order to decrease the laser cost, new concepts, particularly multipass techniques, have been developed in close collaboration with the Lawrence Livermore National Laboratory (USA, CA).

The Optical Pulse Generation System is based on four tunable single-frequency fiber oscillators working at 1.0523, 1.0533, 1.0543, 1.0553 µm ; each wavelength is phase modulated to reduce stimulated Brillouin Scattering in the final optics, and then split in 60 ways to generate one signal for each beamline.

The amplifying section will be composed of 240 4.pass full size cavities fitted for 40 x 40 cm² beamlets. As Pockells cells and polarizors are an issue for such large sizes, we have designed the so-called "L-Turn" scheme (Fig. 10). After input, the pulse goes through different pinholes at each pass. The pick off mirror deflects the beam into the L-Turn after the second pass at a relatively low energy (200 J). The pulse is then injected in the third pinhole for passes 3 and 4 in order to achieve an output energy above 17 kJ. The output pulse then propagates up to the KDP crystals and is focused on the target after third harmonic conversion. The L-Turn concept has been successfully tested on the full scale LLNL beamlet.

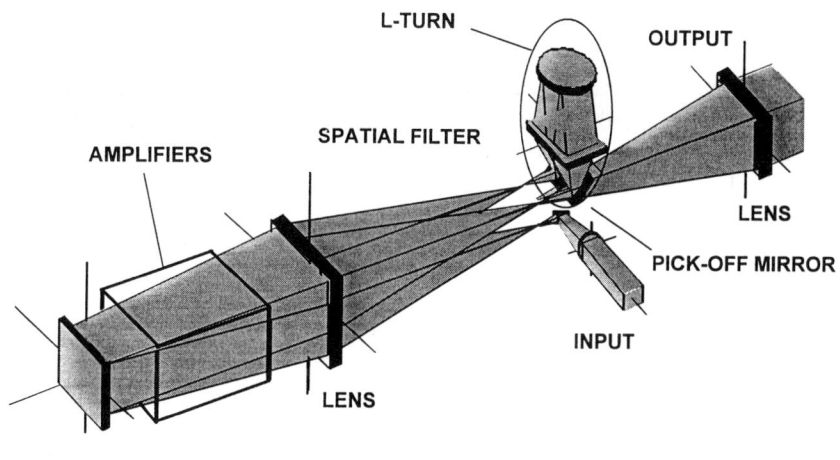

FIGURE 10 : Principle of L-Turn concept

The construction of such a facility raises technology issues : significant improvements have to be obtained in several domains, for example : 1.05 and 0.35µm high damage threshold coatings, large size high quality KDP and DKDP crystals, polishing of square optics. A deep concern is low cost fabrication, as LMJ will need high number of optical components ; industrial companies are already involved in developing new techniques for material construction.

CONCLUSION

Significant progresses in Inertial Confinement Fusion have been obtained at Centre d'Etudes de Limeil-Valenton. High densities have been realized in X-ray driven implosions, owing to the good irradiance uniformity the indirect scheme is proper to provide.

Theoretical works have shown that fuel ignition and moderate target gain would be obtained by this concept with about 1.5 - 2 MJ of laser energy.

The construction of such a laser appears achievable, and a conceptual design is under preparation.

REFERENCES

[1] CEL-V laser researches group, Twenty two years of laser-matter work at CEL-V *Nucl. Fus* 25 n°9 (1985) 1333

[2] Murakami M., Meyer-Ter-Vehn J., *Nucl. Fusion* 31, 1315 (1991)

[3] Marshak R.E., *Phys. Fluids* 1, 24 (1958)

[4] Busquet M., *Phys. Fluids* B5, 4191 (1993)

[5] Kilkenny J.D. et al., *Phys. Plasmas* 1, 1375 (1954)

[6] Takabe H., et al., *Phys. Fluids* 28, 3676 (1985)

[7] Munro D., *Phys. Rev.* A38, 1433 (1988)

[8] D. Veron, G. Thiell and C. Gouedard, *Optics Commun.* **97**, 259 (1993).

[9] C. Rousseaux, B. Meyer and G. Thiell, *"Stimulated Raman backscattering with and without optical fiber smoothing technique in 0.53 µm laser-created plasmas"*, accepted for publication in *Phys. Plasmas* (1995).

[10] M. Andre et al, *Progress in ICF physics at CEL-V* European Conference on Laser Interaction with matter PARIS (1993).

Path To Ignition: US Indirect Target Physics

M. Cray[1] and E. M. Campbell[2]

[1]*Los Alamos National Laboratory, Los Alamos, New Mexico 87545 USA*
[2]*Lawrence Livermore National Laboratory, Livermore, California 94550 USA*

Abstract. The United States ICF Program has been pursuing an aggressive research program in preparation for an ignition demonstration on the National Ignition Facility. Los Alamos and Livermore laboratories have collaborated on resolving indirect drive target physics issues on the Nova laser at Livermore National Laboratory. This combined with detailed modeling of laser heated indirectly driven targets likely to achieve ignition, has provided the basis for planning for the NIF.

A detailed understanding of target physics, laser performance, and target fabrication is required for developing robust ignition targets. We have developed large-scale computational models to simulate complex physics which occurs in an indirectly driven target. For ignition, detailed understanding of hohlraum and implosion physics is required in order to control competing processes at the few percent level. From crucial experiments performed by Los Alamos and Livermore on the Nova laser, a comprehensive indirect drive database has been assembled. Time integrated and time dependent measurements of radiation drive and symmetry coupled with a detailed set of plasma instability measurements have confirmed our ability to predict hohlraum energetics. Implosion physics campaigns are focused on underdstanding detailed capsule hydrodynamics and instability growth. Target fabrication technology is also an active area of research at Los Alamos, Livermore, and General Atomics for NIF. NIF targets require developing technology in cryogenics and manufacturing in such areas as beryllium shell manufacture. Descriptions of our NIF target designs, experimental results, and fabrication technology supporting NIF target performance predictions will be given.

© 1996 American Institute of Physics

1. INTRODUCTION

The U.S. Inertial Confinement Fusion (ICF) Program is preparing to proceed with a next-generation ICF facility to demonstrate ignition: the National Ignition Facility (NIF). NIF is presently conceived as a Nd:glass laser facility with 192 beams capable of delivering 1.8 MJ of 351 nm laser light in a shaped pulse to an indirect-drive target. In indirect drive,[1] the energy of the laser or particle-beam driver is deposited inside an enclosure made of a high atomic-number material (hohlraum) which surrounds the fusion capsule. The hohlraum wall radiates X-rays which ablate the outer capsule wall, driving an implosion. NIF will also have a direct-drive capability using approximately 1.5 MJ. (In direct drive, the laser illuminates the capsule directly.) The NIF will be capable of 500 TW peak power, while maintaining a better than 50μm pointing accuracy and a less than 8% RMS beam to beam power imbalance over a 2 ns interval. The facility will be capable of handling up to a 45 MJ of D-T fusion yield.

Experiments at the Nova laser [2] by both LLNL and LANL personnel are resolving the key issues to ensure success of indirect drive on NIF. These issues are the physics related to capsule-illumination symmetry, laser-plasma instabilities and hydrodynamic instabilities.

With the LASNEX code, [3] state-of-the-art radiation-hydrodynamics modeling has been used to design NIF capsules and to understand to sensitivity of the designs on laser performance, capsule surface finish, hohlraum performance, etc.

The beamlet laser at LLNL is been used to establish the validity of the novel laser technology to be used in NIF. Another important technological area for NIF is target fabrication, particularly the manufacture of sufficiently smooth cryogenic D-T layers. Present capability has advanced to within a factor of two of the roughness presently deemed tolerable for ignition.

2. NIF TARGET DESIGN AND FABRICATION

Various NIF targets have been explored by designers from both LLNL and LANL[4]. All these capsules are imploded by the X rays from the same hohlraum, namely, a gas-filled Au cylindrical hohlraum with a length of 9.5 mm and a diameter of 5.5 mm (see Fig. 1). The hohlraum gas fill is kept by polyimide

($C_{22}H_{10}O_5N_2$) windows. Here we describe briefly two capsule designs for which LASNEX modeling indicates ignition. The modeling uses the integrated technique, where the laser, the hohlraum, and the capsule implosion and burn are all calculated self consistently in a single simulation.

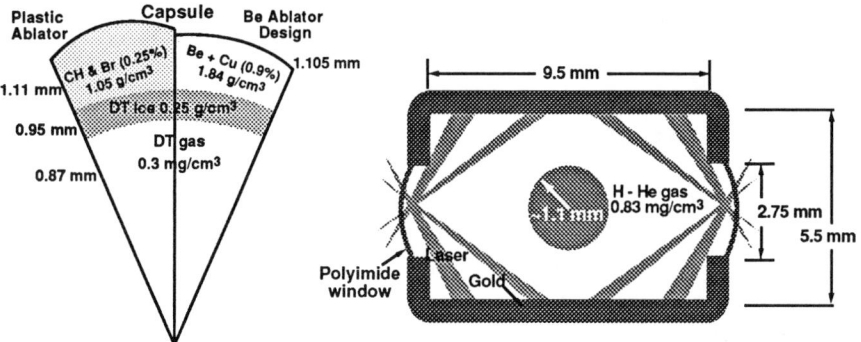

FIGURE 1. The ignition hohlraum for NIF is shown (right). Its size is roughly three times that of typical Nova hohlraums. On the left, details of two ignition capsules are shown.

The first target has a Br-doped CH outer ablator layer. We have used a radiation temperature profile T_r which reaches 300 eV to drive NIF capsules in our simulations (no mix of fuel and ablator due to hydrodynamic instabilities) yield for this target is 12MJ. The second ignition capsule has a Cu-doped outer ablator layer. LTE LASNEX modeling shows that with 1.30 MJ of laser energy (and 330 eV T_r), the clean fusion yield is 7 MJ. Other details of these targets are shown in Fig.1.

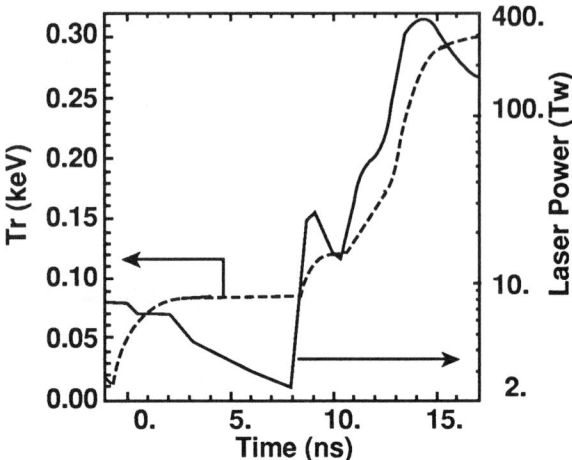

FIGURE 2. Idealized evolution of T_r and required laser power for the CH ignition capsule.

In order to achieve the capule compression and timing of the shocks which creates a hot spot in the capsule which ignites the propagating burn, various processes in hohlraum plasmas must be controlled. These include laser ray propagation, refraction, absorption and X-ray conversion; multi-group X-ray transport; hohlraum plasma dynamics (filling, stagnation, heat conduction), and; capsule evolution and burn. Successful capsule compression and shock timing to ignite a spark in the fuel that initiates the propagating burn requires a highly tailored T_r evolution, shown in Fig. 2 for the plastic-ablator capsule described above. Due to the complex hohlraum processes at work, the laser power evolution is even more complicated (see Fig. 2). The requirement of highly-tailored laser pulses, with differing details for different target, place heavy pulse-shaping demands on NIF.

Emphasis on target design is shifting towards ignition-sensitivity studies of the different target designs. These robustness studies can be divided into four categories:
- capsule studies: capsule surface imperfections, volume imperfections, material imperfections, and coupling of these imperfections with asymmetry of imperfect radiation drive

- hohlraum studies: variations in the hohlraum length, diameter, laser-entrance-hole diameter and lining material, gas-fill density, and hohlraum wall material
- laser studies: determvariations in the laser pointing, laser energy and power, laser pulse shape, and beam balance
- modeling: sensitivity of to the use of particular physics packages and algorithms, and assumptions and neglected physics in the modeling of the target performance.

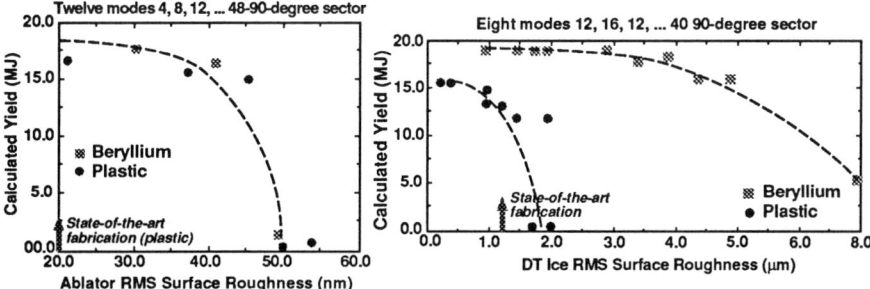

FIGURE 3. The yield from 2-dimensional LASNEX calculations is plotted versus ablator outer roughness (left) . The yield versus interior-surface roughness of the D-T ice is also plotted (right).

A detailed review of the extensive studies under way is beyond the scope of this paper. As an example, a sensitivity study of capsule yield versus ablator-surface roughness is discussed. This roughness is the seed for Raleigh-Taylor instability which could induce unacceptable levels of mix. Both the plastic and beryllium ablators have been studied. The technique used to simulate the instability is "non-linear multimode modeling", where direct 2-D LASNEX simulation of the effect of a realistic surface roughness on a capsule surface is calculated with a minimum of modeling approximations. Results of the study are shown in Fig. 3. The results indicate that a 50 nm RMS outer surface roughness is tolerable to achieve ignition. Current Nova capsules have a surface roughness as low as 20 nm, so targets currently meet this specification. The cliff in target performance for the inner ice surface is calculated to be 1.5 μm for the CH-ablator target. The advantage of the Be target on this score is noteworthy.

An important consideration for NIF targets is the feasibility of fabrication. In particular, most ignition-capsule designs rely on a cryogenic layer of DT which must be very smooth, as discussed above. Fortunately, the melting of the ice due to heating from the radioactive β decay of T and subsequent refreezing provides a path for achievement of the required smoothness[5]. In a toroidal test cell, a D-T layer of thickness 110 μm has been deposited with a surface roughness (sum in quadrature from all modes 1--110) of only 1.2 ± 0.3 μm[6]. Given the requirements discussed above, this represents a very encouraging result.

3. EXPERIMENTAL TARGET-PHYSICS CAMPAIGNS

Present work at Nova has concentrated in completing the "Nova technical contract" originally laid out in 1990 by the review panel from the National Academy of Sciences chartered by the US DOE to review the ICF program. Summary of the status at present status is provided. There are three main issues being presently addressed:

Drive symmetry: Given the relatively high convergence ratio needed for ignition of NIF capsules, a high degree of illumination symmetry is needed. Current estimates indicate the time-averaged symmetry needs to be $\leq$ 1%. Instantaneous asymmetries can be higher than that value. Excellent symmetry control and modeling (with LASNEX) has been demonstrated in both vacuum and lined Nova hohlraums using various diagnostic techniques[7]. However, recent symmetry experiments in gas-filled hohlraums have generally shown anomalously pole-hot capsule illumination. These asymmetries are consistent with a beam-steering effect early in time of about 0.2 mm (compared to a 0.5--0.7 mm beam footprint) as measured by both implosion and re-emission ball techniques[8,7]. It appears that this steering is the result of the deflection of hot spots undergoing mild filamentation[9] by strong plasma flows transverse to the propagating beam[10]. Future experiments will evaluate whether beam smoothing techniques such as finite bandwidth, by virtue of suppressing filamentation, can control this steering effect.

Laser-plasma instabilities: Potentially high levels of parametric laser-plasma instabilities [9] are possible because of the relatively long scale lengths predicted within NIF hohlraums. Laser scattering by unstable hohlraum plasmas are a potential problem for hohlraum energetics (in the case of back scatter) and symmetry (in the case of oblique scatter).

An extensive study of Stimulated Brillouin scatter (SBS), where the laser decays into an ion acoustic wave and scatteed light, has been done in both open targets[11] and hohlraums[12,13] designed to approximate plasma conditions with the longest scale lengths in gas-filled NIF hohlraums. It is found that properly chosen mixtures of ion species (such as He/H for NIF hohlraums) can suppress SBS.

Hydrodynamic stability: The implosion of the ICF capsule is susceptible to the Rayleigh-Taylor (RT) instability during the ablation process and also during the deceleration. Nevertheless, high convergence implosions have been achieved on Nova and successfully modeled[14]. For studying single-mode RT in convergent geometry with good access for time-dependent diagnosis, implosion of foam-filled thin metal cylinders is now used[15]. Implosion of capsules with laser-ablated pits, providing a broad mode-number spectrum, are also used to study convergent RT.

The authors gratefully acknowledge the hard work by the members of the LLNL and LANL ICF programs. This work is supported by the US DOE.

5. REFERENCES

[1] J. H. Nuckolls, L. Wood, A. R. Thiessen, and G. B. Zimmerman, Nature **239**, 139 (1972).

[2] E. M. Campbell, Rev. of Sci. Instrum. **57**, 2101 (1986).

[3] G. Zimmerman and W. Kruer, Comments of Plasma Physics and Controlled Fusion **2**, 85 (1975).

[4] S. W. Haan *et al.*, Phys. of Plasmas 2, 2480 (1995).

[5] J. K. Hoffer and L. R. Foreman, J. of Vac. Sci. Tech., A7, 1161 (1989).

[6] J. K. Hoffer, private communication.

[7] A. A. Hauer, Phys. Plasmas **2**, 2488 (1995).

[8] N. D. Delamater, private communication.

[9] W. L. Kruer, "The Physics of Laser-Plasma Interactions", Addison Wesley, Reading, MA, 1988.

[10] H. A. Rose, E. A. Williams, D. Hinckel, private communications.

[11] B. J. MacGowan, *et al.*, Proc. 15th Int. Conf. Contr. Nucl. Fus. Res., IAEA, Vienna, 1994.

[12] J. C. Fernandez, *et al.*, Phys. Rev. Lett. (1995) submitted.
[13] L. V. Powers, *et al.*, Phys. Rev. Lett. **74,** 2957 (1995).
[14] M. Cable, *et al.*, Phys. Rev. Lett. **73,** 2316 (1994).
[15] W. W. Hsing, Phys. Rev. Lett. (1995) submitted.

Inertial Confinement Fusion Program at CAEP

Hansheng Peng

China Academy of Engineering Physics, P.O.Box 501,
Chengdu, China 610003

Abstract. Proposed by Prof.Ganchang Wang in 1964 and officially started in 1976 the ICF Program at CAEP includes research on every aspect of ICF science and technology: high power and energy laser technology, target fabriction, diagnostics, target physics, and potential applications. Two solid−state lasers have been operated since the middle of the last decade. The most of research on target physics has been focused on indirect−drive approach, covering laser−plasma coupling, parametric instablilities and suprathermal electrons, x−ray conversion and transport, plasma energetics, ablation, hydrodynamic instabilities. Neutron production experiments were conducted successfully on the Shenguang facility ($2 \times 800J$, 1ns, and $1.053 \mu m$) with radiation−driven targets in 1990 .In addition, equation of state experiments have been performed using both laser−and x−ray−driven targets for Fe, Cu, and glass. Furthemore, a much bigger laser facility is now being considered.

INTRODUCTION

Since the sixties scientists have been pursuing the goal of realizing controlled nuclear fusion with powerful laser beams. Worldwide efforts in this field have made significant progresses in developing science and technology for laser fusion. The encouraging accomplishments and the predictable applications have provided a solid fundation and strong arguments for making a decision to take a historical step towards achieving ignition. The enthusiasm and interest of either scientists or governments of countries, such as the United States, France, Japan, China, Russia and others, presages that a new era for ICF has come.

In 1964 China Academy of Engineering Physics started the laser fusion program proposed and led by Prof.Ganchang Wang[1]. Unfortunately, it had been interrupted by the so−called cultural revolution for about ten years before Prof. Wang restarted the program again in 1976. The program includes research on every aspect of ICF science and technology: high power and energy laser technology, target fabrication, diagnostics, target physics, and applications.

In 1986, Chinese government set up a National High Technology Program covering many important fields of science and technology for the long-term development of the country. After a few-year review, ICF was added to the program in 1993 to facilitate the research, which had been conducted for many years at CAEP, with the participation and dedication of the experts nationwide and an additional financial support. CAEP is the main contractor for it. Along with the tide worldwide, CAEP is also making efforts to promote the present ICF program. The new step includes upgrading the Shenguang facility, high power and energy laser technology development, and conceptual design of the next generation laser facility of a few tens kilojoules.

GLASS LASER FACILITIES

Two major Nd:glass laser facilities, referred to as Xingguang and Shenguang, have been built and operated for ICF experiments.

Xingguang[2] is a single beam laser facility located in Institute of Nuclear Physics and Chemistry, CAEP. It was first built in 1984 with an output of 70J in a 1-ns pulse from a 70-mm aperture amplifier at the fundamental frequency and then frequency doubled and tripled. In 1992, the facility was upgraded by adding in two stages of 100- and 150-mm disk amplifiers and the output increased to 250J (1ns, 1.05μm). A frequency converter, installed in 1993, could be operated for both green and blue light. The type II/tpye II crystals are sealed in a common case, giving external efficiencies of 70% and 65% for 2ω and 3ω, respectively. To meet the requiremets of physical experiments on pulse duration the front end are installed with an actively mode-locked and Q-switched oscillator for generating Gaussian pulses of 0.2-0.8ns, and a single-longitudinal-mode and Q-switched oscillator with negative feedback for 1-5ns pulses. There are two target chambers to provide the convenience for different experiments to conduct parallelly. After the upgrade, the facility is referred to as Xingguang II and the specifications are listed in Table 1.

To explore new laser plasma phenomena, the CPA technique has been applied to Xingguang-II to produce a 1.5-ps and 1.5-J pulse output with the rod amplifiers (Φ40mm) as the first step towards a high brightness source.

The Shenguang[2] laser has been the main user's facility for CAEP to

TABLE 1. Xingguang−II Laser Facility

	Target chamber I (Diameter=0.7m)	Target chamber II (Diameter=1.0m)
Energy	40J/1ns (ω) 25J/1ns (2ω) 20J/1ns (3ω)	250J/1ns (ω) 200J/1ns (2ω) 130J/1ns (3ω)
Duration	0.2−0.8, 1−5ns	0.2−0.8, 1−5ns
Contrast ratio	$\geq 10^6$	$\geq 10^6$
Divergence	$\leq$10DL	$\leq$10DL

conduct ICF, X−ray laser and some related physical experiments on since its completion in 1985. It is situated in Shanghai and run by the Joint Laboratory of High Power Laser and Physics managed by Academia Sinica and China Academy of Engineering Physics. The facility has two beams, each yielding 800J at 1.053μm in a 1−ns pulse from the 200−mm aperture amplifier. In 1991, one of the two beams was frequency doubled. A three−year upgrade program for the facility from 1994 through 1996 has been under way. The upgraded Shenguang (referred to as Shenguang −II) will still have two beamlines but eight beamlets splitted at the 40−mm amplifiers stage. The final amplifiers will use segmented Nd:glass disks (200×400×40mm×5/beamlet). Even though the multipass amplifier architecture is preferred for the next generation glass lasers due to the lower cost, double−pass amplifiers will be adopted here to avoid unavailable large −aperture Pockels cells. Second and third harmonic conversion systems will be built in to provide the required wavelengths for ICF and X−ray laser research. Besides, a pulse tailoring technique with temporal and spatial transform will be used.

In addition to the upgrade programs, a conceptual design for a few tens kilojoule level laser facility has been started. Meanwhile, special support has been oriented to innovative laser technology development,

TABLE 2. Shenguang−I and −II Facilities

	Shenguang−I	Shenguang−II
Wavelength	1.053, 0.53μm	1.053, 0.53, 0.35μm
Pulse duration	0.1−1.0ns	20, 100ps, 1−3ns
Beam number	2	2×4
Energy	1.6kJ (1ns, 1.053μm)	4.8−6kJ (1ns, 1.053μm)
Energy imbalance		<10%
Contrast ratio	>10^6	>10^6
Divergence	$\leq$10DL	$\leq$10DL
Pulse shaping	No	Preliminary

such as cost-effective new architecture for high energy lasers and precision techniques for existing facilities. Precision optics and its manufacturing have also received more attention.

ICF TARGET PHYSICS

The primary objective of the program using the above mentioned facilities has been to address the basic physics of laser plasma interaction, hohlraum physics and imploding hydrodynamics. In parallel with the development of high power lasers, we have established a target laboratory to study target materials and develop fabrication technology and characterization techniques. The laboratory has the ability to produce glass capsules $100-300\mu m$ in diameter and $1-5\mu m$ in wall thickness and to fill DT mixture into capsules to a pressure range of $0-5\times10^6 Pa$. Submillimeter indirect drive hohlraum targets can be fabricated, most of which are a cylindrical case with a suspended capsule and some necessary structure inside. Besides, polystyrene capsule fabrication technique, capsule coating technique, and low density materials have also been developed. In the meantime, a great variety of sophisticated diagnostics have been developed and used for temporal, spatial and spectral measurements of laser-produced plasmas. Theoretical studies play a significant role in target physics, and therefore, we have paid special attention to develop codes and models for numerical simulations in accordance with experimental investigations of almost all the important physical issues of interest.

The research on target physics has been focused on addressing indirect-drive related issues, covering laser-plasma coupling, parametric instabilities, suprathermal electrons, x-ray conversion and transport, plasma energetics, ablation and hydrodynamic instabilities. Typical hohlraums are a gold case $600-800\mu m$ in diameter and $1.2-1.3mm$ long with a wall thickness of $20\mu m$ (see Figure 1). For special experiments the wall can be made as thin as $1\mu m$. The standard capsule diameter for

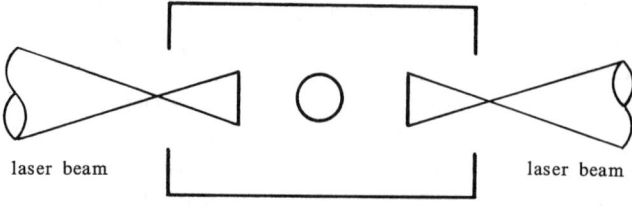

FIGURE 1. Standard hohlraum target on Shenguang-I

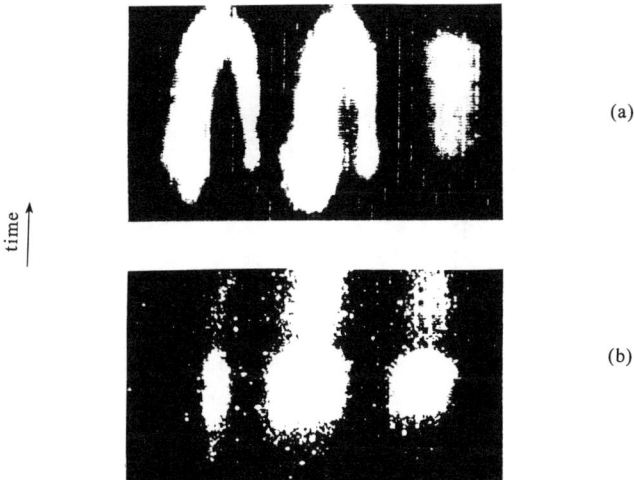

FIGURE 2. Streaked plasma images recorded with 3-pinhole cameras
a. plasmas expanding inwards in cylindrical hohlraum
b. hohlraum diagnostic hole plasma closure process

also produces suprathermal electrons, causing the preheat of the capsule fuel. Observed Raman scattered light spectra range typically from 1.40μm to 2.00μm, corresponding to plasma densities of 0.06Nc to 0.22Nc. However, the long−wavelength cut−off in our case is strongly intensity dependent. When the laser energy increases from, for instance, 50J to 300J, the peak wavelength of backscattered light changes from 1.59μm to 1.72μm, while the spectral width (FWHM) becomes broader from 45nm to 270nm as shown in Figure 3. These results are qualitatively consistent with the Landau damping theory for the short wavelength cut off and the collisional damping for the long wavelength edge[5]. With more laser energy collisional damping rates decrease dramatically due to the higher plasma temperature and, thus, the cut off shifts to denser plasma areas. The averaged Raman light costs about 10% of the laser energy for most of the cylindrical hohlraums in the experiments.

Up−shifted Raman scattering in the range of 0.75μm to 0.95μm, 2ω_0 and 3/2ω_0 have also been observed, whose intensities are a few orders of magnitude lower in comparison with SBS and SRS light. However, the forward Raman scattering leads to very energetic electrons generation and the inferred electron temperature from the recorded x−ray spectra can

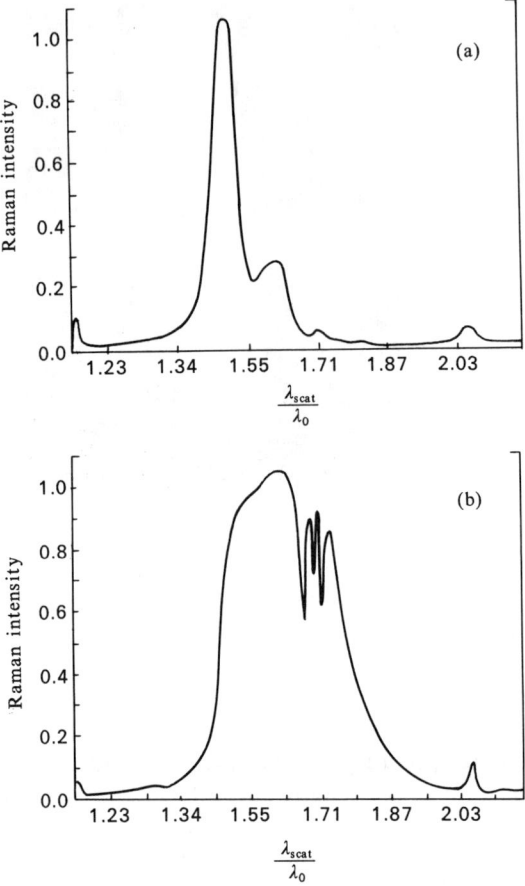

FIGURE 3. The Raman−scattered light spectra from Hohlraums (a.50J, b.280J)

reach 200−300keV.

Suprathermal electrons have also been studied in correlation with Raman scattering by measuring the hard x−ray spectra. Figure 4 presents a typical triple temperature spectrum [6] of x rays emitted from the hohlraums. The intensities of the hard x rays are linearly related to that of the SRS light, causing about 5−10% incident laser energy loss.

The total laser−plasma coupling efficiency for the hohlraums with one micron light is only 50−60% and x−ray measurements show 40~50% of the obsorbed energy could convert to x rays. To make it worse, the converted x rays lose much of their energy to transport and to heat up the hohlraum inner walls. At last, only a few percent of the laser energy is available for the capsule.

imploding experiments is 200μm.

First we experimentally studied laser entrance hole closure effect[3] to see laser injection efficiencies as a function of laser energy, hole size, injection angle and pulse duration. When the power density on the rim of the hole reached 10^{12}W/cm^2, the produced plasma expanded inward closing the hole and the laser energy could not pass through the hole anymore. Time−integrated x−ray pinhole image showed a bright spot in the middle of the hole, where the plasma became denser and hotter due to stagnation and more x rays could be emitted. Also the plasma caused the laser beam refraction, changing the spatial distribution of the laser light on the hohlraum inner walls.

Plasmas inside hohlraums even more concern us because they affect energy transport, excite nonlinear instabilities and, in particular, cause drive asymmetry. Spatially and temporally resolved distributions of plasmas produced by both laser light and x−ray heating at power density of 10^{14}W/cm^2 have been streaked. The results of the initial experiments have shown that the expansion velocity of x−ray produced plasmas reaches $1-2\times 10^7$cm/s and blue light produced plasmas expand at $\sim 3\times 10^7$ cm/s. However, filterd M−band measurements give a reduced velocity of about 6×10^6cm/s for the latter one, while the former one remains the same. Apparently, more experiments are needed to get more information about density and temperature profiles and spectra. Besides, the plasma closure effect of hohlraum diagnostic holes leads to big errors in radiation temperature measurements. Closure velocities for holes of 200μm in diameter in gold walls at a temperature of 120eV were derived to be $(6\pm 2)\times 10^6$cm/s. A radiation temperature correction of 5−23% should be made at different times. Plasma images from some experiments are shown in Figure 2.

Interactions of intense laser light with plasma is a particular rich topic for hohlraums with a plasma scale length in excess of 10^2 and the fundamental frequency laser light. With detectors working in a spectral range of 0.35−2.4μm scattered light has been carefully measured[4].

The Brillouin instability represents the decay of an incident laser light wave into a scattered light wave and an ion−acoustic wave. Because of the small frequency shift, the Brillouin scattering is rather difficult to quantify. For typical hohlraums as mentioned above 25%−30% of incident laser light energy at 1.053μm are back−reflected (shifted plus unshifted). In the experiments, for most of the shots with 400−500J incident energy on the targets, 110−150J are recorded. This is one of the major energy loss processes in our case.

We have made special efforts to study stimulated Raman scattering in hohlraum targets, because this instability not only dissipates energy but

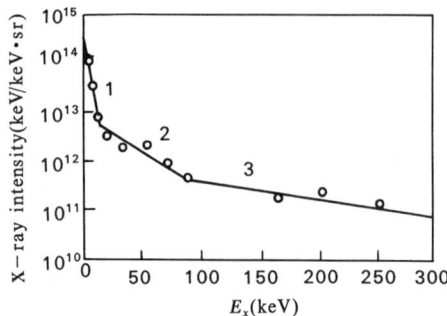

FIGURE 4. The three-temperature x-ray spectrum of Hohlruams. 1.Te=2keV ; 2.Te=44keV ; 3.Te=270keV

To study the x-ray transport properties, we have designed hohlraums divided into two parts: the x-ray conversion section and the capsule implosion section. The primary x-ray spectra have clear band structure and the reemitted is more thermalized. The radiation temperature created by two beams ($2 \times 600J$, ω_0) in millimeter hohlraums is typically 120eV. Based on these data we have carried out indirect drive implosion experiments and obtained detectable fusion neutrons of $10^3 - 10^4$ [7].

In addition to the equation of state experiments using laser beam direct drive with fly-eyes array smoothing technique[8] to get the shock adiabatic curves for Cu in a pressure range of 0.4−0.8TPa, experiments for double-foil targets with a flyer have been conducted and the pressure have reached 3TPa for Al. Meanwhile, radiation-driven EOS experiments with hohlraum have been tried for the first time, obtaining some preliminary data. Figure 5 gives some streaked shock breakout images.

Target physics studies on Shenguang−I laser facility have stopped due to the upgrade and target design and computational simulations have begun for Shenguang−II laser facility. Furthermore, conceptual research of target physics has been conducted in recent years to clarify the

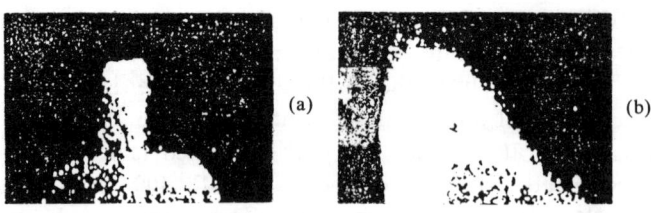

FIGURE 5. Streaked shock breakout of EOS stepped (a) and wedge (b) targets

requirements on the performance of the next bigger laser facility to be built.

X−RAY LASERS

Germanium x−ray lasers experiments with computational simulations have been conducted since 1987 at CAEP [9] using a single slab, double slabs, and quadruple slabs. Deep˙ gain saturation was reached in the experiment for the quadruple configuration and a multilayer soft x−ray mirror with a curvature radius of 8cm in 1992. To improve the laser beam quality we introduced spatial filtering technique by placing the second two−slab set 1m apart axially from the first two slabs and 1.5mrad divergence angle was obtained for 23.2 and 23.6nm lasing lines in 1993. Further improvement of the spatial coherence was realized by using a planar soft x−ray mirror instead of the curved one [10]. The divergence angle was thus decreased to 0.8mrad (Figure 6). The intensity recorded on film with a flat−field spectrometer was 1.7×10^9 photons/μm which was very close to calculated saturation value of 2.0×10^9 photons/μm.

Collisionally excited x−ray lasers have been successfully demonstrated with saturation and small divergence angle in a large wavelength range, but they need very big laser facilities to pump. To increase the pump efficiency, double−pulse pump experiments were conducted in 1993 [11]. The first pulse of 150ps at a power density of $(2-4) \times 10^{13}$ W/cm^2 and the wavelength of 0.53μm was used to burn through a germanium foil target.

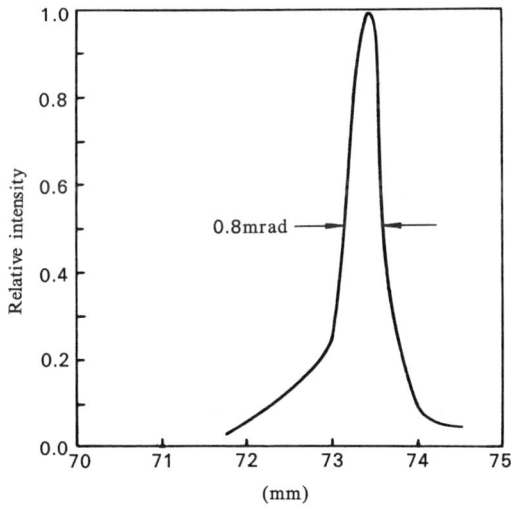

FIGURE 6. Normalized intensity profile at 23.2nm

Having expanded adiabatically for about 350ps to create the required electron density and ion aboundance, the plasma was then heated by the second pump pulse of 160ps at $1.053\mu m$ and a doubled power density of $2\times10^{14}W/cm^2$ to increase the electron temperature and gain coefficiets. In comparison with the single pulse pumped slab target, lasing lines at 19.61 and 23.63nm were much more intensive. Instead, the $J=0-1$ transition $J=0-1$ transition became stronger than the $J=2-1$ transition. Unfortunately, the gain coefficients could not obtained due to the limited shots.

ACKNOWLEDGEMENTS

The author is grateful to his colleagues in Institute of Nuclear Physics and Chemistry, Institute of Applied Physics and Computational Mathematics, and Joint Laboratory of High Power Laser and Physics for their contributions to conducting the programs. The subjects were supported in part by the National Laser Program and National ICF Program.

REFERENCES

1. Wang G.C., Chinese J. of Lasers, Vol.14, No.11, 641(1987).
2. Peng H.S., Laser Optics '93, SPIE Vol.2096, 33.
3. Ding Y.N., 18th ICHSPP, 1988.
4. Mei Q.R., et al., High Power Laser and Particle Beams (in Chinese), Vol.6, No.1, 5(1994).
5. Drake R.P., , et al., Phys Fluids B, Vol.1, No.11, 2219(1989).
6. Qi L.Y., et al., High Power Laser and Particle Beams (in Chinese), Vol.5, No.3, 415(1993).
7. Peng H.S., et al., Laser Interaction and Related Plasma Phenomena, Vol.10, 99, (1991).
8. Gu Y., et al., High Power Laser and Particle Beams (in Chinese), Vol.3, No.1, 263(1991).
9. Peng H.S., Laser Interaction and Related Plasma Phenomena, Vol.11, 263 (1993).
10. Wang S.J., et al., Private communication.
11. Huang W.Z., Zhang G.P., et al., High Power Laser and Particle Beams (in Chinese), Vol.6, No.2, 171(1994).

Experiments on the OMEGA Laser to Validate High-Gain, Direct-Drive Performance on the National Ignition Facility

R. L. McCrory, J. M. Soures, C. P. Verdon, T. R. Boehly, D. K. Bradley, R. S. Craxton, J. A. Delettrez, R. Epstein, P. A. Jaanimagi, S. D. Jacobs, R. L. Keck, J. H. Kelly, T. J. Kessler, H. Kim, J. P. Knauer, R. L. Kremens, S. A. Kumpan, S. A. Letzring, F. J. Marshall, P. W. McKenty, S. F. B. Morse, A. Okishev, W. Seka, R. W. Short, M. D. Skeldon, S. Skupsky, M. Tracy, and B. Yaakobi

Laboratory for Laser Energetics
University of Rochester
250 East River Road
Rochester, NY 14623-1299

INTRODUCTION

The top priority of the U.S. Inertial Confinement Fusion (ICF) program was defined by the 1990 National Academy of Sciences' (NAS) review committee (1) to be: "The expeditious demonstration of ignition and gain..." To attain this objective, the ICF program is directed toward an experimental and theoretical program with the intermediate goal of defining ignition-capsule physics issues and glass-laser driver technology advances necessary to construct a National Ignition Facility (NIF).

The NIF will be a 351-nm (UV), Nd:glass laser with energy capability of 1.8 MJ. The conceptual design report for the NIF is complete and the Key Decision 1 (KD1) to proceed with detailed design and development has been made. The principal objective of the NIF is to demonstrate ignition and modest gain. The principal approach to achieve this objective is indirect drive.

The direct-drive approach to inertial confinement fusion (ICF) could potentially provide higher gain at lower incident energy that indirect drive on the NIF. For this reason, a joint University of Rochester/Laboratory for Laser Energetics (UR/LLE), Lawrence Livermore National Laboratory, and Los Alamos National Laboratory effort is underway to explore the direct-drive approach and potentially use the NIF (properly configured for direct drive) to test high-performance direct-drive targets.

To validate the direct-drive approach, design and construction of the OMEGA

Upgrade facililty (>30-kJ, 60-beam, UV laser) was initiated in 1990 at UR/LLE. In this paper we review the requirements for achieving ignition with direct-drive targets. We then discuss the features of the OMEGA Upgrade designed to address these requirements, and the direct-drive experimental program to address the physics of these targets.

IGNITION REQUIREMENTS

The requirements for achieving ignition and high gain for laser-driven, direct-drive targets include: efficient energy absorption and energy transport, uniform laser energy irradiation, controlled preheat, and hydrodynamically stable capsule designs.

An estimate of the stability "operating" regime can be identified using a simple constant acceleration model (1). The results of such calculations show, that hydrodynamically stable capsule implosions are possible for implosion velocities in the range of 2.5 to 3.5×10^7 cm/s and for $3 < \alpha < 4$ (where α is the Fermi degeneracy parameter—ratio of cold fuel pressure to Fermi degenerate pressure) only if the initial perturbation amplitude (combination of equivalent laser irradiation imprinting and outside surface target imperfections) is less than 1000 Å.

When these limits are included in the direct-drive target designs for NIF, it has been calculated that target gain will be approximately 20, compared to 10 for indirect-drive targets with similar incident laser energy.

OMEGA REQUIREMENTS

It is the objective of the OMEGA experimental program to test these predictions by using "hydrodynamically equivalent" capsules, i.e. capsules whose physical behavior scales to that of capsules appropriate for the NIF. Based on hydrodynamic behavior alone, the laser and capsule parameters would be scaled by the following relations: energy $E \propto R^3$, power $P \propto R^2$ and time $\propto R$, where R is the capsule radius. Comparing the performance of NIF capsules with that of energy-scaled implosions using 30 kJ of 351-nm laser light, one-dimensional simulations show that these capsules have similar: number of R-T e-foldings during both the acceleration and deceleration phases of the implosion; in-flight aspect ratio; hot-spot convergence ratios (C_R); and implosion velocities.

Based on these considerations, the OMEGA specifications were set as follows:

Energy:	up to 30,000 J—on target
Pulse width:	Variable—with a minimum pulse width capability of approximately 500 ps
Pulse shape:	Flexible with contrast ratio >200:1
Number of beams:	60

Irradiation nonuniformity: 1 to 2% rms
Pointing precision/stability: <±15 µm rms
Prepulse control: <10^8 W/cm^2

OMEGA LASER DESIGN

OMEGA is a 60-beam, Nd:glass laser system with traditional master-oscillator—power amplifier architecture. It is designed to achieve high-irradiation uniformity with flexible pulse shaping and is capable of a shot repetition rate of one shot per hour. A schematic view of the OMEGA laser is shown in Fig. 1. The laser and target bay are situated on a single isolated reinforced concrete structure 66 m long and 27 m wide. An top view of the OMEGA layout is shown in Fig 2.

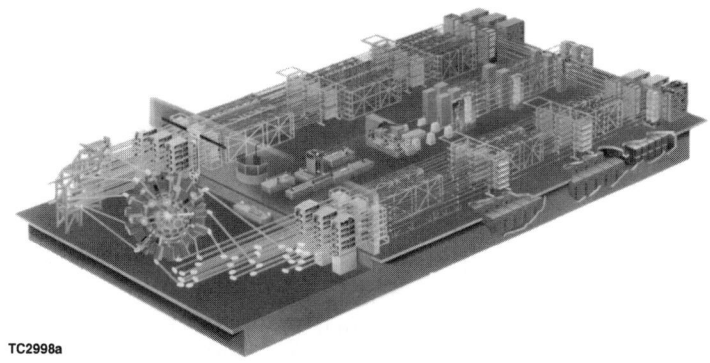

Figure 1. A schematic of the OMEGA laser and target bay facility.

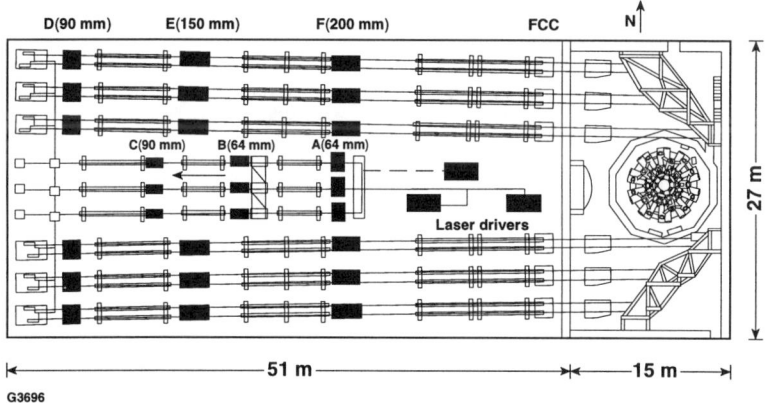

Figure 2. Layout of the OMEGA laser system. The locations of four stages of rod amplifiers (A–D), two stages of disk amplifiers (E,F), and the frequency conversion crystals (FCC's) are indicated.

73

Three separate drivers are provided for OMEGA. There are two drivers that supply the appropriate pulse shape for the foot of the pulse and the peak of the pulse respectively. [For high-dynamic range pulse shaping OMEGA is capable of operating in a copropagating pulse configuration which has previously been described (2).] The third driver can be directed into selected beamlines for use as x-ray backlighting beams.

Referring to Fig. 2, the driver output is split into three equal energy beams and directed to the A amplifiers (64 mm diameter rod amplifiers). The A-stage output is then directed to the B area where each laser beam is split into five equal energy "legs" and is then reamplified by another 64-mm amplifier (B stage). The energy of each leg is then directed to the C amplifiers (90 mm diameter rod). After exiting the C-stage amplifier, the legs are split into four equal energy beams. Two of the beams are directed to the south-side of the laser bay and two to the north side. Image-relaying spatial filters separate all amplifier stages of OMEGA.

The subsequent staging is illustrated in Fig. 3. Each beam is amplified by another 90 mm-diameter rod amplifier and then by 150 mm diameter (E-stage) and 200 mm disc (F-stage) amplifiers sequentially. The nominal output of each beamline is approximately 1000 J. The 200-mm beams are expanded to a diameter of 280 mm by means of the final system spatial filter and directed to the frequency converter crystal (FCC) housed in a special thermally isolated structure located at the output of the system.

The frequency tripling system is comprised of a pair of KDP crystals in a type II-type II polarization mismatch system (3,4). To maintain optimum conversion efficiency, a high-contrast-ratio, high-damage-fluence polarizer is used to keep the fraction of the beam in the wrong polarization below 3×10^{-5}.

To measure all three wavelenghts produced by the FCCs, the harmonic energy detector (HED) system was developed and implemented on OMEGA. The HED system is comprised of an integrating sphere (which accepts a small fraction of the light from a single beam), a fiber to transmit the light, and a spectrometer coupled to

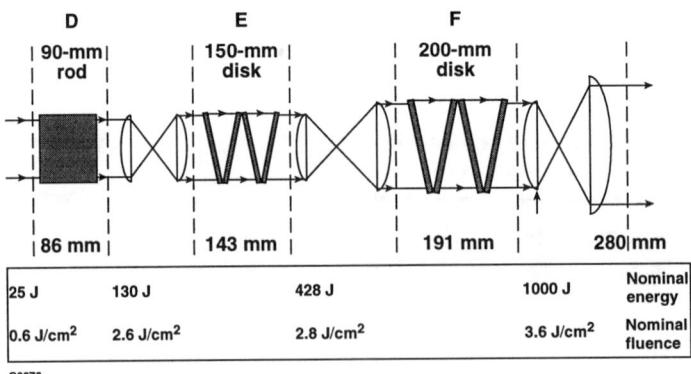

Figure 3. Amplifier staging (from stage D on) of the OMEGA laser system and nominal energy and fluence at key locations throughout the beamline.

a CCD to detect the energy of each wavelength of each of the 60 beams of OMEGA. Figure 4 illustrates the concept of the HED.

Upon passing through the final alignment sensor package (FASP), each OMEGA beam is directed to the target focusing lenses by means of two mirrors. The focusing lens assembly (FLAS) is mounted directly on the OMEGA target chamber—a 3.3-m diameter aluminum sphere. The beam orientation mimics a bucky ball to assure optimum irradiation uniformity.

FUTURE PLANS/EXPERIMENTAL PROGRAM

An experimental program plan has been developed for OMEGA with the mission of proving the validity of high-performance, direct-drive target designs. One of the key issues to be explored on OMEGA is direct-drive hydrodynamic instability control using pulse shaping. Figure 5 illustrates the range of cryogenic capsule implosions (from $\alpha = 1$ to 4) that will be examined using the pulse shapes shown schematically on Fig. 6. The high-contrast, continuously varying pulse shape can produce a nearly Fermi-degenerate implosion (which produces higher densities—but is also more susceptible to failure due to hydrodynamic instabilities). The "picket fence" pulse shape shock heats the ablator and fuel, putting the target on a higher isentrope. This is expected to reduce the hydrodynamic instability growth rate—but also reduce target performance.

To achieve the required irradiation uniformity level, phase conversion and smoothing by spectral dispersion (SSD) are required (5). Using photolithographic techniques, new distributed phase plates have been demonstrated at LLE which

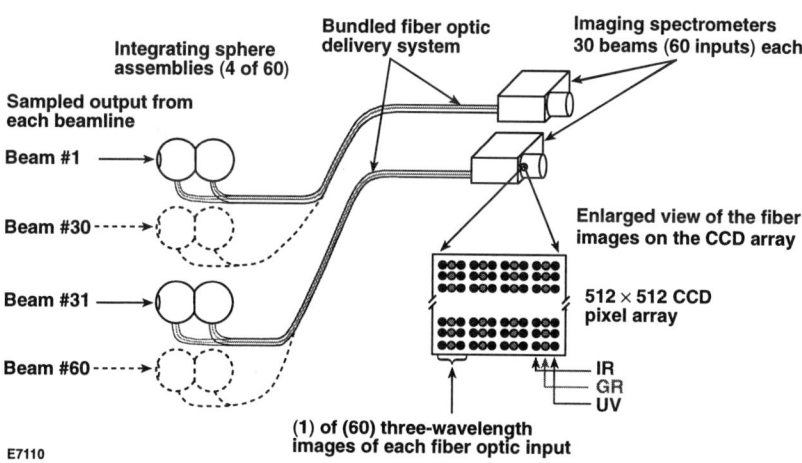

Figure 4. A schematic of the harmonic energy detector system used on OMEGA to determine the 1054-nm, 531-nm, and 351-nm energies.

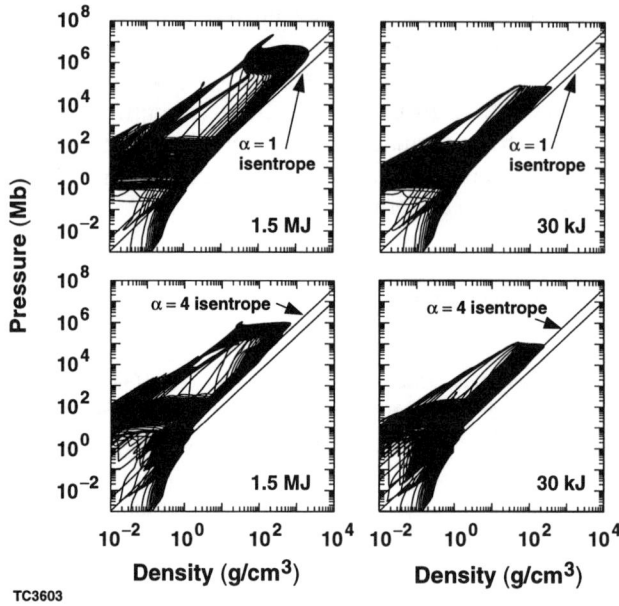

Figure 5. Pressure versus density trajectories of a near-Fermi degenerate ($\alpha = 1$) implosion and an implosion for which $\alpha = 4$.

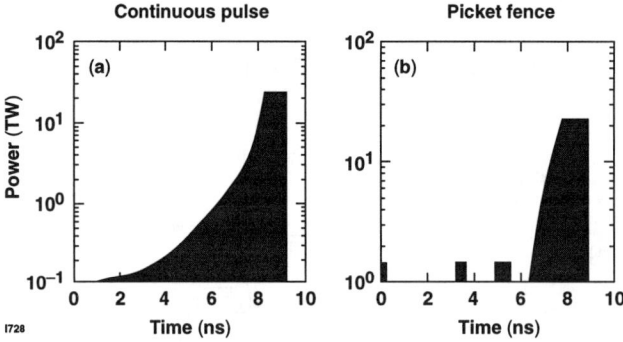

Figure 6. Power versus time for a continuous and picket-fence pulse shape to be used on OMEGA to study the control of Rayleigh-Taylor linear growth rates.

eliminate most of the 20% loss associated with prior technology phase plates. The capability of these new phase plates is shown in Fig. 7. Enhancements to the previously applied 1-D SSD system are also to be implemented on OMEGA. The first of these enhancements is 2-D SSD. This technique is illustrated in Fig. 8. Other future enhancements include the birefringent wedge which can provide beam smoothing through a relative shift of "o" and "e" speckle patterns as shown in Fig. 9. With these techniques the irradiation nonuniformity levels on OMEGA are expected to drop to the 1% level as shown on Fig. 10.

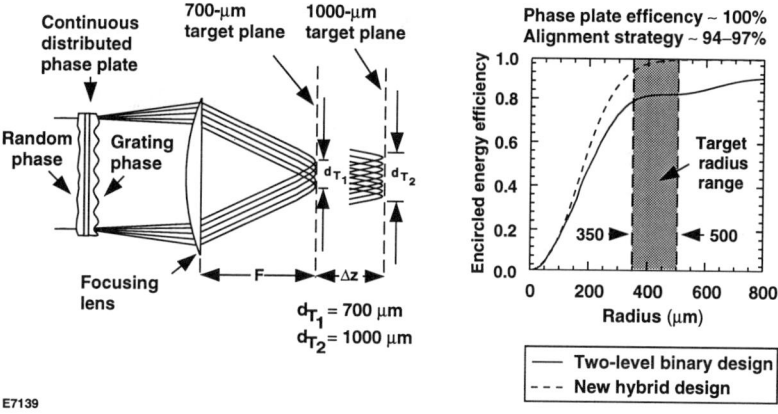

Figure 7. New hybrid design DPP's will accommodate a wider range of target diameters and significantly reduce the energy lost around the target when compared to old binary design DPP's.

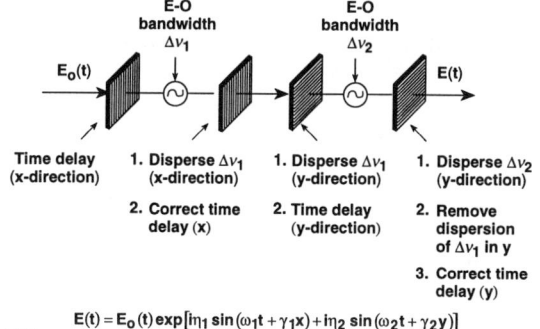

$$E(t) = E_0(t)\exp[i\eta_1 \sin(\omega_1 t + \gamma_1 x) + i\eta_2 \sin(\omega_2 t + \gamma_2 y)]$$

Figure 8. Schematic of a 2-D SSD scheme to be incorporated on the OMEGA laser system.

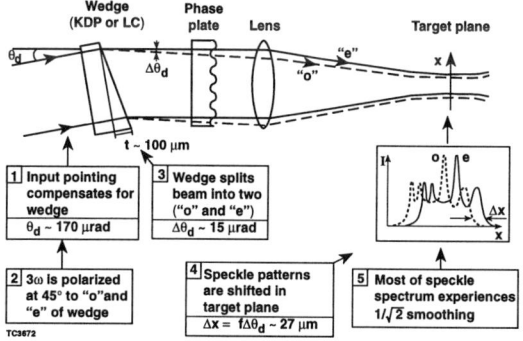

Figure 9. Polarization dispersion will be incorporated on OMEGA to add modes to the far-field resulting in an $\sim\sqrt{2}$ instantaneous reduction in irradiation nonuniformity.

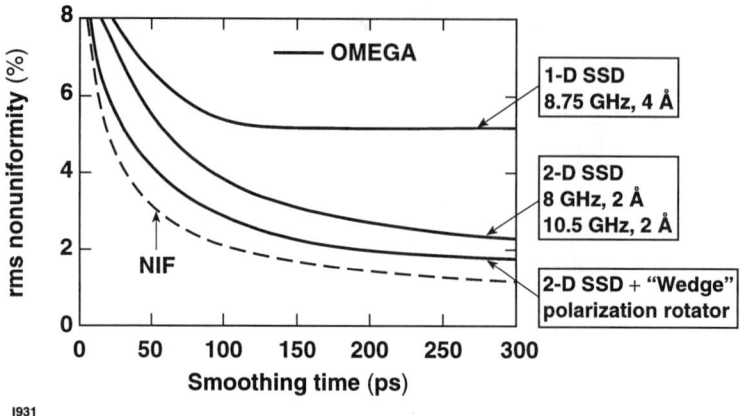

Figure 10. Predicted levels of rms irradiation nonuniformity versus smoothing time for the three irradiation smoothing schemes planned for OMEGA. For comparisons, prediction for the NIF incorporating 2-D SSD and polarization rotators are also included.

To carry out these experiments, extensive plasma diagnostics are required. OMEGA will have a complete suite of nuclear, x-ray, particle, and optical diagnostics. One of the most important of these systems is MEDUSA, A 960-channel neutron detector array that will be used to make fuel areal density and temperature measurements.

The schedule for the experimental and diagnostic development campaigns on OMEGA is shown in Fig. 11. The pinnacle of the direct-drive campaign is HE2—a campaign involving fully cryogenic DT hydrodynamically equivalent direct-drive targets that will begin in late 1998.

SUMMARY

In summary, the OMEGA facility will conduct a full range of direct-drive implosion target experiments. With 60 beams, in excess of 30 kJ energy capability and 1% to 2% irradiation nonuniformity, OMEGA will be capable of validating high-performance hydrodynamically equivalent target performance. An extensive suite of capsule diagnostics and cryogenic capsule fabrication make OMEGA a unique inertial confinement fusion experimental facility.

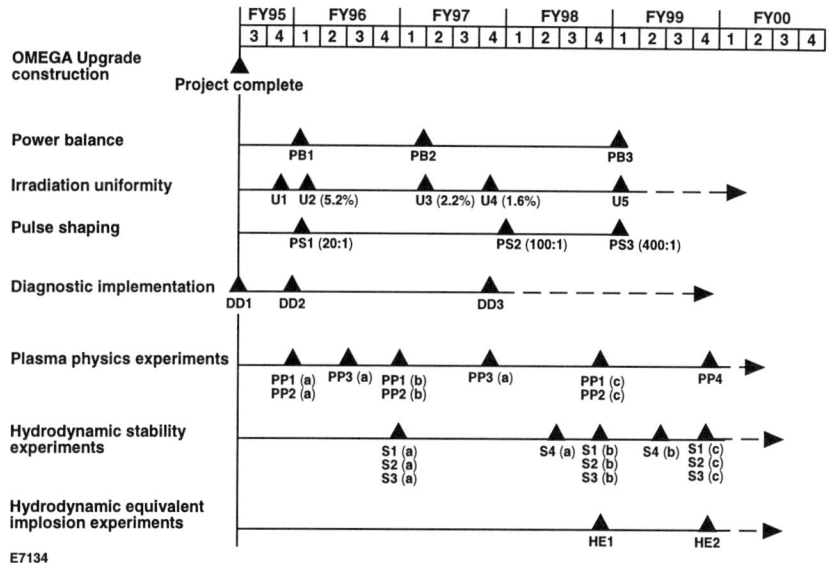

Figure 11. The UR/LLE direct-drive experimental program plan for FY95–FY99, showing completion dates for various program elements.

ACKNOWLEDGEMENTS

This work was supported by the U.S. Department of Energy Office of Inertial Confinement Fusion under Cooperative Agreement No. DE-FC03-92SF19460, the University of Rochester, and the New York State Energy Research and Development Authority. The support of DOE does not constitute an endorsement by DOE of the views expressed in this article.

REFERENCES

1. R.L. McCrory et al., "Direct-drive laser-fusion experimental program at the University of Rochester's Laboratory for Laser Energetics," to be published in *Proceedings of Fifteenth International Conference on Plasma Physics and Controlled Nuclear Fusion Research*, International Atomic Energy Agency, Seville, Spain, 26 September–1 October 1994.
2. 1993 Annual Report, University of Rochester, Laboratory for Laser Energetics, pp. 178–191, (January 1994).
3. W. Seka, S. D. Jacobs, J. E. Rizzo, R. Boni, and R. S. Craxton, *Opt. Commun.* **34**, 469–473 (1980).
4. R. S. Craxton, *IEEE J. Quantum Electron.* **QE-17** (9), 1771–1782 (1981).
5. S. Skupsky, R. W. Short, T. Kessler, R. S. Craxton, S. Letzring, and J. M. Soures, *J. Appl. Phys.* **66**, 3456–3462 (1989).

Status of HIB Fusion in Europe

D.H.H. Hoffmann, J. Jacoby

Physikalisches Institut, Universität Erlangen-Nürnberg, 91058 Erlangen, Germany

Abstract

Inertial Confinement Fusion (ICF) with heavy ion beams has a promising perspective for future energy generation. For the exploration of this concept Europe has a unique potential concerning the technologies involved. In particular, there is an excellent knowledge and experience in RF linear accelerators, storage rings and the handling of intense beams in European laboratories. During the last decade, key issues of ICF in the fields of accelerators, targets and systems have been studied in the framework of national programs. Intense heavy ion beams have been used in recent years to study the properties of matter under extreme conditions.

1 Introduction

Thermonuclear fusion of hydrogen isotopes to helium appears to be a sufficiently clean and abundant long-term energy source of mankind. The climatic effects of CO_2 as recognized today, have led many Governmental Authorities to envisage the replacement of fossil fuel by new energy sources.

In energy strategies for the next century considered in a global frame (UNCDE, Rio 1992) and within the European Community (Decision of the Council of Ministers for Energy and Environment, October 1990) a stabilization or, if possible, a decrease of CO_2 pollution, in particular by the industrialized countries is foreseen. However, both the potential of energy savings and the replacement potential by renewable non-fossile energies are limited.

Research on controlled fusion, initiated already in the fifties, has followed two distinct approaches, a quasi-continous process where the required plasma is confined by strong magnetic fields, magnetic confinement fusion (MCF), and a process based upon successive micro-explosions where the necessary conditions are created in the implosion of a hydrogen filled pellet triggered by high power irradiation, inertial confinement fusion (ICF).

Both laser and particle beams can be used to explore the physics of inertial confinement fusion. With respect to energy production in an reactor, there is a world wide consensus today that heavy ion beams are a more promising to the driver problem [1, 2, 3]. An attractive feature of ICF in general is the seperation of driver and fusion chamber.

The progress of inertial fusion research was severly impaired by constrains of classification. It is only since the end of 1993 that substantial parts of pellet physics results are beeing declassified. Today, the feasibility in principle of inertial confinement fusion is no longer in doubt, but an estimate of technical merits and costs must be considered before a choice can be made for future fusion power generating plants.

2 The European Study Group Initiative

The European Community, recognizing the importance of long-term energy supplies for the highly industrialized countries, is supporting energy research since many years. The means provided for fusion research were concentrated in an important magnetic confinement programme which brought Europe into the frontline of MCF research and was very successful in approaching conditions near thermonuclear burn. The report on the 1990 Fusion Programme Evaluation held under the chairmanship of Prof. Colombo, however, emphazises the necessity of activity also in ICF research, so that Europe will dispose of the necessary scientific and technological basis for important future decisions.

Therefore, it is now proposed to combine the heavy ion related ICF activities existing in Europe, under the auspices of the European Commission into a coordinated feasibility study including also rough cost estimates of an ignition facility, which shall have the potential to attain the threshold of ignition of heavy ion induced ICF. European studies of the heavy ion driver required for the irradiation of the target were conducted in a loose collaboration of a few accelerator laboratories. High power irradiation by heavy ion beams for the necessary pulse duration of 10 ns can be achieved by compression of longer pulses at lower current by a factor of the order of 10^5; those currents can be produced by available sources with the necessary beam quality. The acceleration of the ion beam to the required energy and its compression can be achieved with at least two different driver techniques, of which in Europe the "linac plus storage ring" route was pursued and included in a feasibility study of an ICF facility published in 1985 under the acronym HIBALL. HIBALL is a conceptual design of a fusion reactor driven by heavy ion beams. It has shown that accelerator, reactor and target can be made consistent with the requirements of future energy production.

Clearly, revisiting this concept is essential after 10 years of progress in accelerator technology and target theory. in the new international context establishing a close cooperation with US scientists and even a direct participation of East-European scientists in several aspects of the study could be envisaged. The European Ignition Facility would make use of the body of excellent knowledge and experience in Europe in linear RF accelerators, storage rings and the handling of intense ion beams, to be complementary to the US driver studies underway. Within this study experimental work should be pursued on now existing European ion accelerators so as to provide a sound basic for the choice of critical driver parameters.

3 High energy density in matter driven by heavy ions

Ion beams are well suited to produce high energy density in matter. The energy deposition process is very effective and the range of ions in solid matter is perfectly determined by the ion energy. This enables a precise and homogeneous heating of a target volume. Intense ion beams provide a new tool to study matter at extreme conditions of pressure and temperature, testing the conditions of solar matter in reproducible and controlled laboratory experiments.

At Gesellschaft für Schwerionenforschung (GSI) in Darmstadt and at Universität Erlangen an experimental program is carried out to make use of the unique accelerator facilities here to address key issues of heavy ion driven inertial confinement fusion.

Recent experiments at the heavy ion synchrotron SIS at GSI have been performed using

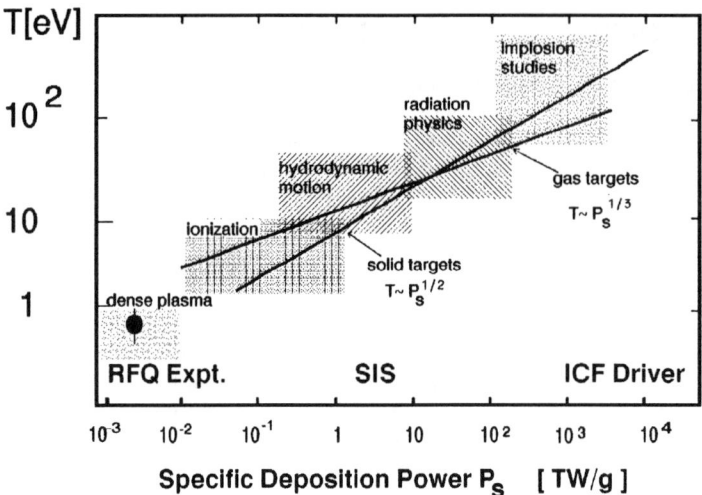

Figure 1: Temperature achieved in a target as a function of the specific deposition power. The experimental data point shown in the figure corresponds to an experiment at low beam energies (at MAXILAC [4]). Increasing the specific depositon power allows access to different regimes in hot dense matter. At high specific deposition power conditions similar to the center of the sun are achieved.

$1\text{-}2 \cdot 10^{10}$ high energy neon ions (300 MeV/u) at an ion pulse duration of about 1 μs. These ion beams have been focused, using conventional quadrupole focusing to a spot size below 1 mm² onto a frozen krypton crystal. To induce hydrodynamic action in solid material, at least the chemical evaporation energy has to be deposited in a target. Rare gas crystals offer the opportunity to induce hydrodynamic motion already at relatively low beam intensities, because here the chemical evaporation energy is one order of magnitude smaller than in metals. Figure 2 shows the ion beam impinging on a krypton crystal and the hydrodynamic reaction produced in this target. For this beam the total energy delivered to the krypton crystal is 1.1 J. This corresponds to a specific energy deposition of 60 J/g.

In the near future the intensity upgrade program at GSI offers the opportunity to increase the number of beam ions within a few years by two orders of magnitude. This will increase the specific deposition power induced by an ion beam up to 1 TW/g.

4 Stopping power

The detailed modelling of ion beam heated matter at high energy density depends strongly on the precise knowledge of the energy deposition process of ions in a plasma. First experimental evidence of a stopping power enhancement in a plasma was observed for deuterons and protons [5, 6]. Later, systematic energy loss measurements in a hydrogen plasma verified the difference in stopping between a highly ionized plasma and cold matter also for heavy ions [7, 8, 9]. Experiments have been carried out at energies between 1 – 6 MeV/u for a large

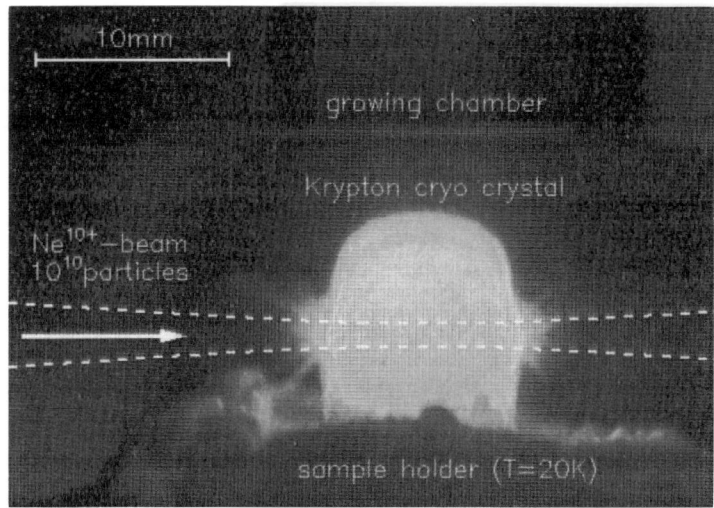

Figure 2: Krypton crystal heated by a 300 MeV/u neon beam. The picture shows the self emission of the krypton crystal, integrated for 100 μs (the dashed line indicates the calculated beam envelope). Hydrodynamic motion of the evaporated krypton ice is observed at the positions where the ion beam enters and leaves the crystal.

variety of ion species from carbon to uranium [10, 11]. An enhancement factor of up to three was demonstrated there. This enhancement was mainly attributed to the higher Coulomb logarithm in a plasma, which depends only on the amount of free electrons available in a plasma. For lower kinetic energies, where the projectile velocity is comparable to the thermal velocity of the plasma electrons, the energy loss is expected to be even more dramatic.

To investigate the stopping power at low beam energies an experiment has been set up at the radio frequency quadrupole-accelerator MAXILAC at GSI [12]. This accelerator delivers singly charged heavy ions at a final energy of 45 keV/u (Kr$^+$, 3.8 MeV). A homogeneous plasma is produced in a 20 cm long cylindrical quartz tube. A capacitor bank (2.6 μF, 5–10 kV) supplies discharge currents of 10–20 kA to the plasma. This current is oscillating with a half period of about 4 μs. At initial hydrogen gas pressures of 0.5 to 2 mbar electron densities of up to $10^{17} cm^{-3}$ are produced in the discharge. The two-stage differential pumping system consists of a powerful roots pump and two turbomolecular pumps at the second pumping stage, each reducing the pressure by about a factor of hundred. Discharge tube, roots pumping, and turbomolecular pumping are separated by small apertures ($\varnothing$ 3 mm × 30 mm), which allow a windowless penetration of the ion beam into the plasma. These small apertures define the beam path through the plasma close to the optical axis of the discharge tube.

The time-of-flight measurement to determine the energy loss is based on the 13.4 MHz radio frequency structure of MAXILAC. The microstructure of the ion beam leads to an ion bunch every 74.8 ns registered by the final stop detector. The arrival time of these ion bunches is digitized at a sampling rate of 1 GHz and recorded during 20 μs. The maximum phase shift registered during the discharge yields 95 ns delay.

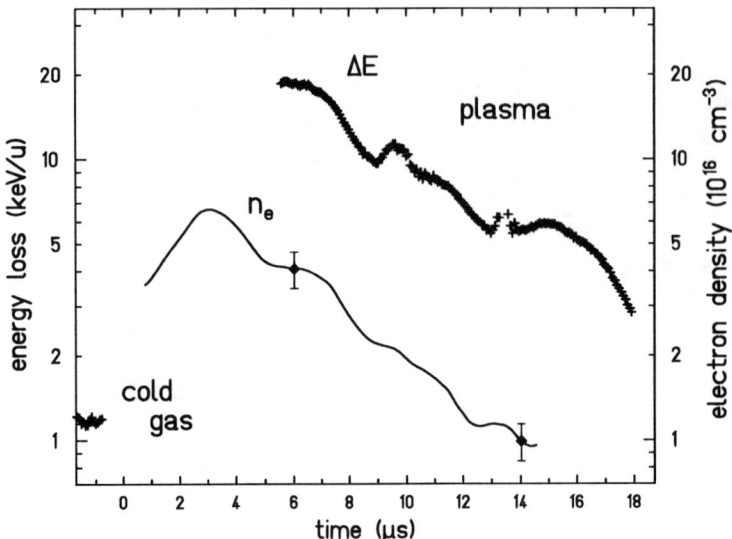

Figure 3: Energy loss of krypton ions in a hydrogen plasma and corresponding electron density in the plasma.

Results of the energy loss measured by the time-of-flight system (crosses) and the corresponding electron density of the plasma (solid line) are shown in Figure 3. Data for the energy loss in cold hydrogen gas are obtained from the time-of-flight signal before the discharge. The start of the discharge has arbitrarily been chosen to be time zero. Due to the strong influence of the discharge to the ion transmission, no stop signals are observed at the beginning of the discharge. The first detected time-of-flight signals of the discharge about 5 µs after ignition correspond to an energy loss of 18 keV/u, or to a relative energy loss of 40 %. The comparison of the energy loss during the plasma discharge with the values obtained for cold gas (1.2 keV/u), shows already a dramatic enhancement of the energy loss in the plasma. For a calculation of the stopping power, the determination of the plasma density is a prerequisite. The plasma parameters were measured by side-on observation of the Balmer emission from the hydrogen plasma. A system of spectrometer, streak-camera, and CCD-camera served as a tool for the time and wavelength resolved detection of the H_α and H_β lines of the plasma. The plasma electron density is derived from the Stark broadening of the H_β emission. The experimental error of the electron density is marked at two positions in Figure 3. For the results shown above, the measured electron density reaches its maximum of $n_e = 7 \cdot 10^{16}$ cm^{-3} after about 3.5 µs heating time, decreasing during the consecutive 10 µs to $n_e = 1 \cdot 10^{16}$ cm^{-3}. During this period the energy loss in the plasma follows nicely the decreasing electron density. Within the experimental accuracy obtained for the electron density, no further temporal dependence on plasma parameters for the energy loss except on plasma density can be deduced. The electron temperature of the plasma is determined by the ratio of H_β to continuum emission, or from the line ratio of H_α to H_β. The temperature maximum is observed about 4 µs after ignition of the discharge, reaching a value well above

3 eV. During the following 11 µs, the temperature decreases to about 1 eV, thus, keeping the degree of ionization in the plasma for the first 14 µs to 99 % and higher.

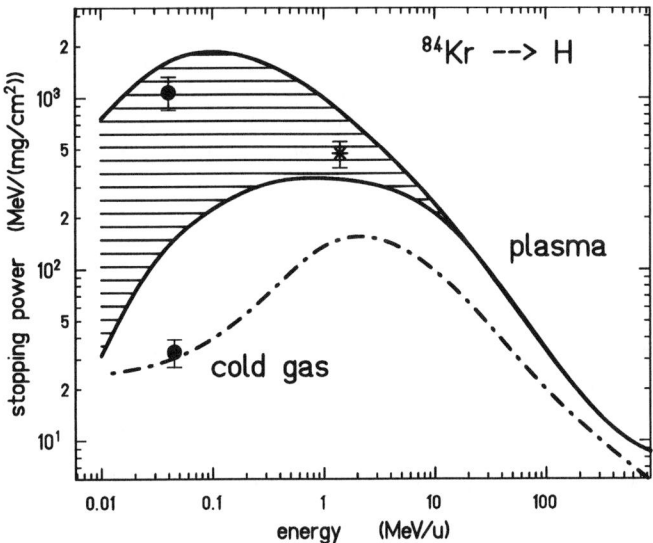

Figure 4: Comparison of experimental results for the stopping power of krypton in a plasma and in cold hydrogen gas with theoretical calculations.

A comparison of the measured stopping power in a plasma and in cold hydrogen gas with theoretical calculations is shown in Figure 4. The solid lines represent theoretical calculations for stopping power in a hydrogen plasma performed with a Monte-Carlo code [10]. Theoretical values for a stopping power in cold hydrogen gas are represented by the dashed–dotted line. The data shown here are within 10 % in agreement of the values with Ref. [13]. For cold hydrogen gas the experimental result yields 33±5 MeV/(mg/cm^2). The dashed area represents the region of stopping power achieved with the maximum charge state expected for a fully ionized plasma (upper curve) and the stopping in a plasma obtained with charge states of cold hydrogen gas (lower curve). For an average ion energy of about 40 keV/u a stopping power in the plasma of 1080±210 MeV/(mg/cm^2) is measured. The experimental value at 1.4 MeV/u was obtained during an earlier plasma stopping experiment performed at the UNILAC accelerator [14].

5 Summary and future activities

It is now proposed to combine the heavy ion related ICF activities existing in Europe, under the auspices of the European Commission into a coordinated feasibility study including also rough cost estimates of an ignition facility, which shall have the potential to attain the threshold of ignition of heavy ion induced ICF. Within this study experimental work will be continued on now existing European ion accelerators so as to provide a basic for the choice of critical driver parameters. Two experiments have been discussed, where already today some of the basic questions of heavy ion fusion have been adressed. Heavy ion beams have been used to heat and evaporate krypton crystals. A specific deposition energy of 60-120 J/g has been induced by a neon beam with $1\text{-}2 \cdot 10^{10}$ ions for a heating time of 1 μs. The produced hydrodynamic reactions in the target are studied. Further experimental efforts are directed to increase the beam currents at the SIS to increase the specific deposition power. In Erlangen a new approach will be tested to produce high energy density in matter using magnetic forces to accelerate the positive and negative ions of a plasma together. A stopping power experiment with MAXILAC demonstrated for the first time the extreme enhancement of stopping power in a plasma close to the theoretical maximum. In comparison to the theoretical cold gas data, an enhancement of a factor 35 was found for this experiment.

References

[1] D. Keefe, Ann. Rev. Nucl. Part. Sci. **32**, 391 (1982
[2] R. Bock, I. Hoffmann, and R. Arnold Nucl. Sci. Appl. **2**, 97 (1984)
[3] C. Deutsch, Ann. Phys. **11**, 1 Paris (1986)
[4] J. Jacoby, D.H.H. Hoffmann, R.W. Müller, K. Mahrt-Olt, R. Arnold, V. Schneider, and J. Maruhn, Phys. Rev. Lett. **65**, 2007 (1990)
[5] F.C. Young, D. Mosher, S.J. Stephanakis, S.A. Goldstein, and T.A. Mehlhorn, Phys. Rev. Lett. **49**, 549 (1982)
[6] J.N. Olsen, T.A. Mehlhorn, J. Maenchen, and D.J. Johnson, J. Appl. Phys. **58**, 2958 (1985)
[7] D.H.H. Hoffmann, K. Weyrich, H. Wahl, Th. Peter, J. Meyer-ter-Vehn, J. Jacoby, R. Bimbot, D. Gardès, M.F. Rivet, M. Dumail, C. Fleurier, A. Sanba, C. Deutsch, G. Maynard, R. Noll, R. Haas, R. Arnold, and S. Maurmann, Z. Phys. **A30**, 339 (1988)
[8] C. Deutsch, G. Maynard, R. Bimbot, D. Gardès, S. Della-Negra, M. Dumail, B. Kubica, A. Richard, C. Fleurier, A. Sanba, D.H.H. Hoffmann, K. Weyrich, and H. Wahl, Nucl. Instrum. Methods Phys. Res., Sect. A **278**, 38 (1989)
[9] K.-G. Dietrich, K. Mahrt-Olt, J. Jacoby, E. Boggasch, M. Winkler, B. Heimrich, and D.H.H. Hoffmann, Laser Part. Beams **8**, 583 (1990)
[10] D.H.H. Hoffmann, K. Weyrich, H. Wahl, D. Gardès, R. Bimbot, and C. Fleurier, Phys. Rev. **A42**, 2313 (1990)
[11] K.-G. Dietrich, D.H.H. Hoffmann, E. Boggasch, J. Jacoby, H. Wahl, M. Elfers, C.R. Haas, V.P. Dubenkov, and A.A. Golubev, Phys. Rev. Lett. **69**, 3623 (1992)
[12] J. Jacoby, D.H.H. Hoffmann, W. Laux, R.W. Müller, H. Wahl, K. Weyrich, E. Boggasch, B. Heimrich, C. Stöckl and H. Wetzler, Phys. Rev. Lett. **74**, 1550 (1995)
[13] L.C. Northcliffe and R.F. Schilling, Nucl. Data Tables **A7** (1970)
[14] K. Weyrich, D.H.H. Hoffmann, J.Jacoby, H. Wahl, R. Noll, R. Haas, H. Kunze, R. Bimbot, D. Gardès, M.-F. Rivet, C. Deutsch, and C. Fleurier, Nucl. Instrum. Methods Phys. Res., Sect. A **278**, 52 (1989)

2. Critical Elements for Ignition

High Convergence, Indirect Drive Inertial Confinement Fusion Experiments at Nova

R. A. Lerche, M. D. Cable, S. P. Hatchett, J. A. Caird,
J. D. Kilkenny, H. N. Kornblum, S. M. Lane, C. Laumann,
T. J. Murphy, J. Murray, M. B. Nelson, D. W. Phillion, H. Powell,
and D. Ress

Lawrence Livermore National Laboratory, P.O. Box 5508, L-473, Livermore, CA 94550

Abstract. High convergence, indirect drive implosion experiments have been done at the Nova Laser Facility. The targets were deuterium and deuterium/tritium filled, glass microballoons driven symmetrically by x rays produced in a surrounding uranium hohlraum. Implosions achieved convergence ratios of 24:1 with fuel densities of 19 g/cm^3; this is equivalent to the range required for the hot spot of ignition scale capsules. The implosions used a shaped drive and were well characterized by a variety of laser and target measurements. The primary measurement was the fuel density using the secondary neutron technique (neutrons from the reaction ^{2}H(^{3}H,n)^{4}He in initially pure deuterium fuel). Laser measurements include power, energy and pointing. Simultaneous measurement of neutron yield, fusion reaction rate, and x-ray images provide additional information about the implosion process. Computer models are in good agreement with measured results.

INTRODUCTION

Experiments recently conducted at the Nova Laser Facility addressed the overall quality of inertial-confinement fusion (ICF) implosions. The experiments were designed to answer two questions: how well do the implosions work and how well can we model them? Many researchers believe high gain ICF will be achieved by hot spot ignition in which a relatively small mass of gaseous fuel at the center of a target heated to 5-10 keV will ignite a higher density, lower temperature gas surrounding the core [1,2]. Because existing lasers are too low in energy to achieve thermonuclear gain, a set of hydrodynamically equivalent implosions are used to demonstrate that important, scalable parameters of ignition capsules are scientifically and technologically achievable.

The implosion experiments described in this talk used gas-filled glass capsules that were symmetrically driven by x rays produced in a surrounding hohlraum or cavity [3]. These implosions simultaneously achieved high convergence ratios (initial capsule radius R_0 to final fuel radius R_f) and implosion configurations of high density glass with hot gas fill that are equivalent to those required for ignition

© 1996 American Institute of Physics

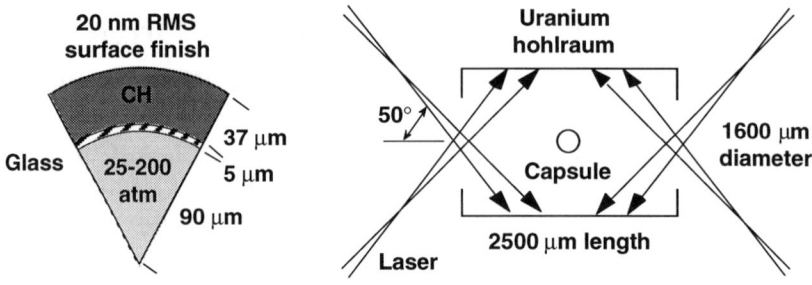

Figure 1. Geometry of glass microballoon and indirect drive hohlraum.

scale targets. The implosions, including nonideal effects, were modeled with detailed computer codes such as LASNEX (a radiation transport and hydrodynamics code). All observable quantities were in good agreement with simulations, demonstrating a good understanding of the energy transport and implosion hydrodynamics.

IMPLOSION EXPERIMENTS

The indirectly driven targets were gas-filled glass microballoons mounted inside uranium hohlraums (see Fig. 1). A relatively small ($R_0 = 90$ μm) capsule was selected to allow secondary neutrons to be used for fuel areal density measurements without saturation of the neutron spectrometer. Deuterium and equimolar deuterium/tritium fill pressure which ranged between 25 and 200 atm was used to vary capsule convergence while maintaining constant drive. The capsules were driven by x rays produced when ten Nova laser beams with a total of 21 kJ of 0.35 μm light were incident on the interior wall of the hohlraum at an intensity of 2×10^{15} W/cm^2. Laser beams were uniformly spaced around the circumference of two rings on the inner surface of the hohlraum.

Capsule implosion is driven by the ablation of the outer surface material of the capsule by x rays. The measured laser power and x-ray drive temperature are shown in Fig. 2. Hohlraum temperature was measured using a time resolved, multichannel, K- and L- edge filtered x-ray spectrometer [4]. Measurements were made looking at both the directly illuminated laser spots and the indirectly illuminated wall. Observed spectra were nearly Planckian. Laser pulse shape was selected to produce an x-ray drive that optimized the pressure-density trajectory of the capsule without producing excessive hydrodynamic instability. Use of a uranium hohlraum and glass shell minimized x-ray preheating of the capsule.

X-ray drive symmetry limits the convergence of these capsules. Drive uniformity for these cylindrically symmetric targets is usually expressed as in terms

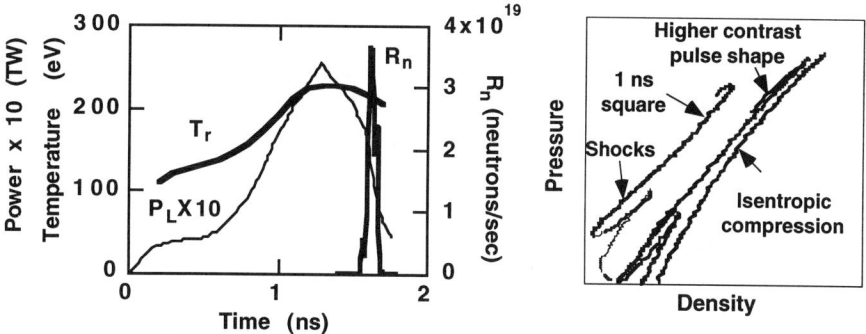

Figure 2. Observed laser power (P_L), hohlraum temperature (T_r), and neutron reaction rate (R_n) for 100 atm DT filled capsule. Graph at right shows how high contrast pulse improves presure-density trajectory. More efficient compression is hydrodynamically less stable. These experiments have moderate growth factors of about 80.

of Legendre polynomials. Symmetry eliminates the odd mode components while higher order even modes are relatively small because each point on the capsule surface views a large region of the hohlraum wall. Capsule dimensions and the laser irradiation point on the hohlraum wall are used to minimize the effect of the first two even moments (P_2 and P_4). Symmetry changes during the laser pulse because both the wall albedo and the effective position of laser energy deposition moves. By design, simulations of this target show a P_2 asymmetry that changes sign during the pulse, averages to a low value, and is at most 8%. Details of hohlraum design are reported by Suter *et al.* [5].

A second source of time-dependent drive asymmetry is imprecise power balance and pointing of the laser beams. During the period of these experiments, the "Precision Nova" project produced substantial improvements in beam-to-beam power balance, beam pointing, and beam synchronization. Power balance in the peak of the pulse was improved to 4% (RMS) and 8% in the foot; previously it had been 10 and 20% respectively. Beam pointing accuracy was improved from ±100 μm to ±30 μm, and beam synchronization was improved from 25 ps to 7 ps (RMS).

Capsule convergence η and final fuel density ρ can be determined from the areal density of the compressed fuel, $"\rho R" = \int_{\text{fuel}} \rho R(r) dr$. If the fuel density is assumed to be uniform, then convergence is given by $\eta = R_0/R_f = (\rho R/\rho_0 R_0)^{1/2}$ and final density by $\rho = \rho R \eta / R_0$. Areal density was determined using the "secondary neutron" technique in which ρR is determined from the ratio of fusion neutrons produced by the secondary reaction $D(T,n)^4 He$ to the number of neutrons produced by the $D(D,n)^3He$ reaction in initially pure deuterium fuel [6]. A "secondary neutron" is produced when a 1.01 MeV triton produced by the $D(D,p)T$ reaction interacts with a deuterium atom. If the tritons do not slow down significantly before they escape from the fuel, then the number of tritons producing neutrons is proportional to the fuel ρR. The shape of the secondary neutron energy spectrum depends on the amount of slowing of the tritons and ranges between 12 to 17 MeV. For fuel in this work, significant triton slowing occurs for fuel densities above a few mg/cm^2 and corrections must be made for cross section energy dependence.

The primary diagnostic for these experiments was a sensitive neutron spectrometer [7]. It measures both the secondary neutron energy spectra and secondary neutron yield. The spectrometer consists of an array of 960 neutron time-of-flight detectors each capable of detecting single neutrons with an energy resolution of 150 keV. To complete a ρR measurement, the primary neutron yield was measured by measuring the activation of an In sample placed near the target.

EXPERIMENTAL RESULTS

In the initial set of experiments, the calculated density was not achieved for low fill pressures (see Fig. 3). Instead of increasing, density decreased at lower fill pressure. After the Precision Nova project was completed and beam balance and pointing improved, the experiments were repeated. This time observed density values increased significantly and matched calculated values. Measured fuel areal densities, which were as high as 16 mg/cm^2, allowed the convergence ratios and fuel densities presented in Fig. 4 to be calculated. The measured values are averages for several implosions (2 at 200 atm, 6 at 100 atm, and 10 at 25 atm); error bars are dominated by the statistics associated with the number of secondary neutrons recorded. For 25 atm capsules, convergence reached 24 with a final fuel density of 19 gm/cm^3. Observed convergence and density values are consistent with simulations when the effects of fuel-pusher mix are included at a level predicted with current models [8] for the capsule surface finish. Fuel-pusher mix produces two important effects: high-Z material moving into the fuel region increases triton slowing and

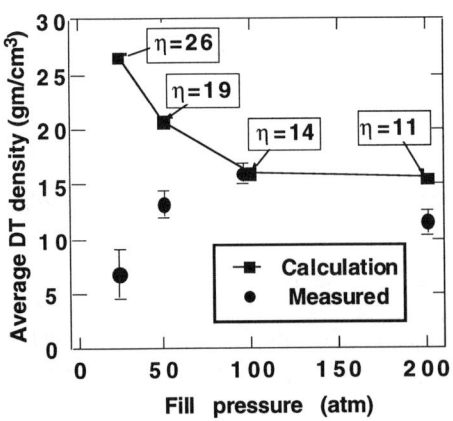

Figure 3. Density measurements made before the completion of "Precision Nova."

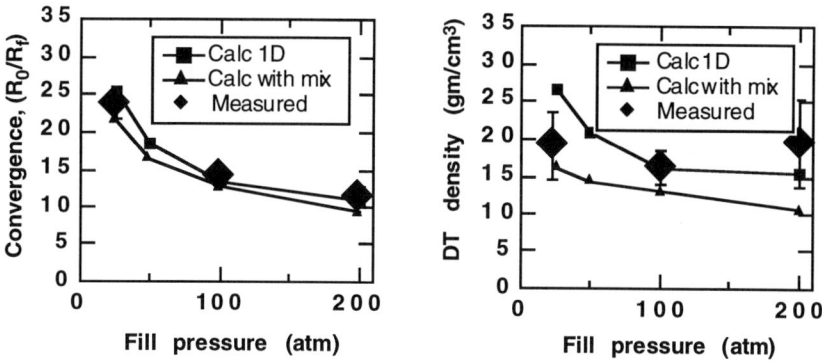

Figure 4. Observed and calulated fuel convergence and density. Measured were made after the completion of Precision Nova. Calculations are for clean 1D and with mix. Density is expressed as an equivalent deuterium-tritium fill even when the capsule is filled with deuterium.

fuel mixing into the shell reduces convergence. The amount of triton slowing can be estimated from the shape of the secondary neutron spectrum. A spectrum obtained by summing the 25 atm capsule data (10 implosions) is shown in Fig. 5 along with a simulated spectrum that includes mix. The energy spectrum is in good agreement with simulation supporting the validity of the mix model.

A number of other experimental measurements were also made for these targets. Neutron yields (2.45 MeV neutrons) for deuterium filled targets were 2×10^7 to 3×10^8, while equimolar deuterium/tritium capsules produced 3×10^8 to 3×10^9 neutrons (14 MeV) with the lower yields observed for the lower fill, higher convergence capsules. Pusher areal density was measured by activation of Rb dopants in the glass shell [9]. Observed glass shell areal densities were 73 ± 16 mg/cm^2 (2 shots at 100 atm) and 60 ± 19 mg/cm^2 (2 shots at 25 atm), while simulations show 54 and 81 mg/cm^2, respectively. Fuel ion temperature was determined from the Doppler broadening of the primary neutron energy spectrum recorded with a time-of-flight detector that is separate and simpler than that used for the secondary neutrons [10]. Fuel temperatures for all fill pressures were 0.9 ± 0.4 keV. Capsule burn duration and burn time relative to the laser pulse were determined with a fast scintillator-streak camera detector capable of 30 ps temporal resolution [11]. Burn for 100 atm capsules occurred at 1600 ± 100 ps and lasted for 50 ± 15 ps (see Fig. 2); simulations show 1603 ps and 33 ps, respectively.

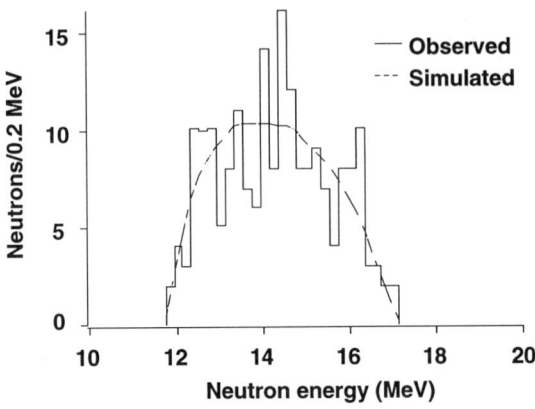

Figure 5. Secondary neutron energy spectrum. Observed data is the sum of ten 25 atm capsules.

These experiments demonstrate the effect of improved "Precision Nova" beam quality and associated laser and target diagnostics on target performance. Through detailed comparisons between measured quantities and simulations, a deeper understanding was obtained for the sensitivity of the implosion process to laser and target parameters. In these experiments we demonstrated our ability to model and control implosion dynamics sufficient enough to achieve convergence levels comparable to those required for the hot spot of an ignition scale capsule.

This work was performed under the auspices of the U. S. Department of Energy by the Lawrence Livermore National Laboratory under Contract No. W-7405-ENG-48.

REFERENCES

1. J. Nuckolls et al., *Nature* (London) **239**, 139–142 (1972).
2. J. D. Lindl, R. L. McCrory, and E. M. Campbell, *Phys. Today* **45**, No. 9, 32–40 (1992).
3. M. D. Cable et al, *Phys. Rev. Lett.* **73**, 2316–2319, (1994).
4. H. N. Kornblum, R. L. Kauffman, and J. A. Smith, *Rev. Sci. Instrum.* **57**, 2179–2181 (1986).
5. L. Suter et al., *Phys. Rev. Lett.* **73**, 2328–2331 (1994).
6. M. D. Cable and S. P. Hatchett, *J. Appl. Phys.* **62**, 2223 (1987).
7. M. B. Nelsom and M. D. Cable, *Rev. Sci. Instrum.* **63**, 4874–4876 (1992).
8. S. Haan, *Phys. Rev. A* **39**, 5812–5825 (1989).
9. S. M. Lane and M. B. Nelson, *Rev. Sci. Instrum.* **61**, 3298–3300 (1990)
10. T. J. Murphy and R. A. Lerche, *Rev. Sci. Instrum.* **63**, 4883–4885 (1992).
11. R. A. Lerche, D. W. Phillion, and G. L. Tietbohl, *Rev. Sci. Instrum.* **66** 933–935 (1995).

Symmetry Experiments in Gas Filled Hohlraums at NOVA

N.D. Delamater*, T.J. Murphy‡, A.A. Hauer*, R.L. Kauffman‡, A.L. Richard**, E.L. Lindman*, G.R. Magelssen*, B.H. Wilde*, L.V. Powers‡, S.M. Pollaine‡, L.J. Suter‡, , R. Chrien*, D.B. Harris*, M.B. Nelson‡, M.D. Cable‡, J.B. Moore‡, K. Gifford+, and R.J. Wallace‡

*Los Alamos National Laboratory, Los Alamos, NM. 87545
‡Lawrence Livermore National Laboratory, Livermore, CA. 94551
+General Atomics, San Diego, CA.
** Centre D'Etudes De Limeil-Valenton, France.

Abstract

Understanding drive symmetry in gas filled hohlraums is currently of interest because the baseline design of the indirect drive ignition target for the planned National Ignition Facility uses a gas filled hohlraum. We report on the results of a series of experiments performed at the Nova laser facility at Lawrence Livermore National Laboratory with the goal of understanding time dependent drive symmetry in gas filled hohlraums. Time dependent symmetry data from implosions in gas filled hohlraums will be discussed. The purpose of filling the hohlraum with gas is to tamp the motion of the high-Z material ablating from the hohlraum walls, reducing the motion of the laser deposition regions and resultant temporal variations in drive symmetry. We have obtained time integrated and time resolved x-ray images of the implosion of plastic deuterium filled capsules, neutron yields, implosion times and spectroscopy of argon emission from the imploded core. Preliminary results show that the gas is effective in impeding the motion of the wall blowoff material, and that the resulting implosion is in qualitative agreement with modeling. These experiments are relevant to those currently being planned for the National Ignition Facility.

Introduction

We report the results of a series of experiments at the Nova laser which measure the symmetry of implosions in gas-filled hohlraums. We are pursuing these experiments because the baseline design of NIF targets use gas-filled hohlraums and we wish to benchmark and verify currently used theoretical modeling codes. Our experiments apply techniques which have been used previously[1,2] in developing the symmetry scaling database for experiments in vacuum. This includes time integrated symmetry measurements and measurements with time-resolved techniques. In this paper, we give details of the experimental measurements, discuss the results and give future directions of our work.

Experimental Design

We measure time integrated symmetry by imaging the x-ray self-emission of the stagnating core of an imploded plastic microballoon. The shape of the stagnated core gives a direct measure of the drive symmetry imposed on the target. Capsules that are typically used in such measurements are filled with 50 atm of deuterium fuel and a small amount (0.1 atm) of Ar which helps make the imploded core visible in x-ray imaging at about 3 keV. The standard capsule which has been used in symmetry experiments is about 540 microns in total diameter with a total shell thickness of 55 microns. The capsule is mounted in a gold hohlraum of lengths varying from about 2000 to 2500 microns. The hohlraum diameter is 1600 microns with a laser entrance hole diameter of 1200 microns. Hohlraums which are gas filled have windows over the laser entrance hole which are made of mylar (0.6µm) or polyimide (0.35µm). The laser pulse used in our experiments is a shaped pulse. Temporally shaped laser pulses are necessary to produce efficient compression. The pulse delivers about 26 kJ of energy in 2.2 ns and has a low power foot with a 1:3 contrast ratio with the peak power. Diagnostics included time gated and time resolved pinhole camera x-ray imagers which provide spatial resolution of about 6 µm and time resolution of 80 ps (picoseconds). The diagnostics view both along the hohlraum axis and perpendicular to it. Other diagnostics were time resolved spectrometers to measure Ar line spectra from the fuel, neutron detectors to measure yield, ion temperature and fuel ρR.

A goal of the experiment was to observe capsule performance (compression, neutron yield, fuel ion temperature) in gas-filled hohlraums and measure differences if any from earlier vacuum results with the same hohlraum and laser pulseshape. We performed a symmetry scaling experiment by measuring the shape of the imploded fuel core as a function of the laser pointing. As has been seen before with vacuum implosions, the shape of the core changes with a characteristic P_2 asymmetry as the beams are moved in and out. The Nova laser beams can be aimed with an accuracy of 25µm. We decided to perform a number of shots with various pointing conditions to map out a symmetry scaling of shape vs pointing for both methane and propane filled hohlraums, providing a scan in gas densities representative of potential conditions in NIF hohlraums. In addition we sought to investigate any effects on symmetry which may be due to the windows over the laser entrance hole so it was decided to shoot some targets with windows but no gas.

Results

Our principal results are that the gas in the hohlraum is effective in impeding the motion of the gold blowoff from the wall, and that the resulting implosion performance of the capsule is not significantly degraded from vacuum results. We have also demonstrated the sensitive control of symmetry with our pointing scans in gas-filled hohlraums just as with the pointing scans in vacuum. The implosion symmetry, however, differs somewhat from vacuum results with similar laser pointing. The drive on the capsule appears to be "pole high" (hotter at the poles than the equator) and this is a surprise as it is not predicted with current calculations[3]. The implosion acts as though the laser beams were shifted about 150 µm outward. We have looked at effects of the windows by comparing targets with thin and thick windows and targets with windows only and no gas. We find that when gas is in the hohlraum, the implosion symmetry is not sensitive to the thickness of the windows. However, in shots with windows but no gas, the thickness of the window has a significant effect on the implosion symmetry. In these cases, the thicker window produces a more oblate implosion.

Figure 1 shows the capsule distortion (major to minor axis ratio) versus laser pointing ("pointing scan") results for methane implosions. The plot also shows vacuum results and it can be seen that the line defining the pointing scan is shifted 150μm inward from the vacuum data. Figure 2 shows the yields for these methane implosions as a function of capsule distortion. It can be seen that yields approaching the vacuum results are attained only at near round implosions and the yield falls off abruptly for oblate shaped implosions (distortion greater than 1). This behavior had not been seen previously. Figure 3 shows the effect of the gas at holding back the gold ablating from the wall. The figure shows the x-ray image at ~3keV in the axial direction as the capsule reaches peak emission (~2.5ns); note that the higher density gas (propane) holds off the gold more effectively than the lower density gas (methane). In vacuum it was seen that the gold had already stagnated on axis by the implosion time.

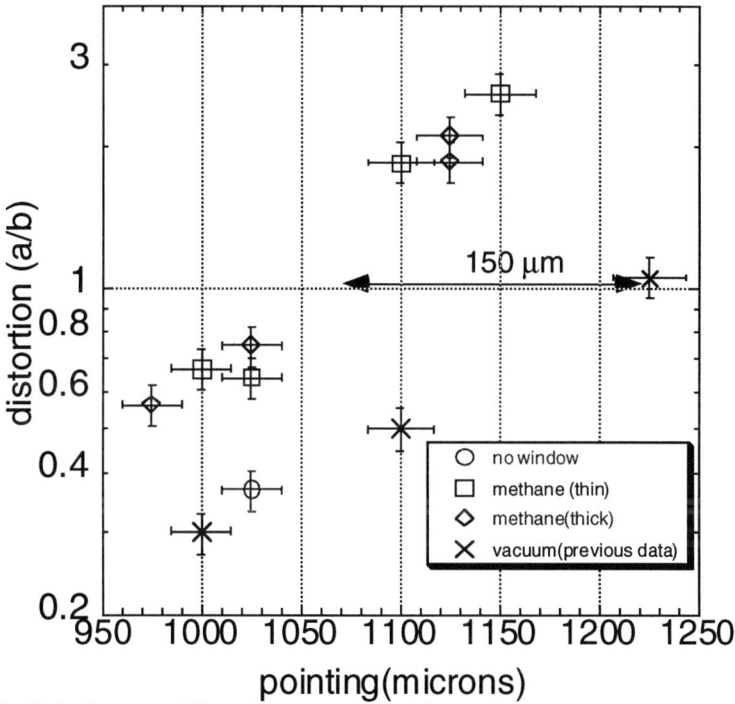

Figure 1. Pointing scan (distortion vs pointing) in methane

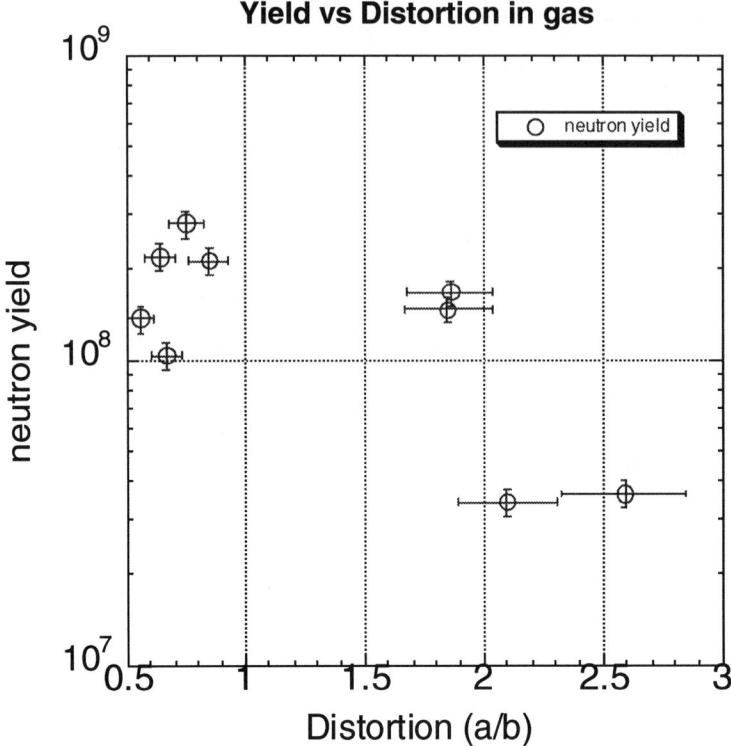

Figure 2. Yield vs distortion in methane

Propane is more effective at impeding motion of gold than methane

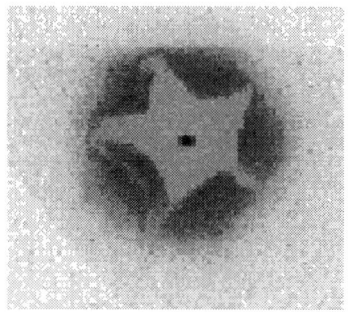 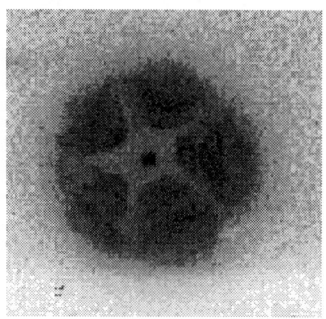

propane
bang time=2.6ns
yld=5.6e7

methane
bang time=2.5ns
yld=1e8

hohlraums used thin (3500A) polyimid windows
Figure 3. Axial images of gas implosions showing wall motion

Summary and Future Work

We have shown that implosion symmetry can be sensitively controlled with implosions in gas, just as in vacuum implosions. We have seen that capsule performance in gas, as measured by yield and compression, is not significantly degraded from the vacuum results. We have also discovered that the drive on a capsule in gas filled hohlruams is somewhat hotter at the poles than expected, indicating a pointing shift of about 150µm in the symmetry scaling as compared to earlier results in vacuum. Explaining this shift is the subject of current work and may be due to such effects as laser beam refraction by the window plasma or thermal conduction. We are also pursuing time resolved symmetry methods in order to determine whether the "refraction effect" is occuring at early or late times.

Acknowledgements

The authors acknowledge the valuable assistance of the Nova laser and target operation staff. We would also like to acknowledge the invaluable role of the Los Alamos and Lawrence Livermore technical staff in the execution of these experiments. This work was performed under the auspices of the U.S. D.O.E. under contract no. W-7405-ENG-36.

References

1. Hauer, A.A., et. al., Rev. Sc. Inst. 66, 672 (1995).
2. Suter, L.J., et. al., Phys. Rev. Letters, 73, 2328 (1994).
3. Lindman, E.L., et. al., "Modeling of Drive Symmetry Experiments in gas filled hohlraums at Nova", to be published in the Conference Proceedings of the 12th International Conference on Laser Ineraction and Related Plasma Phenomena, Osaka, Japan, 1995.

Implosion Experiments with Uniformity-Improved GEKKO XII : Overview

Y. Kato, H. Azechi, H. Takabe, N. Miyanaga, T. Kanabe,
T. Norimatu, H. Nishimura, H. Shiraga, M. Nakai, R. Kodama,
K. A. Tanaka, M. Takagi, M. Nakatsuka, K. Nishihara,
K. Mima, T. Yamanaka, and S. Nakai

Institute of Laser Engineering, Osaka University
2-6 Yamada-oka, Suita, Osaka, 565 Japan

Abstract. We report the results of the recent implosion experiments performed with the uniformity-improved GEKKO XII laser. A high quality plastic shell filled with deuterium gas was irradiated with the power-balanced, smooth-profile laser beams of 526 nm wavelength. The laser pulse was square-shaped with a controlled prepulse in order to reduce the instability growth. The observed neutron yield under these conditions correspond to the 1-D simulation value well within the stagnation phase. These results suggest that the implosion stability is now better controlled in comparison to previous experiments.

INTRODUCTION

Ignition and burn of the fusion fuel in the inertial confinement fusion requires formation of a hot spark at the center of a compressed high-density fuel. Generation of a compressed plasma which has the same structure as the igniting plasma, but with a smaller scale, is an important proof-of-principle experiment for achieving ignition and burn. It will provide the data bases for accurately predicting the conditions for ignition, burn and high gain.

We have achieved previously a compression density of approximately 600 g/cm^3 by irradiation of a spherical plastic hollow shell target directly with 12 laser beams [1]. The experimentally measured values of the implosion velocity (2.5 x 10^7 cm/s), the compression density (540 g/cm^3), and the temperature of the compressed plasma (0.4 - 0.8 keV) were in agreement with the corresponding values for the cold part of the compressed core calculated by the one-dimensional (1-D) simulation code. However, the measured neutron yield of 1 x 10^6 was approximately 3 orders of magnitude smaller in comparison to the 1-D simulation value.

In a series of implosion experiments using gas-filled targets irradiated by a 1-ns Gaussian pulse, that followed this high density compression demonstration, the observed neutron yields corresponded to the 1-D simulation values at the start of the stagnation phase, suggesting that the fusion reaction was significantly reduced in the stagnation phase. These results show that, although formation of the high density cold fuel has been demonstrated, formation of the central hot spark required for ignition of the fusion reaction has not been achieved. Detailed analyses on the implosion process show that the central hot spark will be formed only when the growth of the perturbations due to the target and irradiation non-uniformity is carefully controlled [2].

In order to perform the implosion experiments under the controlled conditions, we have implemented power balancing of GEKKO XII together with beam smoothing [3]. Since July 1994, a series of experiments have been carried out with this uniformity-improved GEKKO XII. In these experiments, the target non-uniformity was characterized in detail and the laser pulse shape was re-designed to reduce the Rayleigh-Taylor (R-T) instability growth rate. The neutron yields that have been observed so far correspond to the 1-D simulation values well within the stagnation phase. These results indicate that the implosion stability is now under better control, an important step toward formation of the central hot spark.

EXPERIMENTAL CONDITIONS

The important parameter in the target design is the growth of the R-T instability. We have made linear analysis for evaluation of the e-folding of the R-T instability (γt) for various implosion conditions based on the 1-D ILESTA simulation. Figure 1 shows γt vs. mode number evaluated for different pulse shapes: a 1.6 ns-width Gaussian pulse, a 1.6 ns-width square pulse, and a 1.6 ns-width square pulse preceded with a 0.2 ns-width prepulse with a time gap of 0.4 ns. The γt allowed in typical ignition designs is 6-8, the shaded region in Fig.1. With a single Gaussian pulse irradiation, γt becomes 16 which is too large for stable implosion over the

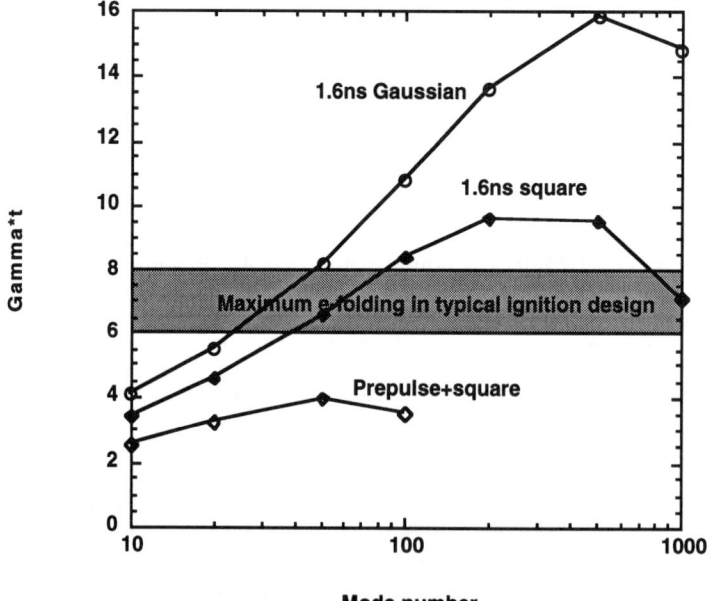

Figure 1. The e-folding gt of the Rayleigh-Taylor instability vs. the mode number for different laser pulse shapes: a 1.6 ns Gaussian pulse, a 1.6 ns square pulse, and a prepulse + a square pulse. The gt in typical ignition target designs is shown by the shaded region.

whole implosion process. Shortening the pulse rise time increases the adiabat, thereby reducing γt to around 10. Adding a prepulse further reduces the R-T growth making the implosion more stable, although it is accompanied with decrease in the compression density. In the present experiment, we have varied γt by changing the time gap between the prepulse and the main pulse.

A polystyrene shell of 520 μm diameter and 8 μm thickness was filled with D_2 gas. The polystyrene shell was used because it has good thickness uniformity of over 99 % and small surface roughness of 2 nm rms over $l \geq 50$. The shell thickness was measured over 4π rad using a reflection microscope spectrometer with 5 nm measurement accuracy. The surface roughness measured with a WYCO interferometer was expanded into spatial frequencies to determine the mode number distribution of the initial perturbation, which was used for evaluation of the instability growth.

Since the polystyrene shell cannot contain a high pressure gas over a long time, the target was filled with D_2 *in situ* and was kept in vacuum under a liquid nitrogen temperature in a cryostat. The cooled target was exposed to an open vacuum just before laser irradiation by removing the shroud of the cryostat at a high speed [4]. The fill pressure was varied from 30 to 5 atm in order to change the convergence ratio of implosion.

The irradiation non-uniformity of high mode numbers (l greater than 10) were reduced by operating GEKKO XII with the partially coherent light (PCL) mode. A divergent laser pulse of a broad-bandwidth (0.6 nm) generated at the front end was amplified through the 12 amplifier chains with a large acceptance angle (32 times diffraction limit) and converted to the second harmonic radiation of 526 nm wavelength. The low-mode non-uniformity (l less than 10) was reduced by using random phase plates of less than 1 % phase error. The low-mode non-uniformity was also minimized by balancing the powers of the 12 laser beams over the whole pulse duration [3]. The power imbalance in this experiment was 3-6 % rms. The time integrated absorption non-uniformity evaluated from the measured beam pattern at the equivalent focal plane is estimated to be 2.5-4.5 % rms. The laser energy at 526 nm was typically 3 kJ in the power-balanced experiment.

The square laser pulse with a prepulse was produced from a fast-rise 0.2 ns single pulse at the front end using a fiber-coupling pulse adder. After amplification, the prepulse and the main pulse had a rise time of 50 ps. The square pulse had a constant power to within 5 % over the 1.6 ns pulse duration and the prepulse had the same peak power. The separation between the prepulse and the main pulse was varied from 0 (no prepulse) to 0.8 ns. This laser pulse was accompanied with a long duration, low intensity foot pulse due to amplified spontaneous emission at the front end, which is not desirable since it may cause imprinting of perturbation before target acceleration. The intensity of the foot pulse was reduced to 2×10^{-4} of the main pulse.

EXPERIMENTAL RESULTS

The irradiation uniformity on the target was evaluated by two independent measurements: the x-ray emission distribution measured with an advanced Kirkpatrick-Baez microscope coupled to an x-ray CCD camera [5], and the spatial intensity distribution of the 263 nm emission which is the second harmonic of the irradiation laser [6]. These measurements indicate that the major non-uniformity arises from smaller laser energy deposition at oblique incidence angles on the target due to smaller absorption rate . The mode-6 non-uniformity due to this factor may amount to 3.5 % in the present configuration.

The stability in the acceleration phase was studied using a tracer layer target where a chlorinated CH shell was coated with a 5-μm thick CH [7]. Addition of a picket pulse in front of the main pulse reduced the emission region of the Cl He-β line in the time-integrated, spectral-resolved x-ray imaging, showing that the prepulse stabilized the acceleration phase.

Figure 2 shows the neutron yield N_y vs. the time gap t_s between the prepulse and the main pulse. The closed circles and the closed squares are the measured neutron yields for the weak (2×10^{-4}) and the medium (3×10^{-3}) intensity foot pulses, respectively. The N_y for the stronger foot pulse intensity was below the detection limit of 10^4. When the foot pulse is controlled to below 2×10^{-4}, The N_y stays almost constant as t_s is increased up to 0.4 ns, above which N_y gradually decreases with t_s. These results were reproduced in repeated series of experiments. Comparison of these results with simulation is presented in the next section.

The ion temperature was deduced from the neutron spectrum measured with a 96-channel neutron spectrometer via the first hit technique [8]. The measured ion temperature was somewhat lower than the value expected for the fuel at the stagnation phase. Since this is the new measurement technique, careful evaluation of the measurement procedure is being made.

Dependence of the neutron yield on the convergence ratio (initial fuel radius/ compressed fuel radius) was studied in the prepulse + square-pulse irradiation with $t_s = 0.4$ ns by varying the D_2 pressure from 30 to 5 atm. The closed circles in

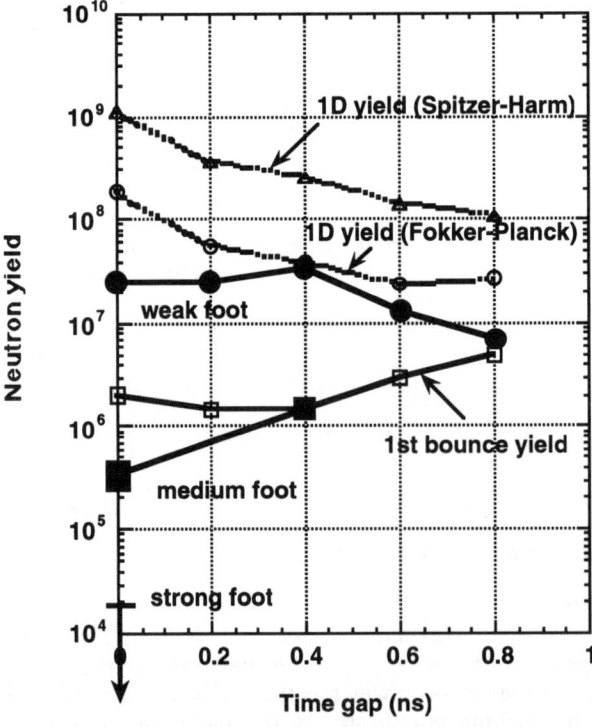

Figure 2. The neutron yield vs. the time gap between the prepulse and the main pulse. The closed circles and squares are the experimental results with weak and medium intensity foot pulses, respectively. The open triangles and squares are the 1-D simulation yields time-integrated over the whole implosion and up to the 1st shock bounce, respectively. A Spitzer-Harm transport model was used in this code. The open circles are the 1-D simulation yields with a Fokker-Planck transport model.

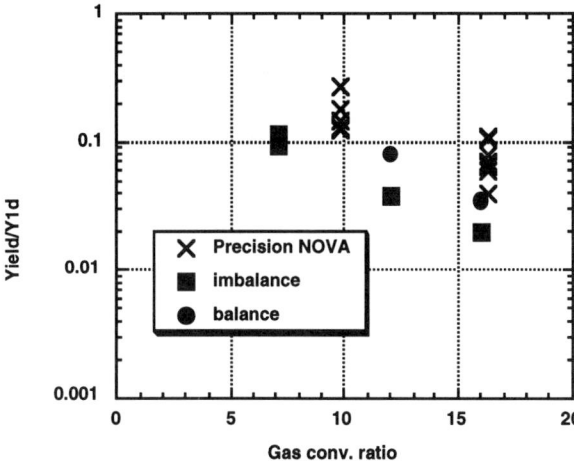

Figure 3. The neutron yield ratio (experimental yield / 1-D simulation yield) vs. the gas convergence ratio. The closed circles and squares are the present experimental results with and without good power balance, respectively. The cross marks show the results reported from LLNL.

Fig. 3 show the neutron yield ratio (observed yield/ 1-D simulation yield) vs. the convergence ratio. The yield ratio decreased gradually from 0.1 to 0.04 as the convergence ratio was increased from 12 to 16. The neutron yield decreased when the mode-1 power balance was offset by an order of ~20 % as shown by closed squares. Although the yield ratio is less than unity, the value of 0.04 for the 16 x convergence is significantly higher than our previous result obtained with a single Gaussian pulse PCL irradiation (0.008 for the 7.4 x convergence). For comparison, the results obtained at LLNL with the Precision NOVA [9] in the indirect-drive experiments are shown by the crosses in Fig. 3, where the radial convergence reported by LLNL (initial target outer radius/ compressed fuel radius) has been renormalized to the gas convergence in the present definition.

DISCUSSION

In Fig. 2, we compare the experimental results of the neutron yield with the simulation values obtained with the 1-D ILESTA code (open triangles). The latter are the time integrated yield for the whole implosion process, whereas the yield integrated up to the start of the stagnation phase (first bounce yield) is shown by the open squares. The neutron yields calculated by the 1-D HIMICO code, where the Fokker-Planck equation is used for the electron transport [10] instead of the flux limited Spitzer-Harm model in ILESTA, are shown by the open circles. In the Fokker-Planck treatment, the neutron yield decreases due to the non-local electron transport which increases the shell adiabat.

Comparing the experimental yields (closed circles) with the standard 1-D Spitzer-Harm calculation (open triangles), the difference between these values decreases as the time gap increases from 0 to 0.4 ns. This indicates that, at the small prepulse time gaps, the implosion is less stable due to increase in γt. At $t_s=0.4$ ns, the experimental yield is approximately 10 % of the 1-D simulation value over the whole implosion and significantly (20 x) higher than the value at the start of stagnation (open squares).

Figure 4 shows the flow diagrams near the stagnation phase and the neutron yields for the two irradiation conditions: single 1-ns Gaussian pulse irradiation (left) and the present, prepulse + square-pulse irradiation (right). The target size

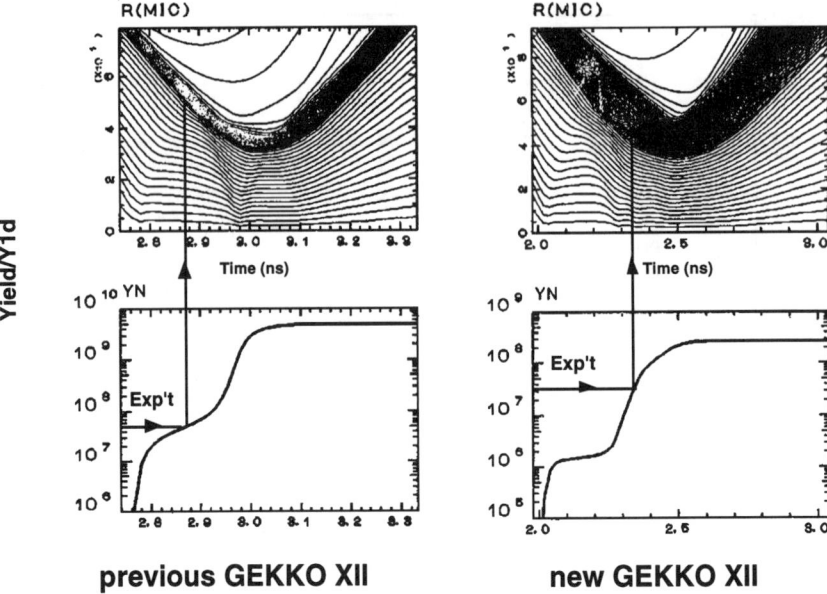

Figure 4. The flow diagrams (upper figures) and the time-integrated neutron yields (lower figures) for the 1-ns Gaussian pulse irradiation (left figures) and the prepulse + main pulse irradiation (right figures), respectively.

and the laser energies are similar in these cases. In the Gaussian pulse irradiation, the observed neutron yield corresponds only up to the first bounce of the shock wave. In comparison, the neutron yield in the present experiment correspond almost close to the second bounce of the shock. These results indicate that the fuel compression and heating continued well during the stagnation phase and the R-T instability is now in better control in comparison to previous experiments.

However the implosion is still away from complete spherical symmetry. The x-ray emission distribution of the compressed core observed with the advanced Kirkpatrick-Baez microscope having a 3-μm resolution shows that the core image is comprised of localized x-ray emission regions that correspond to the 12 illumination patterns of GEKKO XII. Determination of the relationships between the initial and final phases of implosion based on experimental observations and multi-dimensional simulations is necessary for quantitative evaluation for formation of the central hot spark.

SUMMARY

We have performed implosion experiments incorporating the following aspects, using the uniformity-improved GEKKO XII:
1. Reduction of the initial target perturbation by use of the cooled plastic shell target which has small low-mode as well as high-mode non-uniformity.
2. Improvement in irradiation uniformity; partially coherent light for high mode suppression, use of new random phase plates with small phase errors

for low-mode suppression, and power balancing between 12 beams for low mode suppression.
3. Target irradiation with a fast rise-time square pulse preceded by a prepulse to reduce the R-T growth rate.

The major results that have been obtained so far are summarized as follows:
1. The neutron yields observed with the prepulse irradiation correspond to the 1-D simulation value well within the stagnation phase.
2. The neutron yield ratio of 4 % has been achieved for the gas convergence ratio 16.

These results show that the implosion stability is better controlled during the stagnation phase in the present experiment. Further experiments are in progress to characterize the detailed aspects of the implosion process and to achieve 1-D compression at higher stagnation.

ACKNOWLEDGMENT

We would like to thank the staffs at the Institute of Laser Engineering for the technical support to this work.

REFERENCES

1. H. Azechi, T. Jitsuno, T. Kanabe, M. Katayama, K. Mima, N. Miyanaga, M. Nakai, S. Nakai, H. Nakaishi, M. Nakatsuka, A. Nishiguchi, P. A. Norrays, Y. Setsuhara, M. Takagi, M. Yamanaka, and C. Yamanaka, *Laser and Particle Beams* **9**, 193-207 (1991).
2. H. Takabe, K. Mima, K. Nishihara, A. Nishiguchi, M. Murakami, Y. Fukuda, H. Azechi, and S. Nakai, "Numerical Studies on Stability and Mixing in Laser Driven Implosion", presented at 15th Int. Conf. Plasma Phys. Contr. Nuc. Fusion Res., Seville, Spain, Sept. 26-Oct.1, 1994, Paper no. IAEA-CN-60/B-P-9.
3. M. Nakatsuka, T. Kanabe, T. Jitsuno, N. Miyanaga, and S. Nakai, "Improvement of the Irradiation Uniformity of GEKKO XII - Power Balance -, presented at the 12th Int. Conf. Laser Interaction and Related Plasma Phenomena, Osaka, Japan, April 24-28, 1995, Paper no. TuI-1 (in these Proceedings).
4. T. Norimatsu, H. Ito, C. Chen, M. Yasumoto, M. Tsukamoto, K. A. Tanaka, T. Yamanaka, and S. Nakai, *Rev. Sci. Instrum.* **63**, 3378-3383 (1992).
5. R. Kodama, N. Ikeda, and Y. Kato, unpublished.
6. T. Ohnishi, K.A. Tanaka, A. Ozaki, et al., "Measurement of Absorption Distribution by Second Harmonic and X-Ray Images", presented at the 12th Int. Conf. Laser Interaction and Related Plasma Phenomena, Osaka, Japan, April 24-28, 1995, Paper no. TuP-7 (in these Proceedings).
7. H. Nishimura, H. Honda, K. Fujita *et al.*, "Experimental Study on Direct-Drive Implosion Dynamics and Stability with Power-Balanced, Partially Coherent GEKKO XII Green Laser", presented at the 12th Int. Conf. Laser Interaction and Related Plasma Phenomena, Osaka, Japan, April 24-28, 1995, Paper no. TuP-9.
8. R.A. Lerche, S.P. Hatchett, M.D. Cable, M.B. Nelson, *Rev. Sci. Instrum.* **63**, 4877-4879 (1992).
9. M.D. Cable *et al.*, *Phys. Rev. Lett.* **73**, 2316-2319 (1994).
10. A. Nishiguchi, K. Mima, H. Azechi, N. Miyanaga, and S. Nakai, *Phys. Fluid.* **B4**, 417-422 (1992).

Improvement of the Imploded Core Performance with Uniform Gekko-XII Green Laser System

H. Shiraga, M. Heya, M. Nakasuji, H. Azechi, N. Izumi,
T. Yamagajo, M. Saito, T. Urano, Y. Kitagawa, S. Miyamoto,
Y. Kato, H. Takabe, K. Mima, K. Nishihara and S. Nakai

Institute of Laser Engineering, Osaka University
2-6 Yamada-Oka, Suita, Osaka 565 Japan

Abstract. Investigation of the hot spark heating is now possible with a 10-ps temporal resolution. Hot spark heating was improved with the increased stand-off distance and RT stabilized implosion mode. Irradiation nonuniformities have been reduced by precise power balancing of the beams and by the partially coherent light technique. Improved spark heating in the deceleration phase resulted in enhanced neutron production up to the second bounce yield.

INTRODUCTION

High density implosion of the main fuel has been successfully achieved already up to 600 times liquid density(1). On the other hand, another function of the imploded core plasma is to create a high temperature and relatively low density hot spark within the surrounding low temperature and high density main fuel plasma. Thus the kinetic energy of the incoming main fuel or pusher must be efficiently converted to the thermal energy of the hot spark plasma via deceleration process at the final phase of the implosion. So far, the experimentally observed neutron yields except the stagnation-free implosions(2) have been much lower than those predicted by one-dimensional hydrodynamic simulations and more likely been those accumulated up to the beginning of the deceleration, indicating that some discrepancy in spark heating is taking place in the deceleration phase. Therefore it is one of the most important issues in ICF research to improve the spark heating in heavily stagnated implosions.

We have performed recently a series of implosion experiments to improve the target core performance with uniformity-improved GEKKO-XII green laser system. Imploded core dynamics with ultra-fast x-ray imaging and core performance were measured in the experiments. Comparing the results with simulations, we conclude that we could improve the spark heating up to the time when the secondary reflected shock wave hits the pusher.

CORE DYNAMICS

In the experiments, plastic shell targets with diameter of 500-600 µm, thickness of 7 µm were irradiated with Gekko-XII 0.53 µm laser light with 5-6 kJ energy in 1.3-2.5 ns pulse. Targets were filled with 30-40 atm DD gas doped with small amount of Ar. Such implosion can be considered as a model for creating the central hot spark and surrounding cold main fuel structure at the present level of the experiment. DD/Ar gas part is considered as the hot spark with relatively high temperature and low density, while the plastic shell as the main fuel with low temperature and high density. Thus we can investigate the hot spark heating by observing x-rays and neutrons emitted from the imploded DD/Ar region, as illustrated in Figure 1.

Time-resolved Ar x-ray images of the imploded core can give information on structure and history of the spark heating. Since the typical life time of the imploded core plasma is of the order of 100 ps, temporal resolution should be as good as 10 ps. We have recently developed a new technique, Multi-Imaging X-ray Streak Camera(MIXS) (3,4), to obtain ultra fast 2-dimensional x-ray images to meet these requirements. An array of multiple (in this case, 11) identical pinhole images were made on slit photo cathode of an x-ray streak camera with a certain tilting angle. Then, streaked multiple 1-dimensional images each observing different positions of the target at a certain time were broken into pieces and reconstructed to form the original 2-dimensional image, which is now time selected. In the present system, time- and spatial-resolutions were 11 ps and 15 µm, respectively. Observed photon energy was in the range of 2.5-4.9 keV. Typical images are shown elsewhere (4).

Figure 2 shows experimental results of the MIXS observation in two cases of the implosion, one is low intensity laser irradiation and the other is high intensity. Comparing the results of x-ray intensity at the target center with the 1-dimensional simulation, it is found that the spark heating is terminated in the way of the stagnation. In the low intensity case, termination takes place at the time when the reflected shock wave hits the incoming plastic shell, that is the beginning of the deceleration phase. In the high intensity case, experimental x-ray intensity follows the simulation for a longer time by about 70 ps, indicating that the spark heating is improved.

What is the difference between these two data? Estimated Rayleigh-Taylor (RT) growth spectrum in the simulation is essentially the same for both cases. Difference is in the stand-off distance between the laser absorption region and the ablation front. In the high intensity case, it is larger than that in the low intensity

Time- and space- resolved measurements are required.

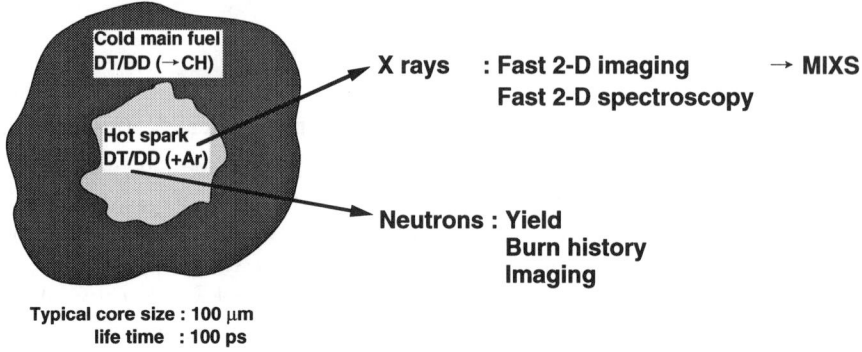

Figure 1. Diagnostics for investigating the spark heating.

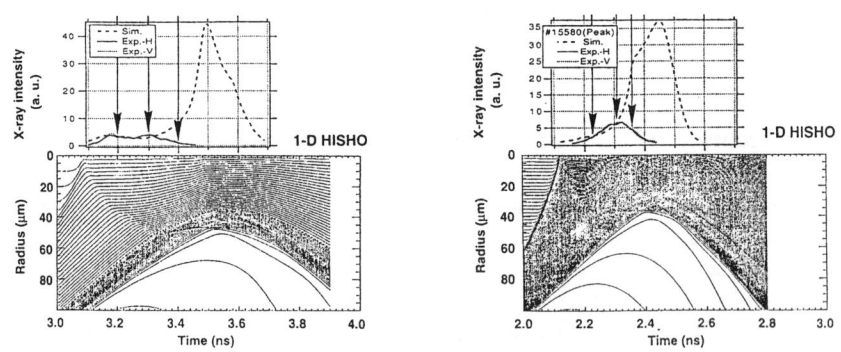

Figure 2. Improvement of the spark heating. (a)Low intensity irradiation, (b)high intensity irradiation. Upper figures are history of observed and calculated x-ray emission intensity at the center of the target. Lower figures are calculated flow diagrams.

case by a factor of about 2, which is due to the higher plasma temperature. From these results, one can conclude that the spark heating can be improved by enhancing the thermal smoothing with larger stand-off distance in the high intensity irradiation. However, the achieved heating is still far below the final state of the simulation. We need much more improvement.

CORE PERFORMANCE

In order to further investigate and improve the spark heating, we have conducted a series of experiments with uniformity improved Gekko-XII laser system. Low-mode drive nonuniformities on the spherical target have been reduced by improving the power balance between twelve laser beams down to about a few %(5). High-mode nonuniformities have been reduced by more than one order of magnitude by introducing partial coherent light technique into the laser system(6). Furthermore, the implosion mode has been controlled to make the implosion more RT stable by varying the laser pulse shape, a 1.6 ns square shaped pulse with a foregoing 0.2 ns single picket fence pulse(6). Similar plastic shell target with 500 µm diameter and 8 µm thickness filled with 30 atm DD gas was irradiated by 12 beams of partially coherent light with energy of 2.5 kJ in total.

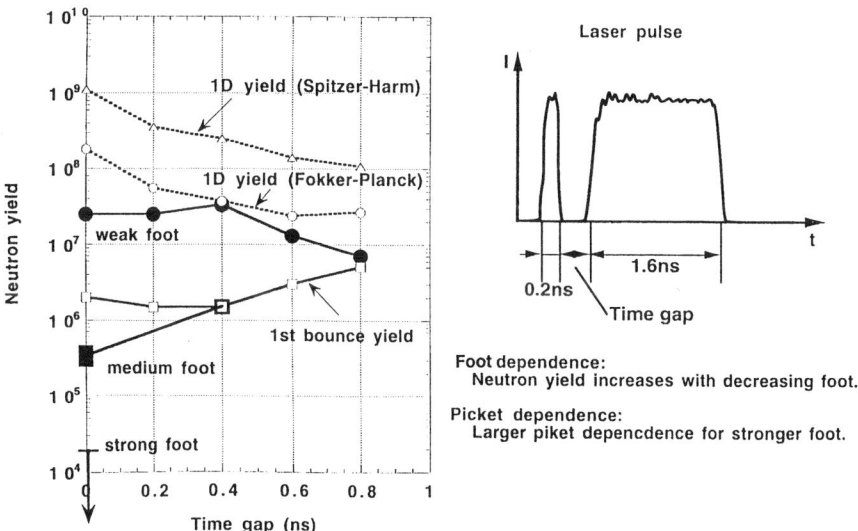

Figure 3. Neutron yield vs. foot-pulse-intensity and picket-time gap.

Figure 3 shows the results of the neutron yield measurement. In the actual experimental condition, laser pulse had a low-intensity foot component, which was found to be deleterious to the implosion performance and was tried to be reduced. When the foot intensity was kept weak (less than 1/820 of the main pulse), neutron production was much improved depending on the time gap between the picket and the main pulses. Comparing with simulation, the best result was obtained at the gap of 0.4 ns. Observed yield has largely exceeded the value at the first shock bounce and was close to the second bounce yield. It is suspected that RT stabilization is not sufficient at smaller time gap and imprinting of the nonuniformity in the picket pulse degrades the performance at the larger gap.

CONCLUSIONS

We have shown that the investigation of the hot spark heating is now possible with a 10-ps temporal resolution. Hot spark heating was improved with the increased stand-off distance and R-T stabilized implosion mode. Reduction of the irradiation nonuniformities by precise power balancing of the beams (low modes) and by adopting the partially coherent light (high modes) has been performed. Improved spark heating in the deceleration phase resulted in enhanced neutron production up to the second bounce yield, approaching 1-D simulation value.

ACKNOWLEDGEMENT

We thank Gekko-XII laser operation group, target fabrication group, plasma diagnostics group and simulation group of the Institute of Laser Engineering for their excellent technical assistance in carrying out the experiments.

REFERENCES

1. Azechi, H., et al., *Laser and Particle Beams*, **9**, 193-207 (1991).
2. Yamanaka, C., et al., *Phys. Rev. Lett.*, **56**, 1575-1578 (1986).
3. Shiraga, H., Heya, M., Fujishima, A., Maegawa, O., Shimada, K., Kato, Y., Yamanaka, T., and Nakai, S., *Review of Scientific Instruments*, **66**, 722-724 (1995).
4. Heya, M., Shiraga, H., Miyanaga, N., Nishihara, K., Tanaka, K., Kato, Y., Yamanaka, T., and Nakai, S., in this Proceedings.
5. Nakatsuka, M., Kanabe, T., Jitsuno, T., Miyanaga, N., and Nakai, S., in this Proceedings.
6. Miyanaga, N., Matsuoka, S., Ando, A., Amano, S., Nakatsuka, M., and Nakai, S., in this Proceedings.

ACCELERATION OF PLANAR FOILS BY THE INDIRECT-DIRECT DRIVE SCHEME

J.J. Honrubia, J.M. Martínez-Val
Institute of Nuclear Fusion, Polytechnical University of Madrid, Spain

J.L. Bocher, G. Faucheux
Centre d'Etudes de Limeil-Valenton, France

Abstract. We have investigated the hydrodynamic response of plastic and aluminum foils accelerated by a pulse formed by an x-ray prepulse followed by the main laser pulse. This illumination scheme, so-called indirect-direct drive scheme, has been proposed as an alternative to the direct and indirect drive. The advantages of such a scheme are that it can contribute to solve the problem of uniformity of the direct drive and, at the same time, it can be much more efficient and use simpler targets than the indirect-drive.

Experiments about this hybrid drive scheme have been performed at Limeil with the PHEBUS facility and the standard experimental set-up and diagnostics. The agreement between experiments and simulations is good for quantities such as the energy of the laser converted into x-rays and the burnthrough time of the converter foil. To simulate the full hydrodynamic evolution of the converter and target foils separated a distance of 1 mm, 2-D effects should be taken into account. The basic goals have been to check the simulation codes developed by the Institute of Nuclear Fusion and to determine the hydrodynamic response of the target foil to the hybrid pulse. These goals have been fulfilled.

INTRODUCTION

One of the major problems associated with the concept of direct drive, is the symmetry of the implosion. It is accepted that, in spite of the fact that the ISI, RPP and SDD techniques reduce to a large extent the problem of uniformization, there is still a real need for improving the implosion uniformization.

Desserberger et al. (1) have shown in an experiment performed at Rutherford Appleton Laboratory that non-uniformities are particularly critical at the beginning of the illumination because they can induce asymmetries in the ablation surface that are not smoothed out along the duration of the laser pulse. They conclude that these non-uniformities can be overcome by preforming a plasma with an x-ray source before the laser pulse. The possibility to use x-ray prepulses to drive ICF pellets was pointed out in reference (2), in which the so-called indirect-direct drive scheme was proposed as an alternative to the direct and indirect drive schemes. The advantages of such a scheme are that it can contribute to solve the problem of uniformity of the direct drive scheme (1) and, at the same time, it can be much more efficient and use simpler targets than the indirect drive schemes. Thus, the indirect-direct drive scheme combines the best features of the direct and indirect drive schemes.

Very recently, Dunne et al. (3) have pointed out that the indirect-direct drive scheme has two drawbacks: (i) the impossibility to control the blow-off plasma of the converter foil that leads to large plasma instabilities (parametric and filamentation) generation, and (ii) strong shock waves are launched into the solid target foil that could produce some preheating of the fuel. For these reasons, Dunne et al. proposed a new scheme based an using an x-ray preheated low density foam layer as a supercritical plasma buffer to thermally smooth laser non-uniformities before reaching the ablation surface.

In this paper, the results of a set of experiments about the indirect-direct drive scheme proposed in 1993 and performed in 1994 at Limeil under the Human Capital and Mobility Programme of the European Commission are presented.

The goals of these experiments have been: (i) to check simulation models and, in particular, radiation transport and Non-LTE opacities developed at the Institute of Nuclear Fusion (INF); (ii) to determine the hydrodynamic response of the target to the hybrid x-ray + laser pulse; and (iii) to investigate the uniformisation of the laser imprint phenomenon.

EXPERIMENTAL SET-UP

Experiments were carried out with PHEBUS laser at 0,35 μm, energies of, roughly, 3kJ and square pulses with FWHM = 1,42 ns. The incident average laser intensities were closed to $8 \cdot 10^{14}$ with 600 μm of focal spot. A typical laser pulse is shown in figure 1.

We have used three types of targets. The first one was gold disks of 2 mm diameter and thickness of 0.097 μm, 0.15 μm, and 0.25 μm. The goal of this first set was to characterize the x-ray prepulse. The second type of targets consists on a gold converter with the former thicknesses separated 1 mm of a target of 10 μm of aluminum or plastic (CH). In the third type of targets, the converter and the target foils are together. Again, the thickness of the converter goes from 0.097 μm to 0.25 μm, and the thickness of the aluminum or CH foil is 10 μm.

The most important diagnostics used in the experimental campaign have been:
- FMS and FXD to study the evolution in space and time of the x-ray emission of plasmas in the soft (250 eV) and hard (a few keV) ranges.
- OMBRIX to perform the radiography of targets. The spectrum of the Molybdenum source is in the range between 2500 and 2800 eV.
- DEMIX and SPART to measure the x-ray spectra in the ranges of 150 eV to 60 keV (DEMIX) and 150 eV to 4 keV (SPART).
- SMART a streak camera with a transmission grating for time evolution measurement of x-ray radiation between 200 eV and 2 keV.
- Diamond diodes to determine the angular distribution of the x-ray emission in the range from 0.4 to 2 keV.
- Stenope images to record the time integrated x-ray emission. It is located at 32 degrees from the axis of the laser beam.

SIMULATION CODE

Simulations have been performed with the SARA-1D code, that includes a non-LTE multigroup radiation transport coupled with two-temperature hydrodynamic and laser deposition by ray tracing (5). Radiation transport equations take into account the angular dependence of the radiation field by means of the S_N approximation, in such a manner that it reproduces accurately the optically thin and thick limits without introducing flux limiters and without ambiguities in the boundary layers. The radiation intensities are implicitly solved by means of multifrequency-grey acceleration of the radiation source.

The opacities and emissivities are computed in line by SARA from the populations obtained from a non-LTE average atom model with populations depending on the principal quantum number. The line profile modelling for the gold converter is based on the smearing of the lines into bands, that takes into account in a simplified way the n, l line splitting and the line shifts due to multiple electronic configurations. This procedure is similar to that used by Tsakiris and Eidmann in reference (6). For the target foil of aluminum or CH, a Gaussian profile is assumed for each line with a phenomenological line width obtained by comparison with more detailed models.

CHARACTERIZATION OF THE GOLD CONVERTER

A first set of shots was considered to characterize the x-ray prepulse produced by the converter as a function of the thickness. The goal of these shots was to measure the burnthrough time of the gold foils, the X-ray conversion efficiency and the time integrated and time dependent spectra emitted by the rear side.

The burnthrough time can be estimated in two ways: (i) from the FWHM of the refracted energy at the front and rear sides of the gold foil, and (ii) from the hard x-ray images of the target with two foils separated 1 mm. The burnthrough times obtained from these two kind of measurements are consistent. In figure 2, the average of the burnthrough times corresponding to the former diagnostics are compared with the results of the simulations. The flux limiter is a free parameter in the SARA code to adjust simulations to the experimental values. The values of the flux-limiter go from 0.01 to 0.03, that are consistent with limiters used in reference [4] to reproduce the rear side conversion efficiency of gold foils.

The comparison between the rear side energy conversion measured by different diagnostics and calculated with the SARA code is shown in figure 3. We can see that simulation follows the same trend, but is 2-3% lower that the experimental results. Similar effect can be shown in the time integrated rear side spectra. The M, N and O bands are present in measurement and simulations, but the intensities of the simulated spectra are lower than the measured ones. The simulation of temporal evolution of the x-ray emission of hard x rays (between 0.4 and 2 keV) by the rear side follows the same trend than that observed in the experiments, increasing the time duration of the emission with the thickness of the converter.

TARGET WITH DOUBLE FOIL SEPARATED 1 mm

In this configuration, 2-D effects are important because the distance between the converter and target foils is larger than the focal spot. The gold converter diffracts a significant fraction of the laser light, in such a manner that from the time integrated x-ray images of the stenope diagnostic, we can measure a diffraction angle of 30 degrees. This effect gives a dilution factor of the laser intensity between the gold and aluminum foils of, roughly, 4. Similarly, the intensity of the x rays emitted by the gold foil is diluted by a factor of 15, assuming a Lambert angular distribution. Taking into account these effects, the hybrid pulses seen by the target foils are shown in table 1.

The x-ray emission by the rear side of the target foil is used to estimate the opacity of the target material. In the case of the plastic foil, the photons with a few keV suffer a small attenuation when they pass through it. On the contrary, in the case of aluminum, this attenuation is important, in such a manner that most of the photons impinging the aluminum foil are absorved. This effect can be analyzed by comparing the rear side emission in the cases of gold foil alone and double gold + aluminum foils. The results are shown in figure 4, in which we can see also the results of simulation of the double foil case. It is worth noting the oposite trend of the single and double foil cases when increasing the thickness of the gold. Simulation results reproduce the trend of the double foil case for thicknesses of 0.097 μm and 0.15 μm, and give larger values than experiments. This can be explained by taking into account that the photons emitted by the rear side can be produced by the gold converter, by the aluminum foil, or by the colliding area between the gold and aluminum plasmas. Then, the overestimation of the extra burst of x rays produced in the collision between these plasmas by the 1-D code can lead to the differences observed in figure 4 between simulations and experiments. These differences increase for thicker gold foils in the same way that the burst of x rays increases with the gold thickness. We have also observed some qualitative differences between the experimental and simulated time integrated spectra emitted by the rear side. These differences can be due to the poor detail of the opacity line used in the simulations.

The hard x-ray images of the space between both foils indicate that in the case of 0.097 μm, the collision between gold and aluminum plasmas produce the compression of the subcritical gold plasma by the expanding aluminum plasma. In the cases of 0.15 μm and 0.25 μm this effect is not shown. Simulations reproduce this effect, as can be shown by comparison of isocontours of hard x-ray emission in x-t diagrams with the x-ray images.

TARGET WITH DOUBLE FOIL TOGETHER

With this configuration we investigate the acceleration of the target foil by the hybrid pulse without the dilution effect shown when the foils are separated 1mm. Moreover, it is interesting to observe the smoothing effects with both foils together.

The hard x-ray image of the FXD diagnostic allows to determine the burnthrough time of the gold foil and to see the effect on the rear side of the acceleration by the hybrid pulse. The ablation pressure associated with the x-ray prepulse is lower than the ablation pressure of the laser drive, then there is a change in the slope of the rear side velocity when the laser interacts with the target foil. This effect is also shown in the simulations.

In figure 5 are represented the rear side velocities measured with the OMBRIX diagnostic and the velocities obtained from simulation. The laser intensities considered in the simulations have been adjusted in such a manner that the code gives the same velocity than OMBRIX in the case of an aluminum foil alone. In this case, the intensity used is 50% of the laser pulse intensity. This reduction can be explained taking into account the losses due to 2-D effects. In figure 5, we can see that simulations follow the same trends that experimental values, and the discrepancies between simulations and experiments.

CONCLUSIONS

The goals of the experimental champaign have been to check the simulation codes of the INF and to determine the hydrodynamic response of low-Z targets to the hybrid pulse. These goals have been completely fulfilled. In particular, simulation results are able to follow the tendencies and to reproduce the order of magnitude of the experimental measurements. However, the full simulation of the experiments, and, specially, the shots of the target with two foils separated 1 mm, needs the use of a 2-D code. Simulations with the MULTI-2D code (7) are been performed.

According with the time integrated x-ray images of the stenope diagnostic, smoothing effects in the target foil are shown. In particular, the smoothing of large scale non uniformities (with a wavelength of, roughly, 100 μm) of the laser beam can be produced in part because of the preconditioning of the target foil by the x-ray prepulse, and, in part, by the diffraction of the laser beam in the converter foil, that acts as a diffussive medium. Then, because of this diffraction effect, to study in detail the uniformization of the laser imprint has not been possible in this champaign.

To continue with this set of experiments, it would be necessary to characterize more precisely the smoothing effect. However, according to the experimental results presented in last ECLIM, the concept of foam buffered indirect-direct drive proposed in reference [3] seems to be promising to solve the problem of the smoothing of laser non-uniformities.

ACKNOWLEDGEMENTS

This work has been supported by the European Commission under the Access to Large Scale Facilities of the Human Capital and Mobility (HCM) Programme, and the scientific network 'Dense Plasmas and Laser Compression Physics' also auspicied by the HCM Programme.

REFERENCES

1. Desselberber, M., Afshar-rad, T., Kattak, F., Viana, S. and Willi, O., *Phys. Rev. Lett.* **68**, 1539 (1992).
2. Eliezer S., Honrubia, J.J., Velarde, G., *Phys. Lett. A* **166**, 249 (1992).
3. Dunne, M., Borghesi, M., Iware, A., Jones, M.W., Taylor, R., and Willi, O., "Experimental Evaluation of X-ray Preheated Foam-Buffer Targets for use in Inertial Confinement Fusion", Book of Abstracts, ECLIM'94, Oxford, 19th-23rd September, (1994).
4. Barbonneau, D., Bocher, J.L., Bayer, C., Decoster, A., Juraszek, D., Perrine, J.P., and Thiell, G., *Laser and Particle Beams* **9**, 2 (1991).
5. Honrubia, J.J., *J. Quant. Spectrosc. Radiat. Transfer*, Vol **49**, No. 5, pp 491-515, (1993).
6. Tsakiris, G.D., and Eidmann, K., *J. Quant. Spectrosc. Radiat. Transfer*, Vol **38**, 353 (1987).
7. Ramis, R., and Meyer-ter-Vehn, J., Report MPQ 174, *Max-Planck-Institut fur Quantenoptik*, Garching, Germany (1992).

TABLE 1.- Hybrid pulse impinging on the target foil. FWHM of the laser pulse 1.42 ns.

	Gold thickness(μm)		
	0.097	0.15	0.25
Duration of x-ray prepulse (ns)	0.37	0.53	0.68
x-ray intensity (W/cm^2)	$1.3 \cdot 10^{13}$	$1.6 \cdot 10^{13}$	$2.5 \cdot 10^{13}$
Laser intensity (W/cm^2)	$2.4 \cdot 10^{14}$	$2.4 \cdot 10^{14}$	$2.4 \cdot 10^{14}$

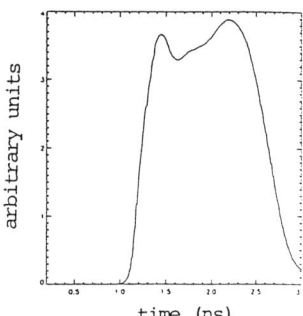

Figure 1.- Laser pulse used in experiments.

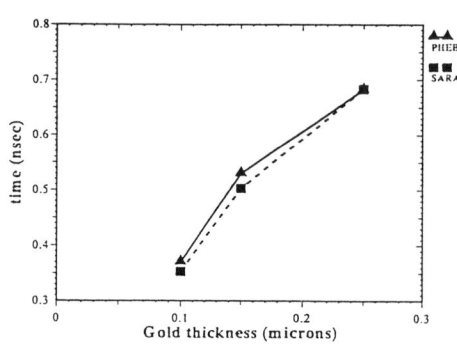

Figure 2.- Comparison of the burnthrough times of gold foils.

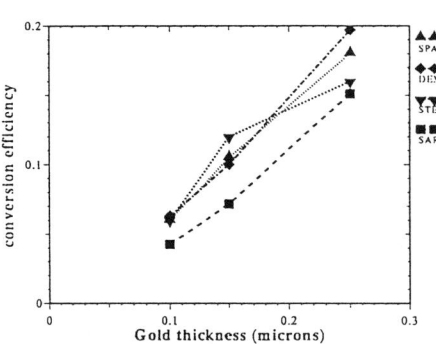

Figure 3.- Rear side x-ray energy conversion of gold foils.

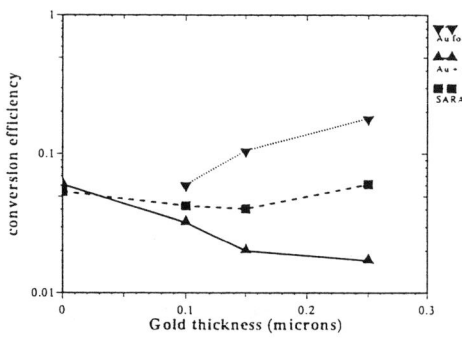

Figure 4.- Rear side x-ray energy conversion of double foil targets compared with single foil targets.

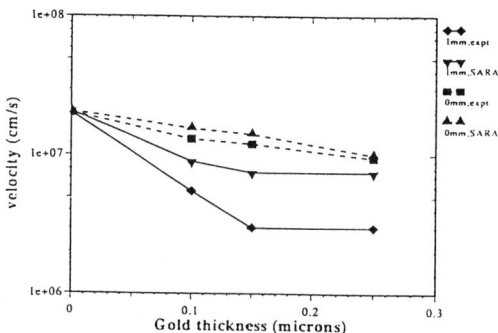

Figure 5.- Comparison between experiments and simulation of the rear side velocity of the double foil targets.

A Nonsupported Pellet for Laser Fusion Scheme

Y.Sakagami, H.Yoshida, T.Mizutani, S.Miyagawa
M.Sekimura and K.Yasufuku

*Department of Electronics and Computer Engineering,
Faculty of Engineering, Gifu University
Yanagido, Gifu 501-11, Japan*

Abstract. For achievement of ignition, one of the most important issue is uniform compression for which a nonsupported pellet is reported in this paper. The magnetic force to a Ni coat on a glass microballoon is clarified by theoretical analysis and model experiments. To introduce horizontal dumper, an LD is verified effective for horizontal dumper. The preliminary test of active dumper is done.

INTRODUCTION

Uniform compression is one of the key issue to the laser fusion scheme. A nonsupported pellet has been desired. Magnetic suspension system is possible to suspend a pellet without any physical contact(1, 2, 3).

FABRICATION OF Ni-GMB

To obtain thin Ni coat on a glass microballoon (Ni-GMB), vacuum evaporation technique was used. The uniform and smooth coat was realized by using a magnetic vibrator which rotated GMBs. The thickness of the Ni coat was measured 70 nm by a X'tal thickness monitor during evaporation. After it, the thickness was confirmed by a multiple-interferometer.

CLARIFICATION OF THE MAGNETIC FORCE

When the glass microballoon (GMB) is in the electric field, electric dipole moment is induced. If the electric field has gradient, the translational force is exerted on the GMB. This force has been analyzed by the electromagnetic theory. The solution is expressed in Eq. (1). The geometry and notations used in the equation is shown in **Fig.1**.

In order to justify the equation we have made next experiment. It is referred from the paper(4) in which bubble dielectrophoresis is treated. Application comes to this GMB dielectrophoresis. **Fig.2** shows the schematic of the apparatus. A

$$\vec{F}_e = \frac{\varepsilon_0}{2} K_e \frac{4\pi}{3}(b^3 - a^3) \vec{\nabla} E^2,$$

$$K_e = \frac{3\varepsilon_1}{\varepsilon_3 + 2\varepsilon_1} \left\{ \frac{3\varepsilon_1(\varepsilon_3 + 2\varepsilon_2)(\varepsilon_2 - \varepsilon_3)}{(\varepsilon_3 + 2\varepsilon_2)(2\varepsilon_1 + \varepsilon_2) + 2\frac{a^3}{b^3}(\varepsilon_2 - \varepsilon_3)(\varepsilon_1 - \varepsilon_2)} + \frac{(\varepsilon_3 - \varepsilon_1)b^3}{b^3 - a^3} \right\}. \quad (1)$$

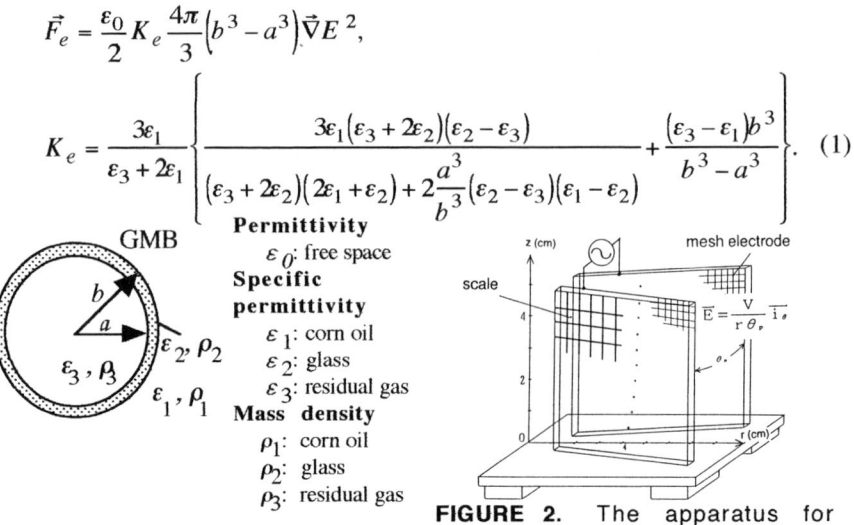

Permittivity
ε_0: free space
Specific permittivity
ε_1: corn oil
ε_2: glass
ε_3: residual gas
Mass density
ρ_1: corn oil
ρ_2: glass
ρ_3: residual gas

FIGURE 1. A GMB with its notations.

FIGURE 2. The apparatus for measurement of GMB dielectrophoresis.

pair of meshes was attached to the surfaces of acrylic resin plates. Between these electrodes alternating voltage was applied. Effective value V of it was 12 kV with the frequency of 60 Hz. The angle θ_p of the electrodes was 15°. The GMB was put into a small hole drilled in a bottom plate. The analytical formula to the trajectory can be formulated as Eq. (2).

$$r = \left[\frac{\varepsilon_0}{2} K_e \frac{8(b^3 - a^3)V^2 z}{\{(b^3 - a^3)\rho_2 + a^3\rho_3 - b^3\rho_1\} g \theta_p^2} + r_i^4 \right]^{1/4} \quad (2)$$

The trajectory of the GMB was observed by a CCD camera-VCR set. The result is shown in **Fig.3**. Dots are experimental. They are quite fitting to the analytical curve when the specific permittivity of the glass is 3.8.

Analytic expression of the dielectric was applied to the ferromagnetic with substitutions $E \rightarrow H$, $\varepsilon \rightarrow \mu$, $b \rightarrow c$, $a \rightarrow b$. The force exerted on the Ni-GMB is expressed by Eq. (3).

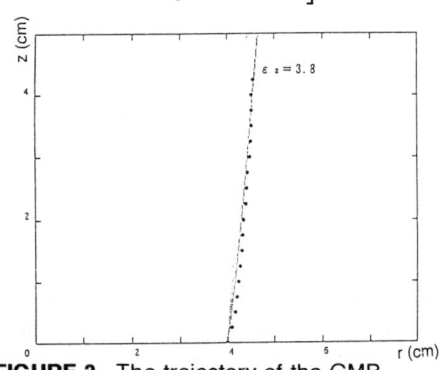

FIGURE 3. The trajectory of the GMB.
$a=352\mu m$, $b=350\mu m$, $\varepsilon_1=3.1$, $\varepsilon_2=3.8$, $\varepsilon_3=1.0$
$\rho_1=0.89 \times 10^3$ kg/m^3, $\rho_2=2.50 \times 10^3$ kg/m^3
$\rho_3 \fallingdotseq 0$ kg/m^3, $r_i=4 \times 10^{-2}$ m

$$\vec{F}_m = \frac{1}{2\mu_0} K \frac{4}{3}\pi(c^3 - b^3)\vec{\nabla}B^2,$$

$$K = \frac{3}{\mu_1(\mu_3 + 2\mu_1)} \times \left\{ \frac{3\mu_1(\mu_3 + 2\mu_2)(\mu_2 - \mu_3)}{(\mu_3 + 2\mu_2)(2\mu_1 + \mu_3) + 2\frac{b^3}{c^3}(\mu_2 - \mu_3)(\mu_1 - \mu_2)} + \frac{(\mu_3 - \mu_1)c^3}{c^3 - b^3} \right\}. \quad (3)$$

In this expression K is susceptibility that is a function of the specific permeability μ and the radii. The structure of the Ni-GMB is shown in **Fig.4**.

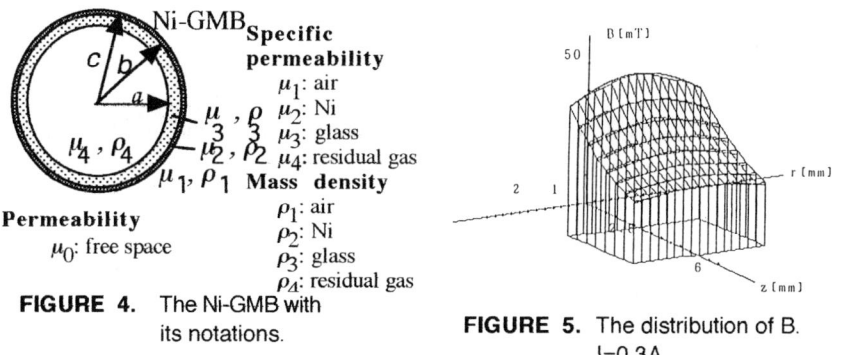

FIGURE 4. The Ni-GMB with its notations.

FIGURE 5. The distribution of B. I=0.3A

The distribution of component of the magnetic flux density Bz was measured by a Hall probe, as a function of z and r, with also the other component Br. The absolute value of $B=\sqrt{B_z^2 + B_r^2}$ is plotted in **Fig.5**. The dependence of B on electromagnet coil current I was also measured. Empirical formula can be made as Eq. (4).

$$B(z,r,I) = \frac{\alpha I}{z\left\{1 + \left(\frac{r}{r_1}\right)^2\right\}}, \quad \alpha = \frac{\beta}{\gamma I + \delta},$$

$$r_1 = \left(\frac{z}{5.7 \times 10^6}\right)^{1/4}, \quad \beta = 4.8 \times 10^{-4}, \gamma = 1.9, \delta = 0.72 \quad (4)$$

By inserting this equation into Eq. (3), one can obtain the relation between z and I. This relation has the meaning such that a Ni-GMB is suspended at the suspension distance z at I. In **Fig.6**, the theoretical curve is shown along with the experimental data. Adequate agreement between them is seen when the specific permeability of the Ni was 15.

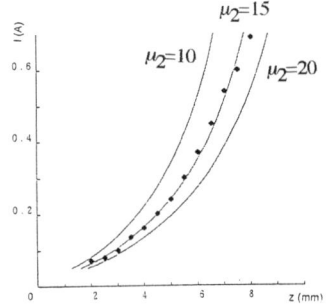

FIGURE 6. The relation between z and I.

HORIZONTAL DUMP BY OPTICAL FORCE

In contrast to the vertical control, no feed back control is necessary horizontally because static equilibrium exists at $r=0$. Nevertheless, when the Ni-GMB is suspended in high vacuum, small disturbance makes it oscillate for long time. For the magnetic force is proportional to the small disturbance as r is much less than r_1. Therefore rapid dumping of the oscillation is wanted. As a method of it, we have tried to introduce dumping using optical forces. Photon force, one of the optical forces, is considered first which acts on the Ni-GMB. It is expressed by Eq. (5).

$$F_p = \int_0^{\pi/2} \left\{ \frac{4P}{cw^2} \exp\left(-\frac{2r_p \sin^2\theta}{w^2}\right) r_p^2 \sin\theta \cos\theta \right\} n^2 R \cos\theta d\theta. \qquad (5)$$

In this equation P is total power of an LD, n the refractive index of surroundings. The beam can be considered parallel experimentally around the focal area with spot radius w. The values of pellet radius r_p, reflective index R of Ni coat and other parameters are given in **Table 1**. Therefore in this equation, the fractional force acting on a small area with the incident angle θ is integrated, resulting in the expressed photon force F_p. Friction force $F_d = 6\pi r_p \eta v$ can be given by Stokes's formula. The magnetic force F_m is given by Eq.(3).

When the photon force F_p is a step function, one can get the analytical solution of the horizontal position as Eq.(6).

$$r(t) = r_0 \left\{ 1 - e^{-\frac{t}{\tau}} \left(\cos\omega t + \frac{1}{\omega\tau} \sin\omega t \right) \right\}, \qquad r_0 = \frac{\mu_0 F_p}{2KV_m}\left(\frac{r_1 z}{\alpha I}\right),$$

$$\frac{1}{\tau} = \frac{3\pi r_p \eta}{m}, \quad \omega = \sqrt{\frac{2KV_m}{\mu_0 m}\left(\frac{\alpha I}{r_1 z}\right)^2 - \frac{9\pi^2 r_p^2 \eta^2}{m^2}} \approx \sqrt{\frac{2KV_m}{\mu_0 m}}\frac{\alpha I}{r_1 z}. \qquad (6)$$

TABLE 1. The Pellet Condition

Item	Symbol	Value	Unit
Pellet radius	r_p	325	μm
Glass thickness		2.3	μm
Ni coat thickness		70	nm
Pellet mass	m	8.3	μg
Reflective index of Ni Coat	R	0.5	
Ni coat volume	V_m	90	fm³

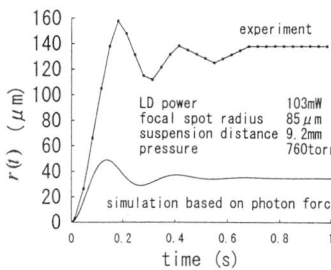

FIGURE 7. The horizontal motion of the GMB.

The power of the LD (λ=810nm), fed through an optical fiber, is given to a small area of the Ni-GMB by using an objective. The horizontal motion of it was detected by a CCD camera-VCR set. The consecutive images were transferred one by one to a PC.

The dependence of $r(t)$ is plotted in **Fig. 7**. In this figure, experimental conditions are given. The analytical solution that simulates the behavior based on photon force is given in this figure. One can see the $r(t)$ which the experiment gives is much larger than that of the simulation. Note that the final shift of 140 μm is obtained by the LD power 103 mW at the pressure 760 Torr.

Fig. 8(a) shows the dependence of the final shift on the LD power at the degree of vacuum 0.2 Torr. As the LD power increases the shift becomes larger. One can notice that the final shift 140 μm is obtained by about 0.5 mW of the LD power. **Fig. 8(b)** shows the final shift as dependence of degree of vacuum at constant LD power 0.35 mW. From this figure, one can see that the LD power becomes more effective to work against the Ni-GMB as the degree of vacuum becomes smaller.

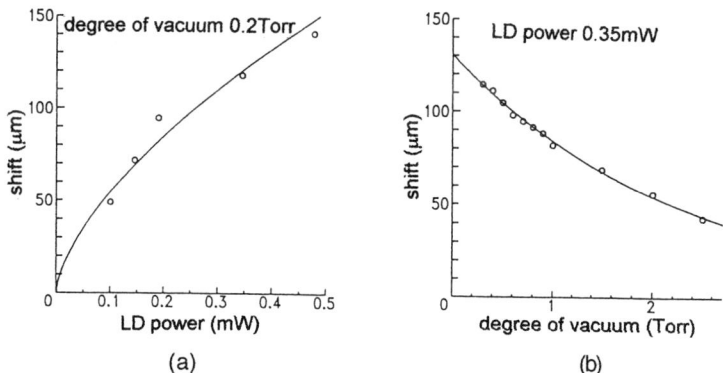

(a)   (b)

FIGURE 8. The dependence of the shift on the LD power and the degree of vacuum.

From the experiments described above, one can see that optical force is effective to be used in the horizontal dumping. The mechanism of the optical force cannot be explained only by the photon force.

In order to realize horizontal dumping, an active dumper was prepared. The magnetically suspended Ni-GMB was monitored by an He-Ne laser. The shadow of the Ni-GMB was imaged on a position sensitive photo sensor. The horizontal position of the Ni-GMB was obtained by the shadow area. The optical force discussed above was controlled. The LD's intensity was varied to dump the Ni-GMB motion. The preliminary test of dumping runs in satisfying condition.

SUMMARY

Uniform and smooth Ni coat on the GMB was realized. The Ni coat thickness was 70 nm. Magnetic force was clarified by theoretical analysis and model experiments of GMB dielectrophoresis. Application to the ferromagnetic Ni-GMB can give the good agreement between the derived formula and the experimentally suspended Ni-GMB data. In order to realize horizontal dump, the LD was verified to be very effective especially as the degree of vacuum decreases. The preliminary test of the active control to the horizontal dump is in satisfying condition.

ACKNOWLEDGMENTS

We are very grateful to the encouragement and support from Professors C.Yamanaka, S.Nakai, T.Yamanaka, Y.Izawa and useful discussion with Drs H.Azechi, T.Norimatsu, M.Takagi at ILE Osaka University. We are very thankful for experimental support to G.Ohashi, Y.Sano, M.Kato, Y.Hayashi, K.Oshita and Y.Suzuki in our laboratory.

REFERENCES

1. Y.Sakagami, H.Yoshida, S.Tanase, T.Mutou and M.Niiyama, ILE Sympo. '94 Annual Collaboration Report, ILE-ACR-94, 79 (1994)
2. H.Yoshida, K.Katakami, Y.Sakagami, H.Nakarai, H.Azechi, and S.Nakai, *Proceedings of the IAEA on Drivers for Inertial Confinement Fusion at Osaka, Japan 15-19 April 1991* , 410 (1992)
3. H.Yoshida, K.Katakami, Y.Sakagami, H.Nakarai, H.Azechi, and S.Nakai, *Laser and Particle Beams* , **11**, 455 (1993)
4. T.B.Jones and G.W.Bliss, *J. Appl. Phys.* , **48**, 1412 (1977)

Dynamics of Gas-Filled Hohlraums

T. J. Orzechowski, R. L. Kauffman, R. K. Kirkwood, H. N. Kornblum, W. K. Levedahl, D. S. Montgomery, L. V. Powers, T. D. Shepard, G. F. Stone, L. J. Suter, R. J. Wallace

Lawrence Livermore National Laboratory
P. O. Box 808
Livermore, CA 94550

J. M. Foster, P. Rosen

Atomic Weapons Establishment
Aldermaston, U. K.

In order to prevent high-Z plasma from filling in the hohlraum in indirect drive experiments, a low-Z material, or tamper is introduced into the hohlraum. This material, when fully ionized is typically less than one-tenth of the critical density for the laser light used to illuminate the hohlraum. This tamper absorbs little of the laser light, thus allowing most of the laser energy to be absorbed in the high-Z material. However, the pressure associated with this tamper is sufficient to keep the hohlraum wall material from moving a significant distance into the interior of the hohlraum. In this paper we discuss measurements of the motion of the interface between the tamper and the high-Z hohlraum material. We also present measurements of the effect the tamper has on the hohlraum temperature.

INTRODUCTION

In the indirect drive approach to inertial confinement fusion, the fuel pellet is imploded by x-ray driven ablation inside a high-Z cavity, or hohlraum (1). This x-ray drive is generated by the interaction of intense laser beams with the hohlraum wall interior. In addition to ablating the fuel pellet surface, this radiation heats and ablates the hohlraum wall causing the hohlraum to fill with a dense, high-Z plasma. This plasma can refract the incident laser beams or cause the absorption region to move a significant distance from the hohlraum wall. Both of these situations can affect the drive symmetry. If the density gets too high, parametric instabilities can scatter the laser light, affecting both symmetry and energy balance.

To prevent the ablated hohlraum wall material from filling the cavity, a low-Z material, or tamper, is used to hold back the high-Z hohlraum wall material. This tamper can be generated from a solid, such as CH, coated on the interior hohlraum wall, or gas such as methane, confined in the hohlraum. These gas-filled hohlraums must be designed with windows to confine the gas. The tamper must allow the laser light to penetrate to the hohlraum wall, but must be of sufficient density and temperature to hold back the ablated high-Z material. A gas filled hohlraum is shown schematically in figure 1. Thin (3500 Å) polyimide ($C_{22}H_{10}N_2O_5$) windows cover the laser entrance holes (LEHs) to provide a gas-tight seal. The laser light rapidly burns through the windows and gas, heating and ionizing the low-Z material as it propagates to the hohlraum wall. In the experiments discussed here, this occurs in about 200 ps.

© 1996 American Institute of Physics

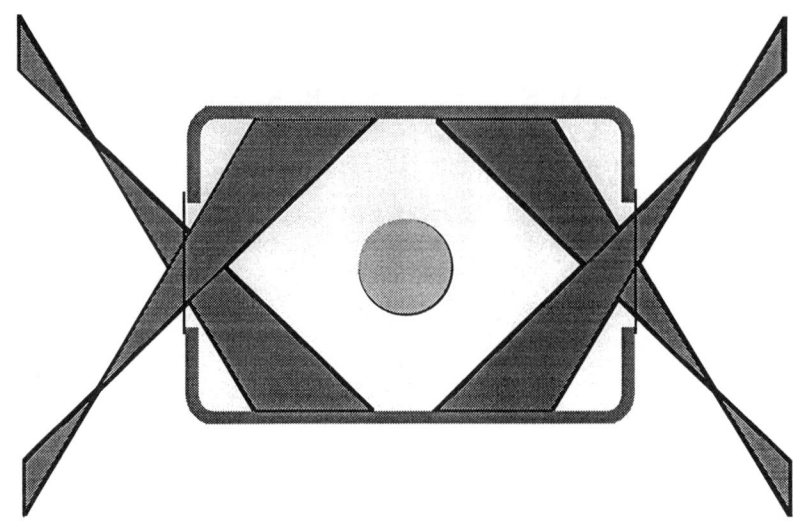

FIGURE 1. Gas-filled hohlraum. The laser entrance holes are covered with polyimide to provide a gas-tight enclosure.

The x-rays generated by the interaction of the laser with the Au hohlraum wall interact with the high-Z wall material via multiple absorption and re-emission to produce a near Planckian photon distribution with a temperature on the order of 200-eV inside the hohlraum. The radiation heat wave (2) propagating through the hohlraum wall ablates the Au causing it to expand toward the center of the hohlraum. The pressure associated with the tamper retards the motion of the high Z plasma.

In this paper we discuss two experiments that study the effect of the tamper on the hohlraum. In the first experiment, we use a modified hohlraum to monitor the motion of the tamper-hohlraum interface as a function time. The hohlraum has 100% LEHs to permit an unobstructed view of the hohlraum wall. A gated imager is used to view the hohlraum along its axis by monitoring the x-ray emission from the high-Z ions in three different energy bands (3). In these experiments, the laser beams are pointed in such a way that they illuminate the midplane circumference with ten-fold symmetry. This makes the illumination more "2-Dimensional" which is easier to model with the 2-D LASNEX (4) code and is a better approximation to the azimuthal illumination scheme intended for the National Ignition Facility (NIF). In the second experiment, we measure the effect of the tamper on the time dependent drive in a hohlraum. In these drive experiments we use a standard scale-1 hohlraum (see below for details). The laser beams illuminate the hohlraum on either side of the midplane with five-fold symmetry, as depicted in figure 1. A ten-channel, absolutely calibrated time-resolving spectrometer, Dante (5), is used to monitor the x-ray emission emanating from a small aperture in the hohlraum wall. This measurement provides the time history of the wall temperature inside the hohlraum, from which we can infer the drive temperature. In these experiments, we measure the Stimulated Brillouin Scattering (SBS) back into the lens of one of the ten Nova laser beams. Monitoring the scattered laser light helps us understand the overall hohlraum energetics.

INTERFACE MOTION EXPERIMENTS

To monitor the motion of the interface between the tamper and the ablated hohlraum wall material, we use a modified hohlraum that has 100% LEHs. Because of the large LEH diameter in these hohlraums, the windows are fabricated from 6500-Å thick polyimide. Both the hohlraum diameter and length are 1600 μm. The laser beams are pointed at the midplane of the hohlraum to provide tenfold symmetry in the azimuth. This provides a more uniform illumination of the wall, which is easier to model with a 2-dimensional code such as LASNEX. This is also a closer approximation to the anticipated illumination of a NIF hohlraum wall. The hohlraum is illuminated with a 2-ns long square pulse with a total energy of approximately 20 kJ. The beam spot at the hohlraum wall is calculated to be about 675 μm long and 460 μm wide (in the azimuthal direction). The laser intensity at the wall is ~3×10^{14} W/cm^2.

The primary diagnostic is the Soft X-ray Framing Camera (SXRFC) that views the hohlraum along its axis. This instrument is filtered to monitor emission around 450-eV, 1200 eV and M-band emission (hν>2 keV). The SXRFC provides images at four different times in each of the three energy channels. Another framing camera views the hohlraum along its axis from the opposite side. This second camera has a fixed sweep length of 500 ps. A multiple pinhole array on the front of the camera provides twelve images in the M-band region during the 500-ps sweep. Figure 2 contains images from the SXRFC. We show here the 450-eV channel at four different times during a pulse for two different hohlraum configura-

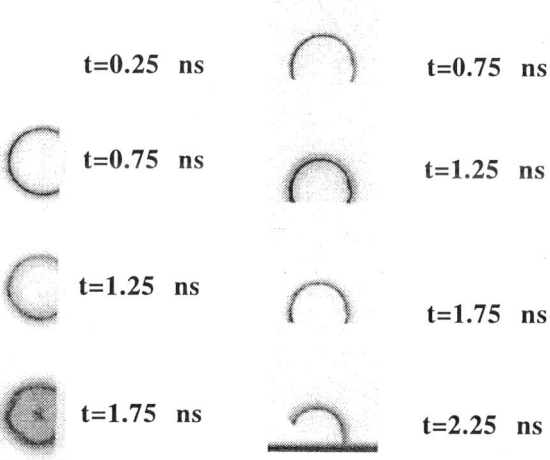

FIGURE 2. Axial images of an empty (left) and a neopentane-filled (right) hohlraum. These data correspond to the 450-eV channel of the framing camera.

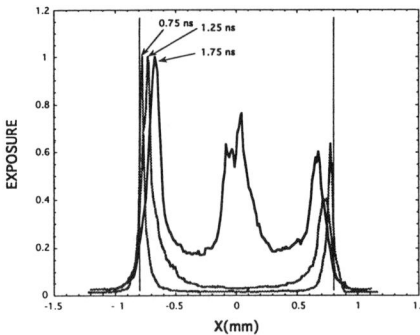 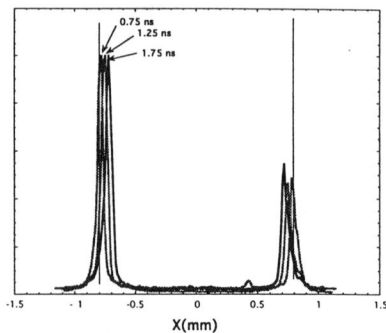

FIGURE 3. Lineouts of the images shown in figure 2. The exposure is in arbitrary units.

tions (windows on the LEHs but no gas fill and 1-atm C_5H_{12} gas fill). Neopentane corresponds to 0.1 n_c when fully ionized. It is quite apparent that for the empty hohlraum the cavity fills with high-Z plasma and that toward the end of the pulse the plasma is stagnating on the axis. In contrast, the tamped hohlraums show the central region free of the emission corresponding to the high-Z material for the duration of the laser pulse. Lineouts of the axial images corresponding to the empty hohlraum and the C_5H_{12} filled hohlraum are shown in figure 3. The most apparent difference between the empty and gas filled hohlraums is the absence of low density gold filling the hohlraum. Not only does the 50% intensity point move in during the pulse, but the peak of the intensity moves in as well. This corresponds to the absorption region moving away from the original location of the wall. The data is in good agreement with LASNEX calculations, except in the M-band region where the code predicts the emission (and hence the location of the high-Z material) to emanate from further inside the hohlraum (i. e. closer to the axis).

We determine the velocity of the interface by somewhat arbitrarily tracking the position of the 50%-intensity point as a function of time. The velocity of the interface in the absence of the tamper is approximately 3×10^7 cm/s while in the presence of the tamper generated from neopentane it is approximately 1×10^7 cm/s. However, we should point out that this method of quantifying the effect of the tamper totally neglects the fact that the low density, high-Z filling of the hohlraum is virtually eliminated by the tamper.

HOHLRAUM TEMPERATURE MEASUREMENTS

In order to determine the effect of the gas tamper on the hohlraum drive, We measured the temperature of a standard, scale-1 hohlraum without a tamper (or windows) and with a tamper generated from methane gas. These hohlraums are 1600-μm in diameter and 2700-μm long with 75% LEHs (1200-μm in diameter). The LEHs on the gas-filled hohlraums are covered with 3500-Å thick polyimide. The laser pulse shape, shown in figure 6, is designed for implosion experiments and incorporates a low-intensity foot followed by a peak in the drive. A ten channel, time resolving x-ray spectrometer views a portion of the interior hohlraum wall through a small aperture (~400-μm diameter) in the side of the hohlraum. The results of these measurements are shown if figure 5 for empty and methane-filled hohlraums. The effect of the tamper on the hohlraum temperature is two-fold: the peak of the drive is reduced about 20-eV and the "foot" of the drive is modified by the energy required to ionize and heat the gas. Preliminary investigations indicate that the gas species has little effect on the drive modification. That is propane and methane reduce the hohlraum temperature by about the same amount.

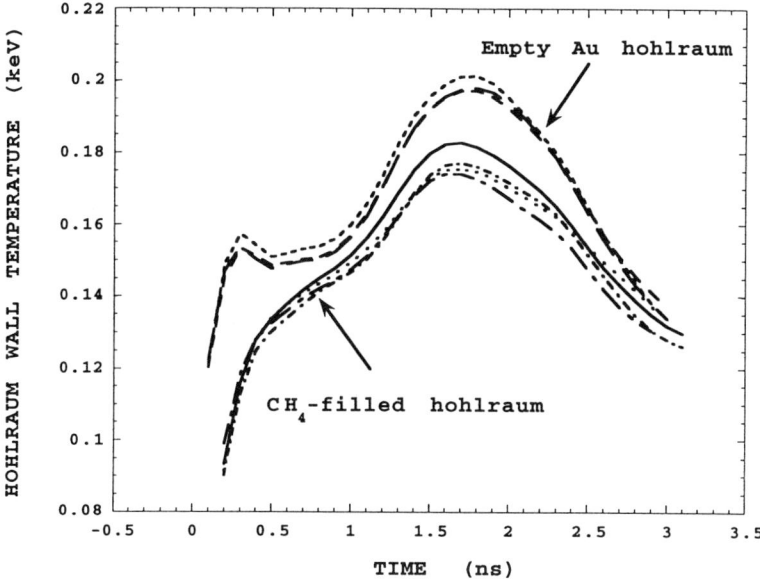

FIGURE 4. Hohlraum wall temperature as a function of time for empty and CH_4-filled hohlraums.

In order to account for the temperature drop, we are measuring the reflected laser light from the hohlraum. We have made extensive measurements of the SBS and are beginning measurements of the SRS. Figure 6 shows the time dependent SBS that reflects back into a specific beam. We plot here the incident laser pulse (for a single beam), the SBS for a gas-filled hohlraum and for an empty hohlraum. Although the average SBS is nearly the same for the gas-filled hohlraum and the empty hohlraum, the temporal characteristics are markedly different. The empty hohlraum shows the SBS signal peaking in the latter part of the drive while the gas-filled hohlraum shows the peak of the SBS signal occurring during the front of the peak of the drive pulse.

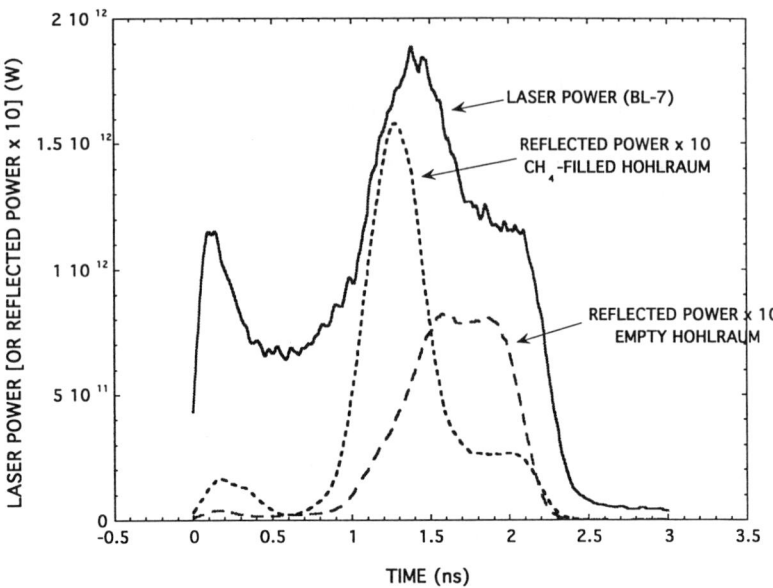

FIGURE 5. Laser pulse shape (power vs time) used in the hohlraum drive measurements. Also shown is the reflected SBS for an empty and a CH_4-filled hohlraum.

CONCLUSION

We have measured the effect of the tamper in a gas-filled hohlraum. The ionized, low-Z plasma is quite effective in preventing the high-Z wall material from filling-in the hohlraum. We observe a temperature drop of about 10% when a tamper is introduced into the hohlraum. We are currently investigating the sources of this energy loss such as laser scattering via SBS and SRS. We also are investigating the properties of the low-Z tamper.

REFERENCES

1. Lindl, J. D., McCrory, R. L., and Campbell, E. M., *Physics Today* **45**, 32 (1992)

2. Marshak, R. E., *Physics of Fluids* **1**, 24 (1958)

3. Ze, F., Kauffman, R. L., Kilkenny, J. D., Wiedwald, J., Bell, P. M., Hanks, R., Stewart, J., Dean, D., Bower, J., and Wallace, R., *Review of Scientific Instruments* **63**, 5124, (1992)

4. Zimmerman, G. B. and Kruer, W. L., *Comments on Plasma Physics and Controlled Fusion* **2**, 51 (1975)

5. Kornblum, H. N., Kauffman, R. L., and Smith, J. A., *Review of Scientific Instruments* **57**, 2179, (1986)

Measurement of Absorption Distribution by Second Harmonic and X-Ray Images

K.A. Tanaka*, T. Ohnishi, A. Ozaki, T. Matsushita, S. Nakaji,
H. Hashimoto, R. Kodama, Y. Oshikane, M. Nakai, M. Tsukamoto,
S.Matsuoka, K. Tsubakimoto, M. Murakami, K. Shimada,
H. Azechi, H. Nishimura, N. Miyanaga, H.Shiraga,
Y. Kato, K. Mima and S. Nakai*

Institute of Laser Engineering, Osaka University, Yamada-Oka 2-6, Suita, Osaka 565 JAPAN
*also at the Department of Electromagnetic Energy Engineering, Osaka University, Yamada-Oka 2-1, Suita, Osaka 565 JAPAN

Abstract. The second harmonic of 527 nm laser light was imaged by a Schwarzschild microscope. Obtained second harmonic images are compared with the x-rays. The second harmonic originates from near the critical density, when laser light is absorbed via. resonance absorption or ion acoustic decay instability. A plastic target coated with 1000Å Al was irradiated with a 527 nm laser beam without any beam smoothing technique. The obtained second harmonic image at 263 nm was compared with an x-ray pinhole picture with a proper filter to observe Al 1.8 keV emission. This comparison was taken as a proof of principle since the over all patterns between the second harmonic and x-ray images show very close correlation. Then the second harmonic images were taken with random phase plate and partially coherent laser light. These images should show the absorption distribution at an early part of the irradiation laser pulse.

INTRODUCTION

In laser inertial confinement fusion, a spherical fuel target should be illuminated and imploded directly by laser light or indirectly by x-rays to a high temperature and high density state. Ablation pressure uniformity is required to be less than a few % in the target to avoid the distortions due to fluid instabilities. In order to fulfil this requirement various laser beam smoothing techniques have been introduced. The final goal is to create a uniform ablation layer or uniform absorption layers.

The second harmonic of 527 nm laser light was imaged by a Schwarzschild microscope. Obtained second harmonic images were compared with the x-rays. It is well known that the second harmonic light originates from resonance absorption with a short-density-scale-length plasma and/or from ion acoustic decay instability. The second harmonic comes from the neighbourhood of plasma cut-off density.

EXPERIMENTS

The experiments were conducted using the NEW GEKKO XII Nd: glass laser system of coherent laser with random phase plate (RPP) or partially coherent laser with

© 1996 American Institute of Physics

RPP. The irradiation configuration used a planer geometry. Laser wavelength was 527 nm. Irradiation intensities were typically 2×10^{14} w/cm^2 with a pulse duration of 0.1 to 1.6 nsec. Targets were plastic planer targets. A half plane of the targets were coated with 1000Å Al for comparison with x-ray pin hole pictures. A Schwarzschild microscope observed the second harmonic image via. a 263 nm ($\Delta\lambda$=18 nm) interference filter using Q-plate films (ILFORD). X-ray pinhole images were recorded directly by x-ray CCD cameras. Figure 1 shows that a comparison of 263 nm and x-ray images for a same shot. The laser pulse width was 1 nsec. Target was a plastic plane coated with Al on the right half of the target. The laser energy was 377 J in 527 nm wavelength, giving a 2.3×10^{14} W/cm^2 intensity for a 500 μm diameter focal spot. Since no beam smoothing technique was introduced, the laser beam had many hot spots within the focal spot. One can see that the x-ray image shows such hot spot patterns only on the right half plane of the focal spot as shown in Fig. 1 (a). The 263 nm picture in Fig. 1 (b) shows very close similarity to the x-ray image on the right half plane, where 1000Å Al was coated to emit the Al Ly-α line(1.8keV).

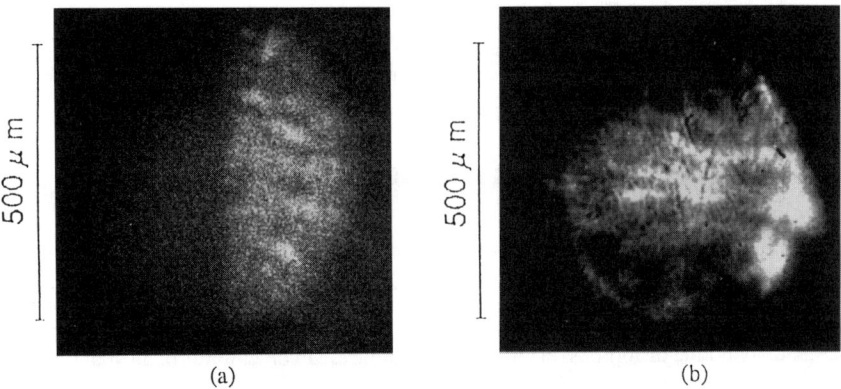

Fig.1 (a) x-ray and 263 nm images

There were many structures due to the laser hot spots. The x-ray pinhole camera had a proper Si filter to transmit mainly the Al line emission. The Al coating was so thin that it could be ablated away after a very short time of the laser irradiation. The Al emission x-ray image should represent a absorption pattern of a very early part of the laser pulse. The left half plane shows almost no structures. This left half plane was a plastic irradiated with the normal laser, which had hot spots as many as the right half plane had. No structures were seen probably because the plastic had a rather high thermal conduction and the emission from the plastic was time integrated within all the laser pulse width of 1 nsec. If the absorption pattern due to the 263 nm image is due to the resonance absorption, one may expect that the absorption pattern should have a very peculiar shape, depending on the laser polarization and incidence angle(1). Figure 1 shows no such pattern. There might be a contribution from ion acoustic decay instability (IADI) to the second harmonic emission, if the laser intensity is above the threshold. However, considering the close similarity between two pictures in Figs. 1 and knowing that the thin aluminium layer is subject to the ablation at an early time of the pulse, the second harmonic emission is dominantly emitted at an early part of the laser, where the density scale length is still rather short. In such a short density scale length plasma, the resonant absorption should be a dominant absorption mechanism. IADI should be important when the density scale length becomes long. The reason why no characteristic resonance pattern is not observed is not clear and is under our discussion. One possible way to interpret is that the hot spots within the focal spot were created due to phase matching and self-focusing. Those hot spots then behave as independent beamlets within the focal spot. Each hot spots excite resonance absorption independently within a very local region near the plasma cut off density, resulting in many independent second harmonic emissions.

In order to show more clearly the close similarity of the 263 nm and x-ray pictures, modal intensities of both pictures are shown in Fig. 2.

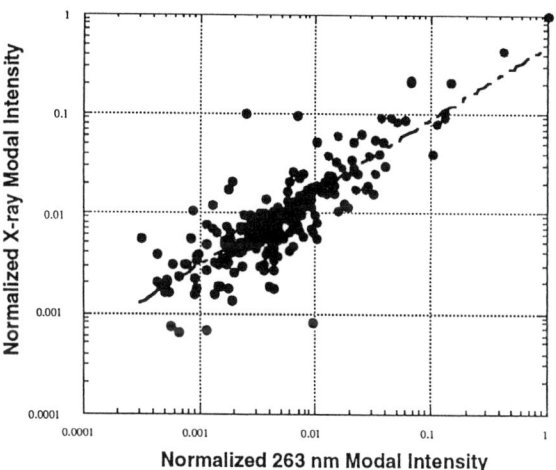

Fig.2 Correlation of x-ray and 263 nm modal intensities

From Figs. 1, a spatial intensity distribution was Fourier transformed to a modal intensity distribution. The Fig. 2 shows that there is a strong correlation with a slope 1 for both the 263 nm and x-ray pictures. In other word, modal intensities of both pictures could be almost linearly related each other. This result was considered as a proof of principle that a 263 nm image could represent an absorption pattern. X-ray images could not show fine scale nonuniformity of the absorption pattern, since x rays could be emitted from over dense, near the cut off, and all the corona regions. In addition x-ray may not be sensitive to monitor an absorption pattern, since a free-free emission rate is proportional to square root of the plasma temperature. Using only specially designed target such as the one used here x-ray pictures could show an absorption distribution. If the absorption energy or intensity is correlated with Al Ly α emissions, it is possible to quantitatively correlate the absorption intensity to 263 nm images. These results should be reported elsewhere.

Figure 3 shows that if a beam smoothing was introduced to a laser beam the intensity distribution of the 263 nm image could be significantly improved. In this shot, the 527 nm laser was beam smoothed with a random phase plate (2) and a large spectral band width (4Å) with a angular distribution enforced (partially coherent light:PCL)(3). The spot size was 250 μm with a 305 J laser energy, giving 4.2×10^{14} W/cm^2 with a 1.6 nsec pulse.

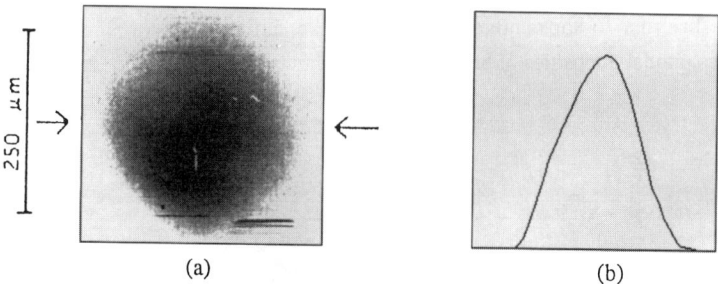

Fig.3 (a) 263 nm image of partially coherent light irradiation and (b) its intensity profile.

Figure 3 (a) shows that the 263 nm pattern and Fig. 3(b) shows that the intensity distribution of the pattern. In Fig. 3(b), the intensity profile shows a very smooth one. In terms of spatial resolution limit of the 263 nm images, another shots were taken with a 527 nm random phased plate and coherent ordinary laser with a pulse width of 1 nsec. The 263 nm image showed many striations due to the random phase speckles of the order of 5-6 μm. Thus at least 5-6 μm spatial resolutions were attained with the Schwarzschild microscope with a 263 nm filter. As seen in Fig. 3(b), the pattern shows almost no structures of the order of 5-6 mm or larger. This indicates that the combination of PCL and the random phase plate gives a laser absorption pattern very smooth.

SUMMARY

We have shown that the second harmonic image could measure an absorption pattern at an early part of the laser pulse, comparing this with an x-ray pinhole picture taken at the same shot. Using a normal laser light without any beam smoothing, there were many hot spot structures in the focal spot. These hot spots were chosen for the comparison. Fourier transformed modal intensities of both pictures show a good correlation, indicating that the 263 nm image could represent an absorption pattern probably dominantly by resonance absorption. Using the combination of PCL and a random phase plate, the 263 nm image showed much improved absorption uniformity.

ACKNOWLEDGEMENTS

We are indebted with many technical supports given by the laser, experimental, and target fabrication groups of ILE.

REFERENCES

1. J.J.Thomson, C.E.Max, J.Erkkila, and J.E.Tull, Phy.Rev.Lett. 37, 1052 (1976).
2. Y.Kato, K. Mima, Y. Kitagawa, M. Nakatsuka, and C. Yamanaka, Phys. Rev. Lett. 53, 1057 (1984).
3. H. Nakano, T. Kanabe, K. Yagi, K. Tsubakimoto, M. Nakatsuka, and Nakai, Optics Commun. 78, 123 (1990).

ρΔR Measurement of Imploded Cryogenic Foam Target by DD-Protons

Y. Kitagawa, K. A. Tanaka, M. Nakai, T. Yamanaka, K. Nishihara, H. Azechi, N. Miyanaga, T. Norimatsu, T. Kanabe, C. Chen, A. Richard, M. Sato, H. Furukawa, and S. Nakai

Institute of Laser Engineering, Osaka University
Yamada-oka, Suita, 565, Japan

Abstract. DD-produced protons are used to measure the imploded fuel-shell areal density ρΔR of a cryogenic foam target. 12 green beams of the GEKKO XII laser directly imploded a cryogenic foamshell target containing a liquid or solid D_2 fuel. The proton emission is detected through a dipole magnet on a CR-39 plate, whose energy shift from 3.02 MeV yielded the maximum ρΔR of 12 mg/cm^2 or the density of 8 g/cm^3, in agreement with that from the secondary neutron method within a factor of 2. The proton energy spread is sensitive to the laser energy imbalance.

TARGET AND LASER

Here is reported measurement of the imploded cryogenic target areal density by using fusion produced protons. The cryogenic target, we used, is a low density plastic foamshell sphere, containing a liquid or solid D_2 fuel[1, 2]. The foamshell is made of trimethylolpropane-trimethacrylate (typically $H_{14}C_{10}O_4$) of 10± 1.3 μm thickness and 220±28 mg/cm^3 density (20% weight of solid plastics) and is coated with 4-μm thick polyvinylphenol ablator both to increase laser absorption and to reduce preheating[3]. The shell radius is within R_0 = 303±24 μm. D_2 density at 21 K is 170 mg/cm^3. The target is sustained vertically with 7μm-thick glass fiber at the center of the vacuum chamber and is cooled down to a presumed temperature in a retractable liquid-He-cooled shroud[4]. Once D_2 gas becomes liquid or solid in the foam layer, then the shroud is retracted and in 10 ms 12 laser beams illuminate the target. At each exit of beams from the GEKKO XII system are a random phase plate and a single KDP crystal for second harmonics. The beams of 32 cm in diameter are focused in tetrahedral configuration on the target by f/3.15 aspheric lenses. The focussing depth $d/R_0 = -5$. The total energy (527 nm) on target is from 3.5 to 12.1 kJ in 2.05±0.25-ns quasi-flat-top pulse with 0.94±0.14-ns rise. Energy imbalance σ between 12 beams is 3.7±1.7% r. m. s. for all the shots.

PROTON SPECTRUM MEASUREMENTS

In the final stage of implosion, a colder but denser main fuel-plastic mixed plasma layer (the second layer) surrounds a central hot core (the first layer). In the hot core, deuterons collide and fuse each other, yielding two groups of fusion products: T (1.01 MeV) + p (3.02 MeV) and He^3 (0.82 MeV) + n (2.45 MeV). In the second layer, tritons collide with deuterons, producing secondary neutrons according to T + D → He^4 (3.6 MeV) + n (14.1 MeV). A yield ratio of secondary neutron to primary neutron gives the second layer (main fuel) ρΔR, where ΔR is a thickness of the second layer[5]. The protons as well run through the second layer plasma, loosing their energies. The resulting proton energy spectrum provides ρΔR. For the first time, we successfully used the proton spectrum to estimate the imploded fuel-shell ρΔR. The obtained ρΔR agrees within a factor of two with that from the secondary neutron method.

A newly developed proton spectrometer uses a ceramic dipole magnet (TDK : REC-22) of 7 kG. An entrance slit into the spectrometer is 0.5 mm wide and 2 mm long and is set 15.3 cm apart from the target. A CR-39 plate of 2 mm width and 40 mm length detects protons of from 1 to 3.5 MeV. Considering particle divergence inside the spectrometer, the effective detection angle is 3.4×10^{-6} str. In Fig. 1(a) is

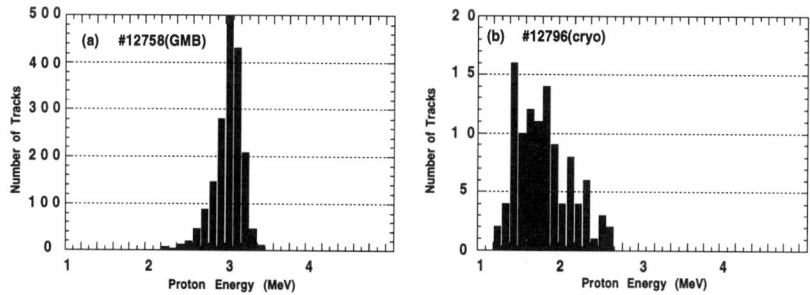

FIGURE 1. Proton spectrum from (a) simple D_2 filled glass microballoon(GMB #12758): (b) cryogenic foamshell target containing gas-liquid D_2 #12796.

shown a proton spectrum from D_2 filled glass microballoon (GMB:#12758) for comparison. The energy peak shift from 2.0 MeV is 0.085 MeV and the energy spread is 0.3 MeV (Gaussian width). The spread is mainly due to Doppler broadening, because a source temperature of 1 keV can yield energy spread of about 0.2 MeV. The GMB energy spread, after substracting the Doppler broadening, gives us an energy resolution of the spectrometer to be 0.22 MeV/3 MeV = 7.4%. The spectra in Fig. 1(b) is from the cryogenic target. By Gaussian fitting the spectra, we estimated both the peak shifts and the spreads. The number of tracks should be 3.4×10^{-6} times Neutron Yield (Ny), assuming that the isotropic emission all over the angle. The protons may be affected not only from the inner plasma fields, but also from the outer field such as a self-generated magnetic field. If the outer fields scatter or distort proton orbits, the number into the spectrometer might be affected. The GMB and cryogenic target tracks are within 50 to 100% of 3.4×10^{-6} Ny, showing that the spectrometer does not appear to have such an effect seriously and the tracks reflect the history inside the targets.

ρR MEASUREMENT BY PROTONS

We calculated the stopping power of 3 MeV proton and 1 MeV triton in a hot dense plasma[6] and obtained proton and triton peak energy shifts ΔE as a function of ρR, which is plotted in Fig. 2 for D_2 plasma of the temperatures Te = Ti = 0.1, 0.3 and 1 keV at the density of 10 g/cm^3. In Fig. 2, maximum proton $\Delta E \approx 3$ MeV provides the upper limit of detectable ρR ≈ 50 mg/cm^2, close to the limit of the secondary neutron method.

In the present mixed ρ/Z, where Z is the plasma charge density. Since the initial density ρ_0 including 80% liquid D_2 (21

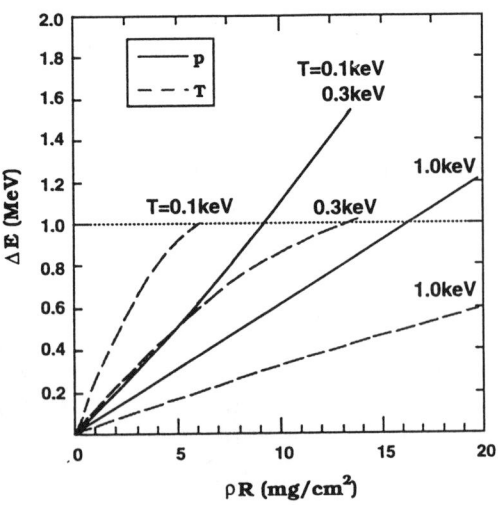

FIGURE 2. Proton and triton energy shift from 3.02 and 1.01 MeV, respectively, calculated versus D_2 plasma areal density ρR. Te = Ti = 0.1, 0.3 and 1 keV for the plasma density = 10 g/cm^3.

K) and 20% foam is 356 mg/cm^3, i. e., 2.1 times the liquid D_2 density, then ρ of the compressed plasma must also be 2.1 times that of pure D_2 plasma. While, supposed charge states of C and O in the foam to be around 5 and 7 from the Thomas-Fermi ionization model[7], Z of the mixed plasma will be twice that of D_2 plasma. which can cancel the ρ increment. Thus without crucial correction we apply Fig. 2 to the present mixed plasma. Fitting the peak shift of each shot to the curve for Te = 0.3 keV in Fig. 2, we obtained ρΔR and plotted it as a function of the initial fuel temperature in Fig. 3. We also plotted ρΔR from the secondary neutron method. The error bar on proton is from the spectral spread and the error bar on neutron is from data scattering between four scintillators. We obtained the maximum fuel ρΔR = 12.7 ± 4 mg/cm^2 for shot #12796. To this shot, the secondary neutron provides 6.3±2.6 mg/cm^2. The proton and neutron data agree within a factor of two each other, except that the error bars are so large. The lower the cryogenic temperature is, the higher the fuel ρΔR should be. Regrettably, cold fuels below the solid temperature, 15 K, could hardly produce either detectable proton or neutron. Even for such cold targets, however, x-ray streak cameras (XSC) and x-ray framing cameras (XFC) caught hot core images as reported[8], yielding hot core areal densities, assuming uniform fuel compression, as plotted in Fig. 3. The x-ray data

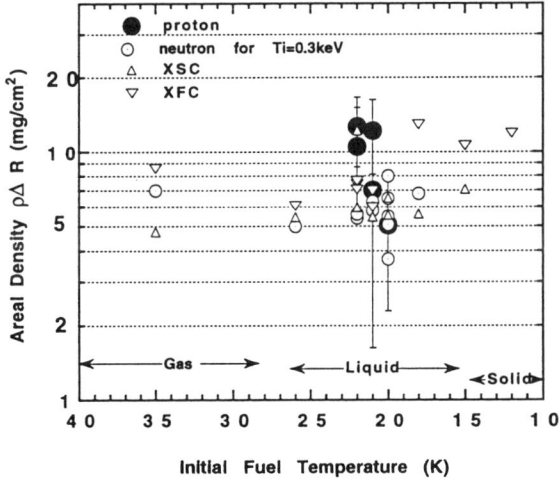

FIGURE 3. ρΔR of second layer (plastic foam and D_2 fuel) from proton energy shift (●) and from secondary neutron method(○) versus initial fuel temperature. Error bar on proton is from spectral broadening and error bar on neutron is from data scattering between four scintillators. △ is core areal density from x-ray streak image and ▽ from x-ray framing image.

suggest that cooling targets down to solid temperature will increase areal densities to around 20 mg/cm^2. X-ray streak images show that the overall implosion is achieved rather spherically and the main shell (second layer) stayed surrounding a hot core (the first layer) at the maximum compression[8]. The maximum x-ray emission takes place at the end of the maximum compression phase at the contact surface between the first and second layers just before it expands. The minimum hot core radius is typically 24 µm, very close to the calculated contact surface radius: R - ΔR/2 ≅ 18 µm. Experimental convergence ratio was 12.4 at the contact surface. Thus, at the maximum compression the second layer seems yet to remain and surround the hot core, then the compression is given by $\rho/\rho_0 = (R/\Delta R)(\rho\Delta R)^{3/2} / (R_0^2 \Delta R_0 \rho_0^3)^{1/2}$, where ρ_0 = 356 mg/cm^3. From the x-ray core emission, the imploded shell radius R ≅ 30 µm[8]. If ΔR is close to the initial shell thickness ΔR_0 = 10 µm, as suggested by the simulation, then R/ΔR ≅ 3. Finally ρΔR of 12.2 ± 4.1 mg/cm^2 for shot #12874 gives ρ = 7.8 ± 1.1 g/cm^3, or ρ/ρ_0 = 22 ± 3. If the shell structure does not remain, but the fuel is compressed uniformly, then R/ΔR ≅ 1, or the density must be roughly a third. The fuel density of GMB(#12758) is provided to be ρ = 50 ± 20 mg/cm^3, or the compression ρ/ρ_0 = 27, assuming an uniform fuel compression. As shown in Fig. 4(a), both DD and DT yields increase as E_L^4, keeping the ratio or ρΔR to be almost constant, as plotted in Fig. 4(b).

The proton ρΔR of cryotargets is 1~2 times the neutron ρΔR, which gap is explained as: the second layer includes not only D$_2$ but also plastics. Though the secondary neutron method gives only the D$_2$ ρΔR, while the

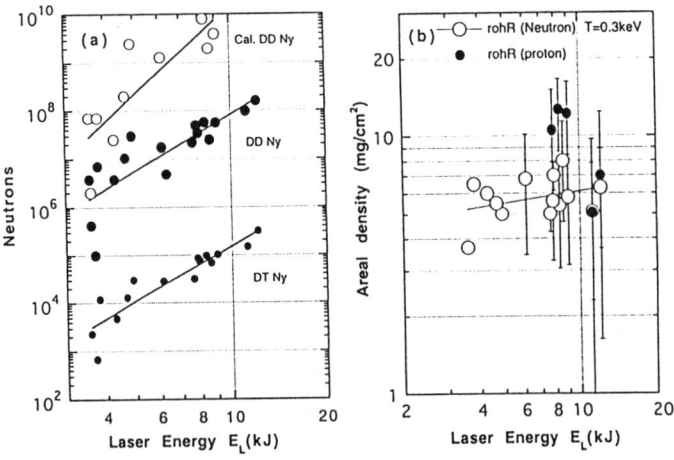

FIGURE 4(a) Primary DD and secondary DT neutron yields as a function of laser energy E_L illuminated on target with maximum DD yield calculated by 1-d hydrocode HISHO. Curves are the optimum fitted ones. **(b)** ρΔR from proton energy shift (●) and from secondary neutron method (○) versus E_L.

proton method gives total $\rho\Delta R$ of the layer, because any charge can reduce proton energies. Since the mass ratio of the total layer to the D_2 fraction is 356 mg/ (0.8 × 170) mg $\cong$ 2.6, consistent with the experiment that the proton $\rho\Delta R$ is 1 ~ 2 times the neutron $\rho\Delta R$.

The spectral spreads of the cryogenic target protons are so large as 0.8±0.2 MeV(HW of Gaussian curve) or 40 to 80 % of ΔE, as shown in Fig. 1(b). If the protons take not only radial but also tangential paths in the layer, a simple path integration over the shell gives 1.3 times effective path length, resulting in about 30 % geometrical spread, which is less than the experiment. The spread seems to be due to other effects such as laser energy imbalance σ, which can distort the implosion symmetry. The velocity imbalance of the imploding shell may bring about an initial shell perturbation at the stagnation. In future power-balanced laser illumination from the New GEKKO will make clearer the cause of the spectral spreads.

CONCLUSIONS

D-D protons are used to measure the imploded fuel-shell areal density $\rho\Delta R$, as $\rho\Delta R = 12\pm 4$ mg/cm^2, giving $\rho = 8 \pm 1$ g/cm^3, or the compression $\rho/\rho_0 = 22 \pm 3$. The results agree with those from the secondary neutron method within a factor of 2. The proton energy spread is sensitive to the laser energy imbalance.

ACKNOWLEDGEMENTS

We acknowledge the GOD and OT groups at the Institute of Laser Engineering for laser operation, the T group for the cryogenic target fabrication and the MT group for plasma diagnostics. Also we thank Mr. R. Ishizaki for stopping power calculation.

REFERENCES

[1] Marshall F. J. et al. Phys. Rev. **A 40**, 2547 (1989).
[2] Nakai M. , and Yamanaka T., Kakuyugo Kenkyu **68** **Suppl.**, 79 (1992) (in Japanese).
[3] Takagi M., Ishihara M., Norimatsu T., et al., J. Vac. Science Tech. **11A**, 2837 (1993).
[4] Norimatsu T., Ito H., Chen C., et al., Rev. Sci. Instrum. **63**, 3378 (1992).
[5] Azechi H., Miyanaga N., Stapf R. O. , et al., Appl. Phys. Lett. **49**, 555 (1986).
[6] Furukawa H. and Nishihara K., Phys. Rev. **A 46**, 6596 (1992).
[7] More R. M., Lawrence Livermore National Laboratory, LLNL-report, UCRL-84991, 1981.
[8] Richard A., Tanaka K. A., Kanabe T., Kitagawa Y., Nakai M., Nishihara K., et al., Phys. Rev. **E 49**, 1520 (1994).

Hot Spark Structure in Laser-Imploded Core Plasmas Observed with 10-ps-Resolved X-ray Imaging

M. HEYA, H. SHIRAGA, N. MIYANAGA, K. NISHIHARA, K. TANAKA, Y. KATO, T. YAMANAKA and S. NAKAI

Institute of Laser Engineering, Osaka University
2-6 Yamada-Oka, Suita, Osaka 565, Japan

Abstract. We have developed a multi-imaging x-ray streak camera (MIXS) technique and applied it to obtain time-resolved x-ray images of the laser-imploded core plasmas. Temporal changes in the core structure of laser-imploded targets were first observed with a temporal- and a spatial-resolution of 10 ps and 15 µm, respectively. It has been found that these structures are dependent on the laser irradiation conditions.

INTRODUCTION

In the present laser-driven implosion experiments, disagreements have been seen between the experimentally observed neutron yields from the core plasmas and those predicted by 1-dimensional simulations.[1] This means that some discrepancies are taking place in the heating process of the hot spark.

We have developed an ultra-high speed x-ray imaging technique, multi-imaging x-ray streak camera with a 10-ps temporal resolution, and have utilized it successfully in the GEKKO-XII implosion experiments. The information on the heating process of the hot spark, especially with regard to temporal changes and spatial structures, has been obtained by measuring the x-ray emission images from Ar doped DD fuel plasmas.

In one experiment, two types of the laser pulse shape were used to change the implosion mode. In another experiment, the laser focusing condition was varied to enhance the low-mode drive nonuniformities. Low-mode structures in the hot spark are seen depending on the laser focusing condition. From these results, we conclude that both the implosion mode and the focusing condition affect the spark heating.

MIXS SYSTEM

A time-resolved two-dimensional x-ray imaging technique using multi-imaging x-ray streak camera (MIXS) technique, first proposed by Choi and co-worker,[2] has been improved and implemented on laser plasmas.[3] In order to obtain time-resolved x-ray images of the hot spark from the core plasmas, we have developed MIXS system with a 10-ps temporal resolution and utilized it successfully in the GEKKO-XII implosion experiments.[4] The multi-imaging x-ray pinhole camera was coupled to an x-ray streak camera, and 2-dimensional images were reconstructed by handling the streaked image data.[4] The image distortion of the system was evaluated and nonuniformities in sensitivity were corrected. Temporal resolution, or the exposure time of each framing image of the present MIXS system is determined by the width due to the electron velocity distribution at the photocathode and the broadening due to the cathode and sampling width. In the present case, the estimated temporal resolution is 11.7 ps. Spatial resolution is determined mainly by the geometrical resolution of the pinhole camera, or the pinhole size. Diffraction effect, spatial broadening in the streak tube including the cathode resolution, and the width of the cathode slit also affects the spatial resolution. The estimated spatial resolution is 15 µm. The observed spectral range was determined by transmission of the filters and response of the Au cathode. In case of a 25-µm-thick Be (a blast shield) and a 5-µm-thick Ti filter, the system had sensitivity in the range of 2.2-4.9 keV and the spectral response was almost flat in this range.

MIXS was applied to the implosion experiment at the GEKKO-XII laser facility. Time-resolved two dimensional x-ray imaging was performed and temporal behavior of the structure in the x-ray image of the imploded core was observed. A spherical ICF target was irradiated by 12 laser beams delivered from the GEKKO-XII system. All energy imbalances between the beams were about less than about $\pm 5\%$. Laser wavelength was 0.53 µm. Each beam was equipped with a random-phase plate[5] to improve the irradiation uniformity. The targets were a deuterated plastic (C_8D_8) shell. Nonuniformity in the shell thickness was within $\pm 1.1\%$.

IMPLOSION EXPERIMENTS AND DISCUSSION

We have conducted a series of experiments and measured the x-ray emission images from the core plasmas, changing the laser pulse shape and the laser focusing condition in order to change the implosion mode and the drive nonuniformity intentionally. In the first experiment, two types of the laser pulse shape; low-intensity / long-pulse (#15291) and high-intensity / short-pulse (#15580) were used. We used identical plastic shell targets at the same focusing condition, which is indicated by a parameter, $d/R = -5$, where d is the distance from the center of the target to the focal position of the laser and R is the initial target radius. The target diameter, shell thickness, laser energy, laser pulse duration (FWHM) and laser intensity were 601 µm, 6.81 µm, 4.29 kJ, 2.45 ns, 1.1×10^{14} W/cm^2 for #15291 and 486 µm, 7.23 µm, 5.66 kJ, 1.34 ns, 3.2×10^{14} W/cm^2 for #15580, respectively. Deuterium and Ar pressure were 40 atm, 0.3 atm for #15291 and 50 atm, 0.1 atm for #15580. Figure 2 shows spatial profiles of x-ray intensity from the core plasmas and radius-time (R-T) flow diagrams from HISHO 1-dimensional simulation for #15291 and #15580, respectively. The relative time used in Figure 2 was determined by adjusting the observed and calculated trajectories of x-ray emission from the imploding shell with a possible of ±20 ps. As for the x-ray intensity, we have adjusted the experimental results to 1-D simulations at the early rising part of the core emission. In the case of low-intensity / long-pulse implosion (#15291), a disagreement of the x-ray emission between the experimental result and the 1-dimensional simulation is found to appear at the time when the reflected shock hits the pusher. In the case of high-intensity / short-pulse implosion (#15580), the disagreement appears somewhat later in the deceleration phase. This means that the heating time of the hot spark was extended to the deceleration phase. What made this improved heating? One possible explanation is the difference in the growth of Raleigh-Taylor instability. However, accumulated growth factors, $\int \gamma \, dt$, according to the acceleration history calculated by 1-D simulations were found identical for both implosions. Therefore, this is not the case. Another possibility is the difference in thermal smoothing effect. The stand-off distance from the laser absorption region to the ablation surface predicted by 1-dimensional simulations is 23 μm and 43 μm, respectively. The irradiation nonuniformities with very small wavelengths can be smoothed out by thermal electrons in plasmas as $\exp(-kD)$, where k is the wave number of the perturbation and D is the stand-off distance. From these estimations, we conclude that in the high-intensity / short-pulse implosion the heating process is improved by the enhanced thermal smoothing of the irradiation nonuniformities at the middle- and high-modes.

Second experiment was conducted at the different focusing conditions, $d/R = -4$ (#15591) and -5 (#15580), with the same high intensity / short pulse. Figure 1 shows an irradiation nonuniformities on a target surface at the mode number of 1-20. It is shown that $d/R = -4$ and $d/R = -5$ create 20% and 13% nonuniformities at the mode number 6, indicating that there is an improvement in the uniformity with this mode at $d/R = -5$. Figure 3(a) and (b) show the MIXS images for the focusing conditions $d/R = -4$ (#15591) and -5 (#15580), respectively. From Figure 3 we can clearly find that the low-mode nonuniformities of the laser-imploded core plasmas for #15580 is more uniform than those for #15591. However, it is found that the laser focusing condition does not affect the bulk heating of the spark significantly. This implies that low-mode nonformities observed in the time-resolved x-ray images did not affect the heating process of the hot spark. Probably the heating process might be affected more by the middle- to high-mode nonuniformities than lower modes.

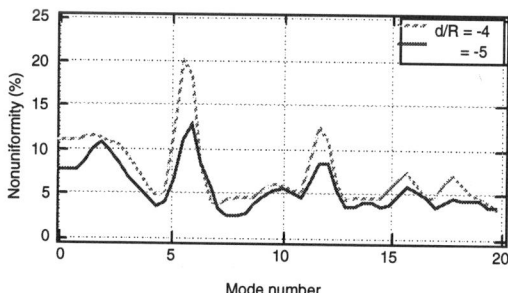

Figure 1. An irradiation nonuniformities on a target surface of 1-20 at the mode number.

SUMMARY

We have developed the MIXS method with a 10-ps temporal resolution and used it successfully in the GEKKO-XII implosion experiments. Core structures were found to be rapidly changing with a 10-ps time scale, while the time-integrated images showed smooth profiles. It was found that both the implosion mode and the focusing condition are affecting the temporal and spatial structure of the core. Still there is a large degradation in spark heating, which should be much improved by reducing irradiation nonuniformities. We are preparing to use beam power balancing and coherence control technique in order to reduce the low- and high-mode nonuniformities, respectively.

REFERENCES

[1] H. Azechi et al, Laser Part. Beams **9**, 193 (1991).
[2] C. Deeney and P. Choi, Rev. Sci. Instrum. **60**, 3558 (1989); P. Choi and R. Aliaga, *ibid*. **61**, 2747 (1990).
[3] O. L. Landen, Rev. Sci. Instrum. **63**, 5075 (1992).
[4] H. Shiraga et al, Rev. Sci. Instrum. **66**, 722 (1995).
[5] Y. Kato et al, Phys. Rev. Lett. **53**, 1057 (1984).

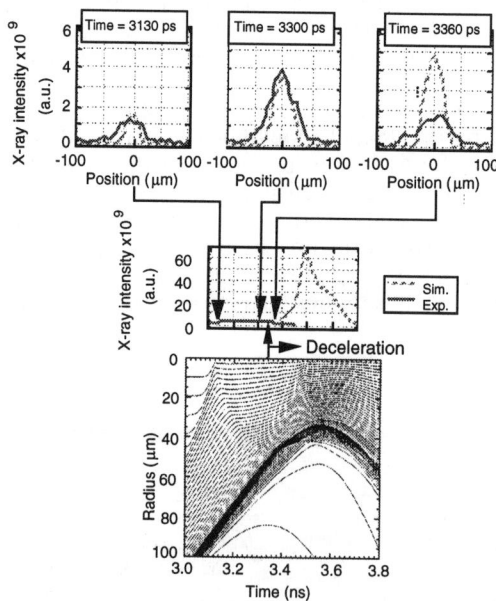

Figure 2(a) low-intensity / long-pulse implosion (#15291)

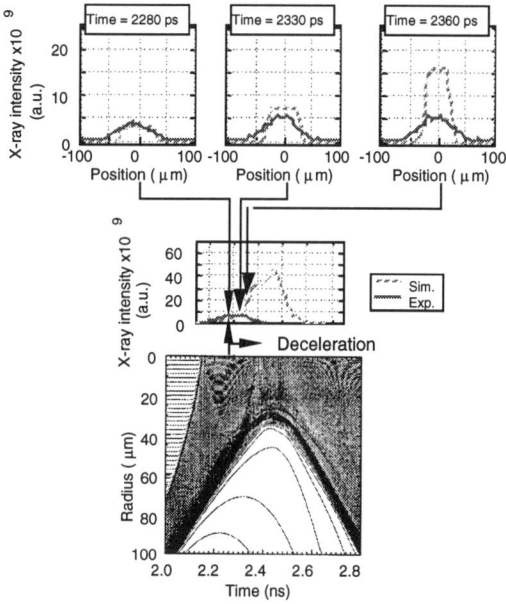

(b) high-intensity / short-pulse implosion (#15580)

Figure 2. Spatial and temporal behavior of the core emission obtained from 2-D images by the MIXS system. Spatial profiles at the equator of the core and temporal profiles at the target center are shown. Also indicated is the corresponding R-T diagram from 1-D simulation.

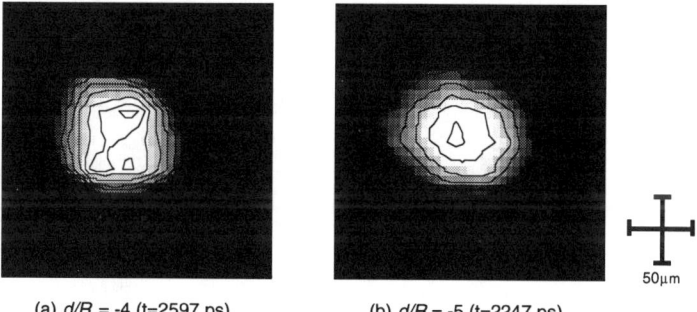

(a) $d/R = -4$ (t=2597 ps) (b) $d/R = -5$ (t=2247 ps)

Figure 3. MIXS images of the imploded core plasmas at the time of the peak x-ray emission for $d/R = -4$ and $d/R = -5$ implosions.

Preliminary Analysis of Core Capsule X-Ray Spectroscopy and Image Results for Medium-to-High Growth Factor Implosions

Greg Pollak*, Norm Delamater*, Nino Landen[†],
Bruce Hammel[†], Chris Keane[†]

*Los Alamos National Laboratory; PO Box 1663;Los Alamos, NM 87545 and
**Lawrence Livermore National Laboratory; 7000 East Avenue; Livermore, CA 94550

Abstract. Recent capsule implosions using indirect drive on NOVA have probed core and near-core capsule T_e, ρ and mix structure using non-trivial pulse shapes (i.e. with a foot). These experiments have been performed using smooth as well as artificially roughened capsules. They have been performed using basically 3 non-trivial pulse-shapes with 3 different types of capsules with correspondingly different growth regimes for Rayleigh-Taylor instabilities. These experiments have employed time-dependent spectroscopy, gated imaging and <u>absolutely calibrated</u> time-integrated imaging as x-ray diagnostics. We compare nominal and "modified" 1D calculations with the spectroscopic and time-integrated image results. We find that the core T_e is less than calculated (not surprising), but also that the T_e of the inner pusher is substantially higher (at least 20%) than predicted, with perhaps some enhanced mix of the PVA layer towards the core.

INTRODUCTION

This paper compares experimental and theoretical results for x-ray diagnostics of imploding indirectly-driven ICF capsules. The 2 diagnostics which are analyzed in some detail here are a streaked, spatially integrated, crystal spectrometer and time-integrated, absolutely calibrated pinhole camera which simultaneously images thru 6 different filters.

In this discussion only 1 of the 6 filters will be analyzed : 4μm of silver. Silver has two important characteristics which merit emphasis. The first is that it will truncate transmission above an L edge of 3.351 kev. This allows the He and H α (2→1) lines of Ar (which some of the capsules are doped with) thru, but not the β (3→1) lines. More importantly, the active film material for the camera is AgBr, and, because of the same Ag L edge, the film undergoes sharp discontinuities and fluctuations in its response to photons above this energy. Below the edge the frequency dependence of the differential response is smooth and linear. In both diagnostics there are Ti patches and a Be blast shield which also need to be taken account of.

© 1996 American Institute of Physics

The theoretical analysis is done in two parts. In the first part, the code LASNEX is run in a 1D spherical mode, including any prescribed mix. Periodically, dumps from LASNEX are written which contain T_e, density, grid, and species fraction information. These dumps are then read by the postprocessor TDG/DCA[1]. This postprocessor does a number of things. The DCA portion calculates non-LTE opacities using a sophisticated detailed configuration accounting procedure (with essentially no approximations), and it calculates a radiation field self-consistently with these opacities and iterates the opacity/photon field calculations to some user specified degree of convergence. The TDG portion of the postprocessor calculates the x-ray image or spectra seen by a detector by doing various line-of-sight source/sink integrals.

Two different pulse shapes and capsules are considered here. One, known as PS 22, is approximately 2.2ns long and has a foot that produces a radiation temperature (T_r) of ~165ev in a NOVA standard holhraum. The peak of the pulse produces a T_r of between 200 and 215ev. The capsule has a pusher thickness of ~55µm of which the inner ~3.5µm are polystyrene (which may be doped with 1% Cl) and a adjacent layer of 2.5µm of PVA (CH_2O). The 2nd pulse shape (PS26) has a lower foot at ~145eV and a higher peak ~220eV. The total laser energy in the UV for the first pulse is ~25KJ→28KJ, and 32KJ→36KJ for the 2nd pulse shape. The 2nd capsule has a total pusher thickness of ~45µm, with 2 inner layers as for the 1st capsule, except that the dopant in the inner layer is Ti (.07%). Generally, the main outer layer of the 2nd capsule is doped with a significant amount (1→2%) of either Br or Ge. In the capsules considered here, the dopant was Br at ~1.8%. The first capsule experiences "moderate" growth of surface roughness-induced instabilities; the 2nd capsule experiences "high" growth-- mostly due to the Br dopant (which reduces ablative and scale length stabilization) and the thinner shell (which increases in-flight aspect ratio)[2].

ANALYSIS

We shall first consider the spectra. Fig. 1a) is a typical spectra from a PS22 capsule near peak compression. The 4 major lines (from left to right) are Heα, Hα, Heβ and Hβ. There are two important points to note concerning the spectra. One is that the line intensities are increasing the higher the line energy (for the Heα, Hα, Heβ lines). This is unusual, and has been traced to two facts: the pusher has a decreasing attenuation with increasing photon frequency; and the β lines are abnormally strong because of n=2→3 collisonal excitation in the high density (Ne $\cong 10^{24}$) environment in the core[1]. The 2nd important point is that there is a non-zero continuum background. Fig. 1b) gives the result of a simulation in which the drive is reduced ~15% compared to the theoretically calculated hohlraum drive. Note that the line ratios are approximately correct, but that there is essentially no continuum background. It is known from the time-dependent data that the background turns on and off approximately in accord with the Ar lines, so the capsule is the source of the background. By looking (calculationally) lower down in photon energy, it was observed that such a continuum was present, but was essentially zero in the spectrometer window.

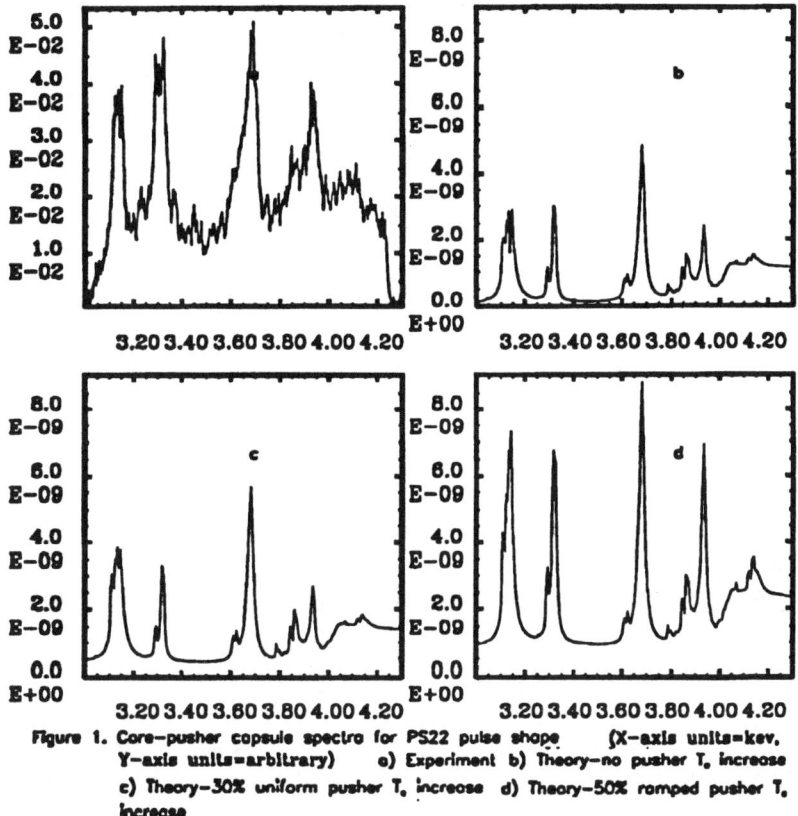

Figure 1. Core-pusher capsule spectra for PS22 pulse shape (X-axis units=kev, Y-axis units=arbitrary) a) Experiment b) Theory-no pusher T_e increase c) Theory-30% uniform pusher T_e increase d) Theory-50% ramped pusher T_e increase

Fig. 1c) gives the calculational result for the same case as in 1b), except the pusher T_e has been artificially enhanced by a multiplicative factor of 1.3. This gives substantially better agreement with Fig. 1a). Fig. 1d) gives the calculated result for a case where the T_e enhancement is ramped in space, being nominal at the center and far out in the pusher, and with a maximum enhancement of 50% at the inner PVA boundary.

Figures 2a), 2b), 2c) and 2d) are equivalent results for PS26. The spectrometer window in these shots is higher in energy than in the corresponding PS22 shots because it was desireable to detect the Heα line from the Ti dopant in the inner pusher (at ~4.75 kev). Here Fig. 2c) gives the results for a uniform 20% enhancement, while Fig. 2d) is for a uniform 30% enhancement. Note the

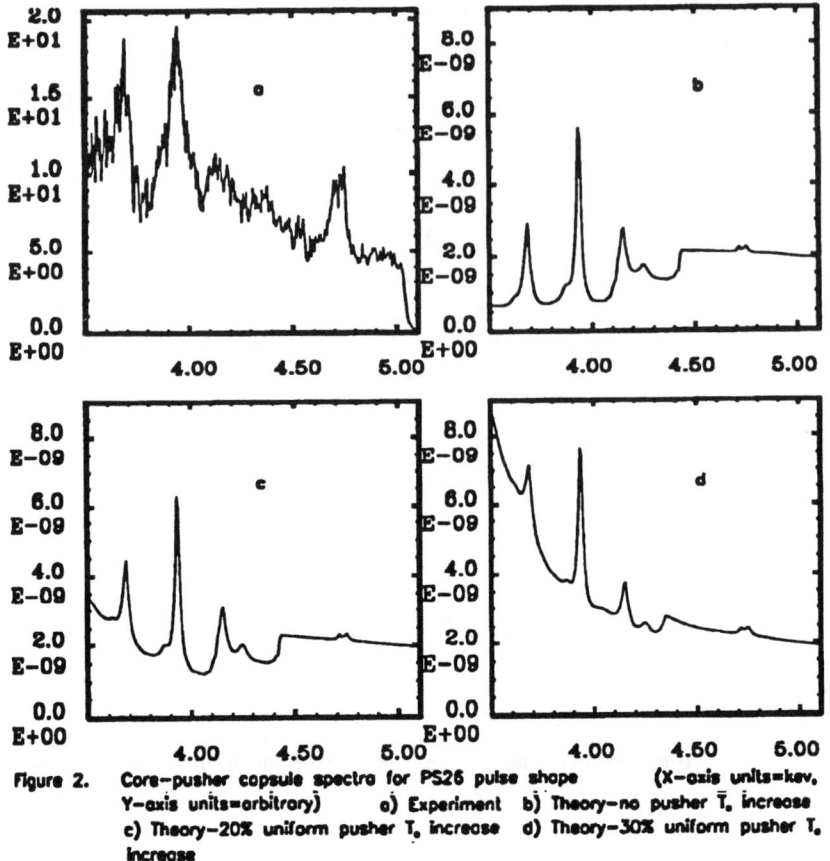

Figure 2. Core-pusher capsule spectra for PS26 pulse shape (X-axis units=kev, Y-axis units=arbitrary) a) Experiment b) Theory-no pusher T_e increase c) Theory-20% uniform pusher T_e increase d) Theory-30% uniform pusher T_e increase

experimental result appears to be bracketed by these 2 calculations. Note the same issues are present for the continuum background as in PS22, except that the background is now quite strong and has an obvious slope.

In order to better understand the nature of this effect, the image diagnostic results for PS22 were modeled. Table 1 gives results for total (time-and space-integrated) photon yield and 2 σ, where σ is the standard deviation of a gaussian fit to the spatial image. Two capsules are analyzed: a completely undoped capsule, and a capsule with Ar in the core. Both experimental and calculated results are given. The calculated results include an unmodified run as well as runs where the pusher T_e has been multiplied by a spatial-and time-dependent multiplier. In one such modified run, the multiplier was .8 on the central core T_e, 1.3 at the outer edge of the PVA layer, 1.0 far out in the pusher, with linear ramps between these points. This is a modest enhancement to the pusher T_e. From the results we can see there is large disagreement between unmodified theory and experiment for both photon yield (~100x) and width (~3x) for the undoped capsule. Modifying the T_e brings the undoped capsule characteristics into better agreement with experiment, but there are still substantial discrepancies. For the doped capsule the discrepancies are less severe for total photon yield (~6x), but are still as bad for the width (~3x). It should be noted that the calculations have been done using theoretically calculated hohlraum drives, and the measured drives are lower, especially for PS26. If the measured drives had been used, the reduction in core T_e needed to match data would have been less, but the increase in pusher T_e would have been higher. This effort to use measured drives in the calculations is currently ongoing.

TABLE I. Key results for image analysis (Ag filter)

	Experiment Pinhole camera		Theory LASNEX TDG/DCA	
	Yield (ergs)	Width (μm)	Yield (ergs)	Width (μm)
No Ar, no Cl	1.0^{-2}	44	9.5^{-5}	15
No Ar, no Cl, increased pusher T_e decreased center T_e	1.0^{-2}	44	4.9^{-4}	22
Ar, no Cl KDCA Ar fine photon bins	1.5^{-2}	36	2.2^{-3}	12
Ar, no Cl-DCA increased pusher T_e decreased center T_e	1.5^{-2}	36	2.7^{-3}	14

Another possibility for the enhanced emission is mix of the oxygen in the PVA layer in towards the core resulting in non-LTE free-bound recombination radiation. Preliminary analysis of this possibility suggests a maximum enhancement of 2x or 3x in photon yield from pre-peak compression mix for the undoped capsule. This is substantial but no-where near what is needed. Late time mix offers some possibility for somewhat more significant emission and is under study.

CONCLUSION

We conclude that there is substantial evidence from both the continuum background in the spectroscopy and from the photon yield and image width that the inner pusher is emitting more strongly than predicted. By zeroing the pusher absorption opacity, the possibility can be ruled out that the pusher has been partially breached and that the enhanced emission is core emission (this can enhance Ar lines but does not change the continuum background much). Mix of the PVA layer and subsequent emission of recombination radiation by the oxygen can probably only account for a small fraction of the peak compression emission. That leaves only enhanced pusher T_e as the source of the excess emission. There are, in-turn, two obvious sources of enhanced T_e: turbulent viscous heating; and non-local heat transport effects (e.g. Fokker-Planck). Both of these are under current investigation.

REFERENCES

1. Pollak, G. D. et al, "Development of a non-LTE spectral postprocessor for Dense Plasma Simulations with applications to spectroscopic diagnostics in spherical Implosions at NOVA", *JQSRT*, 1994, pp. 51, 303.

2. Keane, C. J. et al, "X-Ray Spectroscopic Diagnostics of Mix in High Growth Factor Spherical Implosions", *JQSRT*, to be published.

2D and 3D Ablation Front Hydrodynamic Instability Experiments on Nova

B.A. Remington, M.M. Marinak, S.V. Weber, K.S. Budil,
O.L. Landen, S.W. Haan, J.D. Kilkenny, and R.J. Wallace

Lawrence Livermore National Laboratory, Livermore, CA 94550

Abstract. Single-mode experiments have been conducted on the Nova laser to examine the effect of perturbation shape on ablation front Rayleigh-Taylor growth. The perturbations investigated had the same magnitude wave vector $k=(k_x^2+k_y^2)^{1/2}$ and the same initial amplitude. The shapes corresponded to 2D $\lambda=50$ μm, 3D square $k_x=k_y$, and stretched $k_x=3k_y$ perturbations. We observed that the 3D perturbations grew more than the 2D perturbation. Numerical simulations in 2D and 3D are in agreement, showing the most symmetric modes growing the largest.

Understanding the Rayleigh-Taylor (RT) instability is of critical importance to inertial confinement fusion (ICF) because large RT growth on imploding capsules can degrade implosion performance. In direct drive, nonuniformities in laser illumination imprint nonuniformities onto the capsule pusher ablation front. In indirect drive, residual capsule surface imperfections lead to perturbations at the ablation front. In both cases, the ablation front is RT unstable and perturbations grow. In the linear regime, the perturbation growth is exponential,

$$\eta = \eta_0 e^{\gamma t}, \qquad (1)$$

where η represents perturbation amplitude and the growth rate γ can be written approximately as (1-3)

$$\gamma = [kg/(1+kL)]^{1/2} - \beta k v_a . \qquad (2)$$

Here g is acceleration, $k=2\pi/\lambda$ is perturbation wave vector, L is the density gradient scale length, β is a constant between 1 and 3, and v_a is ablation velocity. When the perturbation spatial amplitude is nonnegligible compared to its wavelength, the RT evolution enters the nonlinear regime. For semi-infinite, incompressible fluids, the bubble approaches its terminal velocity, corresponding to the buoyancy being exactly balanced by kinematic drag (4,5). In this limit the bubble amplitude is just $\eta(t) = \int u_B dt$, with bubble velocity

$$u_B = \alpha(g\lambda)^{1/2}, \qquad (3)$$

© 1996 American Institute of Physics

where $a=0.23$ in 2D and 0.36 in 3D, and we have assumed an Atwood number of 1 for simplicity. Notice that for the same perturbation wavelength (bubble diameter) λ, the bubble velocity is larger in 3D, since the kinematic drag per unit volume is less.

Perturbation growth is sensitive to perturbation shape. The transition to the nonlinear regime occurs approximately when the bubble velocity in the linear regime equals the nonlinear terminal bubble velocity, namely, when $\dot{\eta}_{linear} = \gamma\eta \approx u_B$. Since u_B is larger in 3D than in 2D from Eq. 3, the transition to the nonlinear regime happens later in 3D. Hence, in 3D the linear regime exponential growth phase lasts longer and the asymptotic nonlinear growth rate is higher. Single-mode perturbations therefore are expected to grow larger in 3D than the equivalent perturbation in 2D. This simple qualitative picture is supported by 3rd order perturbation theory (6), recent work with a potential flow model (5), and full numerical simulations (7-9). Until recently, however, experiments in the ICF regime have been lacking due to their complexity. We present here new results of an experimental and computational investigation of 2D versus 3D single-mode perturbation growth at the ablation front in planar, indirectly driven foils (10). Recent advances in target fabrication (11) and diagnostic development (12) have made this experiment possible.

The experimental configuration is shown in Fig. 1a and is described in more detail elsewhere (13). A 750 μm diameter, 60 μm thick CH(Br) planar foil ($C_{50}H_{47}Br_3$, $\rho=1.26$ g/cm^3) is mounted across a diagnostic hole on a 3 mm long, 1.6 mm diameter gold cyclindrical hohlraum. Eight of the 10 Nova laser beams (14) are used to generate a 3.3 ns low-adiabat, shaped drive, as shown in Fig. 1b. Two 3 ns square beams are delayed relative to the drive and focused onto a Sc backlighter disk to generate 4.3 keV He-α x-rays to back-illuminate the accerlating planar foil. Random phase plates with 5 mm diameter hexagonal elements are inserted as the last optic in the two backlighter lasers to generate a smooth 700 μm diameter x-ray spot. Typical timing of the backlighter lasers relative to the drive lasers is illustrated in Fig. 1b. On each laser shot, two-dimensional gated x-ray images were obtained with a new flexible gated x-ray pinhole camera (12). Four pinhole images are obtained for each strip on the MCP, and the interstrip delay was set to 700 ps. Half of the pinholes on each strip were filtered with 12.5 μm of Ti to eliminate higher energy backlighter x-rays such as from He-β and He-γ transitions in Sc.

The foils were made using a new laser ablation technique to make molds in substrates of either kapton or mylar (11). We prepared perturbed foils all with the same magnitude wave vector $k=(k_x^2+k_y^2)^{1/2}$ and nominally the same amplitude. The "2D" foil (1D wave vector $\mathbf{k}=k_x$) was a simple $\lambda=50$ μm sinusoid with initial amplitude $\eta_0=2.5$ μm. One of the "3D" foils [2D wave vector $\mathbf{k}=(k_x,k_y)$] corresponded to a "stretched" $k_x=3k_y$ perturbation, and the other was a square $k_x=k_y$ mode. Characterization of the three foils was done using a contact radiography system, the resulting images of which are shown in Fig. 2a-c, and contact profilometry. The radiographs were converted to spatial amplitudes using

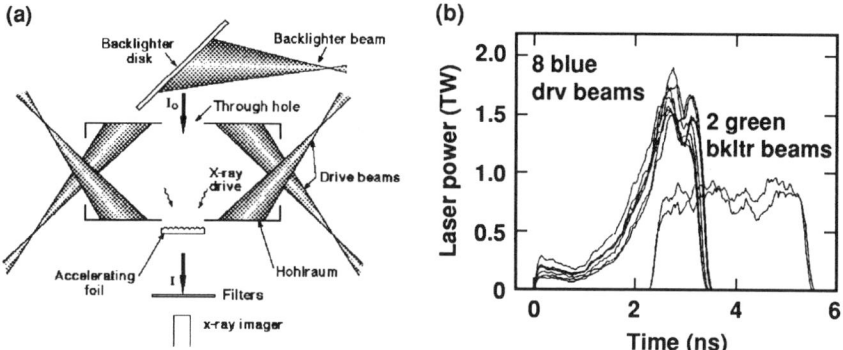

Figure 1. (a) The experimental configuration consists of a Au cylindrical hohlraum with the modulated CH(Br) foil mounted on the wall. The laser beams convert to x rays in the hohlraum, which ablatively accelerate the foil. Two additional laser beams generate backlighter x rays used for in-flight diagnosis of the foil. (b) Power versus time of the eight 0.351 µm wavelength drive laser beams (curves starting at t=0) and the two 0.528 µm wavelength backlighter beams (curves starting at t=2.4 ns).

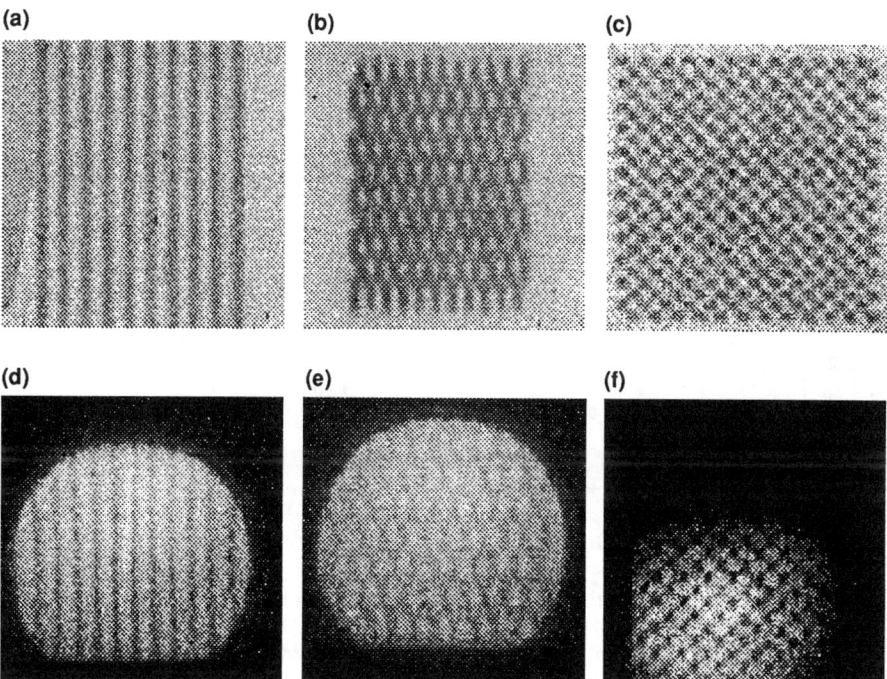

Figure 2. Contact radiographs of foils identical to those used in the Nova experiments are shown in (a)-(c). The perturbations correspond to (a) 2D λ=50 µm, η_o=2.5 µm, (b) 3D k_x=3k_y: λ_x=53 µm, λ_y=158 µm, η_o=2.4 µm, and (c) 3D k_x=k_y: λ_x=λ_y=71 µm, η_o=2.7 µm. The corresponding images from the Nova shots taken at 4.3 ns are shown in (d)-(f).

a step wedge of the same material, CH(Br). Images from the Nova shots at 4.3 ns, which is near peak growth, are shown in Fig. 2d-f. The gated x-ray pinhole camera for these images was run at 8x magnification with 10 μm pinholes, and 150 μm Be filtering. The backlighter was scandium at 4.3 keV.

Each image from the Nova shots is converted to ℓn(exposure) $\propto$ -OD = -$\int \rho \kappa dz$. Hence, modulations in ℓn(exposure) correspond to modulations in foil areal density. The images are Fourier analyzed, and the amplitudes corresponding to the fundamental mode are extracted. A purely experimental demonstration of the effects of dimensionality on perturbation growth depends upon holding conditions between experiments identical. Two of our shots were done back to back on the same day, where the only change made between shots was the target (2D λ=50 μm versus 3D k_x=3k_y). The total laser energy for these two shots was close (16.6 vs 15.8 kJ), and the timing and filtering of the diagnostic were identical. The results for the evolution of the fundamental mode for these two shots is shown in Fig. 3a. The 3D k_x=3k_y perturbation has clearly grown larger late in time in the nonlinear regime, as expected.

Shape effects on perturbation evolution can be examined under <u>identical</u> conditions with computer simulations. This is shown in Fig. 3b using the new 3D radiation-hydrodynamics code HYDRA (10). The perturbations, in order of decreasing peak growth, correspond to k_x=k_y, k_x=2k_y, k_x=3k_y, and 2D λ=50 μm. Our simulations clearly show that the most symmetric perturbations grow the largest, as has been reported by others (6-8). This is qualitatively in agreement with our experimental observations; quantitative comparisons are currently underway.

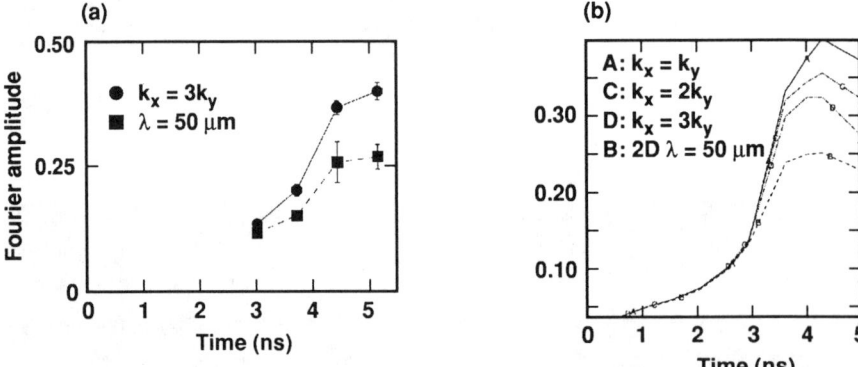

Figure 3. (a) Results of the evolution of the fundamental mode Fourier amplitude of ℓn(exposure) for the 3D k_x=3k_y (circles) and 2D λ=50 μm (squares) perturbations. The connecting lines are meant only to guide the eye. Experimental conditions were kept the same for these two shots to best illustrate the effect of shape on perturbation growth. The diagnostic setup was the same as described in Fig. 2. (b) Predicted Fourier amplitude of ℓn(exposure) from 3D simulations for the evolution of four different perturbation shapes all with the same magnitude $k=(k_x^2+k_y^2)^{1/2}$ wavevector, for drive conditions slightly different from those of (a). The most symmetric (k_x=k_y) mode is seen to grow the largest.

Acknowledgements

We are pleased to ackowledge the assistance of S.G. Glendinning with software, invaluable discussions with D. Shvarts at the inception of this work, and the highly skilled Nova technical staff. *Work performed under the auspices of the U.S. Department of Energy by the Lawrence Livermore National Laboratory under contract number W-7405-ENG-48.

References

1. H. Takabe et al., Phys. Fluids 26, 2299 (1983); ibid 28, 3676 (1985).
2. D.H. Munro, Phys. Rev. A 38, 1433 (1988).
3. M. Tabak et al., Phys. Fluids B 2, 1007 (1990).
4. D. Layzer, Astrophys. J. 122, 1 (1955).
5. U. Alon et al., Phys. Rev. Lett. 72, 2867 (1994); ibid 74, 534 (1995).
6. J.W. Jacobs and I. Catton, J. Fluid Mech. 187, 329 (1988).
7. J.P. Dahlburg et al., Phys. Fluids B 5, 571 (1993).
8. J. Hecht et al., Phys. Fluids 6, 12 (1994).
9. D. Youngs, Lasers and Part. Beams 12, 725 (1994).
10. M.M. Marinak et al., submitted to Phys. Rev. Lett. (1995) and UCRL-JC-120191.
11. R.J. Wallace et al., ICF Quarterly Report 4, 79 (1994).
12. K.S. Budil et al., submitted to Rev. Sci. Instrum (1995).
13. B.A. Remington et al., Phys. Plasmas 2, 241 (1995).
14. J.D. Kilkenny et al., Rev. Sci. Instrum. 63, 4688 (1992).

Experimental Analyses of Rayleigh-Taylor Growth in Cylindrical Implosions

J. Bradley Beck*,[†], Warren H. Hsing*, Nelson M. Hoffman*, and Chan K. Choi[†]

*Los Alamos National Laboratory, Los Alamos, New Mexico 87545 USA and
[†]School of Nuclear Engineering, Purdue University, West Lafayette, Indiana 47907 USA

Abstract. The implosion of an inertial-confinement-fusion (ICF) capsule is most efficient if the imploding flow is exactly spherically symmetric. It is well known, however that real ICF flows are subject to a variety of fluid instabilities that destroy spherical symmetry. An experiment has been devised in cylindrical geometry that allows for the study of instability growth during an implosion. Results from these experiments have shown the effects of "feedthrough" of the instability from the ablation front, the effects of convergence, and the transition from linear to nonlinear growth regimes. The design of the cylinders and the drive configuration will be discussed, after which results are presented. Growth factors greater than ten have been created with these experiments.

INTRODUCTION

A successful demonstration of inertial-confinement-fusion (ICF) will require controlling the growth of the fluid instabilities that arise during the implosion of the fuel-containing capsule. These fluid instabilities can cause mixing of the fuel with high-Z contaminants from the surrounding shell, mixing of the high-entropy fuel "hotspot" with low-entropy main fuel, and the creation of fluid turbulence. All of these effects will degrade the performance of the capsule. For these reasons, the understanding of how unstable modes grow throughout an implosion is extremely important. Of specific interest is the Rayleigh-Taylor instability (1), which occurs when a less dense material accelerates a more dense material.

In an implosion, there are two specific regimes where the Rayleigh-Taylor instability occurs. The first, the ablative acceleration phase, occurs when the hot, low-density ablated material accelerates the remaining shell. Perturbations that exist on the outer surface, either from capsule surface non-uniformities or drive non-uniformities, will grow during this regime. The second regime occurs near the end of the implosion as the hot, low-density fuel decelerates the remaining high-density shell prior to ignition (2). The unstable modes that exist during this phase are due to inner surface non-uniformities along with the "feedthrough" of the outer surface unstable modes (3). There have been many experiments measuring the growth rate due to ablative acceleration either in direct (4) or in indirect drive (5). These experiments have verified the stabilizing effect of mass ablation and density gradients according to predictions. However, few experiments have examined the decelerational phase, where the growth is expected to have classical growth, albeit with some density gradient stabilization (6). Convergent geometry effects will also become important during this phase due to the fact that both the acceleration and wavelength will vary in space and time, along with the physical thickness of the shell (7).

Due to diagnostic difficulties, very few experiments have been conducted to study the Rayleigh-Taylor instability in convergent geometries (8). Inner surface effects are difficult to measure in spherical geometry due to the lack of a direct line-of-sight. While amplitude growth can be inferred from time-dependent spectral measurements in spherical geometry (9), the results are dependent upon the details of the atomic physics model used. For these reasons, we have chosen to use the benefits of cylindrical geometry. Cylindrical geometry has the distinct advantage of creating an implosion which will have the decelerational phase while still allowing for a direct diagnostic line-of-sight. We have chosen a feedthough experiment in cylindrical geometry using an indirect drive created with the NOVA laser at Lawrence Livermore National Laboratory.

EXPERIMENTAL SETUP

The experimental configuration is shown in Figure 1. The cylindrical target was mounted transverse to the hohlraum to allow for a straightforward line-of-sight to the diagnostics, to avoid interference with the laser beams, and to remove the possibility of radiation flow into the ends of the cylinder that would be present in a coaxial configuration. Eight NOVA beams (10) were pointed symmetrically about the cylinder. A low-adiabat drive was used for the main beams at 3ω and a separate beam of 2 ns duration at 2ω was used to irradiate a 2 mm diameter silver (Ag) disk, creating an X-ray backlighter. The main drive consists of a low power foot followed by a ramp to higher power with a peak to foot ratio of about 10. This drive is designed to create a sustained acceleration while minimizing shock heating. A random phase plate (11) was used to smooth the backlighter beam intensity on the Ag disk, creating a resultant laser spot $\sim$ 750 µm in diameter. A 1 mm Be foil was placed between the backlighter and the cylinder to filter out soft x-rays and to prevent reflected 2ω light from being incident upon the inside of the cylinder. On both ends of the cylinder, 400 µm diameter circular apertures made of 1 mm Au were placed concentric with the cylinder axis. The purpose of these apertures was threefold: to prevent any x-rays from the walls of the cylinder from entering the pinhole camera diagnostic, to provide an alignment and parallax diagnostic, and to provide a centering fiducial for each frame on the pinhole camera.

As of the present time, there have been two primary designs of the cylinders. While there have been modifications between the two designs, the basic geometric configuration has remained the same. In both designs, the central region of the cylinder has been tapered inward. The purpose of this is to cause the central region to implode before the outer region, thus minimizing edge effects. In the first design (12), the cylinder was made of polystyrene, with a small 4 µm thick, 160 µm long dichlorostyrene ($C_8H_6Cl_2$) section placed flush on the inner central surface of the cylinder. This material serves as a marker layer; the X-ray backlighter is transparent to the polystyrene cylinder and is opaque to the marker layer. Since the marker layer is on the inside of the cylinder and has no perturbations initially, any perturbations will be indicative of feedthrough of the

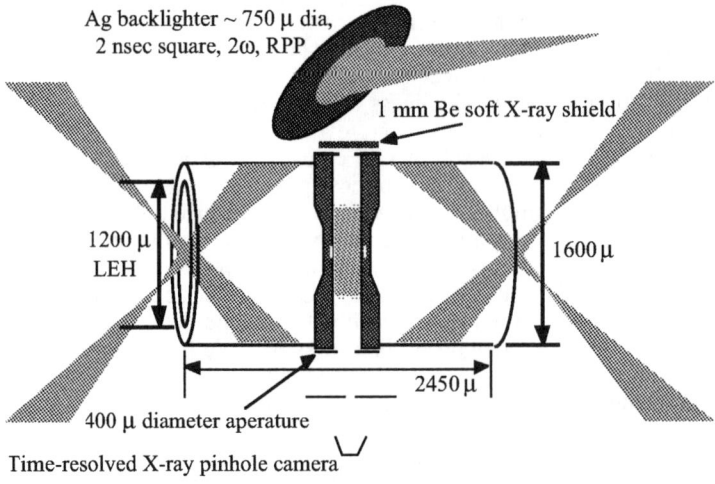

FIGURE 1. Side view of the experimental geometry

initial perturbations onto the inside surface. Over the ends of the cylinder, an additional 5 μm of Au is added. The purpose of this additional Au is twofold: the Au keeps the hohlraum hotter when compared to an uncoated surface, and the Au will further slow down the implosion of the ends, further minimizing edge effects. In both designs, a 60 mg/cc microcellular triacrylate foam ($C_{15}H_{20}O_6$), with a cell size ~ 1-3 μm (13), is placed on the inside of the cylinder to provide a back pressure as the cylinder implodes. In order to minimize opacity to the backlighter, the foam only encompasses the central 800 μm in the initial design. The initial perturbations were placed upon the central 400 μm long region. While a sine wave perturbation (single-mode) is preferable because it does not contain higher harmonics, fabrication is extremely difficult. For that reason, in this design, a pure dodecagon was placed upon this central region in the first design.

The second cylinder design is shown in Figure 2. The first cylinder design proved that this method of investigating instability growth is possible, while the second cylinder was designed to yield higher amounts of growth. In the original design, the use of polystyrene allowed for significant preheating of the central region to occur, decreasing the hydrodynamic efficiency of the implosion. For that reason, part of the polystyrene was replaced with monobromostyrene (C_8H_7Br), effectively reducing the amount of preheat. With the addition of the monobromostyrene, a new layer was created with high opacity to the backlighter. In order to keep the opaque layers separate, the layer was placed upon the outer regions of the cylinder, and was tapered along the length of the cylinder. This creates a constant radial thickness along the length of the cylinder. A layer of polystyrene was then placed between the marker layer and the layer of monobromostyrene. Another minor difference in the two designs is the length of the triacrylate foam. For reasons of edge effects, the foam length was increased by 100 μm. The last difference in the experimental setup involved the manner in which the perturbations were placed upon the surface. In the initial design, a pure dodecagon was cut into the outer portion of the cylinder. With the use of only 8

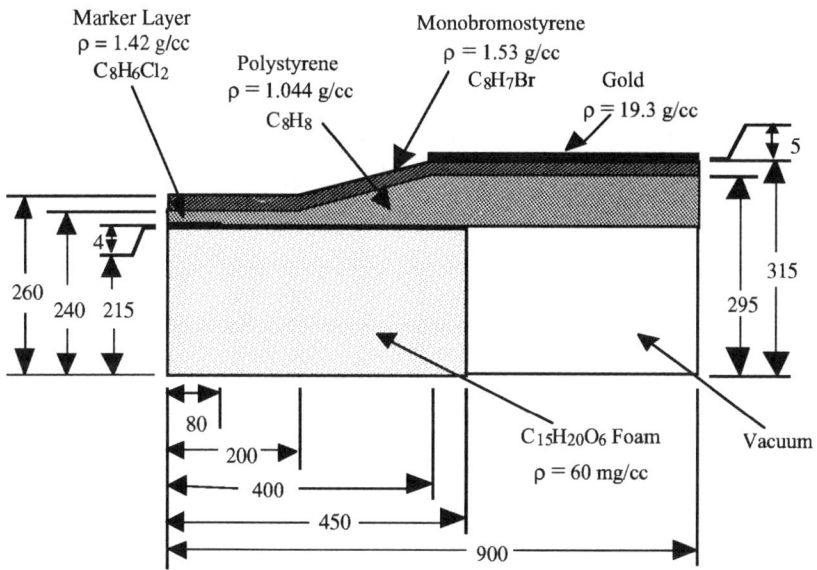

FIGURE 2. R-Z quadrant on new cylindrical design. All dimensions are in μm.

NOVA beams to drive the implosion, a mode 4 asymmetry can be present. Due to the fact that mode 12 is a harmonic of mode 4, the mode of the perturbation to was changed to 10 in the hope that mode-coupling effects might be seen. A second change was required due to the increased amount of growth that occurs in the second design: the initial amplitude of the unstable mode needed to be decreased. This was accomplished with the use of both mode 10 and mode 20 cuts. With the proper cut depths, small initial amplitudes of the mode 10 perturbation can be achieved while reducing the initial second harmonic to zero. This method was used to create an initial perturbation amplitude of 0.75 μm.

EXPERIMENTAL RESULTS

The second cylinder was experimentally shot with the mode 10 perturbation applied. Several time frames from the resulting images are shown in Figure 3. As can be seen, 10 distinct waves can be seen, as can a strong mode 4 asymmetry. This effect is attributed to a drive asymmetry problem which was caused by improper laser beam pointings. Of importance is the fact that the perturbation is growing while the radius is decreasing. The last time picture also shows that we can image the marker layer as it starts to re-expand. This last image shows how much mix has occurred both due to the perturbation and the drive asymmetry.

In each of these images, contours can be taken at the outer positions of the marker layer. The resulting contour data can then be fit to the equation:

$$r(\theta) = \frac{a_0}{2} + \sum_{n=1}^{10} \left(a_n \cos 2n\theta + b_n \sin 2n\theta \right) \tag{1}$$

where $\quad a_m = \dfrac{1}{\pi}\int\limits_{-\pi}^{\pi} r(\theta)\cos m\theta d\theta \quad$ and $\quad b_m = \dfrac{1}{\pi}\int\limits_{-\pi}^{\pi} r(\theta)\sin m\theta d\theta.$ (2)

The data is fit using a least-squared-fit program. All of the even modes are included, although the specific modes of interest include mode 2 and 4 (which are attributed to drive asymmetries) and modes 10 and 20 (which are the fundamental mode and the first harmonic). The results for these modal amplitudes are shown in Table 1.

The results from these images show a growth of both mode 4 and mode 10 wavelengths as the cylinder implodes. The mode 4 growth shows an acceleration around the latter times, where the cylinder is being decelerated by the foam. This shows the importance of a symmetric drive. While the initial magnitude is hard to define, we can note that the mode 4 has become nonlinear by 2.975 ns, as its value of the amplitude divided by the wavelength (a/λ) has increased beyond 0.1. At this point in time, the higher harmonics should start to appear.

The growth of the mode 10 can be determined as an initial magnitude is available. The data thus shows that the amplitude of mode 10 has undergone 11 growth factors at 2.920 ns. At this time the mode has entered the nonlinear regime as the value of a/λ has increased beyond 0.10. This can then be seen as the large increase in the second harmonic at 2.975 ns. The significance of the

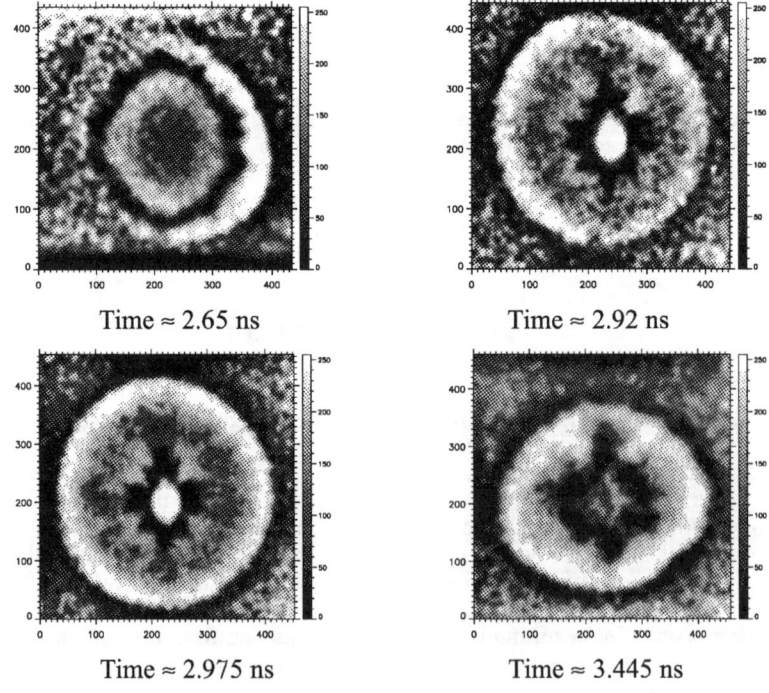

FIGURE 3. A sequence of gated X-ray images of the cylindrical backlit implosion.

TABLE 1. Fourier amplitudes of modes of interest

Time	Outer Radius (μm)	λ_{10} (μm)	a_2 (μm)	a_4 (μm)	a_{10} (μm)	a_{10}/λ_{10}	a_{20} (μm)
t=0.000 ns	260.00	163.30	0.000	0.000	0.750	0.004	0.000
t=2.650 ns	134.40	84.45	4.865	4.872	5.985	0.071	0.376
t=2.920 ns	69.04	43.38	7.017	7.345	8.269	0.191	0.869
t=2.975 ns	64.96	40.82	5.253	11.25	5.756	0.141	2.936

appearance of the second harmonic is that the unstable mode is now entering into the bubble-and-spike regime. Due to the fact that both mode 4 and mode 10 are entering into the nonlinear regime at approximately the same time, the formation of higher harmonics and mode-coupling effects make it extremely difficult to further analyze later times due to the proximity of wavelengths.

ACKNOWLEDGMENTS

We would like to thank S. Rothman, J. Edwards, S. Haan, T. Dittrich, B. Hammel, B. Remington, O. Landen for useful discussions; K. Pfeuffer for 3D CAD modeling to determine proper beam pointings; P. Apen, L. Foreman, P. Gobby, H. Bush, V. Gomez, J. Moore, V. Gurule, for their work on target fabrication; J. Oertel and Los Alamos National Laboratory's P-24 technicians for help with the GXI; the NOVA operations crew for their assistance; A. Hauer for suggesting this experiment; and G. Magelssen for calculations of the hohlraums. This work was supported by the U.S.D.O.E. under contract W-7405-ENG-36.

REFERENCES

1. Taylor, G.I., *Proc. Roy. Soc. A* **201**, 192-196 (1950).
2. Lindl, J.D., and Mead, W.C., *Phys. Rev. Lett.* **34**, 1273-1276 (1975); Bodner, S.E., *Phys. Rev. Lett.* **33**, 761-764 (1974).
3. Haan, S.W., *Phys. Rev. A* **39**, 5812-5825 (1989).
4. Desselberger, M., and Willi, O., *Phys. Fluids B* **5**, 896-909 (1993); Grun, J., et al., *Phys. Rev. Lett.* **58**, 2672-2675 (1987); Nishimura, H., et al., *Phys. Fluids* **31**, 2875-2883 (1988); Fews, A.P., Lamb, M.J., and Savage, M., *Laser Part. Beams* **11**, 257-268 (1993); Glendenning, S.G., et al., *Phys. Rev. Lett.* **69**, 1201-1204 (1992).
5. Remington, B.A., et al., *Phys. Rev. Lett.* **73**, 545-548 (1994).
6. Town, R.P.J., and Bell, A.R., *Phys. Rev. Lett.* **67**, 1863-1866 (1991).
7. Sakagami, H., and Nishihara, K., *Phys. Fluids B* **2**, 2715-2730 (1990).
8. Wark, J.S., Kilkenny, J.D., Cole, A.J., Key, M.H., and Rumsby, P.T., *Appl. Phys. Lett* **48**, 969-971 (1986); Nihimura, H. et al., *Phys. Fluids* **31**, 2875-2883 (1988).
9. Dittrich, T.R., et al., "Diagnosis of Pusher-Fuel Mix in Indirectly Driven Implosions," to be published.
10. Whitlock, R.R., et al., *Phys. Rev. Lett.* **52**, 819-822 (1984).
11. Kato, Y., et al., *Phys. Rev. Lett.* **53**, 1057-1060 (1984).
12. Hsing, W.W., and Hoffman, N.M., "Measurement of Feedthrough of the Rayleigh-Taylor Instability in Cylindrical Implosions," submitted to *Phys. Rev. Lett.*
13. Falconer, J.W., Golnazarians, Baker, M.J., and Sutton, D.W., *J. Vac. Sci. Tech. A* **8**, 968-971 (1990); Apen, P.G., Williams, J.M., *J. Cell. Plastics* **28**, 557-570 (1992).

Multimode Hydrodynamic Stability Calculations for National Ignition Facility Capsules

Nelson M. Hoffman, Douglas C. Wilson, William S. Varnum, William J. Krauser, and Bernhard H. Wilde

Thermonuclear Applications Group
Applied Theoretical and Computational Physics Division
Los Alamos National Laboratory, Los Alamos, New Mexico 87545 USA

Abstract. We examine the hydrodynamic stability of imploding ICF capsules by explicitly calculating the evolution of a realistic surface perturbation far into its nonlinear regime, using a 2D Lagrangian radiation-hydrodynamics code. The perturbation, which consists initially of mesh displacements in the capsule, is represented by the sum of many spherical harmonic modes, having finite amplitudes and realistic spectrum. A 90-degree sector of the capsule is modeled, allowing proper boundary conditions for all modes simultaneously. Because of the large distortion of the mesh that occurs during the calculations, it is necessary to rezone the mesh frequently, by mapping physical variables to a new undistorted mesh. No model-specific parameters are required in this technique. We have used the technique to calculate the yield of several designs for a National Ignition Facility capsule as a function of initial root-mean-square surface roughness σ of the outer ablator surface or the inner cryogenic DT surface. Typically for a capsule we find a "cliff" at a critical value of $\sigma = \sigma_{crit}$ such that the yield of the capsule decreases abruptly for $\sigma > \sigma_{crit}$, indicating a failure to ignite. The values of σ_{crit} we compute are probably upper limits because of the lack of 3D effects, and inaccuracies in Lagrangian modeling of such unstable flows. It is expected that more accurate modeling, perhaps with 3D Eulerian codes, will lead to smaller values for σ_{crit}. We are beginning to carry out studies of the coupling of low-mode radiation flux asymmetries to higher-mode surface perturbations. We report also on sensitivity studies that examine the response of a capsule to small variations in the driving laser's power history. We find that realistic surfaces decrease a capsule's ability to tolerate drive variations.

INTRODUCTION

Hydrodynamic instabilities are processes that amplify small perturbations in the implosion of an ICF capsule, generally leading to a degradation in the performance of the capsule. If perturbations grow large enough, they may prevent fusion ignition by interfering with the compressional heating of the fusion fuel, for example, or by increasing the surface area through which heat is lost from the fuel, or by reducing the effectiveness of α-particle trapping. With even more growth, the perturbations may generate hydrodynamic turbulence, which can transport mass, momentum, and energy in a way that disrupts the desired flow.

© 1996 American Institute of Physics

To have confidence that our predictions of ignition are correct for National Ignition Facility (NIF) capsules, then, it is necessary that we have a sound theoretical and computational understanding of the effect of hydrodynamic instabilities in ICF implosions.

Several approaches have been used over the years to compute the effect of such instabilities. An early technique was to solve the linearized perturbed hydrodynamic equations, including energy flow by radiation and conduction, in a postprocessing mode (1). Later it was realized that the first-order flow could be calculated by a 2D nonlinear radiation-hydrodynamics code simultaneously with the zeroth-order calculation (2-4), obviating the need for a separate linearized perturbation code. We call these first-order single-mode (FOSM) calculations. One may compute the instability amplification factor of single spherical harmonic modes at some unstable interface by either of these approaches, assuming that the modes do not interact. Then one superimposes the modes linearly and determines the RMS deviation of the surface they define, giving a length scale s over which materials are assumed to be mixed across the interface. The process, as outlined in Refs. 4 and 5, accounts for the weak nonlinearity by which neighboring modes influence one another's saturation, using a saturation parameter which must be normalized to experiment.

The mixing length scale s obtained by this approach of linear mode superposition (LMS) is then used as input to a separate 1D calculation in which the capsule performance is finally determined. In recent implementations of this model, a heat-conduction multiplier is used in the 1D calculation to account for the fact that the thermal conductive flux across the unstable interface is enhanced as its area increases due to the growth of the perturbation.

NONLINEAR MULTIMODE CALCULATIONS

Recent evidence from theory and experiment (6,7) points to a picture of late-stage instability in ICF implosions which is at variance with the LMS picture. In the latter, materials are assumed to be thoroughly mixed at an atomic scale across unstable interfaces (5), throughout a region whose thickness is given by the mixing length scale s. The new evidence, however, suggests that unstable interfaces in ICF implosions are deformed into a network of polygonal bubbles surrounded by spike sheaths during the acceleration phase of the implosion. The bubbles occur with a range of sizes, but there is a dominant scale characterizing the largest and therefore most rapidly growing bubbles. The implication is that instability disrupts these flows, not by fine-scale mixing as in the LMS picture, but by large-scale effects such as those mentioned above, provided that the initial perturbation is not too large.

To model the bubble-network picture of hydrodynamic instability, we have performed calculations in which the unstable flow is represented directly with as much detail and as few modeling approximations as possible, given the constraints of finite machine memory and two-dimensionality. We use the same type of 2D radiation-hydrodynamics code that is used for the FOSM calculations described above, but our approach differs in several ways: many spherical harmonic modes (up to 24 thus far) are present in a single calculation; the amplitudes of the modes are realistic, not infinitesimal; and a 90-degree quadrant of the capsule is represented, to allow proper boundary conditions for all modes simultaneously.

Furthermore, unlike the LMS approach, we do not run a separate calculation to model capsule performance, since the yield and other attributes of the capsule are calculated at the same time as the perturbation growth. Thus we avoid assumptions about mode saturation or enhanced heat flow, since we calculate these effects explicitly, and have no need for model parameters that adjust their magnitudes.

Two-dimensionality introduces at least two shortcomings in the calculations, one of which we can compensate for, while the other we cannot. First, a 2D surface is inherently less rough than a 3D surface with the same spherical harmonic spectrum, because of its axial symmetry. This is reflected in the expressions for the RMS deviation of such surfaces. For an isotropically rough 3D surface,

$$\sigma^2_{3D} = \frac{1}{4\pi}\sum_l (2l+1) R^2_{l0}$$

while

$$\sigma^2_{2D} = \frac{1}{4\pi}\sum_l R^2_{l0}$$

for a 2D surface, where R_{l0} is the amplitude of the spherical harmonic Y_l^0. The factor $2l+1$ in the expression for σ_{3D} arises from the azimuthal roughness in a 3D surface, and clearly has the effect of increasing the importance of higher mode numbers in 3D relative to 2D. To try to compensate for this effect in our calculations, we modify the relative amplitude of all modes in the 2D spectrum by the factor $2l+1$, while keeping the total σ constant. This modified spectrum we call a "bandwidth-equivalent-RMS" (BER) spectrum, because it has not only the same total σ as the 3D spectrum which it is supposed to represent, but each mode, or band of modes, makes the same relative contribution to the total σ as in the 3D spectrum.

The other shortcoming arises because of the fact that 2D bubbles grow more slowly than 3D bubbles. This is called the "shape effect" (7), and was pointed out by Layzer (8). The terminal velocity of a bubble is about 50% higher in 3D than in 2D, for large density ratios. There is no obvious way to compensate for the shape effect in a 2D calculation, so a resolution must await 3D calculations. In the meantime, we take the viewpoint that 2D calculations furnish a lower limit to the magnitude of the disruptive effects of a perturbation with a given initial σ_0, and therefore our conclusions about the maximum tolerable $\sigma = \sigma_{crit}$ are actually upper limits to its value.

Another issue raised by our nonlinear multimode calculations is that of Lagrangian mesh distortion. Because the mode amplitudes are finite, the perturbation at an interface can eventually grow to seriously disturb the capsule implosion. In this case the distinction between zeroth-order and first-order flow is lost, distinct bubbles and spikes form in the density structure, and the mesh becomes greatly deformed. The code's Courant time step drops to an extremely small values or the implicit radiation-diffusion algorithm fails, halting the calculation.

To continue the calculation requires rezoning, that is, defining a new, more regular mesh, and mapping the physical variables from the old deformed mesh to the new regular mesh. To define the new mesh, we use a remeshing technique due to Brackbill and Saltzman (9), which allows the mesh to adapt to steep gradients in physical variables, solving an elliptical equation to define the new mesh lines using the old mesh as a boundary condition. Then a 2nd-order-accurate technique due to Scannapieco (10) is used for the remapping of variables. By repeated application of the rezoning procedure we can follow flow patterns which are quite turbulent. It is clear, however, that the calculated performance of a capsule, in particular its yield, can be affected by the exact prescription for rezoning (e.g., the time at which rezoning begins, the frequency with which it is carried out, and the degree to which the new mesh is allowed to approach the exact Brackbill-Saltzman solution). The sensitivity is greatest near "cliffs" in initial surface roughness, i.e., when σ_0 is near σ_{crit}. The effect is to make the location of the cliff uncertain to some degree. We are currently assessing the relative accuracy of the various prescriptions for rezoning, through comparison to experiments and theory. In the meantime, we will simply conclude that rezoning is required to allow these calculations to continue through the time of fusion burn of the capsule, but introduces some uncertainty into the results.

NIF CAPSULE CALCULATIONS

We have applied the nonlinear multimode technique to several NIF designs, in order to specify the maximum tolerable σ_0 of various surfaces in the capsule. Figure 1 . shows the initial perturbation of the inner deuterium-tritium (DT) ice

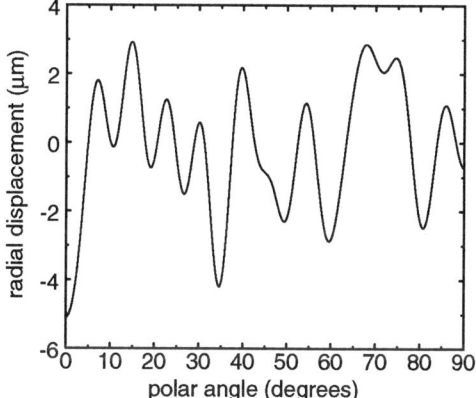

Figure 1. Initial 24-mode BER perturbation on inner DT ice interface, with $\sigma = 1.75$ μm.

interface in one calculation of the "PT" design (11), where the perturbation is the sum of 24 spherical harmonic modes (l=2,4,6,...,48). The modal amplitudes are given by a BER spectrum obtained by scaling a 1D Fourier analysis of cryogenic DT data to a 2D power spectrum, and then applying the BER enhancement discussed earlier.

Calculations using the BER spectrum are still in progress, but we have carried out other studies of the PT capsule with spectra that are simply scaled directly from the 2D power spectrum of DT ice surfaces without the BER enhancement, to determine the effect of spectral shape on our conclusions about the maximum tolerable surface roughness in these studies, we have taken a particular initial spectrum and scaled it up or down to vary σ. As Fig. 2 shows, the

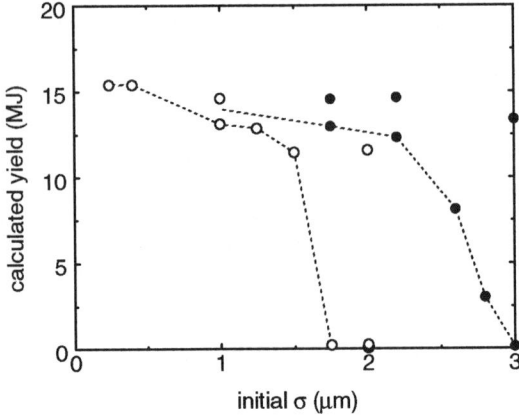

Figure 2. Variation of yield with DT ice inner surface roughness, for an 8-mode perturbation (open circles) and a 24-mode perturbation (filled circles).

yield of the PT capsule is relatively insensitive to σ until it reaches a critical value, σ_{crit}, above which the yield falls dramatically, indicating a failure to ignite. Open circles in Fig. 2 show results for an 8-mode perturbation with relatively high modes only ($l=12,16,20,...,40$), while the filled circles show results for a 24-mode spectrum with low modes included ($l=2,4,6,...,48$). The location of the cliff, $\sigma = \sigma_{crit}$, increases for the spectrum with low modes included, because low modes grow less rapidly than high modes and produce less disruption of the flow. The scatter in the calculated yields near the cliffs is caused by variations in the prescription for rezoning. The dashed lines indicate a lower limit to the calculated performance, which we adopt at present as the most cautious estimate.

Using the technique of direct nonlinear multimode instability simulations, we can now begin to examine the coupling of various asymmetries and departures from ideal driving conditions, simultaneously or taken a few at a time. For example, we can incorporate the low-mode radiation drive asymmetry, determined by an integrated hohlraum calculation, into our instability calculations, to examine its coupling to surface perturbations. Such studies are now in progress. We can perturb two or more surfaces simultaneously, to examine how such perturbations couple and reduce the allowable roughness at either surface. We can study the sensitivity of a capsule to variations in the hohlraum drive history, as is often done in 1D, but now include the non-ideal circumstances of surface roughness and drive asymmetry. Figure 3 shows results of such a study, in which the yield of a NIF capsule is determined as a function of the temperature in the final drive pulse. Open circles show the standard 1D result, which assumes perfect smoothness of all capsule surfaces. The filled circles show the effect of roughness with 20 nm RMS on the outer surface of the ablator, about the limit of our current fabrication capability. Most of the capsule's ability to tolerate low drive is negated by surface roughness. Future studies will include drive asymmetry and perturbations on all surfaces simultaneously, when examining sensitivity to drive variations.

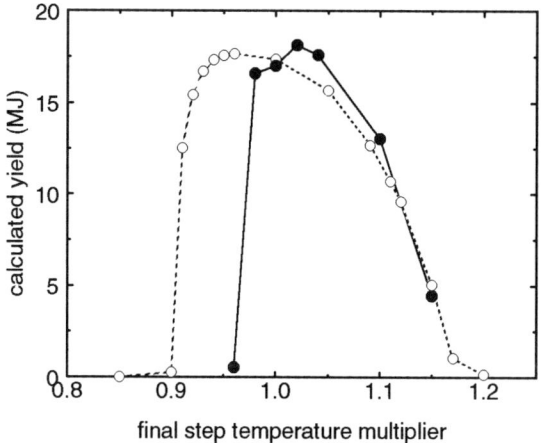

Figure 3. Drive sensitivity profiles for NIF capsule. Open circles show standard 1D result. Filled circles show result of 2D calculations with ablator roughness of 20 nm.

ACKNOWLEDGMENTS

We are grateful to Steve Haan for pointing out the necessity of using bandwidth-equivalent-RMS spectra in 2D calculations, and for suggesting the calculations which led to this work in the first place. We also thank Bill Powers for improvements and advice in the use of the Brackbill-Saltzman remesher. This work was supported under USDOE contract W-7405-ENG-36.

REFERENCES

1. McCrory, R. L., Morse, R. L., and Taggart, K. A., *Nucl. Sci. Eng.* **64**, 163-176 (1977); Scannapieco, A. J., and Cranfill, C. W., "A Derivation of the Physical Equations Solved in the Inertial Confinement Stability Code DOC," Los Alamos Scientific Laboratory report LA-7214-MS (1978).
2. Verdon, C. P., McCrory, R. L., Morse, R. L., Baker, G. R., Meiron, D. I., and Orszag, S. A., *Phys. Fluids* **25**, 1653-1674 (1982).
3. Munro, D. H., *Phys. Fluids B* **1**, 134-141 (1989).
4. Haan, S. W., *Phys. Rev. A* **39**, 5812-5825 (1989).
5. Dittrich, T. R., Hammel, B. A., Keane, C. J., McEachern, R., Turner, R. E., Haan, S. W., and Suter, L. J., *Phys. Rev. Lett.* **73**, 2324-2327 (1994).
6. Remington, B. A., Weber, S. V., Haan, S. W., Kilkenny, J. D., Glendinning, S. G., Wallace, R. J., Goldstein, W. H., Wilson, B. G., and Nash, J. K., *Phys. Fluids. B* **5**, 2589-2595 (1993).
7. Dahlburg, J. P., *Bull. APS* **39**, 1576 (1994); Shvarts, D., *Bull. APS* **39**, 1576 (1994).
8. Layzer, D., *Astrophys. J.* **122**, 1-12 (1955).
9. Brackbill, J. U., and Saltzman, J. S., *J. Comp. Phys.* **46**, 342-368 (1982).
10. Scannapieco, A. J., "Automatic Rezoner for Lagrangian Quadrilateral Hydrocodes," Los Alamos National Laboratory report LA-UR-82-2897 (1982).
11. Haan, S. W., *Bull. APS* **39**, 1635 (1994).

Volume Ignition for Inertial Confinement Fusion Tested by different Stopping Power Models

H. Hora*, S. Eliezer, J. J. Honrubia, R. Höpfl*, J. M. Martinez-Val,
M. Peria, and G. Velarde

*European Network High Energy Density Matter - Polytechnic University of Madrid, Spain,
and (*) Anwendungszentrum der FH Regensburg, Germany*

Abstract. Within the low cost energy alternatives to carbon burning, nuclear fusion based on inertial confinement offers the earliest solution. The existing difficulties with spark (hot spot) ignition are reduced by volume ignition. Two basically different computations with different stopping power models are compared and only minor differences in the fusion gains established. An effective density-radius criterion results in comparably very high values if the strong effect of self-heat is included.

THE POSITION OF VOLUME IGNITION

In view of the need for non-fossil large scale, safe and low-cost energy sources to fight the greenhouse effect and the atmospheric warming catastrophy [1,2], the use of fusion power within the next 15 years is an option. One has to be aware that this cannot come from magnetic confinement fusion where the GW power station is at least 50 years away [3] where the wall erosion is one of the essential problems [4] even if present gains of a little less than one may be increased within the next 10 years. The ITER and JET are pulsating tokamaks and it has been underlined that any hope for a more stationary continuous operation has to go back to the stellarator configuration [5], leaving the ITER as a questionable task for which anyhow the essential support from the USA is under review [6].

For a much less expensive and shorter development of the fusion reactor, inertial confinement with special view to heavy ion beam driving is close to be decided [7]. For the timing, about 20 years for the fusion power reactor are considered [8] though R. Bock`s concept is based on the very complex spark ignition and not on the rather simplified volume ignition [4]. For this easier case even a shorter development time for a crash program may be expected.

The difficulties of spark (hot spot) ignition consist in the fact that the compression of the DT plasma fuel has to be performed in such a special time dependent way that a hot low density core up to a radius r* has to be produced changing into low temperature and high plasma density at r*. While the programming of such profiles is most difficult numerically in one dimension - and the time development e.g. of Planck radiation for indirect driving has to be most refined and accurate - the tree dimensional solution is even more difficult with respect to instabilities and nonuniformity. If at one spot the detonation wave starts but at other directions not, all is in vain.

These difficulties were summarized [9]: "The formation of an ignition DT volume in the center of the implosion (hot spot) is possibly the most difficult point in ICF research as was discussed.... One should recall that, in the mainline approach to high gain ICF, only a small fraction of the fuel has to form the hot spot and the bulk of the fuel is compressed at a low adiabat to much higher density. Ignition has to occur in the hot spot and then propagates through the fuel reservoir. The problem is that the formation of the hot spot is again RTI-unstable. The imploding dense fuel has the tendency to mix with the lower-density hot spot during stagnation, and this may degrade ignition".

Instead of the spark ignition, the volume ignition works without the extravagant density and temperature profiles, without any fusion detonation wave but with natural adiabate compression profiles, [4,10,11] achieved by stagnation-free compression as measured from very smooth x-ray emission [12,13]. Instead of the earlier found low gains, it was discovered [10] that the selfheat by fusion reaction products works as an additional driver of very high amplification such that a simply naturally and adiabatically compressed DT fuel can reach gains close to the high gains known from spark ignition calculations [4,10,11]. Especially it was underlined [14] that "volume ignition sacrifices high gain for lower losses, lower ignition temperature, lower implosion velocity and lower sensitivities of the more robust capsule to small fluctuation and asymmetries in the drive system. The reduction in gain is about a factor 2.5, which is small enough to make the more robust equilibrium ignition an attractive alternative." This difference can even be less down to a factor 1.2 or lower see p. 357 of Ref. [4].

For volume ignition [10-21], sometimes the expression "uniformity ignition" is used [17]. Since uniformity is much more critical to produce the spherical symmetry of the detonation front at spark ignition, this expression seems not to be the best choice. Another expression used instead of volume igniton is "equilibrum ignition" [14]. In view that non-local thermal equilibrium occurs and that this non-equilibrium is a special advantage [16,18] one may prefere "volume ignition" to underline the three-dimensional volume process contrary to the two dimensional detonation wave at spark ignition.

Significant early results on volume ignition were produced by Mima et al [19] as a consequence of the stagnation free high gain fusion experiments. A confirmation is given by the work of Atzeni and by Meyer-ter-Vehn [19].

DIFFERENT STOPPING POWER MODELS

Thanks to the self heat which lets the ion temperature grow up to 100 keV temperature and beyond [10,15-18], it was confirmed that very high fusion gains about of the same values as computed for spark ignition can be expected. The properties of volume ignition were splendidly confirmed by several authors [19,20] where the temporal increase of the temperature by the self heat was exactly as in the first cases [10] growing to the mentioned very high values while very high gains were reached and the temperature and density profiles were typically not that of spark ignition but that of nearly the adiabatic profiles [16,21]. The advantage of volume ignition with the "Wheeler modes" [11] as observed numerically earlier [10], is given in the fact that volume ignition provides the more robust dynamics. This is based on the fact that natural adiabatic compression is used for volume ignition and the extremely complicate radial density and temperature profiles for spark ignition are not necessary for generating a spherical detonation wave.

The earlier calculations [10] tacitly assumed local thermal equilibrium where the characteristic result - contrary to spark ignition with a very sensibly controlled relatively low temperature of the fusion detonation wave - consists in the reheat driven temperature up to 100 keV and more as shown also in the calculations by Anisomov, Oparin and Meyer-ter-Vehn [17]. Similar more detailed computations [16,18,21] including the neutron reheat and the separate development of electron and ion fluids, indicated a further very favorable improvement for the fusion gains by the fact that the reheat and volume ignition process appears within such a short time that only the ions reach their 100 and more keV temperature, while the electrons are remaining at nearly half of this temperature and the black body radiation temperature reaches even a very much lower temperature only. This is so favorable because only the ions need to have the high temperature for the good fusion reactions while electrons and radiation with their lower temperature need much less enthalpy what increases the gain considerably.

The collective stopping power model [22] used the decay of the energy E of the energetic ions of velocity v in a plasma of the temperature T and an electron density n, the electron mass m the electron charge e the charge state Z of the colliding ions (see reviews by Stepanek, [23] and Honrubia [24]).

$$-\frac{dE}{dx} = -\frac{2\pi Z^2 e^4 n}{mv^2} \log \frac{kT m^2 v^4}{4\pi e^4 n Z^2} = \frac{4\pi n Z^2 e^4}{mv^2} ln \frac{b_{max}}{b_{min}} \quad (1)$$

where $b_{max} = \lambda_D$ (Debye length) and $b_{min} = Ze^2/mv^2$ are the maximum and the minimum impact parameters. The dispersion and degeneracy problems were specially analyzed [25]. Using

$$S(E) = -\frac{dE}{dx} \qquad (2)$$

results in the stopping length

$$R = \int_0^A \frac{dE}{S(E)} = \frac{2 e^2 M}{kTm} Ei(\log gA^2) \qquad (3)$$

where M is the ion mass, A the initial ion energy and the integral logarithm is Ei(x). This model for the collision of energetic ions with the whole collective of a Debye sphere is identical with the earlier result of Gabor [26] based on the Debye-Milner theory of plasmas.

Different to this collective interaction, an extented binary collision model was derived by Anisimov and Oparin [27] on which basis the volume ignition of DT pellets was calculated [17]. There is one set of parameters where these calculations can be compared with our results [15,21] using the collective stopping power model. The case is that of Anisimov et al [17] where the plasma parameter H (the density times pellet radius) is 3 g/cm² the temperature is 5 keV and a total fuel depletion by the fusion reaction (called burn efficiency in Fig. 3. of Anisimov et al [17]) is 33%.

This case is the same as in our preceding computation (see Fig.2 of Ref. [15]) for an initial (solid state) volume before compression of 10 mm³ arriving at a compression of 890 times the solid state density. This case arrives at an initial temperature of 5.3 keV and a fuel depletion of 41%.

RESULTS

In view of the completely different computation codes and the not identical models involved one may accept this similarity of the results as a rather good agreement. Going then from the mentioned fixed point of comparison to other values of the results of Hora et al [15] one has to take into account that in the case of Anisimov et al [17], the parameters of initial mass and compression may be different. But following simply the parameter H one finds for H=2 in our case an initial temperature of 7 keV (7 keV), and a depletion of 26% (19%) where the numbers in brackets are the values of Anisimov et al [17]. For a parameter H=1 we have an initial temperature T=9 keV (10 keV) and a fuel depletion of 20% (14%) for comparison. Even then the agreement is rather good as is shown also when putting these results together in Fig.1.

At this point we should explain in more details the parameter H in Fig.1. Using the density n_0 with a radius R at highest compression, the parameter

$$H = n_0 R \qquad (4)$$

has the value of 3.34 at thousand times solid state density. When compressing to this maximum value, the computationally optimized temperature T has to be used which value is achieved from the computation when the plasma sphere is adiabatically compressed with spatially uniform temperature but temporally changing at each time step. While there is a change of the temperature by adiabaticity a stronger increase is given by the alpha reheat and by partial reabsorption of bremsstrahlung [4,10].

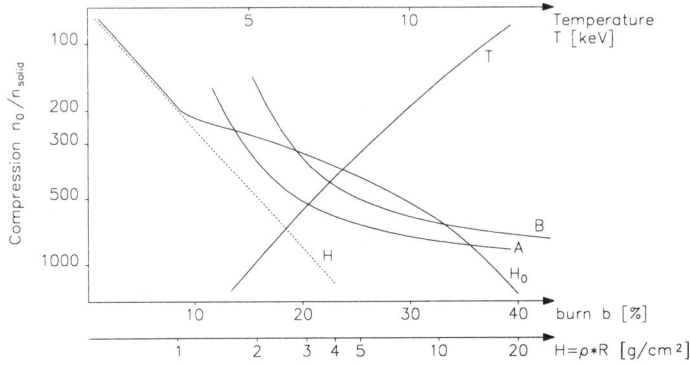

FIGURE 1. Volume ignition calculation for DT depending on the compression density. Initial plasma temperature T, parameter H and the burn b (depletion of fuel due to the reaction) is taken from the case B of an initial solid state DT volume of 10 mm^3 [15] while A denotes the burn b from the results of Anisimov et al [17]. The parameter H is the (naively taken) product of initial density and radius, Eq. (4); when taking into account volume ignition this leads to a higher "effective parameter H_0" given by Eq. (6).

In the cases of the net gain for B in Fig. 1, the alpha reheat is calculated for each time step with varying temperature and density using the collective stopping power model [22,28] as described in the preceding section. At the optimized conditions of volume ignition, the plasma increases its temperature up to or above 100 keV and quickly cools then adiabatically. What numbers of the binary stopping power model for curve A in Fig. 1 were used, is given implicitly by Anisimov et al [17]. In these results A as well as in our calculation B in Fig. 1, thermal equilibrium was used for simplification. The phenomenon of non-local thermal equilibrium is well known for similar conditions as mentioned before [18,21].

From the result that the final fusion gains as shown by the net fuel depletion ("burn" in the terminology of Anisimov et al [17]) of DT in Fig. 1 differ by less than a factor 2 between the cases using individual or collective models for the stopping power, one may conclude that the essential part of the reheating effect is then given at the temperatures of few keV where both models have similar stopping powers as shown before [22,23] and less at the very high temperatures where the stopping powers are very different. This agreement seems to be characteristic for very high burn efficiency.

The H-criterion in Eq. (4) is the simple product of the density and radius of the burning DT volume. As it was shown since 1970 [4], this is algebraically identical with the fusion gain ratio G given by the generated fusion energy to the energy E which was to be transferred into the compressed fuel (i.e. after subtracting the energy needed for ablation)[4].

$$G = \left(\frac{E_0}{E_{BE}}\right)^{\frac{1}{3}} \left(\frac{n_0}{n_S}\right)^{\frac{2}{3}} = const \ n_0 \ R \qquad (5)$$

(E_{BE} ~ break even energy = 6MJ, n_0 = density at highest compression
                                                  n_S = solid state density)

When, however, strong self-heat and volume ignition occurs (at all gains above G=8 for DT), the gain follows the law G_1 [4,29] until saturation by fuel depletion appears for gains above 300

$$G_1 = G \left[\frac{7.2 \cdot 10^{-4} G^3}{\left(1 + 1.44 \cdot 10^{-3} G^3\right)^{\frac{1}{2}} - 1}\right]^{\frac{6}{5}} \qquad (6)$$

Instead of exponent 6/5 one may use 1 for fitting the numerical results [29].

One can then define an effective H_0 which naive value H would correspond to this elevated gain. Calculating this for the results in Fig. 1, one arrives at the curve for H_0. It is remarkable that then the rather low values of H with total fusion gains G_{tot} (related to the incident laser or heavy ion energy) of 100 with 2 MJ driver energy correspond the to effective H-values of 20 and more, just as an effect of the self heat.

ACKNOWLEDGMENT

Assistance by Dipl. Ing. Alois Schönberger and cand. Ing. Elke Anleitner is gratefully acknowledged. This work was supported by the European Network Grant ERB CHRX CT93 C327.

REFERENCES

[1] Gore, Al, *Earth in the Balance* (Penguin, New York, 1993); Hora, H., and Höpfl, R., AIP Conf. Proceed. No. **318**, *Laser Interaction and Realted Plasma Phenomena*, Miley, G. H., ed, (Am. Inst. Phys., New York 1994) p. 625
[2] Velarde, G., Ronen, Y., and Martinez-Val, J. M., eds, *Nuclear Fusion by Inertial Confinement* (CRC Press, Boca Raton Fa. 1993)
[3] Maisonier, C., *Europhys. News* **25**, 167 (1994)
[4] Hora, H., *Plasmas at High Temperature and Density* (Springer, Heidelberg 1991)
[5] Pinkau, K., (interview with N. Lossau) Dialog 30 (No.2) 5 (1995)
[6] Macilwain, C., *Nature* **373**, 375 (1995)
[7] Bock, R., *Europhys. News* **25**, 167 (1994)
[8] Bock, R., *IAEA Techn. Committee Meeting ICF*, Paris, Nov. 1994
[9] Meyer-ter-Vehn, J., *15th IAEA Int. Conf. Plasma Physics and Controlled Nucl. Fusion Res.*, Seville, Sept 1994, invited summary talk on ICF
[10] Hora, H., and Ray, P. S., *Z. Naturforsch.* **33A**, 890 (1978)
[11] Kirkpatrick, R. S., and Wheeler, J. A., *Nucl. Fusion* **21**, 389 (1981)
[12] Yamanaka, C., Nakai, S., Yamanaka, T., et al, *Laser Interaction and Related Plasma Phenomena*, Hora, H., and Miley, G. H., eds (Plenum, New York 1986) Vol.7, p. 395
[13] Nakai, S., *Laser and Particle Beams* **7**, 457 (1989)
[14] Lackner, K. S., Colgate, S. A., Johnson, N. L., Kirkpatrick, R. C., Menikoff, R., and Petschek, A. G., AIP Conf. Proceed. No. 318 *Laser Interaction and Related Plasma Phenomena*, Miley, G. H., ed (Am.Inst.Phys. New York 1994) p. 356
[15] Hora, H., Eliezer, S., Höpfl, R., Martinez-Val, J. M., Velarde, G., Honrubia, J. J., and Piera, M., *IAEA Int. Conf. Plasma Physics and Contrl. Nucl. Fusion Res.*, Seville, Sept 1994, Paper CN-60/B-P-2 (IAEA Vienna 1995) in print
[16] Martinez-Val, J. M., Eliezer, S., and Piera, M., *Laser and Particle Beams* **12**, 618 (1994)
[17] Anisimov, S. I., Oparin, A. M., and Meyer-ter-Vehn, J., 1994 Gesellschaft für Schwerionenforschung Darmstadt, *Annual Report High Energy Density Matter Produced by Heavy Ions 1993*, GSI-94-10 Rpt. June, p.43
[18] He, X. T., and Li, Y. S., AIP Conf. Proceed. No. 318 *Laser Interaction and Related Plasma Phenomena*, Miley, G. H., ed. (Am.Inst. Phys. New York 1994) p. 334
[19] K.Mima. H- Takabe, S.Nakai, Laser and Particle Beams **7**, 245 (1989), Basko, M.M. Nuclear Fusion **30**, 2442 (1990); Basko, M.M. Laser and Part. Beams **11**, 733 (1993); Atzeni, S. Jap.J.Appl.Pys. **34**, 1980 (1995); Meyer-ter-Vehn, J. Workshop by K. Mima at this conference
[20] Tahir, N. A., and Hoffmann, D. H. H., *Fusion Engineering and Design* **24**, 413 (1994)
[21] Martinez-Val, J. M., et al, *Nuovo Cimento* **106**, 1873 (1993)
[22] Ray, P. S.,and Hora, H., *Nucl. Fusion* **15**, 535 (1976)
[23] Stepanek, J., *Laser Interaction and Related Plasma Phenomena*, Schwarz, H., et al eds. (Plenum, New York 1981) Vol 5, p. 341
[24] Honrubia, J. J., in *Nuclear Fusion by Intertial Confienement*, Velarde, G., Ronen, Y., and Martinez-Val, J. M., eds, (CRC Press, Boca Raton Fa.), p. 211
[25] Honrubia, J. J., et al, *Ann.Rpt. ICF DENIM*, Polytech Univ. Madrid 1986, p. 41
[26] Gabor, D. Proc. Royal Soc. (London) **A213**, 73 (1983)
[27] Anisimov, S. I., and Oparin, A. M., *JETP-Letters* **57**, 635 (1993)
[28] Kasotakis, G., et al, *Laser and Particle Beams* **8**, 531 (1989); Pieruschka, P., et al, *Laser and Particle Beams* **10**, 145 (1992); Stening, R. J., et al, *Laser Interaction and Related Plasma Phenomena*, Hora, H., and Miley, G. H., eds. (Plenum, New York 1992) Vol 10, p.347; Khoda-Bakhsh, R., Nucl. Instr. Meth. A330, 268 (1993)
[29] Hora, H., *Z. Naturforsch.*, **42A**, 1239 (1987)

Effects of Nonlocal Heat Transport on Laser Implosion

K.Mima, M.Honda, S.Miyamoto, and S.Kato

Institute of Laser Engineering
Osaka University 2-6, Yamada-oka, Suita, Osaka, 565, Japan

Abstract . A numerical simulation code describing the spherically symmetric implosion hydrodynamics has been developed to investigate the nonlocal heat transport effects on stable high velocity implosion and fast ignition. In the implosion simulation code HIMICO, the Fokker Planck equation for electron transport is solved to describe the nonlocal effects. For high ablation pressure implosion with a pressure higher than 200Mbar, the isentrope is found higher by a factor 2 in the nonlocal transport model than in the Spitzer Harm model.

As for the fast ignition simulation, the neutron yield for the high density compression with 10KJ laser increases to be 20 times by injecting an additional heating pulse of 10KJ with 1psec.

Introduction

Recently, it has been clarified that the turbulent mixing between the cold dense edge plasma and the hot spark is the critical problem for achieving ignition and burn in the direct drive laser fusion(1). In reducing the thickness of the turbulent mixing layer, there have two approaches been considered.

The first approach is the precise control of laser irradiation and target fabrication. As for the laser irradiation system, new techniques for beam smoothing and power balance among laser beams have been developed to reduce the laser intensity fluctuation on a target to be less than 1% in the root mean square average. The random phase plate(2), ISI(3), 1Dand 2D SSD(4), PCL(5)and so on have been proposed as the beam smoothing techniques and the combinations of those methods are used in the implosion experiments.

The second group of the researches are looking for the new implosion schemes which are robust to the implosion asymmetry. They are the high ablation pressure implosion scheme, fast ignition, hybrid implosion scheme and so on. Since in the high ablation pressure implosion (HAPI), the shell acceleration time is shorter than in the low ablation pressure implosion, the e-folding of the Rayleigh-Taylor instability is lower in HAPI. Therefore, the HAPI is more robust for the laser irradiation nonuniformity and the target fabrication errors. In the fast ignition, it is not necessary to form the central hot spark. Therefore, the implosion stability requirement is only for the acceleration phase. As the results the introduction of the fast ignition concept will significantly reduce the uniformity requirements.

In this paper, we show the simulation results on the HAPI and the fast ignition and discuss the preheating problems in the HAPI and the laser plasma coupling issues in the fast ignition.

Fokker-Planck Simulations of HAPI

The e-folding, G of the Rayleigh-Taylor instability for the acceleration phase is estimated by

$$G = \int_0^{t_a} \gamma dt \approx \sqrt{kgt_a^2}$$

where t_a, g and k are acceleration time, acceleration rate and the wave number. When the ablation pressure Pa and the implosion velocity v_{imp} are given, the acceleration rate g and the implosion time t_a are given by

$$g = P_a/M,$$

and

$$gt_a = v_{imp}$$

Therefore, the e-folding for the spherical mode number ℓ, namely $k = \dfrac{\ell}{R}$ is given by

$$G = \sqrt{\ell}\sqrt{\dfrac{Mv_{imp}^2}{RP_a}} = \sqrt{\dfrac{\ell}{A_{inf}}}\sqrt{\dfrac{\rho_a v_{imp}^2}{P_a}},$$

where M is the areal mass density of the imploding shell which is rewritten to be $\rho_a R/A_{inf}$ for the inflight aspect ratio A_{inf}.

When the implosion velocity is 3×10^7 cm/sec and $\rho_a \approx 4 g/cm^3$,

$$G \approx \sqrt{\dfrac{\ell}{A_{inf}}}\sqrt{\dfrac{3600 Mbar}{P_a}}$$

In the HAPI with Pa≥300Mbar, Rayleigh-Taylor e-folding is less than 4 for the modes $\ell \sim A_{inf}$. Therefore, the HAPI is relatively stable and does not require very high irradiation uniformity. However, the HAPI is sensitive to the preheating since the laser irradiation intensity is required to be higher than 10^{15} w/cm^2.

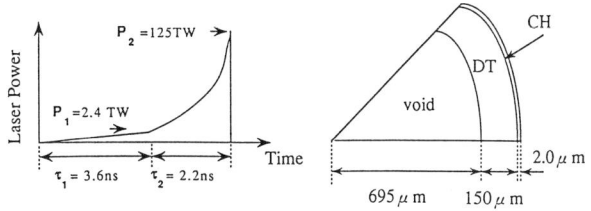

FIGURE 1. Laser pulse shape and target geometry

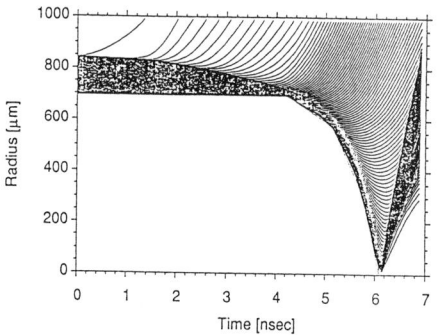

FIGURE2. r-t diagram for HAPl

The typical target design and corresponding laser conditions in this implosion mode are shown in Fig.1. In this target, a frozen DT shell is covered with a thin plastic CH layer and accelerated up to 5×10^7 cm/sec by the pressure higher than 100 Mbar. In order to generate 200Mbar, the absorbed laser intensity is required to be several times 10^{15} w/cm^2. In Fig. 2, the R-t diagram of the implosion is shown, where the in-put laser energy is 50KJ with 0.35 µm wavelength and the gain is approximately 1. In the simulation, we used the implosion hydrodynamic code HIMICO in which the electron heat transport can be described by the two models, namely, the Fokker Planck multi-group diffusion model (6) and the flux limited Spitzer-Härm model. The target design was optimized for the SH model for igniting the fusion burn. The maximum irradiation intensity of the Fig. 2 is 4×10^{15} W/cm^2 and the ablation pressure and the critical surface temperature are 200Mbar and 3kev, respectively.

The isentrope and density distributions during the implosion (at t=5.6sec when the shell shrinks to half of the initial radius) are shown in Fig.3(a) for the FP simulation and in Fig. 3(b) for the SH simulation.Because of the preheating by the Maxwellian tail electron, the plasma density at the ablation front for the FP simulation decreases to be 0.6 times that for the SH simulation. The minimum isentrope of the imploding shell is 4 for the FP simulation and 2 for the SH simulation. I believe that the FP simulation gives the more realistic density and

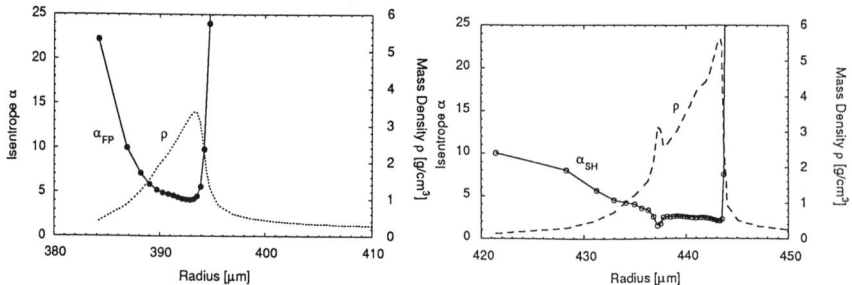

FIGURE3(a). Profiles of the isentrope of an accelerated target shell (Fokker-Planck simulation)
FIGURE3(b). Profiles of the isentrope of an accelerated target shell (Fokker-Planck simulation)

isetrope profiles than the SH simulation. Therefore, it is expected that the ablative stabilization of R-T instability is more effective due to the increase of the ablation flow velocity in the FP simulation than in the SH simulation.

Fast Ignition

Although high density compression by laser implosion has been demonstration by recent experiments, the hot spark formation was not successful in the high density compression experiments. Since the requirements on the laser irradiation uniformity are much higher for the hot spark formation than for the high density compression, a new concept for hot spark formation has been recently proposed(7), which is called the fast ignition. The advantage of this concept is relaxing the requirements on the laser irradiation symmetry. Namely, the hot spark is not required to be formed by the stagnation but by the additional heating with a ultra-intense short pulse laser.

Roughly speaking, the laser pulse parameters for the fast ignition are determined as follows. Since a dense DT plasma of approximately $100g/cm^3$ has to be heated up to 10kev, the deposition energy density is required higher than $10^{11} J/cm^3$. Since the plasma lifetime and radius are a few tens of picosecond and 30μm respectively, the total deposition energy, the heating laser power and the surface laser intensities are higher than 10KJ, $1 \sim 10$ PW and $10^{20} \sim 10^{21}$ w/cm^2 respectively.

A typical simulation example is shown in Fig. 4. A DT cryogenic layer target with a thin plastic ablator is irradiated with a long tailored pulse and a 1 pico second heating pulse. The pulse energy for the compression is 10KJ with 3ω and the heating pulse is assumed 10KJ with 4ω. In this example, the maximum compression occurs at 3.92 nsec and the density and temperature profiles at this time are shown in Fig. 5 (a). The density reaches more than 3000times the solid DT density and the areal density is 0.4 g/cm^2 although the ion temperature is less than 3 keV even in the hot spark. The heating pulse is injected uniformly at 80 psec before

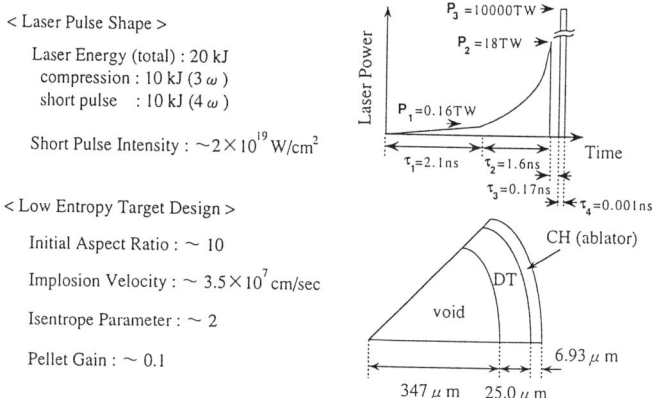

FIGURE 4. Fast Ignition Simulation Parameters for Fokker Planck implosion simulation

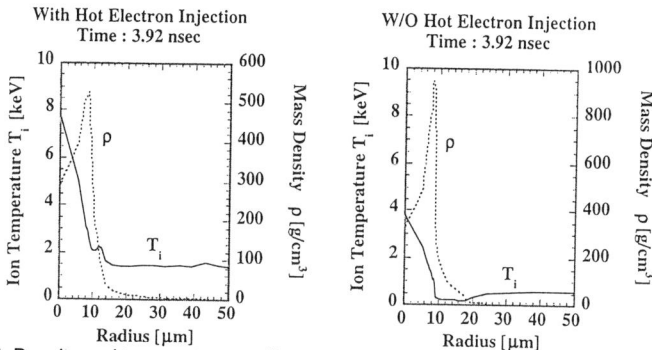

FIGURE 5. Density and temperature profile of the compressed core plasmas both for (a) with and (b) without heating pulse

the maximum compression, where the hot electrons of the temperature of 100keV are generated at the relativistic cut-off surface with the total energy of 5 KJ (50% of the heating pulse energy). The hot electron transport is described by the Fokker Planck code. Note here that the relativistic cut off density is defined by (8)

$$n_{c\,rel.} = \gamma\, n_c,$$
$$\gamma = \sqrt{1 + I\lambda^2 / 3 \times 10^{18}\,\mu m^2 w/cm^2}$$

where n_c, I and λ are cut-off density, laser intensity and laser wavelength respectively. As the laser intensity in the above formula, we used 2×10^{19} w/cm^2. The figure 5(b) shows the density and temperature profiles where the heating pulse is applied. The hot spark temperature increases to be higher than 5keV. Even in the dense cold edge plasma, the temperature is around 3keV, which is 2 times higher than that for the non- heating case. When the heating pulse injection timing is varied,

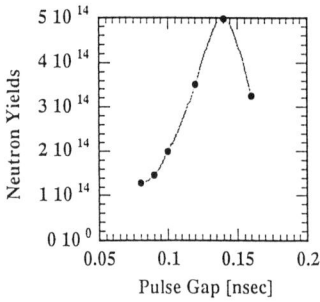

FIGURE6. Heating pulse timing dependence of neutron yield

than that for the non- heating case. When the heating pulse injection timing is varied, the neutron yield changes as shown in Fig. 6. In this time scale, the maximum compression is at 0.22 nsec. This result indicates that the heating pulse timing jitter should be less than 50 psec around the optimum time.

Recently, we investigated the relativistic laser pulse propagation in overdense plasmas. The figure 7 shows how the effective group velocity of the laser pulse increases with $I\lambda^2$. The solid curve and the dotted curve show the theoretical group velocity estimated by

$$v_g/c = \sqrt{1 - \frac{n_0}{n_c}\frac{1}{\gamma}}$$

for $n_0/n_c=5$ and 10 respectively. The solid and open circles in Fig.7 are the 1D particle simulation results for $n_0/n_c=5$ and 10 respectively(9). According to those relativistic effects togather with the hole boring or the self-focusing, the heating pulse with an intensity higher than 10^{20}w/cm^2 will deeply penetrate into the dense plasmas to couple with the implosion core plasmas.

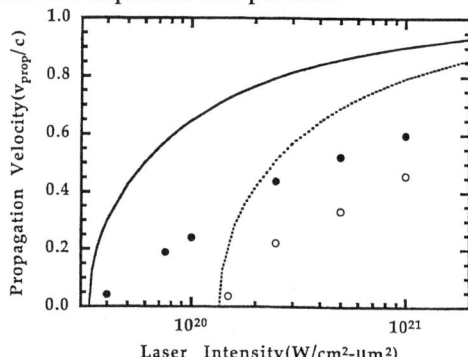

FIGURE7. Intensity dependence of laser propagation velocity

Concluding Remarks

In this paper, a stable implosion and hot spark formation schemes are discussed. Although multi-dimensional simulations have not been carried out, the high ablation pressure implosion(HAPI) and the fast ignition schemes seem promising. Further theoretical and experimental investigation will be continued in future.

REFERENCES

1. Haan, S.W.,Phys.Rev.A. 39(1989)5812
   Mima, K.etal, Plasma Phys. and Cont. Fusion 34 (1992)1775.
   Izawa. Y, Takabe, H. and Mima, K., J. of Plasma and Fusion Res., 70(1994)756.
2. Kato, Y. and Mima, K., Appl. Phys. B29(1982)186, Kato, Y., Mima.K etal, Phys. Rev. Lett. 53 (1984)1057
3. Lehmberg, R., Shmidt,A.and Bodner,S., J. Appl. Phys. 62(1987)2680
4. Skupsky, S.etal, J. Appl. Phys. 66(1989)3456
5. Nakano, H . etal, Optics comm. 78(1990)123
6. Nishiguchi, A., Mima, K.etal, Phys. Fluids, 134, (1992)417
7. Tabak, M., Hammer, J., Glinsky, M.E. etal, Phys. Plasmas 1, (1994)1626
8. Lefebure, E. and G.Bonnand, Phys. Rev. Lett. 74 (1995)2002
9. Sakagami, H.and Mima,K., Private Communication

Integrated Ignition Calculations for Indirectly Driven Targets*

William J. Krauser, Bernhard H. Wilde, Douglas C. Wilson, Paul Bradley, and Fritz Swenson

Los Alamos National Laboratory
Los Alamos, New Mexico 87545 USA

Abstract. We present two-dimensional LASNEX calculations of the hohlraum and ignition capsules proposed for the National Ignition Facility (NIF). Our current hohlraum design is a 2.76 mm radius, 9.49 mm long gold cylinder with 1.39 mm radius laser entrance holes (LEH) which are covered by 1 µm thick polyamide foils. Laser beams with less that 1.4 MJ total energy and less than 400 TW peak power irradiate the cylinder wall from two separate cones entering each LEH. The hohlraum interior is filled with hydrogen-helium gas (50-50 atomic) at a density of 0.83 mg/cm^3 to suppress the inward expansion of the wall. The capsule uses either a 160 µm plastic ablator doped with bromine (the baseline design), or a 155 µm beryllium ablator doped with copper (the beryllium design). The ablator surrounds an 80 µm thick deuterium-tritium (DT) ice layer with an inner radius of 0.87 mm. We will show the results of integrated, two-dimensional calculations of the hohlraum and the capsule. Plasma conditions within the hohlraum will be described. Peak radiation temperatures in the hohlraum are about 300 eV. These calculations proceed through the implosion, ignition, and burn of the DT capsule. Current peak calculated yields are 12 MJ for the baseline design and 6.9 MJ for the capsule with the beryllium ablator, although higher yields should be achievable with improved "tuning" of the laser power levels.

*This work supported under USDOE contract W-7405-ENG-36.

1. Introduction

As currently proposed, the NIF laser is a 192 beam frequency tripled Nd:glass system providing a peak power of 500 TW and a total energy of 1.8 MJ. The baseline NIF ignition target design (1) is indirect drive. In this technique laser light is utilized to heat the interior walls of a high-Z cylindrical enclosure, or hohlraum. The hohlraum serves to isotropize and thermalize the initially highly localized and non-Planckian x-rays created when the laser light is absorbed. A spherical capsule in the hohlraum is irradiated nearly uniformly from almost all directions by the x-rays filling the hohlraum. A highly symmetric ablation-driven implosion of the capsule results.

The baseline capsule is cryogenic with a DT ice shell surrounding a central DT gas region. The ice, in turn, is surrounded by an ablator. The hohlraum radiation temperature peaks at ~ 300 eV and has a carefully shaped pre-pulse which keeps the fuel and the ablator on a low adiabat during the long, high-convergence-ratio (~ 20 - 35) implosion. For the baseline capsule and hohlraum, the peak hohlraum temperature cannot be pushed much higher than 300 eV because the hohlraum will fill too rapidly with plasma blown off from the hohlraum wall, which prevents subsequent laser beam propagation within the hohlraum. Also, laser-plasma instabilities become more severe at the higher intensities needed to drive a hohlraum to higher temperature. At a lower peak temperature, a capsule implosion with a higher convergence ratio (i.e., a thinner ice shell at larger initial radius) would be required to achieve the same fuel density and yield. In order not to quench the ignition, there must be little mixing of the ignition hotspot with the cooler main fuel. The higher the convergence ratio, however, the larger the growth of inevitable initial perturbations on, for example, the ice-ablator interface. If this growth is too large, catastrophic hotspot mixing will occur and the capsule will fail to ignite. The baseline hohlraum and capsule design occupies a prudent compromise location in this parameter space.

2. Modeling requirements

In designing or assessing the performance of hohlraums and ignition targets for the NIF, we need to model a wide variety of physical phenomena that occurs during the course of the implosion. Examples include the refraction, reflection, and absorption of laser light passing through the spatially and temporally varying laser entrance holes; the filling of the hohlraum with plasma; laser-plasma interactions including instability seeding and growth; the spot motion of the laser absorption region; the time dependent emission of x-rays from the laser-heated plasma, their absorption on and re-radiation from the hohlraum walls, and final absorption in the capsule ablator; the spatial and temporal variation of the capsule ablation; the capsule implosion and short-wavelength hydrodynamic instability growth and mixing within the capsule; and thermonuclear burn. No single code exists today which is capable of modeling simultaneously all of these phenomena, some of which require a three-dimensional treatment.

3. The integrated modeling approach

The best we can do currently is to combine the hohlraum and capsule calculations into a single two-dimensional (2-D) radiation-hydrodynamics calculation using the LASNEX code (2) with the laser light as the energy input. These "integrated calculations," each of which requires ~ 100 hours of time on a Cray-YMP, generally use the best available numerical packages for laser ray tracing (3), radiation transport, non-local thermodynamic equilibrium (non-LTE) atomic physics (4) in the hohlraum wall, hydrodynamics (mostly Lagrangian, but with a semi-Eulerian rezoning treatment (5,6) for material in or near the LEH's), and fusion burn. This comprehensive but computationally intensive technique was developed over the last three years and includes many of the effects listed above; nevertheless the calculations are incomplete. For example, the calculation of colliding low density plasmas is performed with Lagrangian hydrodynamics and does not model plasma interpenetration. Laser plasma instabilities such as Brillouin and Raman scattering are not explicitly modeled. However, the integrated calculations only use approximately 1.4 MJ of laser energy, leaving 0.4 MJ of the NIF laser design energy to account for these and other losses. And capsule mixing and yield degradation due to short scale length hydrodynamic instability growth must be modeled separately (7). The techniques and physics packages used in integrated modeling of the NIF have been used in modeling relevant NOVA experimental data (8,9). The excellent agreement between those calculations and the data increases our confidence that the modeling techniques and physics packages are adequate.

4. Hohlraum and ignition target description

There are many possible hohlraum and ignition capsule designs. We have focussed our work on one hohlraum design and on two different capsule designs (see Fig. 1). Our colleagues at Lawrence Livermore Laboratory designed one of the capsules, the baseline capsule, and the basic hohlraum, a 9.49 mm long by 2.76 mm radius gold cylinder, 40 μm thick, with 1.39 mm radius LEH's on either end. We added a 10 μm thick plastic liner around the lip of the LEH to help keep it open. The capsule uses either a 160 μm thick polystyrene (plastic) ablator doped with bromine, or a 155 μm thick beryllium ablator doped with copper. The ablator surrounds an 80 μm thick DT ice layer with an inner radius of 0.87 mm. The ablator dopant serves to reduce x-ray preheat in the ablator and eliminates an unstable density step at the ablator/DT interface.

Our first integrated calculations used a 2.5 μm thick plastic lining on the gold hohlraum walls with the expectation that it would slow the expansion of the gold wall into the (nearly evacuated) hohlraum center. However we found that the plastic stagnated on axis, creating a modest pressure pulse which caused the capsule to implode too early on the pole and prevented ignition. Our solution was to remove the plastic liner and fill the hohlraum with gas at a density ~ 1 mg/cm^3. Since the capsule must be maintained at about 18° K, the gas could be only hydrogen or helium, or a mixture of the two. Our most recent design uses 0.833 mg/cm^3 of a

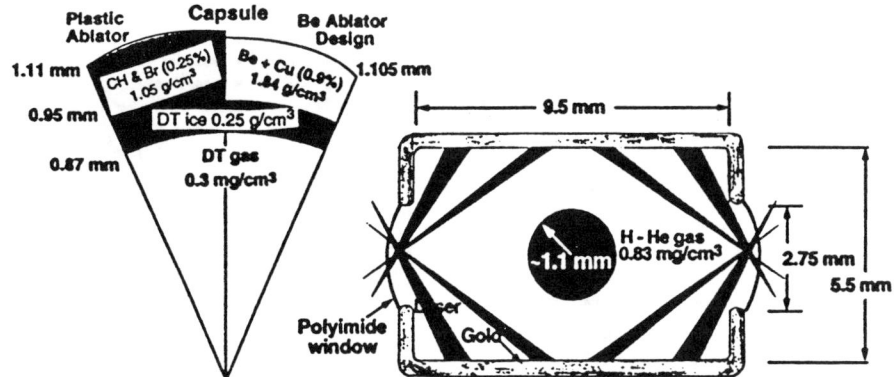

Figure 1. NIF capsules and hohlraum. The gas-filled, gold-walled hohlraum is driven by 192 laser beams (1.8 MJ, 500 TW) arrayed in inner and outer cones.

50-50 (atomic) mixture. The mixture of gases minimizes stimulated Brillouin scattering (SBS). To contain the gas prior to firing the lasers, we covered each LEH with a 1 μm thick polyimide foil. With these changes stagnation pressures were reduced to the point where they no longer degraded the capsule implosion. Fig. 2 shows that throughout the laser pulse the gas fill successfully holds back the gold walls of the hohlraum and maintains hohlraum plasma densities comfortably less than the critical electron density.

As shown in Fig. 1, the laser light coming in each LEH is in two cones, an "inner" cone which illuminates the hohlraum wall near the waist of the capsule, and an "outer" cone which illuminates an area of the wall closer to the LEH. This arrangement was chosen so that the time-dependent asymmetry of the radiation incident on the capsule can be minimized by dynamically varying the relative power in the cones.

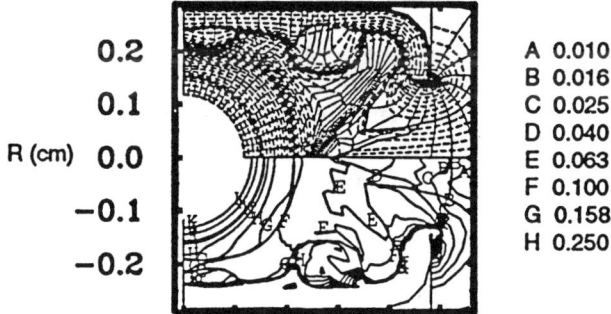

Figure 2. Mesh (upper) and contours (lower) of fraction of critical electron density at time of peak laser power (14.5 ns), from an integrated calculation.

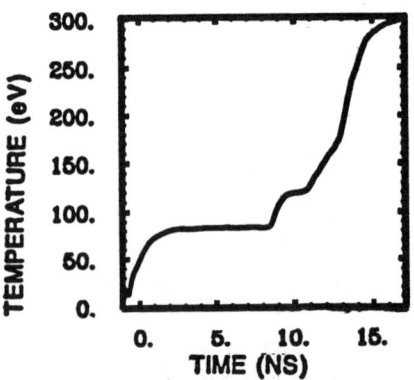
Figure 3. Ideal radiation temperature drive for the plastic capsule.

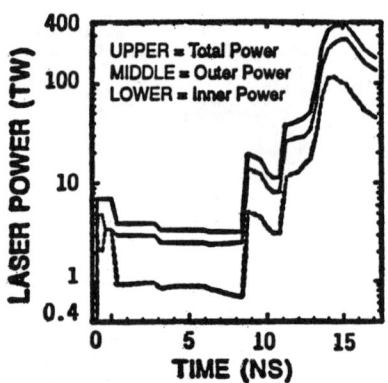
Figure 4. Laser power levels needed to approximate temperature drive and give good capsule symmetry.

5. Achieving ignition in integrated calculations

A 1-D clean calculation of the implosion and burn of the plastic capsule with an ideal radiation temperature drive shown in Fig. 3 gives a yield of 14.7 MJ. This drive, which peaks at 300 eV, creates four shocks in the capsule, with the final shock bringing the ablator up to peak pressure with sufficiently low entropy in the fuel. Since the NIF target will not be in a 1-D environment, integrated modeling calculations must be employed throughout the hohlraum and target design process, as well as in efforts to verify that the overall design has a reasonable chance of achieving ignition, the primary NIF goal. In a 2-D integrated calculation of a particular capsule and hohlraum, the only input variables are the laser pointings and the laser power levels. In iterative calculations, the total laser power time history is adjusted until the hohlraum radiation temperature history reasonably approximates the ideal profile. In general the capsule will fail to ignite at this stage because asymmetries in the drive cause too much distortion in the final configuration of the capsule.

It is customary to quantify the radiation drive asymmetry by expanding the radiative flux incident on the capsule in terms of spherical harmonics, which reduce to Legendre polynomials on the surface of a sphere having cylindrical symmetry. The independent variable in the expansion is the angle θ, which is 0° at the capsule pole and 90° at the capsule equator. Left-right symmetry in the hohlraum ensures that the odd moments P_1, P_3, etc., in the expansion vanish. The P_0 moment is the average flux incident on the capsule and is roughly proportional to the fourth power of the hohlraum radiation temperature. P_2 is a measure of the pole-to-equator asymmetry. A positive P_2, for example, means that the capsule is "pole hot," perhaps due to the outer cone of laser beams having too much power relative to the inner cone thus causing the hohlraum wall near the LEH to appear hotter than the wall near the capsule equator. If this coefficient is mostly positive throughout the implosion, the capsule generally will be "pancaked" at ignition time. Ideally, all but the P_0 moment are zero, so the design problem is to minimize all of the higher moments.

P_2 and P_4 are minimized in the next series of calculations. The relative power levels in the inner and outer laser cones are adjusted dynamically (i.e., the beams are "phased") to minimize P_2 at all times. With two cones (per side), this time dependent control is possible only for the P_2 coefficient. Blowoff of the hohlraum wall causes the position of the laser absorption region on the wall to move radially inward during the illumination. As viewed from the location of the capsule, this "spot motion" makes the hot region on the wall due to the outer cone appear to move steadily toward the LEH. With a judicious pointing of the cones (i.e., the initial spacing between the inner and outer rings of illumination on the hohlraum wall) it is possible to achieve a *time averaged* value near zero for P_4. Direct control, even in a time averaged sense, of the higher moments is not possible. Fig. 4 shows typical laser power levels needed to approximate the desired hohlraum temperature and achieve adequate capsule symmetry in the integrated calculation of the plastic ablator capsule. Fig. 5 shows typical P_2 and P_4 moments.

Minimizing the capsule drive asymmetry in this manner leads to acceptable implosion symmetry followed by ignition and burn. The plastic design yields 12.2 MJ and the beryllium capsule produces 6.9 MJ in integrated calculations. The beryllium design achieves 17.4 MJ in 1-D simulations, and with more tuning of the integrated calculations to decrease the drive asymmetry we believe its yield could be raised substantially.

6. Robustness studies

Two capsules ignite and burn in integrated calculations, but how believable are these results? To better quantify the probability of successful ignition, we have

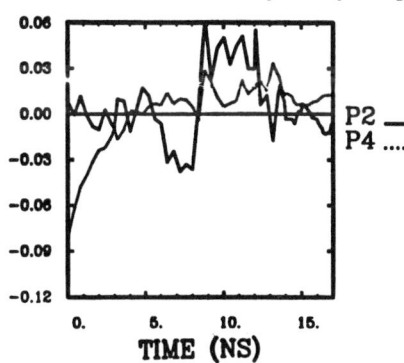

Figure 5. Typical P_2/P_0 and P_4/P_0 moments of the incident radiation flux on the capsule.

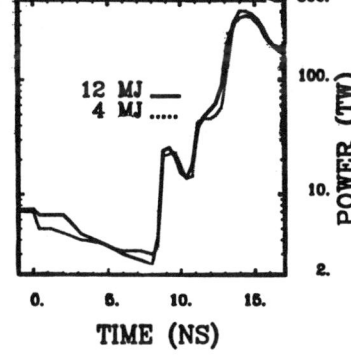

Figure 6. The nearly optimum power profile gives 12 MJ. Non-optimum profile gives 4 MJ in the integrated calculations.

performed numerous robustness studies of the NIF design. We have examined sensitivities to the length, shape, and gas fill density of the hohlraum, and we are assessing whether a very thin plastic lining on the hohlraum wall might offer certain advantages.

For the capsule, we have examined the effects of asymmetric radiation drive and non-ideal radiation temperature drive profile. Fig. 6 shows a laser power profile modestly different from the optimized profile (the laser specifications easily allow one to control the power to this level of accuracy). This non-optimum profile yielded only 4 MJ in the integrated calculation for the plastic capsule.

For the laser we are examining sensitivities to pointing and beam phasing. One pleasantly surprising, totally unanticipated observation was that *no* beam phasing is required for certain optimum pointings. For less than optimum pointings (e.g., moving the outer cone 100 μm toward the LEH), beam phasing *is* required to get adequate burn.

Finally, we are examining sensitivities to some of the modeling approximations in LASNEX: we are investigating different hydro techniques, different electron conductivity approximations, different equations of state and opacities, and LTE vs. non-LTE atomic physics treatments. Our non-LTE atomic physics package attempts to model the time dependence of the level populations in the laser irradiated gold plasma. This enhancement causes less radiation emission from the laser deposition regions and more gold plasma motion.

7. Conclusions

In summary, using our best integrated calculations we confirm Livermore calculations of ignition with a plastic capsule, and we have added an alternate capsule design with a beryllium ablator. Although robustness calculations show the designs are sensitive, they add to our confidence that NIF can achieve ignition. Many additional calculations remain to be performed.

ACKNOWLEDGMENTS

We wish to thank our colleagues at Lawrence Livermore Laboratory, particularly Steve Haan and Steve Pollaine, who have collaborated with us and shared their capsule and hohlraum designs. This work is supported by the USDOE under contracts W-7405-ENG-36 and W-7405-ENG-48.

REFERENCES

1. Haan, S. W. *et al.*, "Design and Modeling of Ignition Targets for the National Ignition Facility," to be published in *Physics of Plasmas* **2** (1995).

2. Zimmerman, G. B., and Kruer, W. L., *Comments Plasma Physics of Controlled Thermonuclear Fusion* **2**, 51 (1975).

3. Friedman, A., Lawrence Livermore Laboratory Laser Program Annual Report No. UCRL-50021-83, 1983, pp. 3—51.

4. Post, D. E. *et al.*, *Atomic Data and Nuclear Data Tables* **20**, 397 (1977); Zimmerman, G. B., and More, R. M., *Journal of Quantitative Spectroscopy and Radiative Transfer* **23**, 517 (1980).

5. Hoffman, N. M. *et al.*, "Nonlinear multimode instability modeling for national ignition facility capsules," presented at 12th International Conference on Laser Interaction and Related Plasma Phenomena, Osaka, Japan, April 24—28, 1995.

6. Brackbill, J., and Saltzman, J., *Journal of Computational Physics* **46**, 342 (1982).

7. Scannapieco, A., Los Alamos National Laboratory Report No. LA-UR-82-2897 (1982).

8. Lindman, E. L. *et al.*, "Modeling of drive-symmetry experiments in gas-filled hohlraums at NOVA," presented at 12th International Conference on Laser Interaction and Related Plasma Phenomena, Osaka, Japan, April 24—28, 1995.

9. Suter, L. J. *et al.*, "Modeling and Interpretation of Nova's Symmetry Scaling Data Base," *Physical Review Letters* **73**, 2328—2331 (1994).

The Computational Optimization of Indirect-Driven ICF Targets

V.A. Lykov, E.N. Avrorin, N.G. Karlykhanov, V.E. Chernyakov, M.Yu. Kozmanov, V.A. Murashkina and Ya.Z. Kandiev

Russian Federal Nuclear Center – VNIITF
P.O. BOX 245, Snezhinsk (Chelyabinsk-70), 456770, Russia.
e-mail: lyk@ch70.chel.su

Abstract. The results of the ICF indirect–driven targets optimization performed by ZARYA/ERA code for a better insight into the requirements imposed on both target designs and hohlraum drive temperature to gain the ignition with laser of minimum power are presented. Two modification of cryogenic shell targets for hohlraum drive temperatures in the range of 0.25–0.38 keV are proposed for the ignition. The 500 TW lasers are needed to perform such investigations.

INTRODUCTION

The review of theoretical and computational works on ICF problem performed at VNIITF was presented in the reports (1,2). The major point of these investigations is the clearing up a requirements imposed on both target designs and lasers to achieve the thermonuclear ignition and high energy gain of targets. ZARYA/ERA code was developed for numerical simulation of ICF targets (3,4). Some results of ZARYA/ERA code simulation of the direct and indirect–driven targets were published in (1–3,5–6).

The indirect–driven target physics has been published in papers (7–11). The actuality of the these research are growing up in view of the NIF Project (12). The keystone of the NIF Project is the creation of Nova Upgrade Laser at LLNL. The NIF Project realization to allow unique possibility for Inertial Fusion Energy basic and applied physics research (7–15) in this century yet.

The results of the ICF indirect–driven targets optimization performed by ZARYA/ERA code for a better insight into the requirements imposed on both target designs and hohlraum drive temperature to gain the ignition are given in this paper.

1. ZARYA/ERA CODE PHYSICAL MODEL

Physical models and numerical methods, realized in 1D ZARYA/ERA code are described in (1-6). The effective temperature (T_r) and multi-group spectral approximations (kinetic or diffusion) are used for the radiation transfer simulations. The atomic statistical model data, Saha equation as well as radiative-collisional kinetic model for ions spices and state populations of the simplest multi-charged ions levels in plasma are used. Energy and momentum transfer by charged fusion reaction products and suprathermal electrons are simulated either by one-group approach with the effective free path length or by multi-group spectral kinetic approach for Landay equations with accounting self-consistent electric fields and reversed currents for charged particles in plasma. Semi-empirical turbulent mixing model (16) and Takabe model (17) are used for simulation of mixing at the hydrodynamic-unstable contact surface and at ablation front of fusion targets. Theoretical absorption model used in ZARYA/ERA code was described in (6). This model takes into account the inverse-bremsstrahlung, the resonance and parametric absorption, the suprathermal electrons generation, the steepening of plasma density profile due to pondermotor forces, the refraction and Brillouin scattering in the target corona and peculiarities of laser light focusing to a targets.

2. RESULTS OF INDIRECT-DRIVEN TARGETS OPTIMIZATION

The tentative results of indirect-driven targets calculations obtained by ZARYA/ERA code have been published in (1,2,5). The results of the indirect-driven targets optimization carried out by ZARYA/ERA code for hohlraum drive temperature in the range of $T_r \cong 0.25$–0.38 keV at target surface are presented bellow.

The simplest type of cryogenic shell-target designs with the initial shell aspect ratio of $A \cong 10$–20 and initial aspect ratio of DT-ice layer of $A_{DT} \cong 25$–50 were taken for the consideration. The large aspect ratio of the target are limited by the hydrodynamics instabilities. The low aspect ratio of shell requires the precision-shape driver pulse that leads to the instabilities development also (13).

The range of $A \cong 10$–20 corresponds to the in-flight aspect ratio of $A_{inf} \cong 50$–75 which is the upper limit by the paper (17) but it do not impose strong requirements on the shape and accuracy of the driver pulse (3). For example, the time–dependence of target absorbed energy power could be $P \sim t^n$. Using paper (7) it can be shown, that linear time–dependence of hohlraum drive temperature corresponds to the power index about of $n \cong 3$. It seems reasonable to say that this index power value is the optimal one for targets given below.

Targets optimization was carried by 1D ZARYA/ERA code (3-6). The time–dependence of the hohlraum drive temperature was used by:

$$T_r(\text{keV}) = \begin{cases} 0.05 + (T_0 - 0.05)t/t_0, & t \leq t_0 \\ T_0, & t_0 < t \leq t_0 + \Delta \\ 0, & t > t_0 + \Delta \end{cases}$$

Analysis of calculations for hohlraum drive temperatures in the range of $T_r = 0.25$–0.38 keV have showed that given–below Case A,B could be optimal for ignition experiments.

Case A The shell with density about of 6–7 g/cm^3, initial diameter of 1.05 mm, initial aspect ratio of $A \cong 15$–20 and DT-ice mass of 10 μg. Hohlraum drive temperature parameters: $T_0 \cong (0.36$–$0.38)$ keV, $t_0 \cong (4$–$5)$ nsec, $\Delta \cong 1$ nsec.

Case B The glass shell with initial diameter of 2.15 mm, initial aspect ratio of $A \cong 12$–14 and DT-ice mass about of 50 μg. Hohlraum drive temperature parameters: $T_0 \cong (0.28$–$0.32)$ keV, $t_0 \cong (9$–$11)$ nsec, $\Delta \cong 2$ nsec.

The results of ZARYA/ERA calculations of targets compression and burn for Cases A,B are presented in the Table 1 and Fig. 1. The results of laser energy and power estimations are presented in the Table 1 also. These estimations were obtained by using of publications (7-10) and additional ZARYA code simulations also.

Probably, that requirements for the accuracy and technical feasibility of considered targets as well as required symmetry of hohlraum drive radiation at the target surface are the same as for published targets designs (7-10,17). In accodance with carried estimates, the homogeneity of irradiation of spherical targets placed at the center of hohlraum designs (7-10) could be sufficient for the ignition of above–described targets. The 500 TW lasers are needed for the ignition and burn targets both in Case A and Case B. But it is rather difficult to predict the potential energy gains decreasing that could be caused by the compression asymmetry.

Table 1. Results of indirect-driven targets compression and burn.

Results of calculations	Case A	Case B
Energy absorbed by target (kJ)	45	140
Ablated target mass (%)	87	90
Compression velosity (cm/nsec)	0.035	0.035
Maximal density at target center (g/cm^3)	180	90
Maximal ion temperature at center (keV)	40	47
DT-fuel burn up (%)	25	22
Thermonuclear energy yield (MJ)	0.84	3.7
Needed laser energy (MJ) (estimation)	0.5	1.3
Needed laser power (TW) (estimation)	500	500
Gain (estimation)	1.6	2.8

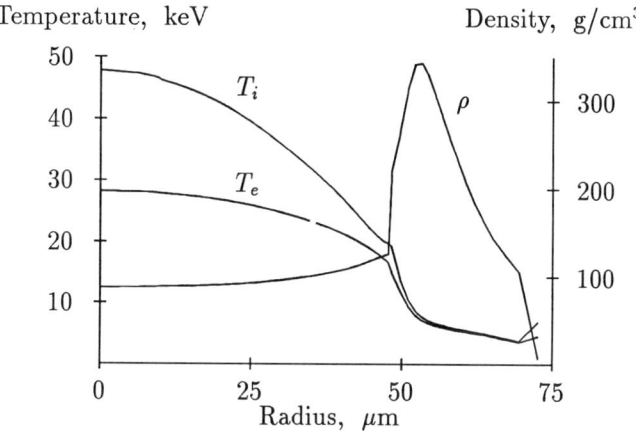

Figure 1: Density, Ion and Electron Temperatures versus radius at the time of intensive thermonuclear burn of the target in Case B

Symmetry of target irradiation and compression could be improved by using hohlraum designs proposed in paper (18). The asymmetry of the target compression to be caused mainly by the first harmonic resulted from the practical-inevitable power disbalance between laser beams in this case. The NIF power balance would be better than 5 % rms between laser cones to obtained the energy gain predicted in papers (9–13) and in this report also.

ACKNOWLEDGMENTS

The authors thank Dr.A.Andriyash for calculations of opacities in the range of low-value hohlraum drive temperatures and Dr.I.Glazyrin for the preparing the paper for publication.

REFERENCES

1. Avrorin,E.N., and Lykov,V.A., "Computations of the Targets for ICF by ZARYA code," in *Proceedings of the International conference on LASERS'90*, STS PRESS. McLEAN, VA., 1991, pp.811–818.

2. Avrorin,E.N., and Lykov,V.A., "Theoretical Works on ICF and High-Z Ions Plasma Physics carried out at VNIITF,"in *AIP Conferece Proceedings No.318. Laser Interaction and Related Plasma Phenomena.* Editor: George H.Miley, AIP PRESS, NY, 1994, pp.268-269.

3. Avrorin,E.N., Zuyev,A.I., Karlykhanov,N.G., Lykov,V.A., and Chernyakov,V.E. *Sov.JETP Lett*, **32**(7), 437–440 (1980).

4. Zyev,A.I. *Zh. Vytchislit. Mat. i Mat.Fiz.*, **32**, 82–96 (1992).

5. Avrorin,E.N., Lykov,V.A., Murashkina,V.A., et al. "Evidence for Reduction of Turbulent Mixing at the Ablation Front in Experiments with Shell Targerts". in *AIP Conferece Proceedings No.318. Laser Interaction and Related Plasma Phenomena.* Editor: George H.Miley, AIP PRESS, NY, 1994, pp.390–399.

6. Lykov,V.A., Chernyakov,V.E., Karlykhanov,N.G., and Chikulaev,A.A., "ZARYA Code Simulation of Physical Phenomena Proceeded Under Compression and Thermonuclear Burn of Spherical Targerts for ICF," presented at 23rd European Conference on Laser Interaction with Matter, Oxford, September 19–23, 1994.

7. Murakami,M. and Meyer-Ter-Vehn,J. *Nuclear Fusion,* **31**, 1315–1341 (1991).

8. Nishimura,H., et al. *Physics of High Power Laser Matter Interactions.* Editors: S.Nakai and G.H.Miley. World Scientific Publishing, 1992, pp.334–341.

9. Lindl,J.D., "Two Decades of Progress Toward ICF Ignition and Burn," presented at 23rd European Conference on Laser Interaction with Matter, Oxford, September 19–23, 1994.

10. Cray,M., "Resolving Key Ignition Issue: Implosion Symmetry," presented at 23rd European Conference on Laser Interaction with Matter, Oxford, September 19–23, 1994.

11. Wilson.,D.C. and Krauser,W.J., "Indirectly Driven Targets for Ignition," presented at 23rd European Conference on Laser Interaction with Matter, Oxford, September 19–23, 1994.

12. Hogan,W., Paisner,J., Lowdermilk,W.H., "National Ignition Facility, Performance and Cost," presented at 23rd European Conference on Laser Interaction with Matter, Oxford, September 19–23, 1994.

13. Nuckolls,J., Wood,L., Thiessen,A. and Zimmerman,G. *Nature*, **239**, 139–142 (1972).

14. Afanasev.,Yu.V., Basov.,N.G. et al., *Pizma Zh. Exp. Teor. Fiz.* **24**, 23–25 (1976).

15. Feoktistov,L.P., Avrorin,E.N., et al., *Kvantovaya Electronica*, **6**, 349–358 (1978).

16. Neuvazhayev,V.E. and Yakovlev,V.G. *VANT. Ser. Theor. Appl. Physics.* **1**, 28–36 (1988).

17. Takabe,H. and Yamamoto,A. *Phys.Rev.*, **44**, 5142–5148 (1991).

18. Phillion,D.W. and Pollaine,S.M. *Physics of Plasmas* **1**, 2963–2975 (1994).

Physics Issues Related to the Confinement of ICF Experiments in the U.S. National Ignition Facility*

M. Tobin, A. Anderson, J. Latkowski, M. Singh,
C. Marshall, and T. Bernat

Lawrence Livermore National Laboratory, P.O. Box 5508, L-481, Livermore, CA 94550

Abstract. ICF experiments planned for the proposed US National Ignition Facility [NIF] will produce emissions of neutrons, x rays, debris, and shrapnel. The NIF Target Area [TA] must acceptably confine these emissions and respond to their effects to allow an efficient rate of experiments, from 600 to possibly 1500 per year, and minimal down time for maintenance. Detailed computer code predictions of emissions are necessary to study their effects and impacts on Target Area operations. Preliminary results show that the rate of debris shield transmission loss [and subsequent periodicity of change-out] due to ablated material deposition is acceptable, neutron effects on optics are manageable, and preliminary safety analyses show a facility rating of low hazard, non-nuclear. Therefore, NIF Target Area design features such as fused silica debris shields, refractory first wall coating, and concrete shielding are effective solutions to confinement of ICF experiment emissions.

Introduction

The repeated confinement of ICF experiments producing yields up to 20 MJ in the proposed US National Ignition Facility introduces unique challenges. The per shot increase in laser energy between NIF and the existing Nova laser facility, for example, is a factor of 40, while the increase in total yield will be 10^5. Two of the unique challenges are understanding the characteristics of target emissions (neutrons, x rays, debris, and shrapnel) and their effects on materials; and designing a facility that will meet both stringent environmental standards and an acceptable experiment rate, both for reasonable capital and operational costs.

Here we summarize the results of primarily radiation-hydrodynamics code calculations that predicted NIF target emissions of x rays, debris, and shrapnel as a function of fusion yield, including considerations of two-dimensional effects such as debris jetting and Lambertian x-ray emission. Predicted responses of first wall materials (a refractory material such as plasma-sprayed boron), the debris shield, and the tip of the target positioner is presented, including characteristics of the heated material (vapor, melt, spall, etc.), as well as the consequences to NIF operations.

We also discuss the work-in-progress in experimentally assessing the degradation in optical transmission of fused silica and KDP due to neutron and γ-ray deposition. The specific impacts of *pulsed* neutron and γ-ray irradiation and spectral dependence is discussed.

NIF has received a preliminary safety rating as non-nuclear, low hazard. Public dose and occupational exposure are restricted through shielding and the use of low activation material (aluminum). From an environment and safety perspective, the facility site can accommodate individual yields ≤ 45 MJ and an annual total yield of ≥ 2000 MJ, enough to encompass the expected demand for access to NIF.

NIF Target Area Design Overview

Prior to a shot, the target area must provide target chamber vacuum, and a stable platform for the target and its diagnostics. During a shot, the impact of the energy introduced into the chamber must be mitigated, and workers and the public protected from prompt radiation. Careful material selection controls residual radioactivity to allow required accessibility. Tritium and other radioactive wastes are collected and disposed. Diagnostic data is also retrieved, and the facility is readied for the next shot.

The 5-m radius NIF target chamber, made from Al 5083, is 10-cm thick. The target positioner inserts targets at the chamber 'waist' and laser beam ports are positioned in two concentric cones (27° and 53°, top and bottom). The final optic, a fused silica debris shield, is 6.75 m from chamber center. [see Fig. 1](1)

The NIF chamber and target area support structures become radioactive following shots producing neutrons. Shielding on the chamber exterior [~40 cm of borated concrete] reduces activation of structures as well as reduces the dose from the radioactive decay of the chamber. Controlled/monitored access to the chamber area is required for hours to several days after shots producing neutrons, depending on the yield. The target chamber support structure is made from aluminum-reinforced, borated concrete to provide stability *and* minimal radiation levels with reasonable cost. NIF's design lifetime is 30 years. A possible annual shot 'mix' is shown in Table 1. Designing to this shot envelope will ensure adequate facility flexibility.

Table 1. Assumed annual NIF target experiments for design purposes

Laser Energy (MJ)	Number of shots	Target Yield (MJ)
1.0	225-600	0
1.5	225-600	0
1.8	150-300	0.1 - 20 MJ
	100-250	0.1 MJ
	10-50	20 MJ

ICF Target Emissions and Their Effects

To repeatedly confine NIF fusion experiments, the pressure pulse from the target must be well contained, debris shields must remain at ≥ 97% transmission by avoiding surface contamination, and neutron activation must be maintained at acceptable levels. A quantitative treatment of the threat to the debris shields requires an understanding of target emissions and materials responses.

X rays

X-ray and debris emissions were calculated for seven fusion NIF target yields [0.45 to 30 MJ] with a 1.8-MJ laser drive using the code LASNEX in one-dimension (2). The target was modeled as shown in Figure 2 but without the cryogenic support system (e.g. hohlraum only). A 'leak source' was used to model the laser entrance hole, revealing some two-dimensional features to the emission, such as the relative fraction of x-ray energy emitted from the hot hohlraum wall as opposed to the laser entrance hole. A Lambertian (cosine) distribution of radiation from the laser entrance holes is assumed. The results are listed in Table 2.

Debris

Debris is shown in Table 3. The effects of debris jetting have been observed on Nova. We postulate debris jetting with a 37° half angle, as this is consistent with other two-dimensional calculations done to examine this issue.(3) The mass involved in the jet is inferred from the mass of inward and outward moving material from the LASNEX calculation. The pulse length is determined by the range in material velocity for a given case.

Shrapnel

Shrapnel generation calculations show that more than five million few-micron average size droplets reach velocities of ~1 km/s for a 20-MJ yield.

X-ray Effects

The target positioner will be protected with a frost-coated plate, where the frost [likely water] will condense from the $\leq 10^{-6}$ Torr chamber vacuum by a cryogenically-cooled panel. Despite several hundred microns of frost removal, no deposition on the debris shield from the positioner is expected since only a non-condensable gas is produced. The first wall surface, likely a boron-based plasma spray material, will be exposed to the x-ray and debris fluences listed in Tables 2 and 3. For a 20-MJ shot, we estimate 120 g of thermal spray coating to be removed, possibly depositing ≤0.2 µm on debris shields. This is done by calculating an energy deposition profile in the material due to the absorption of the x-ray fluences and spectra shown in Table 2. Photon interaction cross sections

Table 2. X-ray fluences and other characteristics of NIF x-ray target emissions.

Shot Energy & Fusion Yield (MJ)	X-ray energy (MJ)	~BB Temp (eV)	Pulse (ns)	Target Positioner Fluence (J/cm^2)	Peak 1st Wall Fluence (J/cm^2)	Peak Debris Shield Fluence (J/cm^2)	#/year
≤ 1.5	≤ 0.59	60% ≤ 260	≤ 10	0	≤ 0.22	≤ 0.12	450-1200
		40% ≤ 34	≤ 61	≤ 190	≤ 0.08	≤ 0.04	
Total				≤ 190	≤ 0.30	≤0.16	
1.8	0.68	60% @ 260	9.9	0	0.26	0.13	150-300
		40% @ 34	61	217	0.09	0.05	
Total				217	0.35	0.18	
100 kJ	0.70	60% @ 260	9.8	0	0.27	0.13	100-250
		40% @ 34	62	223	0.09	0.05	
Total				223	0.36	0.18	
20 MJ	4.23	40% @ 400	6.3	0	1.07	0.52	10-50
		14% @ 150	49	469	0.19	0.10	
		46% @ 47	49	1558	0.62	0.34	
Total				2028	1.88	0.97	

are taken from the LLNL Evaluated Photon Data Library.(4) For conservatism, we assume all material at or above the incipient melt energy density is removed. The target assembly will deposit ~0.5 Å of material on the debris shield each shot. Total deposition in one week *before* a 20-MJ shot is a few Å. Therefore, weekly changeout of each debris shield is planned.

Debris Effects

Although the debris loading is a large total energy fluence, the power is low enough due to long deposition times that thermal conduction effects result in no material being ablated.

Shrapnel Effects

For a single 20-MJ shot, approximately 3% of the area of the inside of the chamber in a band near the centerline is pitted.(3) Although shrapnel will ultimately degrade the NIF debris shields due to pitting, we expect ablated material deposition to be the most stressing area. When debris shields are removed for cleaning, any shrapnel damage can be repaired at that time.

Table 3. Debris fluences and other characteristics of NIF debris target emissions.

Shot Energy & Fusion Yield (MJ)	Debris Energy (MJ)	Average Velocity (cm/µs)	Target Positioner Fluence Mass Pulse	Peak 1st Wall Fluence Mass Pulse	Peak Debris Shield Fluence Mass Pulse	#year
≤ 1.5	≤ 0.40	≤ 8	188 J/cm^2	0.33 J/cm^2	0.18 J/cm^2	450-1200
			104 µg/cm^2	68 ng/cm^2	38 ng/cm^2	
			1.35 µs	67 µs	91 µs	
1.8	0.44	8	207 J/cm^2	0.37 J/cm^2	0.20 J/cm^2	150-300
			104 µg/cm^2	68 ng/cm^2	38 ng/cm^2	
			1.35 µs	67 µs	91 µs	
100 kJ	0.46	8	216 J/cm^2	0.39 J/cm^2	0.21 J/cm^2	100-250
			104 µg/cm^2	68 ng/cm^2	38 ng/cm^2	
			1.35 µs	67 µs	91 µs	
20 MJ	1.85	16	916 J/cm^2	1.46 J/cm^2	0.80 J/cm^2	10-50
			93 µg/cm^2	87 ng/cm^2	48 ng/cm^2	
			1 µs	50 µs	68 µs	

Neutron Effects

The neutron and γ-ray dose to the debris shields is calculated to be ~2 krads due to neutrons and ~1 krad for neutron-induced γ-rays, for a single 20-MJ NIF yield. Therefore, a total NIF lifetime dose could range from ~2 to 9 Mrad. Previous exposures on RTNS-II indicated that SiO$_2$ did not experience any significant degradation until approximately 20 Mrad total dose. We also find that ~300 krads reduces the transmission of fused silica by <1% at 3ω. An experiment to investigate whether the short-pulse NIF fluence might cause more damage than the continuous testing fluences was conducted on the Sandia Pulsed Reactor III. For equal doses of radiation [400 krads γ-ray/160 krads neutron], delivered over a period of 30 minutes for some samples and in four 100 µs pulses over 30 hours for others, the samples exposed to the higher flux actually exhibited less reduction in transmission. Further investigations using the LLNL ^{60}Co source revealed that fused silica with as-manufactured no-oxygen deficiency is very hard to γ-ray irradiation. Fused silica has a potential advantage over fused quartz due to this issue. However, *after* neutron irradiation, fused silica shows susceptibility to γ-ray damage. This is due to oxygen displacements, creating oxygen-deficient centers for color center damage to accumulate. The oxygen deficient color center is 246 nm. Fortunately, the 246 nm line is transmuted to 210 nm by the subsequent γ-ray radiation. KDP appears at least as damage resistant as fused silica, although controlling impurities is now known to be a key issue for making KDP hard to radiation damage. Further investigations will explore the effects of the spectral differences between the SPR-III fission source and NIF where the damage is mainly due to 14 MeV neutron deposition. An important difference between the two spectra is average γ-ray production cross section. The value for 14-MeV is about 10^3 greater than for a fission spectrum. Predictions of the impact of NIF operations on optical performance can be made when the study is completed.

Environment and Safety

The NIF project is considering expanding the annual shot envelope from 385 to 2000 MJ per year. Recent environmental analyses indicate that the increased annual yield would result in a site boundary dose only 15% higher [0.15 mrem vs. 0.13 mrem]. Additionally, increasing the largest single shot yield [for environmental considerations] from 20 to 45 MJ results in a *lower* accident site boundary dose. This dose is substantially due to activated air emissions. The reduction is due to accounting for the elevated release height and mixing with fresh air, aspects ignored in previous analyses. The result is a reduction in dose to an individual at 30m, from 2.7 rem to < 0.025 rem. The significance is the NIF facility non-nuclear rating is based on the dose at 30-m staying below 10 rem. These analyses show that instead of being within a factor of ~3 of a possible nuclear rating, NIF is more than a factor of 100 away.

Acknowledgements

Work performed under the auspices of the U.S. Department of Energy by the Lawrence Livermore National Laboratory under Contract No. W-7405-ENG-48.

References

1. *National Ignition Facility Conceptual Design Report*, LLNL, L-16973-1, UCRL-PROP-117093, May 1994.
2. *LASNEX and Atomic Theory*, LLNL Laser Program Annual Report, Vol 2, UCRL-50021-80, 1980.
3. Tokheim, R., et al., *NIF Target Area Design Support*, SRI International, Final Report, December 1994.
4. Cullen, D., et al., Tables and Graphs of Photon-Interaction Cross Sections Derived from the LLNL Evaluated Photon Data Library (EPDL), UCRL-50400, Vol.6, Rev.4, PartA, Oct. 31, 1989.

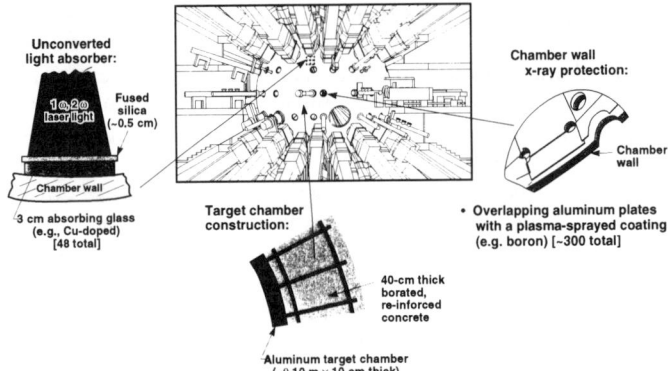

Figure 1. The NIF Target Area chamber will require several specialized components to mitigate the effects of target emissions.

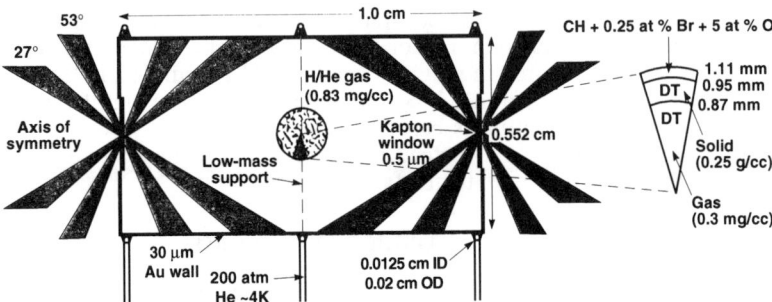

Figure 2. A possible NIF target design that may achieve as much as 20-MJ yield. Target will be inserted from the horizontal on cryogenic capillary tubes.

LARGE-SCALE LASER TARGET DESIGN.
ALTERNATIVE APPROCHES.

Feoktistov L.P., Lebo I.G., Rozanov V.B.
Tishkin V.F.

Lebedev Physics Institute, Moscow, 117924,
Moscow, Leninski pr. 53

Institute of Mathematical Modeling, Moscow
125047, Moscow Miusskaya sq. 4-A

1. INTRODUCTION

It is necessary to reach large gains in laser target (G) to get useful energy in a fusion reactor. Using the fission blanket in reactor allows one to increase the energy gain.

The first russian projects of laser fission-fusion reactor were studed in /1,2/. In /3/ it was suggested to use a two-cascade hybrid reactor scheme, allowing one to reach gain in fission blanket more than 1000. As a result it will be possible to use laser target with G ~ 0.1-1 in such type of a reactor.

In / 4 / we suggested the target design for "ignition" experiment at the laser energy 0,3-0,5 MJ. It was high aspect ratio cryogenic direct driven target. But it is necessary to use a lot of laser beams aruond target to provide with spherical symmetry of laser irradiation. Large surface "will be lost" for fission blanket. In addition the laser pulse should have a sharp time profile.

It is necessary to use about an five-ten times more laser energy to get "ignition" in indirect driven target.

We are studing two alternative approches to the laser target design for hybrid reactor:
   1) the target with inner laser enery input; 2) high aspect
   ratio target for laser with pulse duration about
100 ns. The numerical simulation were made by using Lagrange code "ATLANT".

2. THE TARGET WITH INNER LASER ENERGY INPUT

The scheme of target is shown in Fig. 1a. May be it will be enough to use one or two side laser radiationbeams to "ignite" DT-fuel in such type of the target. We made the series of 1D numerical simulations. Two shell target,- outer gold layer and inner layer - DT-ice. Laser flux heats DT-layer. The results of numerical simulation are shown in Table 1. The laser pulse duration (in full width, half maximum - FWHM) - 0.1 ns. We vary the laser energy, wavelength and parameters of the targets.

The variants 9-12-9-14, laser energy - 300 kJ, variant 9-10 - laser energy - 500 kJ. The most calculations were made by using 2T physical model (electron and ion heat conductivity, without radiative transport). A 3T physical model was used in variant 9-12w (electron, ion and radiation temperature). Table 1 shows neutron yield (Y) and target gain at the moments t=1 and 5 ns, R - outer radius of target, M Au - mass of gold shell, M DT - mass of fuel, E - laser energy

Table 1.

N var.	lamda mkm	E kJ	R cm	M Au mg	M DT mkg	t=1 ns Y 10**16	G	t=5 ns Y 10**16	G
912a	1	300	0.12	49.7	500	1.5	0.14	2.7	0.25
913	1	300	0.12	49.7	318	1.5	0.14	4.0	0.37
914	1	300	0.12	49.7	276	1.5	0.14	4.5	0.4
912b	0.35	300	0.12	49.7	500	0.85	0.08	2.4	0.22
910	0.35	500	0.12	49.7	500	2.3	0.13	6.1	0.34
912w	0.35	300	0.12	49.7	500	0.4	0.04	1.0	0.1

CONCLUSIONS:

1) It is possible to reach $G \sim 0.1$-0.3 at the laser energy 300-500 kJ; 2) laser pulse duration is about 0.1 ns, laser flux > 10^{17} Wt/cm^{2};

Temperature of hot electrons ~ 20-30 keV (see /5/); 3) the problems of 2D geometry, input of laser beams through the hole, plasma instabilities etc. require separate investigation.

3. THE HIGH ACPECT RATIO TARGET FOR LONG TIME LASER PULSE.

It is easier to reach 10% and more of the pump energy conversion into the laser emission by making use a laser facilities with pulse duration of 100 ns. This would allow one to close to energy cycle at $G \sim 0.1$, when the gain of the blanket is over 1000.

The estimation showed, that it was necessary to use very high aspect ratio shell (AS > 500 !) and laser energy > 1 MJ to reach $G \sim 0.1$.

We studed two types of the targets - a) gas-filled shell, b) cryogenic two-cascade target (see Fig. 2a and Fig. 2b).

In case b) we modeled the compression of the following target - outer gold shell with radius R0 and the mass M0, inner target - DTball with with mass M DT and thin gold shell with the radius R1 and mass M1.

We made two sets of the numerical simulations for Nd(lamda=1.06 mkm) and KrF- (lamda=0.27 mkm) lasers. Variants 8-20: gas-filled shell targets, variants 8-22: cryogenic two-cascade targets.

Table 2 shows the results of the simulations:

N var	lamda mkm	M0 mg	R0 cm	M1 mkg	R1 mkm	M DT mkg	t ns	ro g/cm**3	Y 10**16	G
822c	1.06	36.6	0.5005	–	–	293	106	2.4	1.3	0.02
820d	0.27	54.0	0.5009	–	–	280	102.5	6.0	2.2	0.03
822s	1.06	36.6	0.5006	24.5	45	128	106	20.0	3.5	0.06
822v	0.27	54.4	0.5009	24.5	45	126	103.5	14.	20.	0.3

CONCLUSIONS.

1) One can achieve $G \sim 0.1-0.2$, when the laser energy is 2-3 Mj and pulse duration is about 100 ns. One needs a shortwavelength laser for this (a KrF-laser, for example) and the high-aspect-ration target (AS $\sim$ 1000 !!!). 2). The problem of a stable compression is extremely complicated for sich type of the targets. Note only that in order to obtain $G \sim 0.1$ one should not require high compression. It is sufficient to compress a DT-ice ball by 50-100 times (density $\sim$ 10 -20 g/cm**3), and heat it up to 3-5 keV

REFERENCES.

1. Feoktistov L.P., Avrorin E.N. et al. ,*Kvantovaja elektronika*, *1978*, *5*, p.349

2. Basov N.G., Shejndlin A.E. et al. *Izvestija Akademii nauk SSSR,, Energetika i transport,* 1979,N2, 3

3. Feoktistov L.P. Matematicheskoe modelirovanie, 1995,.**7**,N3,p.41,

4. Afanas'ev Yu.V., Volosevich P.P. et al. in *Trudi FIAN*,Moscow,Nauka, 1982, **134**, pp.167-187 (in Russ.)

5. Henderson D.B. *Preprint LA-UR-77-1442*, 1977

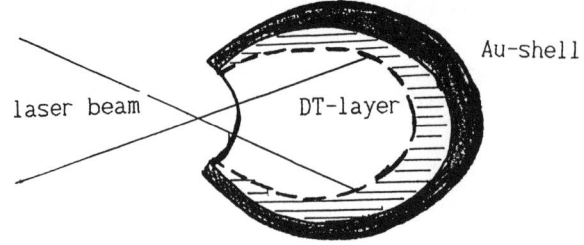

Fig. 1

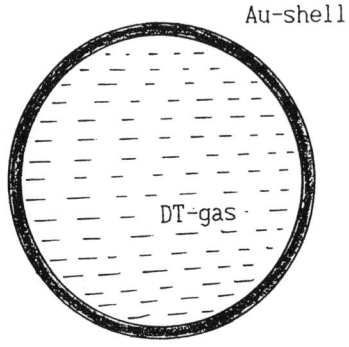

Fig. 2a

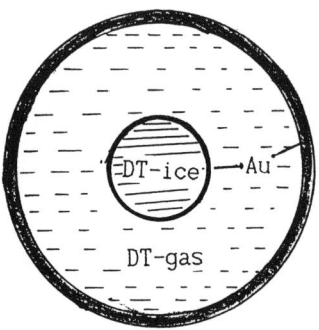

Fig. 2b

Fast Ignitor Concept. Numerical Simulation.

Alexander Pukhov and Jürgen Meyer-ter-Vehn

Max-Planck-Institut für Quantenoptik, 85748 Garching, Germany

Abstract. Two-dimensional (2d3v) PIC simulations related to fast ignition of compressed ICF cores by ultra-intense short laser pulses are presented and interpreted by analytic estimates. Channel boring and forward-peaked relativistic electron generation are considered.

INTRODUCTION

The Fast Ignitor Concept proposed recently by Tabak et al. [1] may drastically lower the ignition requirements for inertial confinement fusion (ICF). The idea is to use target implosion only for fuel compression and to leave ignition to a short and very intense laser pulse (10 ps, 10^{20} W/cm^2) focused on the compressed core. Corresponding short-pulse lasers will become operational soon [1]. The goal is to heat a small portion of the high-density core (typically 1000 times solid density) to ignition temperatures above 5 keV isochorically. The heating is to occur by relativistic electrons on a timescale short enough that hydrodynamic expansion is negligible. Such isochoric hot spot ignition requires considerably less energy than the standard ICF scheme with a hot spot of relatively low density formed isobarically in the center of the core [2].

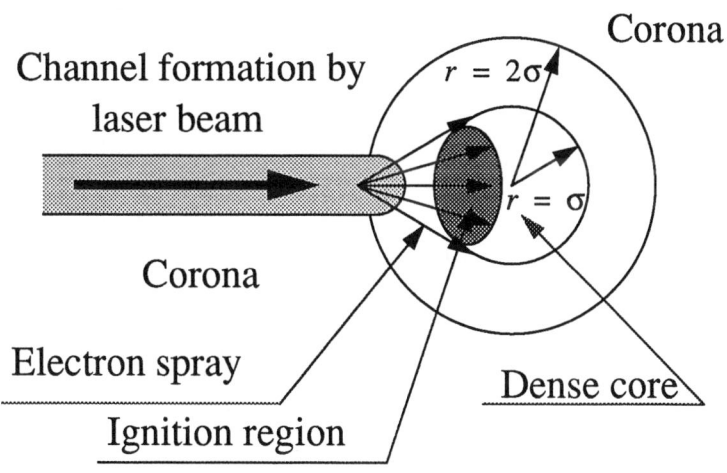

FIGURE 1. Sketch of Fast Ignitor concept.

© 1996 American Institute of Physics

The Fast Ignitor concept, though attractive, is still highly speculative and needs a lot of work to verify the physics elements involved. The imploded core is surrounded by a plasma cloud which is overcritical for the laser light in the inner region, and the laser pulse has first to bore a channel into this plasma cloud through which the light can propagate and come close to the core. The geometry is illustrated in Fig. 1. The effect of channel boring into overcritical plasma has been simulated in the present work (see Figs. 2-4); also analytic estimates for the speed of the channel front are derived below. The fast boring is mainly due to light pressure which is in the order of tens of Gigabars at these intensities, considerably larger than the thermal plasma pressure.

Another critical element is the heating of the fuel core by suprathermal electrons as soon as the light pulse approaches the core. It turns out that laser pulses at intensities of 10^{20} W/cm^2 produce relativistic electron beams sharply peaked in the forward direction. Results on the generation of such *electron sprays* are presented in Figs. 5 and 6. Both channel boring and forward peaked electron emission rely on relativistic physics.

PIC SIMULATIONS WITH A 2D3V CODE

We have performed numerical simulations for the Fast Ignitor concept using the fully relativistic electromagnetic PIC code LPlas2d3v. The main parameters of the code are briefly described in the Appendix.

The simulation geometry is presented in Fig. 2. The laser beam of intensity $I \sim 10^{20}$ W/cm^2 (dimensionless amplitude $a = eE/m\omega c = 5$) is incident from the left boundary on the plasma slab with density rising from 0 to $n_{max} = 10\ n_{cr}$

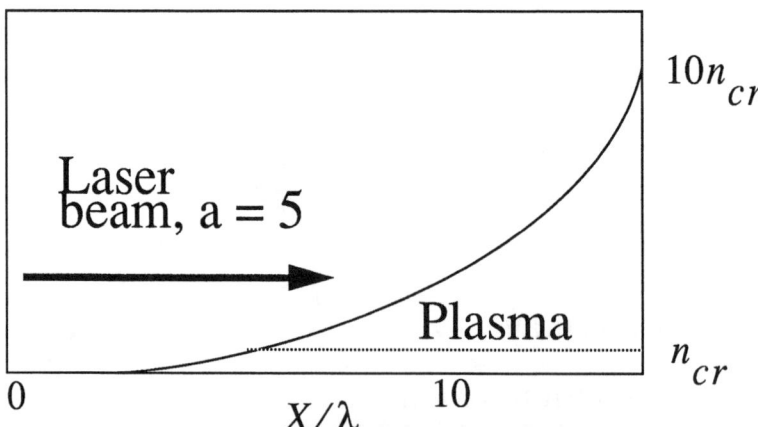

FIGURE 2. Numerical simulation configuration. The laser beam is incident on a plasma slab from the left.

over a distance $\Delta X = 15 \lambda$. The density profile is quadratic, so that the plasma frequency changes linearly. This geometry is close to the real situation with the critical surface inside an ablating plasma corona.

The region of underdense plasma is of high importance for the creation of a well collimated electron spray, because the electrons need some distance to be accelerated to relativistic velocities either by the electromagnetic wave or by another mechanism. The length for direct acceleration of an electron by the electromagnetic wave can be estimated as $l = \lambda\, a^4$.

The results of our PIC simulations are presented in Figs. 3-6. In Figs. 3-4 the electron density distribution is shown at two different times. One can see that a channel is really being bored with a front moving toward the right boundary.

The speed of hole boring can be estimated analytically by balancing the momentum flux of the ion mass flow with the light pressure, neglecting the thermal electron pressure [3]. This readily gives

$$u \approx ac \sqrt{\frac{n_{cr} m_e}{2 n_i M_i}}, \qquad (1)$$

where $n_{cr} = \omega^2 m_e / 4\pi e^2$ is the electron density at the critical surface, $\omega = 2\pi c/\lambda$ is the laser frequency, and n_i is the local ion density.

Having expression (1), it is interesting to estimate the time needed to bore the hole in a real ICF target at the ignition stage. The plasma density profile of the compressed target can be approximated as

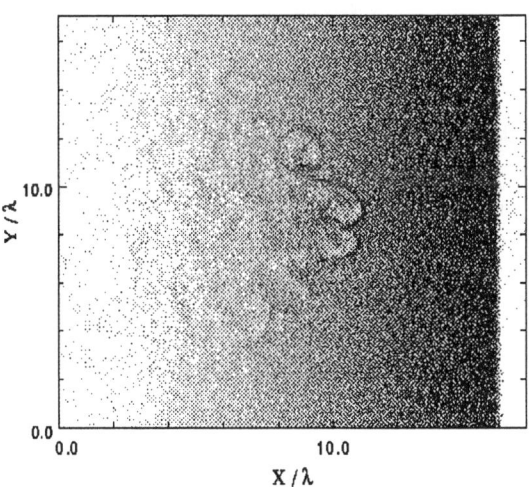

FIGURE 3. Electron density distribution after 20 laser periods.

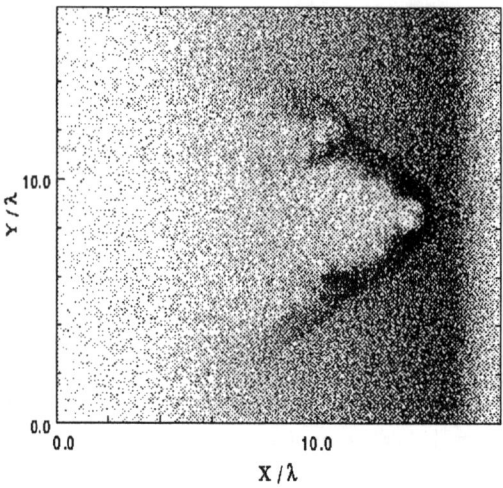

FIGURE 4. Electron density distribution after 70 laser periods.

$$n(r) = n(0) \exp(-(r/\sigma)^2), \qquad (2)$$

where $n(0)/n_{cr} = 10^4 - 10^5$ and $\sigma = 30\ \mu m - 100\ \mu m$. From (1) and (2) we obtain the time τ at which the front reaches radius r:

$$c\tau = \frac{\sigma}{a}\sqrt{\frac{\pi M_i n(0)}{m_e n_{cr}}}\left(erf\left(\frac{r_{cr}}{\sqrt{2}\sigma}\right) - erf\left(\frac{r}{\sqrt{2}\sigma}\right)\right), \qquad (3)$$

here, $erf(x)$ is the error function. Taking $n(0) = 10^5\ n_{cr}$, $M_i = 3600\ m_e$, $\sigma = 50$ μm, $I = 10^{20}$ W/cm^2, we find that it takes about 200 ps to reach the core at $r = \sigma$ It needs only 20 ps to arrive at $r = 2\sigma$, much shorter due to the exponential density slope. Pulse energies of 1 - 10 kJ are required to bore such channels. The lasers capable to produce such intense long pulses have still to be developed.

GENERATION OF RELATIVISTIC ELECTRON SPRAY

The actual ignition of the compressed core is to be accomplished by relativistic electrons generated in the underdense plasma of the laser channel close to its front. A crucial point for the Fast Ignitor concept is that these electrons are emitted sharply peaked into forward direction (see Fig.1).

There are several mechanisms of such electron acceleration. First, it is acceleration of electrons by the superintense electromagnetic wave. This mechanism

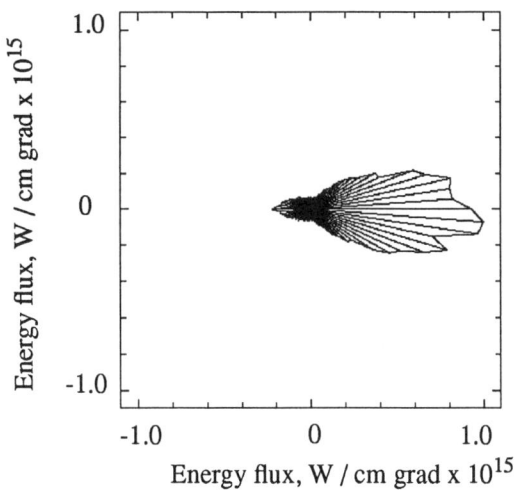

FIGURE 5. "Antenna"-diagramm of electron energy flux.

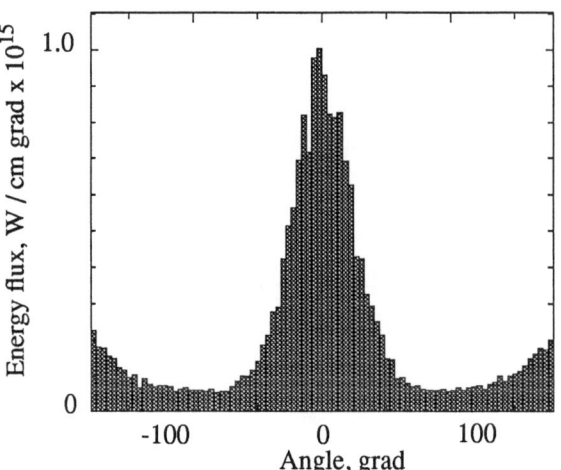

FIGURE 6. Angular distribution of energy flux in the form of histogram.

can produce the longitudinal momenta of electrons up to $p = a^2 mc/2$ in vacuum for ultrarelativistic laser beams ($a \gg 1$). Another mechanism of electron acceleration is acceleration in the wake field of the laser pulse [4]. There are also collective effects which can focus or defocus the spray of suprathermal electrons.

Magnetic field generation leads to pinching of the electron flow [5]. Whereas electrostatic interaction between electrons tends to defocus the beam. An advantage of the PIC simulations is that one has a kinetic description of the plasma and the distribution function of the suprathermal electrons.

The angular distribution of the suprathermal electron flow obtained in our PIC simulations just behind the critical surface is given in Figs. 5-6. Fig. 5 shows the "antenna"-diagram of the energy flux transported by suprathermal electrons, and Fig. 6 has been plotted in the form of a histogram. The angular anisotropy of the electron flow and the formation of a collimated electron spray in the direction toward the target core is readily seen from these pictures. The angular divergence of the electron spray is about $\delta\theta \sim 30°$.

Energies of the suprathermal electrons injected in the forward direction are in range from 1 to 10 MeV.

A more detailed study is now in progress.

APPENDIX

The code LPlas2d3v is a fully relativistic electromagnetic (EM) PIC code. All quantities depend only on two space coordinates (X,Y), but the velocities of the particles as well as the electromagnetic fields have 3 components. The code uses the Fourier-spectral representation for the EM fields. This gives the correct dispersion relation for the EM fields and removes unphysical numerical Cherenkov radiation by relativistic particles that often plagues finite-difference codes. The present PIC simulation used a 512 x 512 spatial grid and 10^6 macro particles both for electrons and ions.

REFERENCES

1. M. Tabak et. al. Physics of Plasmas, Vol.1, p.1626 (1994).
2. J. Meyer-ter-Vehn, Nucl.Fusion, Vol.22, p.561 (1982).
3. S. C. Wilks, W. L. Kruer, M.Tabak, and A.B.Langdon, Phys. Rev. Lett., Vol.69, p.1383 (1992).
4. S. V. Bulanov, F. Pegoraro, and A. M. Pukhov, Phys.Rev.Lett., Vol.74, p.710 (1995).
5. G. A. Askar'yan, S. V. Bulanov, F. Pegoraro, and A. M. Pukhov, JETP Lett., Vol.60, p.251 (1994).

Thermal Smoothing and Hydrodynamical Compensation of the Nonuniformities of Laser Energy Deposition in a Direct-Driven Target

S. Gus'kov, V. Rozanov, I. Lebo, G. Vergunova
P. N. Lebedev Physics Institute of Russian Academy of Sciences, 117924 Moscow, Russia, Leninsky prospect, 53

V. Tishkin, N. Zmitrenko, T. Kozubskaya, I. Popov, V. Nikishin
Institute for Mathematical Modeling of Russian Academy of Sciences, 125047 Moscow, Russia, Miusskaya sq. 4.

Abstract. Design of the LIGHT target based on an electron heat conductivity smoothing and hydrodynamical compensation of the nonuniformities of energy deposition of a small quantity laser beams is discussed. 2D distribution of absorbed laser energy deposition in an undercritical low-Z absorber of LIGHT target under propagation of a super sonic electron heat conductivity waves and 2D implosion of target with ablator having special initially given distributions of mass are presented. Numerical simulations show the flexibility of LIGHT target ignition at the laser energy of (200 - 300) kJ and quantity of beams 6.

1. Introduction

The design of Laser Irradiated Green House Target (LIGHT target) is directed to smooth the nonuniformities of laser beam energy deposition in plasma by the super sonic electron heat conductivity waves and use spatial mass and / or density distributions of a thermonuclear capsule shell[1]. The general design of LIGHT target presents the two co-axial shells: external shell with holes for the laser beams input into the target and an internal shell, which contains a DT fuel. The space between these shells is filled by a low density foam material of light elements (for example, foam plastic) with density close to a critical density of laser-produced plasma. That part of target is an absorber of the laser beams. There are a "volume" absorption of laser radiation and super sonic electron heat conductivity waves propagation leading to thermal smoothing of laser energy deposition in an undercritical extensive plasma of that absorber. The shell of thermonuclear capsule (ablator) may to have a special distribution of mass and / or density for an additional hydrodynamical compensation of the laser energy deposition nonuniformities. Ablator includes the inner layer of heavy element material to absorb the thermal radiation from "hot" absorber.

The results of theoretical analysis and numerical simulations devoted to study the thermal smoothing and hydrodynamical correction of the nonuniformities of laser beam energy deposition and also the prediction of the gain of LIGHT target near a breakeven point are presented in this paper

2. Theoretical Model of a Super Sonic Thermal Smoothing

The criterion for smoothing of the absorber temperature distribution nonuniformities presents the condition of super sonic heat waves propagation a half of the distances between the absorption region of the neighboring laser beams during the time less than the laser pulse duration. That distance is

$$\Delta \simeq (4/k)^{1/2} (1 - \sigma^{1/2}) R,$$

where $\sigma = k R_L^2 / 4 R^2$ is a relative square of the introducing holes; k, a number of the holes, R and R_L, radii of the target and the hole, respectively.

The super sonic electron conductivity wave is an initial stage of a heat wave of the duration is approximately equal to the time of the electron-ion energy exchange:

$$t_s \simeq 310^{-12} \left(\frac{A}{z}\right)^2 \frac{T_a^{3/2}}{\rho_a}, s \qquad (1)$$

where: absorber temperature T_a in keV and density ρ_a in g/cm³, A and z are atomic number and ionization degree of the absorber atoms.

From a self-similar solution for radius and temperature of spherical electron conductivity wave in absorber:

$$R_{\varkappa} = \left[\frac{19}{7} \left(\frac{3}{2\pi}\right)^{5/2} \frac{\varkappa_a (I S_L)^{5/2}}{(B_a \rho_a)^{7/2}}\right]^{2/19} t^{7/19} \qquad (2)$$

$$T_{\varkappa} = \left[\frac{7}{19} \left(\frac{3}{2\pi}\right)^{2/3} \frac{(I S_L)^{2/3} (B_a \rho_a)^{1/3}}{\varkappa_a}\right]^{6/19} t^{-2/19} \qquad (3)$$

the time of the temperature distribution smoothing is :

$$t_{\varkappa} \simeq 7,410^3 \left[\frac{z \rho_a}{(\gamma-1)A}\right] \frac{(1-\sigma^{1/2})^{19/7}(z+4)^{2/7} R^{9/7}}{k^{9/14} I^{5/7} \sigma^{5/7}}, s \qquad (4)$$

220

In the expressions (2)-(4): $æ_a \simeq 10^{19} / (z + 4)$, erg / keV$^{7/2}$ cm s; $B_a = 10^{15} (z + 1) / A(\gamma-1)$, erg/ g keV; laser beam spot S_L in cm^2; laser intensity I in W / cm^2; R in cm; γ is adiabatic constant.

By using (2)-(4) the "smoothing" criterion may be written as

$$R < 3 \cdot 10^{-2} \left[\frac{(\gamma - 1) A}{z \rho_a}\right]^{7/23} \frac{E_L^{7/23} k^{9/46}}{(z + 4)^{2/23}(1 - \sigma^{1/2})^{19/23} (I \sigma)^{2/23}}, cm$$

here E_L is the energy of all the laser beams in J.

For $\rho_a \simeq (3 \cdot 10^{-2} - 3 \cdot 10^{-3})$ g / cm^3, $I \simeq 10^{15}$ W /cm^3, $\lambda \simeq 0.35$ μm, $E_L = 10^5$ J, $k = 6$ and $\sigma = 0.1$ the "smoothing" time $t_æ$, according formula (4) is about 0.3 ns and this is much less than the needed laser pulse duration of 3 ns. On other hand, the absorber temperature, according formula (3), is about of (1 - 3) kev and from formula (1) duration of the super sonic electron conductivity heat wave t_s is (0.5 - 8) ns, and this exceeds the "smoothing" time and close to the laser pulse duration. The upper limit for the LIGHT target radius is 0.18 cm Thus, there are wide possibilities to optimize LIGHT target parameters, since the radius of a thermonuclear capsule lies in the interval of (0.07 - 0.1) cm.

So analysis shows the effective smoothing of laser energy deposition in LIGHT target by the super sonic waves of electron heat conductivity before hydrodynamic perturbations arise.

Main sources of a thermal radiation of the LIGHT target are the ablator of a thermonuclear capsule (that is main source of the thermal radiation for usual direct driven target, also) and the inner layer of the external shell heated by electron conductivity wave from absorber.

Energy of a thermal radiation of LIGHT target absorber is less than (1-2)% of the absorbed laser energy. This low value is due to the low density of the absorber. But a radiation spectrum is enough wide because of a high temperature of this part of the target.

Using the self-similar solution (2)-(3), and considering an expansion of the heated layers in the adiabatic approximation, the average value of a temperature and density of the heated layers of the external shell may be estimated in following form:

$$T_s \sim \left[\frac{\rho_a B_a}{\rho_{so} B_s}\right]^{(\gamma-1)/\gamma} \beta^{2/9} T_æ, \quad \rho_s \sim \left[\frac{\rho_a B_a}{\rho_{so} B_s}\right]^{1/\gamma} \rho_{so} \qquad (5)$$

here $R_æ$ and $T_æ$ are given by expression (2) and (3); index s corresponds to shell matters; $\beta = 4 \, æ_a B_a \rho_a / æ_s B_s \rho_s$.

For $\rho_a \simeq 10^{-2}$ g / cm^3, $\rho_{so} = 1$ g/cm^3, $R = 0.15$ cm, $I \simeq 10^{15}$ W /cm^3,

τ_L = 3 ns, $\lambda \simeq$ 0.35 μm, E_L = 10^5 J formulas (5) give for the thickness of heated layer the value of (8 -12) μm and for the temperature and density of the producing plasma - the values of (150 - 200) eV and (0.05 - 0.1) g/cm^3, respectively. The estimate the thermal radiation energy of such a plasma is (5 - 10)% of the absorbed laser energy. This value closes to level of the own radiation of the ablator of a usual direct driven target.

Spectrum of the heated layer radiation is considerably narrow, than the spectrum of the absorber radiation. To protect the compressed thermonuclear capsule from heating by high energy quantums of the absorber radiation, a layer of heavy matter in the shell of thermonuclear capsule may be used.

3. Numerical Simulations of LIGHT Targets

2D numerical simulations of LIGHT target were carried out by means of code "ATLANT"[2]. Irradiation of the target by 6 laser beams was simulated by the following dependence of laser energy flux on angle:

$$q(\theta,t) = q_o [0.51 + P_4(\cos\theta)] / 0.51$$

here $P_4(\cos\theta)$ is the fourth Legandre function. This formula describes the action of two laser beams which propagate through two holes in the external shell poles and action of another four beams for which the energy flux of radiation uniform distributed along target equator.

For hydrodynamical correction of the nonuniformities of laser energy deposition the distribution of the thermonuclear capsule mass must be closed to distribution of the laser intensity. For 6 laser beams thickness of the ablator is given by following expression

$$r_c = r (1 + \alpha \Delta P_4(\cos\theta) / r \exp[2 |R - R_{b2}| / \Delta]$$

here: α is a parameter, $\Delta = R_{b2} - R_{b1}$ is the ablator thickness, R_b are the boundary radii, r_c and r are the current ablator radii, with and without correction, respectively.

2D numerical simulations were carried out for following laser-target system: laser input energy 100 kJ, radiation wavelength 1.06 μm, pulse duration 2 ns, 6 beams; thermonuclear capsule radius 910 μm, CH-ablator thickness 10 μm, DT-ice layer thickness 10 μm, CH-absorber thickness 590 μm, absorber density 1 μg/cm^3. 2D distribution of the absorber electron temperature and shape of the ablator for moment close to the end of laser pulse in the case of uniform initial distribution of the ablator mass are presented in the Figure. These data demonstrate the high electron temperature of the absorber plasma of (2 - 4) keV and the effective thermal smoothing by electron heat conductivity.

The set of 2D calculations was carried out for the targets with different degree of the ablator mass modulation (different α-parameter). The LIGHT targets with mass modulation demonstrate the considerable lower growth of the ablator thickness perturbation degree $\delta = (R_{b2} - R_{b1}) / (R_{b2} + R_{b1})$ with growth of the convergence ratio $\zeta = R_o / R(t)$ in comparison with usual target, which is same as the LIGHT target thermonuclear capsule without absorber and external shell. For example, in the moment when $\zeta = 7$ for usual target the perturbation degree is close to 1 but for LIGHT targets $\delta \simeq 0.2$ for $\alpha = 0.55$ and $\delta \simeq 0.12$ for $\alpha = 0.8$. Ratio of the 2D and 1D neutron yields for LIGHT target with mass modulation is closed to 0.1.

The detail numerical simulations of the radiation transport in the LIGHT target by code RADIANT[3] in multy-group model shows enough low energy of the thermal radiation in LIGHT target in range of (3 - 7)% from absorbed laser energy. On base that simulations the thickness of the heavy elements protecting layer of LIGHT target ablator was found. For the targets corresponding to laser energy of (100 - 200) kJ, for example, the cooper layers with a thickness (0.5 - 1μm) are suitable.

1D numerical simulations by code DIANA[4] show very high energy characteristics of LIGHT target. The simulations of the cryogenic

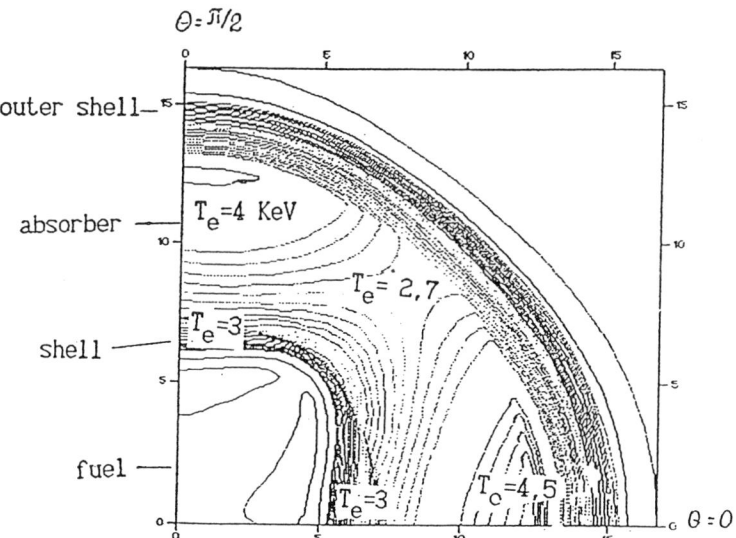

Figure . Distribution of the absorber electron temperature and shape of the LIGHT target ablator for t= 1.95 ns.

targets with the radius from 0.15 up to 0.3 cm and the thickness of the "smoothing" absorber from 0.03 up to 0.08 cm at the laser energy of $(0.1 - 1)$ MJ for $\lambda = (1.06 - 0.35)$ μm and the pulse duration and the pulse duration $(2 - 8)$ ns give the following results: the ablation pressure, $(20 - 40)$ Mbar; the implosion velocity, $(3 - 5)$ 10^7 cm / s; the hydrodynamics efficiency, $(0.12 - 0.2)$. The thermonuclear gain of 2 - 4 was obtained for the laser energy of 0.1 - 0.2 MJ and one of 50 - for the laser energy 1MJ.

4. Conclusion

Presented theoretical results reveals the effectiveness of super sonic electron heat conductivity and hydrodynamics compensation due to special ablator mass distribution to smooth the energy deposition of relatively small number laser beams in direct driven ICF target. That fact and possibility of high gain makes the further research of Laser Irradiated Green House Target interesting and important.

References

1. Gus'kov, S. et al, *Book of Abstracts of 23 rd European Conf. on Laser Interaction with Matter* (Oxford UK), 1994, PA-17.
2. Lebo, I. et al, *J. Russian Laser Research* 15, 136-143 (1994).
3. Vergunova, G. et al, *Book of Abstracts of 23 rd European Conf. on Laser Interaction with Matter* (Oxford UK), 1994, PB-3.
4. Samarskii, A. et al, *Problems of Nuclear Science and Engineering* 2, 38-42 (1983).

Kinetic effects on the electron thermal transport in ignition target design

M.Honda, A.Nishiguchi*, K.Mima, H.Takabe, H.Azechi, S.Nakai

Institute of Laser Engineering, Osaka University, Suita, Osaka, 565, Japan
*Osaka Institute of Technology, Asahi-ku, Osaka, 535, Japan

Abstract. The preheating is one of the most critical issues in laser fusion, because it causes significant reduction of volume compression. The nonlocal heat transport by the high energy tail electrons in ablative plasmas is found to be essential for the preheating under high intensity laser irradiation. In such a situation, electron heat transport is described by the Fokker-Planck (FP) equation in a fluid implosion code, since the Spitzer-Härm (SH) thermal conduction model is not applicable. The numerical simulations of the implosion have been carried out for the fast (high entropy) implosion mode in which the implosion velocity is as high as 6×10^7 cm/sec in order to reduce the required laser energy for ignition. The control of preheating is essentially important for this type of implosion mode. The isentrope of an imploding shell is evaluated to see the preheating level. It is found in the fast implosion mode that the isentrope in the FP simulation code is higher by 2 to 4 times than that in the flux limited SH simulation.

INTRODUCTION

The preheating due to electron heat transport is one of the most crucial issues in laser fusion, because the preheating effects disturb the efficient adiabatic compression in low entropy implosion mode [1-2]. In this paper, we will concentrate on the quantitative estimation of the preheating by nonlocal heat transport of suprathemal electron, but not resonance hot electron caused by laser plasma interaction. This kind of preheating is mainly dominant in fluid acceleration phase with low Z ablative plasma under the high intensity laser irradiation. It is well known in such situation that SH classical diffusive approximation theory [3] for electron heat transport fails, so FP equation should be numerically solved.

The coulomb collision FP simulation for nonlocal electron heat transport had been extensively performed in many previous works [4-9]. These simulations, however, almost had been limited to be applied to the steady ablative corona plasma. For the exact treatment of the preheating due to suprathermal electron heat transport, the FP equation must be numerically solved with hydrodynamic motions. We could be utilize of the one dimensional FP implosion code (HIMICO), coupled with radiative transfer, atomic processes, nuclear reaction and alpha particle transport, and so on. We could also survey the kinetic effects for comparatively

© 1996 American Institute of Physics

short time of numerical calculation. The formulation of FP equation is introduced in Ref. [10].

KINETIC EFFECTS ON THE ELECTRON THERMAL TRANSPORT

First of all, we define the key parameter, that is, isentrope parameter α which is evaluated to see the preheating level. We will consistently use this parameter to present the quantities of preheating in following description. It can be described as,

$$\alpha = (P/P_0) \cdot (\rho_0/\rho)^{\gamma}. \tag{1}$$

where, P and ρ express the total pressure and the mass density of fluid respectively, and suffix "0" means initial value. ρ_0 (= 0.2 g/cm^3 at frozen DT) is initial mass density, and γ (= 5/3 in ideal gas) is the ratio of specific heat. As regards P_0, we assumed the Fermi pressure P_f by electron degeneracy [11],

$$P_f = \frac{1}{20}\left(\frac{3}{\pi}\right)^{2/3} \frac{h^2}{m_e} n_e^{5/3}$$

$$= 3.3\rho(g/cc)^{5/3} \quad [Mbar]. \tag{2}$$

It should be noticed that if we could ideally perform above equation of state (EOS) for zero temperature Fermi pressure around the maximum compression, isentrope becomes unity.

In the case of central ignition, laser energy E_L required to perform ignition and burning is dependent on the third power of isentrope as following,

$$E_L \propto \alpha^3 \eta^{-1} v_{imp}^{-6 \sim -10}. \tag{3}$$

where, α is isentrope, η is coupling efficiency, and v_{imp} is implosion velocity. According to this formula, laser energy is sensible to isentrope parameter, so the control of preheating is essentially significant. Previously the many simulations coupled with SH heat transport have shown it to be 1~3. Coupling efficiency η approximately equals to hydrodynamic efficiency η_H. High gain strategy desire to the lower isentrope and the higher hydrodynamic efficiency [12]. On the other hand, we will remark that the required laser energy is inversely proportional to the maximum tenth power of the implosion velocity with fixed α and η.

Absorbed laser energy is mainly carried to the ablation front by electron thermal transport. The ablative corona plasma composed of polystyrene CH (Z*=3.5) are heated up to 5~10keV by electron-electron elastic collision under the irradiation of intense (~5×10^{15}W/cm^2) laser light. The scaling low of electron temperature at critical surface is given by Manheimer, et al [13], as following,

$$T_{ec} = 5.2(\lambda\ (\mu m)/0.53)^{4/3}(I\ (W/cm^2)/10^{15})^{2/3}\ [keV]. \quad (4)$$

where, λ and I are laser wave length and intensity, respectively. According to this simple estimation for steady state planer ablative flow, the ratio of the thermal electron mean free path to the stand off distance which characterizes decoupling length between ablative corona and cold dense core plasma equals to 0.02~0.05. If peak laser intensity reaches to 5×10^{15}W/cm^2, electron could be heated up to ~6keV around critical density regime. In this case, electron heat transport should be described by FP equation, since SH theory is not applicable.

Suprathermal electrons whose energy is 9 times as much as thermal energy, mainly carry the absorbed laser energy to ablation front, as we can carefully see the formula for the electron heat flux (q);

$$q \propto \frac{1}{Zn_e}\int_0^\infty v^8\left(v\frac{\partial f}{\partial r} + \frac{eE_r}{m_e}\frac{\partial f}{\partial v}\right)dv . \quad (5)$$

where, Z is charge number, n_e is electron density, v is electron velocity, and E_r is electric field for radial direction. The mean free path of the energy carrier electron often exceeds ~100μm. The energy spectrum of electron has bi-Maxwellian around the ablation front. It is found that nonlocal electron heat transport becomes remarkable, and high energy tail electrons preheats the cold fuel.

TARGET AND LASER PARAMETERS FOR NUMERICAL SIMULATION

Now, we shows the optimized target and laser parameter for the full simulation of the implosion hydrodynamics, and briefly explain the reason why those simulation parameter are selected.

Recently, the concept of fast laser implosion mode which positively makes use of a kind of preheating, becomes very attractive as one of the alternative methods in laser fusion [14]. The typical target design and corresponding laser conditions in this implosion mode are shown in Fig.1(a). In this implosion mode, the relatively thick frozen DT shell covered with very thin plastic (polystyrene) CH layer is far accelerated up to $5\sim8\times10^7$ cm/sec by high ablation pressure exceeding 100Mbar, and absorbed laser intensity is required of $10^{15}\sim10^{16}$W/cm^2 at main pulse peaking. Our recent high entropy target design has an advantage over the other hydrodynamic (without additional heating) implosion modes. For instance, this implosion mode closely becomes the stagnation free-mode; the core structure becomes the isochoric-like around the maximum compression, since the fuel is heated up to the ignition temperature due to high implosion velocity, and the fluid instabilities on deceleration phase is likely to be eliminated. According to (3), laser energy required to perform ignition and burning may be also remarkably reduced by high implosion velocity.

Whereas, in order to suppress the Rayleigh-Taylor fluid instability, we might expect the ablative stabilization by the useful of DT ablator in fast implosion mode. And we must keep a low inflight aspect ratio by the positive use of an adequately controlled preheating, so fast implosion mode requires essentially high entropy as a trade off for the high ablation pressure.

We will apply the kinetic effects to this implosion mode, in order to attain the more exact estimation of preheating.

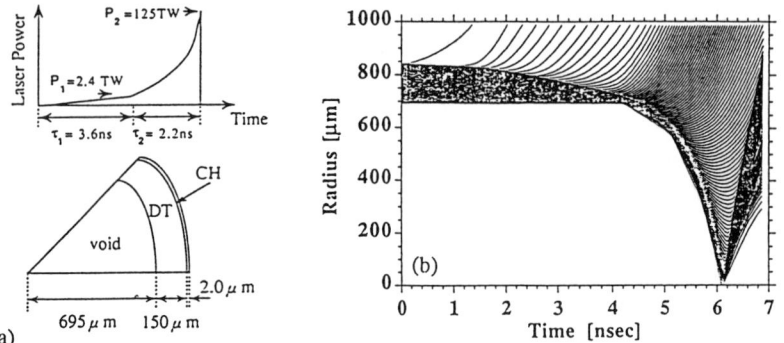

FIGURE 1. (a) Typical target parameter and tailored pulse shape for fast (high entropy) implosion mode. (b) Flow diagram for that simulation parameter.

SIMULATION RESULTS

Now, our interests are concentrated on the problems due to suprathermal electron preheating in fast implosion mode. The typical flow diagram for fast laser implosion mode is shown in Fig.1(b), where, the simulation parameters are the same as shown in Fig 1(a). As previously we said, according to the irradiation of intense laser light (4×10^{15} W/cm^2), the target shell is accelerated up to 5.5×10^7 cm/sec, and electron temperature at critical surface and ablation pressure reach to 3keV and 200Mbar, respectively. The isentrope and density profiles at in-flight phase (t=5.6nsec), when the shell is converted to half of initial radius, for the same simulation conditions are shown in Fig.2(a), and the corresponding simulation results by classical SH heat transport are also shown in Fig.2(b). In the SH simulation, we employ the flux limiting factor (f) of 0.1 in this formula,

$$q = \min(q_{SH}, f q_{free}) \tag{6}$$

where, q_{free} means the free streaming value for electron heat flux given by $q_{free} = n_e v_{th} T_e$. The mass averaged isentrope over the part of main fuel are 4 (FP) and 2 (SH) respectively, and we can also find the shell swellings in the FP case obviously. Isentrope is very high around the inner radius side, because high entropy hot spark has to be formed in this part for the central ignition mode. It is shown in Fig.3(a) that the mass averaged isentrope is dependent on time evolution. Since main pulse power from 3.6 nsec to 5.0 nsec is in proportional to the square of the time and target shell is converted with time evolution, laser intensity becomes very high. Hence, kinetic effect is remarkable in such acceleration phase. It should be noticed that the isentrope in the FP simulation is higher by 2 to 4 times than that in the SH simulation, and although no hot electrons due to resonance absorption exist, the nonlocal heat transport due to suprathermal electrons are enough to cause the preheating. If hot electrons induced by laser-plasma interaction and radiation induced by high Z materials exist, the more preheating may be allowed.

At last, in Fig.3(b), we presents an interesting results: so called "target size effects". This effects means that accompanied with the increasing of the thickness of the target shell, the plasma stopping of suprathermal electrons increases. As a result of this effects, the preheating becomes not effective as the laser energy increases. We also introduce the peak inflight aspect ratio in order to clear the shell swelling. We obviously found an expansion of target shell by an increasing of the preheating due to nonlocal electron heat transport. According to this target size effects, the peak inflight aspect ratio is certainly suppressed under 100, so high entropy implosion mode may be expected to be very stable. The results also shows that the preheating due to nonlocal effects would not be significant in low entropy and high gain target design.

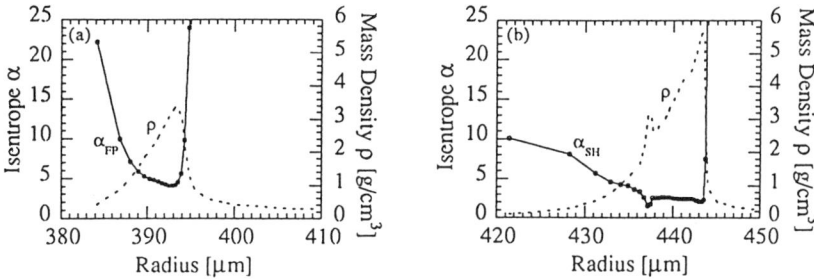

FIGURE 2.(a) The spatial profiles of isentrope parameter by the FP simulation, and **(b)** the SH simulation, respectively. Dotted lines show the background mass density.

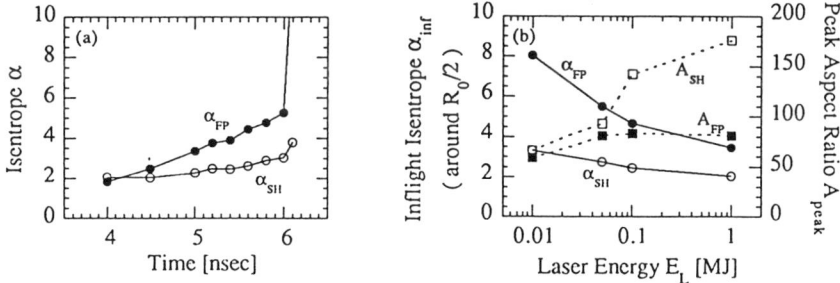

FIGURE 3.(a) The time evolution of the mass averaged isentrope until the laser irradiation is cut. Suffix "FP" and "SH" mean Fokker-Planck and Spitzer-Härm case respectively. **(b)** Laser driver energy versus inflight isentrope and peak inflight aspect ratio. The inflight isentrope means mass averaged one at time when the target shell is converged to the half of the initial radius.

CONCLUDING REMARKS

We have performed the simulation in order to investigate the preheating by suprathermal electrons, using Fokker-Planck (FP) code coupled with hydrodynamic phenomena. Particularly, we have chosen the fast (high entropy type) implosion target design, and applied consistently this implosion mode to the survey of preheating due to nonlocal electron heat transport, because the Spitzer-Härm (SH) theory of heat flow is not applicable in this implosion mode. We also introduced the isentrope parameter which quantitatively estimates the preheating.

A comparison of the preheating due to kinetic (FP) and classical (SH) heat transport for the same simulation parameter was made. An increasing of preheating due to nonlocal electron heat transport and the shell expansion were obviously indicated. We also cleared the "target size effects", which means that the isentrope parameter has relation to laser energy, so that the stopping power of suprathermal electron increase, accompanied with the increasing of target size. This effects tells us that suprathermal electron preheating would not be significant in low isentrope and high gain experiments by MJ class laser driver in future.

REFERENCES

[1] Kidder, R.E., Nucl. Fusion, **21**, 145 (1981).
[2] Meyer-ter-Vehn, J., Nucl. Fusion, **22**, 561 (1982).
[3] Spitzer, L., and Härm, R., Phys. Rev, **89**, 997 (1953).
[4] Bell, A.R., Evans, R.G., and Nicholas, D.J., Phys. Rev. Lett, **46**, 243 (1981).
[5] Matte, J.P., and Virmont, J., Phys. Rev. Lett, **49**, 1936 (1982).
[6] Albritton, J.R., Phys. Rev. Lett, **50**, 2078 (1983).
[7] Fuciani, J.F., Mora, P., and Virmont, J., Phys. Rev. Lett. **51**, 1664 (1983).
[8] Kho, H.T., Bond, D.J., Haines.M.G., Phys. Rev. A**28**, 156 (1983).
[9] Bell, A.R., Phys. Fluids, **28**, 2007 (1985).
[10] Nishiguchi, A., Mima, K., Azechi, H., Miyanaga, N., and Nakai, S., Phys.Fluids,B**4**, 417 (1992).
[11] Duderstadt, J.J., and Moses, G.A., Inertial Confinement Fusion (JOHN WILEY & SONS, 1982).
[12] Gardner, J.H., and Bodner, S.E., Phys. Fluid, **29**, 2672 (1986).
[13] Manheimer, W.M., Colombant, D.G., and Gardner, J.H., Phys. Fluids, **25**, 1644 (1982).
[14] Takabe, H., Hashimoto, T., Mima, K., and Nakai, S., in the present Proceedings, IAEA-TCM, Osaka, April (1991).

Modeling of Initial Imprinting Caused by Laser-Intensity Nonuniformities in Ablative Plasmas

Y.Shimuta and K.Nishihara

Osaka University, 2-6 Yamada-oka Suita Osaka, 565 Japan

Abstract. A theoretical model that studies smoothing of non-uniform ablation pressure is presented. The "cloudy day" model that is widely used to estimate the smoothing considers only thermal wave in homogeneous plasmas. However, sound wave caused by temperature perturbation and steep ablation structure are considered in presented model. Previous work of Manheimer et al. tries to include these. However, we newly 1) consider ablation structure with finite density ratio between sonic point and ablation surface; 2) solve eigen value problem to initialize the variables consistently with fluid equations in the shock compression region; 3) impose different boundary condition at the sonic point; and 4) consider higher mode. A nonlinear relation between the perturbed ablation pressure ratio at sonic point to ablation surface, and transverse wave number of perturbation is shown. As the wavenumber becomes larger, the smoothing becomes more efficient.

INTRODUCTION

In order to achieve laser fusion, there are several critical elements that one must control. The symmetry of the implosion is one of these. Asymmetry is caused by target roughness and non-uniformity of laser irradiance. The initial perturbations caused by laser nonuniformity are called initial imprint. These can cause shell break because the perturbations grow exponentially at acceleration phase due to Rayleigh Taylor Instability. It is thus important to estimate what degree of asymmetry is tolerable. For high gain implosion, the asymmetry of the ablation pressure must be less than a few percent; i.e., $\frac{\delta P}{P} \leq 1-3\%$ [1]. However, it does not mean that the laser absorbtion nununiformity must satisfy the condition $\frac{\delta I}{I} \leq 1-3\%$, because there is physical mechanism that smoothes out the perturbed pressure in the ablation layer. This smoothing mechanism has been experimentally observed [2,3]. In present paper, the perturbed ablation pressure at sonic point caused by nonuniform laser energy absorption is related to that of ablation surface. The "cloudy day" model has been used to estimate the smoothing effect in the ablation layer[1, 3, 4]. However the fluid motion in the ablation layer is not considered in the model as References [5] and [6] point out. Our model and References [5 ,6] consider the fluid motion in the ablation layer. Moreover some new methods, for example initialization of the variables by solving an eigen value problem, are included in present model.

Two dimensional steady-state fluid equations system is used to estimate the smoothing of nonuniform pressure in the ablation layer[5]. Then the basic equations are given by

$$\tilde{\nabla} \cdot \tilde{\rho}\tilde{v} = 0, \tag{1}$$

$$\tilde{\rho}\tilde{v} \cdot \tilde{\nabla}\tilde{v} = -\tilde{\nabla}\tilde{\rho}\tilde{T}, \tag{2}$$

$$\frac{3}{2}\tilde{\rho}\tilde{v} \cdot \tilde{\nabla}\tilde{T} = -\tilde{\rho}\tilde{T}\tilde{\nabla} \cdot \tilde{v} + \tilde{\nabla}\left(\tilde{T}^{\frac{5}{2}}\tilde{\nabla}\tilde{T}\right), \tag{3}$$

where $\tilde{\rho}$, $\tilde{v}$, $\tilde{T}$ are normalized density, velocity and temperature with respect to the values at isothermal sonic point of statinary solutions ρ_s, v_s and T_s. Independent variable x is normalized with the mean free path at the sonic point, $x_0 = \dfrac{K_{sp}A^{\frac{3}{2}}T_s^2}{\rho_s}$ where $A = \dfrac{M_i}{(Z+1)}$ for plasmas whose ion mass and charge are M_i and Ze, and K_{sp} is constant given by Spizer.

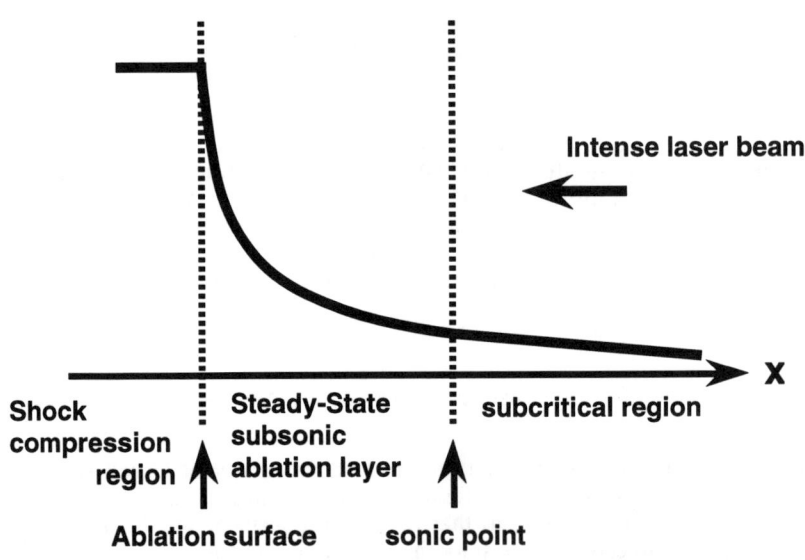

FIGURE 1. Laser driven ablative flow

ABLATION STRUCTURE

When a uniform intense laser illuminates a fuel target, one-dimensional ablative flow occurs(Fig. 1). The laser is on the right, flow is in the positive x direction. There exit three distinct regions in the laser driven ablative flow. Subcritical region which is apploximately described as isothermal rarefaction region is farthest to the right. Shock compression region is farthest to the left. There exists a steady-state subsonic ablation layer between them. The fluid flow in the steady state subsonic ablation layer[7] is given by

$$\tilde{\rho}_0 \tilde{v}_{x0} = 1, \tag{4}$$

$$\tilde{\rho}\tilde{v}_{x0}^2 + \tilde{\rho}_0 \tilde{T}_0 = 2, \tag{5}$$

$$\frac{5}{2}\tilde{\rho}_0 \tilde{v}_{x0} \tilde{T}_0 + \frac{1}{2}\tilde{\rho}_0 \tilde{v}_{x0}^2 \tilde{v}_{x0} - \tilde{T}_0^{\frac{5}{2}} \frac{d\tilde{T}_0}{d\tilde{x}} = 5\beta - 2\beta^2, \tag{6}$$

where β is the density ratio between the sonic point and ablation surface. Note that finite β allows ablation flow through the ablation surface and supresses unphysical behavior of the perturbation near the ablation surface. These unperturbed one dimensional solutions are used as the 0th order values in the following sections.

A MODEL FOR THE SMOOTHING OF NONUNIFORM PRESSURE IN THE ABLATION LAYER

In order to consider the fluid motion caused by the temperature perturbation at the sonic point, Eqs. (1)-(3) are linearlized in small perturbation to give

$$\frac{\partial \tilde{J}_1}{\partial \tilde{x}} + \tilde{k}_y \tilde{\rho}_0 i\tilde{v}_{y1} = 0, \tag{7}$$

$$\frac{\partial \tilde{P}_1}{\partial \tilde{x}} + \tilde{k}_y \tilde{\rho}_0 \tilde{v}_{x0} i\tilde{v}_{y1} = 0, \tag{8}$$

$$\tilde{\rho}_0 \tilde{v}_{x0} \frac{\partial i\tilde{v}_{y1}}{\partial \tilde{x}} + \tilde{k}_y \frac{-\tilde{T}_0 \tilde{P}_1 + 2\tilde{v}_{x0}\tilde{T}_0 \tilde{J}_1 + \tilde{\rho}_0 \tilde{v}_{x0}^2 \tilde{T}_1}{\tilde{T}_0 - \tilde{v}_{x0}^2} = 0, \tag{9}$$

$$\frac{d\tilde{E}_{x1}}{d\tilde{x}} + \frac{5}{2}\tilde{\rho}_0 \tilde{T}_0 \tilde{k}_y i\tilde{v}_{y1} + \frac{1}{2}\tilde{\rho}_0 \tilde{v}_{x0}^2 \tilde{k}_y i\tilde{v}_{y1} + \tilde{T}_0^{\frac{5}{2}} \tilde{k}_y^2 \tilde{T}_1 = 0, \tag{10}$$

where perturbed x component of mass flux $\tilde{J}_1$, xx component of momentum tensor $\tilde{P}_1$, and x component of total energy flux are given by

$$\tilde{J}_1 = \tilde{\rho}_0 \tilde{v}_{x1} + \tilde{\rho}_1 \tilde{v}_{x0}, \tag{11}$$

$$\tilde{P}_1 = 2\tilde{\rho}_0 \tilde{v}_{x0} \tilde{v}_{x1} + \tilde{v}_{x0}{}^2 \tilde{\rho}_1 + \tilde{\rho}_0 \tilde{T}_1 + \tilde{\rho}_1 \tilde{T}_0, \tag{12}$$

$$\tilde{E}_{x1} = \frac{-\tilde{v}_{x0}{}^3 \tilde{P}_1 + \left(\frac{5}{2}\tilde{T}_0{}^2 - \tilde{v}_{x0}{}^2 \tilde{T}_0 + \frac{1}{2}\tilde{v}_{x0}{}^4\right)\tilde{J}_1 + \tilde{\rho}_0 \tilde{v}_{x0}{}^3 \tilde{T}_1}{\tilde{T}_0 - \tilde{v}_{x0}{}^2}$$

$$+ \left(\frac{5}{2}\tilde{\rho}_0 \tilde{v}_{x0} - \frac{5}{2}\tilde{T}_0{}^{\frac{3}{2}} \frac{\partial \tilde{T}_0}{\partial \tilde{x}}\right) \tilde{T}_1 - \tilde{T}_0{}^{\frac{5}{2}} \frac{\partial \tilde{T}_1}{\partial \tilde{x}} \tag{13}$$

When the spatial derivertives of the 0th solutions is small enough, exponential profiles of the perturbed variables can be assumed. Then $\frac{\partial}{\partial \tilde{x}}$ in Eqs. (7)-(10), (13) are replaced by an eigen value λ. The eigen values are calculated in the shock compression region because the spatial derivertives of the 0th solutions are nearly equal to zero in this region so that the conditions that ensure exponential profiles are easily satisfied. Two positive eigen values are obtained for any transverse wave numbers. One is identified as a thermal mode because transvers themal conduction of the mode is always larger than that of the other. The other is sound mode because the energy flux associated with fluid motion always dominates the thermal conduction. These two modes are combined imposing the following boundary condititon at sonic point.

$$\left(\tilde{P}_{1s} + \tilde{P}_{1t}\right) - 2\left(\tilde{J}_{1s} + \tilde{J}_{1t}\right) - \left(\tilde{T}_{1s} + \tilde{T}_{1t}\right) = 0 \tag{14}$$

where subscript s indicates sound mode, and t thermal one. The boundary condition Eq. (14) is obtained to aboid that the spatial derivatives $\frac{\partial iv_{y1}}{\partial \tilde{x}}$ in Eq.(9) and $\frac{\partial T_1}{\partial \tilde{x}}$ in Eq. (13) diverge at the sonic point.

Imposing Eq. (14) at the sonic point, the mass and mommentum fluxes associated with the sound and thermal modes have opposite signs each other. The mass flux $\tilde{J}_{1s}$ has negative sign, $\tilde{J}_{1t}$ positive, when total perturbed temperature $\tilde{T}_{1\,total} = \tilde{T}_{1s} + \tilde{T}_{1t}$ is assumed to be positive at sonic point. The magnitude of the mass flux associated with the sound mode $\tilde{J}_{1s}$ is larger than that of thermal

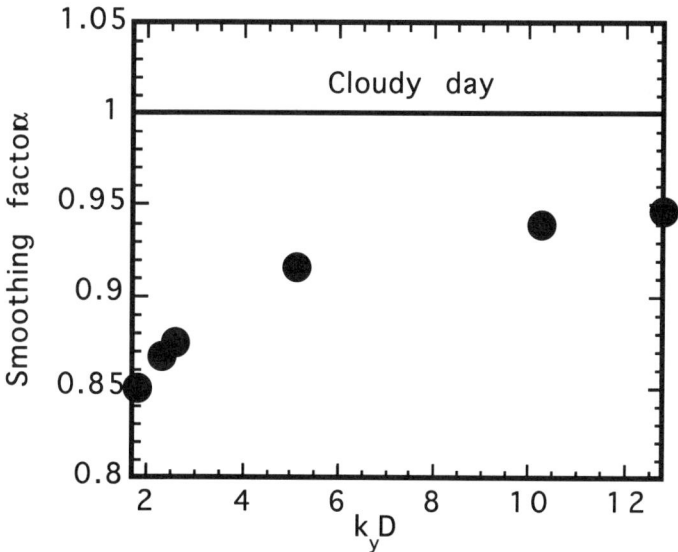

Figure 2. A plot of smoothing factor as function of $k_y D$

one $\tilde{J}_{1t}$ near the sonic point. The total pressure $\tilde{P}_{1total} = \tilde{P}_{1s} + \tilde{P}_{1t}$ has negative sign there. However $\tilde{P}_{1total}$ changes its sign where $\tilde{P}_{1t}$ becomes larger than $\tilde{P}_{1s}$ in the middle of the ablation layer. Then $\tilde{P}_{1total}$ become negative again just inside the shock compression region because the thermal mode does not penetrate the region.

Finally we estimate the smoothing factor α as a function of transverse wave number normalized with the distance from the ablation surface to the sonic point, $k_y D$. The smoothing factor α is defined as

$$\Gamma = \frac{\left(\tilde{P}_{1total} \text{ at the ablation surface}\right)}{\left(\tilde{P}_{1total} \text{ at sonic point}\right)} = \exp(-\alpha k_y D) \qquad (15)$$

The results are shown in Fig. 2. In the cloudy day model the smoothing factor is the constant of 1. The smoothing factor is somewhat smaller than that of the cloudy day model, especially for small wavenumbers.

CONCLUDING REMARK

A model that studies smoothing for a non-uniform ablation pressure is presented. The smoothing effect plays an important role in surpressing the initial imprinting. Fluid motion in the ablation layer must be considered to estimate the smoothing effect precisely. In present model, we have solved the eigen value problem to initialize the variables consistently with fluid equations; and considered the ablation structure with finite density ratio between the sonic point and the ablation surface. It is shown that the smoothing becomes more effecient as the transverse wave number of perturbations becomes larger.

ACKNOWLEDGMENT

Many useful discussions with Professors K. Mima, and Dr. K.A.Tanaka are acknowledged.

REFERENCE

1) Gardner, J. H. and Bodner, S. E., *Phys. Rev. Lett.* **47**, 1137, (1981).
2) Obenschain, S. P., Grun, J., Ripin, B. H., and McLean, E. A., *Phys. Rev. Lett.* **46**, 1402 (1981).
3) Weber, S. V., Kilkenny, J. D., Gendinning, S. G., Powell, H. T., Bell, P. M., Wallance, R.J., Dixit, S. N., Knauer, J. P., Henesian, M. A., ICF Quart. Report. LLNL 2 (1992).
4) Emery, M. H., Gardner, J. H., Lehmberg, R. H., and Obenschain, S. P., *Phys. Fluids.* **B3**, 2640 (1991).
5) Manheimer, W.M., Colombant, D. G., and Gardner, J. H., *Phys Fluids,* **25**, 1644(1982).
6) Meyer, B., Saleres, A., and Decoster, A. "Effect of Non-Uniform Illumination in Laser Driven Planar Target," *in Proceedings of the Conference on 7th Laser Interaction and Related Plasma Phenomena*, 1985, PP 495.
7) Takabe, H., Nishihara, K., Taniuti, T., *J. Phys. Soc. Jpn.* **45**, 2001, (1978).

Ablation Effects in Weakly Nonlinear Stage of the Ablative Rayleigh-Taylor Instability

Susumu Hasegawa and Katsunobu Nishihara

Institute of Laser Engineering, Osaka University, Suita, Osaka 565, Japan

Abstract. Weakly nonlinear stage of the ablative Rayleigh-Taylor instability has been studied by perturbation theory. Mode coupling of linear growing waves with wavenumbers k_A and k_B drives new excited waves with wavenumbers k_0 (= $k_A \pm k_B$, $2 k_A$, $2 k_B$). We have investigated time evolution of the excited waves and found that the ablation effect plays an important role even in nonlinear stage to reduce amplitude of the excited waves. Differences between an ablation surface and a classical contact surface have been discussed. Dependence of the excited wave amplitude on the wavenumber k_0, the ablation velocity v_a, and the effective gravity g is also investigated.

I. Introduction

The Rayleigh-Taylor (RT) instability occurs in the acceleration of a fluid by one of lower density, whereupon the two tend to interchange positions. This instability may represent an obstacle to inertial-confinement fusion by destroying the symmetry of high-aspect-ratio imploding shells. The presence of the ablation in the inertial confinement fusion context introduces important modifications to the RT instability evolution that can not be accounted for in the classical RT theory. The principal differences between the ablation surface and the classical contact surface are the presence of mass and heat flows across the surface. These effects can lead to a reduction in a growth rate and even to stabilization at large wave numbers. Takabe et al.[2] found that the linear growth rate $\gamma(k)$ is expressed $\gamma(k)=0.9(kg)^{1/2}-\beta k v_a$ (β=3–4) where v_a is an ablation velocity.

If an initial perturbation amplitude is much smaller than its wavelength, the perturbation grows exponentially in time. As the amplitude of the RT instability becomes larger, nonlinear effects become important. Mode coupling of the RT unstable surface waves has been investigated by using ablatively stabilized growth rates in coupled equation obtained for the classical incompressible fluids[3]. In this paper, we study mode coupling of two waves for the ablative RT instability by perturbation theory (2nd order theory). We assume that there are two linear growing waves with wavenumbers k_A and k_B at the ablation front. Mode coupling between two waves excites new waves with wavenumbers $k_0 = k_A \pm k_B$, $2k_A$, $2k_B$. The growth of the wave is known to saturate when the amplitude becomes comparable to its wavelength. The amplitude of the wave with the longer wavelength becomes larger than that with the shorter wavelength when they begin to saturate. Thus a wave with a longer wavelength may be dangerous in the achievement of inertial confinement fusion, especially a wave whose wavelength is comparable to the target thickness. We thus investigate time evolution of an excited wave with a longer wavelength, that is, a wave with the wavenumber k_0 ($\equiv k_A - k_B$) and $k_0 \Delta R \approx 1$, where ΔR is an effective target thickness.

In Sec.II we apply perturbation theory to the basic equations and derive a set of equations up to the 2nd order, and in Sec.III we investigate time evolution of the excited wave with the wavenumber k_0 and clarify the difference between an ablative RT instability and a classical RT instability. In Sec.IV, the dependence of the amplitude of the excited wave on the wavenumber k_0, the ablation velocity v_a, and the effective gravity g is also investigated. It will be shown that the ablation effect plays an important role even in nonlinear stage to reduce amplitudes of the excited waves.

© 1996 American Institute of Physics

II. Basic Equations

We adopt one-fluid one temperature equations as basic equations[2]. Since the ablation structure is mainly determined by electron thermal conduction, it should be included in the equations. In a frame moving with an ablation front being accelerated at constant acceleration g, they are given in the normalized form

$$\frac{\partial}{\partial t}\rho + \nabla \cdot (\rho v) = 0 \quad , \quad (1) \qquad \rho(\frac{\partial}{\partial t}v + v \cdot \nabla v) = -\nabla \rho T + \rho g \quad , \quad (2)$$

$$\frac{3}{2}(\frac{\partial}{\partial t}T + v \cdot \nabla T) = -\rho T \nabla \cdot v + \nabla \cdot (K_0 T^{5/2} \nabla T) \quad , \quad (3)$$

where ρ, v, T are the density, velocity, and temperature of the ablating plasmas respectively. They are normalized by the corresponding values at the isothermal sonic point of the stationary state, and we normalize the length by the electron mean free path ℓ_s at the sonic point. In plane (x-y) geometry, we expand the variables up to the 2nd order

$$U = U_0 + \varepsilon U_1 + \varepsilon^2 U_2 + \cdots \quad , \quad (4)$$

where the column vectors $U_i = {}^t(\rho_i, v_{xi}, D_i, T_i)$ (i = 0, 1, 2) consist of the i th order variables and D_i is defined as $\partial v_{yi}/\partial y$. Subscript 0 denotes an steady ablation profile, 1 denotes linear growing mode profile, 2 denotes new excited profile. By submitting them into the basic equations, we get the set of equations expressed vector and matrix form as follows

$$0\text{ th} \qquad A\ U_0 = 0 \qquad , \quad (5)$$

$$1\text{ st} \qquad \frac{\partial}{\partial t}U_1 + A\ U_1 = 0 \qquad , \quad (6)$$

$$2\text{ nd} \qquad \frac{\partial}{\partial t}U_2 + A\ U_2 = S_2 \qquad , \quad (7)$$

where A, S_2 are the operator matrix and the column vector, respectively.

Equation (5) is obtained by assuming a stationary ablation profile. Equation (6) is the linear equation for the column vector U_1. Equation(7) is the nonlinear equation, namely the column vector S_2 contains the product of the 1st order variables and excites the new wave with the wavenumber k_0. The details of the 0th and 1st order profiles are discussed in Takabe et al.[2] and Hasegawa et al.[4].

III. Time Evolution of Waves Excited by Mode Coupling

We consider a case that there are two linear growing waves whose wavenumbers and linear growth rates are k_A and k_B, γ_A and γ_B, respectively. Mode coupling of the linear growing waves excites new waves whose wavenumbers are $k_A \pm k_B$, $2k_A$, $2k_B$. We solve the initial value problem of the 2nd order equation. Figure 1 shows the time evolution of the 2nd order velocity perturbation (v_{x2}) profile ($k_0 = k_A - k_B = 50$) which is zero initially and excited by the mode coupling of the waves with the wavenumbers $k_A = 160$ and $k_B = 110$. In Fig.1, we used the normalized ablation velocity $v_a = 0.01$ and the gravity force normalized by c_s^2/ℓ_s, $g=1$, where c_s is the sound speed at 1st sonic point. The position of the ablation front is 0.40625 and that of the 1st sonic point is 0. As clearly seen in Fig.1, in the beginning the 2nd order velocity perturbation is convected towards both the 1st and the 2nd sonic points with the convective velocities ($v_{x0} \pm c_s$), where v_{x0} and c_s are the local ablation velocity and sound speed. But later the 2nd order velocity perturbation grows exponentially at the adjacent of the ablation front. The observed convection of the excited perturbation for the ablative RT instability does not exist in the classical surface. In the ablative RT instability, the unstable wave consist of three different modes, i.e., sound mode, thermal conduction mode and surface mode (entropy mode), and the velocity perturbation excited by the mode coupling is not an eigen mode of the stationary state. The curl of the 2nd order velocity perturbations is not equal to zero, and the perturbation is shed as vortexes.

The dotted line in Fig.2 shows the maximum 2nd order velocity perturbation at the adjacency of the ablation front which is initially equal to zero. In the beginning the 2nd order velocity perturbation spends time in growing exponentially as discussed above. But later the 2nd order velocity perturbation grows exponentially with the growth rate of $\gamma_A+\gamma_B$. The solid line in Fig.2 shows the 2nd order velocity perturbation which initially has a finite value. In the beginning the 2nd order velocity perturbation grows exponentially with the linear growth rate γ_0 corresponding to the wavenumber k_0, but later it grows with the growth rate $\gamma_A+\gamma_B$. As clearly seen in Fig.2, the amplitudes of the waves become the same in later time for both cases. Namely they do not depend on their initial amplitudes. If the inequality $(\gamma_A+\gamma_B \gg \gamma_0)$ is valid and the initial amplitude of the perturbation with the wavenumber $k_0=k_A-k_B$ is not extremely large, the amplitude is mainly determined by mode coupling.

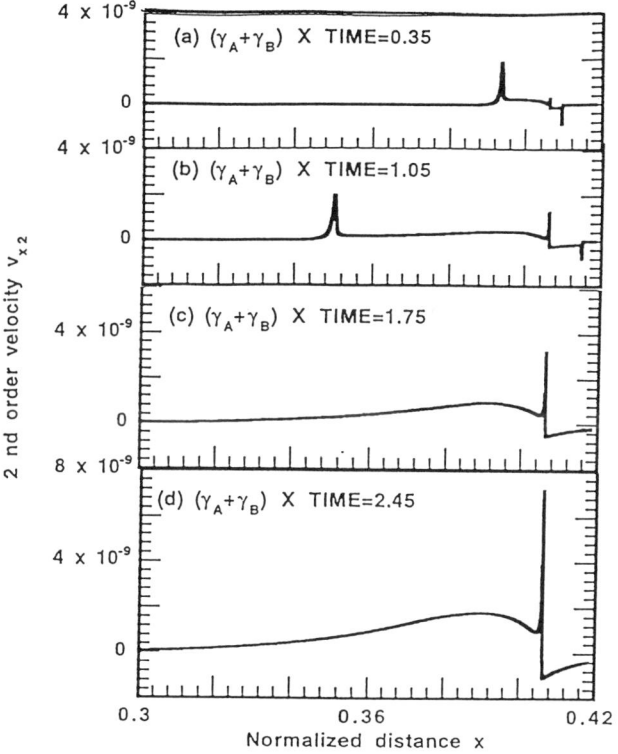

FIGURE1. Time evolution of the 2nd order velocity profile.

Figure 3 shows the various physical variables at the adjacency of the ablation front when the normalized time (that is $(\gamma_A+\gamma_B) \times$ TIME) is equal to 8.6. The solid line shows the 2nd order velocity, the broken line shows the 2nd order mass density, and the dotted line shows the 2nd order mass flux, $J_2=\rho_2 v_{x0}+\rho_0 v_{x2}+\rho_1 v_{x1}$. In the ablative RT instability the fluid velocity perturbation is not equal to the time derivative of the surface displacement because of the ablation. The deformation of the ablation surface can be determined by the mass flux perturbation across the surface. The mass flux perturbation divided by the stationary mass density gives the time derivative of the local displacement in the ablative RT instability. It should be noted that the absolute value of the 2nd order mass flux has a maximum at the ablation surface.

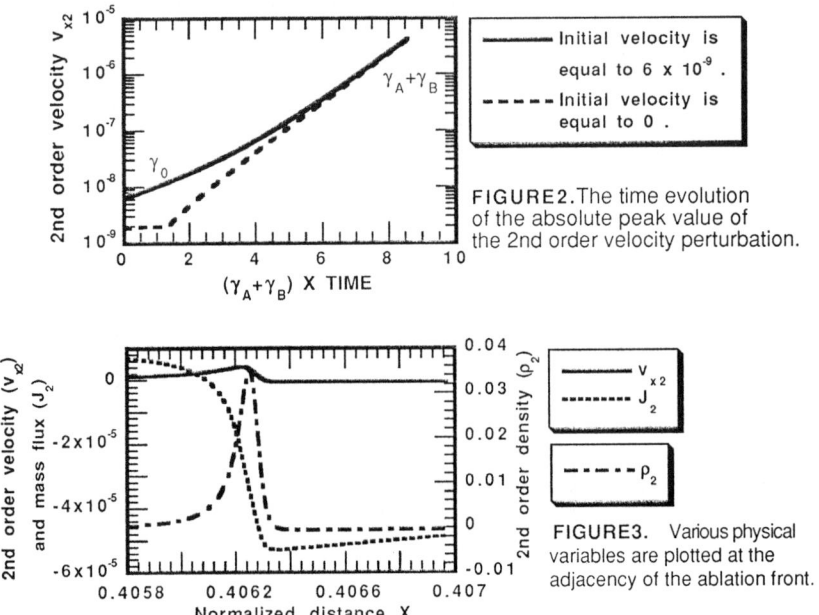

FIGURE 2. The time evolution of the absolute peak value of the 2nd order velocity perturbation.

FIGURE 3. Various physical variables are plotted at the adjacency of the ablation front.

IV. Discussion and Conclusion

As shown in the previous section, two linear growing modes at the adjacent of the ablation front whose wavenumbers are k_A and k_B and growth rates are γ_A and γ_B respectively excite new mode whose wavenumber is k_0 ($= k_A-k_B$) and growth rate is $\gamma_A+\gamma_B$. We investigated the dependence of the amplitude of the excited wave on the wavenumber k_0, the ablation velocity v_a, and the effective gravity g by changing their values. As discussed in Sec.I, we choose the wavenumbers k_A and k_B so that they give the corresponding growth rates $\gamma_A(k_A)$ and $\gamma_B(k_B)$ around the maximum growth rate, and also $k_0(=k_A-k_B)\Delta R \approx 1$.

Figure 4 shows the linear growth rates as a function of the normalized wavenumber $k_y \Delta R$ for the four different cases. The practical target thickness ΔR is defined by the distance of the points where the stationary density has 1/e of its peak value. Figure 5 shows the zeroth order normalized densities versus the normalized distance x for the corresponding cases. In both figures, the solid line shows the case for the ablation velocity $v_a=0.01$, the effective gravity $g=1$, the effective target thickness $\Delta R=0.015$. The broken line shows that for $v_a= 0.01$, $g=0.5$, $\Delta R = 0.030$. The dotted line shows that for $v_a=0.02$, $g=0.5$, $\Delta R= 0.058$. The dotted-dotted line shows that for

v_a=0.04, g=0.5, ΔR=0.112. In Fig.4, open circles are the wavenumbers k_A and closed circles are k_B. The lines in Fig.4 are plotted from Takabe formula and the symbols (○ and ●) show the actual simulation points. Thus the acceleration g is different between the solid line and the other lines. The ablation velocities v_a are different among the broken line, the dotted line and the dotted-dotted line. From the zeroth order equation of continuity, the normalized ablation velocity is equal to the inverse of the normalized maximum density (see Fig.5).

As discussed in the previous section, the deformation of the ablation surface is determined by the mass flux perturbation. In Fig.6, the maximum 2nd order mass fluxes divided by the product of the 1st order mass fluxes are shown as a function of the normalized wavenumber for the four cases. Symbols in Fig.6 correspond to the different four cases; open circles (v_a=0.01, g=1, ΔR=0.015), closed squares (v_a=0.01, g=0.5, ΔR=0.030), crosses (v_a=0.01, g=0.5, ΔR=0.030) and triangles (v_a=0.04, g=0.5, ΔR=0.112). As shown in Fig.6, the 2nd order mass flux increases with the increase of the wavenumber k_0. This dependence of the excited wave amplitude on the wavenumber k_0 has also been obtained for the classical RT instability[3]. It should be also noted that the acceleration g affects little on the amplitude as also obtained for the classical RT instability.

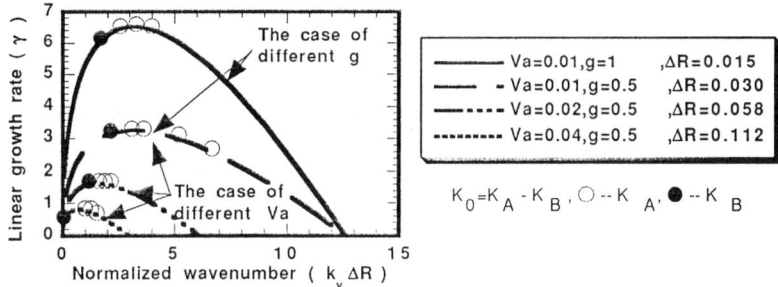

FIGURE 4. Linear growth rates versus normalized wavenumber $k_y\Delta R$ for the various cases, where ΔR is the practical target thickness.

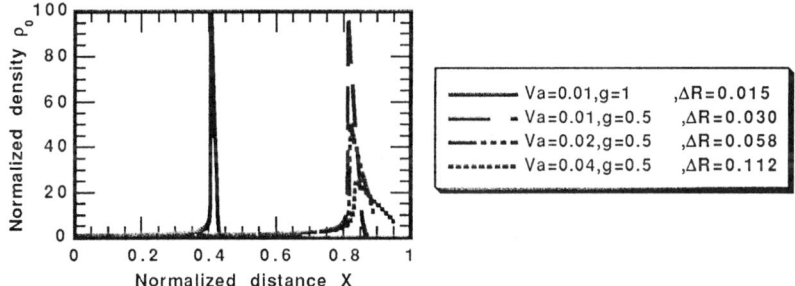

FIGURE 5. This figure shows the zeroth order densities versus the normalized distance x for the various cases.

As clearly seen in Fig.6, a larger ablation velocity leads to a smaller 2nd order mass flux, that is not observed in the mode coupling theory for the classical RT instability.

Figure 7 is plotted in order to clarify the dependence of the excited wave amplitude on ablation velocity v_a. In Fig.7, the maximum 2nd order mass fluxes divided by the product of the 1st order mass fluxes are plotted as a function of the ablation velocity for the fixed wavenumber $k_0 \Delta R$=1. Their values for the wavenumber $k_0 \Delta R$=1 are obtained by the interpolation of the values in Fig.6. The 2nd order mass flux decreases with the increase of the ablation velocity v_a. Thus the ablation effect plays an important role even in weakly nonlinear stage to reduce the amplitude of the excited waves.

A mode coupling theory has been developed for the ablative Rayleigh-Taylor instability. A part of the 2nd order perturbation excited by the mode coupling of two waves is found to be convected out from the ablation surface. It takes a time that the perturbation remaining near the ablation surface grows exponentially with the sum of the linear growth rates of the two waves. A larger ablation velocity leads to a smaller 2nd order perturbation.

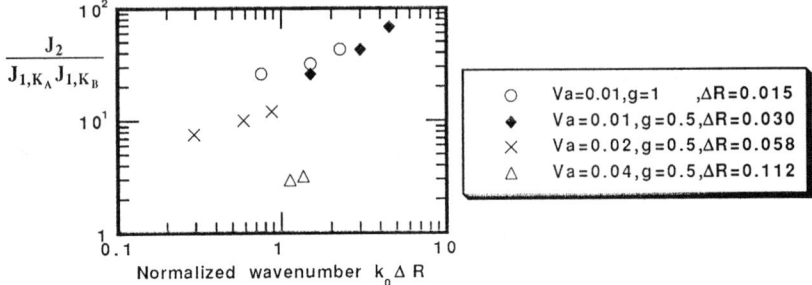

FIGURE 6. The vertical line shows the maximum 2nd order mass fluxes divided by the product of the 1st order mass fluxes, and the horizontal line shows the normalized wavenumber $k_0 \Delta R$.

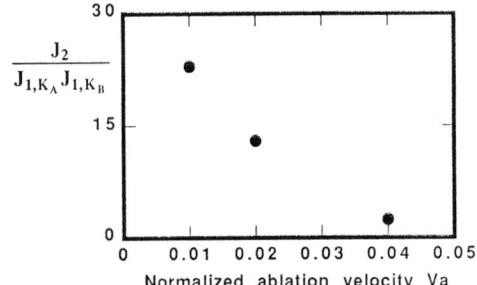

FIGURE 7. The vertical line shows the maximum 2nd order mass fluxes divided by the product of the 1st order mass fluxes, and the horizontal line show the normalized ablation velocity v_a.

ACKNOWLEDGEMENTS

We would like to express thanks to Professor S.Nakai and Professor K.Mima for their interests and encouragements.

REFERENCES

1. Kilkenny, J.D. *Phys.Plasmas* **1**, 1379, 1994 and their references.
2. Takabe, H. et al. *Phys.Fluids* **26**, 2299., 1983; Takabe, H. et al. *Phys. Fluids* **28**, 3676, 1985.
3. Haan, S.W. *Phys.Fluids* **B3**, 2349, 1991.
4. Hasegawa, S. et al submitted to *Phys.Plasmas*

A New Instability of a Contact Surface Driven by a Nonuniform Shock Wave

R.Ishizaki, K.Nishihara, H.Sakagami*, and M.Murakami

Institute of Laser Engineering, Osaka University, Suita, Osaka 565 Japan.
*Himeji Institute of Technology, Himeji, Hyogo 671-22 Japan

Abstract. Propagation of a nonuniform shock wave driven by a ripple piston has been investigated both analytically and computationally. Oscillating surface ripple of the shock wave is found to decay exponentially with its propagation. The oscillation period and the decay time observed in simulations agree very well with the linear theory. A new instability of a contact surface is shown to be driven by the ripple shock wave. Namely a uniform contact surface becomes unstable after the nonuniform shock passing through it. The growth rates depend on the phase of the oscillating shock wave at the time when the shock hits the contact surface. Physical mechanism of the instability is qualitatively discussed. Nonlinear evolution of the instability is also investigated. Properties of the instability are quite different from those of the Richtmyer-Meshkov instability in both linear and nonlinear phases.

INTRODUCTION

In inertial confinement fusion, fuel is required to be compressed up to 1000 times solid density. An asymmetric implosion associated with hydrodynamic instabilities disturbs uniform high density compression and reduces fusion reaction yield to significantly lower values than those predicted by one-dimensional simulations. Nonuniform laser irradiation leads to nonuniform ablation and thus a nonuniform shock wave. It is important to investigate the propagation of the nonuniform shock wave, since the nonuniform shock wave seeds density perturbations that grow due to the Rayleigh-Taylor instability in acceleration phase, even if a target surface is initially uniform.

It is well known as the Richtmyer-Meshkov [1,2,3](R-M) instability that a nonuniform contact surface becomes unstable when a (uniform) shock wave propagates through the contact surface. In this paper we investigate stability of a uniform contact surface when the nonuniform shock propagates through the contact surface. It will be shown that the uniform contact surface becomes unstable and the growth rates depend on the phase of the oscillating shock waves at the time when the shock hits the contact surface. Linear and nonlinear evolutions of the instability is found to be quite different from the R-M instability.

PROPAGATION OF RIPPLE SHOCK WAVE

We consider a shock wave driven by a ripple piston as shown in Fig.1, in which ρ, p, u denote density, pressure, and velocity, respectively, in a reference frame moving with the piston. The subscripts 0 and 1 denote the values ahead and behind the shock, and v_s, a_0, and $a(t)$ denote the shock speed, the amplitudes of the surface ripple of the piston, and the shock, respectively.

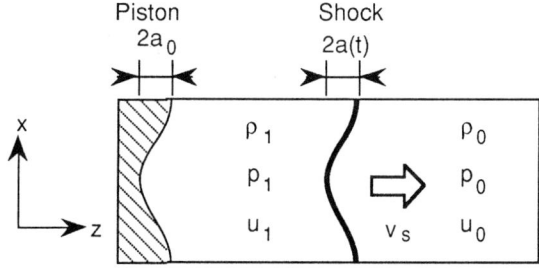

FIGURE 1. A ripple piston drives a nonuniform shock wave.

In the linear theory [1], the pressure perturbation in the shock compressed region satisfies the wave equation,

$$\frac{\partial^2 p_1^1}{\partial t^2} = c_1^2 \frac{\partial^2 p_1^1}{\partial z^2} - c_1^2 k^2 p_1^1 \qquad (1)$$

where c_1 is the sound speed in the shock compressed region, and k is the wave number of the perturbation, exp(ikx). From the shock Hugoniot condition, the amplitude of the shock surface ripple[4] is given by

$$\dot{a}(t) = \frac{1}{2} \frac{1}{\rho_1^0 - \rho_0^0} \left(\frac{1}{v_s} - \frac{v_s}{Kc_1^2} \right) p_1^1(v_s t, t) \qquad (2)$$

where K is the dimensionless slope of the Hugoniot in the P-V plane evaluated behind the shock. If we denote its Hugoniot in P-V plane as P(V), then we have

$$K = -\frac{1}{(\rho_1^0 c_1^0)^2} \left[\frac{dP(V)}{dV} \right]_{V=1/\rho_1^0} \qquad (3)$$

Figure 2 shows the amplitude of the ripple shock surface as a function of the normalized time $v_s t/\lambda$, where λ is the wavelength of the perturbation. The parameters used are ρ_0=0.01g/cm3, p_0=1Mbar, the Mach number M=2, a_0/λ=0.01 and λ=100µm. Open circles show simulation results obtained by using the two-dimensional Eulerian fluid code IMPACT-2D[5]. Solid line is the theoretical value obtained from Eq.(1) and (2). They are in good agreement. It is clearly seen that the ripple shock front oscillates as it propagates[4]. When the displacement of the shock surface is maximum, the transverse velocity perturbation u_x^1 is also maximum and the longitudinal velocity perturbation, $u_z^1 \sim \dot{a}$, pressure perturbation, and density perturbation at the shock surface become zero. When the displacement of the shock surface is zero, u_x^1=0 and the other perturbation become their maximum values. The oscillation of the velocity perturbations, u_x^1 and u_z^1, will be also seen in Fig.3(b). The oscillating amplitudes of the ripple shock surface decay exponentially as it propagates out from the piston. This decay can not be obtained from the CCW approximation, which is shown by a dotted line in Fig.2. The decay was also observed in experiments[6].

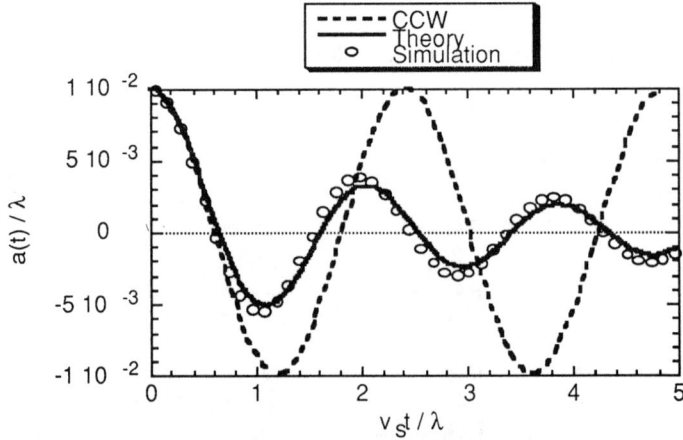

FIGURE 2. Displacement of the shock surface decays exponentially as it propagates. Parameters used are ρ_0=0.01g/cm³, p_0=1Mbar, the Mach number M=2, a_0/λ=0.01 and λ=100mm.

INSTABILITY OF CONTACT SURFACE DRIVEN BY NONUNIFORM SHOCK

In this section we consider that the nonuniform shock wave propagates through a uniform contact surface. The shock wave shown in Fig.2 hits the contact surface around the normalized time $v_{st}/\lambda=4$. As discussed in the previous section, the shock surface oscillates as shown in Fig.3(a), in which the solid lines indicate the shock surface and the arrows the velocity perturbations at the shock surface. When the displacement of the shock surface is zero, the velocity perturbations $|u_z^1|\sim|\dot{a}|$, has its maximum value and $u_x^1=0$. The velocity perturbations u_z^1 and u_x^1 at the shock surface which are normalized by the contact surface velocity u_c are shown in Fig.3(b) as a function of the normalized time. In Fig.3, (1), (2), and (3) correspond the cases that $u_z^1\sim a$ has a minimum value, zero, and a maximum value, respectively.

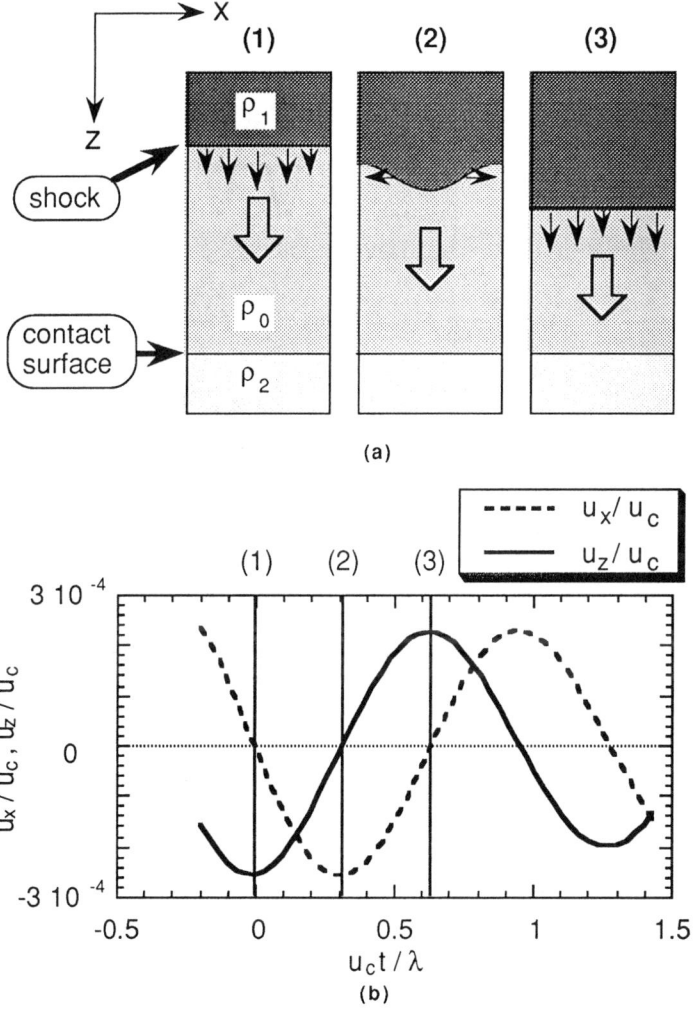

FIGURE 3. Perturbations of a ripple shock front oscillate. (1), (2), and (3) correspond, respectively, the cases that $u_z^1\sim a$ has a minimum value, zero, and a maximum value in both (a) and (b). (a) Images of the shock phase. (b) velocity perturbations of the shock front.

The growth rates of the contact surface perturbation induced by the propagation of the nonuniform shock wave are examined for three different cases. Namely the shock hits the contact surface at the three different oscillating phases of the shock corresponding to (1), (2), and (3) in Fig.3. The velocity perturbations of the contact surface, $\dot{a}_c/u_c$, are shown in Fig.4, in which (1), (2), and (3) correspond to those in Fig.3. For all cases the velocity perturbations of the contact surface oscillate with the same period as the velocity perturbation behind the shock oscillates. However, their averaged values, which are indicated by the solid lines in Fig.4, are different for three cases. Namely, for the case (1), the averaged perturbation velocity of the contact surface, $<\dot{a}_c>$, is negative, for the case (2), $<\dot{a}_c> \approx 0$, and for the case (3), $<\dot{a}_c>$ is positive. The growth rates of the new instability driven by the nonuniform shock thus depend on the phase of the oscillating shock at the time when the shock hits the contact surface. The dependence of the growth rates on the shock phase is found to be the same even for the density jump of the contact surface $\rho_2/\rho_0=3$.

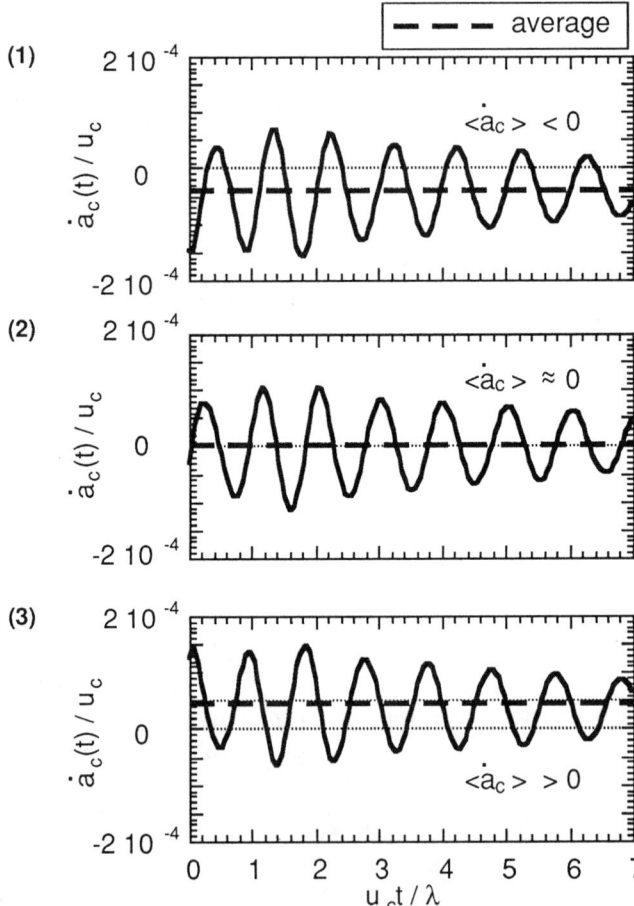

FIGURE 4. Growth rates of the instability are different for the phases of the ripple shock wave. A = -0.5: A = $(\rho_2-\rho_0)/(\rho_2+\rho_0)$

The physical mechanism of the instability can be understood as follows. As discussion in the previous section, the velocity perturbation behind the nonuniform shock driven by the rippled piston is characterized as the damping oscillation. The contact surface is, therefore, accelerated continuously with the acceleration of the damping oscillation, in the form

$$g(t) = \Delta g \exp(-\gamma t)\sin(\omega t + \phi) \qquad (4)$$

The growth rate of the instability is thus estimated by integrating the acceleration

$$\dot{a}_c(t) = \int_0^t g(t)\, dt$$

$$\Rightarrow \frac{\Delta g}{\gamma^2 + \omega^2}(\gamma \sin\phi + \omega\cos\phi) \quad (t \to \infty) \qquad (5)$$

$$\Rightarrow \frac{\Delta g}{\omega}\cos\phi \quad (\omega \gg \gamma)$$

where $a_c(t)$ is the displacement of the contact surface. Equation (5) clearly indicates that the growth rate depends on the phase of the oscillating shock at the time when the shock hits the contact surface. Δg in Eqs.(4), (5) can be estimated to be $\Delta g = du_z^1/dt$. The new instability can be thus understood as a gravitational instability with the acceleration of the damping oscillation.

The dependence of the growth rate on the phase of the oscillating shock is investigated for the various Atwood numbers of the contact surface $A=(\rho_2-\rho_0)/(\rho_2+\rho_0)$, ± 0.25, ± 0.5, ± 0.75 as shown in Fig.5. The dotted line in Fig.5 is the theoretical values obtained from Eq.(5), where Δg and ω are estimated from Fig.3(b). The growth rates observed in simulations agree fairly well with the theoretically estimated values.

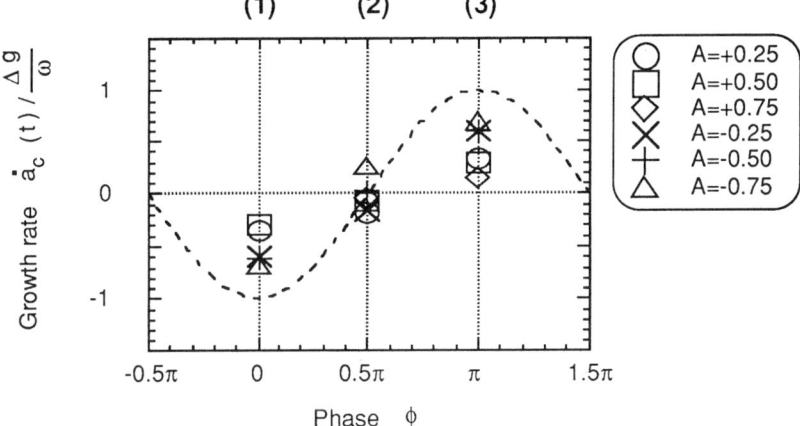

FIGURE 5. A gravity force causing the instability is an acceleration of a damping oscillation. A dotted line shows theoretical value. Parameters is the same as ones in Fig.2 except Atwood number.

Nonlinear evolution of the instability is also investigated, in which the amplitude of the ripple piston is large, $a_0/\lambda=0.2$. The contrast in Fig.6 indicates the iso-density contour in the case that the density jump of the contact surface is $\rho_2/\rho_0=1/3$. The black area in the top represents the ripple piston of which surface shape is given by a trigonometric function. The boundary between black and gray in the lower part indicates the unstable contact surface. The shock hits the contact surface at the normalized time $v_{st}/\lambda=1.6$ and 2.1 in Fig.6(2) and (3), respectively. The phases of the oscillating shock in Fig.(2) and (3) correspond to those of (2) and (3) in Fig.3(a).

In the nonlinear regime, the ripple shock surface is not a sinusoidal any more. In the linear regime, there is no average growth rate of the contact surface when the shock hits the contact surface at the phase that the displacement has its maximum value. However, the mashroom shape of the contact surface appears in the nonlinear regime as shown in Fig.6(2). In Fig.6(3), the z-component of the velocity perturbation at the shock front has large values near the edge, when the shock hits the contact surface. This large z-component of the velocity perturbation leads to the deformation of the contact surface as shown in Fig.6(3) after the shock propagates through it.

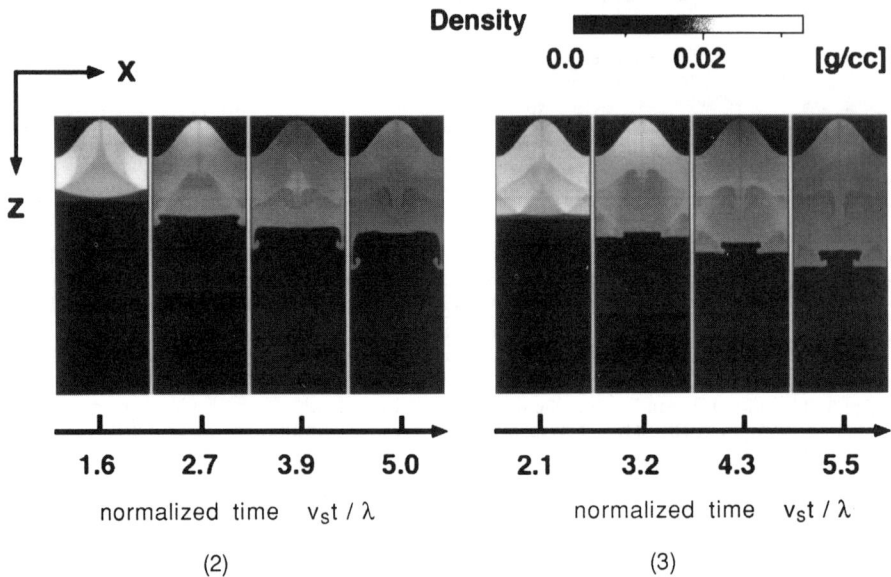

FIGURE 6. Nonlinear evolution of the instability, iso-density contour. Numbers (2) and (3) correspond to those in Fig.3 to 5. A=-0.5 (A transmitted shock decelerates). (2) A displacement of shock front is minimum. (3) A displacement of shock front is maximum.

CONCLUSIONS

We have found a new instability of the uniform contact surface when the nonuniform shock driven by a ripple piston propagates through the contact surface. Both linear and nonlinear evolutions of the instability are quite different from the R-M instability. The physical mechanism of the instability can be explained as a gravitational instability driven by an acceleration of the damping oscillation. The growth rate depends on the phase of the oscillating shock at the time when the shock hits the contact surface.

REFERENCES

1. Richtmyer,R.D., "Taylor instability in shock acceleration of compressible fluids," *Comm. Pure Appl. Math*, 13, 297 (1960)
2. Meshkov,E.E., "Instability of the interface of two gasses accelerated by a shock wave," *Fluid Dyn.* 4, 101 (1969) [Izv. Akad. Nauk SSSR, Mekh. Zhidk. Gaza 5, 151 (1969)]
3. Sakagami,H.,Polionov,A.V.,Nishihara,K., "Nonlinear evolution of Richtmyer-Meshkov instability," in *Proceedings of the Conference on Japan U.S. Seminar on Physics of High Power Laser Matter Interactions* Kyoto, Japan, March 9-13, 1992.
4. Briscoe,M.G.,Kovitz,A.A., "Experimental and theoretical study of the stability of plane shock waves reflected normally from perturbed flat walls," *J.Fluid Mech.* 31, 529 (1968)
5. Sakagami,H.,Nishihara,K., "Rayleigh-Taylor instability on the pusher-fuel contact surface of stagnating targets," *Phys.Fluid.* B2, 2715 (1990)
6. Endo,T.,Shigemori,K.,Azechi,H.,Nishiguchi,A.,Mima,K.,Sato,M.,Nakai,M.,Nakaji,S.,Miyanaga,N.,Matsuoka,S.,Ando,A.,Tanaka,K.A.,Nakai,S.,"Dynamic behavior of rippled shock waves and subsequently-induced areal-density-perturbation growth in laser-irradiated foils," *Phys.Rev.Lett.* 74, 3608 (1995)

Two Dimensional Simulation of Turbulent Mixing in Stagnation Dynamics of Implosion

A. Sunahara, H. Takeuchi, H. Takabe

Institute of Laser Engineering, Osaka University
2-6 Yamada-oka, Suita Osaka 565 Japan

abstract: We have studied the evolution of Richtmyer-Meshkov instability and turbulent mixing in the deceleration phase of implosion process. It is found that the spectrum of the perturbation of the interface between the main fuel and hot spark regions is shifted to longer wavelength by the inverse-cascade process. It is also found that bouncing of a shock wave in the spark region enhances the inverse-cascade process. As for the nonlinear electron thermal conduction effect on the mixing, it is found that it does not affect drastically the growth of the mixing layer width in the deceleration phase, but enhances the cooling of the hot spark region in the final stage. It is concluded that the cooling of the hot spark region by the thermal conduction is enhanced by the growth of the mixing layer width and degrades the implosion performance.

I. Introduction

In inertial confinement fusion, it is required to achieve a high density compression of about 10^3 times the solid density. In order to achieve a high convergence of a target, spherically symmetric hydrodynamic motion is required. Due to nonuniformities in irradiated laser intensity and/or non-sphericity of the target, hydrodynamic instabilities are generated to degrade an implosion performance. In the implosion experiments carried out so far, it has been found that the experimental neutron yields are much below than the values predicted with one dimensional implosion codes[1,2,3]. This has been thought to be due to the hydrodynamic instabilities[4]. The hydrodynamic instabilities consist of two componets, shorter wavelength modes which lead to the turbulent mixing at the interface of the main fuel and hot spark in the deceleration phase, and longer ones which lead to deformation of the hot core structure. Both components cause the lowering of the implosion performance.

In studying the hydrodynamic instabilities in the implosion process, the Rayleigh-Taylor[5] and Richtmyer-Meshkov[6] instabilities become important in both of the acceleration and deceleration phases. The Rayleigh-Taylor instability is induced when a dense fluid is accelerated by a light fluid. In contrast, the Richtmyer-Meshkov instability is induced when a layer with density profile varying in space is impulsively accelerated by a shock wave. In the present paper, we investigated the development of the hydrodynamic instability induced by the shock wave in the deceleration phase with a two dimensional simulation code. In the deceleration phase, a reflected strong shock wave generated at the center of the target collides a stratified boundary interface between the main fuel and the hot spark. If the interface is not uniform, vortexes are generated by the baroclinic effect and the interface deformation grows in time. Finally, it causes the turbulent mixing which decreases the implosion performance through the cooling of the hot spark region.

The basic properties of this instability have been studied by Zabusky et.al[7], and we have reported the fundamental process of the interaction of a shock wave and stratified boundary interface in Ref.[8] by relative to ICF topics. In this paper, we deal with the evolution of the perturbation spectrum, mixing layer width, and nonlinear electron thermal conduction effect in the deceleration phase of implosion. In order to carry out a two dimensional numerical simulation, we have used the 2nd order Godunov scheme described in Sec. II. In Sec. III, we study the initial condition dependence in turbulence spectrum. In sec. IV, multiple shock effect on evolution of the mixing layer is studied. Effect of nonlinear electron thermal conduction on the evolution of the mixing layer

© 1996 American Institute of Physics

is studied in Sec. V. Section VI is devoted to a brief summary.

II. Simulation Condition

We have carried out two dimensional simulations by using the 2-nd order Godunov scheme[9]. We have assumed an one-temperature, invicid, ideal fluid with a specific heat of 5/3. A part of a spherical target to be imploded is shown schematically in Fig. 1. In the deceleration phase, the shock wave generated at the center of the target decelerates the main fuel converging forward the center. After the first shock wave collides the main fuel, the shock wave is separated into a transmitted shock wave and a reflected shock wave. The reflected shock wave collides the main fuel again in later time. As the shock collides the interface, the region inside the main fuel, i.e., the hot spark region is heated. In the present simulation, we set the initial conditions so that the hot spark becomes high temperature (3.5keV) and low density (200g/cm^3).

In order to set the initial condition of density, velocity, and pressure to simulate the stagnation dynamics in the high gain target, we have used the ILESTA_1D code[1]. This initial condition of our simulation corresponds to the timing just before the collision of the first shock wave with the main fuel. It is assumed in the simulation that the contact surface propagates to the right, and the shock wave propagates to the left. At the top and bottom boundaries, we have assumed slip boundary condition in all cases. That is, the flow velocity normal to the boundaries are set to be equal to zero. Boundary conditions at the left and right are altered accordingly to difference of situation.

III. Initial condition dependence

In the stagnation phase, which starts when the shock wave reflected from the target center collides the interface of the main fuel and hot spark region, the perturbation amplitude of the interface grows.

The perturbation of the interface is thought to stem from the target non-sphericity and/or laser irradiation non-uniformities, and is initiated at the contact surface by the feedthrough process in the acceleration phase. In order to investigate the dependence of the growth of the perturbation on the initial perturbation spectrum, we compare two types of simulation results shown in Fig. 2. One is the case when only the longer wavelength modes with the wavenumbers of 0.5-$7.5*2\pi/100\mu m$ are set for the initial condition, while the other is the case when the mode with high wavenumbers of 8-$15*2\pi/100\mu m$ are set for the initial condition. In these simulation, free boundary is assumed at the left and right boundaries by setting all the physical variables are continuous there, so that a shock wave is not reflected and passes through the density-stratified boundary. The simulation region consists of 64*128 numerical zones with the square zone of $0.78\mu m * 0.78\mu m$.

In Fig. 3, the density contours at 700 psec are shown, where small scale eddies evolve to large ones in both cases. This is called inverse-cascade process and shorter wavelength modes of the initial perturbation are translated to the longer wavelength modes. It is found that the perturbation spectrum, therefore, the evolution of mixing layer seems to be not very different at 700ps in these two cases. This can be confirmed by comparing the density fluctuation spectra obtained by Fourier transfer of the density perturbation. The resultant spectra at t=700ps are shown in Fig. 4. In addition, velocity power spectra are shown in Fig. 5. The amplitudes are different, but the spectra are almost same. It is concluded that the longer wavelength modes are generated by the inverse-cascade process regardless of the initial perturbation spectrum. As the result, the circulation of the two cases plotted in Fig. 6 tend to be same value in later time.

IV. Multiple shock effect in stagnation phase.

In the deceleration phase, the main fuel is stagnated by the collision of the multiple shock wave in a short time. After the shock wave collides the interface between the main fuel and the hot spark, the first shock wave is separated to a transmitted shock wave and reflected shock wave. The reflected shock wave bounces in the spark region, and the spark is heated by the shock wave. We compare the evolution of the mixing layer in the multiple shock case with in the single shock one described in Sec. III to study the effect of the multiple shock.

The simulation region is 256*256 (zones) with the size of each zone 0.39μm * 0.39μm. At the right boundary, reflection condition is assumed.

The velocity power spectrum at 200 ps after the first shock wave collides the interface of the main fuel is shown in Fig. 7 (a), and the velocity power spectrum at 300ps just after the second shock collides the main fuel is shown in Fig. 7 (b). Both cases are same at 200ps, but are obviously different at 300ps and the longer wavelength modes grow extremely in the multiple shock case. This is explained as follows. After the first shock collides the interface, the longer wavelength modes are so dominant by the inverse-cascade process described in Sec. III that shorter modes decrease when the second shock collides the interface. The shorter wavelength modes are not generated by the second interaction of shock and density stratified boundary interface.

In Fig. 8, the time evolution of the mixing layer width is shown for both cases of the multiple shock wave and the single shock wave. In the single shock wave, the mixing layer width grows linearly with time. In the multiple shock wave case, on the other hand, the mixing layer width decreases once by the compression of the fluids when the shock wave collides with the interface. However, it grows linearly at higher rate than before the collision. These results shows that the evolution of the longer wavelength modes is enhanced after every collision of the shock wave.

V. Thermal conduction effect in stagnation phase

In this section, we describe nonlinear electron thermal conduction effect on the evolution of mixing layer width in the stagnation phase. We have studied the evolution of the perturbation amplitude by including the nonlinear electron thermal conduction effect, and compared with the results without the thermal conduction effect. The simulation condition is the same as in Sec. III. It is noted that we assumed that the temperature is twice the simulation result in evaluating the nonlinear electron thermal conduction coefficient. This is because in the spherical high gain target, the temperature of the hot spark is about 5-6keV, while in our simulation it is about 3.5keV due to no spherical convergence effect. This assumption is reasonable in evaluating the time scale of thermal conduction effect in the implosion process.

In Fig. 9, the density contours at 500ps for the both cases without thermal conduction and with thermal conduction are shown respectively. It is found that the shorter wavelength modes are stabilized by the thermal conduction effect. However, the evolution of mixing layer width is not strongly affected. The time evolution of the mixing layer width is shown in Fig. 10. It is found that the thermal conduction does not affect to the longer wavelength modes. In the stagnation process, the longer modes are dominant by the inverse-cascade process, so that the thermal conduction does not affect the stagnation phase so much. At the earlier time when nonuniformities starts to grow the temperature of that spark is not so high that the stabilization by the thermal conduction is not expected. However, the thermal conduction help to cool the hot spark region and may degrade the fusion performance. The temperature profile at 400ps which is maximum value for each case, is shown in Fig. 11. It is found that hot saprk region is cooled by the non-linear electron thermal conduction. In the experiments carried out with a multi-beam irradiation system, it has been found that the experimental neutron yields are much below than the values predicted with one-dimensional implosion codes. This has been thought to be due to the coupling of the evolution of mixing layer with the cooling of the fuel by the thermal conduction. It can be concluded from the present simulation that the cooling of the thermal conduction is enhanced by the evolution of the mixing layer width.

VI. Summary and Discussion

We have investigated the turbulent mixing and the effect of the thermal conduction on the mixing by using the 2-D simulation code. We have studied the dependence of the mixing layer evolution of the initial perturbation spectrum. We compared two both cases where the simulations starts with only longer and shorter wavelength spectrum. In the simulations, it is seen that due to inverse-cascade process, the final spectrum were shifted to the longer wavelength mode in the both case. Small eddies collides coalescence each other, to grow as relatively large-scale eddies.

In present paper, all the results are obtained in two dimensions. We have to consider the difference in three dimensional phenomena, i.e.; the stretch of the eddies would occur in the direction of the vorticity vector, and physical process would be modified from the inverse-cascade process as seen in present results. Furthermore, we picked up a part of the target and used a two dimensional plane geometry. We didn't treat a convergence effect, which is thought to affect to the growth rate of the perturbation amplitude. In the next step, we have carry out the simulation with a convergence effect.

We have also studied the couple of multiple shock waves. It is found that the multiple shock wave collision enhances the inverse-cascade process. As for the thermal conduction effect on the turbulent mixing process, it is shown that the cool of the hot spark region due to the thermal conduction is as the mixing layer grows. In the stagnation phase, the longer mode spectrum is dominant by the inverse-cascade process, so that the growth of the perturbation amplitude was not so much affected by the thermal conduction. This cooling of the hot spark region by the thermal conduction can be thought to be associated to the decrease of the neutron yield in the experiments.

References

[1] H. Takabe et al., Phys. Fluids 31 (1988) 2884.
[2] M. C. Richardson, P. W. Mckenty, F. J. Marshall, C. P. Verdon, J. M. Soures,
   R. L. McCrory, O. Varnouin, R. S. Craxton, J. Deletrez, R. L. Hutchison,
   P. A. Jaanimagi, R. Keck, T. Kessler, H. Kim, S. A. Letzring, D. M. Roback,
   W. Seka, S. Skapsky, and B. Yaakobi, in Laser Interaction and Related Plasma
   Phenomena (Plenum,New York,1986) vol.7, p.179.
[3] E. M. Campbell, J. T. Hunt, E. S. Biss, D. R. Speck, and R. P. Drake, Rev. Sci,
   Instrum. 57 (1986) 2101.
[4] Y.Isayama et al., Journal of Plasma and Fusion Reserch 70 No.7(1994) 756
   in Japanese.
[5] S. Chandrasekhar, "Hydrodynamic and Hydromagnetic Stability"
   (Oxford, London) 1961.
[6] R. D. Richtmyer. Commun.Pure. Appl. Math. 13 (1960) 297.
[7] N. J. Zabusky et al., Proc.of 18th International Symp. on SHOCK
   WAVES, Sendai JAPAN, 1 (1991) pp.19.
[8] A. Sunahara, H. Takeuchi, and H. Takabe, to be published.
[9] J. C. Liu et al., "Code description Document TSUNAMI 1.0 A program for
   predicting Gas Dynamics in Inertial confinement Fusion Reactors",
   Department of Nuclear Engineering University of California, Berkeley (1992).

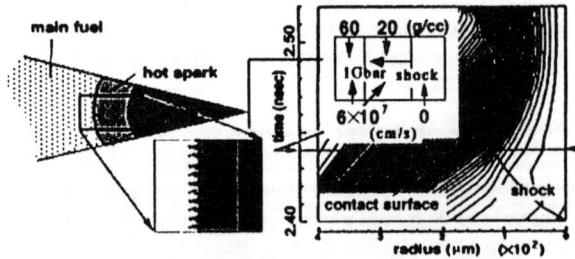

Fig. 1.

A part of a spherical target is shown. The initial condition corresponds to the timing just before the collision of the first shock wave with the main fuel

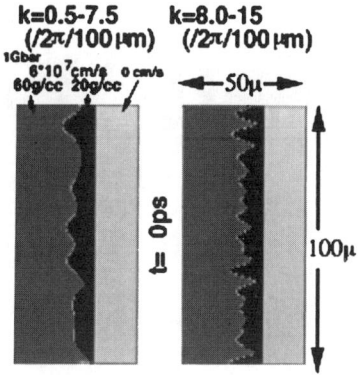

Fig. 2.
Two types of the initial condition are shown.
(a) has the modes with the wavenumbers 0.5-7.5 ($2\pi/100\,\mu m$)
(b) has the modes with the wavenumbers 8-15 ($2\pi/100\,\mu m$)

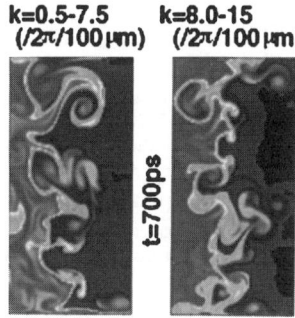

Fig. 3.
The density contours at 700 ps are shown. Both types are not very different.

Fig. 4. The density spectra at 700 ps are shown.

Fig. 5. The velocity power spectra at 700 ps are shown.

Fig. 6. The circulation of the two cases are shown.

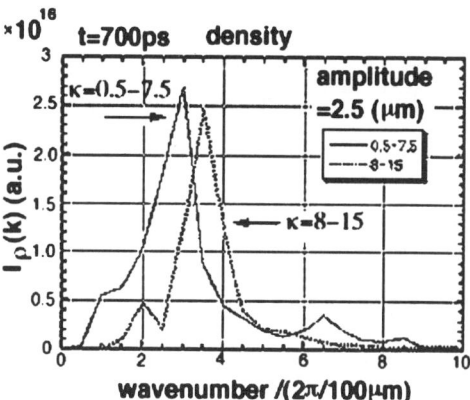

Fig. 4.

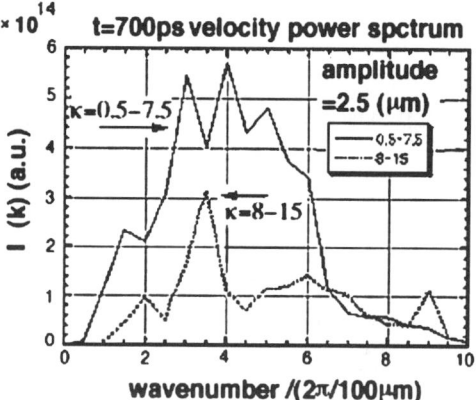

Fig. 5.

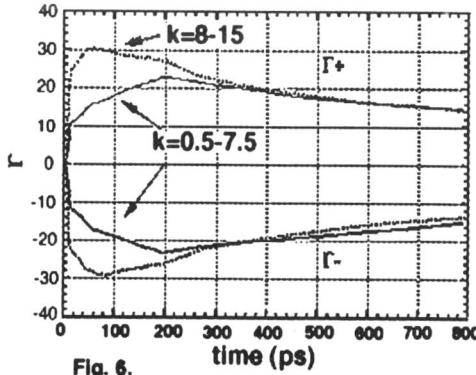

Fig. 6.

253

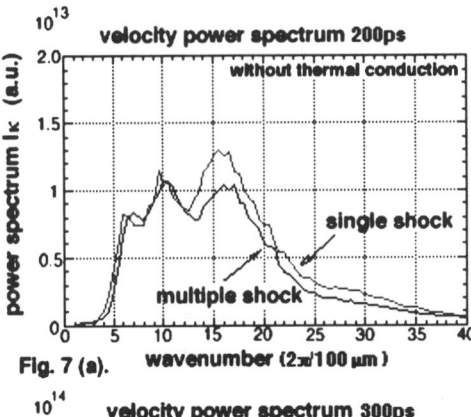

Fig. 7 (a).

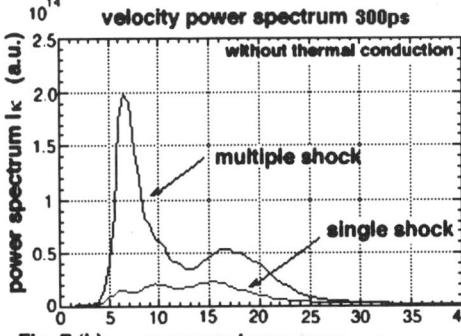

Fig. 7 (b).

Fig. 7. The velocity power spectra are shown. (a) is at 200 ps after the first collision of the shock wave with the stratified inter face. (b) is at the time after the second collision for the multiple shock case.

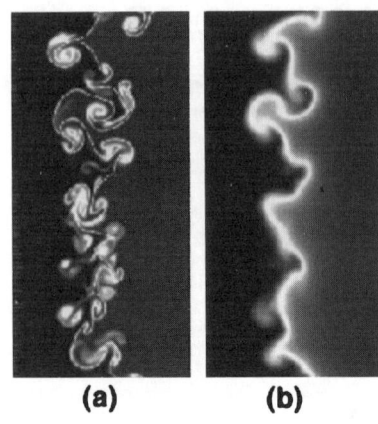

Fig. 9. The density contours at 500 ps are shown. (a) is w/o thermal conduction, (b) is with thermal conduction.

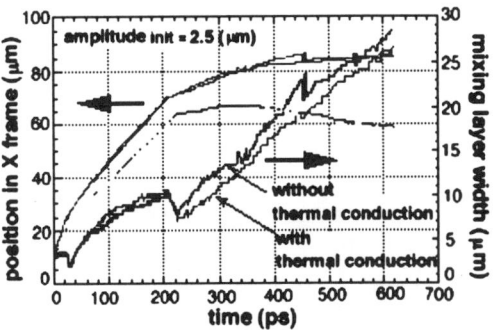

Fig. 10. The time evolution of the mixing layer of both cases are shown.

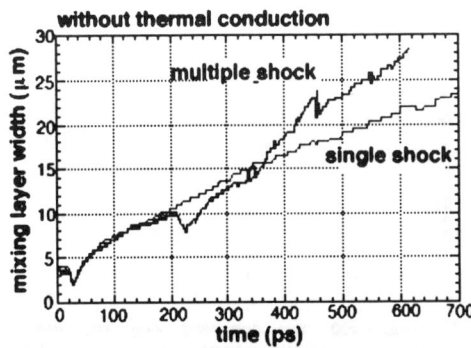

Fig. 8. The time evolution of the mixing layer width is shown.

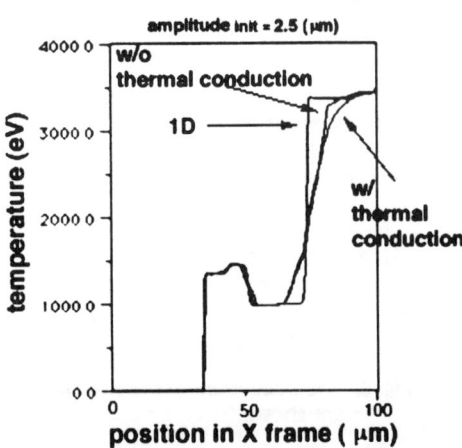

Fig. 11. The maximum temperature profile is shown. All the cases are at 400 ps.

The Design and Characterization of Toroidal-shaped NOVA Hohlraums that Simulate National Ignition Facility Plasma Conditions for Plasma Instability Experiments*

B. H. Wilde†, J. C. Fernandez†, W. W. Hsing†,
J. A. Cobble†N. D. Delamater†, B. H. Failor‡
W. J. Krauser†, and E. L. Lindman†

† Los Alamos National Laboratory, Los Alamos, New Mexico 87545
‡ Physics International, Los Angeles, CA

Abstract. Special Nova hohlraums have been designed to simulate the plasma conditions calculated for various NIF hohlraum point designs. These hohlraums attempt to maximize the laser pathlength for parametric instability measurements. A toroidal-shaped hohlraum with a diameter of 3200 microns and a length of 1600 microns allows a laser pathlength of about 2 mm. Filling the hohlraum with 1 atmosphere of neopentane gas gives an electron temperature of 3 keV and electron density near 0.1 of critical. Detailed LASNEX calculations for these hohlraums and comparisons to the NIF point design will be presented. Comparisons between data and calculations that characterize the plasma conditions (electron, radiation, and ion temperatures, electron density, etc) in these Nova hohlraums will also be shown.

*This work supported under USDOE contract W-7405-ENG-36

INTRODUCTION

Computer simulations of the indirect-drive point designs for the National Ignition Facility (NIF) using the LASNEX code (1,2) show that the plasma conditions inside of these hohlraums are such that Laser Plasma Instabilities (LPI's) may have a deleterious effects on the performance of the ignition target. In particular, stimulated Brilloiun scattering (SBS) and stimulated Raman scattering (SRS) could scatter a sizeable fraction of the laser energy back out of the target and therefore that energy would not be available for driving the capsule. In addition, unpredictable scattering and absorption inside of the hohlraum would adversely affect capsule implosion symmetry. We have designed a special target for the NOVA laser to investigate LPI's under conditions that *approach* those expected for the NIF point design. We first describe the plasma conditions in this special toroidal hohlraum design and compare them to the NIF design. Next the

© 1996 American Institute of Physics

characterization of this hohlraum will be discussed and compared to calculations. The SBS and SRS measurements in this hohlraum are discussed in Refs. 3-6.

TOROIDAL HOHLRAUM DESIGN AND THE NIF

Figure 1 shows same-scale LASNEX(7) computational meshes and laser rays (at peak power) of the NIF point design, the typical NOVA Scale 1 hohlraum, and our toroidal-shaped hohlraum. The hohlraums are made of gold and are filled with gas to mitigate hohlraum filling and hydrodynamic coupling to the capsule in the NIF. The NIF hohlraum is 5.7 mm in diameter and 9.6 mm in length, while the Scale 1 NOVA hohlraum measures 1.6 mm in diameter and about 2.3 mm in length. As can be seen from the figure, the latter hohlraum has scale lengths of less than a mm along the laser path in the gas. To study LPI's it is desirable to have scale lengths approaching those for the NIF target. However, since NOVA's maximum energy is only about 1/50th that of the NIF (1.8 MJ), actual NIF scale lengths near 10% of critical density cannot be achieved on NOVA with the required electron temperatures of 3 to 5 keV.

The toroidal hohlraum for NOVA is 3.2 mm in diameter and has a length of 1.6 mm giving a scale length of about 2 mm. There are 5 NOVA beams on each side with some of them crossing at the central vertical axis of the hohlraum. Normal incidence at the wall is intended to simplify future modeling of the SBS. The laser typically delivers 30 kJ at 3ω (351 nm wavelength) in the pulse shape as shown in Fig. 1d. For comparison, Fig. 1d also shows the so called PS22 pulse shape used in the Scale 1 symmetry experiments and the NIF pulse shape for the point design. For the interaction experiments the probe beam would typically have a 1 ns square pulse delayed a few hundred picoseconds with respect to the main heater beams and with different RPP's (Random Phase Plates) for spatial smoothing, one or four colors, SSD (Smoothing by Spectral Dispersion), f/4.3 or f/8, etc. (3-6). These toroidal hohlraums have been fielded with a 2 μm thick CH liner, 4 mg/cm^3 CH$_2$ foam, and with 1 Atm of neopentane (C_5H_{12}), deuterated neopentane (C_5D_{12}), and carbon dioxide (CO_2). The gas is contained by placing a 0.25 μm thick Si_3N_4 window across the laser entrance holes (LEH'S).

At room temperature 1 Atm of C_5H_{12}, C_5D_{12}, and CO_2 have mass densities of 3, 3.4, and 1.8 mg/cm^3 respectively, and when totally ionized give electron densities that are 11, 11, and 6% of the critical density at 3ω (9×10^{21}/cm^3). Recently, we have also fielded these hohlraums with 5 Atm of H and He-H (3% of critical). He-H at 3% critical is the present gas fill of choice for the NIF point design. However, to approach the near 10% critical density achieved in the NIF target at peak laser power, we hope to field 10 Atm of He-H(at 6%). Cryogenic liquid nitrogen targets are also being considered to get even higher densities. Since most of the interaction and the plasma characterization experiments were fielded with neopentane, the rest of this paper will concentrate on neopentane. However, Refs. 3-6 discuss LPI measurements for various gases.

Both the NIF and the toroidal hohlraum modelling were done with the same physics using the two-dimensional radiation-hydrodynamics LASNEX code (7).

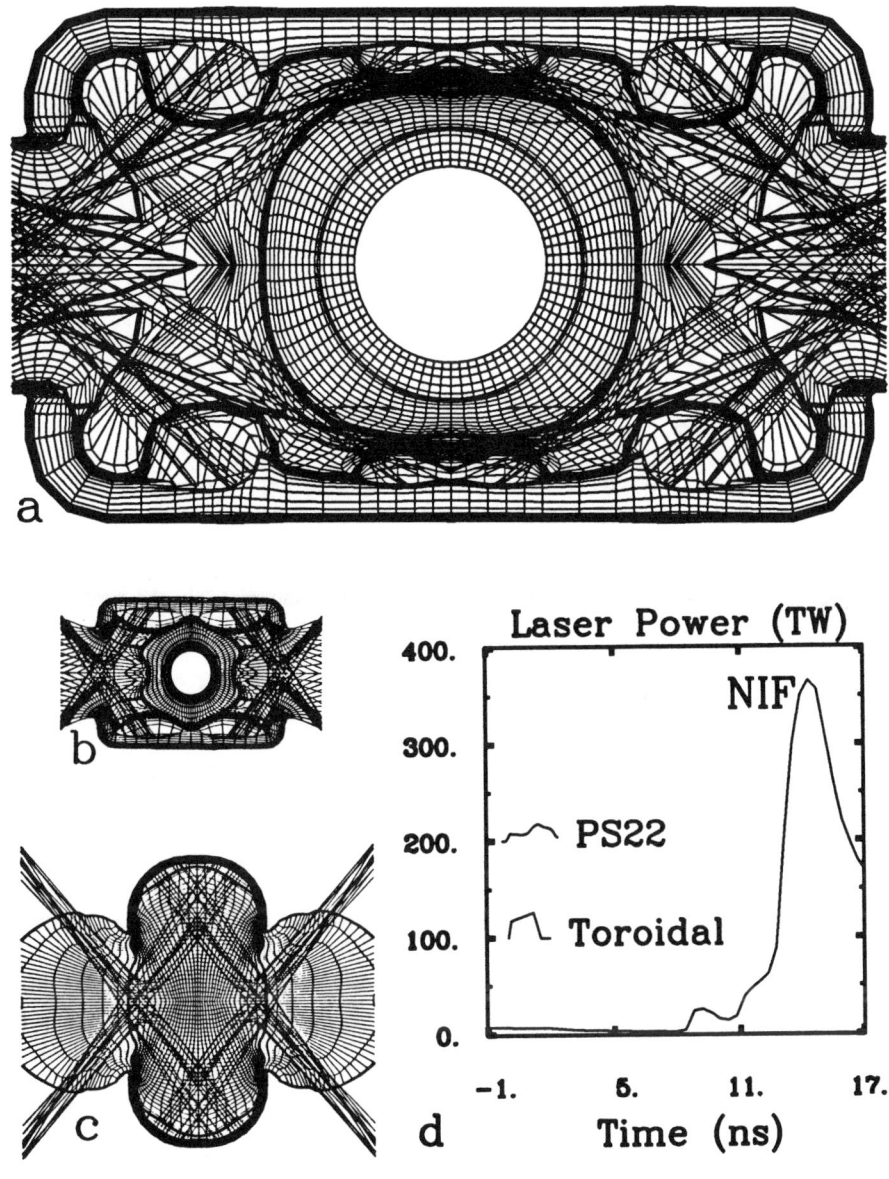

Figure 1. LASNEX meshes and laser rays at peak laser power for a) the NIF (0.03n_c He-H gas), b) the standard Scale 1 NOVA (PS22 pulse shape, 0.03 n_c methane gas), and c) the toroidal NOVA hohlraums. The laser power for these 3 hohlraums is shown in d), where the NOVA pulses are offset for clarity. The 3D laser beams are approximated by 2D cones.

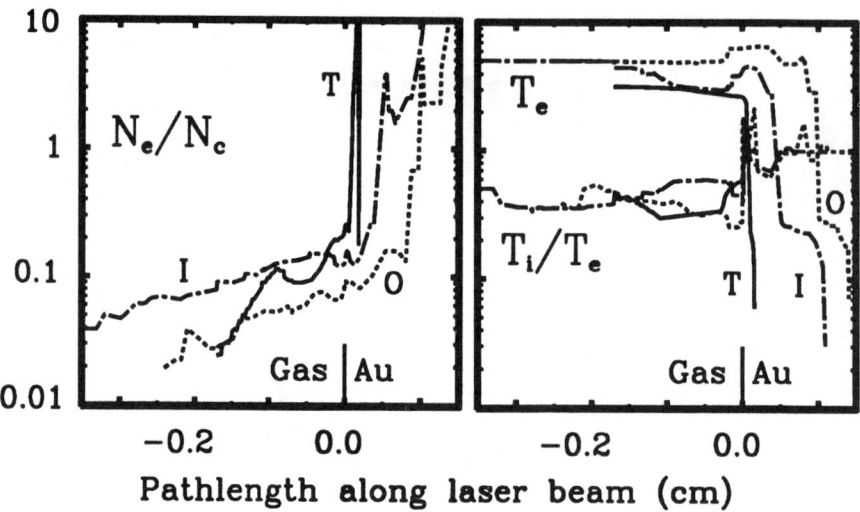

FIGURE 2. LASNEX plasma conditions along the pathlength of the laser beams for the inner (I=dot-dash) and outer (O=dash) cones of the NIF and the toroidal (T=solid) hohlraums. Since presumably, the gas is the amplifying medium for LPI's, the curves have been lined up such that the gas-gold interface is at the origin.

To obtain the correct gold conversion efficiency and albedo requires non-LTE (Local Thermodynamic Equilibrium) atomic physics and the best available radiation transport.

Although, the NIF He-H fill starts at 3% critical, at peak laser power (14.5ns), the gold wall and the CH capsule ablator have blown off to give near 10% critical densities in the gas. Figure 2 shows plasma conditions as calculated with LASNEX for both the inner and outer cones of the NIF at peak laser power as a function of distance along the laser path. Figure 2 also shows the same conditions in the NOVA toroidal hohlraum at 1 ns slightly before the peak of the laser pulse used for this hohlraum. The increasing power ramp as shown in Fig. 1d keeps the plasma conditions relatively constant for about 400 picoseconds. In the gas near the gold wall, the toroidal hohlraum has electron temperatures near 3 keV, similar to the inner cone of the NIF, but lower than the 5 keV of the outer cone. The electron density in the toroidal hohlraum is near 10% of critical, between that of the inner and outer cone of the NIF. As expected, the scale length in the gas of the toroidal hohlraum is shorter than in the NIF but more than twice as long as in the Scale 1 gas-filled symmetry NOVA hohlraums. The ratio of ion temperature to electron temperature is the same in the outer cone of the NIF, but is lower than in the inner cone of the NIF. However, this should only help to quench SBS in the inner cone of the NIF compared to the toroidal hohlraum through ion Landau damping (8). The radiation temperature is about 300 eV in the NIF and 200 eV in the toroidal hohlraum and is almost constant along the beam paths.

PLASMA CHARACTERIZATION

In order to perform relevant laser plasma instability experiments, it is necessary to confirm the plasma conditions calculated by the LASNEX simulations with experiments. For the toroidal hohlraum we have measured the radiation, electron, and ion temperatures and the electron density, in addition to laser propagation times through the cold gas.

LASNEX calculates that it takes about 100 ps for the laser to burn through the Si_3N_4 windows that are across the LEH's. This is confirmed by temporally and spectrally resolved SBS measurements which show strong SBS signals for about 100 ps when the density of the Si_3N_4 is above $n_c/10$.

To measure the propagation time of the laser through the cold neopentane gas from the Si_3N_4 window to the gold wall, a 2000 A° thick aluminum patch was used to replace the gold wall where one of the beams would hit the wall. Since the Z of the Si and Al are nearly the same, the stripping times of the two elements are comparable. By viewing the emission of the Si He-α line of the Si_3N_4 window and the Al He-α line of the aluminum patch in the same line of sight with a temporally streaked spectrometer, one measures a propagation time to the wall of 360 ps in good agreement with the calculated 350 ps.

The radiation temperature is measured through a hole cut in the side of the hohlraum (covered with 0.5 µm mylar) looking at the gold wall with the broadband DANTE x-ray spectrometer (9). Figure 3 shows the time dependant temperatures deconvolved from the 8 XRD filter-fluorescer channels for three shots with 3mg/cm^3 neopentane gas, normalized to 30 kJ energy in the laser pulse. After

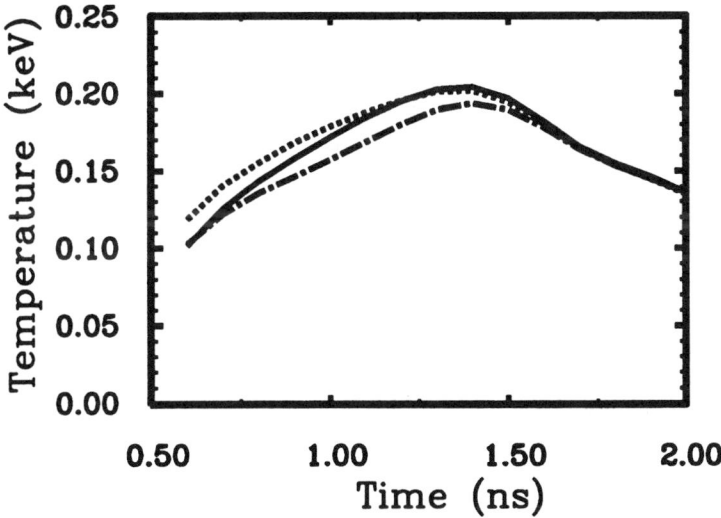

FIGURE 3. DANTE measured radiation temperatures for three almost identical toroidal hohlraums normalized to 30 kJ laser energy.

corrections for the gold albedo and the mylar transmission are applied, the LASNEX calculations agree very well with the data, i.e. the peak, rise time and shape of the radiation temperature agree.

To measure the electron temperature, a Ti/Cr (50/50) coated (2400 A° thick) carbon fiber (9 µm diameter) is placed approximately at the center of the hohlraum along its vertical axis. The isoelectronic line ratios of the Ti and Cr He-α are a function of the electron temperature (10) and are measured with a time and space resolved spectrometer. The fiber expansion as a function of time is monitored with a time resolved x-ray pinhole camera (WAX) that views the fiber through the laser entrance hole. Since the fiber is positioned in a three dimensional configuration, it can only be approximately modelled with the 2D LASNEX code. Figure 4a shows a LASNEX interface and mesh plot at 1 ns with the fiber modelled as a torus in a position that would intercept the laser beams. Models have also been calculated that have the toroidal fiber positioned closer to the axis outside of the laser beams or as a cylindrical fiber on the symmetry axis. Figure 4b shows that the fiber expansion is relatively insensitive to these various 2D approximations to the actual 3D geometry and is in good agreement with the fiber expansion measured by the x-ray images. Figure 4c shows the electron temperature along the fiber from the hohlraum center towards the gold wall as calculated with LASNEX at three times and as measured from the TI/Cr line ratios. Since the experimental errors are about 0.5 keV, the agreement with calculations is quite good.

By filling the hohlraum with deuterated neopentane (C_5D_{12}), and using the time of flight spreading of the DD generated neutrons, a peak ion temperature of 2 keV has been inferred, in good agreement with the LASNEX simulation. Even though the peak temperature does not occur until 2 ns, after the LPI measurements are over, this result provides more confidence in our simulations.

Spectrally and temporally resolved SRS measurements indicate the presence of densities near 10% critical. SDOSS (5,6) measurements also show the presence of 25% critical density (via two plasmon decay at 3/2ω) 800 µm off axis in the midplane of the hohlraum where the laser beams cross. This disagrees with the LASNEX calculations and is not yet understood.

CONCLUSIONS

We have designed a hohlraum for NOVA that has plasma conditions that approach those of the NIF point design with electron temperatures of 3 keV and electron densities of 10% of critical for scale lengths of about 2 mm. These plasma conditions have been verified experimentally and agree well with LASNEX computer simulations. Additional measurements that provide energy balance and laser propagation information also agree with the simulations. Various LPI experiments have been and are being fielded in this toroidal geometry(3-6).

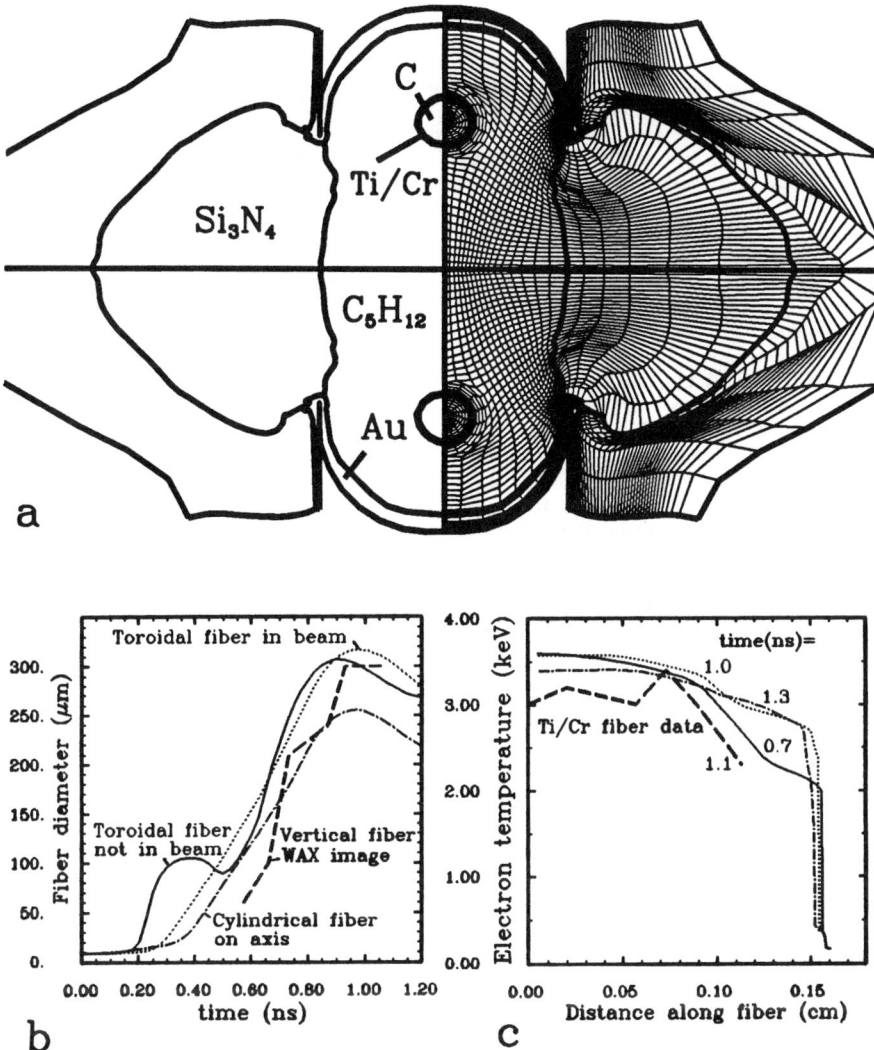

FIGURE 4. LASNEX interface and mesh plots at 1 ns for a toroidal Ti/Cr coated C fiber that is placed in the path of the laser beams (a). The fiber expansion as calculated with various 2D approximations and compared to the x-ray image data (b). The measured electron temperature along the vertical fiber compared to the LASNEX calculations (c).

REFERENCES

1. W. Haan, S. M. Pollaine, J. D. Lindl, L. J. Suter, R. L. Berger, L. V. Powers, W. E. Alley, P. A. Amendt, J. A. Futterman, W. K. Levedahl, M. D. Rosen, D. P. Rowley, R. . Sacks, A. I. Shestakov, L Strobel, G. L. Tabak, G. S. V. Weber, G. B. Zimmerman, W. J Krauser,D. C. Wilson, D. C. Coggesshall, D. B. Harris, N. M. Hoffman, and B. H. Wilde, "Design and Modeling of Ignition Targets for the National Ignition Facility", invited talk.,12th International Conference on "Laser Interaction and Related Plasma Phenomena", April 1995, paper TuII-1, this conference.
2. W. J. Krauser, B. H. Wilde, D. C. Wilson, P. Bradley, F. Swenson, and R. A. Kopp, "Integrated Ignition Calculations for Indirectly Driven Targets", 12th International Conference on "Laser Interaction and Related Plasma Phenomena", April 1995, paper MoP-1, this conference..
3. J. C. Fernandez, J. A. Cobble, P. L. Gobby, E. L. Lindman, D. S. Montgomery, H. A. Rose, B. H. Wilde, and M. D. Wilke, "Brillouin Backscatter and Seeding Mechanisms in NOVA Hohlraums," 12th International Conference on "Laser Interaction and Related Plasma Phenomena", April 1995, paper TuP-5, this conference..
4. J. C. Fernandez, J. A. Cobble, B. H. Failor, W. W. Hsing, H. A. Rose, B. H. Wilde, K. S. Bradley, P. L. Gobby, H. N. Kornblum, and D. S. Montgomery "Dependence of Stimulated Brillouin Scattering on Laser Intensity, Laser F Number and Ion Species in Hohlraum Plasmas (U)" Submitted to Physical Review Letters, 1995.
5. J. A. Cobble, S. Evans, J. A. Fernandez, J. Oertel, R. Watt, B. H. Wilde, "SDOSS: a Spatially Discriminating, Optical Streaked Spectrograph," 12th International conference on "Laser Interaction and Related Plasma Phenomena", April 1995, paper WdII-3, this conference..
6. J. A. Cobble, J. C. Fernandez, B. H. Wilde, S. Evans, J. Jimerson, J. Oertel, D. S. Montgomery, and C. C. Gomez, "Simultaneous Temporal, Spectral, and Spatial resolution of Laser Scatter from Parametric Plasma Instabilities (U)", Submitted to Review of Scientific Instruments, 1995.
7. G. B. Zimmerman and W. L. Kruer, Comments Plasma Phys. Controlled Fusion **2**, 51 (1975).
8. H. X. Vu, J. M. Wallace, and B. Bezzerides, Phys. of Plasmas (1994), submitted.
9. H. N. Kornblum, R. L. Kaufman, and J. A. Smith, Rev. of Sci. Instr. **57**, 2179 (1986).
10. R. S. Marjoribanks, M. C. Richardson, P. A. Jaaninagi, and R. Epstein, Phys. Rev **A46**, 1747, (1992).

Modeling of Drive-Symmetry Experiments in Gas-Filled Hohlraums at Nova

E. L. Lindman*, N. D. Delamater*, A. A. Hauer*, R. L. Kaufman†, G. R. Magelssen*, T. J. Murphy†, S. M. Pollaine†, L. V. Powers†, L. J. Suter†, and B. H. Wilde*

*Los Alamos National Laboratory, Los Alamos, NM 87545 USA and
†Lawrence Livermore National Laboratory, Livermore, CA 94551, USA

Abstract. Experiments on capsule implosions in gas-filled hohlraums have been carried out on the Nova Laser at Lawrence Livermore National Laboratory. Observed capsule shapes are more oblate than predicted using modeling methods which agree well with experiments in evacuated hohlraums. Improvements in modeling required to calculate these experiments and additional experiments are being pursued.

INTRODUCTION

Modeling and execution of drive-symmetry experiments in gas-filled hohlraums at Nova are currently being pursued to verify the accuracy of the design tools which we use to predict target performance for the National Ignition Facility (NIF) (1). Gas-filled hohlraums are currently preferred for the NIF because gas fill reduces the wall motion and avoids hydro-coupling asymmetries. LASNEX calculations suggest that the desired results can be achieved with approximately 1 mg/cm^3 of helium or an equivalent equimolar mixture of helium and hydrogen. When fully ionized this initial gas fill has an electron density of $n_e = 0.033 n_{crit}$. Subsequent blowoff from the walls and capsule raise the electron density in some regions to values approaching quarter critical.

NOVA EXPERIMENTS AND MODELING

The capsule-implosion-symmetry experiments and modeling were carried out (2) in scale-1 hohlraums (gold cylinders with diameters of 1650 μm and lengths between 2000 and 2800 μm) with standard capsules (plastic spheres with an innder diameter of 440 μm and an outer diameter of 540 μm. The laser entrance holes (LEH) at the ends of the cylindrical hohlraum have diameters of 75% or 100% of the hohlraum diameter. The gold hohlraums were unlined and filled with 1 atm of either neopentane, propane or methane held in by a 0.65 μm thick Mylar window or a 0.35 μm thick polyimide window. Since one atmosphere of methane has a mass density closest to the mass density of the gas to be used in the NIF ignition design, most of our experimental data has been obtained with methane.

The symmetry of imploded capsules was varied by changing the laser pointing. The pointing is described by giving the distance from the center of the hohlraum along the axis to the point of intersection with the central ray of the incoming laser beam. As the laser pointing was varied (from 975 to 1325 μm) different hohlraum lengths (from 2000 to 2700 μm) were used to insure that the laser beams always intersected the hohlraum axis and entered the hohlraum at the center of the LEH. The experiments were driven with shaped laser pulses consisting of a 1 to 2 ns low-power foot followed by a 1 ns main pulse with a contrast ratio between 2 and 3 and a total energy of 25 to 30 kJ.

A "standard" capsule has an inner radius of 220 μm and a total wall thickness of 54 μm made up of an inner layer of polystyrene of thickness 3.1 μm, an intermediate layer of PVA of thickness 3.3 μm and an outer layer of CH of thickness 48 μm. X-ray images of imploded standard capsules were used to examine the time-integrated symmetry over the whole laser pulse.

LASNEX calculations of these experiments are being carried out as well. These integrated calculations include the propagation of the laser through the window and gas and the energy deposition in the hohlraum walls and its reradiation as x-rays. Also included are the dynamics of the window, gas, hohlraum walls and capsule as shown in Fig. 1. Loss and redistribution of the energy by stimulated scattering instabilities and other plasma effects are, however, not included.

DATA AND ANALYSIS

Fig. 2a shows Nova data and LASNEX calculations for a methane pointing scan using standard capsules. The discrepancy between Nova data and LASNEX modeling is quite apparent. Similar disagreements were not observed in previous experiments with evacuated hohlraums. Mechanisms and missing physics which might account for the discrepancy include whole beam deflection, nonlocal electron thermal transport, and stimulated scattering of the energy towards either the capsule or the inside of the end caps of the hohlraum. Before discussing work on these effects it is useful to assess the effects

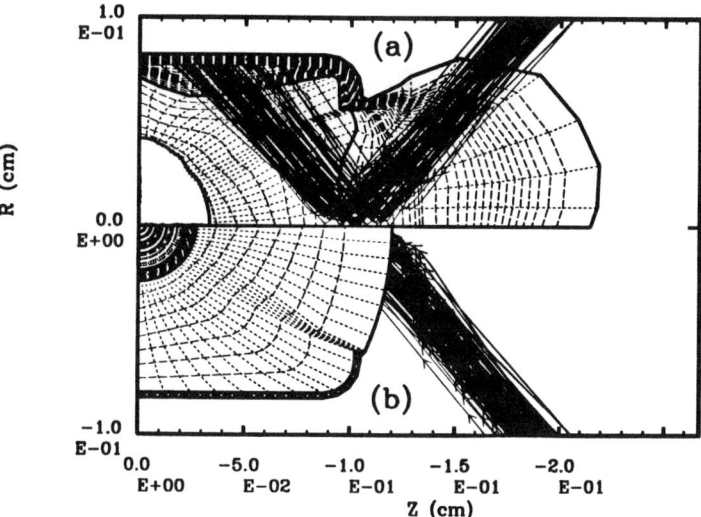

Figure 1: Lagrangian mesh at time = 0.4 ns compared with the initial mesh at −0.3ns. Initially the window and gas are opaque to the laser beams as shown on the bottom. At time = 0.4 ns, shown on the top, the laser has burned through the window and gas and is heating the gold wall which is expanding.

of the windows which hold the gas in the hohlraum prior to the shot. When they are vaporized and ionized, they contribute plasma to the hohlraum in addition to that coming from the gas. Consequently there has been considerable attention paid to their role in these experiments.

Window Plasma

One might conclude from the data in Fig. 2a that the effects of the windows are not important. Data for targets with both mylar and polyimide windows are included. The mass of a mylar window is twice the mass of a polyimide window, and, for the shortest hohlraum, the mass of the methane gas is greater than the mass of two polyimide windows by a factor of 1.5 and is less than the mass of two mylar window by a factor of 0.75. Little difference, however, is seen in the data from the different windows suggesting that the results are not dominated by the extra plasma coming from the windows.

The opposite conclusion might be drawn from Fig. 2b. In it distortion versus gas fill are plotted to show the dependence on plasma density. The data from an evacuated hohlraum with no windows, from another filled with one atmosphere of methane and from others filled with one atmosphere of

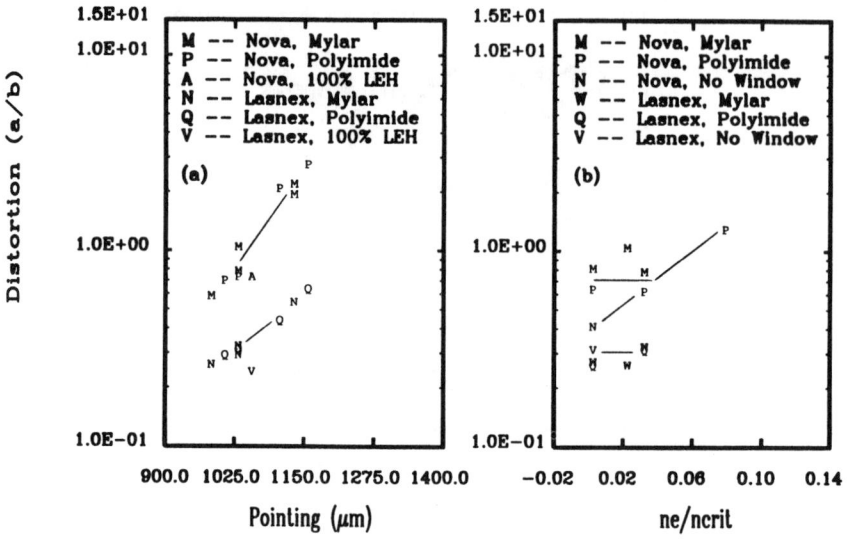

Figure 2: Imploded-capsule distortion versus pointing and density. Capsule distortion is computed by dividing the length of the semimajor axis transverse to the cylindrical axis by the length of the semimajor axis along the axis. The experimental distortions, plotted as a function of pointing for 1 atmosphere of methane on the left, and as a function of gas density at similar pointing on the right, in general have higher values than the corresponding calculated distortions.

propane fall on a straight line. Also plotted at zero gas density is data from an evacuated hohlraum with polyimide windows. The capsule distortion for latter case is similar to that obtained from the methane shot suggesting that the extra plasma from the window has an effect comparable to the methane fill.

These conflicting results may be caused by the different dynamics of the window plasmas which occur when gas is present or not. In any case the more interesting question is: What are the additional effects occurring in the experiments because of the presence of the extra plasma independent of its origin?

In contrast to the experimental data in Fig. 2b, the calculations show a very weak dependence on the extra plasma in the hohlraums. A discrepancy is also seen in Fig. 2a where the computed distortions, plotted versus pointing at fixed density, are quite different from the experimental data. This discrepancy was first observed in earlier experiments and modeling and was reported at the 23rd European Conference on Laser Interaction with Matter (3). In subsequent experiments and modeling we have verified the existence of the discrepancy and also examined some of the mechanisms and physics, missing from the calculations, which might account for the observed discrepancies.

Thermal Conduction

Electron thermal transport is computed using flux-limited diffusion in our modeling. A flux limiter of 0.05 has given better agreement with hohlraum and disk experiments in the past and has been used in these calculations. To test the possibility that the flux limiter was adversely affecting the agreement with experiment in these calculations, a set was performed in which the flux limiter was varied. The results of these calculations shown in Fig. 3b showed improved agreement of the calculated distortions of the imploded capsules with experiment. A calculation in which the flux limiter was set to 20 was used to remove it from the calculation. And the thermal diffusion transport coefficient was multiplied by 3 to test for errors in the coefficient. As shown in Fig. 3a these changes did move the results of the calculations closer to the experimental data, but agreement was not achieved. These results support the contention that thermal transport is not the source of the discrepancy, but do not rule it out completely.

Whole Beam Deflection

A deflection of the laser beams away from the capsule, that is not included in the modeling, could account for the discrepancy. When this occurs the bright spot on the wall of the hohlraum heated by the laser should be displaced from its expected position. Attempts to experimentally measure a shift in the spot position have been carried out, and a shift between 0 and 100 μm away from the capsule has been observed.

The beam deflection required to obtain agreement between modeling and experiment can be assessed computationally by doing calculations in which the laser beams are brought in at different angles. The results from such a sequence of calculations are shown in Fig. 3b. As the angle of incidence is changed and the spot is moved away from the capsule, the imploded-capsule distortion changes as expected. An angular deflection of 10° which corresponds to a spot shift of 200 μm appears to be required to match the experimental data.

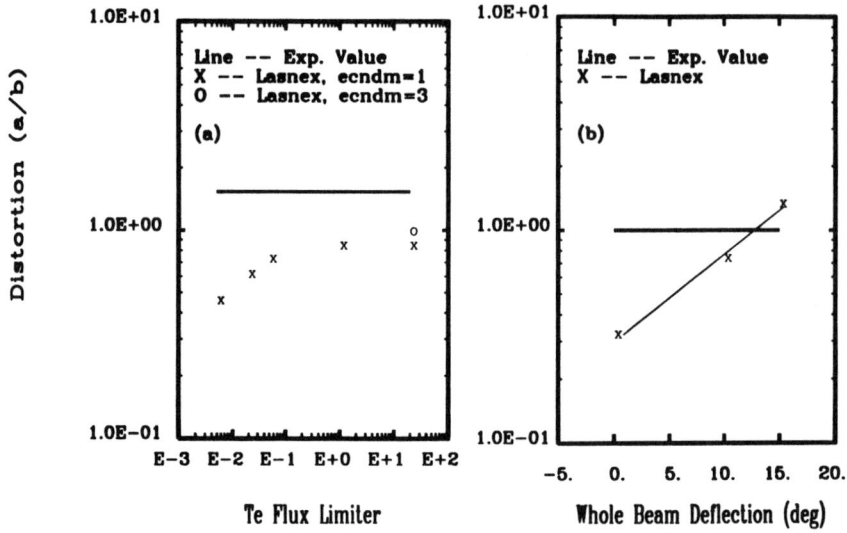

Figure 3: Computed distortion vs flux limiter and beam deflection. The computed distortions approach the experimental value as the flux limiter is moved closer to free streaming, and the diffusion coefficient is made stronger. A whole beam deflection of 10° or spot shift of 200 μm is required for agreement of calculations with experiment.

ACKNOWLEDGEMENTS

We would like to thank the LASNEX support team, target fabrication and the Nova operations staff for their expert contributions to this work. The work was performed under the auspices of the U. S. Department of Energy under Contract No. W–7405–ENG–36 at the Los Alamos National Laboratory and under Contract No. W–7405–ENG–48 at the Lawrence Livermore National Laboratory.

REFERENCES

1. S. W. Haan, et al., "Design and Modeling of Ignition Targets for the National Ignition Facility," To be published in *Physics of Fluids*.
2. N. D. Delamater, et al., "Symmetry experiments in gas filled hohlraums at Nova," To be published in the *Conference Proceedings of the 12th International Conference on Laser Interaction and Related Plasma Phenomena, Osaka, Japan, April 1995*.
3. E. L. Lindman, et al., "Effects of plasma physics on capsule implosions in gas-filled hohlraums," To be published in the *Conference Proceedings of the 23rd European Conference on Laser Interaction with Matter, Oxford, U. K., September 1994*.

Relevance of the U.S. National Ignition Facility for Driver and Target Options to Next-Step Inertial Fusion Test Facilities

B. Grant Logan

Lawrence Livermore National Laboratory
Livermore, CA 94551, USA

Achievement of inertial fusion ignition and energy gain in the proposed U.S. National Ignition Facility (1), shown in Figure 1, is a prerequisite for decisions to build next-step U.S. inertial fusion facilities for either high yield (>200 MJ) or high pulse-rate (>5 Hz) (2,3). There are a variety of target and driver options for such next-step inertial fusion test facilities, and this paper discusses possible ways that the NIF, using a 1.8-MJ glass laser in both direct and indirect-drive configurations, can provide target physics data relevant to several next-step facility options. Figure 2 illustrates several possible next-step facility options for the U.S., and how the NIF contributes to the target and driver decisions for those options. Next-step facility options illustrated in Figure 2 include the Engineering Test Facility (ETF) (4), which needs several-Hz pulse rates for testing relevant to Inertial Fusion Energy (IFE) development. An option for high yield, called the Laboratory Microfusion Facility (LMF) (5), does not require such high pulse rates, but may still benefit from driver technologies capable of much higher shot rates than possible with glass lasers. A high-pulse-rate driver could also be used for a combined ETF/LMF facility, driving multiple target chambers with a common driver (6). Driver technologies that could support high pulse rates for next-step options include heavy-ion (HI) and light-ion (LI) accelerators, diode-pumped solid-state lasers (DPSSLs), and krypton-flouride (KrF) gas lasers.

A U.S. workshop (7,8) last year found that the NIF can contribute data important to IFE in fundamental target physics relevant to both ion and laser drivers, and in direct- as well as indirect-drive configurations. The NIF could be used to provide important data for IFE in generic areas of target chamber damage and materials responses, neutron activation and heating, tritium recovery and safety management, and in performance tests of prototypical IFE targets and injection systems. In the study of ignition in both direct and indirect drive, the

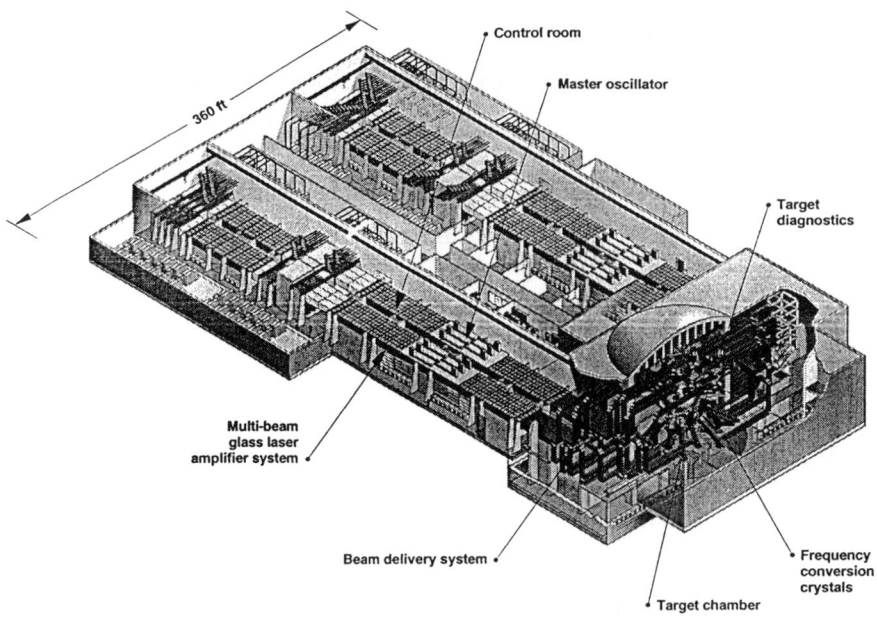

FIGURE 1. The U.S. National Ignition Facility. The NIF will provide 192 beams of 0.34-μm-wavelength light for a variety of inertial fusion ignition and gain experiments, delivering a total energy of 1.8 MJ, at a peak power of 500 TW.

NIF would explore generic ICF fuel-capsule implosion physics common to all driver and target options for next-step facilities. Below, we point out specific ways in which the NIF could be used to study target physics specifically relevant to the above-mentioned driver options for such next-step facilities, as well as how the NIF laser system itself could be relevant to the DPSSL option.

Figure 3 shows how the 192 NIF laser beams can be configured for both indirect drive (a) and direct drive (b) geometries. The direct-drive configuration is achieved by moving 24 of the 48 final optics assemblies (each assembly containing a 2 × 2 array of beams) into ports closer to the equator of the NIF target chamber. The indirect-drive configuration (a) provides illumination of two ends of hohlraums in a vertical-axis orientation from beams arranged around two pairs of cones on the top and bottom of the target chamber. This configuration will be used to test ignition in indirect-drive laser targets, where the beams enter the hohlraum through holes or thin windows (the upper target shown next to Figure 3a). This configuration can also be used to test IFE-model targets simulating heavy-ion (HI) targets (the lower target shown in Figure 3a). In this "HI-simulation" target, the laser beams are absorbed in "gas bags" at each end filled with 0.1-critical neopentane. This hydrocarbon gas can be doped with a high-Z

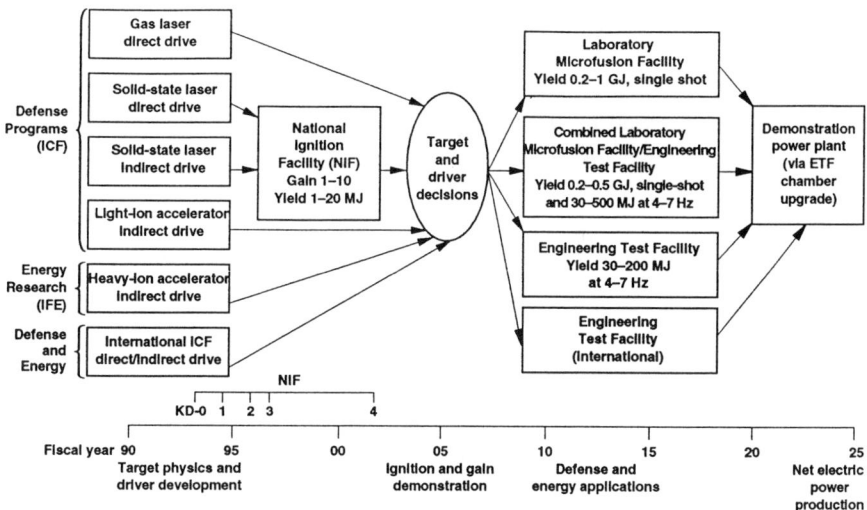

FIGURE 2. The role of the NIF in providing key data for a variety of possible options for next-step inertial fusion test facilities in the post-2005 time frame.

gas like xenon to provide similar radiation optical depth as that provided by the lead dopant used in the beryllium converters of the HIF target. Recent success in volume-heating dense gas-bag targets (9) on the NOVA laser facility at LLNL supports the suggestion that such gas radiators could be used to simulate HI converters. Because of x-ray loss out the ends of the gas-bag radiators, the HI-simulation target shown in Figure 3a would have lower hohlraum-to-capsule coupling efficiency than would the actual IFE target of Figure 4a, which has high-Z radiation cases enclosing the ends of the radiators (through which the heavy ions can pass). Thus, the HI-simulation targets would not be expected to achieve ignition, but would be useful for studying hohlraum drive symmetry relevant to HI-IFE targets. Since the soft x-ray intensity radiating out of the beam converter sources on the poles, viewed from the capsule, would be higher than the x-ray flux from the equator of the hohlraum, "shine shields" such as shown in Figure 3a (lower target) and Figure 4a are used to improve HIF capsule x-ray drive symmetry. Recent NOVA experiments with shine-shields inserted into indirect-drive laser targets have shown that such shine shields can be used to control the x-ray drive symmetry on the capsule (10).

Recent studies (11) have found that high-Z plasma blow-off from the HI-target hohlraum walls into the channels between the shine shields and the beryllium converters can significantly impede the transport of soft x rays through those channels, and thereby affect the capsule implosion symmetry. These types of HIF-model targets can and should be first tested in existing facilities like NOVA at ~1-ns pulse lengths. But such targets could also be tested in NIF at longer

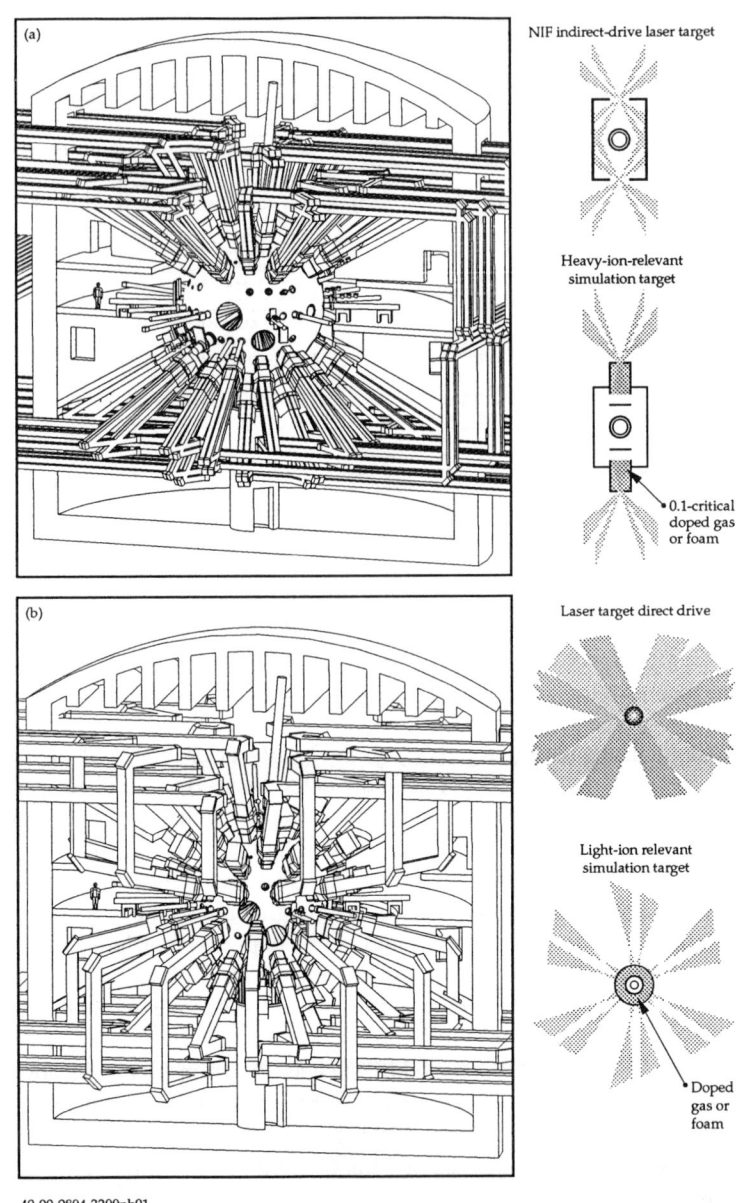

FIGURE 3. The NIF target area shown in two beam-illumination configurations for (a) indirect-drive and (b) direct-drive geometries. Sample targets that may be tested in each configuration are shown to the right of each NIF beam configuration. The light-ion simulation target (lower right of 3b) would use the NIF direct-drive beam configuration for 12-point illumination symmetry, but simulates an indirect-drive light-ion target design.

case to allow laser beam absorption, such "LI-simulation" targets would have lower coupling efficiency from laser energy to the capsule than in actual LI targets (Figure 4b), which would have an outer radiation case. Thus, as in the "HI-simulation" targets, such "LI-simulation" targets would not be used for ignition, but to study relevant issues of x-ray transport and capsule drive symmetry.

Besides target physics relevant to each driver option for next-step facilities, the NIF target chamber could also be used to provide data relevant to wall protection and chamber recovery. For all next-step facility options, the NIF could provide data on chamber wall and final optic materials damage to soft x-ray an target debris emissions from direct- or indirect-drive targets appropriate to each option. For next-step facility target chamber concepts using high-Z gas for attenuation of target x rays and debris, the NIF chamber (13) might be adapted to allow a gas fill (such as a few torr of Argon or Xenon) to test the degree of protection such gases could provide to final optics and in-chamber structural surfaces. The decay of gas ionization, shock-reverberations, and cool-down times subsequent to each shot in the NIF would be useful to determine the chamber recovery time to ambient conditions, which would be important to the maximum pulse-repetition rates that may be allowed in next-step facilities. Experiments with a series of no-yield targets injected into the NIF chamber (such as several foil disks injected 0.2 s apart) might be used to simulate multipulse chamber recovery dynamics in a burst of shots (13). This type of experiment may be possible since all 192 NIF beams have independent front-end pulse timing, allowing laser chain sections to be stagger-fired.

In addition to providing data relevant to target physics and target chamber clearing and materials damage, operation of the NIF's 192-beam glass laser will itself assist the development of any laser driver in terms of broad considerations such as the following:
1. Deployment of hundreds of beams that must be aligned, synchronized, balanced for uniform delivery of energy, and transported through a switchyard to interface with a target chamber.
2. Arrangement of the beams into cones or other patterns subtending large solid angles upon entrance through the envelope of the target chamber.
3. Transport of UV light, which requires the development of high-damage-threshold optics and coatings.
4. Use of the modularity of the laser system in the formation of a suitable pulse shape for a high-gain target.
5. Development of bandwidth (i.e., a band of wavelengths) and the investigation of the effect of bandwidth on target performance.

The NIF laser development and operation can have a special relevance to a DPSSL option for a next-step ICF facility. A major difference is the NIF uses Nd:Glass side-pumped with flash lamps, while a DPSSL uses a special crystal, Yb^{3+}-doped $Sr_5(PO_4)_3F$ (called Yb:S-FAP), end-pumped with laser diodes.

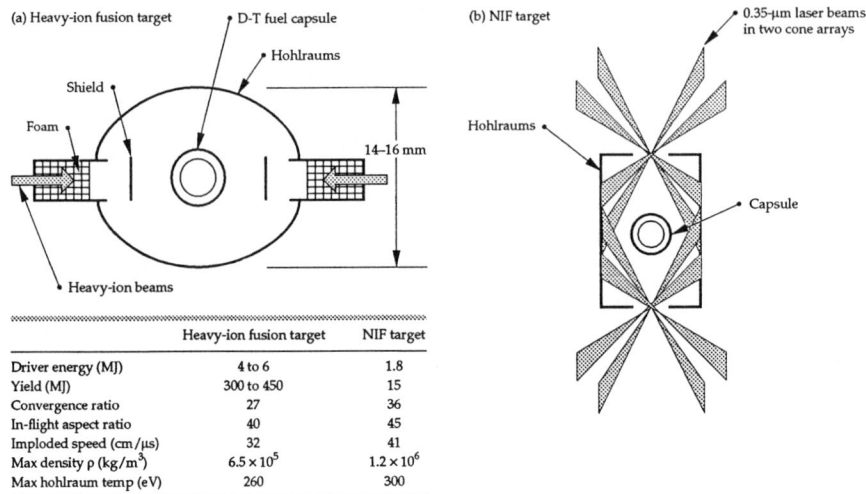

FIGURE 4. Schematic target designs for (a) a heavy-ion driver and (b) a light-ion driver. In both designs, the ion beams penetrate a thin, high-Z radiation case completely surrounding the capsule.

pulse lengths up to 10–20 ns more relevant to the HIF designs, where wall plasma blow-off can be more important. Such experiments in NIF can test ways to mitigate the effects of hohlraum wall blow-off for HIF targets by optimizing the hohlraum and shine shields geometry, and by filling the hohlraum with various densities of helium gas.

The NIF direct-drive beam configuration (Figure 3b) would be used to test both laser direct-drive (the upper target shown next to 3b), and spherical foam-overcoated capsules to simulate indirect-drive LI targets (the lower target shown next to 3b). The direct-drive targets would use 48 spherically-symmetric illumination points, each illumination point overlaped with 2×2 beamlets at different wavelengths to provide the bandwidth needed for 2-dimensional SSD beam smoothing. Direct-drive capsule ignition and gain performance could be studied for a range of capsule aspect ratios, surface finishes, and pulse shapes (various adiabats a) relevant to direct drive with DPSSL and KrF laser options for next-step facilities. By re-aiming the 48 NIF 2×2 beams to 12 spherically symmetric illumination points on a thick-foam overcoated capsule, the NIF can be used to study target physics issues important to LI targets relevant to the LMF (12). At some depth into the foam past the outer laser absorption zone, such LI simulation targets could be used to study the symmetrization from 12 illumination points by x-ray transport through the foam, as well as to simulate internal pulse shaping in the capsule ablator. Because there would be no outer high-Z radiation

Apart from using a different solid-state gain medium and method of pumping, a recent DPSSL design (14) shares many features in common with the NIF laser: Both use four-pass regenerative amplifiers operating near 1.05-µm wavelength, with an optical switch composed of a Pockels cell and a polarizer. Both use similar beam transport equipment and both use harmonic conversion to 3w in KD*P crystals. Both must interface with high-gain targets using hundreds of laser beams configured around a target chamber. The NIF laser operation can verify the functionality of a DPSSL driver in several areas: front-end electronics, four-pass regenerative laser amplifier operation, efficient conversion from 1w to 3w, computerized control equipment for aligning, synchronizing and energy-balancing many beams, final-optic interfaces to the target chamber, damage thresholds for optics and coatings, pulse shaping, beam transport in the target chamber with residual gas, focusability of 3w beams on target, and target physics in both direct- and indirect-drive configurations. Individual DPSSL beamlets have less bandwidth than the NIF glass laser, but the NIF will also test beam smoothing with the use of multiple colors in overlaping beams, which a many-beam DPSSL can also be designed to do.

ACKNOWLEDGMENTS

Work performed under the auspices of the U.S. Department of Energy by the Lawrence Livermore National Laboratory under Contract No. W-7405-ENG-48.

REFERENCES

1. "National Ignition Facility Conceptual Design Report—Executive Summary," LLNL report UCRL-PROP-117093 ES, U.S. National Ignition Facility Document NIF-LLNL-94-113, L-16973-ES, August 1994.
2. National Research Council, *Second Review of the Department of Energy's Inertial Confinement Fusion Program, Final Report*, (National Academy Press, Washington, D.C., 1990).
3. Fusion Policy Advisory Committee, *Review of the U.S. Fusion Program, Final Report*, DOE Office of Energy Research, Washington, D.C., 20585 (1990).
4. "Findings and Recommendations for the Heavy-Ion Fusion Program," *Report of the Fusion Energy Advisory Committee (FEAC) to DOE Energy Research Director William Happer* (April 1993).
5. *The Laboratory Microfusion Capability Study Phase II Report*, DOE Office of Inertial Confinement Fusion DOE/DP-0017, May 1993.
6. C. L. Olson, R. O. Bangerter, B. G. Logan, and J. D. Lindl, "ICF Driver Strategies for IFE," in *Proc. of the IAEA Technical Committee Meeting on Drivers for ICF*, Paris, France, Nov. 14-18, 1994.

7. B. G. Logan, A.T. Anderson, M.T. Tobin, V. E. Schrock, W. R. Meier, R. O. Bangerter, R. E. Tokheim, M. A. Abdou, and K. R. Schultz, "Utility of the U.S. National Ignition Facility for Development of Inertial Fusion Energy," in *Proc. of 15th IAEA Int. Conf. on Plasma Physics and Controlled Fusion Research*, Seville, Spain, Sep. 26–Oct. 1, 1994, paper IAEA-CN-60/B-P-15.
8. B. G. Logan, M. T. Tobin, and W. R. Meier, general editors, *The Role of the U.S. National Ignition Facility in the Development of Inertial Fusion Energy*, Lawrence Livermore National Laboratory Report UCRL-ID 119383, Dec. 15, 1994.
9. B. J. MacGowan, "The Study of Parametric Instabilities in NIF-Scale Plasmas on Nova," *Proc. of 15th IAEA Int. Conf. on Plasma Phys. and Controlled Fusion Research* (Seville, Spain, Sep. 25–Oct. 1, 1994), IAEA-CN-60/B-P16 (1994).
10. T. J. Murphy and P. Amendt, "X-ray Flux Symmetry Control in Advanced Nova Hohlraums," LLNL *ICF Quarterly Report* **4** (3), 101–107 (1994).
11. D. D.-M. Ho, J. A. Harte, and M. Tabak, "Radiation-Driven Targets for Heavy-Ion Fusion," *Proc. of 15th IAEA Int. Conf. on Plasma Phys. and Controlled Fusion Research* (Seville, Spain, Sep. 6–Oct. 1, 1994), IAEA-CN-60/B-P-13 (1994).
12. *Inertial Confinement Fusion Five-Year Program Plan FY 1992–FY 1996*, DOE Office of Research and Inertial Confinement Fusion, Washington D.C, 20585, Dec. 31, 1991, revised March 1993.
13. B. G. Logan, general editor, "White Paper: Laser System and Target Area Design Needs for Inertial Fusion Energy Experiments in the National Ignition Facility," NIF Project Document number NIF-LLNL-95-171, L-19313-1, WBS 1.11.5, available from the NIF Project Office, Jeff Paisner, mail stop L-488, Lawrence Livermore National Laboratory, P.O. Box 808, Livermore, California, 94551 (March 31, 1995).
14. D. Orth, S. A. Payne, and W. F. Krupke, *A Diode-Punped Solid State Laser Driver for Inertial Fusion Energy*, LLNL report UCRL-JC-116173 Aug. 30, 1994 (Submitted to *Nuclear Fusion*).

A New Drive for ICF: Externally Guided Implosions

J.M. Martínez-Val, M. Piera and P.M. Velarde

Institute of Nuclear Fusion
Madrid, Polytechnic University, Madrid 28006, Spain

Abstract. A new drive method to implode and ignite fuel micropellets is propose. It is based on the acceleration of two pellets inside a tube. The pellets are originally placed on the tube mouths and collide in the center of the tube at the end of the acceleration process. In order to increase the fuel areal density during the implosion, the tube cross section decreases from the mouths inward. There is a small cavity in the center of the tube where the collision takes place. The cavity radius expands as the inner pressure goes up, but the short inertial confinement provided by the tube material can be strong enough to trigger a fusion burst. Numerical simulations on this concept are presented in this paper.

INTRODUCTION

One of the main difficulties of any type of ICF concept is to achieve a perfect spherical implosion with a limited number of driver beams. The option of indirect drive has the drawback of a small energy efficiency because of the huge fraction of driver energy that does not arrive as thermal radiation onto the target itself.

The system presented in this paper is devised to try to ignite a fusion microburst with a minimum number of driver beams, namely two. Figure 1 is a sketch of the type of target used in this system (1,2). The fuel pellet and its coating are placed in the ends of a tube made of lead or a similar high density material. Each pellet is illuminated by a beam. The ablator thickness and density must be fitted to the type of driver. Particle beams will need a much heavier coating than laser beams.

The coating ablation will produce the fuel acceleration inside the tube. Although the hydrodynamic regime will be highly turbulent, an almost planar acceleration of the fuel is predicted by our numerical simulations. Nevertheless, the central part of the fuel pellet becomes faster than the outer ring not only because of friction forces but because the conical shape of the tube cross section.

The main objective in this scheme is to obtain a ultrafast stream of fuel flowing inwards. Both fuel jets collide in a small cavity in the center of the tube. As the collision takes place, the kinetic energy is converted in internal energy and the inner pressure goes up. The tube suffers a very fast expansion (and will be blown into pieces at the end). It is difficult, for not to say impossible, to reach very high densities with this scheme, because of the lack of isotropic confinement. Even so, burnup fractions about 10% could be achieved if the fuel is properly accelerated.

© 1996 American Institute of Physics

In the following paragraph, a theory of the central collision is described in order to asess the potentiality of the concept.

THE PHYSICS OF THE CENTRAL COLLISION

Main features of this scheme can be derived by using the mass, momentum and energy conservation in the cavity as the fuel moves in after the acceleration process.

The values of the variables at the entrance sections (throttles) will be identified by the subscript e. Values (averaged) inside the cavity will not have any subindex.

The radius R of the cavity will not remain constant. It will increase according to the material speed of the tube inner face for a cavity pressure P

$$\frac{dR}{dt} = v_t = \left(\frac{P}{\beta}\rho_t\right)^{1/2} \tag{1}$$

where the Hugoniot parameter $\beta \approx 1.4$. In order to be conservative in our estimate, ρ_t, will be kept constant (for instance, $11 g/cm^3$ for lead) although it would be an increasing function with P (because compression effects).

If V stands for the cavity volume and S for the cross section of the entrance holes, the balance of mass can be written as

$$\frac{d}{dt}\int_V \rho dV = \int_S \rho_e v_e dS \tag{2}$$

if γ stands for the fraction of solid angle covered by the entrance holes, we can write

$$V\frac{d\rho}{dt} + \rho\frac{dV}{dt} = \rho_e v_e 4\pi R^2 \gamma \tag{3}$$

The evolution of V is known by Eq. (1) and the evolution of the density corresponds to

$$\frac{d\rho}{dt} = \frac{3}{R}(\gamma \rho_e v_e - \rho v_t) \tag{4}$$

The balance of energy can be written as

$$\frac{d}{dt}\int_V \rho E dV = \int_S \rho_e E_e v_e dS + \int_S \frac{\rho_e v_e^3}{2} dS - \int_C P_t v_t dC \qquad (5)$$

where C represents the surface of the cavity covered by the tube.

The first term of the r.h.s. can usually be neglected (because $E_e < v_e^2/2$). The l.h.s. can be split in the following derivatives

$$\frac{d}{dt}\int_V \rho E dV = V\frac{d(\rho E)}{dt} + \rho E \frac{dV}{dt} \qquad (6)$$

and it can be written for the internal energy

$$\frac{dE}{dt} = \frac{3\gamma}{\rho R}\rho_e v_e \left(E_e + \frac{v_e^2}{2} - E\right) - \frac{15 P v_t}{2\rho R} \qquad (7)$$

It is worth pointing our that the last term of the r.h.s. embodies two phenomena: the expansion of the cavity (by increasing V) and the energy lost by making the expansion (last term of the balance in Eq. (5)). In order to write them, use has been made of the relation $P = 2\rho E/3$

Two additional Eqs. are needed to determine ρ_e and v_e, and they correspond to the continuity and momentum conservation principles in the entrance holes. In a first degree of approximation (without using space derivatives) we can write

$$\rho_i v_i = \rho_e v_e \qquad (8)$$

$$P_i + \rho_i v_i^2 = P + \rho_e v_e^2 \qquad (9)$$

The entrance speed is thus

$$v_e = v_i - \frac{P - P_i}{\rho_i v_i} \qquad (10)$$

From this Eq., it can be seen that v_e becomes negative if $P > P_i + \rho_i v_i^2$

Of course, $v_e < 0$ means that no more fuel can enter the cavity. It is important to remember that the cavity pressure P depends on the density and specific energy of the fuel filling the cavity.

The mass of fuel in the cavity is depends on R(t) and ρ(t) and the temperature will be proportional to the specific energy, T = cE where c=8.68 x 10^{-16} if c.g.s. units are used for E and keV for T.

The foregoing equations show the following features of the central collision dynamics: the fuel in the cavity will initially have a specific energy E (at very low density) close to the value $v_i^2/2$ where v_i is the final implosion speed. As the fuel fills the cavity, v_e will go down ($v_e < v_i$) and so will E. Besides this effect, the density will not exceed a maximum value (see Eq. (4)) that mainly depends on γ, $ρ_i$ and v_i, and is also inversely proportional to E. Although (dE/dt) will in general be negative, its absolute value decreases as ρR increases (see Eqs. 7) what is useful for keeping E as high as possible.

This type of dynamics could be called "tampered impact fusion" because it mainly relies on the incoming speed of the fuel. The tampering effect provided by the guide-tube is essential to have a confinement time long enough for a fusion burst to succeed.

A fundamental requirement for ignition to be triggered is that ρR has to reach a value higher than 0.3 g/cm² before the fuel becomes too cool. Otherwise, the reheat effect by alpha particles (let alone the 14.1 MeV neutrons) is not effective.

The foregoing hydrodynamic equations did not take into account the reheating effect produced by the fusion product energy deposition. This can be taken into account by including an additional term in the energy balance equation, that will thus affect the cavity pressure P, the expanding speed v_t and therefore the density as well as the entrance speed v_e.

Per fusion reaction, the alpha-particle reheat will be

$$E_\alpha(keV) = 3500 \frac{\rho R}{0.1} \tag{11}$$

The neutron reheat can be computed by taking into account the neutron m.f.p. (4.7 g/cm²) and the elastic kerma factor which can be expressed (3) as

$$E_n(keV) = 3000 \frac{\rho R}{4.7} \tag{12}$$

with the limitation that E_n <14100 keV and Eα < 3500 keV

Hence, the rate of specific energy deposition will be (in c.g.s.)

$$E' = 1.6 \times 10^{-9} n^2 \langle \sigma v \rangle (E_\alpha + E_n)/\rho \tag{13}$$

and the energy balance equation will be from Eq. (7)

$$\frac{dE}{dt} = E' + \frac{3\gamma}{\rho R}\rho_e v_e \left(E_e + \frac{v_e^2}{2} - E\right) - \frac{15 P v_t}{2\rho R} \qquad (14)$$

The hydrodynamic evolution of the target is thus severely modified if the fusion burst takes place. The yield of the fusion burst can be computed by

$$E_y = 1.6 \times 10^{-6} \times 17.6 \int_0^t \langle \sigma v \rangle n^2 V dt \qquad (15)$$

and the intrinsic gain (IG) would be

$$IG = \frac{2 E_y}{M v_i^2} \qquad (16)$$

Figures 2-3 show the hydrodynamic evolution of a case where the fusion burst has been triggered. It corresponds to the following specifications: implosion speed $=1.4 \times 10^8$ cm/s, implosion density of the fuel at the cavity holes $=30$ g/cm^3, initial radius $=400$ μm and geometry parameter $\gamma=0.5$. At the beginning, the temperature corresponds to the kinetic energy of the fuel entering the cavity ~ 8.5 keV) but the density is practically zero because the filling of the cavity has just started. Because of the cavity expansion due to the inner pressure, the fuel in the cavity cools down a little bit as its areal density goes up. Fusion power becomes significant as the fuel density increases, and the energy deposition by fusion products also increases as the areal density of the fuel rises. About 0.25 ns, the temperature reaches 10 keV and the density is high enough (25 g/cm^3) to trigger a fusion burst. 20 ps later the inner pressure is so high that no more fuel can come in the cavity. As a matter of fact, at 0.28 ns the fuel is being expelled through the cavity holes at the same time as the cavity explodes at a very high speed (2.5 x 10^8 cm/s).

From this point on, the mechanical disassembly takes place, but the fusion burst still goes for a while, although it practically stops at 0.6 ns. At that time, the temperature still is very high (~50 keV) but the density is rather low (below 2 g/cm^3 inside the cavity). The radius of the cavity is then 1 mm, but a lot of fuel has already escaped through the holes (at that time, the outgoing speed is about 2 x 10^8 cm/s).

In order to drive this case, 30 mg of DT has been used (imploded along the conical pipes joining the cavity). The kinetic energy of the fuel at 1.4 x 10^8 cm/s was 3 x 10^{14} erg (30 MJ).

On the other hand the total energy yield was about 1000 MJ (10^{16} erg) what means an intrinsic gain IG=1000/30=33. The burnup fraction of the fuel has been 10% of the total fuel (16% of the fuel that filled the cavity). This value can be improved if higher implosion densities are achieved (see Eq. (16)). Of course, the actual gain is smaller because the hydrodynamic efficiency of the fuel acceleration, In a laser case, it can be $\eta_H=10\%$, which means that the actual gain G=3.3. For particle beam driving, η_H can be higher (~20%) and G~6.6. Although this value is very modest, it must be remembered the simplicity of the scheme and its unsensitivity to non-uniformities both in the target and in the beams.

REFERENCES

1. Martínez-Val, J.M. and Piera, M., "An externally guided target for Inertial Fusion", *Fusion Technology* (submitted, 1995).
2. Velarde, G. et al. Paper B-2-II-6, 15th IAEA Conference on Plasma Physics and Controlled Nuclear Fusion Research, Sevilla, Spain, Oct. 94.
3. Martínez-Val, J.M. *Fusion Technology* **17**, 476 (1990).

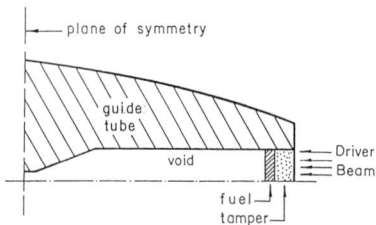

Fig. 1. A sketch of an externally guided ICF target

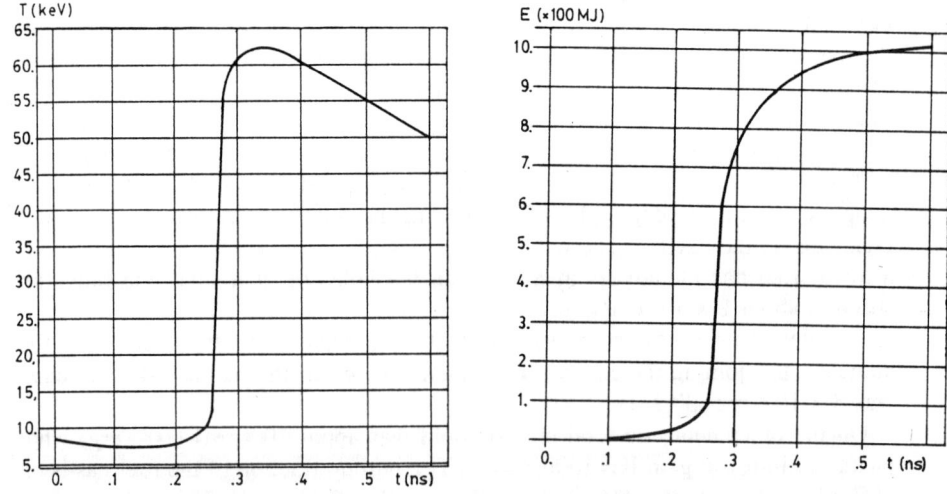

Fig. 2-3. Temperature and fusion yield evolution of a case described in the text.

Implosion and Ignition of Stagnation-Free Targets

J.M. Martínez-Val, G. Velarde, S. Eliezer*,
H. Hora*, J.J. Honrubia, E. Mínguez, M. Perlado
M. Piera and P. Velarde

Institute of Nuclear Fusion
Madrid, Polytechnic University, Madrid 28006, Spain

Abstract. Stagnation-free targets can be produced by tailoring the implosion process in order to get an untamped fuel sphere just before void closure. This type of targets have the advantage of being insensitive to pusher-fuel mixing during the deceleration phase. On the contrary, the confinement time is reduced to a minimum value because there is not any tampering effect and the fuel disassembles as the internal overpressure sends a strong shockwave outwards. Nevertheless, if the implosion is suitably tuned and the density is high enough, a fusion burst can take place and high energy gains can be achieved.

INTRODUCTION

Most of the target designs for Inertial Fusion (1) are based on a hollow microsphere of deuterium-tritium (DT) coated by a suitable material. The coating has two objectives: first, to absorb the energy of the driving beams; second, to push the fuel inwards and to tamper it while the maximum compression is achieved. For that purpose, a multilayered coating is devised, with at least two outer shells: the pusher and the ablator.

Problems related to target mixing and implosion instability have been pointed out in the performance of these targets (2). There seem to be two phases where instabilities can make the implosion to fail. First, the early times of the acceleration, because of non-uniformities in the ablation process. Second, the deceleration period, when the fuel (after void closure) has to stop the heavier pusher. This situation is unstable (in a Rayleigh-Taylor sense) and a finger of the pusher can penetrate the fuel and displace it from the center, so avoiding the actual fuel compression. There are some experimental evidences (3,4) that fuel displacement by the pusher really happens and quenches the ignition burst.

Stagnation-free targets (5) offer an alternative to the standard way of driving the implosion. Stagnation-free targets are also named pusherless targets (6) because the fuel becomes uncoated by the end of the implosion. In laser-direct-drive, experiments (7) have been performed by shock-multiplexing. In particle-beam (direct) drive (5), pusherless fuels can be obtained by tailoring the beam energy

© 1996 American Institute of Physics

deposition into the coating, in order to evaporate it by the end of the implosion. In such a way, neither pusher-fuel mixing non fuel displacement by a pusher finger can happen. Therefore, the stagnation phase, characterized by a confinement time along which several shock-waves sweep back and forth the fuel, is not possible. The fuel begins to disassemble as the first shock-wave moves outwards.

THE PHYSICS OF STAGNATION-FREE TARGET

The confinement time of stagnation-free targets is very short, of the order of

$$t_c = \frac{R}{3c_s} \quad (1)$$

where R is the compressed fuel radius and c_s the sound speed. For DT plasma fully ionized (and taking into account both the electron and ion pressure) the sound speed can be estimated as

$$c_s \sim 3.5 \times 10^7 \, T^{1/2} \quad (2)$$

where T is in keV and c_s in cm/s.

In order to succeed in triggering and propagating ignition, the confinement time must be larger than the heating time and the burnup time. The heating time t_h is defined as the time needed to heat the plasma from the temperature hydrodynamically achieved at the end of the implosion up to 10 keV (for which the fusion reactivity is already very high, namely 10^{-16} cm^3/s). The burnup time t_b is defined as the time needed to get a significant burnup fraction ($\geq 25\%$) in a heated plasma ($<\sigma v>$ greater than 10^{-16} cm^3/s).

These times can be estimated by the following equations (5)

$$t_h = \int_T^{10} \frac{6 \, n \, k \, dT}{<\sigma v> \, n^2 \, E_f} \quad (3)$$

where E_f is the fusion-products energy deposition (in keV). For targets with an areal density of $\rho R \geq 4$ g/cm^2, it can be written

$$t_h \sim 0.3 \times 10^{17} \, n^{-1} \, T^{-3} \quad (4)$$

if the reactivity $<\sigma v>$ is fitted to $10^{-20} \, T^4$, between 1 and 10 keV.

The burnup time can be estimated by the evolution of the burnup fraction ϕ

$$\frac{d\phi}{dt} = \frac{<\sigma v> \, n^2}{n} \quad (5)$$

$$t_b \leq 0.25 \times 10^{16} \, n^{-1} \qquad (6)$$

For standard values of ICF compressions ($n \geq 0.24 \times 10^{26}$) t_b is much shorter than t_h. Hence, a criterion for ignition in stagnation free targets is $t_c > t_h$, that can be written in terms of the fuel mass M, the compression factor $f(=\rho/\rho_o)$ and the temperature achieved at the end of the implosion. We take into account that in cgs units, for a DT plasma

$$R = 1.05 \left(\frac{M}{f}\right)^{1/3} \qquad (7)$$

and thus

$$t_c = \frac{R}{3c_s} = 10^{-8} \, M^{1/3} \, f^{-1/3} \, T^{1/2} \qquad (8)$$

and an index of quality can be defined to assess the criterion for ignition

$$i = \frac{t_c}{t_h} = \frac{10^{-8} \, M^{1/3} \, f^{-1/3} \, T^{1/2}}{1.2 \times 10^{-6} \, f^{-1} \, T^{-3}} \qquad (9)$$

$$i = 0.008 \, M^{1/3} \, f^{2/3} \, T^{5/2} \qquad (10)$$

For Volume Ignition to succeed, this index has to be higher than 1. This equation represents a scaling law very useful to define the condition for V.I. For instance, if M= 0.001g (1 mg), and a compression factor 5.000 is assumed ($f^{2/3} = 290$) the ignition temperature has to be higher than 1.8 keV. However, if the compression factor is only 100 ($f^{2/3} = 100$) the ignition temperature has to be higher than 2.8 keV. It is obvious from the index equation that the role of achieving high compression densities is not negligible. It is worth remembering that the energy gain of volume ignition is inversely proportional to the ignition temperature, and therefore is extremely interesting to reach high densities because they produce high gains.

The gain equation can be established on the basis of the specific fusion yield, which is 335 MJ/mg, and the specific internal energy of the plasma at the end of the implosion, which is 115 T kJ/mg. If a burnup fraction ϕ in the fusion burst and an implosion efficiency η are achieved, the gain can be written as

$$Gain = \eta \phi \, 2.9 \, 10^3 \, T^{-1} \qquad (11)$$

For instance, with T = 2 KeV, η=10% and ϕ =33%, the gain would be 48, a value high enough to be interesting for fusion energy applications.

HOW STAGNATION-FREE IMPLOSIONS CAN BE DRIVEN?

Spark ignition profiles are based on the assumption that a sort of unstable profile can be maintained during the stagnation time, with a high temperature, low density plasma region in the central core and a lower temperature, high density plasma (plus a pusher) in the outer shells.

Volume ignition do present a much more uniform and stable situation, with the T, ρ and P profiles looking pretty much the same, as can be seen in figure 1. The main drawback of this situation comes from the fact that the confinement time is very short because there is not any extra confinement due to an outer pusher. The confinement time is just the while provided by the inertial forces of the imploding fuel. Once the outer fuel is stopped by the outgoing first shock wave after void closure, the mechanical disassembly starts, unlike in tampered targets, where the outer pusher will theoretically refrain the fuel from expanding (although the fuel and the pusher can become mixed up at once).

In order to produce a stagnation-free compressed fuel it is necessary to uncoat it just before the end of the implosion. Hence, the ablation process must be tuned to fulfill this objective, which has been demonstrated by experiments in laser-driven implosions (8). In these cases, it is extremely important to drive the implosion by shock multiplexing in such a way that all shocks arrive in the center almost at the same time. The shock train is produced by the ablation pressure, and this one is produced by the ablation mass rate. Hence, a coherent way of implosion drive can be established to get an optimally accelerate pusherless fuel at the end of the flight.

In particle beam drive there is no experiment evidence because of the lack of facilities. However, numerical simulations (5, 9) with heavy-ion direct drive point out that a pusherless implosion can be performed (provided the beam energy deposition is spherically uniform). The main parameter that governs the implosion performance is the ion energy deposition rate in the coating. In this case, a much thicker ablator than in the laser case is needed. For 1 mg DT class targets driven by Bi ions of some GeV, more than 100 mg of lithium are needed in the coating. The evaporation process is less intense than in the laser case, which is a positive effect for not to preheat the fuel and not to create very strong shock waves that would accelerate too much the innermost shell of the fuel.

Compression efficiency optimization

If the implosion process is not properly controlled and the shock waves arrive to the inner face of the fuel too early, the fuel becomes hydrodynamically decoupled. This means that the innermost part of the fuel will implode at a much higher speed than the bulk of it (10). Why is the hydro decoupling so negative for the compression performance? Because the inner fuel has a specific kinetic energy ($v^2/2$) much higher than the outer fuel. When the void closes, the inner fuel ($v^2/2$) is converted into internal energy E and, correspondingly, the pressure rises (A measurement of the pressure is $2\rho E/3$, i.e., the pressure is proportional to $\rho\, v^2$). The sudden rise in pressure in the fuel center after void closure sends a shock wave outward at the same time that the rest of the fuel shells begin to suffer the deceleration phase in a propagating way. If the fuel was hydro decoupled before void closure, the outer fuel will be stopped at a radius very far from the center, which means a poor compression factor. It is true that all the kinetic energy has been converted to internal energy, with a fraction of radiation losses (the outer temperature is so low and the fuel optical thickness so high that the losses rarely reach 5% of the internal energy during the deceleration period). However, the final density is very low if the fuel was hydro decoupled and a small f means a small index of quality. Therefore, in order to produce efficient stagnation-free implosions is mandatory to drive a hydro coupled implosion. This is possible both in the laser case, where very thin coatings are needed. It also possible in the heavy ion case, where thick coatings receiving a much smaller dose rate do present a different hydrodynamic regime leading to a similar final result: an almost uncoated sphere of fuel compressed to very high densities. If the temperature is big enough according to the criterion established before, volume ignition can take place (11).

NUMERICAL SIMULATION OF VOLUME IGNITION TARGETS

The NORCLA code (12) has been used to simulate the implosion, compression ignition and burn of stagnation-free targets. NORCLA is a one-dimensional, single-fluid, three-temperatures code that solves the hydrodynamic equations with energy transport by diffusion approximation and calculates the fusion reaction rates and the energy deposition of alphas and neutrons.

Figure 1 depicts the result of a pure hydrodynamic calculation of the implosion of a target initially made of two shells: the inner one is a DT layer (r_{in}=1.5539 cm, r_{out}=1.5694, ρ_0=0.21 g/cm^3) and the outer one is a coating of lithium (ρ=0.51 g/cm3, Δr=0.25 cm) driven by a heavy ion pulse (perfectly focused of 10 GeV Bi$^+$ ions with a constant power of 5000 TW for 40 ns (it must be taken into account that the DT mass is in this case 100 mg).

It is seen that the fuel is almost uncoated just before void closure, the density of the lithium being much smaller than the DT density. The void closes at 49.71 ns and the maximum ρR (minimum radius) is achieved at 50.91 ns. The profile of the temperature does not show a hot spark in the center, but a typical shape of a hot black body. The density profile is very flat with a maximum in the center. With these conditions at the end of the implosion, a huge fusion burst would be triggered with an energy gain higher than 60.

ACKNOWLEDGMENTS

This work was done under the auspicies of the European Union Scientific network "High Energy Density Matter".

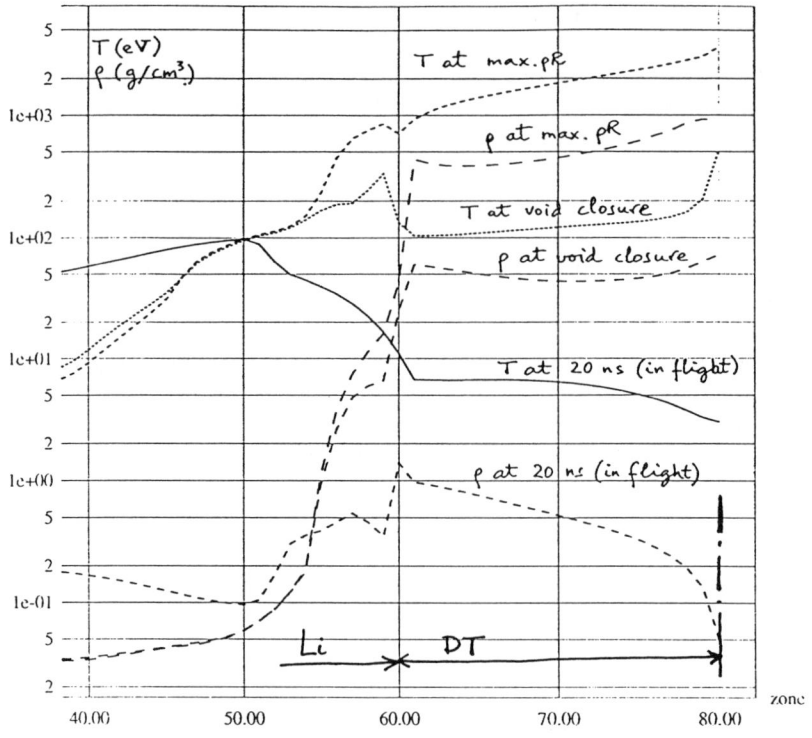

Fig. 1. Temperature and density profiles in V.I.

REFERENCES

1. Martínez-Val, J.M., Velarde, G. and Ronen, Y., *Nuclear Fusion by Inertial Confinement*, G. Velarde, Y. Ronen and J.M. Martínez-Val, Eds. CRC Press, Boca Raton, FL, 1993, ch. 1. pp. 1-43.
2. Gamaly, E., *Nuclear Fusion by Inertial Confinement*, G. Velarde, Y. Ronen and J.M. Martínez-Val, Eds. CRC Press, Boca Raton, FL, 1993, ch. 13. pp. 321-351.
3. Bayer, C. et al., *Nuclear Fusion*, **24**, 1984, p. 573.
4. Yamanaka, T. et al., "Target implosions using deuterium cryogenic system", *IAEA 15th Conf. Plasma Physics and Controlled Nuclear Fusion*, paper B-2-II-5, Seville (Spain) 1994.
5. Martínez-Val, J.M., Eliezer, S. and Piera, M., *Laser and Particle Beams*, **12**, 1994, p. 681.
6. Mima, K., Takabe, H. and Nakai, S., *Laser and Particle Beams*, **7**, 1989, p. 249.
7. Yamanaka, C., Naki, S., *Nature*, **319**, 1986, p. 757.
8. Nakai, S., *Laser and Particle Beams*, **7**, 1989, p. 467.
9. Martínez-Val, J.M. et al., *Il Nuovo Cimento*, **106 A**, 1993, p. 1873.
10. Velarde, G. et al., *Particle Accelerators*, **37**, 1992, p. 537.
11. Eliezer, S. and Hora, H. *The physics of directly driven targets*, Chapter 2 of book in Ref. 1.
12. Velarde, G. et al., *Laser and Particle Beams*, **7**, 1989, p. 305.

Effect of Fusion Reaction Products Heating on The Volume Ignition of DT and D^3He Fuel Pellets

R. Khoda-Bakhsh

Department of physics, University of Urmia,

P.O.Box 165 Urmia-IRAN

Abstract: Laser fusion simulations are carried out for the DT and D-^{3}He pellets by using a hydrodynamic code including heating from all charged reaction pruducts and neutron. It is shown that, the inclusion of the side reactions and heating from all reactions products in the fuel pellets has an appreciable affect on the plasma temperature, the ICF drive energy requirement, fusion gain and the ignition condilions. The total input energy is decreased, the burn efficiency and total gain are increased compared to the results of simple volume ignition calculations.

INTRODUCTION

Since it has been internationally decided to reduce carbon dioxide emission before the year 2005 in order to aroid the greenhouse catastrophy of the earth's atmosphere, and since there is an urgent need of energy especialy in the developing countries, there is now a strong demand for alternative energy sources. While the established low cost energy production by light water nuclear fission reactors could be a solution for a period of transition (limited by resources of the light uranium isotope), fusion energy is of interest for large scale energy production while renewable sources will be able to provide a part of the increased energy demand.

The advantages of using advanced fuels for fusion have been recognized for a number of years. We can look forward to this long - rage goal, which would enable fusion to fulfill its ultimate potential as a very clean and efficient energy supply. The classical deuterium-tritiunm (D-T) mixture is the premier candidate for a fusion fuel on account of its outstanding energy [1] gain, $D+T \rightarrow n(14.1 \text{ Mev}) + {}^4He$ (3.5 Mev). But, because of its meutron production, it certainly can not be the ideal long-term option. It is preferred to avoid the use of radioactive tritium and the undesired radioactivity generated by the neutrons (most of which will be absorbed and used for breeding in the lithium surrounding the reactor). An ideal reaction is the H-^{11}B reaction, in which less radioactivity is emitted per unit of energy produced [1] than the 2 parts per million of uranium given off in the dust that is released from burning coal. It is very difficult, however, to make reaction H-^{11}B occur and the results currently available suggest that inertial confinement fusion (ICF) operation with this fuel is imposible in the near future.

Another " clean" reaction that can be considered is the D-^{3}He reaction [2]:

$$D + {}^3He \rightarrow P(14.7 \text{ Mev}) + {}^4He \text{ (3.6 Mev)} \quad (2)$$

Because of the absence of uncharged reaction products (neutrons) this obviously is a desirable reaction. There remains the problem of the competing reactions

$$D + D \rightarrow n(2.4 \text{ Mev}) + {}^3He \text{ (0,8 Mev)} \quad (3)$$

and

$$D + D \rightarrow P(3.0 \text{ Mev}) + T \text{ (1.0 Mev)} \quad (4)$$

which represent a source of radioactivity.

It has been shown that reaction [2] can dominate up to 95% by application of spin polarisation [3] and hence can be still considered as relatively clean. A detailed inclusion of these complex influncess for the sake of completeness has recently been done for both reactions of type [1] and [2] using a detailed volume ignition modele. Therefore, the problem of the secondary reaction due to two different D-D nuclear reactions, taking place at the same time in a D-^{3}He and D-T pellets [4,5], has been solved. The results of calculations for D-^{3}He show that the effect of secondary reactions cannot be ignored. Inclusion of the D-D reactions in the D-^{3}He pellet has an appreciable effect on the driver energy requiremts for both ignition conditions and fusion gain.

Although the n/D-3He ratio depends on volume and density, this ratio is small and mearly the same for different pellet volumes and pellet concentrations for conditions where pellet ignition occurs. Also, the inclusion increases the fusion gain and further reduce the reaction ratio [5]. Also one of the main advantages of the inclusion of the D-D reactions in the D-T pellet is that the exact ratio $R = \dfrac{DD}{DT}$ can be calculated. Since the fuel depletion and the reaction ratio is known, the true value of the neutron flux can be used in any DT reactor design. Before considering any approach for incluctions of other possible nuclear reactions in the D-T and D-^{3}He pellets we point out here that laser engineering at Osaka University [7-9] demonastrated that the best fusion gains from laser-irradiated pellets result when central shocks are aroided, and an ideal volume compression is achieved. The advantages of volume compression and volume ignition of laser-irradiated fusion pellets against spark ignition have been reviewed in the literature [10-12]. Thus, the experimental success of volume compression led us to use the simple original model of volume ignition [13] with possible improvement in the model for D-T and D-^{3}He fuel pellets fusion calculations.

Volume compression and ignition to clean fusion have been applied to the D-^{3}He reaction [4,10,11,14,15] and the D-T reaction [5,10-12,15] but the role and relative magnitude of all secondary reactions at the same time ; ie D-D and T-T in the D-T reacton , D-D and ^{3}He-^{3}He in the D-^{3}He reaction., with heating by the fusion neutrons have been ignored. After having studied the volume ignition of D-D fuel [16], with inclusion of the D-D reactions in the D-^{3}He reaction (4) and the D-T reaction (5) with encouraging success, we are now able to carry out fusion gain calculations for D - ^{3}He and D-T pellets with all the secondary reactions and the neutron heating included.

In our simulation the effects of all charged reaction products from the D-T , D-D and T-T reactions and neutrons from the D-T reaction for D-T fuel pellet and from the D-^{3}He, D-D and ^{3}He-^{3}He reactions for D-^{3}He fuel pellet, fuel depletion and bremsstrahlung losses for both pellets have been included.

From our calculations, compared with (D-^{3}He) + (D-D) case (4) the inclusion of the (^{3}He-^{3}He) + (D-D) reactions in the D-^{3}He fuel pellet results in , up to 10% reduction in the input energy, and increase in the fusion gain by about 10% maximume plasma temperature is also increased by about 2%.

Unlike the D-^{3}He case, the inclusion of the D-D and T-T reactions in the D-T fuel pellet with respect to different fusion parameters is not considerable , but taking together all reactions allows us to consider a fusion reactor operating with realistic parameters in the volume ignitions mode and then ignore effect of , if any , the T-T reaction. Although our calculation for a D-T feul pellet with neutron interaction included is in its preliminary state, but the first results show that the neutrons which are produced in a D-T plasma with energy 14.1 Mev can not be ignored in the volume ignition mode.

ADIABATIC VOLUME COMPRESSION AND VOLUME IGNITION CALCULATIONS

The calculation of the nuclear fusion gain G from laser compressed D-T or D -^{3}He plasmas

$$G = \frac{\text{nuclear reaction energy}}{\text{input energy } E_o} \quad (4)$$

is very sensitively depending on the parameters chosen as initial volume V_o, temperature T_o, and density n_o. A further question of essential importance is the model chosen for the slowing down rate of the charged reaction products within the reacting plasma, causing reheat and eventvally ignition and self-burning. A further parameter to be inclused is the radiation loss by bremsstrahlung and the depletion of the nuclear fuel by the reaction istself, while the expansion of the plasma, determined by inertial confinement and adiabatic cooling follows the hydrodynamic equations [13]. For the fusion gain calculation in this model [9-16] we consider a spherical plasma of volume V_o and radius R_o and an initial density n_o. By unspecified mechanisms, the plasma has been heated by a laser (or particle beam) of energy E_o to an initial temperature T_o. We calculate the fusion reaction energy during the subsequent expansion and adiabatic cooling.

If the energy produced per fusion reaction of one nucleus i with another nucleus j is E_{ij}, the very simplified gain, quation 4, is given by

$$G = \frac{E_{ij}}{E_o} \int dt \int d^3r \, \frac{n^2_i}{A} <\delta v> \quad (5)$$

Where n_i is the ion density with a constant $A = 4$ for binary reactions (other wise, $A = 2$). The velocity-averaged fusion cross section $<\delta v>$ for a thermalized plasma of temperature T and an average mass m of the nuclei is given by

$$<\delta v> = \frac{\sqrt{m}}{2[\pi(KT)^{3/2}]^{1/2}} \times \int \frac{m}{2} v^2 \delta(v) \exp(\frac{mv^2}{KT}) dv^2 \quad (6)$$

Thus, from Eqs. [5] and [6], gains depending on the initial volum, density, and input energy can be calculated. Using Eq. [5] and including fuel depletion, bremsstrahlung loss, the reheating due to charged reaction products and the effect of secondary D-D reactions, the gain for D-^{3}He and D-T reactions [4,5] has been calculated previously for D-^{3}He and D-T pellets, but because of complexity of calculation the effect of other side reactions has been meglected.

In the light of dramatic changes following the D-D reaction in the D-^{3}He gain efficiency predictions (4), we felt that it would be worthwile to carry out fusion - gain calculations for D-^{3}He and D-T pellets using our recently extended code (4) to fully include all possible side reactions. Thus, besid the D-D reactions we included the ^{3}He-^{3}He reaction.

$$^3He + {}^3He \rightarrow P + P + {}^4He + 12.86 \text{ Mev} \qquad (7)$$

in the (D-^{3}He) + (D-D) reactions and the T-T reaction

$$T + T \rightarrow n + n + {}^4He + 11.33 \text{ Mev} \qquad (8)$$

in the (D-T) + (D-D) reactions by the simultaneous calculation of Eq. (5) for the pellets.

In general our model algorithm includes all nuclear reactons, all direct temperature changing elements, the adiabatic cooling, the temperature change due to bremsstrahlung losses and all charged reactions products reheating of all nuclear reactions.

As in our previous calculations [4,5,14-16] the same method [17] was used for reabsorption of the bremstrahlung, and again the collective model for stoping power [13] was adopted in calculating reheat by the charged fusion products.

RESULTS

In order to test the results with regard to our recent simulation, Fig 1 shows the time dependence of the plasma temperature T for the case of an initial density of 3×10^3 cm^3. Five cases are given in Fig.1 where the input laser energy is 172.3, 189.0, 207.3, 227.4 and 249.5 GJ respectively. The plasma temperature dependence of the D-^{3}He and D-3He + D-D reactions is not the same. For cases with the D-^{3}He reaction alone, the fuel does not ignite. The exact values of the input parameters and the results of calculations of gains, maximum plasma temperatures, and fuel depletions of the D-^{3}He and the D-^{3}He + D-D reactions are given in the figure captions.

In contrast to the D-^{3}He fusion reaction our application of volume compression and ignition to the D-T pellet including the D-D reactions shows that the effect of the inclusion of the D-D reactions on the fusion gain is small that it will have no appreciable affect in any fusion reactor. However, the inclusion of the secondary reactions in the calculation enables us to calculate the n/D-^{3}He and n/D-T neutron to reaction ratio, and therefore true neutron flux, which is an important factor in any reactor design. The ratio not only has an imput energy dependence but also has a pellet volume and density dependence.

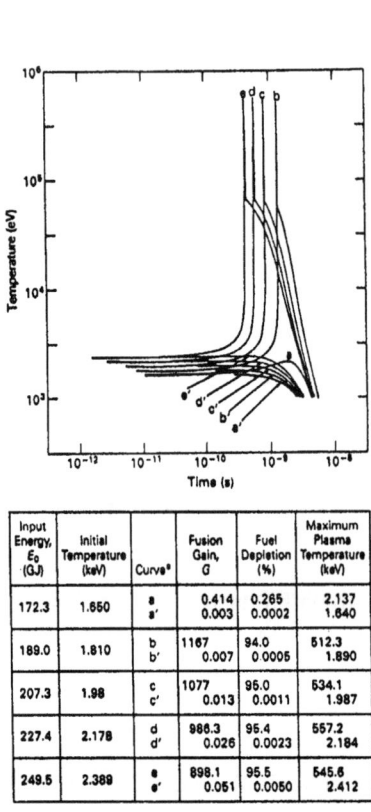

Figure 1- Dependence of the temperature of a D-³He pellet on time at initial compression to 3 ×10³ times the solid state and an initial volume of 10⁻¹ cm³.

Figure 2- Dependence of the temperature of a D-³He pellet on time at initial compression to 4 ×10³ times the solid state and an initial volume of 10⁻¹ cm³.

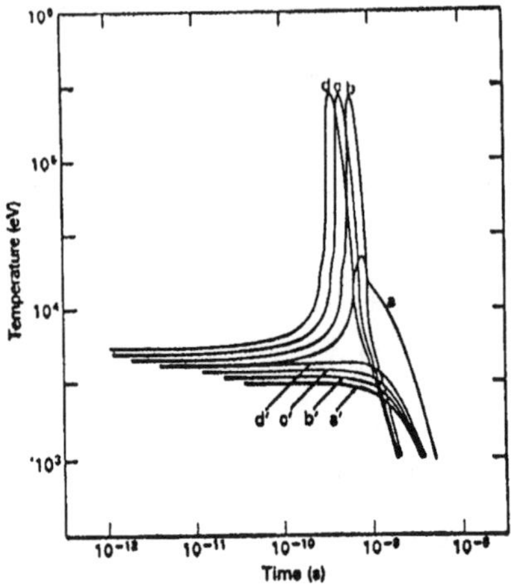

An example of our computations of fusion gain and the time dependence of the plasma temperature for a D-3He fuel pellet with an initial density of 4×10^3 times the solid state density n_s is shown in Figure 2. Figure 3 shows result of similar computations for a D-T fuel with an initial density of 5×10^3 n_s and initial volume of 10^{-1} cm³.

From Fig.2 the inclusion of the D-D reaction in the D-³He also has an appreciable effect on the plasma temperature and fusion gain but the inclusion of the ³He-³He reaction as well, has a small effect on the fusion parameters. The exact values of the input parameters and the results of calculations of different parameters of the D-³He , D-3He + D-D and D-³He + D-D + ³He-³He are also given in the figure captions.

Figure 3 which is the results of the same computations, as for Fig 2, for the D-T pellet. This Figure shows that the effet of the D-D + T-T reactions included in the D-T pellet and the D-D reactions included in the D-T pellet is the same and so small that we can ignor all side reactions in the calculation. But a prelimerary calculation of different fusion parameteres for the case of inclusion of the fusion neutrons of the D-T pellet indicates that the reheating of the neutrons should not be ignord.

A detailed inclusion of these complex influences is on its way and the work on this topic is in progress at Urmia - University.

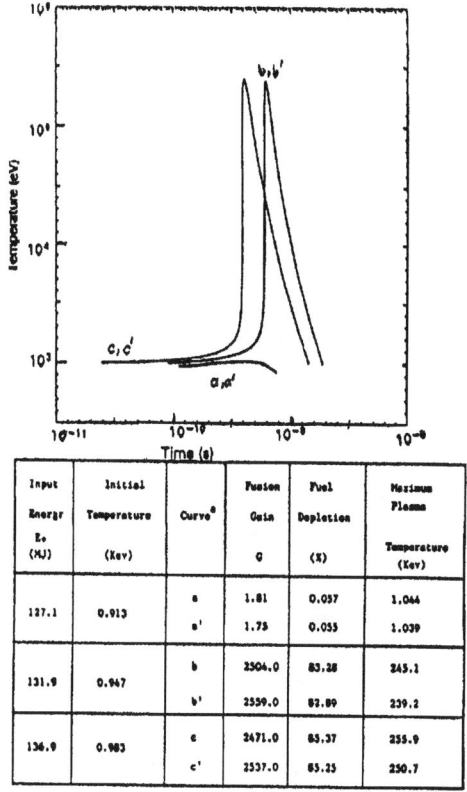

Input Energy E_o (MJ)	Initial Temperature (Kev)	Curve*	Fusion Gain G	Fuel Depletion (%)	Maximum Plasma Temperature (Kev)
127.1	0.913	a	1.81	0.037	1.044
		a'	1.75	0.035	1.039
131.9	0.947	b	2504.0	83.28	245.1
		b'	2559.0	82.89	239.2
136.9	0.983	c	2471.0	83.37	255.9
		c'	2537.0	83.25	250.7

*curves without primes indicate calculations with the (D-D)+(T-T) reactions included; curves with primes indicate calculations with D-D reactions included.

Figure 3- Dependence of the temperature of a D-T pellet on time at initial compression of 5×10^3 time the solid state and an initial volume of 10^{-1} cm^3.

CONCLUSION

The effects of the side reactions due to different nuclear reactions, taking plase at the same time in a D-^{3}He and in a D-T pellet are now known. The results of our calculation for the D-^{3}He pellet show that the inclusion of the ^{3}He-^{3}He reaction in the D-^{3}He + D-D reactions reduces the imput energy E_o up to 10% and inceases fusion gain and maximum plasma temperature up to 10% and 2% respectively. Our calculations show that there is no distingushable difference between gains with and without the inclusion of the D-D and D-D + T-T reactions in the D-T pellet. This is because a small amount of fusion energy is carried out by the charged reactions products. Our preliminary results show that, the effects of the fusion neutrons interaction with the D-T plasma in the D-T reaction should be taken into account.

ACKNOWLEDGEMENTS

I am grateful to Prof. H. Hora for helping me in different ways.

REFERENCES

[1] Waver, T.A., Laser Report, 9,12,1(1973);see also Best,R.W.B., Nucl. Instrum. Methods, 144 , 210(1977).
[2] Miley, G.H., Laser Interaction and Related Phenomena,Vol.5.P.313, Schwarz, H. et al., Eds., Plenum Press. New York (1981).
[3] Dabiri, A.E., Nucl. Instrum. Methods, A271, 71(1988); see also Cicchitelli, L. et al., Laser Part. Beams. 2, 467 (1984).
[4] Khoda-Bakhsh, R., Hora, H.and Miley, G.H., Fusion Technology, Vol.24, 28(1993).
[5] Khoda-Bakhsh, R., Nucl. Instrum. Methods Phys. Res. A330, 268(1993).
[6] Goel,B and Heeringa, W., Nucl. Fusion 28, 355(1988).
[7] Yamanaka, C. and Nakai, S., Naturs, 319, 757 (1986).
[8] Yamanaka, C., et al., Phys. Rev Lett., 56, 1575 (1986).
[9] Nakai, S., Laser Part. Beams, 7, 467 (1989).
[10] Cicchitelli, L., Eliezer, S. et al., Laser Part. Beams, 6, 163 (1988).
[11] Kasotakis, G., Cicchitelli, L., Hora, H.,and Stening, R.J.,Laser Part. Beams,7, 511(1989).
[12] Hora, H., Z.Naturforsch., 42A, 1239 (1987); see also Basko, M.M., Nucl.Fusion,30, 2443 (1990); see also Tan,N.H.andLiu,R.,ChineseJ.Laser, 17, 658(1990); seealso Colgate, S. and Pbtschek, A.,Presented at 6th Int.Conf.Emerging Nuclear Energy Systems, Monterey. California, June 16-21, 1991; see also Eliezer, S.and Hora, H., Nuclear Fusion by Imertial confinement, Elarde, G., Ronen, Y. and Martinezval, j., Eds, R C Press, Bocs Raton, Florida (1992).
[13] Hora, H. and Ray, P.S., Naturforsch., 33A, 890 (1978); see also Ray, P. S.,and Hora, H., Laser Interaction and Related plasma phenomena, Vol. 4B, P. 108, Schwarz,H., et al., Eds., Plenum Press, New york (1977).
[14] Pieruschka, P. Cicchitelli, L., Khoda-Bakhsh, R., et al., Laser Particle Beams, 10,1,154 (1992) .
[15] Stening, R.J., Khoda-Bakhsh, R. et al,. Laser Interaction and Related plasma Phenomena, Vol.10, Miley , G.H. and Hora, H., Eds, Plenum Press, New york 347 (1992).
[16] Khoda-Bakhsh, R., Hora, H., Miley, G.H., et al., Fusion Technology, 22, 50 (1992).
[17] Hora, H., Physics of Laser Driven Plasma, John wiley and Sons, New York(1981); also Hora, H., Plasma at, High Temperature and Density, Springer, Heidelberg (1991).

Implosion dynamics of a hot core

M.Murakami, M.Shimoide, and K.Nishihara

Institute of Laser Engineering, Osaka University, Suita ,Osaka 565 Japan

Abstract A simple model of a hot spot implosion is developed, where the key parameters are the areal density $\rho_s R_s$, the central temperature T_s, and the implosion velocity v_∞. The dynamics is dominated by the mechanical compressing work and the thermal conduction loss. An original self-similar solution is then found, describing the fluid motion in terms of $\rho_s R_s/T_s^2$ and $v_\infty/\sqrt{T_s}$.

I. INTRODUCTION

To achieve ignition and high gain in inertial confinement fusion (ICF), a spherical pellet must implode efficiently and symmetrically. Within the framework of the spark ignition,[1,2] a hot spot, with the central temperature $T_s \approx 5$ keV and the areal mass density $\rho_s R_s \approx 0.3$ g/cm^2, should be formed through the stagnation (deceleration) process to assure the self-heating by alpha particles. In this article, we derive an original self-similar solution, describing the hot spot dynamics, where the energy balance is determined by mechanical compressing work (pdV work) and thermal conduction loss. Of course, the self-similar analysis does not apply to the whole implosion process. It should be here noted that, for more detailed description, the alpha-heating and the bremsstrahlung radiation loss should be additionally taken into account; however, numerical simulations show that these two physical factors do not significantly effect the dynamics until the last moment before ignition.

II. THE SELF-SIMILAR SOLUTION
A. Formalism

The fluid in the hot spot is governed by the equations of continuity, momentum and energy, respectively given in a spherical system by

$$\frac{d\rho}{dt} + \frac{\rho}{r^2}\frac{\partial}{\partial r}(r^2 u) = 0 \tag{1}$$

$$\rho\frac{du}{dt} = -\frac{\partial p}{\partial r} \tag{2}$$

$$\frac{3}{2}\rho\frac{d}{dt}(\Gamma T) + \frac{p}{r^2}\frac{\partial}{\partial r}(r^2 u) = \frac{\kappa}{r^2}\frac{\partial}{\partial r}\left(r^2 T^{5/2}\frac{\partial T}{\partial r}\right) \tag{3}$$

where the density r, temperature T, pressure p, and flow velocity u are functions of time t and position r. In Eq.(3), $\Gamma = 7.7 \times 10^{14}$ [erg/(g·keV)] and $\kappa = 4.7 \times 10^{19}$ [erg/(s·cm·keV$^{7/2}$)] are the constants for the equation of state, $p = \Gamma\rho T$, and the Spitzer conductivity, respectively, where the ionization state $Z = 1$ and the coulomb logarithm $\ln\Lambda \approx 2$ are postulated for a DT plasma. Introducing Lagrangian variables, R, facilitates the analysis of the system. A class of self-similar motions is characterized by the condition\

$$r = Rf(t) \tag{4}$$

The fluid velocity is then given by

$$u = R\,df/dt \tag{5}$$

In previous work,[3-7] the adiabatic relation, $p \propto \rho^\gamma$ with the specific heat ratio g instead of Eq.(3), was used to close the hydrodynamic equations. In this case, it is enough to introduce one function f(t). However, to solve the present system, Eqs.(1)-(3), we need to introduce another time-dependent function h(t) in the form,

$$T = T_0(R)h(t) \tag{6}$$

© 1996 American Institute of Physics

where $T_0(R)$ is the temperature profile at $t = 0$. The functions f and h obey the initial conditions,
$$f(0) = 1, \quad (df/dt)_{t=0} = 0, \quad h(0) = 1 \tag{7}$$
Relations (4) and (6) imply that each fluid element varies with time in precisely the same way. With the density profile $\rho_0(R)$ at $t = 0$, the continuity equation yields
$$\rho = \rho_0(R)/f(t)^3 \tag{8}$$
Equations (2) and (3) are then reduced with the help of Eqs.(4)-(8) to
$$\frac{f}{h}\frac{d^2f}{dt^2} = -\frac{\Gamma}{\rho_0 R}\frac{d}{dR}(\rho_0 T_0) = \frac{1}{t_H^2} \tag{9}$$
$$\frac{1}{fh^{5/2}}\left(\frac{3}{2h}\frac{dh}{dt} + \frac{3}{f}\frac{df}{dt}\right) = \frac{\kappa}{\Gamma\rho_0 T_0 R^2}\frac{d}{dR}\left(R^2 T_0^{5/2}\frac{dT_0}{dR}\right) = -\frac{1}{t_C} \tag{10}$$
where t_H and t_C (>0) are separation constants, which physically represent characteristic times for the hydrodynamics and the thermal conduction, respectively. For further simplification, we here introduce the normalization,
$$\tilde{T} \equiv \frac{T}{T_s}, \quad \tilde{\rho} \equiv \frac{\rho}{\rho_s}, \quad \tilde{t} \equiv \frac{t}{t_H}, \quad \tilde{R} \equiv \frac{R}{t_H\sqrt{\Gamma T_s}} \tag{11}$$
where $T_s \equiv T_0(0)$ and $\rho_s \equiv \rho_0(0)$ are central temperature and density at $t = 0$. Equations (9) and (10) are then rewritten:
$$f\ddot{f}/h = 1 \tag{12}$$
$$-(\tilde{\rho}_0\tilde{T}_0)'/(\tilde{\rho}_0\tilde{R}) = 1 \tag{13}$$
$$\frac{1}{fh^{5/2}}\left(\frac{3\dot{h}}{2h} + \frac{3\dot{f}}{f}\right) = -c_a \tag{14}$$
$$(\tilde{R}^2\tilde{T}_0^{5/2}\tilde{T}_0')'/(\tilde{\rho}_0\tilde{T}_0\tilde{R}^2) = -c_b \tag{15}$$
where the dot and the prime denote the derivatives with respect to $\tilde{t}$ and $\tilde{R}$, respectively. The system is thus parametrized by two dimensionless variables,

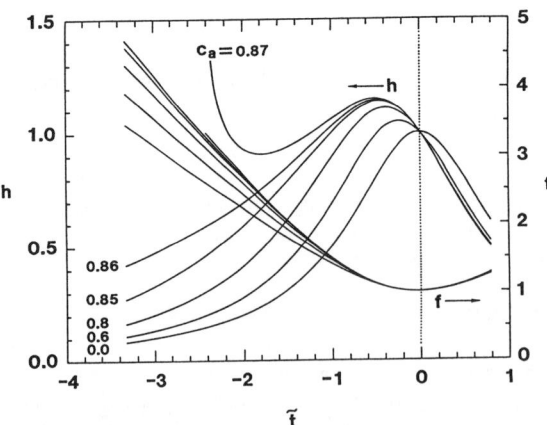

Fig.1 Functions f and h, which correspond to temporal evolution of the radius and the temperature, respectively. The parameter c_a accounts for the ratio of the thermal conduction loss to the pdV work. The values of c_a assigned to h also apply to f just in the same order as for h.

$$c_a \equiv \frac{t_H}{t_C}, \quad c_b \equiv \frac{\Gamma^2 \rho_s t_H^2}{\kappa T_s^{3/2} t_C} \quad (16)$$

As it will be presented in detail below, c_a dominates the temporal evolution of f and h, while c_b governs the spatial profiles of ρ_0 and T_0.

B. Temporal evolution of the background

Figure 1 shows the temporal evolution of f and h, numerically integrated for various values of c_a with the help of Eqs.(7), (12) and (14). It should be noted that $c_a = 0$ corresponds to adiabatic compression, and $f = (1+\tilde{t}^2)^{1/2}$ and $h = f^{-2}$ are then analytically obtained. In this case, the system is symmetric with respect to time, and the temperature reaches its maximum at $t = 0$. For $c_a \neq 0$, however, the peak temperature is achieved before the maximum compression at $t = 0$. This feature can be observed also in numerical simulations. Thus, the system is not symmetric any more for $c_a \neq 0$ due to energy dissipation via thermal conduction. Furthermore, for a relatively small c_a, $h \to 0$ holds in the limit of $t \to -\infty$. However, since the system is strongly nonlinear, there exists the critical value $c_a^{crt} = 0.8613$; for $c_a > c_a^{crt}$ the numerical integration diverges, i.e., $h \to \infty$, and leads to a non-physical solution.

The temporal evolution presented in Fig.1 is interpreted in Fig.2 in a T-ρR diagram, which gives a more comprehensive picture of an implosion. It shows T/T_0 vs $(\rho R)/(\rho R)_0$, namely h vs f^{-2} [compare Eqs.(4), (6) and (8)], for an arbitrary fluid element, where the subscript "0" corresponds to $t = 0$. Lindl[8] showed the scaling in the T-ρR diagram under two limited conditions. First, for a relatively early stage of an implosion, the pdV work dominates the energy balance, and the core compresses adiabatically, following the relation $T \propto \rho R$. Second, in the successive stage close to maximum compression, when pdV work balances thermal conduction loss, the scaling $T \propto (v_\infty \rho R)^{2/5}$ is obtained assuming that the interface velocity is kept constant at v_∞. As can be seen in Fig.2, these features are reproduced within the present model evolving continuously on the way to the maximum compression. This is because the velocity decreases self-consistently along with the compression.

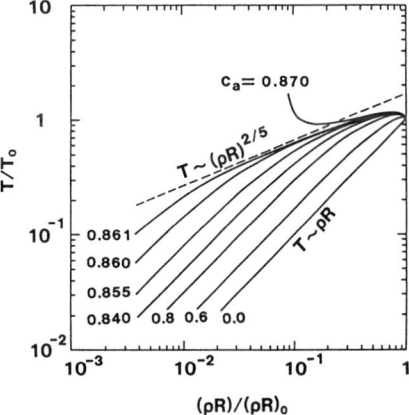

Fig.2 The slope in T-ρR diagram reflects the ratio of the thermal conduction loss to the pdV work; $T \propto \rho R$ for an adiabatic compression, while $T \propto (\rho R)^{2/5}$ for a compression where the two physical effects balance.

C. Spatial profile of the background

The spatial profiles of the temperature $\tilde{T}_0(\tilde{R})$, density $\tilde{\rho}_0(\tilde{R})$ and pressure $\tilde{p}_0(\tilde{R}) = \tilde{\rho}_0(\tilde{R})\tilde{T}_0(\tilde{R})$, which are determined by the external parameter c_b, are numerically obtained from Eqs.(13) and (15) under the boundary conditions,

$$\tilde{T}_0(0) = 1, \quad \tilde{\rho}_0(0) = 1, \quad \tilde{p}'_0(0) = 0 \tag{17}$$

Decreasing the ratio of the characteristic times t_C/t_H, c_b becomes larger, and this means physically that the energy balance is affected more by the thermal conduction, and less by the pdV work; this reflects in an approximately isobaric profile for $c_b \gg 1$. By assuming such an isobaric profile, $\tilde{p}_0(\tilde{R}) \approx 1$, the approximate temperature profile is obtained from Eq.(15):

$$\tilde{T}_0(\tilde{R}) \approx \left(1 - \frac{7}{12} c_b \tilde{R}^2\right)^{2/7} \tag{18}$$

Then, the hot spot radius $\tilde{R}_s$, where the temperature goes to zero, is straightforwardly given by

$$\tilde{R}_s = \sqrt{12/7c_b} \tag{19}$$

Figure 3 shows the profiles numerically obtained for $c_b = 12$. At $\tilde{R} = \tilde{R}_s$, the numerical integration gives $\tilde{T}_0 \to 0$ and $\tilde{\rho}_0 \to \infty$, while keeping $\tilde{p}_c \approx 1 - 6/5c_b$ (the subscript "c" denotes the contact surface) at a finite value. It should be noted that the above approximations, Eqs.(18)-(20), are found to be in good agreement with the exact numerical results, if $c_b \gg 1$. The left part of Fig.3 represents qualitatively the cold main fuel.

In summary, the dynamics of the hot spot implosion have been studied in terms of the original self-similar solution taking the conduction loss into account. It is characterized by the two governing parameters, $\rho_s R_s / T_s^2$ and $v_\infty / \sqrt{T_s}$. It should be also noted that, with the present model, such a crucial condition as the minimum implosion velocity to reach the required compressed state $(\rho_s R_s, T_s)$ can be derived.[9]

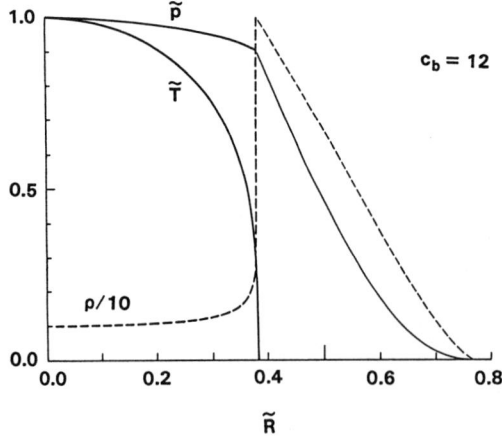

Fig.3 Spatial profiles for temperature $\tilde{T}$, pressure $\tilde{p}$, and density $\tilde{\rho}$. The left half represents an exact numerical result obtained from the self-similar solution, whereas the right half illustrates the profiles of cold main fuel qualitatively.

REFERENCES

1. J.Meyer-ter-Vehn, Nucl.Fusion **22**, 561(1982).
2. M.M.Basko, Nucl.Fusion **30**, 2443(1990).
3. F.Hattori, H.Takabe, and K.Mima, Phys.Fluids **29**, 1719(1986).
4. I.B.Bernstein and D.L.Book, Astrophys.J.**225**, 633(1978).
5. D.L.Book and I.B.Bernstein, J.Plasma Phys.**23**, 521(1980).
6. H.Sakagami and K.Nishihara, Phys.Fluids B **2**, 2715(1990).
7. M.M.Basko, Nucl.Fusion **35**, 87(1995).
8. J.D.Lindl, in *Inertial Confinement Fusion* edited by A.Caruso and E.Sindoni (Editrice Compositori, Bologna, 1988), p.617.
9. M.Murakami, M.Shimoide, and K.Nishihara, to be published in Phys.Plasmas, Sep.issue (1995).

Kinetic Model for DT Ignition and Burn in ICF Targets

S.I. Anisimov, A.M. Oparin, and J. Meyer-ter-Vehn

Max-Planck-Institut fuer Quantenoptik, D-85748 Garching, Germany
L.D. Landau Institute for Theoretical Physics, 117940 Moscow, Russia

Ignition and burn of DT targets is studied taking into account kinetic effects. Kinetic equations describing the interaction of the high-energy reaction products with target plasma are solved using the particle-in-cell (PIC) code for collisional plasma. Volume and spark ignition configurations are simulated for initial temperatures and $<\rho R>$ values of practical interest and target masses between 0.1 and 10 mg. Optically thick configurations igniting at temperatures below 5 keV are considered. Burn of the targets with reduced tritium content is simulated. It was shown that, for 25% tritium concentration, the energy output is reduced only by 15%.

INTRODUCTION

Theoretical studies of various schemes of inertial confinement fusion (ICF) include a self-consistent description of hydrodynamics and thermonuclear burn of ICF-targets. The simplest approach to this problem is based on the approximation of local energy release. This approach is valid if the mean free path of the reaction products is much smaller than the scale length of hydrodynamic motion. For ICF-target burn, this condition is not fulfilled in many cases of practical interest. The mean free path of fast charged particles produced in the fusion reactions (p, He^3, He^4, T - nuclei in the case of DT - fuel) is usually comparable to the target size. Several approximations have been proposed to take into account the nonlocal energy release. The diffusion approximation [1,2], the approximation of a monoenergetic ion beam [3,4], and the Monte-Carlo method [4,5] were used to solve the transport equation for fast ions. The latter method has a broader range of applicability. However, it is based on a simplified description of charged particle interaction with the plasma [5]. More realistic models for ICF-target burn can be constructed on the basis of kinetic equations for the fast reaction products. This approach, however, calls for an efficient method for solving kinetic eqations for collisional plasma. Note that standard methods of numerical solution of the kinetic equations using meshes in coordinate a velocity space include cumbersome calculations and require great computer resources. In the present paper, we employ the particle method worked out in [6,7] to construct a hybrid model, in which the bulk plasma is described by hydrodynamics equtions, whereas a kinetic description is applied to the motion of fast particles.

MODEL

The interaction of charged particles with plasma can be described by the Fokker-Planck equation (FPE). A detailed analysis of the accuracy and limits of applicability of this equation is given in [8]. In the present paper, FPE for the group of fast ions generated in the fusion reactions is solved numerically using the collisional particle-in-cell (PIC) method developed in [6,7]. With this approach, the FPE is replaced by a set of equations for the motion of macroparticles. These equations have the form of Langevin equations,

$$dv_i^\alpha / dt = h_i^\alpha(\mathbf{v}^\alpha) + g_{ik}^\alpha(\mathbf{v}^\alpha)\xi_k(t) \quad (1)$$

where the index α specifies the particle species, $i, k = 1, 2, 3$, $\xi_k(t)$ is the vector of white noise, and $h_i^\alpha(\mathbf{v}^\alpha)$, $g_{ik}^\alpha(\mathbf{v}^\alpha)$ are the functions defined in Ref. 6. Summation is assumed in (1) over recurring subscript k. Note that the equations (1) are very convenient for treating the collisional kinetics of a small group of fast particles moving in the background of the bulk of the plasma. In this case, the fast particles do not directly interact with one another, and the particles of the bulk plasma are in Maxwellian equilibrium. Equations of motion of macroparticles (1) can, therefore, be reduced to:

$$\frac{dv_i^\alpha}{dt} = 2\pi\lambda \left(\frac{e_\alpha}{m_\alpha}\right)^2 \frac{u_i^\alpha}{(u^\alpha)^3}\left\{R + S - Q_\alpha - 2\left[2R(S-R)\right]^{1/2}\right\} +$$

$$+2\frac{|e_\alpha|}{m_\alpha}\left(\frac{\pi\lambda}{u^\alpha}\right)^{1/2}\left\{(S-R)^{1/2}\xi_i - \left[(S-R)^{1/2} - (2R)^{1/2}\right]\frac{u_i^\alpha(u_k^\alpha \cdot \xi_k)}{(u^\alpha)^2}\right\} \quad (2)$$

where

$$R = \sum_\beta e_\beta^2 n_\beta F_\beta \quad , \quad S = \sum_\beta e_\beta^2 n_\beta W_\beta \quad , \quad Q_\alpha = 2\sum_\beta e_\beta^2 n_\beta W_\beta w_\beta^2\left(1+\frac{2m_\alpha}{m_\beta}\right)$$

$$u_i^\alpha = v_i^\alpha - U_i \quad , \quad w_i^\beta = u_i^\beta\left(m_\beta / 2T_\beta\right)^{1/2} \quad , \quad w_\beta = |\mathbf{w}^\beta| \quad , \quad F_\beta = F(w_\beta)$$

$$W_\beta = W(w_\beta) \quad , \quad F(z) = (2/\sqrt{\pi})\int_0^z e^{-x^2}dx \quad , \quad W(z) = (F(z) - zF'(z))/2z^2$$

n_β is the number density of particles of species β, **U** is the local mass velocity of the bulk plasma. The set of equations (2) should be supplemented by the equations for particle coordinates

$$\frac{dr_i^\alpha}{dt} = v_i^\alpha \quad (3)$$

The method of numerical solution of Eqs. (2), (3) is described in detail in [7].

The bulk plasma is described by hydrodynamical equations, that are solved numerically using two-temperature Lagrangean hydrocode, including multigroup radiation transport. The motion of bulk plasma is assumed to be spherically symmetric. For each time step and Lagrangean cell, the number of high-energy nuclei generated in the thermonuclear reactions is determined. The following thermonuclear reactions are taken into account

$$D+T \to He^4 (3.518\,MeV) + n (14.071\,MeV)$$
$$D+D \to T (1.007\,MeV) + p (3.021\,MeV)$$
$$D+D \to He^3 (0.817\,MeV) + n (2.451\,MeV)$$
$$D+He^3 \to He^4 (3.675\,MeV) + p (14.70\,MeV)$$

The interaction of fast nuclei with the background plasma is simulated by solving a set of 3D Langevin equations for the macroparticles. The energy and momentum transferred to the bulk plasma are calculated. As in the usual PIC-method, the number of macroparticles is much less than the actual number of fast particles. Typically, 1500 macroparticles were included in the calculations.

RESULTS OF SIMULATION

1. Volume ignition

Let us discuss first the ignition and burn of initially uniform DT-plasma spheres. This is often referred to as volume ignition [9]. Results of calculations based on the above described kinetic model are shown in Fig. 1. The burn fraction of the fuel is plotted as a function of the confinement parameter $H = \int \rho(r,0)\,dr$ for different fuel masses and initial temperatures. One can see that, for the initial temperatures above 5 keV, an appreciable burn fraction is reached at $H > 1$ g/cm^2. The burn fraction depends only slightly on the target mass and initial temperature. This is consistent with a simple formula proposed in [4]

$$\Phi = H/(H+H_B), \quad H_B = 6.3\,g/cm^2 \qquad (4)$$

which is applicable, according to [4], in the range: $H > 1$ g/cm^2, $20\,keV < T_i < 70\,keV$. At $H < 1$ g/cm^2, the burn fraction Φ depends significantly on the initial temperature, and Eq. (4) loses its meaning. A detailed comparison of the approximation (4) with the results of kinetic simulation is given in [10].

For initial temperatures lower than 4.6 keV, Bremsstrahlung losses exceed the α-particle heating, and the ignition occurs only at sufficiently high values of confinement parameter (opticlly thick configurations). As is seen in Fig. 1, at 4 keV, the fuel ignites, if H exceeds some threshold value dependent on the fuel density.

It is interesting to compare the present results with less sophisticated models often used in burn calculations. In Fig. 2, the evolution of electron and ion temperatures at the center of burning DT sphere with $M = 1$ mg, $H = 1.5$ g/cm^2, and $T_i = 10$ keV is shown. Time dependence of the temperature was calculated using the following models: (1) local energy deposition, (2) one-group α-particle diffusion, and (3) the present kinetic model. It is seen from Fig. 2 that local deposition strongly overestimates the burn temperature, while the diffusion approximation gives the results not too different from the kinetic calculations. The burn fraction calculated using the above three models is presented in Fig. 3 as a function of H. The local energy deposition overestimates Φ at low H and underestimates at higher H. The latter fact may be attributed to faster disintegration of the target in the case of local deposition due to higher electron and ion temperatures. Results for one-group α-particle diffusion agree well with kinetic transport at low H, but fall low for higher H.

2. Optically thick configurations

Let us now discuss more in detail the ignition at low temperature. It is seen from Fig. 1 that for $T_i = 4$ keV and target mass of 10 mg no ignition occurs in the H region displayed. However, configurations with smaller masses and higher densities do ignite at 4 keV provided that H is high enough. This ignition regime has been described before in [11-13]. It corresponds to optically thick configurations in which the radiation emitted by the plasma is partially absorbed in the target, and radiation loses are reduced. A simple estimate shows that the Planck mean free path for a target with $T_i = 4$ keV, and $M = 0.1$ mg becomes comparable to the target size when $H \approx 2\,\text{g/cm}^2$. This value agrees well with the results of simulation presented in Fig.1.

A detailed study of the ignition of optically thick fuel configurations was carried out using one-group diffusion approximation for α- particle transport. Results are given in Fig. 4. It is seen that the fuel ignites even at 1.5 keV provided that $H > 10$ g/cm². At $T = 2$ keV the fractional burnup closely follows Eq. (4) with $H_B \approx 13$ g/cm². Calculations show that, at low initial temperatures, a considerable time is needed to heat up the fuel to the temperature of about 5 keV when the burn reaction rate sharply increases. This phenomenon is similar to well known "period of induction" in thermal explosion [14]. For example, at $T = 2$ keV, $M = 1$ mg, and $H = 5$ g/cm², it takes about 50 ps before the ion temperature reaches 5 keV and ignition occurs.

3. Spark ignition

For reactor applications, the energy gain of ICF-target should be sufficiently high. It is well known that the maximum gain can be reached using spark ignition. In spark configurations, only a small central part of the target is heated to ignition conditions, while the outer part is left at much lower temperature to reduce the total amount of energy invested. Ignition occurs in the spark region, and then thermonuclear combustion wave propagates into the cold fuel. Detailed simulation of spark ignition was carried out using the stochastic transport method. In Fig. 5, the results of calculations performed for isobaric fuel configuration with a mass of 1 mg are shown. The mass of hot spark region is taken equal to 0.1 mg, the temperature of cold fuel around the spark is 1 keV. The results of simulation are presented in Fig. 5 in the form of burn trajectories in the plane (H_s, T_s), where H_s is the spark confinement parameter and T_i is the average temperature of the spark. There are two sorts of trajectories depending on the initial values of the parameters H_s and T_s : (1) the trajectories corresponding to igniting configurations characterized by rising temperatures and confinement parameters, and (2) the trajectories corresponding to decaying configurations with decreasing H_s and T_s. The boundary between these two cases (shown by two dashed lines in Fig. 5) which can be described by the equation

$$H_s T_s \approx 2.7 \quad \text{keV} \cdot \text{g/cm}^2 \tag{5}$$

is not very sharp. There are intermediate situations in the vicinity of the boundary that have been denoted in [15] as half ignition.

Ignition conditions resulting from the present calculations, are slightly less optimistic than the majority of spark ignition criteria proposed in the literature (see, e.g., [16 - 18]). According to our calculations, the values $T_i = 5$ keV and $H_s = 0.4$ g/cm^2, quoted in [16], correspond to decaying configuration; the values $T_i = 10$ keV and $H_s = 0.3$ g/cm^2, suggested in [17], correspond to half ignition. Note that there is a possibility to lower the ignition threshold obtained in the present work by optimizing fractional spark mass.

4. Reduced tritium content

Since tritium is a radioactive isotop, it is desirable to minimize its content in ICF-targets. A reduction of tritium content leads, generally, to lower fractional burnup. The dependence of energy gain on the tritium fraction in DT fuel was studied in [19-21] using simplified burn models. We made similar calculations using the above described kinetic approach. Results of calculations for 1 mg uniform DT sphere are shown in Fig. 6. The burn fraction is presented as a function of tritium fraction $f = n_T / (n_D + n_T)$ for different values of H. The results are normalized to the burn fraction at $f = 0.5$. Solid lines correspond to local energy release approximation; triangles and squares present the results of kinetic simulation for different values of confinement parameter. It is seen that the burn fraction reaches its maximum at $f = f_{max} \approx 0.4$. In the vicinity of the maximum, the burn fraction depends only slightly on changes in f. At $f = 0.25$, the burn fraction is only by 15% lower than at $f = 0.5$. This behavior can be understood if we remind that the burn rate is proportional to $n_D n_T \propto f(1-f)/(2+f)^2$. This dependence, shown in Fig. 6 by the dotted line, is in a qualitative agreement with the results of burn simulation.

ACKNOWLEDGMENTS

The authors would like to thank S. Atzeni for helpful discussions. S.A. and A.O. are grateful to Max-Planck-Institute for Quantum Optics for hospitality during their stay at MPQ in 1993-1994.

REFERENCES

1. S.Atzeni, Plasma Phys. and Controll. Fusion, **29**, 1535, 1987
2. R.L.McCrory, C.P.Verdon, in *Inertial Confinement Fusion*, Ed. by A.Caruso and E.Sindoni, Proc. Intern. School of Plasma Physics "Piero Caldirola", Varenna, Sept. 6 - 16, 1988, Editrice Compositori, Bologna, 1988, p. 83
3. K.Brueckner, S.Jorna, Rev. Mod. Phys., **46**, 32, 1974
4. G.S.Fraley, E.J.Linnebur, R.J.Mason, R.L.Morse, Phys. Fluids, **17**, 474, 1974
5. Ya.Z.Kandiev, V.B.Kryuchenkov, V.V.Plokhoi, Sov. J. Plasma Phys., **5**, 98, 1971
6. M.F.Ivanov, V.F.Shvets, Sov. Phys.- Doklady, **23**, 130, 1978
7. M.F.Ivanov, V.F.Shvets, USSR Comput.Math. and Math. Phys., **20**, 145, 1980
8. B.A.Trubnikov, in *Voprosy teorii plasmy*, Ed. by M.A.Leontovich, Moscow, 1963, v. 1, p.98 (Russian)
9. H.Hora, S.Eliezer, J.J.Honrubia, R.Hoepfl, J.M.Martinez-Val, M.Piera, and

G.Velarde, 12th Intern. Conf. Laser Interaction and Related Plasma Phenomena, Osaka, April, 1995
10. S.I.Anisimov, A.M.Oparin, JETP Lett., **57**, 634, 1993
11. A.Caruso, Plasma Physics, **16**, 683, 1974
12. L.Cichitelli, S.Eliezer, M.P.Goldsworthy, F.Green, H.Hora, P.S.Ray, R.J.Stening, H.Szichman, Laser Part. Beams, **6**, 163, 1988
13. M.Basko, Nuclear Fusion, **30**, 2443, 1990
14. D.A.Frank-Kamenetskii, *Diffusion and Heat Transfer in Chemical Kunetics*, Plenum Press, New York, 1969
15. A.M.Oparin, S.I.Anisimov, J. Meyer-ter-Vehn, Nuclear Fusion (to be published)
16. J.Meyer-ter-Vehn, Nuclear Fusion, **22**, 561, 1982
17. J.D.Lindl, in *Inertial Confinement Fusion*, Ed. by A.Caruso and E.Sindoni, Proc. Intern. School of Plasma Physics "Piero Caldirola", Varenna, Sept. 6 - 16, 1988, Editrice Compositori, Bologna, 1988, p. 617
18. S.Atzeni, A.Caruso, Phys. Lett., **85A**, 345, 1981
19. S.Kawata, H.Nakashima, National Institute for Fusion Sciences Report NIFS-113, Nagoya, October, 1991
20. S.Kawata, H.Nakashima, Laser Part. Beams, **10**, 479, 1992
21. N.A.Tahir and D.H.H.Hoffman, Fus. Eng. Design, **24**, 413, 1994

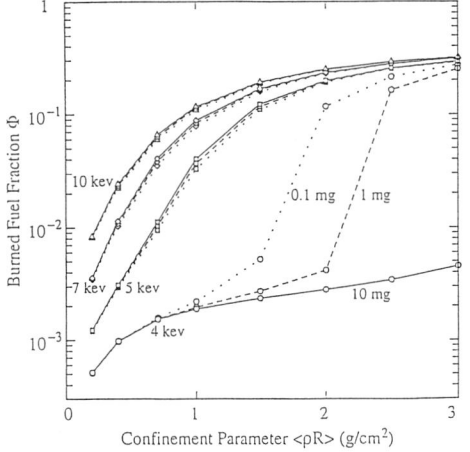

Fig. 1. Burn fraction of DT fuel versus initial confinement parameter $<\rho R>$ for different temperatures and fuel masses. The results refer to initially uniform fuel sphere.

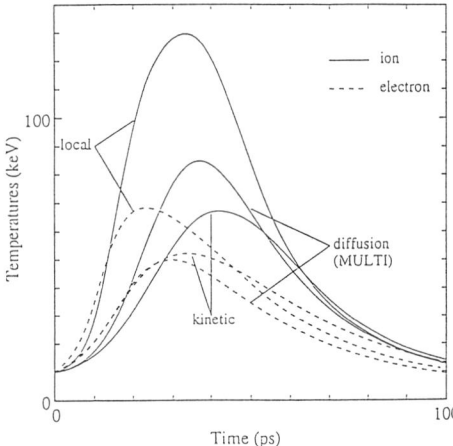

Fig. 2. Evolution of central electron and ion temperature calculated for 1 mg, 10 keV target using different models: local deposition, one-group diffusion, and kinetic simulation.

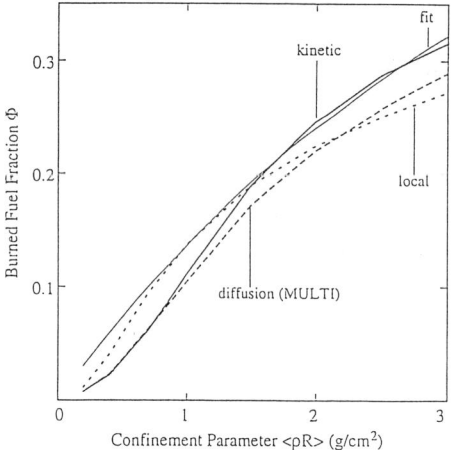

Fig. 3. Fractional burnup calculated for 1 mg, 10 keV target using different methods: kinetic simulation (thick solid line), one-group diffusion (long dashes), and local deposition (short dashes). The thin solid line represents the fit formula (4) with $H_B = 6.3 \ g/cm^2$.

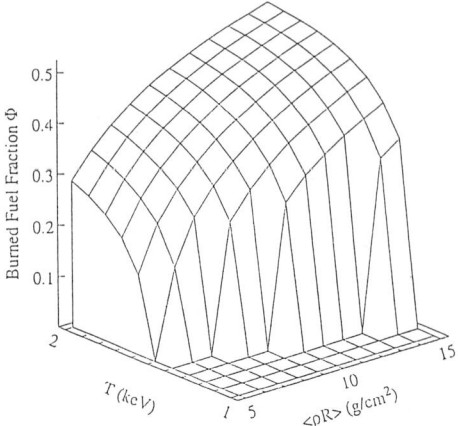

Fig. 4. Burn fraction versus initial confinement parameter and temperature for optically thick fuel spheres.

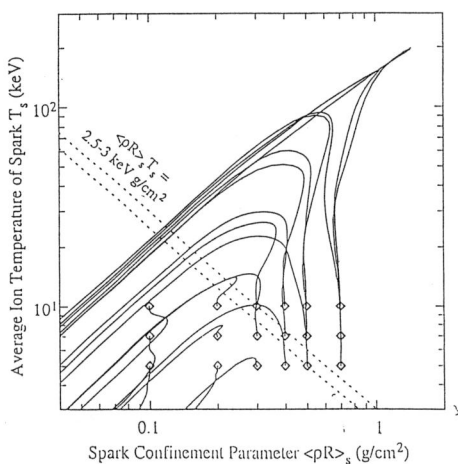

Fig. 5. Burn trajectories of isobaric spark configurations plotted in the plane of average ion temperature and confinement parameter of the spark. The diamonds give the initial values.

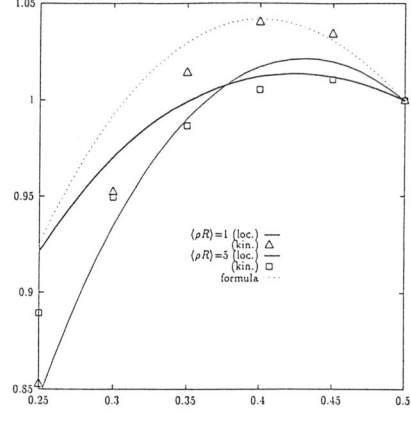

Fig. 6. Normalized burn fraction versus tritium fraction in the initial fuel. The lines refer to local energy release, the squares and triangles represent the results of kinetic simulation.

Emission of Highly Energetic Neutrons from Laser-imploded D-T Pellets and Its Applicability to Pellet Diagnosis

Y.Tabaru, Y.Nakao, H.Nakashima*, K.Kudo

Department of Nuclear Engineering, Kyushu University, Hakozaki, Fukuoka 812, Japan
*Department of Energy Conversion Engineering, Kyushu University, Kasuga, Fukuoka 816, Japan

Abstract. Because of suprathermal fusion reaction and rapid pellet expansion, the highly energetic neutrons (E_n>14MeV) can be emitted from DT pellet. On the basis of coupled transport/hydrodynamic calculation, the suprathermal fusion reaction is examined and the energy spectrum of neutrons bursting from DT pellet is calculated. The pellet ρR diagnosis by the detection of suprathermal fusion neutrons and the method for detecting these neutrons are discussed.

1. INTRODUCTION

Detecting highly energetic neutrons produced by suprathermal fusion is expected as an useful method for ρR diagnosis in upcoming ICF experiments with relatively high ρR value and high neutron yield.[1],[2],[3] Here, the suprathermal fusion means the fusion reaction that recoil ions created in neutron elastic scattering undergo with background ions. Since recoil ions introduce their kinetic energy into the reaction, highly energetic neutrons ($E_n \sim 30$MeV) can be emitted from the reactions.

So far, the relation between the number of suprathermal fusions per primary 14-MeV neutron, *i.e.* suprathermal fusion probability, and the ρR value of pellets was estimated.[1],[3] The energy spectrum of neutrons emitted from the pellets was also derived.[3] These previous studies, however, assumed compressed but stationary pellet states. In actual ICF pellet, the energy spectrum of burst neutrons becomes more broadened, because the fusion reactions occur during rapid expansion of imploded DT pellets.

In this study, we calculate the realistic energy spectrum of neutrons emitted from laser-imploded DT pellets, on the basis of coupled transport/hydrodynamic calculation method[4]. We also carry out calculations of neutron spectra for various-sized pellets. Using these spectra, we discuss the most suitable method for each pellet to detect the highly energetic neutrons and estimate the relation between high energy components of energy spectra of burst neutrons and the ρR values of laser-imploded pellets.

2. METHOD OF CALCULATION

2.1. Transport Process

Analysis of the suprathermal fusions in laser-imploded DT pellets requires the transport calculations for the neutrons and charged particles. We use simultaneous neutron/recoil-ion transport calculation model formulated by Nakao, et al.[4] In this model, the transport equations are described using the modified Eulerian coordinates[5], where the energy and angular variables are defined in terms of the velocity relative to the medium. This treatment is suitable for the particle transport calculation in a moving medium (e.g. burning ICF plasma).

The transport equation for neutrons is the Boltzmann equation. The source neutrons come not only from thermal fusions but also from suprathermal ones. Here, the doppler-broadening of source neutrons is neglected. Neutron interactions we consider are elastic scattering and break-up reaction of deuteron, i.e. D(n,2n)p. The cross sections for these processes are given by Seagrave, et al.[6] The fundamental equation to describe the transport of recoil ions and alpha particles is Boltzmann-Fokker-Planck equation. Coulomb scattering and DT suprathermal fusion are taken into account as the basic interaction processes between these energetic ions and background species. The cross section data for DT fusion reactions are taken from Duane[7] and averaged over the velocity distribution of background ions.

We coupled these transport routines for neutrons and charged particles to one-dimensional hydrodynamics code, MEDUSA.[8]

2.2. Implosion and Burn Simulation

At first, we carry out the implosion simulation using one-dimensional hydrodynamics code ILESTA-1D[9]. In the implosion simulation, neutron interactions are neglected. The temperatures, density and hydrodynamic velocity distributions of the medium obtained around the final stage of implosion are used as the input data for subsequent burn simulation carried out with the MEDUSA code. In the burn simulation, we examine the suprathermal fusion reactions and calculate the burst neutron spectrum from laser-imploded DT pellet, on the basis of coupled

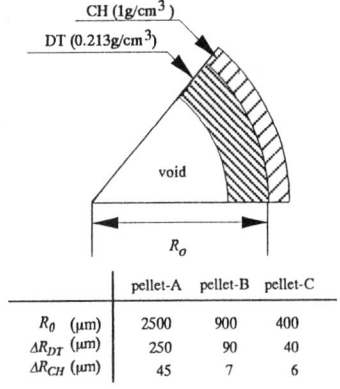

Fig.1. Initial configuration of pellet

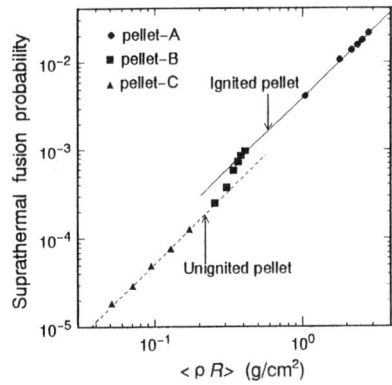

Fig.2. Dependence of suprathermal fusion probability on the $<\rho R>$ values

transport-hydrodynamic calculation method.

3. RESULTS AND DISCUSSION

We examine the suprathermal fusion reactions for three kinds of pellets ((A)reactor grade, (B)ignition experiment grade, (C)current implosion experiment grade) as shown in **Fig.1**. Examination is made of the dependence of suprathermal fusion reactions on the ρR value of pellets for the various values of laser energy E_L (E_L=2.35∼3.35MJ for pellet-A, E_L=50∼110kJ for pellet-B, E_L=3∼8kJ for pellet-C).

3.1. The Dependence of Suprathermal Fusion Reactions on $<\rho R>$ Values of Pellets

The ρR value of burning DT pellet changes momentarily. Therefore, we use the burn-averaged ρR ($<\rho R>$) defined as follows:

$$<\rho R> = \frac{\int_0^\tau \rho R(t) R_F(t)\, dt}{\int_0^\tau R_F(t)\, dt} \qquad (1)$$

where R_F is the rate of fusion reactions in the pellet and τ is the implosion-burn time.

Figure 2 shows the suprathermal fusion probability as a function of the pellet

$<\rho R>$. Here, "ignited pellet" means the pellet whose target gain is more than 1. The probability strongly depends on $<\rho R>$ values of the pellets. We can also observe that the probability in the ignited pellets is larger than in the unignited pellets. In the ignited pellets, the plasma temperature becomes higher than in the unignited pellets, because of the self-heating by the fusion-born particles. The increased temperature reduces the stopping power and hence increase the range of recoil ions. This results in higher suprathermal fusion probabilities.

3.2. Burst Neutron Spectra and $<\rho R>$ Diagnosis

We can not directly measure the suprathermal fusion probability. In practice, the $<\rho R>$ value of pellet should be determined by measuring the ratio of highly energetic neutrons emitted only from suprathermal fusions to total burst neutrons. In this section, we examine the burst neutron spectra including neutrons produced by suprathermal fusions and discuss the most suitable neutron activation reaction for detecting highly energetic neutrons from each pellet. We show the relation between the fraction of energetic population of burst neutrons and $<\rho R>$ value of pellet.

Figure 3 represents the energy spectrum of burst neutrons from pellet-A, reactor-grade pellet, that is imploded by 3.35MJ laser. The dashed line shows the spectrum of neutrons produced by suprathermal fusions. The burst neutrons contains high energy components whose maximum energy reaches about 40MeV. These highly energetic neutrons are produced when the suprathermal fusion reactions occur during rapid expansion of DT pellet and emit neutrons in the same direction as the medium expansion velocity. On the other hand, the maximum energy of neutrons emitted from thermal fusions is about 21MeV. Therefore, we

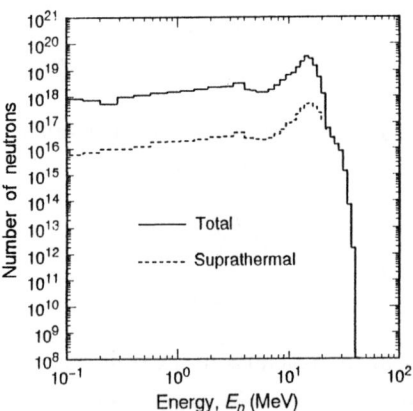

Fig.3 Time-integrated energy spectrum of neutrons bursting from pellet-A(E_L=3.35MJ)

must use a neutron activation reaction whose threshold energy is more than 21MeV, for the purpose of detecting highly energetic neutrons emitted only from the suprathermal fusion reactions. Here, we propose to detect highly energetic neutrons as above by the activation reaction,

$$^{23}\text{Na} \xrightarrow[E_n \geq 24.5\text{MeV}]{(n,3n)} {}^{21}\text{Na} \xrightarrow[t^{1/2}=22.5\text{s}]{\beta^+\text{-decay}} {}^{21}\text{Ne}. \quad (2)$$

The relation between the ratio of the number of highly energetic neutrons $Y_{24.5}$ to the number of total neutrons Y_t and $<\rho R>$ the of pellet are examined and plotted in **Fig.4**. This figure shows that it may be possible to know $<\rho R>$ of the pellet by detecting of β-decay of ^{21}Na and estimating the ratio $Y_{24.5}/Y_t$.

Figure 5 represents the burst neutron spectrum from pellet-B, ignition-grade pellet, imploded by 110kJ laser. The maximum energy of neutrons emitted from suprathermal fusions and thermal ones are 33MeV and 16.5MeV, respectively, because the expansion velocity of pellet is smaller than in pellet-A. Therefore, the condition concerned with the threshold energy is not severe compared with the case of pellet-A. In these ignition-grade pellets, however, we should pay attention to the error being accompanied with the detection of suprathermal fusion neutrons, because neutron yield is expected to be not significant ($Y_t=10^{14}\sim10^{17}$). We estimate the error on the assumption that the activation cylinder cover a solid angle of 10^{-1}. It is found that we can detect the suprathermal fusion neutrons with the uncertainty below 10% using the activation reaction as follows:

$$^{12}\text{C} \xrightarrow[E_n \geq 20.3\text{MeV}]{(n,2n)} {}^{11}\text{C} \xrightarrow[t^{1/2}=20.38\text{m}]{\beta^+\text{-decay}} {}^{10}\text{B}. \quad (3)$$

In the implosion-grade pellet as pellet-C, both neutron yield ($Y_t=10^8\sim10^{13}$) and suprathermal fusion probability ($10^{-4}\sim10^{-5}$) are too small to detect suprathermal

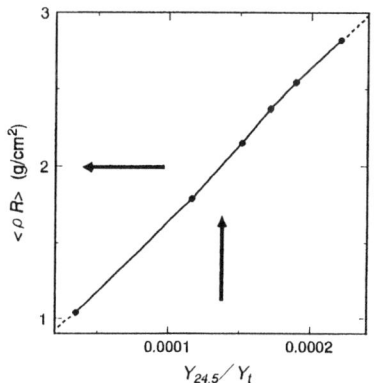
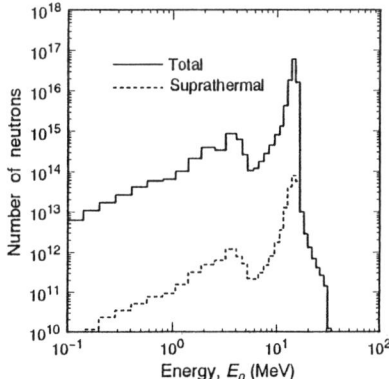

Fig.4. The relation between the fraction of highly energetic neutrons and $<\rho R>$

Fig.5. The burst neutron spectrum from pellet-B (E_L=110kJ)

Table 1. Summary of results

Pellet size	$<\rho R>$ (g/cm²)	Maximum energy of neutron (MeV)		Method for detecting highly energetic neutrons
		(a) suprathermal	(b) thermal	
Reactor	1.0 ~ 2.8	40	21	^{23}Na (n, 3n) (E_{th}=24.5MeV)
Ignition	0.25 ~ 0.4	33	16.5	^{12}C (n, 2n) (E_{th}=20.3MeV)
Implosion	0.05 ~ 0.17	31	15.5	

fusion neutrons accurately. Therefore, it is difficult that we diagnose $<\rho R>$ of the pellet by detecting highly energetic neutrons from these small-sized pellets.

Table 1 summarizes the analysis of suprathermal fusion reactions in laser-imploded pellets and the method of detecting highly energetic neutrons emitted from suprathermal fusion for each pellet.

4. SUMMARY

We have examined the suprathermal fusion reactions in laser-imploded DT pellets and calculated the burst neutron spectra from pellets, on the basis of the coupled transport/hydrodynamic calculation.

It was found that the highly energetic neutrons whose maximum energy reaches about 40MeV could be emitted from laser-imploded DT pellets. We also showed that it may be possible to know $<\rho R>$ values by detecting the highly energetic neutrons produced by suprathermal fusion reactions using the activation cylinders of ^{23}Na and ^{12}C for reactor-grade and the ignition-grade pellets, respectively.

REFERENCES

(1) Welch,D.R., et al., *Rev. Sci. Instrum.*, **59**, 610-615(1988).
(2) Azechi,H., et al., *Laser and Particle Beams*, **9**, 119-134(1991).
(3) Nakao,Y., et al., *Fusion Technol.*, **20**, 824-828(1991).
(4) Nakao,Y., et al., *J. Nucl. Sci. Technol.*, **30**, 18-30(1993).
(5) Wienke,B.R., *Phys. Fluids.*, **17**, 1135-1138(1974).
(6) Seagrave,J.D., et al., *Ann. Phys.*, **74**, 250-299(1972).
(7) Duane,B.H., BNWL-1685, Richmond, WA(1972).
(8) Christiansen,J.P., et al., *Compute. Phys. Commun.*, **7**, 271-287(1974).
(9) Takabe,H. and Mima,K., *ILE Research Report, ILE-8713*(1987).

Neutron Heating in Ignition and Burn Phases of Laser-imploded DT Pellets

Y. Nakao, T. Johzaki, H. Nakashima*, A. Oda, K. Kudo

Department of Nuclear Engineering, Kyushu University, Hakozaki, Fukuoka 812, Japan
*Department of Energy Conversion Engineering, Kyushu University, Kasuga, Fukuoka 816, Japan

Abstract. On the basis of coupled neutronic/hydrodynamic calculations, we examine the neutron heating effects in the ignition and burn phases of the laser imploded D-T pellets. The neutrons deposit their energy all around the pellet, since the mean free path of neutron is long. The fraction of neutron energies deposited to the central spark region in the ignition phase is too small to reduce the threshold energy of laser for ignition. In the burn phase, the neutron heating decreases the maximum compression ratio and accelerates the plasma expansion. The inclusion of neutron heating then decreases the pellet gain from the value in the case of neglecting the neutron heating.

1. INTRODUCTION

In D-T laser fusion reactors, the fuel pellets should be compressed to areal density of more than $3g/cm^2$. The mean free path of 14-MeV neutrons times the fuel density is roughly $4.7g/cm^2$. The neutrons produced in the pellet hence interact with background species and deposit a certain fraction of their energy in the plasma.

Previous burn calculations considering the transport of fusion neutrons revealed that the neutron heating affects considerably the burn characteristics such as fuel gain for reactor-grade ICF pellets.[1],[2] In these burn calculations, however, the neutron heating process was treated in an approximated way. For example, the energy transport by recoil ions arising from neutron collisions was not taken into account. This process is of particular importance, because without it, the temperature dependence of energy deposition can not be accounted for.

Recently the present authors derived a simultaneous neutron/recoil-ion transport model for accurately calculating neutron heating in ICF plasmas, and examined the neutron heating effect on the basis of coupled neutronic/hydrodynamic calculations.[3],[4] In this calculation, however, the implosion process was not included; isobaric state was assumed as the final stage

© 1996 American Institute of Physics

of compression. The calculation including implosion process is indispensable for an accurate evaluation of the neutron heating effect in the actual ICF pellet.

In this paper, we examine the neutron heating effect in the ignition and burn phases of laser–imploded D–T pellets by making the implosion and burn simulations for reactor–grade laser fusion pellets.

2. METHOD OF CALCULATION

2.1. Simultaneous Neutron/recoil–ion Transport Model

In this model, the energy of fusion neutrons is deposited to plasma ions and electrons through the Coulombic slowing down of recoil ions. The time-dependent transport equations used for the transport calculations (*i.e.* the Boltzmann equation for neutrons and the Boltzmann–Fokker–Planck equation for charged particles) are written in terms of the modified Eulerian coordinates originally proposed by Wienke[5]. In this coordinates, the energy and angular variables are defined in terms of the velocity relative to the medium, *i.e.* $E=mq^2/2$, $q=|v-\dot{r}|$, where v and $\dot{r}$ are respectively the velocities of the neutron and the medium in a stationary coordinate. This treatment enables us to account for the motion of background plasmas and consistently couple the transport and hydrodynamics processes.

2.2. Implosion and Burn Simulation

At first, we carry out the implosion simulations using ILESTA–1D code[6]. In this code, the radiation transport, which is important in the implosion process, is treated by the multigroup flux–limited diffusion model with non local thermodynamic equilibrium (*i.e.* non LTE) model. The transport of fusion–born alpha particles is treated by the multigroup flux–limited diffusion model. The Cowan model and the modified Thomas–Fermi model are adopted for ion and electron equations of state, respectively. Neutrons are assumed to freely escape from the pellet. We use the radial distributions of density, ion and electron temperatures, and hydrodynamic velocity around the final stage of implosion as the input data for the subsequent burn simulations.

In the burn calculations, we use MEDUSA–Q code[4] that includes time-dependent transport routines. The transports of neutrons, recoil ions and alpha particles are taken into account. The ideal gas model is adopted for the ion and electron equations of state. The radiation is assumed to freely escape from the pellet.

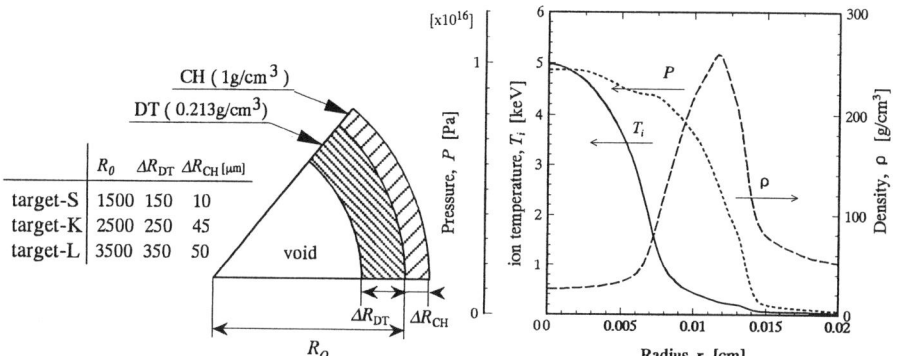

Fig.1 Initial configuration and size of traget models

Fig.2 Typical profiles of ion temperature, density and pressure arround final stage of implosion (target-L, E_L=4.8MJ, t=49.972ns)

3. RESULTS AND DISCUSSIONS

3.1. Implosion Simulation

Figure 1 illustrates the initial configuration and size of the pellet designed with reference to the conceptual ICF reactor design `KOYO'[7]. The initial aspect ratio A_o (*i.e.* $\Delta R_{DT}/R_o$) is fixed as 10. The thickness of CH ablator is adjusted to almost ablate away at the end of implosion. We assume that there is about 1 atm DT gas in the central void.

We have made the implosion simulations for various values of laser energy E_L. Tailored pulse consisting of linearly ramped prepulse followed by main pulse rising with $\{ (t-\tau_1)/\tau_2 \}^2$ is irradiated on the pellet, where τ_1 and τ_2 are the duration times for prepulse and main pulse, respectively. Short wavelength (0.35μm) and f/10 lens with $d/R = -1$ focusing are assumed. The intensity of the prepulse is kept as 10^{13}W/cm^2 on the pellet surface at $t = \tau_1$, while the peak power density of the main pulse with the pulse duration of τ_2 is about 100 times that of the prepulse. The prepulse and main pulse durations are optimized to obtain the maximum fusion output energy for every E_L.

Figure 2 illustrates the profiles of the density, ion temperature, and pressure around the final stage of implosion. It is found that the "spark ignition"[8] consisting of the central hot spark and the surrounding main fuel with higher density and lower temperature has been realized.

3.2. Burn Simulation

Figure 3 shows the pellet gain Q (*i.e.* the ratio of fusion output energy E_f to

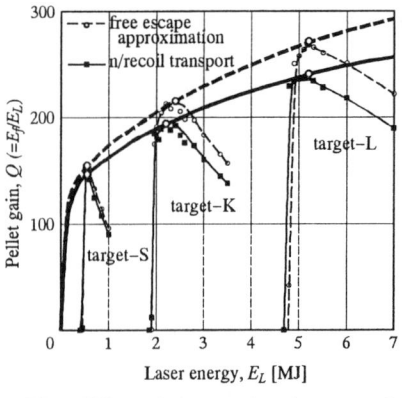
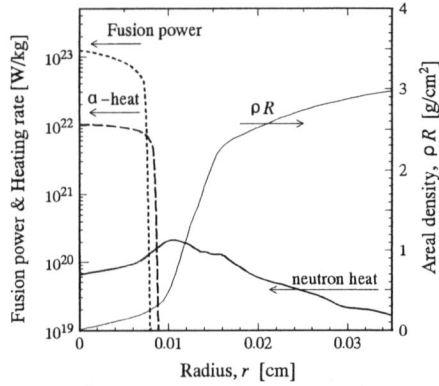

Fig.3 Effect of neutron heating on pellet gain as a function of laser energy

Fig.4 Spatial profiles of energy transfer to pellet by neutrons and α-particles at ignition phase (target-L, E_L=4.8MJ, t=38.680ns)

E_L) with each pellet as a function of E_L. It is found that the inclusion of neutron heating slightly reduces the threshold energy of laser E_{Lth}, i.e. E_L required to obtain $Q > 1$, in the case of target-L alone. On the other hand, the inclusion of the neutron heating decreases the maximum pellet gain Q_{max} from the value obtained by neglecting the neutron heating in all cases. In order to reveal the reason for the reductions in E_{Lth} and Q_{max}, we examine the neutron heating effects in the ignition and burn phases of the pellet.

Ignition phase

In the case of the spark ignition scheme, the pellet can be ignited when the deposited energy by fusion-products in the spark region exceeds the lost energy from this region by the bremsstrahlung and the electron thermal conduction. If the neutrons deposit a large fraction of their energy in the spark region, the spark temperature required for ignition becomes lower, which implies the reduction in E_{Lth}.

The spatial profiles of neutron heating rate at the ignition phase is shown in **Fig. 4** for the case of target-L and E_L=4.8MJ. (For comparative purposes, also is shown the heating rate from α particles.) The neutrons deposit their energy all over the pellet in contrast with that α particles deposit their energy only around spark region. The neutron heating rate in the whole pellet at the ignition phase is roughly comparable to the α-heating rate. Limited in the spark region, however, the ratio of neutron heating rate to α-heating rate is only 2 - 5 % because the areal density of the spark region which is about 0.3g/cm² is too small as compared with the mean free path of 14-MeV neutron times the fuel density. The reduction in E_{Lth}, therefore, is not more than several percent. In the cases of other smaller pellets (*i.e.* target–K,S), this effect no longer appears since the spark ρR is smaller.

Burn phase

The flow diagram around the burn phase for target–K and E_L=3.0MJ is shown in **Fig. 5**. Thermonuclear reaction wave generated by the ignition of the central spark region is propagated to the main fuel mainly by α–heating. During this process, the surrounding main fuel is still being compressed with inward velocity. In this phase, fusion neutrons rapidly heat all over the pellet, especially the main fuel, so that including the neutron heating advances the propagation of reaction wave to the main fuel compared with the case of neglecting the neutron heating. Due to the neutron heating, the temperature of the main fuel increases, which increases the pressure. These effects reduce the attainable density of main fuel at the time of maximum compression. (In the case of target–K and E_L=3.00MJ, the neutron heating decreases the maximum density of the main fuel by about 60%.) The reaction rate of the main fuel hence decreases.

The ignition of the dense main fuel drives the shock wave. The reaction wave is propagated outwards further by this shock wave. In the case of including the neutron heating, the shock wave becomes weaker because of the decrease in the compression of main fuel. The propagation velocity of shock then becomes smaller. The ions in front of shock are however heated up by the neutrons, which accelerates the expansion of the outer part of fuel, so that the confinement time is shortened. Thus, the inclusion of the neutron heating decreases the burn fraction or pellet gain from the value obtained by neglecting the neutron heating. This

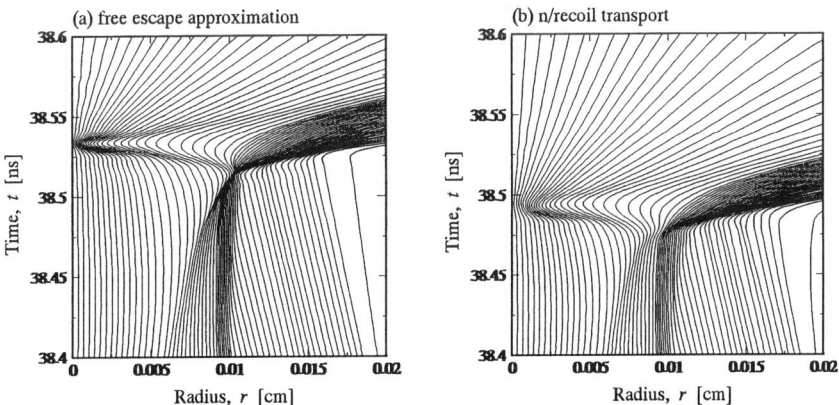

Fig.5 Flow diagram in ignition and burn phases (target-K, E_L=3.00MJ)

effect of neutron heating becomes more remarkable in larger-sized pellets.

4. CONCLUDING REMARKS

We have examined the neutron heating effects in the ignition and burn phases of the laser-imploded D-T pellets on basis of coupled neutronic/hydrodynamic calculations. The simultaneous neutron/recoil-ion transport model has been used for calculating the neutron heating rate in the pellet.

It was found that the neutron heating is not very effective in reducing the threshold energy of laser for the ignition. This is because the neutrons scarcely deposit their energy in the spark region during the ignition phase.

We also found that the inclusion of neutron heating decreases the maximum pellet gain because of the decreased compression of main fuel from the faster propagation of reaction wave to the main fuel and of the rapid expansion from the increased pellet temperature. This effect becomes more remarkable in larger-sized pellets.

ACKNOWLEDGEMENT

The authors would like to thank Dr.Takabe for permission to use ILESTA-1D code for the implosion simulations.

REFERENCES

(1) GOEL,B., HÖBEL,W. : *Plasma Physics and Controlled Nuclear Fusion Research 1984 (Proc. 10th Int. Conf. London, 1984)*, Vienna, IAEA, 1985, Vol.3, pp.345-351.
(2) MARTINEZ-VAL,J.M. : *Fusion Technol.,* **17**, 476-483(1990).
(3) NAKAO,Y., et.al. : *J.Nucl. Sci. Technol.,* **30**, 18-30(1993).
(4) NAKAO,Y., et.al. : *ibid.,* **30**, 1207-1213(1993).
(5) WIENKE,B.R. : *Phys. Fluids,* **17**, 1135-1138(1974).
(6) TAKABE,H., MIMA,K. : *ILE Research Report, ILE*-8713P (1987).
(7) TAKABE,H., et.al. : *Emerging Nuclear Energy Systems 1993 (Proc. 7th Int. Conf. Chiba, 1993)*, Singapore, World Scientific, 1994, pp.101-105.
(8) MEYER-TER-VEHN,J. : *Nucl. Fusion,* **22**, 561-565(1982).

A Moderate-Gain ICF Model From LTE Ignition to non-LTE Burn

X.T.He, Y.S.Li and M.Yu

Institute of Applied Physics and Computational Mathematics,
P.O.Box 8009, Beijing 100088, China

Abstract A moderate - gain indirect-driven ICF model from LTE volume ignition to non-LTE burn is presented, a high compressed glass tamper, which consists of a capsule glass shell that has not been ablated, is required to enhance the DT burn. The numerical simulation shows that the energy yield of 18.1MJ from DT fuel of 0.2mg is achieved when a shaped laser beam of about 1.25MJ is invested in a hohlraum, a enhanced factor of 2.3 for the DT fuel burn up fraction is obtained.

1. INTRODUCTION

At present one concerned about ignition and moderate-gain(<20,say) burn in inertial confinement fusion (ICF) research on National Ignition Facility.

For an indirect-driven ICF, the investigation for above purpose has been shown (1),(2) more recently. The typical ablators of the capsules used on the above work are made up of *Br*-doped *CH* or *Cu*-doped *Be*, both them are in a same inner radius of 0.95mm and 0.16mm thickness. The deuterium-tritium (DT) ice layer neighboring ablator is of a mass of 0.2 mg, and DT gas with density 0.3mg/cm³ is filled in a central region of the capsule. Laser energy of 1.6 MJ in a pulse of 17 ns is delivered into a hohlraum to generate a spatial uniform radiation filed with peak temperature about 3.5-4.0 *MK(MK $\equiv 10^6 K$)* surrounding the capsule, the clean yield is calculated about 5 MJ. It is easy to estimate from the radiative ablation depth that the ablator of 0.16mm thickness is essentially exhausted during the laser pulse, the central hot spot ignition in extremely high density with a marginal ignition temperature 55 MK obtained from the balance between the emission rate of the spontaneous bremsstrahlung and the heating rate of the thermonuclear reaction seems to be required, this is so called non-local -thermal-equilibrium (non-LTE) ignition model.

In this paper, we suggest an alternative model that the local thermal equilibrium (LTE) ignition, which is in a lower ignition temperature comparing to non-LTE ignition, is required, and after short sustaining LTE burn the DT system rapidly transits to the non-LTE burn as discussed in our previous work (3). The capsule configuration is similar to that of the non-LTE ignition model as shown in (1) and (2). However, the shell of the capsule is thickened so that a portion of this shell is ablated to provide an imploding energy and the rest, which has not been ablated in a laser duration, is a high compressed "cold" layer and acts first as a pusher and later as a tamper to confine the high pressure DT plasma and further to enhance the fuel depletion. The material of the shell is carefully chosen so that it is optically thick until LTE ignition is coming, and optically thin during non-LTE burn. Here, the material of the shell is required to be a low Z one or doped a few high Z elements. Such a system is essentially robust, i.e., the central hot ignition is absent, in-flight aspect ratio is far less than 25, and the requirement of the imploding symmetry may be relaxed to a certain extent.

2. PHYSICAL CONSIDERATIONS

The LTE volume ignition in the lower temperature, and the enhanced confinement effect on DT plasma using the properly thickened shell in lower Z material are the essential points of our model. The physical considerations are discussed below.

a. LTE ignition

The LTE ignition occurs in a condition that the heating rate of the thermonuclear reaction balances with the losing rate of the radiative heat conduction that the flux goes into the shell of the capsule from DT fuel , then the transition from LTE to non-LTE happened in burn process. The threshold values of the LTE ignition of

DT system under the optimized implosion conditions are given in (3), that is, the temperature about 20-25MK and areal density $10^{-4}T^{3/2} \ll \rho R < 1$, where fluid density ρ usually is about 20-50 g/cm^3, R in centimeter is the radius of the compressed DT fuel, temperature T is in unit of MK.

Under these conditions, the internal energy for the DT fuel of 0.2 mg is about 0.05MJ. If laser energy of one megajoule is delivered into hohlraum, as long as 5% laser energy is invested to DT fuel, it may trigger a sustained DT burn. For moderate-gain ICF, the target possesses only DT fuel of a few tenths of 1 mg, therefore, the LTE volume ignition may reduce the requirement of laser energy as well.

b. Shell design

As above mentioned, the shell of the capsule must be required to have the properties that it is optically thick for LTE ignition and optically thin for non-LTE burn. Our work has shown that the glass shell or doped a few high Z elements is able to approximately meet this requirement.

At LTE ignition, if the imploding kinetic energy is converted into the internal energy of the high compressed "cold" layer (called tamper) and DT fuel, the energy conservation reads:

$$\tfrac{1}{2} M_1 v_0^2 = M_1 \alpha E_S + m C_v T_{ig} \tag{1}$$

where v_0 is an imploding velocity, α is a modified factor close to one due to radiation heat-conduction from DT to tamper, M_1 and E_S are mass and internal energy for the tamper respectively, m, C_v, T_{ig} are mass, specific heat, ignition temperature for DT fuel, respectively.

We write $m = \dfrac{4\pi(\rho R)_{ig}^3}{3 P_{ig}^2} C_{ig}^4$, where the pressure $P_{ig} = C_{ig}^2 \rho_{ig}$, C_{ig} sound speed for DT fuel. For high compressed "cold" glass tamper, pressure $P_S \propto \rho_S^{5/3}$, internal energy of the unit mass $E_S \propto \rho_S^{2/3}$, where ρ_S is the tamper density.

Assume that the mass of the tamper is an approximate invariant in a main implosion stage to invest energy to the DT fuel, and regard M_1, E_S as a function of $P_s = P_{ig}$; we discuss the minimal energy for the LTE volume ignition, or the minimal mass of the tamper as v_0 is given.

Fix C_{ig} and $(\rho R)_{ig}^3$, and let $\frac{dM_1}{dP_{ig}} = 0$, from Eq.(1) and equation of state for glass, we have $\alpha E_s = \frac{5}{12} v_0^2$, and thus $\frac{m}{M_1} = \frac{v_0^2}{12 C_v T_{ig}}$. If the fuel mass is $m=0.2mg$ and $v_0 = 2.5 \times 10^7$ cm/s, we obtain the tamper mass $M_1 \approx 9mg$.

The ablated mass of the shell is roughly estimated below. The ablative depth from asymptotic similar solution(4) for glass shell is

$$X_R(cm) \approx 0.5 \times 10^{-3} \rho^{-1.2} T^{2.9} \sqrt{t} \qquad (2)$$

where the glass density ρ is in g/cm^3, the ablation temperature T in MK, the ablation duration t in ns.

Taking the effective boundary temperature $T \approx 3.5 MK$, normal glass density $\rho = 2.5 g/cm^3$, the ablative duration t=6ns, for the outer radius of 1.55mm, the ablated mass $M_2 \approx 4\pi R^2 \rho X_R \approx 11mg$. Thus, the total shell mass $M = M_1 + M_2 \approx 20mg$, which is 100 fold mass of the DT fuel.

C. Enhancement of fuel depletion

It is well known that the DT fuel burn up fraction in the case of exhausted ablator and the ion temperature about 200 MK is approximately (5)

$$\eta_0 = \frac{A_0}{A_0+6} \qquad (3)$$

where $A_0 = \rho R$ is an areal density for DT fuel.

However, when there exists the high compressed "cold" tamper, the burning duration τ is approximately determined by the relation $\frac{1}{2}\frac{P_{DT}}{(\rho\Delta R)}\tau^2 = \frac{R}{3}$, and $\frac{\tau}{\tau_o} = \sqrt{\frac{6(\rho\Delta R)_t}{(\rho\Delta R)_{DT}}}$, where $(\rho\Delta R)_t$ and $(\rho R)_t$ are the areal density for DT fuel and tamper respectively, and P_{DT} is the pressure of the DT fuel. As a result, Eq.(3) should be modified by

$$\eta = \frac{A}{A+6} \qquad (4)$$

where $A = A_0\xi$, $\xi \equiv \sqrt{\frac{6(\rho\Delta R)_t}{(\rho\Delta R)_{DT}}}$. In general, $\xi > 1$ as shown in numerical simulation.(6)

As a result, when there exists the high compressed tamper, the DT depletion may considerably enhanced.

3. NUMERICAL RESULTS AND DISCUSSIONS

We consider a target for demonstration of the moderate-gain ICF from LTE volume ignition to non-LTE burn.

For calculation convenience, the hohlraum is simulated into a spherical cavity with a gold case, the gas of $He + H_2$ with density of $1\ mg/cm^3$ is filled inside.

The capsule in the hohlraum is constructed of a glass shell with outer radius of 1.55mm and 0.48mm thickness, a cryogenic DT fuel layer of 0.2mg and DT gas with density $10^{-4}\ g/cm^3$ filled in the central region.

The one-dimensional and three-temperature code CFJ is used to compute the system evolution. A laser beam of $1.25MJ$ with about 6 ns prepulse and 1.5 ns main pulse is invested in the hohlraum, and a radiation filed with peak temperature $4.50MK$ is produced surrounding the capsule to drive an imploding process.

The computed results are as follows. The LTE volume ignition occurs at time t=11.95ns, the internal energy of the DT fuel is $0.043MJ$, thus, the averaged temperature $T_i \approx T_e \approx T_\gamma = 23MK$, where T_i, T_e and T_γ are the temperature for ion, electron and x-ray, respectively; $(\rho R)_{DT} \approx 0.41 g/cm^2$ with the radius of $0.082mm$. For the tamper, with $(\rho\Delta R)_t \approx 1.19 g/cm^2$; About 14mg of the glass shell has been ablated and the rest as a tamper is about 11mg. The in-flight aspect ratio is about 2.7 which is far less than 25-35, as a result, the hydrodynamic instability is expected to be unimportant.

After short LTE burn, at 12.1ns the maximal areal density of $(\rho R)_{DT} \approx 0.8 g/cm^2$ reaches and the non-LTE burn begins; at 12.36 ns the maximal ion temperature $T_{i max} \approx 804 MK$ and $T_e \approx 278 MK$ and $T_\gamma \approx 33.8 MK$ arrive.

Finally, $18.1MJ$ energy yield (gain G=14.5) is released, thus, the DT burn up fraction is $\eta \approx 0.27$. This is very well expressed by Eq.(4) using the maximal areal density $(\rho R)_{DT} \approx 0.8\, g/cm^2$ for DT fuel and simultaneously $(\rho R)_t \approx 1.38\, g/cm^2$ is taken in account, we obtained $\xi = 3.2$, thus, η is enhanced to a factor of 2.3 due to confinement effect from tamper.

This work was supported in part by the foundation of the National Hi-Tec ICF Committee and by the National Natural Science foundation in China.

REFERENCES

1. M.Cray, "Inertial confinement fusion at Los Alamos National Laboratory", invited talk at *China-Us-Japan Workshop on Laser Plasma and Drivers*, October 17-21, 1994, Beijing, China.
2. J.D.Lindl, "Review of indirect drive ICF", Presented to *The Am.Phys.Society and Div. of Plasma Society*, November 7-11,1994, Minneapolis, MN, USA.
3. X.T.He and Y.S.Li, "Physical processes of volume ignition and thermonuclear burn for high-gain ICF", *Laser Interaction and Relation Plasma Phenomena*, AIP Conference Proceedings 318, G.H.Miley, ed., AIP Press, New York, 1994, pp.334-344.
4. X.T.He, T.Q.Chang and M.Yu, "X-ray conversion in high gain radiation drive ICF", *Laser Interaction and Related Plasma Phenomena*, H.Hora and G.H.Miley, eds.,(Plenum Press, New York), **9**,553(1991).
5. R.E.Kidder, "Energy gain of laser-compressed pellets: a simple model calculation", *Nuclear Fusion* **163**,405(1976).
6. Y.S.Li,X.T.He,Y.M.Gao and M.Yu, "The Kinetic Process Analysis of Volume Thermonuclear Ignition in High Gain Inertial Confinement Fusion", *High power Laser and Particle Beams* (to be published in 1995).

Practical Large-Size-Pellet ICF

Shigeo Kawata and Katsushige Kurawaki

*Department of Electrical Engineering, Nagaoka University of Technology,
Nagaoka 940-21, Japan*

Abstract. We discuss a concept of a large-size-pellet inertial confinement fusion (ICF) in which a fuel pellet contains the deuterium-tritium fuel of about one gram. In the large-size-pellet ICF the fuel is compressed to about 100 times the solid density in order to realize $\rho R > 1 \sim 3 g/cm^2$. Because of the low compression ratio, constraints required for the uniform fuel implosion are relaxed compared with those for a high-compression implosion of a small pellet. This concept of the large-size-pellet implosion may present another approach to ICF. We present a simple estimation for the concept and one-dimensional numerical analyses for the large-size-pellet implosion. The simulation results show that sufficient fusion energy output can be obtained in the large-size-pellet ICF. A simple linear estimation for the Rayleigh-Taylor instability presents that the fuel pellet may be robust against the instability.

INTRODUCTION

The researches have been concentrated on a small-size-pellet DT inertial confinement fusion ICF. (1-7) The small pellet is compressed to about 1000 times the solid density to extract the fusion energy in a short time with a high burn rate and to save the input driver energy. It is well known that the nonuniformity should be suppressed less than a few % in order to attain the high compression. (1,5) One of other approaches to ICF may be a low-compression one, which is presented and discussed in this paper. In the low-compression scheme, a fuel should contain a larger DT mass compared with that in a high-compression pellet in order to attain $\rho R > 1 \sim 3 g/cm^2$; for example, about one gram of the DT fuel may be required. Because of the low compression ratio and the large size of pellet, the low-compression scheme relaxes the uniformity requirement and the driver-energy-focusing one, which is quite severe in ICF approaches except laser fusion. However this scheme has the disadvantage of larger input driver energy, which should be about 100 MJ for the one-gram DT fuel pellet. We believe that the advantages in the low-compression scheme prevail over the disadvantage. In this paper first we present the concept of the large-size-pellet ICF. Then numerical analyses for the large-size-pellet implosion are performed by a one-dimensional hydrodynamic simulation code. A simple estimation for the Rayleigh-Taylor(R-T) instability growth is also performed.

The purpose of this paper is to point out that a large pellet is robust against the nonuniform implosion compared with the small pellet, and that sufficient fusion energy output may be obtained in the large-pellet impact ICF. As presented above, ICF researches have been concentrated on the small-pellet scheme. Therefore it may be useful for the future ICF researches to focus our attention on the large-size pellet implosion also.

LARGE PELLET AND UNIFORMITY

One of the main objectives in the large-size-pellet ICF is to reduce the fuel density compression ratio to about 100. In the large-size-pellet ICF a fuel radius is about 1cm depending on a pellet structure and on the total fuel mass being contained in the capsule. Because of the low compression, the uniformity requirement can be relaxed.

The input energy being required to compress and heat the fuel is estimated by the sum of the heating energy $W_h = 38.3 M_{DTh} T_i$ J and the energy (pressure work) to compress the fuel. Here M_{DTh} is a fraction of the fuel (M_{DT} mg) heated up to T_i eV by the input energy, and its radius is assumed to be 3 times the radius at which a produced α particle stops. In the calculation for the pressure work, electrons are assumed to be in the perfect-degenerate gas state. The pressure work is $W_P = 114(\kappa^{2/3} - 1)M_{DT}$ J, where κ is the density compression ratio. In this section we employ the following typical parameter values in DT ICF: $\rho R = 3 g/cm^2$, the implosion efficiency is 5%, the driver efficiency $\eta_d = 20\%$, $M = 1.1$, $f_a = 0.1$, and $T_i = 5$keV. Finally we obtain the relation between the compression ratio and the input energy, as shown in Fig.1; the numerical values beside the line denote the total DT mass. For the large pellet containing about 1 gram of the DT, the required density compression ratio is rather low and is less than 100.

The implosion-uniformity requirement is estimated as follows: first the nonuniformity of implosion speed δv_{imp} is estimated. From the approximate relations of $R_0 = v_{imp} t_f$ and $\delta R = \delta v_{imp} t_f$, $\delta v_{imp}/v_{imp} = \delta R/R_0 = (\delta R/R)(R/R_0) = (\delta R/R)\kappa^{-1/3}$. Here R_0 is the initial radius, R the final radius of the pellet and κ the volume compression ratio. On the other hand ρR is proportional to R^{-2}, and $\delta(\rho R)/(\rho R) = (1 + \delta R/R)^{-2}$. Then $\delta R/R = \{\delta(\rho R)/(\rho R)\}^{-1/2} - 1$. Combining these relations, we obtain $\Delta = \delta v_{imp}/v_{imp} = \{\{\delta(\rho R)/(\rho R)\}^{-1/2} - 1\}/\kappa^{1/3}$. From this result $\Delta_{low-\rho}/\Delta_{high-\rho} = (\kappa_{high-\rho}/\kappa_{low-\rho})^{1/3}$. Here $\Delta_{high-\rho}$ shows the nonuniformity

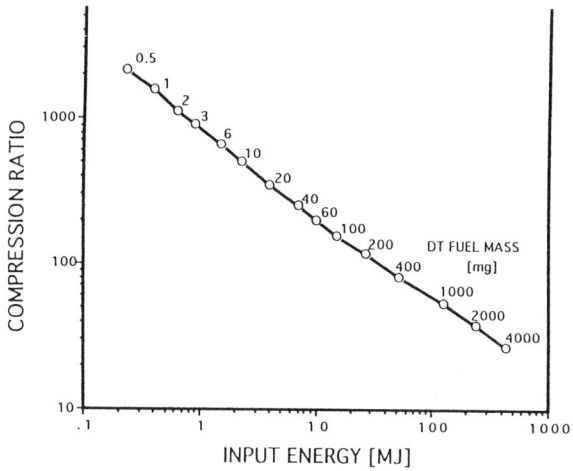

FIGURE 1. Density compression ratio versus the input energy.

in the conventional high-compression scheme and $\Delta_{low-\rho}$ presents one in the large-pellet ICF, for the fixed allowable $\delta(\rho R)/(\rho R)$. For the high-compression scheme the volume compression ratio $\kappa_{high-\rho}$ is about 10000, and for the large-pellet ICF $\kappa_{low-\rho}$ may be a few hundreds to 500 in our pellet employed in this paper. Therefore the factor of $\Delta_{low-\rho}/\Delta_{high-\rho}$ is about 2.7~4. Consequently the nonuniformity tolerance in the large-size-pellet ICF is 2.7~4 times larger than that in the high-compression ICF. The nonuniformity of about 5~8% may still give sufficient fusion energy output in the large-pellet ICF. This result confirmed that the large pellet is rather stable against the nonuniformity compared with the conventional small pellet.

HYDRODYNAMIC SIMULATION AND STABILITY

Our simulation was performed by a one-dimensional hydrodynamic Lagrangian code "Mt.FUJI", which has been developed originally for simulations of ion-beam small-pellet implosion (6,7) and other purposes. In our code we employ the three-temperature model (2,3). We include the DT reactions and the fuel depletion by the reactions. In our model we include the α particle heating and disregard the neutron heating. The flux-limited diffusion model is used for the radiation transportation and the α particle one. We also use the SESAME library as the equation of state.

A DT gas is located at our pellet center. We assume that the DT gas density is 1/10000 times the solid density. A mixture of DT(95%) and Pb(5%) covers the main DT fuel. The pellet is hit by a solid Pb pusher/tamper which is accelerated to 120 km/s. The Pb pusher/tamper may be accelerated by the

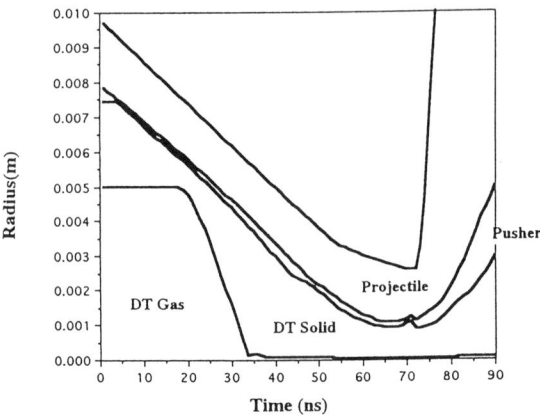

FIGURE 2. Radius-time diagram in a large-pellet implosion.

illumination of laser or ion beams. Hereafter we call the Pb pusher/tamper a projectile.

The parameter values employed and a part of results obtanied are as follows: the DT gas layer radius is 5.00 mm, the DT solid layer thickness 2.43 mm, The total DT fuel mass 0.250 g, the pusher thickness 0.500 mm, the Pb projectile thickness 1.86 mm, the input energy 150 MJ, the maximum DT ion temperature 11.6 keV, the maximum ρR 6.84 g/cm^2, the maximum compression ratio of the DT fuel 383, the fusion energy output 64.2 GJ, and the pellet gain 428. Figure 2 presents the radius-time diagram. By a strong shock wave launched by the projectile impact, first the inner DT gas reacts slightly at about 34 nsec. At this time the main DT is not yet ignited. After the further compression, the DT gas is ignited and burned at the time($\sim$67 nsec) of the maximum compression. The main DT fuel is also heated up by the alpha particles produced in the DT gas. Then the main DT fuel is ignited and burned. Figure 3 shows the spatial profiles of the ion temperature (T_i) and the mass density (ρ) at 70nsec; the characteristic feature of the profiles is the spatial flatness of T_i and ρ in the DT fuel. Between $\sim$60nsec and $\sim$70nsec, the typical values of T_i and ρ in the DT fuel are 3keV and 300 times the solid mass density (ρ_s), respectively. Hereafter we call this $\sim$10nsec the confinement time τ_c. In the DT fuel, the radiation mean free path l_R is about 6.6cm for $T_i = 3$keV and $\rho = 300 \times \rho_s$. The DT outer radius R_{out} is about 0.095cm during τ_c. So the radiation temperature is flat in the DT fuel during τ_c. The ion-electron energy exchange time interval τ_{ie} and the electron-radiation one τ_{er} are very short compared with τ_c. Consequently the ion temperature should be also rather flat in the DT fuel as shown in Fig.3. The sound speed C_s is about 3.4×10^7cm/sec for 3keV. The sound wave propagates in the entire DT fuel in $\tau_f \simeq R_{out}/C_s \simeq 2.8$ nsec: during τ_f the information of the fluid behavior is transported in the

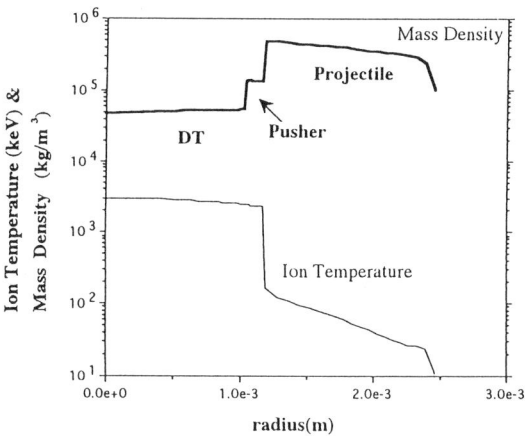

FIFURE 3. Profiles of ion temperature and mass density at 70 nsec.

entire DT fuel. Because of $\tau_c > \tau_f$, the density jump between the gas and solid DT should be also smoothed out during τ_c by the propagation of the compression/expansion wave generated by the density jump. The flatness of ρ and T_i profiles are characteristic of the large size pellet: for a conventional small pellet, $\tau_c < \tau_f$. However nonuniform the compression way of the gas DT is, the initial DT gas should be compressed to almost the density of the main solid DT. The compression/expansion wave propagates in both the inner gas DT and the outer solid DT during τ_c. In this particular simulation result, the implosion efficiency is about 3.5%. The burn time duration is longer than 2 nsec, that is rather long compared with that in a conventional small pellet; this comes from the large DT mass and the tamping effect of the projectile on the DT fuel confinement.

We also performed a simple linear estimation for the R-T instability growth. In order to evaluate the growth, we computed $\gamma\tau \simeq \int_0^t dt \sqrt{Agk}$. Here A is the Atwood number, g the acceleration, τ the time interval during which the R-T instability takes place, and k the wave number which is estimated by $2\pi/$(the thickness of the layer concerned). In our simulation the interface between the solid DT and the pusher is in danger of the R-T instability. The instability is found during the deceleration phase by the reflected strong shock wave (mode 1) and the stagnation phase (mode 2). The thickness of the pusher layer changes in these two phases: the pusher thickness is about 350 μm during the deceleration phase (mode 1) (about 40ns~48ns) by the reflected shock and about 150 μm during the stagnation phase (mode 2) (about 49ns~70ns). Therefore we computed $\gamma\tau$ separately, because the instability modes should be different with each other. $\gamma\tau_s$ is about 3.8 for the mode 2. $\gamma\tau_d$ is about 2.0 for the mode 1.

The estimation result suggests that the R-T instability growth may be suppressed more, if we shape the projectile density in space in order to make the first shock wave weak: inside the projectile the mass density may be shaped from the ablator density at the projectile inner surface to the solid Pb density at its outer surface. We also performed the simulation by using the shaped projectile. In this case the pellet has the same structure as the pellet employed above. Sufficient fusion energy output is obtained in this case also. We have again two unstable phases (modes 1 and 2): $\gamma\tau_d$ is about 1.2 for the mode 1, although $\gamma\tau_s$ is about 3.7 the mode 2. The initial shock is weak and reaches the pellet center late compared with that in the former case. Consequently the reflected shock is weak, and the implosion is more stable in this case. In addition, the pellet gain does not decrease greatly. These estimation results present that the large-size pellet may not be destroyed by the R-T instability.

CONCLUSIONS

The results of this work demonstrate that the large-pellet ICF scheme may supply sufficient fusion energy output. The results of the researches also show that the large pellet may be robust against the nonuniformity compared with the small pellet. We also performed the linear estimation for the R-T instability growth based on the one-dimensional simulation results. The linear analyses present that the large-pellet implosion may be rather stable. We also believe that the large-pellet ICF is another way to the future energy source.

ACKNOWLEDGEMENT

This work is supported partly by the Scientific Research Fund of the Ministry of Education, Science and Culture, and in part by the Institute of Laser Engineering, Osaka University, Japan.

REFERENCES

1. Lindl, J.D., McCrory, R.L. and Campbell, E.M., *Phys. Today* **9**, 32 (1992).
2. Fraley, G.S., Linnebur, E.J., Mason, R.J. and Morse, R.L., *Phys. Fluids* **17**, 474 (1974).
3. Tahir, N.A., Long, K.A. and Laing, E.W., *J. Appl. Phys.* **60**, 898 (1986).
4. Tahir, N.A. and Arnold, R.C., *Phys. Lett.* **A160**, 385 (1991).
5. Kawata, S. and Niu, K., *Jpn.J.Appl.Phys.* **53**, 3416 (1984).
6. Kawata, S. and Niu, K., *Res. Report Inst. Plasma Physics, Nagoya, Japan*, **IPPJ-742**, 149 (1985).
7. Tamba, M., Nagata, N., Kawata, S. and Niu, K., *Laser Part. Beams*, **1**, 121 (1983).

DEVELOPMENT OF HIGH YIELD LOW LEVEL TRITIUM TARGETS FOR INERTIAL FUSION REACTOR SYSTEMS

N. A. Tahir and D. H. H. Hoffmann

Gesellschaft für Schwerionenforschung, Postfach 11 05 52,
64220 Darmstadt, Germany

ABSTRACT

This paper shows that it is not necessary to use an equimolar DT mixture in an inertial fusion target, but instead one may use a fuel composition with a substantially reduced tritium content. For example one may design a target that contains 75% deuterium atoms and 25% tritium atoms and yet it may generate sufficient output energy to run the reactor. This could lead to an overall reduction in the tritium inventory. Moreover the tritium fractional burn in reduced tritium targets in much higher than that in equimolar targets. This means that in the case of low level tritium targets a much smaller tritium mass will be left unburnt that would need to be evacuated from the reactor chamber together with other target debris after every implosion run.

INTRODUCTION

A deuterium-tritium fusion process has a very large crossection even at a relatively low temperature of 5 keV. It is therefore very likely that the first generation of fusion reactors will be based on a scheme that extracts energy from DT fusion. However there will be two very important problems associated with such a system. Firstly the activation caused by the neutrons and secondly the large tritium inventory needed to run the reactor. The tritium is a radioactive substance and is very difficult to contain. Any accident related to release of tritium would lead to environmental and health hazards. It is therefore highly desirable to minimize the tritium inventory in the reactor system. The first problem could be handled by developing low activation materials. In this paper we show how one may treat the second problem by developing inertial fusion targets that use a substantially reduced tritium content. This would reduce the the mass of tritium used per day in the reactor that would eventually lead to a reduction in the overall tritium inventory. It has also been found that the tritium fractional burn in such low level tritium targets is very large as compared to the equimolar DT targets. This means that after the implosion run, a much smaller tritium mass will be left in the reactor chamber that would have to be evacuated. As a result of this the process of chamber evacuation will be much more clean in case of the reduced tritium targets.

© 1996 American Institute of Physics

REDUCTION OF TRITIUM CONTENT IN THE FUEL

The number of DT reactions taking place in the fuel per unit volume per unit time is given by

$$R_{DT} = (\sigma v)_{DT} \, f_D \, f_T \, n_i^2 \qquad (1)$$

where $(\sigma v)_{DT}$ is the DT reaction rate and is a function of the ion temperature only. Also f_D and f_T denote the fraction of deuterium and tritium ions respectively while n_i is the total ion number density. Table 1 shows the values of the product $f_D f_T$ for different combinations of f_D and f_T. It is seen that in the equimolar case $f_D = 0.5$ and $f_T = 0.5$ so that their product becomes 0.25. In case of $f_D = 0.6$ and $f_T = 0.4$ the product $f_D f_T = 0.24$. This means that if we use a fuel composition that contains 60% deuterium atoms and 40% tritium atoms, the tritium content will be reduced by 20% whereas the output energy will be reduced by 4% only. Similarly one may consider other combinations. For further details see[1].

SIMULATIONS OF LOW LEVEL TRITIUM TARGETS

The idea of using reduced tritium targets discussed in the previous section has been used to design indirect drive, reactor-size inertial fusion targets. The initial conditions of a typical target are shown in Fig.1 and the input radiation pulse is shown in Fig.2. The precise target parameters are listed in Table 2 while the input pulse parameters are given in Table 3. Previously[2] we simulated the compression and burn of this target using an equimolar DT fuel mixture. A parameter study of the target gain as a function of input pulse parameters has also been reported[3]. These simulations have been carried out using a one-dimensional, three-temperature, Lagrangian copmuter code MEDUSA-KAT[4]. In the present paper we report compression and burn simulations of this target using the same pulse parameters, but employing different fuel compositions. The results are presented in Table 4 and Fig.3. In Table 4 we give thermonuclear energy output whereas in Fig.3 we plot energy output normalized to the equimolar case. It is seen that the simulation results are much more optimistic than those predicted by simple considerations using Eq.1. This is because in the latter case we assume that the temperature and the density remains constant when we change the fuel compositions. This however is not the case in full target implosion simulations. When the tritium content of the fuel is reduced, the rate of energy production in the hot spot is reduced. As a consequence the hot spot temperature and hence the pressure increases at a slower rate. The bulk of the fuel around the hot spot that still is imploding sees a lower pressure at the center and therefore compresses the hot spot to a higher density. Also the burn temperature in low trtium case is lower than the equimolar case that increases the confinement time. These two effects compensate for the reduction in energy caused due to the decrease in the tritium content. For further details see[5].

Table 1: Reduction in tritium content as a function of deuterium tritium fraction

No.	f_D	f_T	$f_D f_T$	Reduction in $f_D \cdot f_T$	Reduction in Tritium
1	0.5	0.5	0.25		
2	0.6	0.4	0.24	4 %	20 %
3	0.7	0.3	0.21	16 %	40 %
4	0.8	0.2	0.16	36 %	60 %
5	0.9	0.1	0.09	64 %	80 %

Table 2: Target initial conditions

Inner fuel radius	R_1 = 3.02 mm
Outer fuel radius	R_2 = 3.18 mm
Outer radiation shield radius	R_3 = 3.21 mm
Outer ablator radius	R_4 = 3.41 mm
Hohlraum radius	R_5 = 6.91 mm
Hohlraum wall thickness	$R_6 - R_5$ = 0.015 mm
DT fuel mass	4.45 mg
DT fuel density	0.224 g/cm^3
Radiation shield mass (90% Li + 10% Au)	5.8 mg
Radiation shield density	2.0 g/cm^3
Ablator mass (solid C)	61.0 mg
Ablator density	2.20 g/cm^3
Hohlraum casing mass (solid Au)	170 mg
Hohlraum casing density	19.0 g/cm^3

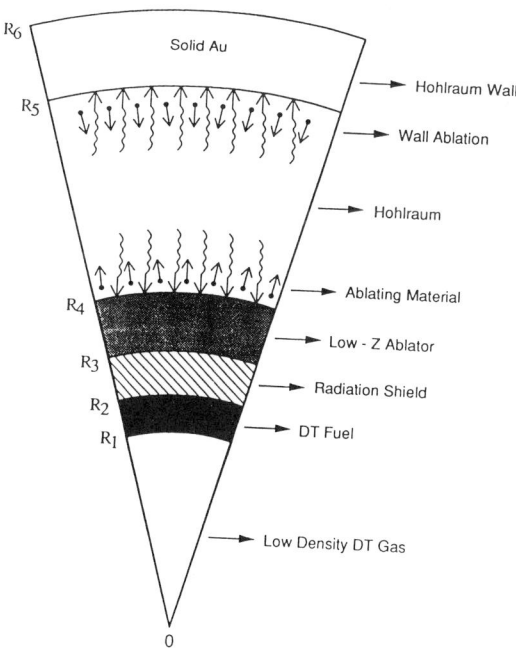

Fig.1 : Target initial conditions

Table 3: Input pulse parameters

Prepulse Temperature	$T_1 = 110$ eV
Prepulse power	$P_1 = 15$ TW/cm²
Maximum temperature	$T_2 = 300$ eV
Maximum power	$P_2 = 830$ TW/cm²
Prepulse duration	$\tau_1 = 24$ ns
Main pulse duration	$\tau_2 = 6$ ns
Shape factor	$p_1 = 1.75$
Shape factor	$p_2 = 20.0$

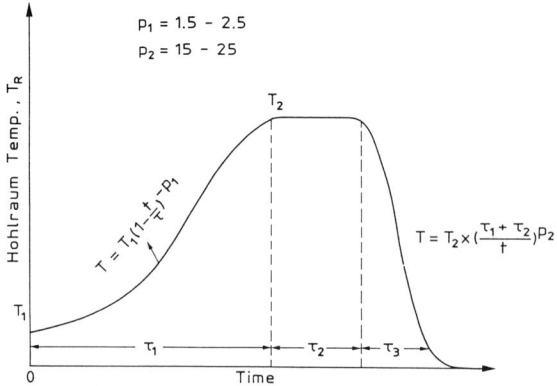

Fig. 2: Input pulse shape

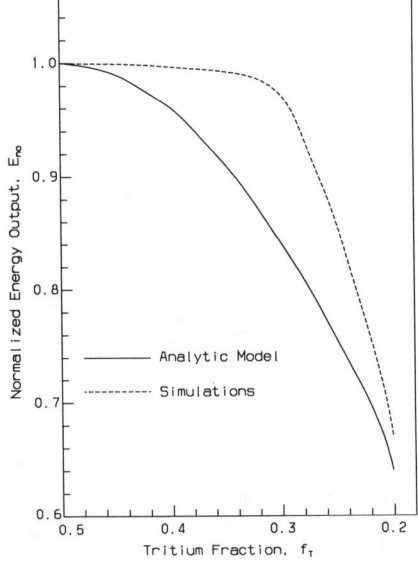

Fig.3 : Normalised energy output vs tritium fraction in fuel

Table 4: Energy output as a function of tritium fraction

fD	fT	Eos [MJ] Simulation Results	EOA [MJ] by Eq. 1
0.50	0.50	700	
0.60	0.40	699	672
0.65	0.35	691	637
0.70	0.30	686	588
0.75	0.25	606	525
0.80	0.20	467	448

FRACTIONAL BURN IN LOW TRITIUM TARGETS

Another very attractive feature of the low tritium targets is that the tritium fractional burn is much higher than that in the equimolar targets. As a result of this a much smaller mass of tritium is left unburnt among the target debris that has to be evacuated after each implosion run. This would make the chamber evacuation process much cleaner in case of low tritium targets. This is seen from Table 5 where we present tritium burn parameters for different fuel compositions.

Table 5: Tritium burn parameters (initial fuel mass 4.45 mg)

No.	Tritium Fraction f_T	Initial Tritium Mass (mg)	Final Tritium Mass (mg)	Tritium Fractional Burn (%)
1	0.50	2.67	1.43	46
2	0.40	2.22	0.99	55
3	0.35	1.99	0.78	61
4	0.30	1.74	0.52	70
5	0.25	1.49	0.42	72

CONCLUSIONS

We have shown that one should not use an equimolar DT mixture in an inertial fusion target, but the target may contain a substantially reduced tritium content. For example one may design a target with 75% deuterium and 25% tritium and yet the energy output may be sufficient to run the reactor. This could eventually lead to a significant reduction in the overall tritium inventory. Thes results are not only applicable to inertial fusion but are also valid for magnetic confinement fusion. Another important point is that a much higher fraction of tritium is consumed in low tritium targets. This implies that a much smaller mass of tritium is left unburnt in the reactor chamber after the target implosion. As a consequence the process of chamber evacuation will be much more cleaner in case of low tritium targets. .

REFERENCES

1. N. A. Tahir and D. H. H. Hoffmann, Fusion Engn. Design 24, 413 (1994).
2. N. A. Tahir et al., Phys. Lett. 172A, 162 (1991).
3. N. A. Tahir et al., Plasma Phys. Controlled Fusion 37, 447 (1995).
4. N. A. Tahir et al., J. Appl. Phys. 60, 898 (1986).
5. N. A. Tahir and D. H. H. Hoffmann, Fusion Technl. Sept. 1995, To be published

DT-Fuel Concentration in an ICF Pellet

Shigeo Kawata

Department of Electrical Engineering, Nagaoka University of Technology,
Nagaoka 940-21, Japan

Abstract. This paper shows that a 40% mixing inhomogeneity of the deuterium (D) and tritium (T) concentrations in a D-T pellet still gives a sufficient fusion energy output in D-T inertial confinement fusion (ICF). Numerical and analytical studies support this new result as long as the D and T total amounts are equal and the implosion process is stable. This result means that fusion energy output and burn process are rather insensitive to the inhomogeneous fuel mixing in the D-T ICF pellet.

INTRODUCTION

In deuterium(D)-tritium(T) inertial confinement fusion (ICF) a fuel pellet is compressed to about a thousand times the solid density. The fuel compression ratio is extremely high. Therefore, the requirement for implosion uniformity is quite stringent (1): the tolerable nonuniformity during the implosion is only a few percent. The nonuniformity arises from the driver energy illumination nonuniformity, the pellet structure nonuniformity and the Rayleigh-Taylor (R-T) instability. In addition, other causes may be the imbalance of the contents of the D and T fuels (2), and also the inhomogeneous mixing of the initial D-T fuel concentrations (3). The fuel-content imbalance in the D-T pellets has already been studied (2): a 30 % content imbalance still gives sufficient fusion energy output.

In this paper, we study the effect of inhomogeneous mixing of the D-T fuel concentrations on the fusion energy output. In a D-T pellet, we may have the inhomogeneously-mixed frozen D-T fuel in a pellet fabrication process, although the total amounts of D and T are equal. We may control the total amounts of D and T fuels in a DT pellet. Therefore we may have pressure and particle-number-density uniformities in a pellet after pellet fabrication. However it may be difficult to control the homogeneous mixing of D and T in the pellet. The inhomogeneous mixing of D and T fuels may influence the DT burn and the fusion energy output, and induces the mass-density nonuniformity. In this paper we study on the inhomogeneous mixing effect on the fusion energy output. The mass-density nonuniformity is discussed briefly at the end of this paper. A one-dimensional numerical simulation presents that a 40 % mixing inhomogeneity of the D and T concentrations still gives a sufficient fusion energy output, as long as the total amounts of the D and T are equal. This result means that fusion energy output and burn process are rather insensitive to the inhomogeneous fuel mixing in the D-T ICF pellet.

In general, the density-radius product ρR is about $3\sim 5 \text{g/cm}^2$ in the D-T ICF. Therefore, the fuel reaction ratio is about $30\sim 40\%$. The rest of the fuel remains inert but plays the important roles of confining the fuel against the expansion and increasing the reaction rate. Consequently, the fusion energy output may be insensitive to the mixing inhomogeneity of the D and T concentrations, as long as the pellet contains a sufficient amount of the D and T fuels.

INHOMOGENEOUSLY-MIXED D AND T

In order to study the effect of inhomogeneous mixing of the D and T concentrations on fusion energy output, we performed a one-dimensional computer simulation which includes the equations for the D-T, D-D and D-H_e^3 reactions (2). In the simulation, the three-temperature fluid model (4) is employed and the fuel-burn process is computed (5). The diffusion equation is solved for alpha particle transport, and neutrons escape freely. Electron and radiation heat conductions (4,6) are also included by the flux-limited diffusion model. The SESAME library is used for the equation of state. As the initial condition for the zeroth-order unperturbed profile, we employ the constant-pressure model (7), in which the temperature of the inner hot spot is 10 keV, that of the outer part is 3.33keV, the density n_{0h} is 333 times the solid density at the hot spot, and the density n_{0out} is 1,000 times the solid density at the outer part. In our simulation, the initial ρR for the hot spot is 0.5g/cm^2 and is fixed, when the total ρR value is changed in the parameter studies. Our D-T pellet has no tamper. The density perturbations for the D and T are imposed to the zeroth-order density profile $n_0(r)$, so that the D and T total amounts are equal:

$$n_D = \frac{n_0(r)}{2} + \delta \frac{n_{0out}}{2} P_N(r/r_{out}) \qquad (1)$$

$$n_T = \frac{n_0(r)}{2} - \delta \frac{n_{0out}}{2} P_N(r/r_{out}) \qquad (2)$$

Here δ is the perturbation amplitude, $N = 2n+2, (n = 1, 2, 3, ...)$, and $P_N(R)$ is the Legendre function which satisfies the following equation:

$$\int_0^1 R^2 P_N(R) dR = 0 \qquad (3)$$

Because of this property of the Legendre function, we can guarantee that the D and T total amounts are equal. This initial density profile may be obtained during the implosion process, in which we may have some instabilities. During a D-T-pellet fabrication, a pellet may have the inhomogeneous mixing of the D and T fuels, although the D and T total amounts can be controlled to be equal. As a result of some instabilities and implosion nonuniformities, the pellet may have the mixing inhomogeneities. Our initial condition is an example of the inhomogeneities.

Figure 1 shows the D and T initial density profiles employed for $\delta = 0.3$, $\rho R = 4 \text{g/cm}^2$ and $N = 8$. Figure 2 presents the relations between the normalized fusion energy output and δ for $\rho R = 3$ g/cm^2. Figure 2 shows that a 40 % mixing inhomogeneity of the D and T concentrations in the D-T pellet still gives a sufficient fusion energy output.

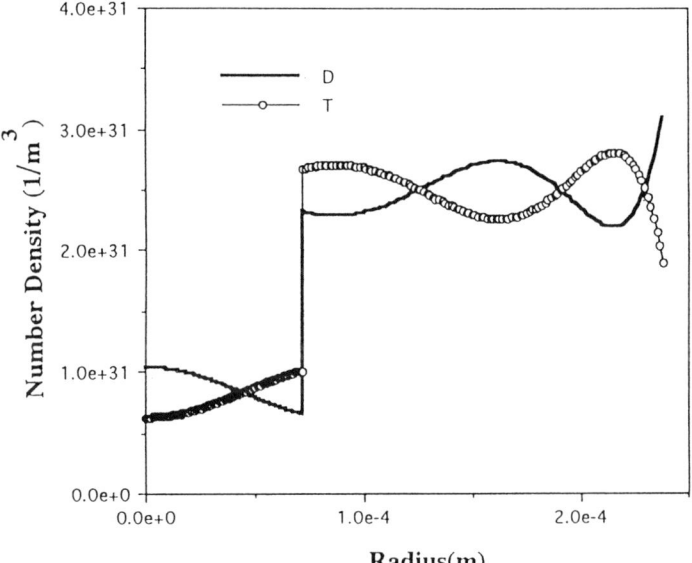

FIGURE 1. The initial profiles for D and T density with the perturbation for ρR=4 g/cm2, δ=0.3, and the mode number N=8.

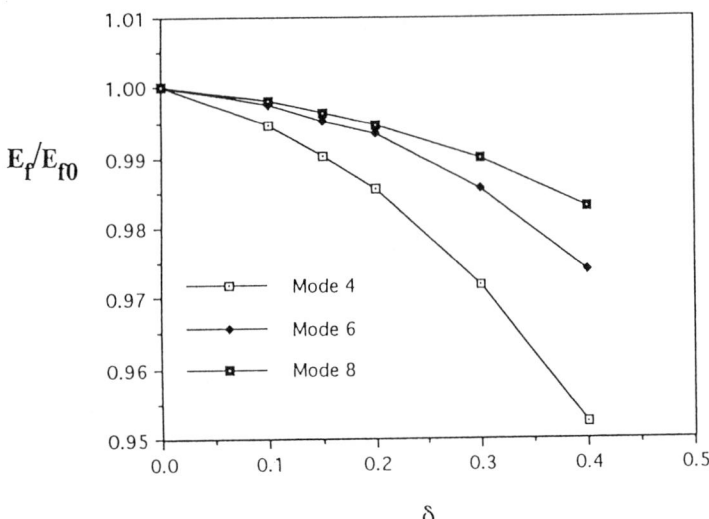

FIGURE 2. The relation between the normalized fusion energy output and δ for ρR=3g/cm².

In order to estimate the inhomogeneous mixing effect of the D and T concentrations on fusion energy output, we employ the following simplified reaction equations for the D and T:

$$dn_T/dt \simeq -n_D n_T <\sigma v>_{DT} + n_D^2 <\sigma v>_{DD}/4$$
$$\simeq n_T^2(-<\sigma v>_{DT} + <\sigma v>_{DD}/4) \quad (4)$$

$$dn_D/dt \simeq -n_D n_T <\sigma v>_{DT} - n_D^2 <\sigma v>_{DD}$$
$$\simeq -n_D^2(<\sigma v>_{DT} + <\sigma v>_{DD}) \quad (5)$$

Here $<\sigma v>$ shows the fusion reaction rate and t the time. In this analysis, we assume that the temperature is constant in time and $\delta \ll n_{0out}$. For simplicity of analysis, the density profile is assumed to be constant (n_{0out}) in space. Then we can estimate the fusion energy output E_f as follows:

$$E_f \simeq \int dt \int dr 4\pi r^2 e_f n_D n_T <\sigma v>_{DT}$$
$$\simeq 4\pi \tau e_f <\sigma v>_{DT} \int dr r^2 n_D n_T \quad (6)$$
$$\propto \int dr r^2 n_D n_T$$

Here e_f is the fusion energy by one D-T reaction, and τ the confinement time. A further calculation gives us the following expression:

$$\int dr\, r^2 n_D n_T \simeq \int dr \frac{r^2}{F_1}\left(\frac{n_{0out}^2}{4}\right)\left[1 + \Delta F_2 - \Delta^2\left(\frac{F_3}{F_1}\right)\right] \quad (7)$$

Here $\Delta = \delta \times P_N(r/r_{out})$ and

$$\begin{aligned}
F_1 &= AB, \\
A &= 1 + (n_{0out}\tau/2)(<\sigma v>_{DT} - <\sigma v>_{DD}/4), \\
B &= 1 + (n_{0out}\tau/2)(<\sigma v>_{DT} + <\sigma v>_{DD}), \\
F_2 &= (n_{0out}\tau/2)(<\sigma v>_{DT} - <\sigma v>_{DD}/4)/A \\
&\quad -(n_{0out}\tau/2)(<\sigma v>_{DT} + <\sigma v>_{DD})/B, \\
F_3 &= (n_{0out}^2\tau^2/4)(<\sigma v>_{DT} - <\sigma v>_{DD}/4) \\
&\quad (<\sigma v>_{DT} + <\sigma v>_{DD})/F_1.
\end{aligned} \quad (8)$$

The first term of the righthand side of Eq.(7) comes from the D-T reaction by the zeroth-order of the D-T density(n_{0out}). From Eq.(3), the second term of the righthand side of Eq.(7) vanishes approximately, if we assume that the reaction rate $<\sigma v>$ does not depend on r. This estimation result shows that the second order of the mixing inhomogeneity δ influences the fusion energy output, as long as the D and T total amounts are equal. Therefore, the fusion energy output is insensitive to $\delta(<< 1)$, and the decrease in the fusion energy output is expected to be small. This supports the simulation results.

DISCUSSIONS

In this paper, we presented a numerical and analytical study of the effect of inhomogeneous mixing of the D and T concentrations on fusion energy output in a DT-ICF fuel pellet. We found that a 40% mixing inhomogeneity of the D-T fuel concentrations in a D-T fuel pellet can still give a sufficient fusion energy comparable to that of a pellet which contains deuterium and tritium homogeneously. This means that the D-T fusion energy output is rather insensitive to mixing inhomogeneities of the D-T fuel concentrations, as long as the total D and T amounts are equal in the D-T pellet. This result comes from the following fact: during the burn, only a fraction of the fuel is burned and the remainder, usually about 60 to 70%, remains inert. Consequently, the fusion energy output and the burn process seem to be insensitive to the inhomogeneous mixing of the D and T concentrations.

In our one-dimensional analyses, we did not include instability effects on the energy output. If the pellet has mixing inhomogeneities during implosion and burn phases, the pellet has a mass-density fluctuation, and may suffer the *R-T* and other instabilities. This should be studied in detail in the near future. Here we discuss on this point briefly. When a DT pellet has the initial DT mixing inhomogeneity

$$n_D = \frac{n_0}{2}(1+\Delta), n_T = \frac{n_0}{2}(1-\Delta), \qquad (9)$$

the pellet has the following mass-density nonuniformity:

$$\rho = \rho 0[1 + \frac{m_D - m_T}{m_D + m_T}\Delta] = \rho_0[1 - \Delta/5] \qquad (10)$$

The tolerable nonuniformity may be a few percent (1). If we impose this nonuniformity requirement, the tolerable DT mixing inhomogeneity may be about 5 times a few percent, that is, $5 \sim 15\%$.

ACKNOWLEDGEMENT

This work is supported partly by the Scientific Research Fund of the Ministry of Education and Culture in Japan, and in part by the Institute of Laser Engineering, Osaka University.

REFERENCES

1. Lindl, J. D., McCrory, R. L. and Campbell, E. M., *Phys. Today* **9**, 32, (1992); Kawata, S. & Niu, K., *J. Phys. Soc. Jpn.* **53**, 3416, (1984).
2. Kawata, S. and Nakashima, H., *Laser and Particle Beams* **10**, 479, (1992).
3. Kawata, S., *Laser and Particle Beams* in print, (1995).
4. Tahir, N.A. et al., *J. Appl. Phys.* **60**, 898, (1986).
5. Fraley, G.S., et al., *Phys. Fluids* **17**, 474, (1974).
6. Zel'Dovich, Ya.B. & Raizer, Y.P., "*Physics of Schock Wave and High Temperature Hydrodynamic Phenomena*", New York: Academic Press,1966.
7. Meyer-ter-Vehn, J., *Nuclear Fusion* **22**, 56, (1982).

3. Laser-Matter Interaction Physics

Options for Laser Compression of Matter to Study Dense-Plasma Phases at Low Entropy, Including Metallization of Hydrogen.

J. Meyer-ter-Vehn*, A. Oparin*, T. Aoki[+],

*Max-Planck-Institut fuer Quantenoptik, D-85748 Garching, Germany, and [+]Tokyo Institute of Technology, 4259 Nagatsuta, Midori-Ku, Yokohama 227, Japan

Abstract. The potential of high-power lasers for detailed studies of strongly coupled plasmas at low entropy is discussed, emphasizing multiple-shock techniques. Some outstanding features like metallization in solids and related ionization phase transitions in the fluid phase - predicted theoretically, but not yet observed experimentally - are reviewed. Planar multiple shock compression of solid hydrogen is decribed, using reverberating shocks between massive liners and, alternatively, a stepped pressure pulse acting from one side. In the latter case, shock splitting and a rarefaction shock show up at the metallic phase transition.

INTRODUCTION

The potential of high power lasers for a detailed study of phase diagrams of matter at temperatures between 0.1 - 10 eV and close to solid density ρ_0, say in the range ρ/ρ_0 = 0.1 -10, has found little attention, so far. This is surprising because lasers with 1 kJ, 1 - 10 ns pulses should be well suited to scan this region, and a considerable number of such lasers actually exists.

While single-shock Hugoniots probing the phase diagram along a single trajectory have been investigated with lasers (see e.g. Cauble 1993, Loewer et al. 1994), the way to perform a two-dimensional scan is to tune entropy either by following a series of expansion isentropes of material shocked to different entropy levels or by low-entropy compression using multiple shocks and shaped pulses (Meyer-ter-Vehn and Oparin 1995). Whereas the first method allows to access densities $\rho \leq \rho_0$, the second one may explore low temperature matter above solid density. Both techniques have been used in dynamic high-pressure experiments driven by explosives (Fortov 1982) and gas guns (Nellis 1992); they still need development when used with lasers.

Also ultrashort-pulse (femtosecond) lasers are of considerable interest in this context. They can heat small probes of solids to 1 - 10 eV isochorically, i.e. so fast that hydrodynamic motion is negligible during the heating process and the density remains essentially unchanged. The expansion isentropes then cut into the liquid-gas two-phase region. This offers new possibilities to explore critical points. The critical points of most solid materials could not be accessed experimentally so far, except for some alkaline metals and Hg (Hensel 1990). The new lasers are likely to change this situation. For example, liquid carbon was produced recently with femtosecond laser pulses (Reitze, Ahn, Downer 1992). Another interesting point is that the short times may allow to study also the kinetics of these phase transformations.

The laser experiments proposed here are relevant to both fundamental physics of condensed matter as well as applications in material science and in geo- and astrophysics. The considered phases prevail in the interior of the earth and the giant planets (e.g. Jupiter), in brown dwarfs, as well as in the envelopes of white dwarfs and neutron stars.

PHASE TRANSITIONS

One of the outstanding features is metallization of non-metallic solids due to successive pressure ionization. This phenomenon has been observed in the laboratory in a few favourite cases, using static compression in diamond anvil cells (e.g. solid xenon, Reichlin et al. 1989 and Goettel et al. 1989). In general, multimegabar pressures are required. The important question whether metallization sets in as a first-order phase transition or more gradually is an open question for most materials. Concerning the solid state, one may expect that the transformation occurs in a series of complex steps involving also changes in the lattice structure.

As to the pressurized liquid state above the melting curve, of major interest for planet interiors, there is increasing evidence now from independent dense-plasma calculations that first-order phase transitions occur in this state related also to pressure ionization. Such transitions have been observed, intermixed with the normal liquid gas transition, in materials with low ionization threshold like Cs and Hg (Hensel et al. 1990). They are actually well described now theoretically (Nagel et al. 1994). For materials with higher ionization potentials, the calculated critical points of the ionization transitions are located at higher pressures than the gas-liquid transitions.

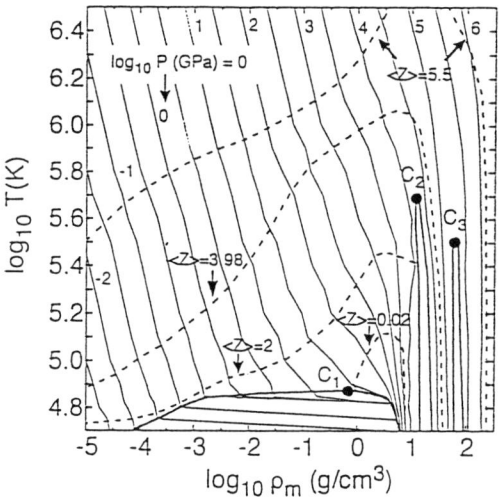

FIGURE 1 Contour lines of pressure P and average ionization <Z> in ^{12}C matter from Ogata, Kitamura, and Ichimaru (1995).

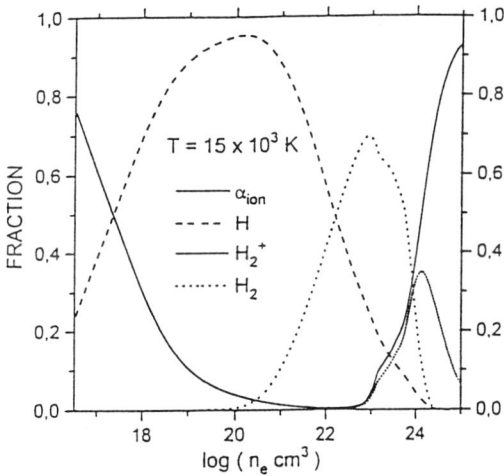

FIGURE 2 Composition of 15000 K hydrogen plasma. The fraction of free electrons (solid line), atoms H, dimers H_2, and molecular ions H_2^+ is shown vs. number density (from Reinholz, Redmer, and Nagel, 1995).

Just to give an example, recent results on carbon by Ogata, Kitamura, and Ichimaru (1995) are shown in Fig. 1. Calculated contours of pressure P and average ionization <Z> are plotted in the temperature-density plane. Three two-phase regions with critical points C_1, C_2, and C_3 at temperatures in the range $(50 - 500) \times 10^3$ K are seen. These first-order phase transitions occur in the fluid phase and involve sharp changes in the degree of ionization. When passing through these transitions from left to right at about 50000 K, the atomic states change from 2+ to 0+, from a mixture of 0+ and 2+ to 4+, and from 4+ to 6+, respectively. Of course, this involves sharp changes in the conductivity.

Results of another calculation by Reinholz, Redmer, and Nagel (1995) for dense hydrogen fluid are presented in Fig. 2. Changes in the composition as a function of density are displayed at 15×10^3 K. As the density approaches 10^{23} atoms/cm^3, the fraction of free electrons (solid line) drops almost to zero and then sharply comes up again due to pressure ionization. The composition is rapidly changing between 10^{23} and 10^{24} cm^{-3}, and one should notice the occurence of dimers H_2 and molecular ions H_2^+. Again, these calculations predict the existence of a first-order plasma phase transition with a critical point at $T_c = 16.5 \times 10^3$ K, $p_c = 570$ kbar, and $\rho_c = 0.42$ g/cm^3. It had been predicted before by a number of other groups with somewhat different critical data (see e.g. Robnik and Kundt 1983, Ebeling and Richert 1985, Saumon and Chabrier 1992).

So far, these phase transitions have not been observed in experiments. Since their critical points are located at relatively high temperatures, it should be straight-forward to access them by double and triple shocks. One should notice that lasers are not only suited to generate these phases thermodynamically, but are also superior tools to diagnose them, since light reflection, absorption and scattering is highly sensitive to the density of electrons and to the species of ions, atoms, molecules, clusters, etc. present.

Skepticism concerning the existence of such plasma phase transitions refers to the chemical model on which most of the predictions are based. This model makes a sharp distinction between bound and free particles and identifies the bound units as chemical species interacting via generalized mass-action laws. Such a picture is bound to fail when approaching the solid regime at lower temperatures. The physical picture dealing just with nuclei, electrons, and their interactions then becomes more appropriate. The transition region is difficult to treat theoretically, and presently large uncertainties exist.

METALLIC HYDROGEN

Considerable experimental efforts have been devoted to the metallization of hydrogen in its solid state, using diamond anvil techniques (Mao and Hemley 1994). At low pressures, hydrogen crystallizes as an insulating molecular solid, but is expected to transform into a metallic atomic phase at a pressure of about 3 Mbar and a density 10 times the normal solid density. The atomic phase represents the simplest metal possible, and producing it in the laboratory has been described as a key problem in modern physics and astrophysics (Ginzburg 1978).

Meanwhile diamond anvils have achieved pressures around 2 Mbar, revealing a number of phase changes in hydrogen at Megabar pressures, but all within the molecular solid. Two-stage gas guns produced 1.2 Mbar at 3000 K in hydrogen by multiple shock compression, and conductivity measurements indicated that the transition from non-conducting molecular hydrogen to the metallic atomic phase was approached, but not yet reached (Nellis et al. 1992, Weir et al. 1993). Early pioneering experiments on dynamic compression of hydrogen by Grigoriev et al. (1972) and by Hawke et al. (1978) reported evidence for the metallic phase; however, it was not generally accepted as convincing.

In Fig. 3, some isotherms of hydrogen are plotted in the pressure-density plane, including the T=0 isotherm as the lowest one. They display the first-order transition to the metallic phase at 2 - 3 Mbar and 1 g/cm^3. The corresponding hydrogen equation of state (EOS)

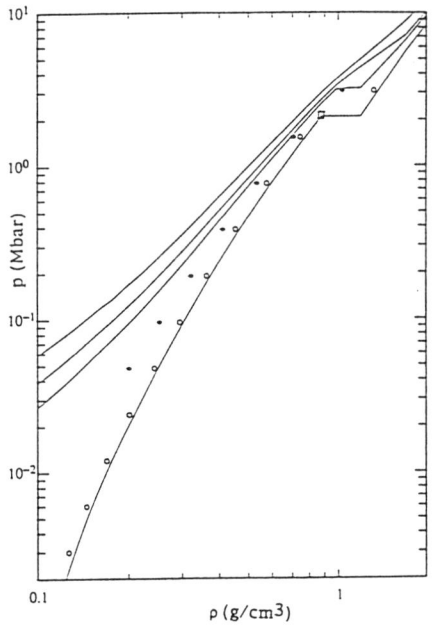

FIGURE 3 Hydrogen isotherms at 0 K, 3500 K, 5486 K, and 8600 K from below, based on SESAME EOS. Open dots refer to a 11 shock sequence described in text., the full dots to a 7 shock sequence leading to higher temperatures.

is taken from the SESAME EOS tables; it was constructed combining different models for the three phases: molecular solid, atomic solid, and fluid (Kerley 1972). The calculations reported below are based on this EOS. The difficulty in hydrogen is to achieve the tenfold compression at sufficiently low entropy. The dots in Fig. 3 give the compression stages in a specific multiple-shock scheme described further below. Multiple shock generation with the probe layer between impacting liners is discussed before.

REVERBERATING SHOCKS BETWEEN LINERS

A convenient method to achieve multiple shock compression is to sandwich the probe layer between massive planar liners and give the liners on one or both sides some initial kinetic energy such that they act as moving pistons. They generate reverberating shocks in the probe layer and lead to very smooth low-entropy compression. The compressed layer can be diagnosed optically in the transverse direction. The flyer plates are accelerated by laser ablation; of course, this acceleration has to occur in a very uniform manner.

In order to give some quantitative idea, we have simulated compression of a 200 μm thick layer of frozen (initially 10K) hydrogen by gold liners each 100 μm thick. Results have been published before (Meyer-ter-Vehn and Oparin 1995). It turns out that double-sided symmetric impact requires a flyer velocity of 2.5 km/s to reach the metallic state in hydrogen. The compression history is seen in Fig.4 in a perspective plot, showing the density evolution

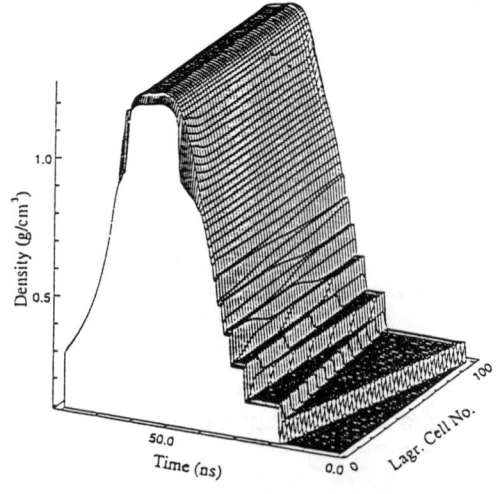

FIGURE 4 Hydrogen density evolution for symmetric impact of two 2.5 km/s flyers striking from opposite sides.

versus spatial mass coordinate and time. Because of spatial symmetry, only one half of the layer is shown; the flyer plate acting from the rear side (at cell 100) is not shown. One sees how the initial sequence of reverberating shocks is finally merging into an adiabatic compression regime, producing a very uniform volume of hydrogen with a density of almost 1.3 g/cm^3 and a temperature of 1500 K at maximum compression. It consists of metallic hydrogen and lives for about 10 ns, a long time for optical diagnostics.

Single-sided impact requires a flyer velocity of about 5 km/s to generate hydrogen with 1.2 g/cm^3 at 3500 K. Also cylindrical compression, using cylindrically symmetric drive, has been considered (Aoki and Meyer-ter-Vehn 1994). Cylindrical implosion has the advantage that the radial convergence required for 10 times compression is only 3-4; also, a liner velocity of only 1 km/s turned out sufficient to generate the metallic state at temperatures between 1000 - 2000 K. However, convergence effects lead to less uniform distributions. Let us briefly mention transverse rarefaction. With a sound velocity of about 25 km/s at highest compression, we estimate the rarefaction front runs about 300 - 500 µm into the sample during the high-density phase. Therefore its lateral extension has to be 1 - 2 mm initially to insure the formation of a highly compressed core. Tamping of lateral outflow may be considered; however, the tamping material needs to remain transparent to allow for optical observation of the compressed sample.

COMPRESSION BY STEPPED PRESSURE PULSE

Let us finally discuss a different option for multiple shock compression using a carefully tuned, stepped pressure pulse. This is certainly more difficult to achieve experimentally, but may be possible with state-of-the-art laser techniques. The advantage is that the sample layer is driven from one side only leaving the other side completely open for observation (no tamping). Here, the scheme is described in an idealized form, replacing the laser drive by pressure boundary conditions. The pressure pulse is shown in the upper part of Fig. 5. In a real experiment, of course, the pressure is generated by laser ablation, requiring an additional ablator and pusher layer on the drive side.

In the standard case, we start with 3 kbar applied to 10 K frozen hydrogen and then double this pressure 10 times to end up with about 3 Mbar, which is somewhat above the transition to the atomic phase. The flow diagram is shown in the lower part of Fig. 5. The first shock needs the time t_0 to pass the layer of thickness x_0 and

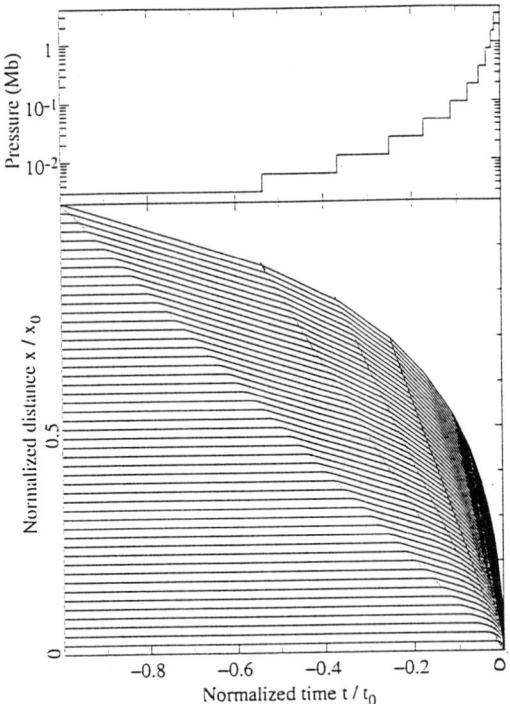

FIGURE 5 Stepped pressure pulse (upper part) and corresponding flow diagram (lower part).

arrives at the rear surface at time t=0; for x_0 =200 µm, we calulate t_0 =61 ns. The first pressure doubling to 6 kbar is timed in such a way that the second shock arrives at the rear surface also at t=0, and this timing is repeated for all successive shocks. For each step, we solve the Hugoniot equations to find the next shock velocity and from this the jump time. The pressure-density points behind the 11 shocks are plotted in Fig. 3 as open circles. As long as the EOS behaves regularly, each shock is faster than the preceding one and can catch up with it.

Figure 6 shows the density evolution close to t=0 when the shocks arrive simultaneously at the rear surface. In Fig. 6a, we stopped pressure doubling after the tenth shock at 1.54 Mbar, somewhat below the phase transition. The density behind the tenth shock is 0.74 g/cm3 (about nine-fold compression) and the temperature is 1154 K. After combined shock break-out from the rear surface, a normal rarefaction wave is seen running back through the compressed foil.

Let us now explore what happens when adding another factor 2 pressure jump, reaching 3.07 Mbar beyond the transition. The density plot is shown in Fig. 6b. One observes that the corresponding shock splits into two fronts. The lower front still cuts through molecular hydrogen, and it is fast enough to reach the rear surface at t=0 (the launching of the 11th pressure jump was timed this way). The state behind it corresponds to the open square in Fig. 4. The upper front transforms the molecular phase into the atomic metal phase with a density jump of about 40% (see Fig. 3). This

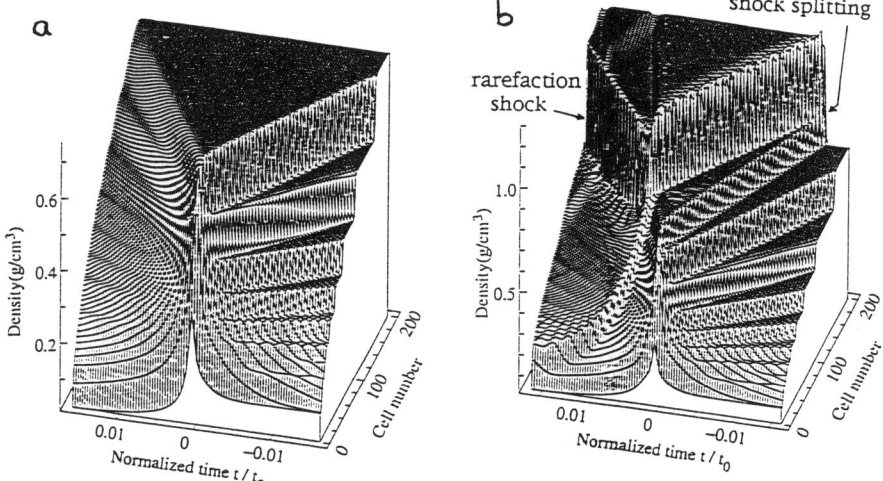

FIGURE 6 Density evolution corresponding to Fig. 5 showing simultaneous shock arrival on the rear surface at t=0. Left side (a): ten shocks with highest pressure 1.54 Mbar; right side (b): one additional shock raising the pressure to 3.1 Mbar; notice shock splitting and rarefaction shock. related to phase transition.

second front moves slower than the first one, cannot reach the shock focus and finally collides with the rarefaction wave. After this collision, rarefaction of the atomic metal phase occurs in form of a rarefaction shock.

Both phenomena - shock splitting and rarefaction shock - are unique signatures of anomalous EOS behavior such as first-order phase transitions (Zeldovich and Raizer 1966). Both features should be easy to detect experimentally. These features have been observed only in a few cases, and the observation in the hydrogen compression experiment would be of great interest in itself. It could be used as a diagnostic for the existence of the metallic phase transition.

ACKNOWLEDGEMENTS

The authors thank S. Ichimaru and R. Redmer for the permission to reproduce results prior to publication. They acknowledge helpful discussions with S. Anisimov and V. Fortov on high pressure equation of states and with S. Bodner and R. Sigel on laser experiments in this context.

REFERENCES

Aoki, T., and Meyer-ter-Vehn, J., 1994, Phys. Plasmas 1, 1962.
Cauble, R., Phillion, D. W., Hoover, T.J., Holmes, N.C., Kilkenny, J.D., and Lee, R.W., 1993, Phys. Rev. Lett. 70, 2102.
Ebeling, W., Richert, W., 1985, Phys. Stat. Sol. B128, 467.
Fortov, V., 1982, Sov. Phys. Uspekhi 138, 361.
Ginzburg, V., 1978, Key Problems in Physics and Astrophysics (Mir, Moscow).
Goettel, K.A., Eggert, J.H., Silvera, I.F., Moss, W.C., 1989, Phys. Rev. Lett. 62, 665.
Grigorev, F.V., Kormer, S.B., Mikhailova, O.L., Tolochko, A.P., and Urlin, V.D., 1972, JETP Letters 16, 201
Haronska, P., Kremp, D., Schlanges, M., 1987, Wiss. Z. Univ. Rostock, 36, 98.
Hawke, P.S., Burgess, T.J., Duerre, D.E., Huebel, J.G., Keeler, R.N., Klapper, H., Wallace, W.C., 1978, Phys. Rev. Lett. 41, 994.
Hensel, F, 1990, J. Phys. Cond. Matt. 2, SA 33.
Kerley, G.I., 1972, A Theoretical Equation of State for Deuterium, Los Alamos Scientific Laboratory, Report LA-4776, UC34.
Loewer, Th., Sigel, R., Eidmann, K., Foeldes, I.B., Hueller, S., Massen, J., Tsakiris, G.D., Witkowski, S., Preuss, W., Nishimura, H., Shiraga, H., Kato, Y., Nakai, S., and Endo, T., 1994, Phys. Rev. Lett. 72, 3186.
Mao, H., and Hemley, R.J., 1994, Rev.Mod. Phys. 66, 671.
Meyer-ter-Vehn, J., A. Oparin, 1995, in *Elementary Processes in Dense Plasmas:*, A Conference Volume" edited by S. Ichimaru and S. Ogata (Addison-Wesley, Reading, MA, 1995), p. 283.
Nagel, S., Roepke, G., Redmer, R., Hensel, F., 1994, J. Phys: Condens. Matter 6, 2137.
Nellis, W.J., Mitchell, A.C., McCandless, P.C., D.J. Erskine, D.J., Weir, S.T., 1992, Phys. Rev. Lett. 68, 2937.
Ogata, S. Kitamura, H., and Ichimaru, S., 1995, in *Elementary Processes in Dense Plasmas:*, A Conference Volume" edited by S. Ichimaru and S. Ogata (Addison-Wesley, Reading, MA, 1995), pp. 145-164.
Reichlin, R., Brister, K.E., McMahan, A.K., Ross, M., Martin, S., Vohra, Y.K., Ruoff, A.L., 1989, Phys. Rev. Lett. 62, 669.
Reinholz, H., Redmer, R., Nagel, S., 1995, submitted to Phys Rev E.
Reitze, D.H., Ahn, H., Downer. M.C., 1992, Phys. Rev. B45, 2677.
Robnik, M., Kundt, W., 1983, Astron. Astrophys. 120, 227.
Saumon, D., and Chabrier, G., 1992, Phys. Rev. A46, 2084.
Weir, S.T., Nellis, W.J., and Mitchell, A.C., 1993, {\it High Pressure Science and Technology - 1993}, edited by S.C. Schmidt, J.W. Shaner, G.A. Samara, and M. Ross (American Institue of Physics, New York), in press.
Zeldovich, Ya.B. Raizer, Yu.P. 1967, *Physics of Shock Waves and High Temperature Hydrodynamic Phenomena*, AP, New York.

SIMULATION OF ASTROPHYSICAL PLASMA DYNAMICS IN THE LASER EXPERIMENTS

Zakharov Yu.P., Antonov V.M., Melekhov A.V., Nikitin S.A., Ponomarenko A.G., Posukh V.G., Stoyanovsky V.O., Shaikhislamov I.F.

Institute of Laser Physics, Novosibirsk, 630090, RUSSIA

Abstract. The properties of laser-produced plasma are similar in many features to the ones of various space plasma releases of explosive nature. It allows to investigate the processes of an interaction of those plasmas with background media by the means of laboratory simulation. In the given paper such simulation experiments with quasispherical plasma clouds at KI-1 laser facility are presented. The experiments are related to dynamics of collisionless expansion of exploding space plasmas in the surrounding magnetized media at various Alfven-Much numbers. A non-uniform media as well has been simulated. The results on generation of shock and wistler waves in background, on formation of diamagnetic cavity and flute instability of plasma clouds are described. Data on a dynamics of laser-produced plasma in a dipole magnetic field are presented also. Results are analized in connection with dynamics of Supernova remnants in interstellar medium, Barium releases in Earth magnetosphere and possible disturbancies of the latter caused by high-energy explosions in near space.

INTRODUCTION

Now in astrophysics (1) and geophysics (2) one can see a great interest in plasma collisionless effects during processes of an interaction between expanded plasma cloud and ionized magnetized background. The explanation is that at the high initial expansion velocities (Vo ~ 1000 km/s) and low background densities (n_* ~ 1 cc) the observed space plasmas interaction can't be explained by usual interparticles collisions. The main goal of our simulation experiments is to study collisionless processes (3,4) of the cloud-background interaction and their effect on dynamics or scales D of the cloud propagation and transformation of the exploding plasma energy. The research program (3-5) includes the majority of the most interesting for astrophysics cases of the cloud interactions, in particular, with the vacuum magnetic field as well as with the magnetized and unmagnetized background plasma and finally the cloud formation conditions themselves (for instance, Supernova envelope structure).

The idea of utilizing laser-produced-plasma to simulate explosive-type astrophysical phenomena was suggested by Dawson (6) as long ago as 1964 and soon the first experiments (7) of this kind were carried out. But only in result of many years' investigations (in plasma physics, space physics and the physics of laser-produced plasma itself), it was possible to move from a general acceptance of the unique similarity between Laser Plasma Cloud (LPC) and some astrophysical objects to laboratory experiments (3,5) on LPC simulation of the collisionless deceleration of Supernova remnants (SNR). The bases of such kind of simulation experiments lies (3,6) upon the property of laser to generate from point and instantly ($\tau_r \ll D/V_o$) a contactless, spherically expanding plasma cloud with high $Z \geq 10$ and velocities $V_o \geq 10^8$ cm/s, which may reach to SNR one's. But simulation of collisionless interaction of plasma clouds with magnetized space media requires (3,4,18) to fulfil an additional important condition, that is the sufficient magnetization of LPC ions $\varepsilon = R_H / D \leq 1$ (with directed Larmor R_H = mcV/ZeB). To achieve such simulation conditions a special set of experiments (4,10,11) at KI-1 on production of plasma clouds with moderate velocity ($V_o \sim 10^7$ cm/s) and large total mass $M \sim 10 \div 100$ μg has been conducted (as LPC deceleration scales $D \propto M^{1/3}$).

KI-1 FACILITY AND LASER-PRODUCED PLASMA CLOUDS

The multi-purpose simulation facility KI-1 (4) is designed to study a wide range of non-stationary processes in space plasma. It includes a large-scale (∅1.2 m x 5 m), high-vacuum (10^{-7} Torr) interaction chamber (Fig. 1), a CO_2-lasers and CO_2-amplifier with output energy $Q_r \approx 1$ kJ. The chamber is equipped with sources of both an axial magnetic field $B_0 \leq 1$ kG and of an impulse dipole field ($\mu \approx 10^7$ G·cm^3). There is also a quasi-stationary source of highly ionized

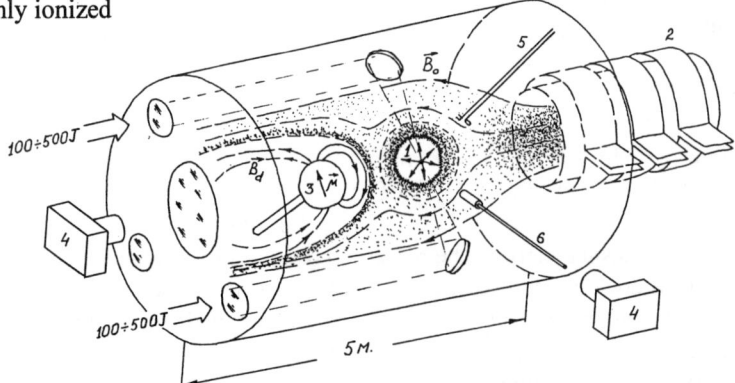

FIGURE 1. KI-1 simulation facility: 1 - LPC, 2 - source of background plasma, magnetic dipole, 4 - gated imager, 5 - Langmuir and dot magnetic probes, 6 - ion collector.

background plasma to fill ~1 m³ of chamber volume by density $n_* \leq 10^{14}$ cm^{-3}. Design of laser system allows for the irradiation of one or several targets by 4 beams, both simultaneously or serially, with a controlled delay of up to several microseconds between irradiations. In addition a method (4) of controlling the form and duration of the CO_2-amplifier pulse($\tau_r = 0.05 \div 3$ µs) has been developed, allowing wide variations in the expansion geometry, energy E_0, and the density profile $n(r)$ of the LPC.

The parameters of plasmas allow to study the effects of collisionless interaction up to Alfven-Much numbers $M_a = V_0 \sqrt{4\pi n_* m_*} / B_0 \geq 100$ or ion magnetization up to $\varepsilon \leq 0.3$. To solve the problem of production of spherical LPC of large masses we used earlier the two-sided irradiation of Nylon $(C_6H_{11}ON)_n$ filament target with diameter $\varnothing_t 0.3$ mm $\ll \varnothing_r$ (focal spot), both in a single-pulse (4) or two-pulse mode. In the latter case (10) with the help of additional prepulse (for evaporating target only) we can obtain a record value of quasispherical LPC energy up to $E_0 = 500$ J (and $M = 100$ µg). Latter on we developed a new, rather effective method (11) of LPC-production by means of two-sided and single-pulse (~1µs) irradiation of massive spherical target with $\varnothing_t = 3 \div 4$ mm $\leq (0.5 \div 0.7)\varnothing_r$. Recently a suitable degree of symmetry (Δn, $\Delta V_0 <$ 30% in all planes, Fig.2) has been achieved along with fulfilment of condition of instantaneous LPC generation ($\tau_r V_0 \sim \varnothing_t \ll D$) at $\tau_r \sim 0.1$µs. A detailed analyses of ion fluxes j = enV , measured by Langmuir probes in vacuum (j_{p0}) at various direction in relation to laser beam (where $\Theta = 0°$) showed, that the initial density distribution of such LPC can be described as follows:

$$n(\theta) \sim (1+0.8\cos^2\theta)\exp\left[-(R/t-V_0(\theta))^2/V_T^2(\theta)\right]/t^3,$$

where $V_0(\theta) = 5.5 + 4.5\sqrt{\cos(\theta)}$; $V_T = 6.5 - 1.5\sqrt{\cos(\theta)}$ (in 10^6 cm/s). For last case irradiation of Nylon target ($\varnothing_t$ 4mm, supported by 0.1 mm metallic wire) there is a cone $\approx 30°$ around wire of essentially reduced n. In experiments presented such LPCs were generated at $q \sim 10^{10}$ W/cm² and consisted of H^+, C^{+3}, C^{+4} - ions with $V_0 = 1.7 \cdot 10^7$ cm/s and $E_0 = 15$ J.

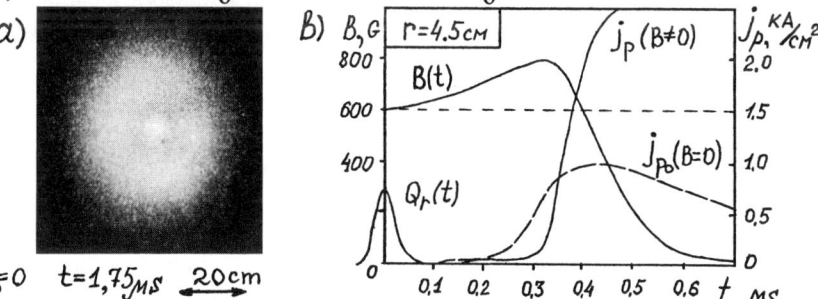

FIGURE 2. Configuration (a) of LPC and dynamics (b) of its interaction with vacuum magnetic field (in "Cavity" experiment at $\theta \approx 90°$).

COLLISIONLESS INTERACTION AND SIMULATION RESULTS

The conditions of simulation experiments at KI-1 facility were deduced from analysis of collisionsless interaction processes and according to corresponding similarity criteria. In the presence of magnetic field the main criteria are an Alfven-Much number (M_a) which determines a kind of interaction taking place and Larmor criterium (ion magnetization - ε) which affects its intensity. Namely at $M_a \ll 1$ a spherical cloud with energy $E_0 = 0.3 MV_0^2$ should be decelerated at radius $D \sim R_* = (3E_0/B_0^2)^{1/3}$ by the action of magnetic field (13,17), whereas at $M_a \gg 1$ by the background plasma (12,18) at $D \sim R_* = (3M/4\pi n_* m_*)^{1/3}$. In the nonuniform background with the density scale of inhomogeneity - L_* or field scale -L_B, a dynamics of cloud expansion is influenced by criterium $\gamma = D/L$ as well.

At KI-1 experiments (4,5,8,14-18) collisionless cloud-background interaction has been investigated in the range $M_a = 0 \div 30$, $\varepsilon = 0.3 \div 10$, $\gamma_* = R/L_* \leq 3 \div 5$ and $0.3 \leq \gamma_B = R/L_B \leq 10 \div 100$ (in a dipole field). Also, a cloud's dynamics in the vicinity of plasma/magnetic cavity created previously in the background by additional similar plasma cloud has been investigated at $M_a \gg 1$.

The formation of diamagnetic cavity plays a crucial role in the physics of collisionles cloud - background interacion. It has been shown for the first time in our experiments (4,5,17,18) that the efficiency of cavity formation and plasma cloud dynamics as well depends on Larmor criterium ε for both regimes of interaction. That is, in background case ($M_a \gg 1$) when cavity radius reaches R_* and $\varepsilon_* = R_H/R_* \leq 1$ a spherical cloud can generate on the boundary a curl electric field that appears sufficient (3,4,12,18) to accelerate the background ions up to velocity $V_* \sim V_0$ (at $R_{H*} \leq R_*$), as though in the "snow-plough" model. In a "vacuum" case ($M_a \ll 1$) cavity reaches R_B and cloud decelerates (17) at R_B only if $\varepsilon_B = R_H/R_B \leq 1$, when observed LHD-turbulence (with $\nu_{eff} = 0.3\omega_{ce}$) still doesn't destroy a skin-layer of cavity on deceleration times ($\sim R_B/V_0$).

Those and other experimental results on plasmas interactions have allowed us to simulate a main characteristics of a number of explosion phenomena in space plasma and to found out some of their peculiarities that up to now can't investigated by other methods, such as astronomical observations, active expeiments in space or computer simulations.

Collisionless, non-turbulent processes of SNR deceleration in uniform and non-uniform interstellar plasma was simulated (5) for the first time at KI-1 facility owing to realization of the necessary $\varepsilon_* < 1$ condition for $M_a \gg 1$ interaction (in contrast to parameters (19,9) of "HELIOS" and "PHAROS" experiments without results). The comparison of our simulation deceleration law of cloud (or shock) front R(t) with corresponding data of the historical SNR allow us to speculate about initial, non-"Sedov" phase of their deceleration (Fig. 3a). Evolution of the Ba-clouds in magnetosphere at $M_a \ll 1$ was simulated in

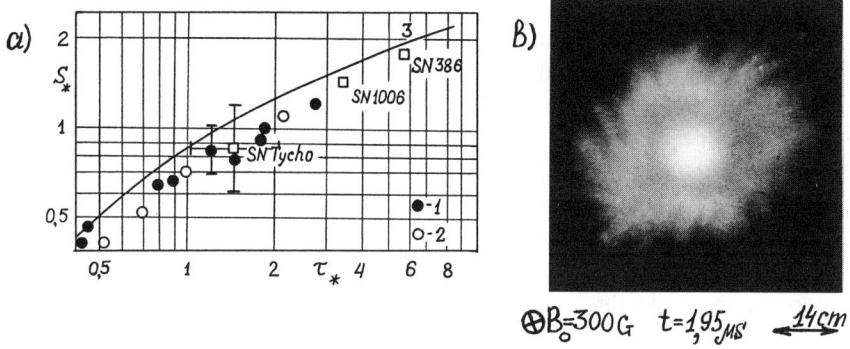

FIGURE 3. Results on simulation of collisionless SNR-deceleration (a) and CRRES/G-10 Barium release (b): 1 - probe, 2 - imager data of KI-1, 3 - improved "snow-plough" (5); $\frac{S_*}{\tau_*} \approx 1.2 \sqrt{1+x^3} dx$

experiments without backgound plasma (4,5,17), where it was predicted for the AMPTE- or CRRES-type releases (with $\varepsilon_B \leq 1$) the development of flute-type instability (Fig. 3b) with mode-number k= $8\pi/\varepsilon_B$. This dependence was confirmed later in computer and laboratory (20) experiments with symmetrical plasma expansion. Beside it we revealed a new effect of flute'stabilization by background in Trans-Alfvenic regime.

Another very interesting for simulation phenomena are observed under widespreaded in Universe Conditions of plasma cloud'expansion in a dipole-like magnetic field, such as hypotetical γ-bursts during asteroids fall on neutron star or "Starfish"- like event in Earth's magnetosphere (connected with possible anti-asteroid explosions in near space). To develop a program of simulation those phenomena we have considered the dynamics of exploding plasma in a vacuum dipole magnetic field within the scope of on ideal MHD-approximation (14). The character of the diamagnetic plasma evolution, shape and scales of its extension region are defined by an energy-scaling parameter $\kappa = 3E_o R_e^3/\mu^2$ where R_e is

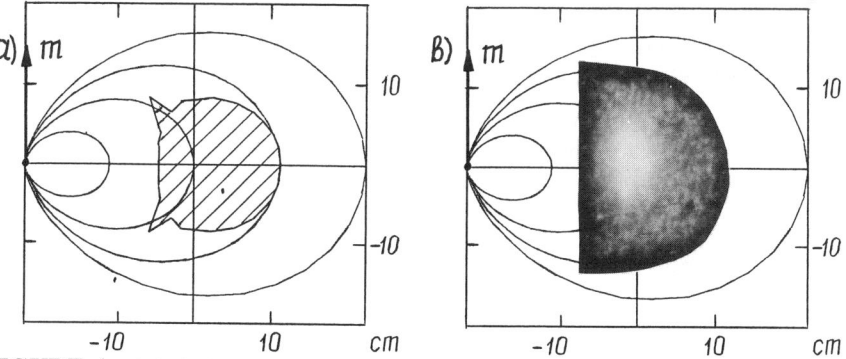

FIGURE 4. Calculation results (14) on an ideal plasma dynamics in a dipole field - 4a and corresponding photograph of LPC - 4b at the end of deceleration stage (t = 0.7 μs, $\kappa \approx 0.04$).

distance from the dipole to the point of explosion. In the particular case of the equatorial explosion a plasma "blow-out" across the field lines should occure at $\kappa \geq \kappa_c = 1/10$, but for $\kappa \ll \kappa_c$ the conditions can realize for the "quasitrapping" the plasma (Fig. 4) on a scale $\sim R_e \kappa^{1/3} \approx R_B$. In such preliminary simulation experiments at $\epsilon_B \sim 1$ in vacuum we've observed a plasma structure's formation like radiation belts, moving diamagnetic cavity and plasma precipitation into poles (8) and a total increasing in factor 4 - 5 of a dipole field compressed between a background plasma flow and LPC exploded inside of magnetopause.

We thank to Mrs. C. Elliott for assistance in preparing manuscript. The researh described in this publication was made possible in part by Grant No. NQ3000 from the International Science Foundation.

REFERENCES

1. Spicer D.S., Clark R.W., Maran S.P., *Astrophys.J.*, **356**, 549-71 (1990).
2. Kazeminezhad F., Dawson J.M., Bingham R., *J. Geophys. Res.*, **98 A**, 9493-9501 (1993).
3. Zakharov Yu.P., Ponomarenko A.G., *Laser Interaction With Matter*, Novosibirsk: Inst. Pure and Appl. Mech., 1980, pp. 4-39/In Russian.
4. Zakharov Yu.P., Orishich A.M., Ponomarenko A.G., *Laser-Produced Plasma and Simulation of Non-Stationary Space Processes*, Novosibirsk: Inst.Pure and Appl.Mech.,1988/ In Russian.
5. Zakharov Yu.P., Orishich A.M., Ponomarenko A.G. et al, "Laboratory simulation of collisionless coupling between Supernova remnants and magnetised interstellar medium", in *Proceed Int. Workshop on Plasma Astrophys,* Sukhumi (ESA SP-251), 1986, pp.37-40.
6. Dawson J.M. *Phys. Fl.* **7**, 981-7 (1964).
7. Tsuchimori N., Yamanaka T., Yamanaka Ch.,*Jpn. J. Appl. Phys.*,**7**, 84 (1968).
8. Zakharov Yu.P., Orishich A.M., Ponomarenko A.G. et al, "Laboratory study of collisionless interaction processes between Supernova-like ejectas and magnetized background under conditions of laser-produced-plasma experiments",in *Proc.Int.Conf.on Plasma Phys.*, Innsbruck Pt. III (Europhys. Conf.Abs.Ser.v.16C), 1992, pp. 1689-92.
9. Ripin B.H., Huba J.D., Mc Lean E.A. et al, *Phys. Fl.* **B5**, 3491-506 (1993).
10. Antonov V.M., Zakharov Yu.P., Maksimov V.V. et.al, *Thermal Phys. of High-Temperature.* **23**, 649-52 (1985) / In Russian.
11. Avdyeva A.A., Zakharov Yu.P., Maksimov V.V. et al, *J. Appl.Mech. and Techn. Phys.* **30**, 892-5 (1989).
12. Golubev A.I., Soloviev A.A., Terekhin V.A., *Ibid*, **5**, 33-42 (1978)/In Russian.
13. Raizer Yu.P., *Ibid,* **6**, 19-28 (1963)/In Russian.
14. Nikitin S.A., Ponomarenko A.G., *Ibid* , **34**, 745-751, (1993).
15. Zakharov Yu.P., Orishich A.M., Snytnikov V.N., Shaikhislamov I.F., *Ibid,* **35**, 481-6 (1994).
16. Shaikhislamov I.F., Antonov V.M., Zakharov Y.P. et al, *Ibid*, **36**, (1995).
17. Zakharov Yu.P., Orishich A.M, Ponomarenko A.G. et al, *Sov.J.Plasma Phys.***12**,674(1986).
18. Antonov V.M., Bashurin V.P., Golubev A.I. et al, *Sov.Phys.-Doklady*, **289**, 72-5 (1986)
19. Borovsky J.E., Pongratz M.B., Russell-Dupre R.A., et al, *Ibid,* **280**, 802-8 (1984).
20. Dimonte G., Wiley L.G., *Phys. Rev. Lett* **67**, 1755-8 (1991).

Anomalous Spectral Signatures of High-Intensity Stimulated Raman Backscattering

Miloš M. Škorić[1], Moma S. Jovanović[1,2] and Milan R. Rajković[1]

[1] *Vinča Institute of Nuclear Sciences, P.O.B. 522, 11001 Belgrade, Yugoslavia*
[2] *Department of Physics, University of Niš, P.O.B. 91, 18001 Niš, Yugoslavia*

Abstract. The nonlinear three-wave interaction model of stimulated Raman backscattering in an underdense plasma layer, with a relativistic detuning taken into account, is simulated in space-time. A mechanism is discovered based on a systematic transition to a chaotic state via quasiperiodic and intermittent regimes, that predicts a progressive growth in the backscatter complexity with an increase in laser intensity. Anomalous signatures of the backscatter, such as bursting and incoherence, together with broad, modulated and blue-shifted spectra are readily observed. Consistency with recent data on high-intensity sub-picosecond laser plasma experiments is established.

In recent years, a solid evidence has emerged that stimulated Raman scattering (SRS) often becomes absolute in laser fusion plasmas, which has brought the problem of nonlinear saturation into a focus of many studies [1,2]. Several possible mechanisms have been suggested based on models of varying complexity [3], that point to a nonlinear physics of the electron plasma wave (EPW) playing a decisive role in determining the nature of saturated states. Moreover, an inherent, strongly nonlinear feature of a transition from regular (coherent) to chaotic (turbulent) dynamics has been anticipated [4,5].

Currently, a large amount of effort is put into studies of fascinating new physics that is observed as one moves into a regime for collective effects in high-intensity ultra-short pulse laser interaction with a plasma [6]. The growing availability of table-top, multi-terawatt, short-pulse lasers has resulted in experiments which begin to reveal the peculiarities of laser-plasma instabilities in the high-intensity sub-picosecond regime.

A Livermore-UCLA collaboration has observed a short-pulse version of the classic SRS instability [7]. The backscattered light displays novel spectral signatures that strongly depend on laser intensity, such as broad and modulated spectrum that spreads to the blue side of the incident light wavelength λ_0 [7-9], that obviously differs from conventional picture of SRS which downshifts the incident frequency ω_0 by approximately the electron plasma frequency ω_{pe}. Such a behaviour is not predictable by a standard nonlinear parametric theory of weakly- or even strongly-coupled SRS. An inclusion of additional effects due to nonlinear dynamics in SRS is needed if one expects to explain, at least qualitatively, the observed spectra.

In this paper, we concentrate on anomalous spectral signatures of high-intensity backward SRS regimes in our one-dimensional model of an underdense, uniform, weakly-collisional plasma layer. We propose the nonlinear SRS model of weakly coupled three-wave interaction, including quadratic coupling of slowly-varying complex amplitudes, that accounts for pump depletion and relativistic nonlinear detuning, a well-known effect due to relativistic electron mass variation [2,10,11].

Assuming that resonant matching conditions among the interacting waves are fulfilled ($\omega_0=\omega_1+\omega_2$, $k_0=-k_1+k_2$), a spatiotemporal evolution of the complex amplitudes of the laser pump (a_0), backscattered wave (a_1) and EPW (a_2) is described by the following set of partial differential equations [1,4]:

$$\frac{\partial a_0}{\partial t} + v_0 \frac{\partial a_0}{\partial x} = -\Omega_0 a_1 a_2, \tag{1a}$$

$$\frac{\partial a_1}{\partial t} - v_1 \frac{\partial a_1}{\partial x} = \Omega_1 a_0 a_2^*, \tag{1b}$$

$$\frac{\partial a_2}{\partial t} + v_2 \frac{\partial a_2}{\partial x} + \frac{\Gamma}{2} a_2 + i\sigma |a_2|^2 a_2 = \Omega_2 a_0 a_1^*, \tag{1c}$$

where the normalized amplitudes are related to the corresponding physical fields through the expressions:

$$a_0(x,t) = \frac{E_0(x,t)}{\mathcal{E}_0}, \quad a_1(x,t) = \frac{E_1(x,t)}{\mathcal{E}_0}, \quad a_2(x,t) = -i\frac{\delta n_e(x,t)}{n_0}. \tag{2}$$

with $\mathcal{E}_0 = E_0(x=0)$. The group velocities in (1) are given by:

$$v_0 = \frac{c^2 k_0}{\omega_0}, \quad v_1 = \frac{c^2 k_1}{\omega_1}, \quad v_2 = \frac{3 k_2 v_T^2}{\omega_2}, \quad \text{where} \quad v_T = \left(\frac{\kappa T_e}{m_e}\right)^{1/2}, \tag{3}$$

while Ω_i-coefficients, EPW damping rate Γ and relativistic detuning parameter σ are in units of frequency:

$$\Omega_0 = \frac{\omega_{pe}^2}{4\omega_1}, \quad \Omega_1 = \frac{\omega_{pe}^2}{4\omega_0}, \quad \Omega_2 = \frac{\varepsilon_0 \omega_{pe}^2 k_2^2 \mathcal{E}_0^2}{4 m_e n_0 \omega_0 \omega_1 \omega_2}, \quad \sigma = \frac{3\omega_{pe}^3}{16 c^2 k_2^2}. \tag{4}$$

The EPW equation (1c) contains a nonlinear term which corresponds to the relativistic detuning and is indispensable when concerning the SRS induced by ultra-intense laser radiation.

The system (1) was solved in space-time by means of a numerical scheme, specially constructed in a way that removes difficulties arising from nonzero source boundary conditions for the pump and backscattered waves at the opposite plasma ends:

$$a_0(x=0,t)=1, \quad a_1(x=L,t)=\epsilon_1, \quad a_2(x=0,t)=\epsilon_2, \tag{5}$$

where ϵ_1 and ϵ_2 are noise levels of the backscattered and plasma waves, respectively.

The following plasma parameters were fixed in the simulations:

$n_0=0.001\,n_{cr}$, $L=100\,c/\omega_0$, $\kappa T_e=300\,eV$, $\Gamma=10^{-5}\omega_{pe}$, $\epsilon_1=10^{-3}$, $\epsilon_2=0$.

We chose these parameters as relevant to current high intensity, short–pulse laser – underdense plasma experiments [7–9]. Only recently, for such conditions, high intensity particle-in-cell simulations of Wilks et al. have revealed that the Raman backscatter instability effectively heats the background electrons significantly above the initially cold underdense plasma temperature [12].

We perform the numerical runs by varying the control parameter — the pump strength β_0, given as the ratio of the electron quiver velocity in the laser pump to the speed of light: $\beta_0 = e\mathcal{E}_0/m_e\omega_0 c$. As revealed by these authors, as the pump strength grows, increasing complexity of the wave patterns in space–time is observed [4,5]. A generic route to a chaotic state was discovered via intermittency following a quasiperiodic scenario.

For completeness, we briefly outline the generic bifurcation sequence, while more details could be found elsewhere [4,5]. For low pump values, the system (1) attains the steady state ("fixed point"), while the reflectivity saturates at low levels. Note that the term reflectivity refers to the fraction of incident laser intensity that is backscattered through the SRS process. For larger pump values, beyond a certain threshold, the steady state bifurcates to a single–periodic one ("limit cycle"). Further, we note that a single–periodic state can make a transition to a more complex regime featuring reflectivity pulsations and modulated backscatter spectra. Once the bifurcation threshold to quasiperiodicity is passed, typical motion on two frequencies ("2–torus") is revealed. In addition to the single period frequency, a second frequency, lower than the fundamental, begins to make an appearance together with higher–order modulation peaks.

Further, it was discovered that a quasiperiodic state makes a transition to chaos [4]. By increasing the pump strength, the reflectivity starts to resemble the temporal intermittency series (Fig. 1a, for $\beta_0 = 0.1$). Although still predominantly quasiperiodic, the power spectrum begins to show appearance of broad–band noise indicative of chaos (Fig. 2a). We note in passing that power spectra presented are calculated for full fields rather than in the am-

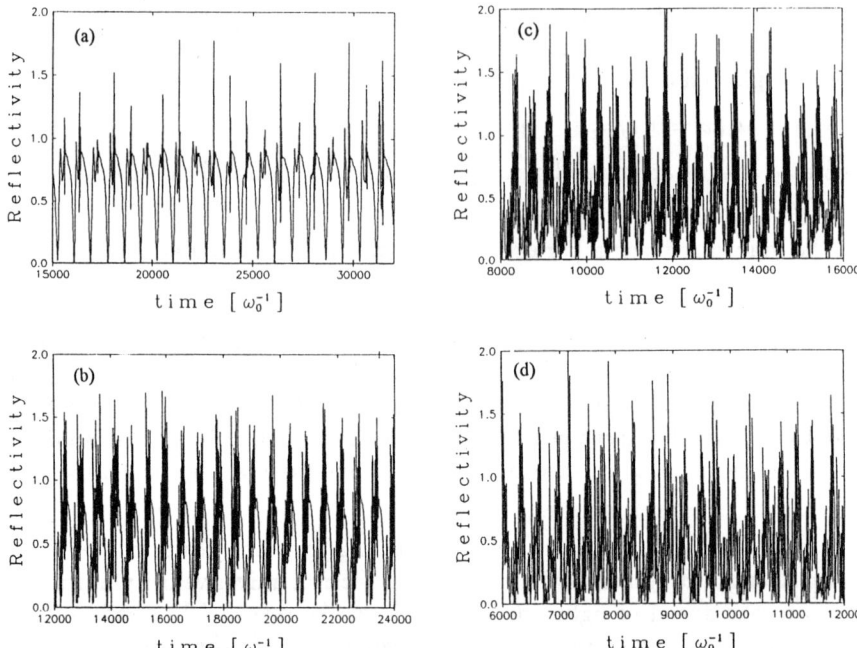

FIGURE 1. Time series of the stimulated Raman backscattering reflectivity from a uniform plasma layer, for parameters: $n_0=0.001\,n_{cr}$, $L=100\,c/\omega_0$, $\kappa T_e=300\,eV$, $\Gamma=10^{-5}\omega_{pe}$, $\epsilon_1=10^{-3}$, $\epsilon_2=0$, and for the values of the pump strength: a) $\beta_0=0.1$, b) $\beta_0=0.2$, c) $\beta_0=0.3$, d) $\beta_0=0.4$.

plitude representation. For higher intensities ($\beta_0=0.2$ and 0.3) intermittency patterns (Figs. 1b and 1c), where quasiperiodic oscillations are interrupted by chaotic bursting, display pronounced spectral broadening that spreads to the blue side of the incident wavelength (Figs. 2b and 2c). Moving deeper into the chaotic regime ($\beta_0 = 0.4$), broad continuous spectrum which loses signatures of coherent modulation is found (Figs. 1d and 2d).

At this point, we briefly refer to recent Livermore experiment on high–intensity sub-picosecond laser interaction with low density gaseous plasmas. According to C. Darrow et al. [7,9], deep modulations on backscattered spectra have been observed at variety of densities and intensities. The major features of their modulated spectra are the large frequency extent, the irreproducibility and the growth in spectrum mainly to the blue side of ordinary SRS, with much of the data taken below the threshold for strongly coupled SRS. We feel that consistency between above experimental facts and features predicted by our SRS model [4,5] seems remarkable. More recent relativistic 1D–PIC simulations [9] appear to further support the above conclusion.

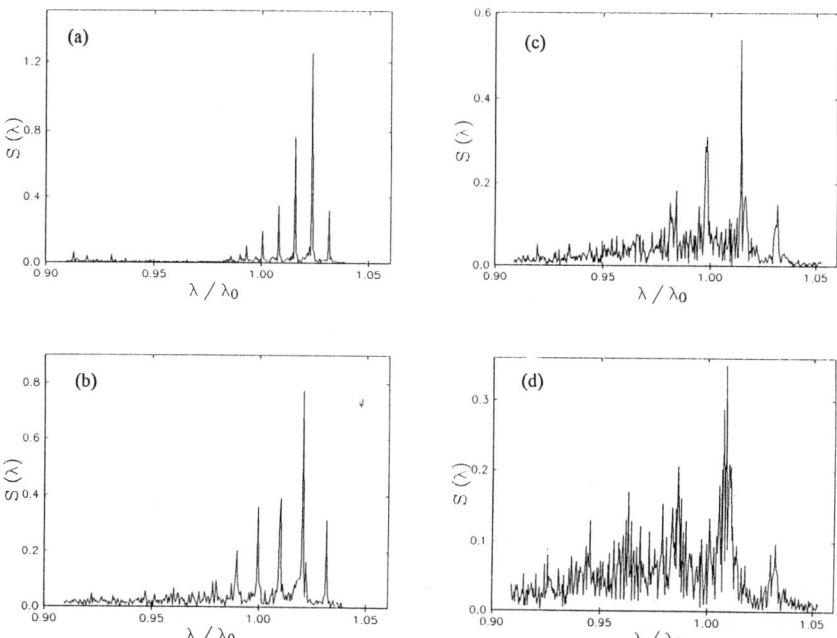

FIGURE 2. Stimulated Raman backscattering spectra for the parameters of Fig. 2.

In summary, we have presented anomalous spectral signatures of high-intensity backward SRS regimes in an underdense, uniform, weakly-collisional plasma layer. The proposed nonlinear model of three-wave interaction that accounts for pump depletion and relativistic detuning of the EPW predicts a progressive increase in backscatter complexity as the laser intensity grows. By increasing a laser pump, a generic route [13] via steady-state, periodic and quasiperiodic regimes, with an intermittent transition to a spatio-temporal chaos was discovered [4,5]. The picture in a physical space reveals patches of turbulence inside the coherent state; the continuous transition amounts to a progressive increase of turbulence through the increase of the pump strength. Anomalous features, such as spiky burst-like reflectivities, modulated, broad and blue-shifted backscatter spectra are obtained, showing a consistency with experimentally observed anomalous spectral properties.

ACKNOWLEDGEMENTS

Stimulating discussions with K. Mima, E.A. Williams and S.C. Wilks are acknowledged.

REFERENCES

1. Forslund, D.W., Kindel, J.M., Lindman, E., *Phys. Fluids* **18**, 1002 (1975).
2. Kruer, W.L., *"The Physics of Laser Plasma Interactions"*, Addisson-Wesley, New York, 1988.
3. Villeneuve, D.M., Baker, K.L., Drake, R.P., Sleaford, B., LaFontaine, B., Estabrook, K., Prasad, M.K., *Phys. Rev. Lett.* **71**, 368 (1993).
4. Škorić, M.M., Jovanović, M.S., in *"Laser Interaction and Related Plasma Phenomena"* vol. 11, ed. by Miley, G.H., AIP, New York, 1994, pp. 380-389.
5. Škorić, M.M., Jovanović, M.S., Rajković, M.R., in *"Dynamical Systems and Chaos"*, ed. by N. Aoki, World Scientific, 1995.; and *Phys. Rev. E* (submitted)
6. Joshi, C.J., Corkum, P.B., *Phys. Today* **1**, 36 (1995).
7. Darrow, C.B., Coverdale, C., Perry, M.D., Mori, W.B., Clayton, C., Marsh, K., Joshi, C., *Phys. Rev. Lett.* **69**, 442 (1992).
8. Najmudin, Z., Dangor, A.E., Malka, V., Marques, J-R., Modena, A., Norreys, P., Taday, P., *Rutherford Appleton Laboratory Annual Report*, RAL-94-042, 23 (1994).; Najmudin, Z., *private communication*
9. Coverdale, C.A., Darrow, C.B., Decker, C.D., Naumova, N., Bulanov, S., Mori, W.B., Tzeng, K.Z., submitted to *Phys. Rev. Lett.*; Darrow, C.B., *private communication*
10. Kruer, W.L., *Physica Scripta* **T30**, 5 (1990).; Bingham, R., *ibid.* p. 24
11. Johnston, T.W., Bertrand, P., Ghizzo, A., Shoucri, M., Fijalkow, M.E., Feix, M.R., *Phys. Fluids* **B4**, 2523 (1992) and references therein
12. Wilks, S.C., Kruer, W.L., Williams, E.A., Amendt, P., Eder, D.C., *Phys. Plasmas* **2**, 274 (1995).
13. Manneville, P., *"Dissipative Structures and Weak Turbulence"*, Academic, San-Diego, 1990.

Investigation of the Dynamic Fracture Process at Ultrahigh Strain Rate Caused by Laser-Induced Shock Waves in Solid Targets

V.I. Vovchenko, I.K. Krasyuk, P.P. Pashinin, and A.Yu. Semenov

General Physics Institute
Russian Academy of Sciences
Vavilov Street 38, 117942 Moscow, Russia

Abstract. This work consists of three main parts. In the first part dependence of amplitude of ablation pressure from intensity of laser radiation in a specified range is experimentally investigated. In the second part measurements of spall strength of a researched material from deformation rate up to $3 \cdot 10^7 \, s^{-1}$ are performed. In the third part measurements of mechanical work, expended on separation of spall layer, are carried out.

In the present work results of research of some characteristics of destruction process of metal targets are indicated at large rates of deformation, Fig. 1. In experiments Nd-glass laser installation with a length of a wave of radiation 1.06 μm was used. The laser pulses had the form, closed to triangular, with a duration of a pulse on the basis from 10 up to 80 ns. The laser radiation was focused on the

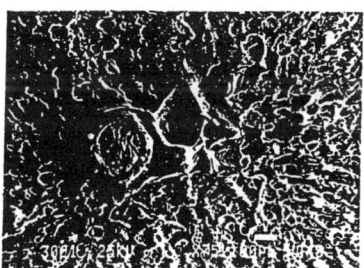 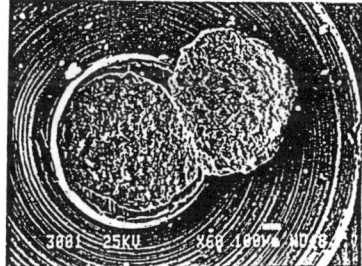

FIGURE 1. The results of the laser-induced spallation process. The laser irradiated surface of the target is showed on the right hand, back surface of the target is showed on the left hand.

© 1996 American Institute of Physics

target in a spot with a diameter 1–1.2 mm. The targets in the majority of cases had thickness from 230 up to 400 μm. These are made of the aluminum alloy AMg6M. As has appeared in these conditions spallation process arose at laser radiation intensity in a range 10^{10}–10^{12} W/cm^2. The researches are based on joint application experimental methods and methods of numerical modeling.

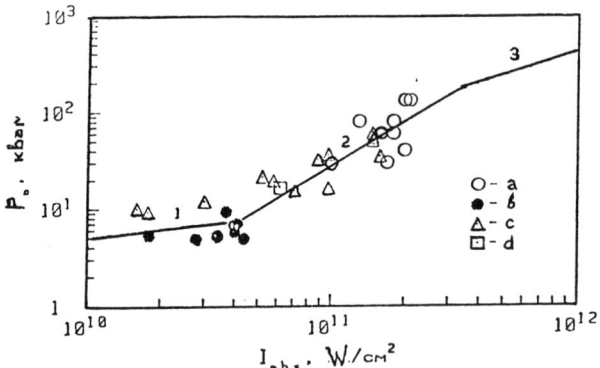

FIGURE 2. Experimental dependence of ablative pressure P_a versus intensity of absorbed laser radiation I_{abs}.

The results of the first part of this work are shown on Fig. 2. For definition of ablative pressure below-listed methods, designated on drawing by letters are used (a-d).

(a) The method is based on measurement of a time of propagation of shock waves in aluminum targets by thickness 100, 150, and 200 μm. In this method a moment of arrival of a shock wave front on a back surface of a target was fixed. Then, amplitude of ablative pressure on a face surface of a target was calculated with the help of a numerical code on the basis of equations of hydrodynamics and real equation of a state of the aluminum.

(b) The method is based on registration of the profile of laser acceleration of aluminum foils with thickness 100 μm with the using of VISAR type velocity interferometer. Then, the effective significance of acting pressure amplitude was calculated on the basis of the Newton second law. Thus, the change of foil mass was not taken into account at laser interaction.

(c) The method is based on discrete registration of the characteristics of braking polymer foils by thickness from 5 up to 12 μm in a cylindrical channel, filled in by air or xenon. The amplitude of acting pressure was calculated on the basis of the foil movement law, the equations determined from joint integration of gas and foil movements (with the account by reduction of its mass in ablation process).

(d) The method is based on continuous registration with the using of Doppler effect of polymer foils laser acceleration with thicknesses from 1 up to 5 μm in a deuterium filled conic channel. The processing of received experimental data by a method of the least squares results in a relation

$$P_a(\text{kbar}) = (28\pm3)(I_{abs} \cdot 10^{-11}, \text{W} \cdot \text{cm}^{-2})^{1.5\pm0.2}.$$

Here P_a is the ablation pressure, I_{abs} is the intensity of absorbed laser radiation. On Fig. 2 this relation is shown by curve 2. There is the concurrence with data of paper (1) in the interval 10^{10}–$4 \cdot 10^{10}$ W·cm^{-2} (curve 1 on Fig. 2). Curve 3 was received in paper (2).

In the second part of present work spall layer thickness is measured in each experience. Then the value of material deformation rate and the value of the maximum strain (negative) pressure were calculated in a spall plane by using numerical modeling method. For this purpose a special numerical code is developed on the basis of equations of hydrodynamics in view of the real equation of state for the researched material and opportunity of the description of rarefaction waves with negative pressure. The results of research are shown on Fig. 3. In some cases the specified values were determined by a conventional method with use of a velocity profile structure of a target free surface. The structures of velocity were measured with the help of VISAR. The analysis shows that these data complement each other and characterize the spallation phenomenon more complete.

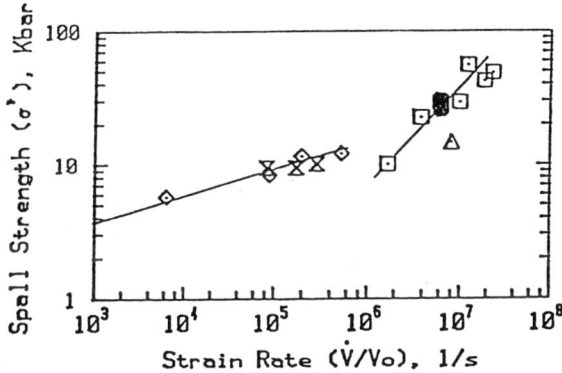

FIGURE 3. The experimental dependence of the spall strength versus the strain rate: (□) new laser data, (Δ) laser data of D.L.Paisley et al., (■) light ion beam data of K.Baumung et al., and (◊) explosive data of G.I.Kanel.

In the third part of the present work measurements of spall layer velocity are perform by means of the electrocontact method. On the basis of these measurements, the kinetic energies values of separated layers were determined. Kinetic energies of aluminum plates glued on a back surface of a target (artificial spall layer) were measured by similar way. The thickness of these layers corresponded to thickness of the real spall layers. The difference between kinetic energies of real and artificial layers was accepted as the value of the destruction work. It occurs that measured values of fracture work of the spall layer is considerably less than it in the case of explosive experiments. The results of these measurements are shown on Fig. 4. It is obviously that sharp decrease of destruction work occurs for a investigated material at deformation rate $10^6 \, s^{-1}$. This result is received for the first time and its explanation requires further researches.

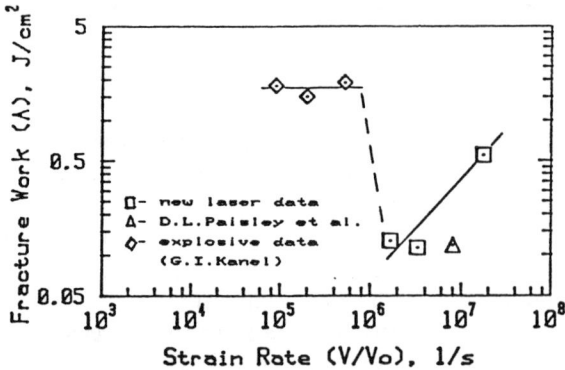

FIGURE 4. The experimental dependence of the fracture work from the strain rate. The marking are the same as on Fig 3.

In addition, influence of the target form on some characteristics of the spallation phenomenon was investigated. In these experiments simple and stepped targets were used. It was established that complete separation of the layer occurs at considerably smaller laser intensity for the stepped targets. Moreover, in the case of stepped targets separating plates had as a rule small spherical bend. The size of the bend did not exceed 10 %. In case of flat targets occurred extension of a separated layer and gap in its middle. Thus, the edges of this layer were not separated from the target. For qualitative understanding of spallation phenomena two-dimensional numerical modeling was carried out. Results of modeling show that conditions for spallation occur on the layer periphery, and then, they will be realized and in a frontal plane of the target.

CONCLUSION

The ablation pressure characteristics were investigated by means of the dynamic methods. For aluminum alloy AMg6M the following results were obtained.

(i) The value of spall strength was studied for the strain rate up to $3 \cdot 10^7 \, s^{-1}$.

(ii) It was found that the value of fracture work at separation of the spall layer significantly decreases when the strain rate is equal to $10^6 s^{-1}$.

This work was supported by the Russian Foundation for Basic Research, Grant No. 94-02-03413-a.

REFERENCES

1. Combis P., Cazalis B., David J., et al. "Low-fluence laser-target coupling." *Laser and Particle Beams*. 1991. 9. No. 2. P. 403–420.
2. Dahamani F., Kerdia T. "Measurements and laser-wavelength dependence of mass-ablation rate and ablation pressure in planar layered targets." *Ibid*. No. 3. p.769–778.

Extremely Cold Plasma Production by Optical-Field-Induced Ionization of a Preformed Plasma

Yutaka Nagata*, Katsumi Midorikawa, Minoru Obara*, and Koichi Toyoda

The Institute of Physical and Chemical Research, RIKEN
Wako-shi, Saitama 351-01, Japan
*Department of Electrical Engineering, Keio University,
3-14-1 Hiyosi, Kohoku-ku, Yokohama 223, Japan

Abstract. We report the production of fully stripped lithium ions with extremely cold electrons by optical-field-induced ionization (OFI). Such cold electrons were created by irradiating a preformed plasma (singly-ionized lithium ions) with a 0.5-ps KrF laser at an intensity of 10^{17} W/cm^2. The measured electron temperature of less than 1 eV was much lower than that expected from an above threshold ionization model. In order to explain this low temperature, a two-component plasma model is presented. The model indicates that preformed free electrons existing prior to the field ionization play an important role to achieve such cold electron temperature. An averaged electron temperature may be decreased by the rapid electron thermal conduction cooling between the OFI plasma and preformed plasma.

1. INTRODUCTION

Recently optical-field-induced ionization (OFI) has been investigated as an appropriate candidate to produced cold and dense multiply ionized plasma that have a potential for generating soft-x-ray laser [1-5]. An ultrashort high intensity laser can produce a plasma consisting of fully stripped ions and cold free electrons on a time scale much shorter than the recombination time. In such a plasma, a rapid recombination cascade of the electrons can lead to a population inversion with respect to the ground state of hydrogen-like ions. This scheme has some favorable characteristics. The first advantage is that lasing to a ground state will make a soft-x-ray laser wavelength shorter compared with that between excited states. The second advantage is that a high-laser intensity required to produce desired ionization states is achievable with a table-top high power laser with ultrashort pulse widths. The requirement of the low pump energy will also allow soft-x-ray lasers in high repetitive operation, which is critical for most of the promising applications.

To realize an OFI soft x-ray laser, it is necessary to minimize undesirable electron heating originating from the high intensity laser matter interaction, such as inverse bremsstrahlung and above-threshold-ionization (ATI) . When an electron density of the plasma is less than 10^{18} cm^{-3}, a subpicosecond pulse width is short enough to avoid inverse bremsstrahlung. ATI heating, however, can not be avoided in any case. Even if we use short wavelength pump laser pulse, such as a KrF laser, the ATI heating prevent to achieve a population inversion with respect to the ground state when neutral gases are used as an initial medium. Therefore, we proposed a new scheme using a Li$^+$ preformed plasma as an initial medium[4]. The new scheme is unique modification of the original OFI scheme proposed previously [1-3]. This modified scheme is shown in Fig. 1, together with relevant energy levels of Li. In this scheme, singly-ionized lithium ions which are initially prepared by a nanosecond KrF excimer laser are further ionized to a fully stripped state by a subpicosecond high-intensity KrF laser. Soft-x-ray lasing on the Lyman-α transition in H-like Li at 13.5 nm is expected by rapid recombination following OFI. By using this scheme, we obtained extremely cold electron temperature of less than 1 eV, while the ionization energy of Li^{2+} is 122 eV[5].

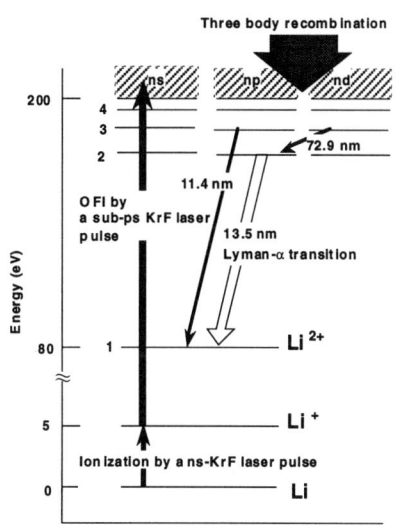

Fig.1 Schematic energy diagram of Li. The preformed plasma production, the 13.5-nm Lymanα soft x-ray laser transition followed by OFI are indicated.

In this proceeding, we discuss the roles of a preformed plasma for the realization of an optical-field-induced ionization soft-x-ray laser.

Calculated energy of electrons produced by OFI

A kinetic energy of electrons produced by OFI in neutral gases is predicted theoretically by quasistatic model [6,7]. By using this model, we calculated the kinetic energy of electrons produced by OFI of a preformed Li$^+$ plasma. The calculated average energy as a function of the laser wavelength is shown in Fig. 2. The initial electron density is 2×10^{17} cm^{-3}. The peak laser intensity and pulse width are 10^{17} W/cm^2 and 0.5 ps, respectively. The kinetic energy of electrons varies with the square of wavelength, because the ionization rate is independent of the wavelength while the electron quiver energy depends on the square of the wavelength. A KrF laser operating with a wavelength of 248 nm will produce electrons having an average kinetic energy of 26 eV under these conditions, which

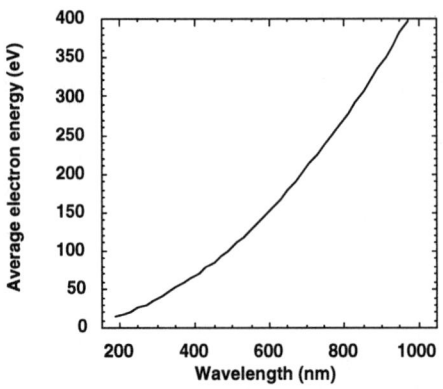

Fig.2 Average electron energy as a function of the laser wavelength for Li $^+$.

is much smaller than an ionization potential of Li^{2+} (122 eV). This value, however, is not low enough to produce population inversion with respect to the ground state by rapid recombination.

Calculated energy distribution of electrons produce by OFI is shown in Fig. 3. Several parameters for the calculation are the same as those used in Fig. 2. The calculated electron energy distribution becomes continuous although the ATI electron distribution form different ionization stage varies. It is noted that the calculated electron energy distribution is different from the Maxwellian distribution. The electron distribution decreases exponentially at the energy between 0 and 5 eV. At the energy larger than 5 eV, the distribution rather slowly decreases. Maxwellian distributions with an average kinetic energy of 1 eV and 26 eV are also shown in the figure for comparison. As shown in Fig. 3, kinetic energy distribution of OFI plasma has larger population in the lower energy region, compared with maxwellian one.

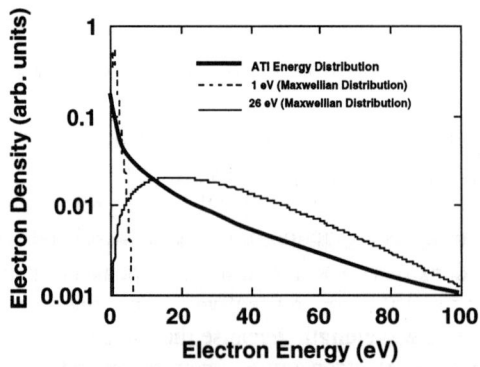

Fig.3 Electron energy distribution for Li^+.

Two component plasma model

In order to explain the role of a preformed plasma for production of a cold plasma, a two-component plasma model has been developed. The preformed plasma component is homogeneously distributed before the subpicosecond KrF laser irradiation and its electron temperature is assumed to be 1 eV. In this cold bath of the preformed plasma, a hot electrons in the OFI plasma column with a diameter of 10 µm is instantaneously produced by OFI. Since preformed electrons are not heated by irradiation of the subpicosecond laser pulse, both the hot and cold electron components are assumed to coexist in an OFI plasma column [5].

In such a plasma, thermalization may cause electron collisional heating of the cold component. When we assumed that energy distribution of electrons is spatially homogeneous and a diameter of the OFI plasma column is sufficiently larger than mean free path of the hotter electrons, thermalization equilibrates the two components. The temperature change of the cold electrons and an equilibration time, t_{eq}, are given by Spitzer, and are described by Eqs.(1) and (2), respectively [8].

$$T_{ecold}(t) = T_{eav}\left(1 - \exp\left(\frac{-t}{t_{eq}}\right)\right) + T_{ecold} \exp\left(\frac{-t}{t_{eq}}\right) \tag{1}$$

$$t_{eq} = 0.137 \frac{(T_{ecold} + T_{ehot})}{(n_{ecold} + n_{ehot})\ln \Lambda} \tag{2}$$

In our case, a preformed plasma has a cold electron density $n_{ecold} = 2 \times 10^{17}$ cm^{-3} and an electron temperature $T_{ecold} = 1$ eV while the OFI plasma has a hot electron density $n_{ehot} = 4 \times 10^{17}$ cm^{-3} and electron temperature $T_{ehot} = 26$ eV. Substituting these parameters into Eqs.(1) and (2), We evaluated a cold-electron heating rate of approximately 1 eV/ps. The heating rate of the cold electron depends weakly on the hot electron temperature. Consequently, this rapid electron heating due to collisional thermalization lead to reduction of gain.

In order to explain the experimental results, we should consider the diameter of the OFI plasma column, because it is shorter than a mean free path of hotter electrons. In this case, electron thermal conduction may decrease the electron temperature of the OFI plasma after the passage of the subpicosecond KrF laser pulse. Neglecting the further ionization after the laser pulse, the temperature change is described by Eq. (3).

$$\frac{\partial}{\partial t}\left(\frac{3}{2}kT_e n_e\right) = \frac{\partial}{\partial x}\left(\kappa \frac{\partial k T_e}{\partial x}\right) \tag{3}$$

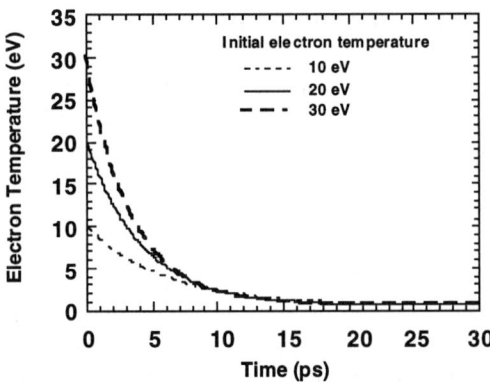

Fig.4 Temporal evolution of the average temperature of electrons produced by OFI in a cold bath of the preformed electrons

where κ is the electron thermal conductivity. For strong temperature gradients or collisionless regime, electrons are assumed to free-stream rather than diffuse. At an electron density of 6×10^{17} cm^{-3} with a temperature approximately 10 eV, the electron-electron collision frequency and the electron velocity are estimated to be 5 ps and 2×10^6 m/s, respectively. Therefore, electrons having an energy of more than 10 eV can escape from the focal volume without collision. In this collisionless regime, the heat flow is simply limited by the maximum flux. Then Eq.(3) reduces to Eq. (4) [9].

$$\frac{3}{2} n_e \frac{\partial}{\partial t}(k\, T_e) = \frac{\partial}{\partial x}(f\, n_e\, k\, T_e\, v_e) \qquad (4)$$

where f is the so-called flux limit and is 0.6 , v_e is the electron thermal velocity. By integrating Eq.(4) , the temporal change of the OFI electron temperature was obtained and is shown in Fig. 4. The temperature decreases exponentially and approaches to 1 eV within 15 ps. This decay time roughly coincides with a delay between the appearance of peak gain and the subpicosecond laser irradiation. Consequently, both the preformed electrons and the cold temperature component of OFI electrons seem to contribute to the rapid three-body recombination in the focal volume.

Summary. Production of extremely cold plasma by optical-field-induced ionization of a preformed plasma has been investigated. Compared with the previous approaches using neutral atoms , the use of an initially ionized medium plays an important role for realizing the cold electron temperature. A preformed plasma supplies a huge cold bath in which a high-temperature plasma column is produced by OFI. A high energy component of electrons produced by OFI would escape from the focal volume without interaction before the thermalization occurs, while the initial electrons produced prior to the OFI would survive and the temperature of these electrons is kept low enough during the OFI.

References

1. Burnett, N. H., and Enright, G. D., "Population inversion in the recombination of optically-ionized plasmas," IEEE J. Quantum Electron. **26**, 1797-1808 (1990).
2. Amendt, P., Eder, D. C., and Wilks, S. C., "X-ray lasing by optical-field-induced ionization, " Phys. Rev. Lett. **66**, 2589-2592 (1991).
3. Eder, D. C., Amendt, P., and Wilks, S. C., "Optical-field-ionized plasma x-ray lasers," Phys. Rev. A **45**, 6761-6772 (1992).
4. Nagata, Y., Midorikawa, K., Kubodera, S., Obara, M., Tashiro,. H., and Toyoda, K., "Soft-x-ray amplification of the Lyman-α transition by optical-field-induced ionization, "Phys. Rev. Lett. **71**, 3774-3777 (1993).
5. Nagata, Y., Midorikawa, K., Kubodera, S., Obara, M., Tashiro,. H., and Toyoda, K., "Production of an extremely cold plasma by optical-field-induced ionization," Phys. Rev. A **51**, 1415-1419 (1995).
6. Burnett, N. H., and Corkum, P. B., "Cold-plasma production for recombination extreme-ultraviolet lasers by optical-field-induced ionization," J. Opt. Soc. Am. B **6**, 1195-1199 (1989).
7. Pulsifer, P., Apruzese, J. P., Davis, J., and Kepple, P.,"Residual energy and its effect on gain in a Lyman-α laser," Phys. Rev. A **49**, 2598-2965 (1994).
8. Spitzer, L., *Physics of Fully Ionized Gases*, New York: Interscience publishers, 1962. ch. 5. pp. 120-136.
9. Kruer, W. L., *The Physics of Laser Plasma Interactions*, Addison-Wesley Publishing Company, Inc., 1988. ch. 12. pp.144-147.

SDOSS: a Spatially Discriminating, Optical Streaked Spectrograph[*]

J. A. Cobble, S. C. Evans, J. C. Fernández, J. A. Oertel, R. G. Watt, B. H. Wilde

Los Alamos National Laboratory, Los Alamos, NM 87545

SDOSS is employed to study broadband laser scattering encompassing SBS, SRS, and the 3/2-ω signature of two plasmon decay for ns-scale laser-plasma experiments with 351- or 527-nm drive. It uses a Cassegrain telescope to image scattered light from a laser plasma onto a field stop. The telescope magnification and the stop aperture provide spatial discrimination of target plane scatter. A UV lens relays the image to a 0.25-m spectrograph which is lens coupled to a streak camera with an S-1 photocathode. The streak output is imaged onto a CCD camera. In its 512x480 pixel array, the CCD covers a spectral range from 200 to 800 nm with 4-nm resolution and can be adjusted to look from 350 to 1060 nm. The sweep speed is variable with full window values of 30, 12, 6 ns, and faster. An optical fiducial provides a spectral and temporal marker.

On the Livermore Nova laser, SDOSS has been used to determine spatial density in gas-filled hohlraums from SRS signals. At Trident in Los Alamos, it has been employed for similar measurements with long scale length plasmas in SBS and SRS seeding experiments. It has proven to be a versatile tool for studying the physics of laser-generated plasmas.

Optical spectroscopy of scattered laser light from inertial confinement fusion (ICF) targets is important to verify the economy of laser coupling to the target and to diagnose the target performance. The former helps quantify laser energy converted to plasma waves, which can scatter laser light from the target and can generate hot electrons that might preheat the fusion capsule. The latter determines the plasma conditions in the target thus providing the opportunity to improve target design and behavior. An example of an optical diagnostic is the subject of this paper: SDOSS, a spatially discriminating optical streaked spectrograph, which obtains time dependent scattered laser spectra from discrete regions of the target.

The major novelty of SDOSS (1) is perhaps its spatial discrimination, which derives from a Cassegrain telescope imaging target light onto an intermediate focal plane. A field stop is located in this plane where an aperture limits the light which passes on to a collection lens for injecting scattered light into the spectrograph. The ~9X magnification of the Cassegrain and the aperture diameter determine the field of view which at present affords a spatial discrimination of 200 µm.

---

[*] This work supported by the US DOE.

The output signal from the spectrograph is lens coupled to a streak camera and subsequently lens coupled again into a CCD camera for storage by a computer. The resulting data are images of the time-versus-wavelength streak for light coming from the selected portion of the target. Figure 1 illustrates the optics of SDOSS, all of which are compatible with ultraviolet (UV) light. The streak photocathode is a

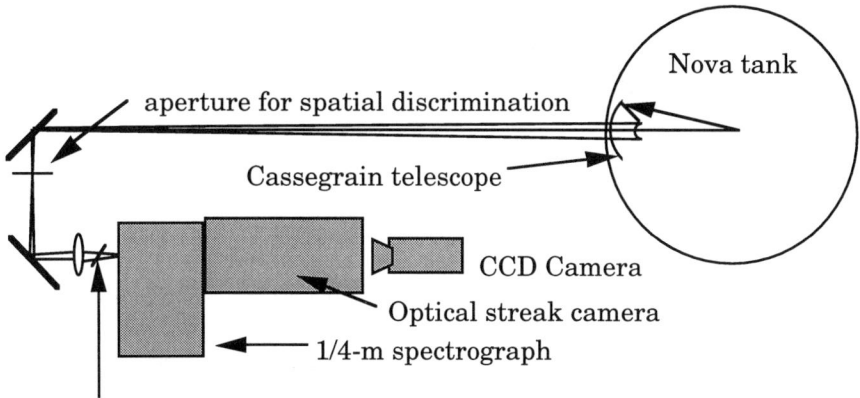

FIGURE 1. Line drawing of SDOSS on the NOVA laser chamber

UV-extended S-1 (2), which gives it sensitivity from 200 to greater than 1000 nm. The 0.25-m spectrograph and the CCD pixel size combine to give a system spectral resolution of 4.5 nm. With a 6-ns window on the streak, the temporal resolution is about 100 ps.

Another novel feature of the system is the broad spectral coverage. With 351-nm laser drive, data extending from 200 to over 700 nm are recorded. This allows simultaneous recording of three primary parametric instabilities (3): stimulated Brillouin scattering (SBS) near 351 nm, stimulated Raman scattering (SRS) between 351 and 702 nm, and two plasmon decay (TPD) with its signature near 234 nm. SDOSS has been run in this mode at NOVA (4). On the Trident laser (5), which operates at 527 nm, a simple grating rotation in the spectrograph permits wavelength coverage from 400 to 1100 nm. At Trident, the Cassegrain telescope was not available, and glass lenses were used to focus the scattered light. Therefore, the two-plasmon signal at 351 nm was not accessible in this case.

Thus, SDOSS has been employed on two laser facilities, NOVA and Trident, where it has demonstrated its capability for resolving plasma parameters and has helped in the understanding of target issues (6 and 7). This is shown next.

On NOVA, SDOSS was used on gas-filled hohlraums (8). Nine of the ten NOVA beams are used to heat the hohlraum as shown in Fig. 2. After ~400 ps, a 1-ns probe beam of intensity 6×10^{14} W/cm^2 is turned on. As indicated in the figure, SDOSS sees the overlap of the probe with a single heater beam at a distance of 840 µm from the hohlraum axis. On other targets, the window has been moved left or right to observe probe sidescatter at different distances from the axis. The sidescatter angle is 104°.

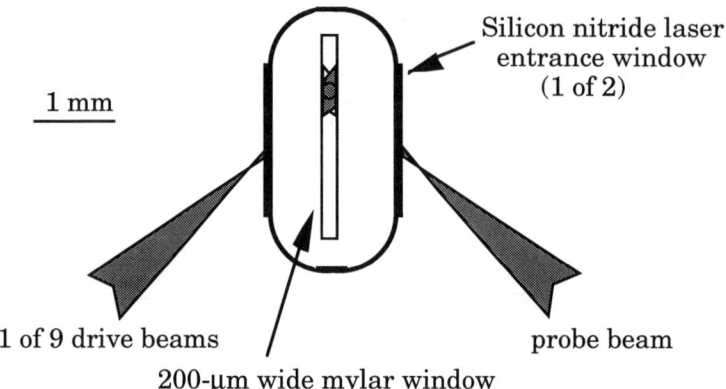

FIGURE 2. Side view of a gas-filled NOVA hohlraum from SDOSS -- The circle shows the 200-μm SDOSS field of view. Five beams enter the target on each end.

The nominal gas fill is one atmosphere of neopentane. This gas, when photoionized by the laser and by x rays, corresponds to a plasma density of 12% of the critical density ($\pi mc^2/e^2\lambda^2$) for the 351-nm drive. However, shocks in the plasma are launched when the laser entrance windows and the interior gold wall are illuminated. These cause a variation in plasma density within the hohlraum. In order to estimate the density, we examine the SRS spectrum. SRS arises when the incident laser light pumps an electron plasma wave and is shifted to longer wavelengths. The spectrum tells from what density the scattered light comes, and since SDOSS has spatial resolution corresponding to its line of sight through the probe beam, we have a point measurement of density in this case at 840±100 μm.

A sample scattered spectrum from a CCD image is shown in Fig. 3. This is raw data with no correction for photocathode sensitivity or spectrometer calibration. The wavelength scale is calibrated with a Hg lamp, which shows that the dispersion is linear. The spectrum is filtered with two high reflectivity rejection filters: one which reduces the 351-nm light by approximately 100 times and another which does the same at 527 nm. (The latter is necessary for when NOVA operates with green drive.)

Signals from three parametric instabilities are indicated by the spectrum. The strongest when corrected for sensitivities is the 351-nm SBS. Next is the TPD signal at 234 nm. This is a surprise since TPD originates at quarter critical density and is strong evidence of a factor of two enhancement of the plasma density somewhere within the small volume of the hohlraum from which light is scattered. The third instability is SRS with a spectrum extending from 460 to 700 nm. The gap in the spectrum from 500 to 550 nm is due to the green rejection filter. (In spite of this filter, the specular reflection of unconverted, unfocused green light can be seen in the streak. This has been eliminated in other data with shine shields to prevent stray light from striking the outside of the target. Then, the green filter can be removed to look at SRS in this spectral region.) Finally, the 700-nm feature is believed to be SRS at near quarter critical. While possibly some of the signal is due to second order SBS, the shape of the profile is different, and the relative intensity

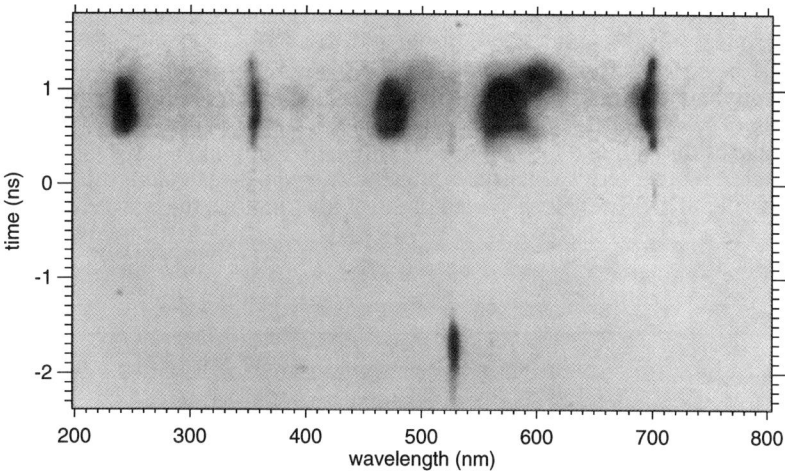

FIGURE 3. Seen through a diagnostic slot at the intersection of two beams inside the hohlraum, simultaneous spectra of SBS, SRS, and TPD are observed by SDOSS.

of first and second order Hg I 365 in the calibration denies the likelihood that this is second order 351. The breadth of the SRS therefore suggests scattering from many densities within the scattering volume and presents a picture of the complicated dynamics in the interior of these hohlraum targets.

The utility of SDOSS is also shown by backscatter experiments on Trident. In this case, resolution is dictated by the laser spot size of 125 μm rather than by a field stop aperture. The scattering angle as shown in Fig. 4 is 180° along the probe beam and the wavelength 527 nm. The laser irradiance is about 1×10^{15} W/cm^2.

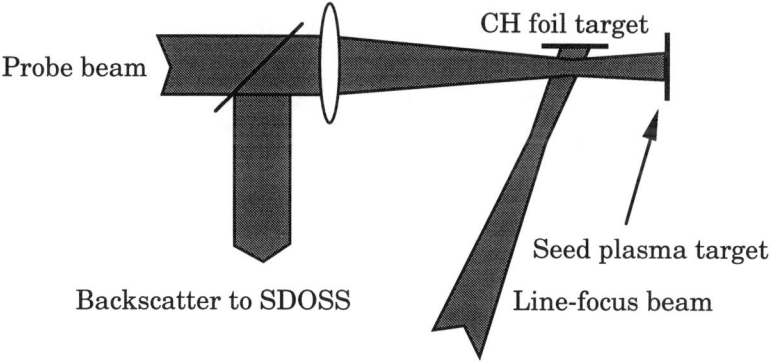

FIGURE 4. SDOSS looks at 180° backscatter of the probe beam on this complex Trident target.

SDOSS was used to characterize two parts of the target for an instability seeding experiment (6). The seed part of the target was a CH-coated gold disk on which the probe beam was normally incident. The second part of the target, more than a hydrodynamic expansion distance away, was a CH foil illuminated by the line focus of a second laser beam. The probe beam passed within 200 μm of the surface of this foil and tangent to it. With the CH foil alone, SDOSS verified the existence of a 5%-critical plasma within the probed plasma volume. Figure 5 below shows the SDOSS data for the seed plasma only. In time, the wavelength of peak

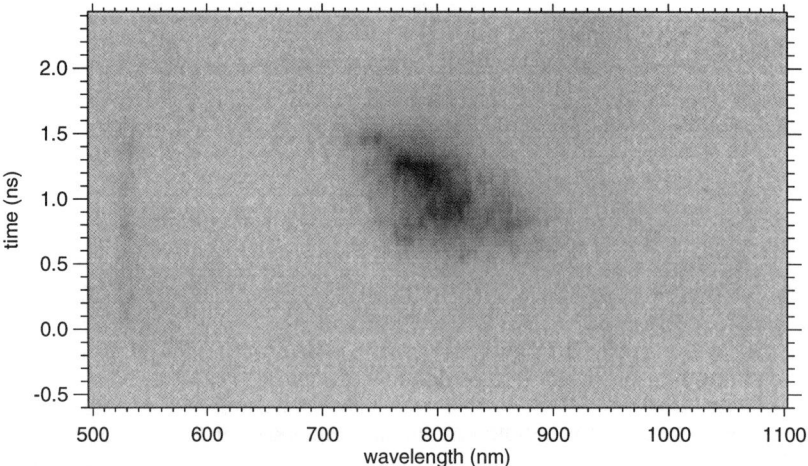

FIGURE 5. The electron density of peak SRS gain is seen to fall off with time for a Au disk target. The 527-nm SBS is attenuated by a high reflectivity green rejection filter.

SRS falls off from 825 to 740 nm. This corresponds to the expansion of the plasma and shows that the density at which SRS gain is highest is blowing down as the plasma expands. In 1 ns, this density drops from about 12% to 7% critical.

These measurements were vital to proper control of this experiment. On both NOVA and Trident, SDOSS is demonstrated to be a versatile diagnostic for recording optical spectra and understanding the plasma physics of complex ICF experiments.

REFERENCES

1. Cobble, J. A., Fernández, J. C., Wilde, B. H., Evans, S., Jimerson, J., Oertel, J., Montgomery, D. S., Gomez, C. C., "Simultaneous temporal, spectral, and spatial resolution of laser scatter from parametric plasma instabilities", accepted for publication in *Rev. Sci. Instru.*, Jan., 1995.

2. Turner, E. B. ,"Optical and Ultraviolet Techniques", Huddlestone, R. H., Leonard, S. L., eds., *Plasma Diagnostic Techniques,* New York: Academic Press, 1965, ch. 7, p. 351.

3. Kruer, W. L., The Physics of Laser Plasma Interactions, Redwood City, CA: Wesley Publishing Co., 1988, chs. 6-8, pp. 57-94.

4. Hunt, J. T., Speck, D. R., *Optical Engineering* **28**, 461 (1989).

5. Moncur, N. K., Johnson, R. P., Watt, R. G., Gibson, R. B., "TRIDENT: a versatile high-power Nd:glass laser facility for ICF experiments", accepted for publication in *Appl. Optics,* 1995.

6. Fernández, J. C., Cobble, J. A., Gobby, P. L., Lindman, E. L., Montgomery, D. A., Rose, H. A., Wilde, B. H., Wilke, M. A., "Brillouin backscatter and seeding mechanisms in NOVA hohlraums", paper TuP-5, this conference.

7. Wilde, B. H., Fernández, J. C., Hsing, W. W., Cobble, J. A., Delamater, N. D., Failor, B. H., Hockaday, R. G., "The design and characterization of toroidal-shaped NOVA hohlraums that simulate national ignition facility plasma conditions for plasma instability experiments", paper TuP-4, this conference.

8. Fernández, J. C., Cobble, J. A., Failor, B. H., Hsing, W. W., Rose, H. A., Wilde, B. H., Bradley, K. S., Gobby, P., Kornblum, H., Montgomery, D. A., "Dependence of stimulated Brillouin scattering on laser intensity, laser f number and ion species in hohlraum plasmas", submitted to *Phys. Rev. Lett.,* Jan., 1995.

Two-temperature Model of Atoms in Dense Plasmas

Yoichiro Furutani and Atsushi Fukuyama

Department of Electrical and Electronic Engineering,
Okayama University, Okayama 700, Japan

Abstract. To analyze the internal structure of a compressed atom or ion immersed in dense plasmas, many theoretical schemes, such as orthodox Hartree-Fock-Slater method, time-honored statistical theory of atoms and a more systematic density functional theory, have been devised and developped. In this article, we propose the two-temperature model of atoms, composed of an ion core in its ground state ($T = 0$) and a clad electron gas at finite temperatures ($T > 0$), the latter being held in thermal equilibrium with a surrounding dense plasma. We analyze the model with recourse to the Thomas-Fermi-Dirac-Weizsäcker statistical theory and present several results. Throughout this article, the atomic units are used, unless otherwise specified.

Basic Formulation

We give a brief survey of the sophisticated Thomas-Fermi- Dirac-Weizsäcker (TFDW) statistical model. First of all, the free energy of the clad electron gas can be expressed as

$$F[n_f] = F_{KG}[n_f] + F_{XC}[n_f]$$
$$- \int_{\Omega_>} d\vec{r} \frac{Z}{r} n_f(r) + \int_{\Omega_>} d\vec{r} \frac{Q_b}{r} n_f(r) + \frac{1}{2} \int \int_{\Omega_>} d\vec{r} d\vec{r}' \frac{n_f(r) n_f(r')}{|\vec{r} - \vec{r}'|} \quad (1)$$

where Z/r is the Coulomb potential produced by the nucleus located at the origin of the spherical coordinate system. Q_b is a total number of electrons in the ion core and n_f the free electron density. Since, in our two-temperature model (TTM), the ion core and the clad electron gas are not in thermal equilibrium, it seems legitimate to introduce the internal energy $E[n]$ to the cold electron gas of the ion core[1]. It is given by

$$E[n_b] = E_{KG}[n_b] + E_{XC}[n_b]$$
$$- \int_{\Omega_<} d\vec{r} \frac{Z}{r} n_b(r) + \int_{\Omega_<} d\vec{r} V_0 n_b(r) + \frac{1}{2} \int \int_{\Omega_<} d\vec{r} d\vec{r}' \frac{n_b(r) n_b(r')}{|\vec{r} - \vec{r}'|} \quad (2)$$

where V_0 is a constant potential produced in the ion core by the clad electrons located at $r > r_c$, with r_c the radius of the ion core. n_b is the bound electron density. The suffixes K, G, X and C denote the kinetic, Perrot's gradient correction to the kinetic energy[2], exchange and correlation. $\Omega_<$ and $\Omega_>$ stand for a volume of the ion core and of the clad electron gas. According to the definition of the chemical potential,

$$\frac{\delta E[n_b]}{\delta n_b} = \mu_b \quad \text{and} \quad \frac{\delta F[n_f]}{\delta n_f} = \mu_f \ . \tag{3}$$

These relations yield

$$\frac{1}{2}(3\pi^2 n_b)^{2/3} + \frac{\delta \mathcal{E}_{XC}[n_b]}{\delta n_b} + \frac{1}{72}\left\{\left[\frac{d}{dr}\ln n_b(r)\right]^2 - 2\frac{\nabla^2 n_b}{n_b}\right\}$$

$$= \mathcal{U}_<(r) + \mu_b, \quad T = 0 \tag{4}$$

and

$$\frac{\delta \mathcal{F}_K}{\delta n_f} + \frac{\delta \mathcal{F}_{XC}[n_f]}{\delta n_f} - \left\{\left(h - n\frac{\delta h}{\delta n}\right)\left[\frac{d}{dr}\ln n_f(r)\right]^2 + 2h\frac{\nabla^2 n_f}{n_f}\right\}$$

$$= \mathcal{U}_>(r) + \mu_f, \quad T > 0 \tag{5}$$

where we defined the energy density $\mathcal{E}$ and $\mathcal{F}$, such that

$$E[n_b] = \int_{\Omega_<} d\vec{r}\,\mathcal{E}[n_b] \quad \text{and} \quad F[n_f] = \int_{\Omega_>} d\vec{r}\,\mathcal{F}[n_f] \tag{6}$$

While equation (4) was derived by virtue of the boundary condition $n'_b(r_c) = 0$ (the dash denotes the derivative with respect to r), equation (5) was obtained with the use of the two boundary conditions $n'_f(r_c) = n'_f(R) = 0$. R is the outer radius. The Hartree potential prevailing in each region is defined, respectively, by

$$\mathcal{U}_<(r) = \frac{Z}{r} - \int_{\Omega_<} d\vec{r}' \frac{n_b(r')}{|\vec{r}-\vec{r}'|} - V_0 \tag{7a}$$

and

$$\mathcal{U}_>(r) = \frac{Z - Q_b}{r} - \int_{\Omega_>} d\vec{r}' \frac{n_f(r')}{|\vec{r}-\vec{r}'|} \tag{7b}$$

where

$$V_0 = 4\pi \int_{r_c}^{R} dr'\, r'\, n_f(r') \tag{8}$$

We can readily verify the continuity of the two potentials at $r = r_c$, by expanding $1/|\vec{r}-\vec{r}'|$ in terms of the Legendre polynomial. A set of equations (4) and (5), coupled with the Poisson's equation which holds in each region, can be solved upon introduction of the screening function $\phi(r)$ defined as

$$\frac{Z}{r}\phi_{>,<}(r) = \mathcal{U}_{>,<}(r) + \mu_{f,b} + \frac{1}{2\pi^2} \qquad (9)$$

The resulting two nonlinear equations of fourth order are expressed as[3]

$$\phi_<^{(4)}(r) = \frac{144\pi rn}{Z}\left(\frac{1}{72}\left[\frac{d}{dr}\ln n_b(r)\right]^2 + \frac{1}{2}(3\pi^2 n_b)^{2/3} + \frac{\delta\mathcal{E}_{XC}[n_b]}{\delta n_b} - \frac{Z}{r}\phi_< + \frac{1}{2\pi^2}\right)$$

$$0 \leq r \leq r_c \qquad (10)$$

$$\phi_>^{(4)}(r) = \frac{2\pi rn}{Z}\left(\left(h - n_f\frac{\delta h}{\delta n_f}\right)\left[\frac{d}{dr}\ln n_f(r)\right]^2 + \frac{\delta\mathcal{F}_K}{\delta n_f} + \frac{\delta\mathcal{F}_{XC}}{\delta n_f} - \frac{Z}{r}\phi_> + \frac{1}{2\pi^2}\right)$$

$$r_c \leq r \leq R \qquad (11)$$

Here, the gradient correction factor due to Perrot[2] can be given by

$$h[n] = -\frac{\sqrt{2}}{24}\pi^2 n\beta^{3/2}\frac{\delta}{\delta\alpha}\left[\frac{1}{I_{-1/2}(\alpha)}\right], \quad \beta = \frac{1}{k_B T} \qquad (12)$$

in terms of the Fermi function

$$I_{-1/2}(\alpha) = \int_0^\infty dx \frac{x^{-1/2}}{e^{x-\alpha}+1}, \quad \alpha = \beta\mu \qquad (13)$$

On the other hand, the kinetic energy density $\mathcal{F}_K$ can be expressed as

$$\mathcal{F}_K[n] = \frac{\sqrt{2}}{\pi^2}\beta^{-5/2}\left(-\frac{2}{3}I_{3/2}(\alpha) + \alpha I_{1/2}(\alpha)\right) \qquad (14)$$

This expression includes a contribution of the entropy part. As for the exchange-correlation energy density $\mathcal{F}_{XC}[n]$, we borrow the fitting formula of Pade type, due to Tanaka, Mitake and Ichimaru (TMI)[4], inclusive of both exchange and correlation parts, which was derived with reference to the local field correction due to Singwi, Tosi, Land and Sjölander[5]. Note that $\mathcal{E}_{XC}[n_b]$ can be obtained from TMI's $\mathcal{F}_{XC}[n_f]$ by letting $T \to 0$. Instead of reproducing their expression here, we invite the reader to refer to their article.

Boundary Conditions and Numerical Scheme

A set of equations (10) and (11) constitutes the eigenvalue problem, in which the boundary conditions play a crucial role. In the present TTM, it is necessary to take into account additional boundary conditions at the interface ($r = r_c$), aside from those usually imposed at the atomic surface ($r = R$) in the uniform temperature model. Since we have two TFDW equations and have to determine two radii, we should impose *ten* boundary conditions as a whole. Let us enumerate them below. At the origin, we impose

$$\phi_<(0) = 1 \tag{15}$$

and
$$n'_b(0) = -18Z n_b(0) \tag{16}$$

which was derived from the quantum-mechanical argument by Steiner[6]. We use the Tu's approximate expression for $n_b(0)$[7] as a clever initial guess of the iterative calculus. At the interface

$$n'_b(r_c) = n'_f(r_c) = 0 \ , \tag{17}$$

already given in the preceding section, and

$$\phi_<(r_c) - \phi_>(r_c) = \frac{r_c}{Z}(\mu_b - \mu_f) \tag{18}$$

$$\phi_<(r_c) - r_c \phi'_<(r_c) = \phi_>(r_c) - r_c \phi'_>(r_c) = 1 - \frac{Q_b}{Z} \tag{19a,b}$$

At the atomic surface,

$$n_f(R) = \bar{n} \quad \text{and} \quad n'_f(R) = 0 \tag{20}$$

and
$$\mathcal{U}_>(R) = 0 \tag{21}$$

where $\bar{n}$ is the electron density of a surrounding plasma. Equation (21) holds only when we deal with a neutral atom. Because of a discontinuity of the temperature at the interface, there is a jump in μ, or what is the same, in ϕ. This does not mean a flow of electrons across the interface, if we assume that the ion core is impenetrable. The above jump can be eliminated, upon introduction of a new screening function $w(r)$ defined as

$$\phi_<(r) = w(r) \quad \text{and} \quad \phi_>(r) = w(r) - (\mu_b - \mu_f)\frac{r}{Z} \ , \tag{22}$$

which is continuous across the interface. Equations (10) and (11) were solved with the aid of the finite difference method, by converting them into a set of algebraic equations. Details of the method was already explained in our previous work[8], so that we skip them here.

Results and Discussion

In numerical analysis, the external parameters are the nuclear charge Z, the electron density $\bar{n}$ and the temperature T of a surrounding plasma. Given $\bar{n}$ and T, μ_f is determined from inversion of the relation: $I_{1/2}(\beta\mu_f) = \pi^2 \beta^{3/2} \bar{n}/\sqrt{2}$. We chose $Z = 20$(Ca) in conjunction with our previous work[3]. In Table 1 we

Table 1 Results for various quantities, characterizing the TTM of Ca atom. In the Table, $(-n)$ stands for 10^{-n}.

T(eV)	$\bar{n}$	r_c	R	$n(0)$	Q_b	μ_b	$(\Delta\mu)_c$
100	0.1	0.7736	2.739	5.279	9.131	-3.643	0.1120
	1.0	0.7905	1.313	5.284	11.85	8.324	$2.872(-2)$
	5.0	0.7049	0.8169	5.297	15.90	27.90	$4.613(-3)$

$n_b(r_c)$	$\phi_<(r_c)$	$(\Delta\phi)_c$
0.4523	$-1.997(-2)$	$-4.333(-3)$
1.341	0.1351	$-1.135(-3)$
5.046	0.4140	$-1.630(-4)$

T(eV)	$\bar{n}$	r_c	R	$n(0)$	Q_b	μ_b	$(\Delta\mu)_c$
200	0.1	0.4965	3.049	5.267	5.399	-12.89	0.1339
	1.0	0.5353	1.377	5.283	7.559	7.614	0.0403
	5.0	0.5439	0.8348	5.298	10.68	34.31	0.0130

$n_b(r_c)$	$\phi_<(r_c)$	$(\Delta\phi)_c$
0.8827	$-6.891(-2)$	$-3.325(-3)$
2.341	$9.774(-2)$	$-1.080(-3)$
5.898	0.3232	$-3.500(-4)$

tabulate the two radii r_s, R, the electron density at the origin $n(0)$ in units of 10^4, the total number of electrons inside the ion core Q_b, the chemical potential μ_b of the ion core, $(\Delta\mu)_c [= \mu_b(r_c) - \mu_f(r_c)]$, the electron density and the screening function at the interface, $n_b(r_c)$ and $\phi_<(r_c)$, and $(\Delta\phi)_c [= \phi_<(r_c) - \phi_>(r_c)]$. Since we dealt with a neutral Ca atom, the total number of electrons in the clad gas can be evaluated as $Z - Q_b$. From the table we can observe several reasonable results: (1) R increases when T rises. This may be attributable to an increase of the exchange-correlation potential (which is negative) and to a reduction of the effect of gradient correction. These affect a dilute part of the electron density profile. (2) r_c decreases with increasing T. This means that more "bound" electrons are stripped out of the ion core. When we fix T and vary $\bar{n}$, the following properties should be mentioned: (3) R decreases with increasing $\bar{n}$, but the ratio r_c/R increases. This is explained by the effect of compression by a surrounding plasma which affects mainly the clad electron gas.

Due to a lack of space, we could not present several figures which will be published in a separate paper.

Conclusion

So far, we constructed the TTM from a purely intuitive reasoning that deep-lying atomic electrons should not be affected by the thermal effect of a surrounding hot, dense plasma. The point is to know how long the ion core can be held out of thermal equilibrium with the clad electron gas. Though the TFDW model can not answer this question, we may ask whether a duration of the stationary state be governed by a life time of low-lying excited states[9], energy levels of which lying within $k_B T$ from the bottom of the continuum.

To ensure the accuracy of the present numerical analysis, our TTM code was so devised as to be capable of analyzing the one (uniform)-temperature model, worked out a decade ago with the use of the Runge-Kutta-Gill method[3]. A typical quantity for comparison is $n(0)$. A discrepancy of results obtained by the two methods was small, with digression of less than 0.01%. We showed that there exist a consistent solution to a set of the coupled TFDW and Poisson equations for the TTM, but could not make sure yet how many boundary conditions are necessary and sufficient to obtain a self-consistent solution. A complete DFT analysis of the TTM is in progress.

References

1. Perrot, F., private communication (1994)
2. Perrot, F., Phys. Rev. A20, 586-594 (1979)
3. Furutani, Y., Shigesada, M., and Totsuji, H., J. Phys. Soc. Jpn. 55, 2653-2670 (1986)
4. Tanaka, S., Mitake, S., and Ichimaru, S., Phys. Rev. A32, 1896-1899 (1985)
5. Singwi, K.S., Tosi, M.P., Land, R.H., and Sjölander, A., Phys. Rev. 176, 589-599 (1968)
6. Steiner, E., J. Chem. Phys. 39, 2365-2366 (1963)
7. Tu, K., J. Math. Phys. 32, 2250-2253 (1991)
8. Furutani, Y., Ohashi, K., Shimizu, M., and Fukuyama, A., J. Phys. Soc. Jpn. 62, 3413-3424 (1993)
9. Nara, S., private communication (1995)

Photoionization in laser-produced hot dense plasmas

H. Furukawa
Institute for Laser Technology
Suita, Osaka, 565, Japan

Abstract The cross-sections and the rate coefficients of photoionization in hot dense plasmas are estimated quantitatively by using the atomic model including electron-electron, electron-ion and ion-ion correlations [H. Furukawa and K. Nishihara, Phys. Rev. A 46 (1992) 6596]. And they are compared with those obtained by other models. The rate coefficient of photoionization obtained by the present calculation is roughly 400 times of one of hydrogenlike atom approximation in the case of the plasma parameters of atomic number Z=4, ion number density $n_i = 5 \times 10^{22} \text{cm}^{-3}$ and plasma temperature T=50eV at the photon temperature T_{ph}=10eV.

INTRODUCTION

The investigation of atomic processes in hot dense plasmas is very interesting and important for the success of inertial confinement fusion. Particle correlation effects greatly affect on atomic processes in hot dense plasmas. The author developed an atomic model including particle correlation effects exactly within the framework of quantal hypernetted-chain (QHNC) approximation [1] in the previous work [2]. In this paper, the cross-sections and the rate coefficients of photoionization in hot dense plasmas are described. The present calculations are compared with those by other models in order to clear the particle correlation effects on those.

ATOMIC MODELS

In order to estimate the effects of particle correlations on photoionization in hot dense plasmas quantitatively, and to clear which effect is dominant on photoionization in hot dense plasmas, I show six types of atomic models and compare them with the results obtained by each model.

First I show the full correlation atomic model indicated "Full" in this paper, second the TCP quantal finite-temperature jellium model indicated "QJL" in this paper, third the no ion-ion correlation model indicated "NOIC" in this paper, fourth the jellium vacancy model indicated "JVC" in this paper , fifth the jellium vacancy Thomas-Fermi model indicated "JTF" in this paper, sixth the ion sphere model indicated "ISPM" in this paper.

Full correlation atomic model [2]

First, I show a full correlation atomic model "Full" presented in the previous work [2]. The word "full correlation" stands all of electron-electron, electron-ion and ion-ion correlations. The effective potentials, $V_{fe-fe}(r)$, $V_{e-i}(r)$ and $V_{i-i}(r)$, which determine the particle distribution functions are calculated through the Ornstein-Zernike (OZ) relation within a framework of QHNC [1].

Note that the suffix fe means the free electron, e means the total (free and bound) electron and i means the ion. They are given by the Eqs. (2.1)-(2.3) in ref. 2.

The free electron number density and the bound electron number density around the test ion $n_{e-i}(r)$ is obtained by solving the Schrödinger equation with the electron-ion effective potential $V_{e-i}(r)$ [2].

TCP quantal finite-temperature jellium model [2]

When I replace the direct correlation function between the free electrons to those by jellium model in the present model, the OZ relation can be reduced as the same as that in "TCP (two component plasma) quantal finite-temperature jellium (QJL) model" defined in ref. [2]. This approximation means that the electron-ion and the ion-ion correlation functions are solved self-consistently in this model. The correlation functions between the free electrons are given by jellium model.

No ion-ion correlation model

When I add one assumption, ions are positive charge which is distributed uniformly at the background, to TCP quantal finite-temperature jellium model, the model is called "No ion-ion correlation (NOIC) model" in this paper.

Jellium vacancy model

When I add one assumption, ions are nothing in the ion sphere and exist out of the ion sphere uniformly to TCP quantal finite-temperature jellium model, the model is called "Jellium vacancy (JVC) model" in this paper.

Jellium vacancy Thomas-Fermi model

In this model, the free electron distribution function around the test ion is calculated by Thomas-Fermi approximation [%]. In the jellium vacancy model, when I replace $c_{fe-fe}(k)$ to $-4\pi/k^2$ and $c_{fe-i}(k)$ to $4\pi Z^*/k^2$, the electron-ion effective potential V_{e-i} is related to the charge distribution functions through the Poisson equation. This model is called "Jellium vacancy Thomas-Fermi (JTF) model" in this paper.

Ion sphere model [3]

As another model, the ion sphere model exits. In this model, there is an nuclear charge Ze at the center of the ion sphere, radius a and the electrons of the number of Z of which distribute in the ion sphere. The effective potential V_{e-i} is calculated through the Poisson equation with the boundary condition at r=0 and r=a. The free electron distribution function around the test ion is calculated by Thomas-Fermi approximation. Note that this model is the same as one in ref. 3 except the exchange correlation potential within the non relativistic regime. This model doesn't include the relativistic effects because they are not so impotent in the case of the low Z.

FORMATION OF THE CROSS-SECTION AND
THE RATE COEFFICIENT OF PHOTOIONIZATION [4]

The cross-section for the process $\varepsilon_{n\ell} + h\nu \to \varepsilon_{\alpha\ell}$ is

$$\sigma(\nu) = 2\pi^2 \frac{e^2}{\hbar c} \frac{df_{n\ell,\alpha}}{d\varepsilon_\alpha} a_B^2 \qquad (1)$$

where a_B is the Bohr radius and $\hbar$ is the Plank constant divided by 2π. In the average atom approximation, the oscillator strength density can be written as [4]

$$\frac{df_{n\ell,\alpha}}{d\varepsilon_\alpha} = \frac{2}{3} h\nu N_{n\ell} \times B_\alpha \times \left\{ |D(n\,\ell, \alpha\,\ell+1)|^2 \frac{\ell+1}{2\ell+1} \frac{dn^{\ell+1}}{d\varepsilon_\alpha} + |D(n\,\ell, \alpha\,\ell-1)|^2 \frac{\ell}{2\ell+1} \frac{dn^{\ell-1}}{d\varepsilon_\alpha} \right\} \qquad (2)$$

where $D(n\ell,\alpha\ell')$ stands for the dipole integral

$$D(n\,\ell, \alpha\,\ell') = \int_0^{R_0} \chi_{n\ell}(r)\, \chi_{\alpha\ell'}(r)\, r\, dr \qquad (3)$$

and $\chi_{\alpha\ell'}(r)$ is the radial part of the ℓ'-th partial wave normalized over the radius R_0. Note that the orthgonality of wavefunctions of bound states and free states is satisfied by Schmidt method. $N_{n\ell}$ stands for the number of electrons in the shell of principal and angular momentum quantum numbers $n\ell$. B_α is regarded as the availability factor [4]. The term $dn/d\varepsilon$ is the number density of states obtained by Using the Bohr-Sommerfeld quantization rule.

Note in Jellium vacancy Thomas-Fermi model "JTF" and ion sphere model "ISPM". The radius R_0 cannot be determined selfconsistently. I set the radius $R_0=10 a_e$ and $R_0=a$ to calculate the photoionization cross-section within the framework of "JTF" and $R_0=a$ to calculate the photoionization cross-section within the framework of "ISPM". Where a_e is the electron sphere radius. The indicator "JTF" in the next section means that I set $R_0=10 a_e$ to calculate the photoionization cross-section within the framework of "JTF". While the indicator "JTFa" in the next section means that I set $R_0=a$ to calculate the photoionization cross-section within the framework of "JTF".

In the case of hydrogenlike atom approximation "H-like" [5], that is written as

$$\sigma(\nu) = \frac{8\pi}{3\sqrt{3}} \frac{Z^4}{n^5} \frac{e^2}{\hbar c} \frac{g_{fb}}{(h\nu)^3} a_B^2 \qquad (4)$$

where g_{fb} is Gaunt factor and I set $g_{fb}=0.93$.

The rate coefficients are related to the cross-sections as

$$A(T_{ph}) = \int U(\nu) \frac{c}{h} \frac{\sigma(\nu)}{N_{n,\ell}} \frac{d\nu}{\nu} \qquad (5)$$

$U(\nu)$ is the distribution function of photon in cgs unit and this is assumed to be a Plankian in this paper.

RESULTS AND DISCUSSIONS

I perform the calculations of photoionization cross-sections using the models mentioned in previous sections for 3 cases. In case 1, the plasma parameters are $Z=3$, $n_i=5\times10^{22}\text{cm}^{-3}$ and $T=50\text{eV}$, in case 2, $Z=4$, $n_i=5\times10^{22}\text{cm}^{-3}$ and $T=50\text{eV}$, in case 3, $Z=3$, $n_i=2\times10^{23}\text{cm}^{-3}$ and $T=50\text{eV}$. In this paper, I describe detail for the case 2.

In Fig. 1, the photoionization cross-sections for the case 2 are shown. In Fig. 1 (a), those by "Full", "ISPM" and "H-like" are shown with the range of photon energy $h\nu=0$–30 with $\Delta h\nu=0.05$. As shown in Fig. 1 (a), each threshold of those by "Full" and "ISPM" is smaller than that by "H-like" because of the continuum lowering effects. The peak values of those by "Full" and "ISPM" exist near the thresholds in Fig. 1 (a). At the range of $h\nu>15$, the cross-sections by "Full" and "ISPM" reach to the finite values in spite of that by "H-like" reduces to zero.

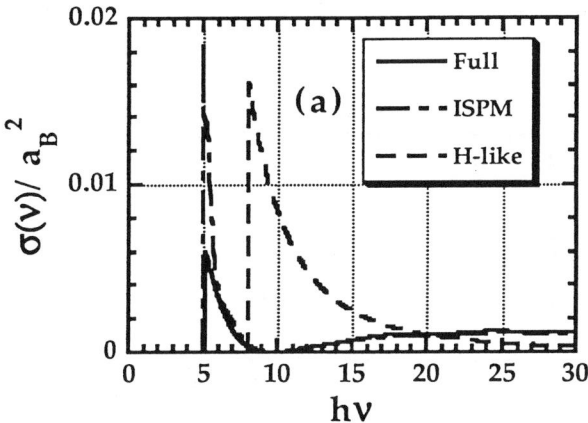

Figure1 (a) The photoionization cross-sections for the case 2 by "Full", "ISPM" and "H-like" with the range of photon energy hv=0-30 with Δhv=0.05.The solid line shows those by "Full", the solid-dotted line shows those by "ISPM"and dashed line shows those by "H-like". The horizontal axis represents the photon energy.

Fig. 1 (b) shows the photoionization cross-sections with the range of $h\nu=4.7$-5.7 with $\Delta h\nu=0.002$. One can see that the photoionization cross-sections by "NOIC" has the sharp peak in Fig. 1 (b). As shown in Figs. 1 (b), ion-ion correlation effects are dominant on the photoionization cross-section. And it is shown that the photoionization cross-sections have the fine structure as a function of phton energy.

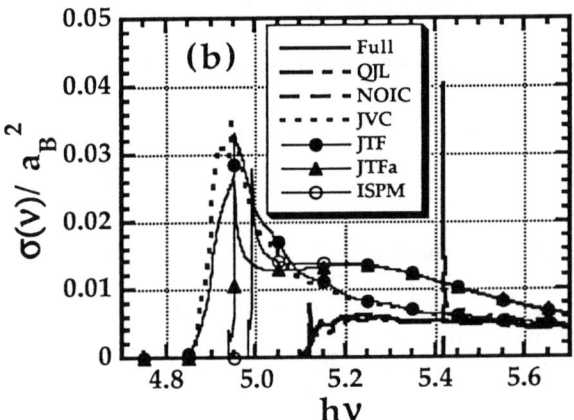

Figure 1 (b) The photoionization cross-sections with the range of hv=4.7-5.7 with Δhv=0.002.

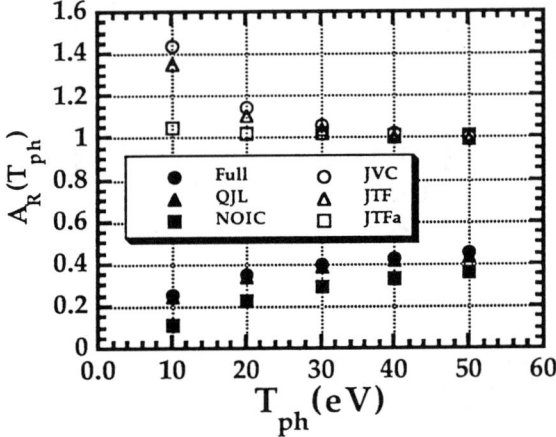

Figure 2 The ratio of the rate coefficients of photoionization by each model to those by "ISPM" as a function of the photon temperature T_{ph} for the case 2.

The rate coefficients of photoionization are estimated as a function of the photon temperature T_{ph}. Fig. 2 shows the ratio of the rate coefficients of photoionization by any model to those by "ISPM" as a function of the photon temperature defined as follows. As shown in Fig. 2, the differences of the values A_R reflect the differences of the treatment of ion-ion correlation effects. The differences of the normalization conditions of wave functions of free states also affect on the rate coefficients at $T_{ph}=10eV$. The accuracy of the estimation of the electronic states of an atom in hot dense plasmas greatly affects on the rate coefficients of photoionization. Fig. 3 shows the rate coefficients of photoionization by "Full", "ISPM" and "H-like" in cgs units. As shown in Fig. 3, the rate coefficient by "Full" is roughly 400 times of that by "H-like" at $T_{ph}=10eV$ and roughly 0.5 times of that by "H-like" at $T_{ph}=50eV$. As mentioned above, particle correlation effects greatly affect on the rate coefficients of photoionization.

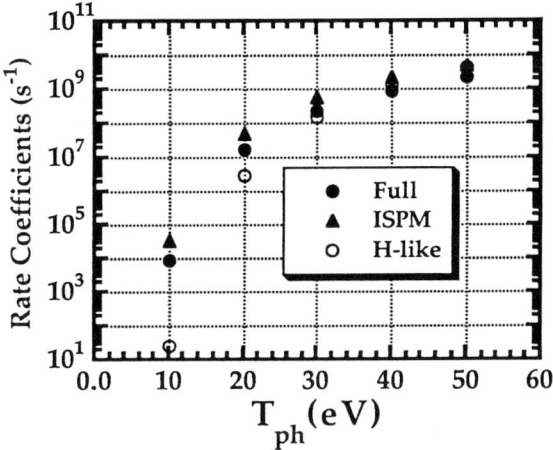

Figure 3 The rate coefficients of photoionization by "Full", "ISPM" and "H-like" in cgs units for the case 2.

REFERENCES

1. J. Chihara, Prog. Theor. Phys. 72 (1984) 940 and therein.
2. H. Furukawa and K. Nishihara, Phys. Rev. A46 (1992) 6596.
3. Balazs F. Rozsnyai, Phys. Rev. A5 (1972) 1137.
4. Balazs F. Rozsnyai, J. Quant. Spectrosc. Radiat. Transfer 13 (1973) 1285.
5). Donald H. Menzel and Chaim L. Pekeris :
   Absorption coefficients and hydrogen line intensities,
   The Royal Astronomical Society 96.

Pressure Ionization of Dense Plasmas in Spherical Ion-Cell Model with Spin-Orbit Interactions

Kenichi Ishikawa*, Thomas Blenski†, Hiroyuki Takahashi*, Tetsuo Iguchi*, and Masaharu Nakazawa*

*Department of Quantum Engineering and Systems Science, Faculty of Engineering, University of Tokyo, 7-3-1 Hongo, Bunkyo-ku, Tokyo 113, Japan
†Institut de Génie Atomique, Département de Physique, Ecole Polytechnique Fédérale de Lausanne, CH 1015 Lausanne, Switzerland

Abstract. We study the continuity of pressure of dense plasmas in pressure ionization in case where spin-orbit interactions are taken into account in calculations. Pressure is calculated using a stress-tensor pressure formula in the relativistically-corrected self-consistent field spherical ion-cell model (average-atom model). It appears that calculated pressure and electronic density distribution change continuously in pressure ionization if we take narrow shape resonances into account properly. This observation stresses the need of a coherent description of bound and free electrons. We also compare the results by the stress-tensor pressure formula with those by other pressure formulas. It appears that different pressure formulas give rather discrepant results in some cases.

1. INTRODUCTION

The self-consistent field spherical ion-cell model (1) is often an important part of equation-of-state or opacity calculations of dense plasmas. One of the most typical phenomena encountered in dense plasmas is pressure ionization of bound states. Several authors (1) have studied it and have shown analytically that pressure ionization induces, in principle, no discontinuity in any thermodynamic quantity. It is known that continuum resonances appear accompanying pressure ionization. R. More (1) discussed this problem and stressed the role of narrow shape resonances. T. Blenski and K. Ishikawa (2) calculated pressure with the relativistically corrected spherical ion-cell model and showed that appropriate calculations actually preserved the continuity of pressure. Their model, however, neglected spin-orbit interactions, which may be important for high-Z elements.

In the present study we include spin-orbit interactions in the spherical ion-cell model and calculate plasma pressure with a stress-tensor pressure formula. We report some results of our numerical calculations and discuss the continuity of calculated pressure. We also compare the stress-tensor pressure formula with other pressure formulas.

© 1996 American Institute of Physics

2. SPHERICAL ION-CELL MODEL

We calculate the electron density by summing over bound and free states,

$$n(\mathbf{r}) = \sum_i |\psi_i(\mathbf{r})|^2 f(\varepsilon_i, \mu), \quad f(\varepsilon, \mu) = \left[\exp\{(\varepsilon - \mu)/T\} + 1\right]^{-1}. \tag{2.1}$$

Wave functions $\psi_i(\mathbf{r}) = (y_{el}/r) Y_{lm}(\theta, \varphi)$ are calculated from the radial Schrödinger equation:

$$\left[-\frac{\hbar^2}{2m}\left(\frac{d^2}{dr^2} - \frac{l(l+1)}{r^2}\right) - eV_{sc} + V_{mv} + V_{Darwin} + V_{so}\right] y(r) = \varepsilon y(r). \tag{2.2}$$

The self-consistent potential $-eV_{sc}(\mathbf{r})$ is the sum of the exchange-correlation potential (3) and the electrostatic potential determined through the Poisson equation with the boundary condition that it smoothly tends to zero at the radius $r_0 = [3/(4\pi\rho)]^{1/3}$ of the Wigner-Seitz sphere, where ρ is number density of atoms. Equation (2.2) has three terms for relativistic corrections [Pauli theory(4)] i.e. the mass-velocity term V_{mv}, the Darwin term V_{Darwin}, and the spin-orbit term V_{so} (2):

$$V_{mv} = -\frac{(\varepsilon + eV_{sc})^2}{2E_0}, \quad V_{so} = -\frac{\hbar^2}{2m} \frac{1}{2E_0 + \varepsilon + eV_{sc}} e \frac{dV_{sc}}{dr} \frac{X}{r},$$

$$V_{Darwin} = \frac{\hbar^2}{2m} \frac{1}{2E_0 + \varepsilon + eV_{sc}} e \frac{dV_{sc}}{dr} \left(\frac{d}{dr} - \frac{1}{r}\right), \tag{2.3}$$

where $X (= l, -l-1)$ is an eigenvalue of the operator $2\mathbf{L} \cdot \mathbf{s}$, and E_0 is the rest-mass energy of an electron. We neglect the Darwin term for $l \geq 1$, where its effect is small. We retain, however, the spin-orbit term. The chemical potential μ is obtained by the neutrality condition of the ion-sphere. The computer program based on this model in the present study is originally written by one of the authors (T. B.) at EPFL.

3. PRESSURE FORMULAS

Pressure formulas are classified into three categories. The first category of formula is derived thermodynamically, i.e., the volume derivative of the free energy F. If we neglect relativistic corrections, F per ion-sphere is given by

$$F = \sum_{\varepsilon, l} \varepsilon f_{el} - TS - \int_0^{r_0} \left[(1/2)V_{ee} + V_{xc}\right] n(r) 4\pi r^2 dr + F_{xc},$$

$$S = -k_B \sum_{\varepsilon, l} 2(2l+1) \int_0^{r_0} \left[f_{el} \ln f_{el} + (1 - f_{el}) \ln(1 - f_{el})\right] y_{el}^2(r) dr, \tag{3.1}$$

where V_{ee} denotes the electrostatic potential minus the nuclear potential, F_{xc} (3) the exchange-correlation contribution to the free energy, and k_B Boltzmann constant. The second category is derived from the virial theorem and written as follows when we neglect the relativistic corrections:

$$P_{virial} = \frac{\hbar^2}{2m}\left[\sum_{\varepsilon,l}\frac{2(2l+1)}{4\pi}f_{el}\left\{\left(\frac{d}{dr}\frac{y_{el}}{r}\right)^2 - \frac{y_{el}}{r}\frac{d^2}{dr^2}\left(\frac{y_{el}}{r}\right)\right\} - \frac{1}{2r}\frac{dn}{dr}\right]_{r_0} + P_{xc}, (3.2)$$

where P_{xc} (2) is the contribution of the exchange-correlation potential. The third category of formula is obtained from the quantum-mechanical stress tensor:

$$P_{stress} = \frac{\hbar^2}{2m}\sum_s\frac{1}{4\pi}\frac{1}{1+\varepsilon_s/2E_0}f_s\left[\left(\frac{d}{dr}\frac{y_s}{r}\right)^2 - \frac{y_s}{r}\frac{d^2}{dr^2}\left(\frac{y_s}{r}\right) - X_s\frac{y_s^2}{r^4}\right]_{r_0} + P_{xc}.(3.3)$$

This formula contains relativistic corrections. The factor $(1+\varepsilon_s/2E_0)^{-1}$ and the term with X_s should be omitted for non-relativistic calculations. The first category has the disadvantage that it requires numerical differentiation that involves the subtraction of two large numbers.

4. RESULTS AND DISCUSSIONS

4. 1. Continuity of pressure in pressure ionization

Let us take as examples the pressure ionizations of $3p_{3/2}$ and $3p_{1/2}$ bound levels of Ar at $T = 2.72$ eV. In this case we observe the pressure ionization of $3p_{3/2}$ ($X = 1$) level between $\rho = 4.22\rho_0$ and $4.24\rho_0$, and $3p_{1/2}$ ($X = -2$) between $4.28\rho_0$ and $4.30\rho_0$ where ρ_0 denotes the liquid density (1.4 g/cm³).

In Fig. 1 we show pressure vs. density of Ar. We used the stress-tensor pressure formula, Eq. (3.3), with relativistic corrections. In Fig. 1 are also presented the bound and free electron contributions to the total pressure. We see that the bound-electron pressure jumps up discontinuously in pressure ionization of $3p_{3/2}$ level. It comes from the fact that the pressure associated with $3p_{3/2}$ level is significantly negative. When $3p_{3/2}$ level is ionized, the free-electron number increases by 2.4. A narrow resonance appears in the continuum after a pressure ionization. Even when the resonance width is small, the electronic density of states is highly concentrated in the resonance energy region. If we take the resonance into account properly, the sudden increase of free electron number is absorbed by the resonance. The resonance contribution to the total pressure is negative and this manifests itself in the discontinuous decrease in free electron pressure and cancels the abrupt change of bound electron pressure exactly to give a completely continuous total pressure curve as in Fig. 1. Since the resonance is very narrow just after pressure ionization (for instance, its half width is 1.8 meV for

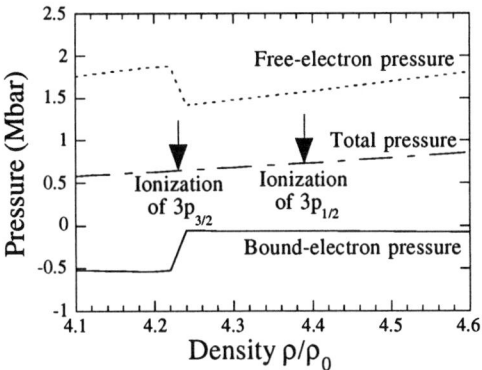

FIGURE 1. The total pressure and bound- and free-electron contributions to it of Ar at $T = 2.72$ eV.

$\rho = 4.24\rho_0$), the mesh of energy-integration should be properly modified in order to evaluate accurately the electronic density of states around the resonance energy.

We see also from Fig. 1 that the bound- and free-electron pressure change continuously in the pressure ionization of $3p_{1/2}$ level unlike in that of $3p_{3/2}$ level. It is because the pressure of a bound level with $X = -l - 1$ approaches zero as its energy eigenvalue approaches zero. This can be proved directly from Eq. (3.3) and the analytic form of wave functions of bound levels at $r \geq r_0$.

Figures 2 and 3 present the electron densities of Ar atom at $T = 2.72$ and $\rho = 4.22\rho_0$ and $\rho = 4.30\rho_0$, respectively, i.e., before and after the pressure ionization of $3p$ levels. We see that before the pressure ionization (Fig. 2) the bound electron density extends rather far outside the Wigner-Seitz sphere. After it (Fig. 3) the bound electron density at r_0 is small. This change of the bound electron

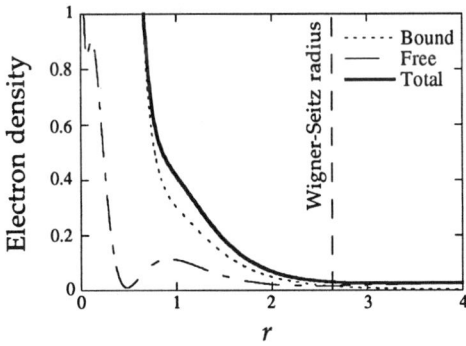

FIGURE 2. The bound, free and total electron densities of Ar atom at $T = 2.72$ eV and $\rho = 4.22\rho_0$. The unit of length is Bohr radius.

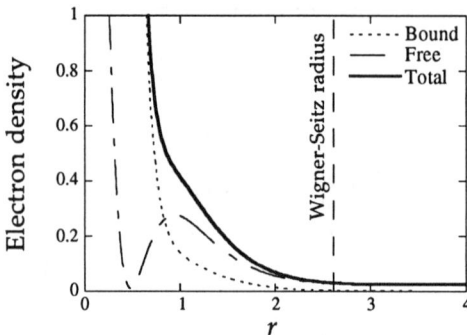

FIGURE 3. The bound, free and total electron densities of Ar atom at $T = 2.72$ eV and $\rho = 4.30\rho_0$. The unit of length is Bohr radius.

density is compensated by the free electron density which includes the effect of the resonance, and the total electron density changes continuously in spite of the pressure ionization.

If we define the degree Z^* of ionization as the free-electron number inside the Wigner-Seitz sphere, it changes discontinuously in pressure ionization. It changes, for instance, from 4.9 to 6.1 when $3p_{1/2}$ level is ionized. Such a naive definition, therefore, comes into doubt when we take account of the above-mentioned continuity of electron density, and Z^* should be related to the total electron density and to the location of electrons. One of such new approaches has been proposed by F. Perrot. (5). We see also from Figs. 2 and 3 that free-electron density is rather high even near the nucleus where bound-electron density is very high.

When we neglect all the relativistic corrections, calculated pressure of Ar at $T = 2.72$ eV and $\rho = 4.60\rho_0$ is 0.900 Mbar. The calculation including relativistic corrections except the spin-orbit term gives 0.856 Mbar, while that including also the spin-orbit term gives 0.854 Mbar. This tells us that the main part of relativistic corrections comes from the mass-velocity term and that the contribution of the spin-orbit term is much smaller. This can be understood by the fact that the term containing X_s in Eq. (3.3) for $X_s = l$ and for $X_s = -l - 1$ nearly cancels. The spin-orbit contribution is, therefore, negligible in equation-of-state calculations.

4. 2. Comparison between Different Pressure Formulas

Three kinds of pressure formulas given in Sec. 3 would give exactly the same result if a plasma model were completely consistent. In our spherical ion-cell model, however, they sometimes give rather discrepant results as we will see.

Equations (3.2) and (3.3) with the relativistic corrections neglected give the same result only when $dn/dr = 0$ at $r = r_0$. Generally it is not the case in the present model as can be seen from Figs. 2 and 3. R. More (1) has shown that the virial theorem gives the same pressure formula as that from the stress tensor if we

TABLE 1. Pressures calculated with the thermodynamic (P_{thermo}), virial theorem (P_{virial}), and stress tensor (P_{stress}) pressure formulas at shown temperatures T. Case A is for Ar at 8.4 g/cm^3 and cases B - E for Zn at 7.5 g/cm^3. Pressures are in Mbar. The numbers in brackets denote multiplicative powers of ten.

	Case A	Case B	Case C	Case D	Case E
T (eV)	2.72	50	100	150	500
P_{thermo}	4.19	35.4	107	176	962
P_{virial}	3.38	35.3	106	198	1.14 [3]
P_{stress}	2.03	35.2	106	198	1.14 [3]

redefine the kinetic energy. This issue, however, may be resolved by the comparison with the results of the thermodynamic way of pressure calculations.

In Table 1 are presented calculated pressures for several cases with the three kinds of pressure formulas. In case A the formula from the virial theorem gives the result very different from that from the stress tensor in spite of the fact that dn/dr looks small at $r = r_0$ in Fig. 3. In cases B - E, on the other hand, the two formulas give almost the same results. Since the difference between the two explicitly depends only on the electronic density gradient, not on electron energy eigenvalue, it does not change much as temperature increases and its contribution becomes relatively less important. It appears that the thermodynamic pressure formula gives results rather different from both the other two formulas in cases A, D and E. This may be because of errors in numerical differentiation or because of the defect of the present model itself. More detailed work on the issue is now going.

5. CONCLUSIONS

We have studied the pressure ionization and the continuity of plasma pressure calculated with the spherical ion-cell model in case where spin-orbit interactions are taken into account in calculations. It appears that pressure-vs.-density curves are continuous and smooth despite pressure ionization if electronic density of states concentrated in narrow resonances are properly evaluated. The correction by spin-orbit interactions in pressure calculations is negligibly small.

We have compared results by the stress-tensor pressure formula with other formulas. At relatively high temperature, it gives almost the same result as the formula obtained from the virial theorem. In some cases, however, discrepancies among results obtained by the three formulas are significant.

REFERENCES

1. More, R. M., *Adv. Atom. Molec. Phys.* **21**, 305—356 (1985), and references therein.
2. Blenski, T., and Ishikawa, K., to be published in *Phys. Rev. E*.
3. Ichimaru, S., Iyetomi, H., and Tanaka, S., *Phys. Rep.* **149**, 92—205 (1987).
4. Bethe, H. A., and Salpeter, E. E., *Quantum Mechanics of One- and Two-Electron Atoms*, Berlin: Springer-Verlag, 1957, secs. 12—13, pp. 54—63.
5. Perrot, F., *Phys. Rev. A* **42**, 4871—4883 (1990).

Time dependent ionization of carbon atom and ions in hot dense plasmas

T. Kato, T. Fujimoto*, T. Kawachi*,
J. Dubau** and U. Safronova***

National Institute for Fusion Science, Nagoya 464-01, Japan
*Department of Engineering Science, Kyoto University, Kyoto 606, Japan
**Observatory of Paris, 92195-Meudon, France
***Institute of Spectroscopy, Troizk, 142092, Russia

Abstract. We have derived the effective ionization and recombination rate coefficients in hot dense plasmas with the use of collisional radiative model of the carbon atom and ions. For this purpose the new atomic data of dielectronic recombination rate coefficients for excited states of C^{+0} and C^{+1} have been calculated. Using the effective ionization and recombination rate coefficients we calculate the time dependent ionization in hot dense plasmas as well as the ionization equilibrium.

INTRODUCTION

In order to understand non - equilibrium ionization in hot dense plasma, it is necessary to include not only ionization and recombination processes between the ground states but also inner - shell ionization and excitation of the lower ionized ion as well as the recombination to the excited states from the higher charged ion. Then we have to solve the rate equations for the populations of all the ions at the same time, in principle. It is necessary to include many highly excited states for the case of dense plasma. We have to solve an enormous set of equations for the level populations. To simplify the situation, as a first step, we tried to derive the effective ionization and recombination rate coefficients as a function of the electron temperature and density. In this paper we consider the effects of the electron collisional process on the ionization and recombination using collisional radiative models of carbon atom and ions.

We apply these effective rate coefficients to the ionization balance in high density plasma.

ATOMIC DATA FOR CARBON ATOM AND IONS

For the ionization rate coefficient $S(C^Z(j) \rightarrow C^{Z+1}(i))$, direct ionization (outer shell, inner shell) and excitation autoionization are included, where $C^Z(j)$ indicates the j-th level of the ion z. For the recombination rate coefficient, $a(C^{Z+1}(i) \rightarrow C^Z(j))$, radiative recombination a_r, dielectronic recombination a_d, and three body recombination a_t are included. For the excitation rate coefficient $R^Z(i,j)$, outer shell excitation and inner shell excitation should be included. In our collisional radiative model, we include the dielectronic recombination rate coefficients from the ground state of each ion to the excited states of the next lower ion. Three body recombinations are included as the inverse process of the ionization from excited states to the excited states.

Dielectronic recombination data poses a problem since this process includes many transitions and it is difficult to estimate the rates to the excited states. For this purpose, we have calculated the dielectronic recombination rate coefficients to excited states of carbon atom and L-shell ions. The dielectronic recombination rate coefficient from the ground state 1 to an excited state nl' is expressed as,

$$a_d(1, nl') = (h^2/2\pi m k T_e)^{3/2} \Sigma_j (g_{z-1}(j, nl)/2g_z(1))$$
$$\times (A_a(j, nl - 1)A_r(j, nl - 1, nl')/(\Sigma A_a + \Sigma A_r)) \exp(-E_s/kT_e), \quad (1)$$

where A_r and A_a are the radiative transition probability and the autoionization rate, respectively. In Eq.(1) $g_{z-1}(j,nl)$ is the statistical weight of the dielectronic state (j,nl) and E_s is the energy of the (j,nl) state measured from the ionization limit of the z-1 ion. Here n is the principal quantum number and l is the angular momentum. The values of A_a and A_r are calculated with AUTOLSJ method by Dubau et al[1] for $C^0(i)$ [2] and with Cowan's program for $C^{+1}(i)$ by Safronova [3]. As an example, $a_d(1,nl')$ for C^{+1} is shown in Fig.1(a). The total dielectronic recombination rate coefficients are calculated as

$$a_d^{tot} = \Sigma_{n,l'} a_d(1, nl'). \quad (2)$$

The characteristic of the dielectronic recombination of L-shell ions is that the n-distributions of $a_d(1,nl')$ spreads over wide n range as shown in Fig.1(b). The recombination rate to the n = 2 levels are only 1% of the total recombination rate. We have taken into account the levels up to n = 500 in order to derive the total dielectronic recombination rate coefficients.

As for the excitation cross sections of carbon atom, the data have been calculated recently by R-matrix method[4]. These data will be compared with the data calculated DW method.

EFFECTIVE IONIZATION AND RECOMBINATION RATE COEFFICIENTS

We have constructed a collisional radiative model (CRM) of carbon atom including the sub-levels up to n = 4[5] and derived the effective ionization and recombination rate coefficients for carbon atom. For C^{+1} and C^{+2} ions, CRM have been made including sub-levels up to n = 3[6] and the effective rate coefficients are obtained. For Li-like[7], He-like[8] and H-like[9] ions CRM have been already constructed.

The definition of the effective ionization and recombination rate coefficients, S_{eff} and a_{eff}, are given as

$$S_{eff} \equiv \Sigma S(i)n(i)/\Sigma n(i) \quad (3)$$

$$a_{eff} \equiv \Sigma_i (A_r(i,1) + R(i,1))n(i) + a_t(1)n_e + a_r(1) + a_d(1) \quad (4)$$

The density dependencies of S_{eff} and a_{eff} are given in Fig.2. The effective ionization rate coefficient increases for increasing density due to ladder-like excitation ionization [10] until the saturation due to collisions as shown in Fig.2(a). The effective recombination rate coefficients decrease very rapidly for increasing density by more than one order of magnitude due to the decrease of the dielectronic recombination for L-shell ions. But at very high densities where the three body recombination is dominant the recombination rate coefficients increase proportionally to the density as shown in Fig.2(b).

IONIZATION EQUILIBRIUM

The ionization equilibrium of different densities using the derived S_{eff} and a_{eff} are calculated. The ionization degree advances when the electron density increases because S_{eff} increases and a_{eff} decreases. But when the electron density exceeds 10^{20} cm^{-3}, recombination through the three body recombination is more effective compared to the ionization as shown in Fig.2, and then the ionization degree goes down towards the high density. The results are shown in Fig. 3 for n_e = 10, 10^{18} and 10^{20} cm^{-3}.

TIME DEPENDENT IONIZATION

We have calculated the time evolution of carbon ions using S_{eff} and a_{eff} for the different densities assuming only the neutral carbon as the initial condition. The ionization advances more effectively at high densities but, as discussed for ionization equilibrium, the ionization does not advance for the densities higher than 10^{20} cm^{-3}. The results are shown in Fig. 4 for $T_e = 5$eV.

DISCUSSION

We have derived the effective ionization and recombination rate coefficients of carbon atom and ions and studied the equilibrium and non equilibrium ionization states in hot dense plasmas. This procedure might be correct only in high density plasmas where most of the levels are near LTE. For intermediate density plasmas where the metastable states are important, this approximation is not correct. Then the time dependent calculation will need a lot of computer time. It would be interesting to find the reasonable approximate way to derive the time dependent ionization for the intermediate density plasmas.

The collisional processes in singly excited states are included in our calculations. But the following processes are not yet included. The dielectronic capture ladder like excitation - ionization increases S_{eff} and the effective excitation rate coefficients, R_{eff} [11]. The l-mixing in dielectronic states by microfield and collisions increases a_d. At very high densities, the atomic structure and the atomic constants are varied such as the lowering of the ionization potential. These effects advance the ionization degree in plasmas. The radiative transport which depends on the plasma condition very much would be also important.

REFERENCES

1. Dubau J, Gabriel A H, Loulergue M, Steeman-Clark L and Volonte S, Mon. Not. R. Astron. Soc. 195, 705(1981)
2. J. Dubau and T. Kato, NIFS-DATA-21 (1994)
3. U. Safronova and T. Kato, submitted to Physica Scripta (1995)
4. K M Dunseath, V M Burke et al, W C Fon, V M Burke,P G Burke et al, JET Order No: JP2/11566 (1993)
5. S. Sasaki, Y. Ohkouchi, S. Takakmura and T. Kato, J. Phys. Soc. Japan, 68, No.8 (1994)
6. T. Kato, in preparation (1994)
7. T. Kawachi and T. Fujimoto, Phys.Rev.A, 51,1428 (1995)
8. T. Fujimoto and T. Kato, Phys. Rev. A 30,379 (1984)
9. N.N.Ljepojevic, R.J.Hutcheon and P. MacWhirter, J. Phys. B, 17,3057(1984)
10. T. Fujimoto, J. Phys.Soc. Jpn, 47,265 and 273(1979), 49,1561 and 1569(1980)
11. T. Fujimoto and T. Kato, Phys. Review Letters, 48, 1022,(1982), Phys. Rev. A 35,3024-3026 (1987), Phys. Rev. A 32, 1663-1668 (1985)

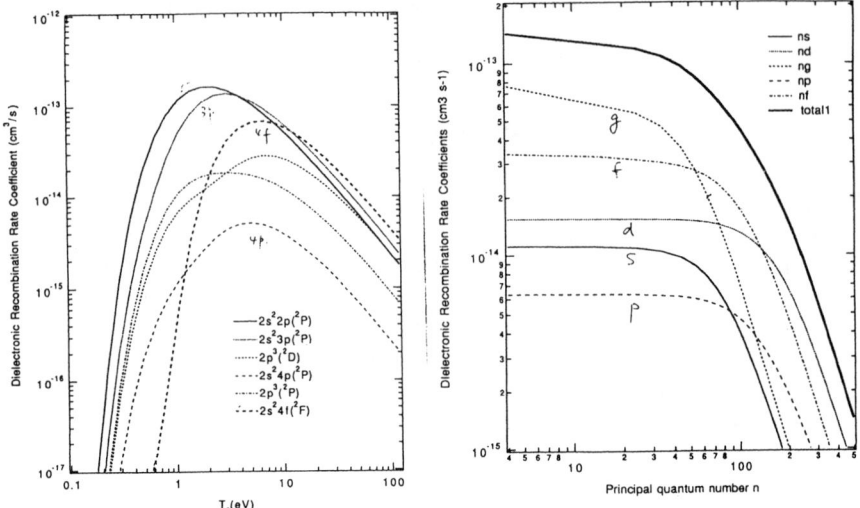

Fig.1
Dielectronic recombination rate coefficients to the excited states of C^{+1} from C^{+2}
(a) as a function of Temperature   (b) n - distribution of $a_d(2s^2 nl)$

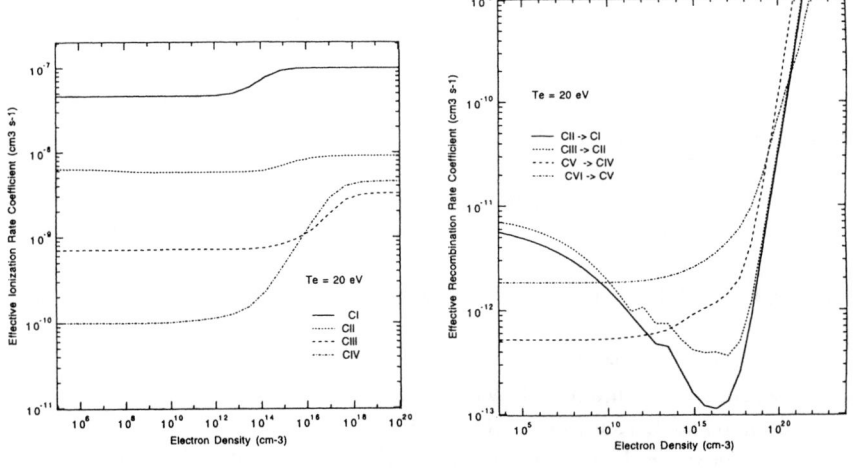

Fig.2 (a)
Effective ionization rate coefficients as a function of electron density

(b)
Effective recombination rate coefficients as a function of electron density

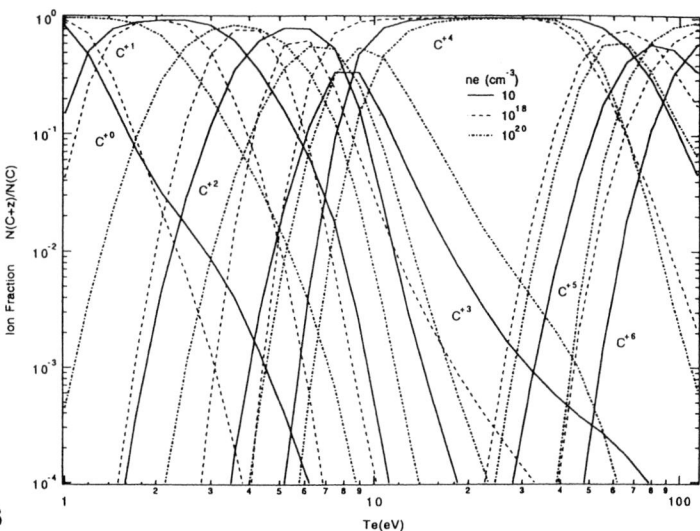

Fig.3
Ionization equilibrium of carbon atom and ions for $n_e = 10$, 10^{18} and 10^{20} cm^{-3}.

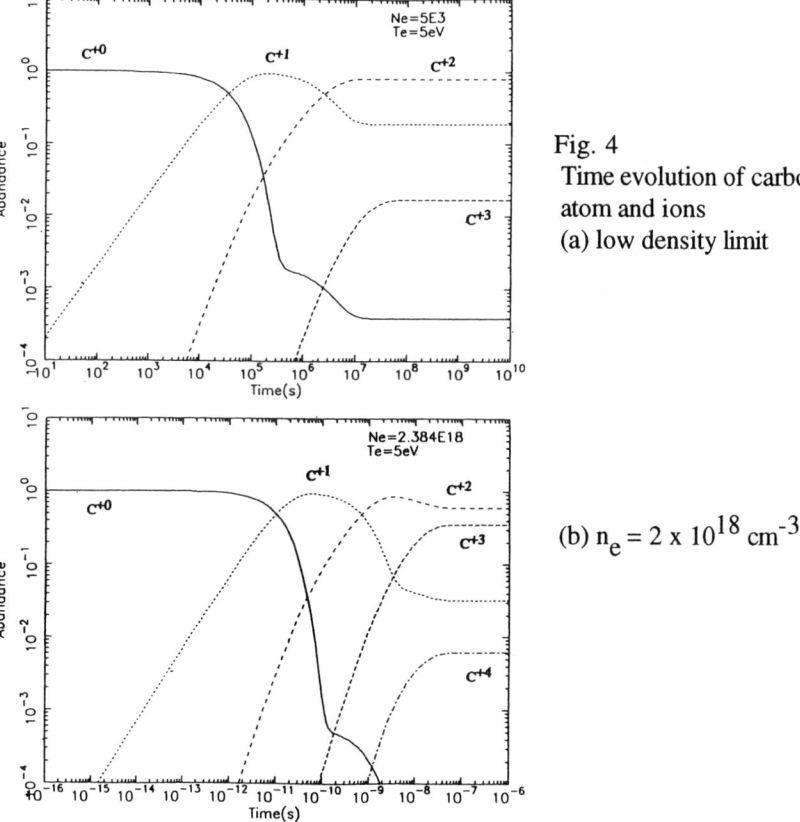

Fig. 4
Time evolution of carbon atom and ions
(a) low density limit

(b) $n_e = 2 \times 10^{18}$ cm^{-3}

A Relativistic WKB Method for Bound States of an Electron in Screened Coulomb Potentials

Keiji KOSAKA*

Department of Physics, Okayama University, Okayama 700, Japan

An overview of a relativistic WKB method for bound states [K. Kosaka and K. Yonei, J. Phys. Soc. Jpn. **60** (1991) 850, 4081] is given. As an application, one-electron states of highly ionized Au atoms are calculated using the effective potential determined by the relativistic and the nonrelativistic TFDW model proposed by Tomishima [Progr. Theor. Phys. **42** (1969) 437] and by Yonei and Tomishima [J. Phys. Soc. Jpn. **20** (1965) 1051], respectively.

I. INTRODUCTION

With the advance of heavy ion physics, revived interest has arisen in the relativistic calculation of atomic structure. In particular, much attention has been paid to the electronic states of highly ionized atoms, which are closely related to atomic processes in plasmas in nuclear fusion reactors.

To exploit a simple method of the relativistic calculation, we have recently developed a relativistic WKB method (RWKB) for bound states of an electron in screened Coulomb potentials [1-3]. The method is based on the idea that the wavefunction can be approximated by a modification of a comparison function. The same idea was used by Goldberg et al. [4] for continuum problems.[1] In ref. [1] (referred to as I hereafter), we gave the formulation for the RWKB method, and in ref. [2] (referred to as II hereafter), we made a detailed examination of the method. Furthermore we applied it to compressed atoms [3].

The method overcomes a difficulty about the eigenvalue condition of the relativistic WKB method proposed by Good [6]. Although the eigenvalue condition given by Good gives the exact eigenvalue for an electron in a pure Coulomb potential, it does not give the eigenvalue specified by certain pairs of quantum numbers ($1s_{1/2}$, $2p_{3/2}$, $3d_{5/2}$, etc.) for an electron in a screened Coulomb potential. In contrast, our eigenvalue condition gives the eigenvalue for any bound state in a screened Coulomb potential. Furthermore, a detailed numerical examination indicates that the method yields a good approximation to both eigenvalues and wavefunctions.

---

*e-mail address: kgk@mp.okayama-u.ac.jp

[1] They also treated the bound state problem independently of us [5].

In this paper, we calculate the atomic structure of highly ionized atoms by applying this method to an effective potential determined by the relativistic Thomas-Fermi-Dirac-Weizäcker (RTFDW) potential [7]. In particular, one-electron states of the Au ions are calculated for a variety of ionization degrees. Nonrelativistic calculation is also carried out for comparison.

II. SUMMARY OF THE RELATIVISTIC WKB APPROXIMATION

Consider an electron moving in a screened Coulomb potential $V(r) = Zs(r)e/r$, whose screening function $s(r)$ is a smooth, continuous, and decreasing function of r satisfying

$$s(r) = \begin{cases} 1 & (r = 0) \\ s_0 = \text{constant} & (r \geq r_0) \end{cases} \qquad (1)$$

r_0 being a certain radial distance specified by the problem. To obtain approximate solutions of the Dirac equation for this potential, we developed the RWKB method in I, which can be summarized as follows.

According to Goldberg and Pratt [8], the radial part of the Dirac equation for this potential can be transformed into a Schrödinger-like equation of the following form:

$$\frac{d^2 y}{dr^2} + k_0^2(r) y = 0 , \qquad (2a)$$

$$k_0^2(r) = p^2 + \frac{2Z\alpha\epsilon}{\lambda_c r} \omega(r) - \frac{\gamma(r)(\gamma(r) - 1)}{r^2} , \qquad (2b)$$

$$p = \frac{\sqrt{\epsilon^2 - 1}}{\lambda_c} , \qquad (2c)$$

where $\alpha = e^2/\hbar c$ and $\lambda_c = \hbar/mc$ are the fine structure constant and the Compton wave length, respectively, while ϵ is the ratio of the energy eigenvalue W to mc^2, i.e. $\epsilon = W/mc^2$. The functions $\omega(r)$ and $\gamma(r)$ are some given functions which are determined from $s(r)$, ϵ, and the quantum number κ defined by $\kappa = \mp(j + 1/2)$ as $j = \ell \pm 1/2$. For further information of these functions, the reader should refer to ref. [8] or ref. [4].

To find solutions of (2a), we assume $y(r)$ to be expressed as a modification of the so-called comparison function $Y(r)$ in the following form:

$$y(r) = A(r) Y(R(r)) . \qquad (3)$$

In the present work, the comparison function is taken to be a solution of the transformed Dirac equation (2a) for the pure Coulomb potential. Then eq. (2a) reduces to the following set of equations for $A(r)$ and $R(r)$:

$$\begin{cases} A''(r)/A(r) + \left[k_0{}^2(r) - R'^2(r)K_0{}^2(R)\right] = 0 , & (4a) \\ A^2(r)R'(r) = \text{constant} , & (4b) \end{cases}$$

where K_0 is k_0 when $s(r) = 1$.

As an approximation, we neglect in eq. (4a) the quantity[2]

$$\left[\frac{A''}{A} + \frac{1}{4r^2} - \frac{R'^2}{4R^2}\right] , \qquad (5)$$

and obtain

$$\varphi(r) \equiv \int_{r_a}^{r} k(\tau)\, d\tau = \int_{R_a}^{R} K(\tau)\, d\tau \equiv \Phi(R(r)) , \qquad (6)$$

where

$$k(r) = \sqrt{k_0{}^2 - \frac{1}{4r^2}} , \quad K(R) = \sqrt{K_0{}^2 - \frac{1}{4R^2}} . \qquad (7)$$

The lower limits of integrals should be chosen such that

$$k(r_a) = 0 \quad \text{and} \quad R_a = R(r_a) . \qquad (8)$$

As discussed in I, R_a necessarily satisfies $K(R_a) = 0$. The equation (6) immediately leads to

$$R(r) = \Phi^{-1}[\varphi(r)] . \qquad (9)$$

Note, however, that this $R(r)$ includes the parameter ϵ which is still undetermined in the bound state problem. The eigenvalue ϵ for a bound state can be determined from the condition

$$I(\epsilon) = \int_{r_1}^{r_2} k(r)\, dr = \pi\left[n - \left|\kappa + \frac{1}{2}\right|\right] , \qquad (10)$$

where r_1 and r_2 $(> r_1)$ are two roots of $k(r) = 0$, whose existence is usually expected for bound states, while n being the principal quantum number. Using the eigenvalue ϵ determined by (10), the solution $R(r)$ now can be calculated according to eq. (9), which in turn determines $A(r)$ via eq. (4b). In this way one can obtain the approximate wavefunction $y(r) = A(r)Y(R(r))$.

We define the effective potential acting on each electron as follows:

---

[2] This quantity can be expressed more compactly as $e^{-2x}\tilde{A}^{-1}d^2\tilde{A}/dx^2$, where $x = \ln r$ and $\tilde{A}(x) = (R/r)^{1/2}A(r)$. The change of variables from (r, A) to $(x, \tilde{A})$ has been done to take the Langer correction into account. See eqs. (2.17)–(2.20) of I.

$$V_{\text{eff}}(r) = \begin{cases} V(r) + \dfrac{4}{3}\kappa_a \rho^{1/3}, & \text{if the right hand side} > \dfrac{(Z-N+1)e}{r}, \\ \dfrac{(Z-N+1)e}{r}, & \text{otherwise}. \end{cases} \quad (11)$$

The electron density ρ and the electrostatic potential V are determined by the RTFDW model proposed by Tomishima [7]. For the nonrelativistic calculation, we use the nonrelativistic counterpart of this [9]. The term $(4/3)\kappa_a \rho^{1/3}$ is the Kohn-Sham exchange potential, where $\kappa_a = 3(3/\pi)^{1/3} e^2/4$.

III. RESULT AND DISCUSSION

TABLE I. One-electron orbital energies for Au ions (with opposite sign and in a.u.). The rest energy is subtracted. N is the electron number, while the nuclear charge $Z = 79$. The italic numbers are the energies of unoccupied states.

Level	N=79	N=59	N=39	N=19
$1s_{1/2}$	2912.40	2955.92	3016.69	3119.61
$2s_{1/2}$	518.702	539.575	585.120	676.312
$2p_{1/2}$	503.784	520.538	565.187	657.008
$2p_{3/2}$	435.103	452.999	497.594	588.562
$3s_{1/2}$	124.004	142.083	184.749	262.028
$3p_{1/2}$	115.544	132.988	175.647	254.063
$3p_{3/2}$	100.501	118.096	160.439	237.093
$3d_{3/2}$	84.5189	102.217	145.108	225.054
$3d_{5/2}$	81.2887	99.0088	141.804	*221.206*
$4s_{1/2}$	28.5424	45.7697	82.6364	*138.581*
$4p_{1/2}$	24.7152	41.7879	78.6369	*135.215*
$4p_{3/2}$	21.0461	38.0604	74.1263	*128.612*
$4d_{3/2}$	14.3041	31.3300	67.5614	*123.879*
$4d_{5/2}$	13.6110	30.6114	*66.6153*	*122.342*
$4f_{5/2}$	4.90081	21.9119	*58.6034*	*118.138*
$4f_{7/2}$	4.75066	21.7463	*58.3385*	*117.575*
$5s_{1/2}$	4.84459	*20.0921*	*47.4857*	*85.5942*
$5p_{1/2}$	3.49053	*18.5207*	*45.5936*	*83.9194*
$5p_{3/2}$	2.77267	*17.4907*	*43.6416*	*80.6756*
$5d_{3/2}$	0.791552	*14.9272*	*40.5820*	*78.3647*
$5d_{5/2}$	0.705466	*14.7341*	*40.1582*	*77.5868*
$6s_{1/2}$	0.370421	*11.6692*	*30.9240*	*58.0694*

Along the line described above, we have calculated one-electron levels of atoms for a variety of ionization degree. To save the space, however, we confine ourselves to showing only a few calculated results as an example. For the other results, we will report elsewhere.

First, we give orbital energies for Au ions in Table I. The data of the table clearly indicate that the orbital energy of each level decreases as the ionization proceeds. This was expected from the outset

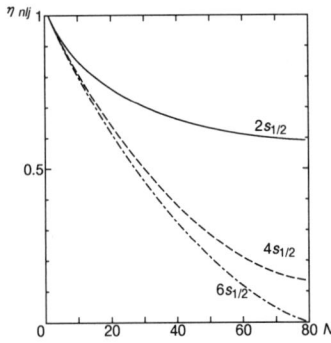

Fig. 1. Plot of $\eta_{nlj}(N)$ against N.

because the ionization diminishes the screening of the nuclear charge due to electrons. To visualize the situation, we give in Fig. 1 plots of N-dependence of the orbital energies. The quantity η_{nlj} means the ratio of the orbital energy $E_{nlj}(N)$ to the corresponding energy of the hydrogen-like ion $E_{nlj}(1)$, i.e., $\eta_{nlj} = E_{nlj}(N)/E_{nlj}(1)$. Here it should be understood that the rest energy mc^2 is subtracted in the expression E_{nlj}. One can observe that for each level η_{nlj} decreases monotonically with the increase of the electron number, reflecting the effect of the screening. The plots also tells us a significant growth of screening effect for the outer electrons.

TABLE II. Comparison between relativistic and nonrelativistic one-electron energies of Au atoms (with opposite sign and in a.u.). As to the relativistic energies, the rest energy is subtracted. Also an average is taken between $j = l \pm 1/2$ sub-states: $E_{nl} = [(l+1)E(n,l,j=l+1/2) + l\, E(n,l,j=l-1/2)]/(2l+1)$. PD means the percentage deviation of relativistic energy from the nonrelativistic: $PD = 100(E_R - E_{NR})/|E_{NR}|$.

Level	$N = 79$			$N = 19$		
	$-E_R$	$-E_{NR}$	PD	$-E_R$	$-E_{NR}$	PD
1s	2912.40	2687.59	−7.72	3119.65	2841.22	−8.93
2s	518.702	449.682	−13.31	676.318	598.195	−11.55
2p	457.997	433.028	−5.45	611.378	582.851	−4.67
3s	124.004	106.569	−14.06	262.030	239.010	−8.79
3p	105.516	98.5722	−6.58	242.750	232.475	−4.23
3d	82.5809	83.3885	0.98	222.745	221.032	−0.77
4s	28.5425	24.0103	−15.88	138.582	129.035	−6.89
4p	22.2692	20.5290	−7.81	130.814	126.217	−3.51
4d	13.8883	14.0537	1.19	122.957	121.691	−1.03
4f	4.81502	5.30762	10.23	117.817	117.384	−0.37
5s	4.84459	3.85699	−20.39	85.5945	80.7573	−5.65
5p	3.01196	2.67901	−11.05	81.7569	79.3321	−2.97

5d	0.739901	0.767475	3.73	77.8980	77.1179	−1.00
6s	0.370421	0.276029	−25.48	58.0696	55.2894	−4.79

In Table II, on the other hand, a comparison is made between relativistic and nonrelativistic orbital energies of Au. This table indicates that in a neutral atom, the relativistic one-electron energy of s- and p-states is lower than that of nonrelativistic, while the energy of d- and f-states is higher than the nonrelativistic. The cause of this can be ascribed to the fact that the effective potential in the relativistic case is different from that in the nonrelativistic case. Note that the peculiar behavior of the energy in d- and f-states disappears in highly ionized atom.

We conclude this report by the following remarks: A detailed comparison shows that the present result is in good agreement with the SCF result, in spite of its very simple procedure [1]. The present method, therefore, provides us with information which is useful to see though essential features of electronic states of an atom under a variety of circumstances in plasmas.

ACKNOWLEDGMENTS

The author would like to express his sincere thanks to Professor K. Yonei for helpful discussions.

REFERENCES

[1] K. Kosaka and K. Yonei: J. Phys. Soc. Jpn. **60**, 850 (1991).
[2] K. Kosaka and K. Yonei: J. Phys. Soc. Jpn. **60**, 4081 (1991).
[3] K. Yonei and K. Kosaka: Rep. Res. Lab. for Surface Science, Okayama University, **7**, 135 (1994)
[4] I. B. Goldberg, J. Stein, Akiva Ron, and R. H. Pratt: Phys. Rev. A **39**, 506 (1989).
[5] I. B. Goldberg, J. Stein, Akiva Ron, and R. H. Pratt: Phys. Rev. A **42**, 2501 (1990).
[6] R. H. Good, Jr.: Phys. Rev. **90**, 131 (1953).
[7] Y. Tomishima: Progr. Theor. Phys. **42**, 437 (1969).
[8] I. B. Goldberg and R. H. Pratt: J. Math Phys. **28**, 1351 (1987).
[9] K. Yonei and Y. Tomishima: J. Phys. Soc. Jpn. **20**, 1051 (1965).

Self-consistent Continuum Lowering Model

K. J. LaGattuta

Los Alamos National Laboratory, MS-B226, Los Alamos, NM 87545, USA

Introduction

The properties of an ion embedded in a background plasma has been a subject of considerable interest, in both atomic physics and plasma physics, for many decades. Although significant progress in understanding has been made, much is still only poorly understood, especially at high density and low temperature, where the microscopic environment changes rapidly as a function of space and slowly as a function of time.

At low to moderate density and high temperature, mean-field theories are rather successful. Under these conditions, plasma dynamics are routinely simulated by state-of-the-art multidimensional radhydro codes which employ some form of atomic model as a driver. Since such a driver is called at every time-step in every hydro zone, it must execute very rapidly. Hence, it must simplify the physics to the maximum extent possible, consistent with a desired accuracy.

The Average-Ion Model (AIM)(1) was created to fulfill this function. It sets up and solves a system of classical rate equations describing the evolution between time-steps of bound ionic level populations for a representative, or average-ion, with fractional positive charge and "a fractional number of electrons in each level."(2) Levels are connected with each other, and with a continuum, through rate coefficients describing bound-bound, bound-free, and free-bound processes. The populations can be " .... computed self-consistently with the free electron density, and the bound-free rates."(2) The effect of the background plasma on binding energies, and on the bound portion of the density-of-states (BDOS), is treated with differing degrees of sophistication within each of the various manifestations of the AIM.

Most forms of the AIM are rudimentary in that only principal quantum numbers (PQN) label bound levels, and the "screened hydrogenic approximation" is invoked; i.e., the one-electron Schrodinger equation is never solved explicitly(3). The effect of the background plasma on bound levels is especially simple. Binding energies are all reduced, relative to the isolated average-ion, by an amount equal to the "continuum lowering parameter", ΔE_{CL}. Typically, the AIM sets

$$\Delta E_{CL} = 3Z^*/2R \tag{1}$$

where Z^* is the average-ion positive charge, and R is the corresponding "ion-sphere radius", $4\pi R^3/3 = 1/n_i$. Here, n_i is equal to the average-ion density, related to the free electron density by $n_e = Z^* n_i$, all in atomic units (a.u.). In this approximation, the energy of level n is

$$E_n = -Z_n^2/2n^2 + \Delta E_{CL} \tag{2}$$

where Z_n is the (hydrogenically) screened positive charge seen by the bound electron. At very low density $(R \to \infty)$ $\Delta E_{CL} \to 0$, as it should for an isolated ion.

The BDOS is often computed from

$$g(n) = 2n^2/(1 + (A\, r_n/R)^3) \tag{3}$$

for a level with $PQN = n$, where $r_n = n^2/Z_n$ is an effective bound orbital radius, and A is a constant(3), chosen to be $A = (40\pi/3)^{1/3}$. Thus, the BDOS is reduced rather slowly, as a function of R, for fixed n, and rather slowly as a function of n, for fixed R. At very low density $(R \to \infty)$ $g(n) \to 2n^2$, as it should for an isolated ion.

It has been suggested(4) that continuum lowering arises primarily from plasma electron screening of the bound level, whereas reductions in the BDOS arise largely from the disruptive influence of neighboring ions(5). Although this point of view may have some intuitive appeal, it is difficult to justify rigorously. Explicit solutions of the time-independent Schrodinger equation (TISE) have been performed(6,7) for an electron moving around a point positive ion, while interacting with a background potential which models seperately the effects of plasma electrons and neighbor ions. In these, and most similar calculations, it appears to be impossible to distinguish those elements of the model potential which are most effective at inducing changes in ΔE_{CL} from those which are most effective at inducing changes in the BDOS. These two changes appear to be inextricably linked.

Having said this, it is worth noting that, whereas eqn 1 has some significant theoretical basis(4), especially at high density, and has been discussed by many authors, eqn 3 is not on so firm a footing. The origin of eqn 3 may be traced to a single work(5), where it seems to have arisen by fitting limited data to an arbitrary functional form. Recent explicit solutions of the TISE have shown that this functional form is generally inadequate(6,7). Changes in the BDOS with R and n have been found to be much more abrupt than are allowed for by eqn 3. However, in this context, the influence of long-lived resonances on the behavior of actual plasma systems may be of some relevance(8).

It is also now generally recognized that eqns 1 and 2 lead to an unnecessarily crude estimate of the binding energy, even at high density. For instance, they imply that changes in binding energy due to the background plasma are independent of n, a notion that is inconsistent with explicit solutions of the TISE, and with current ideas concerning the "plasma polarization shift" (4).

Indications of inadequacies in eqn 1 have existed for some time. For example, it has been found necessary, for some applications, to multiply the right-hand-side of eqn 1 by a factor ranging from 1 to 1/4, in order to obtain acceptable results. The necessity for this multiplier has been traced to the screened hydrogenic approximation, which often leads to under-binding of energy levels(9). However, it may also be that the functional form of eqn 1 is too limiting, and an alternate due to Stewart and Pyatt(10) has been suggested(11).

A New Approach

In has been noted, in the course of AIM-based numerical calculations, that the ion-sphere (IS) form of the CL (eqn 1) can yield implausibly high states of ionization, for moderately dense plasmas. By contrast, the Stewart-Pyatt (SP) form of the CL(8) seems to give a more realistic result. This difference is simply due to the fact that, generally, $\Delta E_{CL}^{(SP)} << \Delta E_{CL}^{(IS)}$, except in the very high density, and/or very low temperature limits. The IS form of the CL being somewhat simpler, and perhaps also being somewhat more palatable on physical grounds, we have suggested the following simple fixup: The IS form of the CL will be retained in kinetics modeling, with no prefactors multiplying it. However, the Boltzmann factor multiplying the BDOS will not be allowed to vanish abruptly as the total energy of a level turns positive. Rather, this factor will persist, decreasing continuously as the CL increases (albeit decreasing at an exponential rate, if the temperature is low). We refer to this approach as the "smoothed" ion-sphere (SIS).

This change in the ususal formulation has two effects. First, discontinuities in predicted ionization balance, and other equilibrium properties, disappear. In fact, such discontinuities are unphysical. Second, predicted ionization balance shifts toward a more recombined state. In effect, one is treating states of low positive energy as if they were bound. The boundary between high-lying bound state and low-energy continuum state is thereby blurred. This is probably physically reasonable. Finally, it is perhaps fortuitous that ionization balance predicted according to this new procedure is generally quite similar to that predicted by the SP form of the CL, with no consideration being taken of positive energy states. We refer to this "normal" approach as the NSP. See the following Fig. 1, displaying data taken with an LTE model.

Obtaining Self-Consistent CL and BDOS Values

We have invoked the following procedure, for a range of background plasma temperature, density, and average ion charge Z^*, and determined the bound electronic structure of an embedded impurity ion of nuclear charge Z.

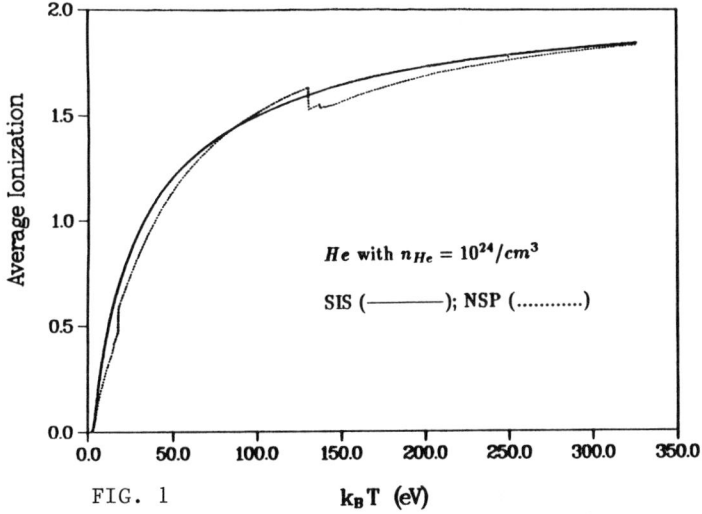

FIG. 1

[1] We first solve the nonlinear Poisson's equation for the total electrostatic potential seen by the background charges in the vicinity of the impurity. In this, we include an estimate of the electronic charge density due to the bound (impurity) component, using unscreened Hartee-Fock wavefunctions.

[2] The electrostatic potential seen by the background charges determines their spatial distribution, in the vicinity of the impurity. With this distribution fixed, we next solve Poisson's equation for the electrostatic potential seen by the bound electrons.

[3] Using the electrostatic potential determined in step 2, we solve the screened Hartree-fock equations for the bound electron wavefunctions and energies. Thus, we find the first approximation to the CL and BDOS, for the impurity ion.

[4] Returning to step 1, the entire procedure is repeated until a converged calculation is obtained. In this way we determine self-consistent values for the CL and BDOS, for a given background plasma temperature, density, and Z^*, for the selected impurity ion.

[5] A new form of the continuum lowering parameter and the BDOS will be developed by fitting our computed results to a simple, yet appropriate, functional form. Preliminary results have already been obtained(6), and new forms for the BDOS have been suggested(5,6), as well as for the continuum lowering parameter(12).

It is conceivable that the effects of resonances(7) are being included in many current AIM calculations by the apparently unphysical form chosen for the BDOS. As already noted, recent solutions of the TISE show that eqn 3 is too smooth, the actual BDOS changing much more abruptly with R and n. However, if the contribution of resonances to the *continuum* density-of-states is included in the BDOS, then this abruptness could be reduced. We will investigate this question further by solving the TISE for the low-lying continua, and computing phase shifts. Of course, it remains to be seen whether resonances play an essential role in any real application. This should be considered to be a research area.

Refinements to this procedure will be accomplished by means of selected comparisons of our results with the predictions of INFERNO(13). The procedure (steps 1-5) has been described earlier by several authors(10,14).

The Poisson equation, for the electrostatic potential $V(r)$, in the neighborhood of an impurity ion of nuclear charge Z, in a background plasma characterized by an average ion charge Z^*, temperature T, and average ion and free electron densities satisfying $<n_e> = Z^* <n_i>$, is

$$\frac{1}{r}\frac{d^2}{dr^2}[rV(r)] = 4\pi[n_e(r) - Z^* n_i(r) + |\psi(r)|^2] \tag{4}$$

The probability density for the bound electron component is $|\psi(r)|^2$.

The spatial dependence of the background ion density is given by

$$n_i(r) = <n_i> e^{(-Z^* V(r)/k_B T)} \tag{5}$$

while the spatial dependence of the free electron density is described by

$$n_e(r) = <n_e> [2(y/\pi)^{1/2} + e^y(1 - erf(\sqrt{y}))] \tag{6}$$

with

$$y \equiv V(r)/k_B \tag{7}$$

all in *a.u.*, as per ref. 14. Examples appear in Figs. 2 and 3.

References

1 B. Roznyai, JQSRT 15, 695 (1975).
2 W. Lokke and W. Grasberger, UCRL-52276 (1977).
3 G. Pollak, LA-UR-90-2423 (1990).
4 J. Albritton and D. Liberman, JQSRT 51, 9 (1994).
5 G. Zimmerman and R. More, JQSRT 23, 517 (1980).
6 D. Liberman and J. Albritton, JQSRT 51, 197 (1994).
7 K. LaGattuta, unpublished data (1994).
8 R. More, in Adv. Atom. and Molec. Phys. 21, 305 (1985).
9 D. Liberman, private communication (1994).
10 J. Stewart and K. Pyatt, Astrophys. J. 144, 1203 (1966).
11 C. Cranfill and G. Pollak, private communication (1994).
12 C. Cranfill and M. Clover, private communication (1994).
13 D. Liberman, JQSRT 27, 335 (1982).
14 S. Skupsky, Phys. Rev. A21, 1316 (1980).

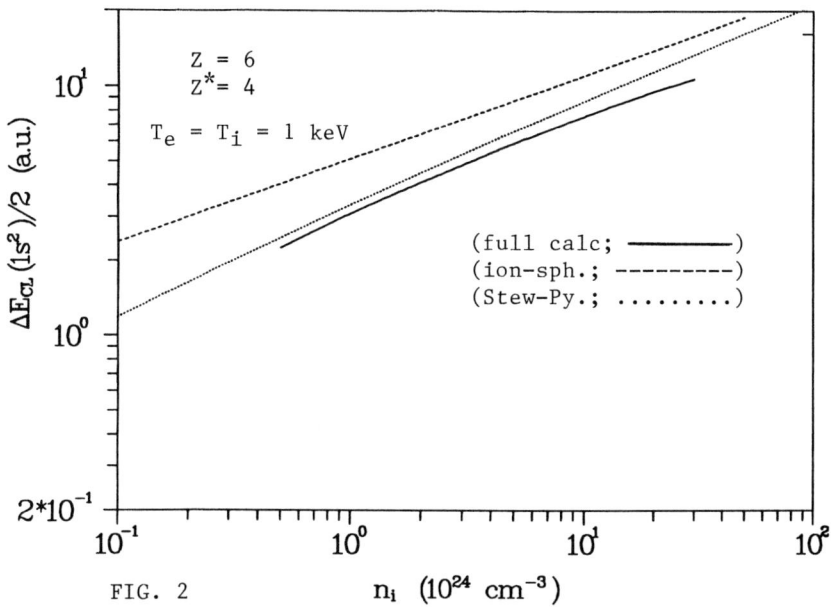

FIG. 2

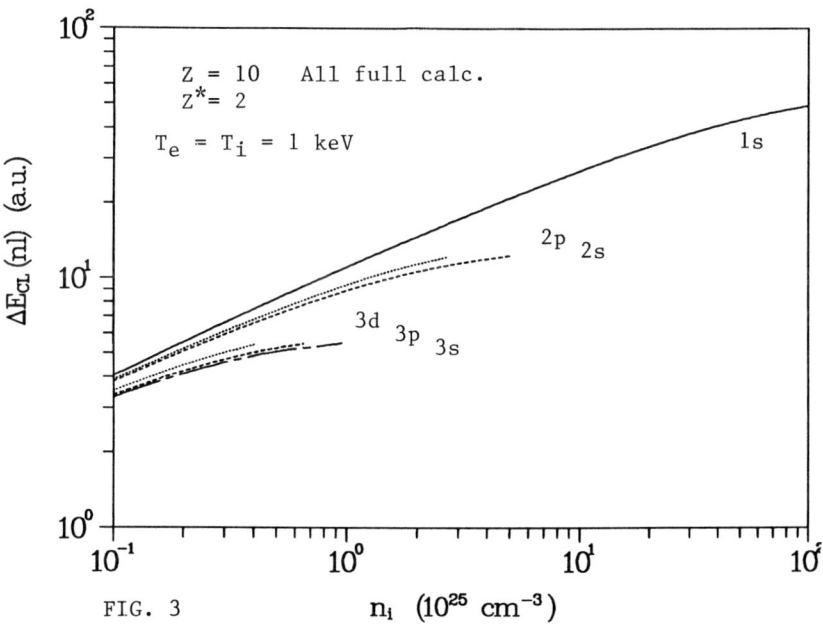

FIG. 3

Improved Screened Hydrogenic Model

Takeshi Nishikawa

Department of Electrical and Electronic Engineering
Okayama University, 3-1-1 Tsushima-naka, Okayama 700, Japan

Abstract. Screened Hydrogenic Model is widely used for energy level calculation in hydrodynamic code of inertial confinement fusion because Screened Hydrogenic Model is simple algebraic calculation. More's Screened Hydrogenic Model and his screening constants are usually used to calculate opacity and equation of state. By the use of his model, energy level can be consistently calculated with ion's total energy. But his model take into account the principal quantum number dependence only and cannot reproduce hydrogenic energy levels. As the precise experiment about opacity measurement is performed, it becomes clear that his model is not enough to use for opacity calculation. In this paper, his model is improved in the framework of Screened Hydrogenic Model. The improved model can reproduce the hydrogenic energy levels and include azimuthal quantum number dependence and the effect from another quantum state (a kind of inner quantum number). Screening constants are fitted by spectroscopic data and sophisticate calculations. By the use of improved model, energy levels are calculated more accurately for low-z ions.

INTRODUCTION

Screened hydrogenic model (SHM) is usually used to calculate spectral opacity of the hydrodynamic code of inertial confinement fusion. To calculate spectral opacity, atomic data of energy level, oscillator strength, line broadening width, and etc. are required for the calculation in the local thermodynamic equilibrium (LTE), and the data of electron collisional processes are required in the non-local thermodynamic equilibrium (NLTE). For the purpose to calculate spectral opacity in the LTE of high quality, atomic data of precise energy levels are required because dipole transition rate is proportional to the square of the transition energy, though the oscillator strength to the power of one. SHM is simple calculation and can easily be extended to highly excited states. The highly excited states are sometimes important for opacity calculations and detailed theoretical atomic code cannot calculate the highly excited states. SHM is developed by Slater and improved by Mayer, More, and etc. But for the purpose to calculate spectral opacity of high quality, their models is not suitable. Therefore, the SHM is improved for opacity calculations in this paper.

SCREENDE HYDROGENIC MODEL (SHM)

Screened hydrogenic model is developed by J.C.Slater in 1930 to calculate wave function of non-hydrogenic ions based upon hydrogenic wave function (1). For the case of energy level calculations, H.Mayer had calculated screening constants from hydrogenic wave function analytically (2). In his article, the screening constant is defined as

$$\sigma_{nm} = \frac{F^0(n,m)}{-\frac{dE_n}{dZ}},$$

and effective charge of n-th shell Z_n is defined as

$$Z_n = Z - \sum_{m \neq n} \sigma_{nm} P_m - \sigma_{nn} P_n \left(1 - \frac{1}{g_n}\right),$$

and effective charge of n-th shell E_n is defined as

$$E_n = -\frac{Z_n^2}{2n^2}.$$

Until just recently, Mayer's screening constants had been used for opacity calculations (3). In 1970's, experiments on high-z materials are performed by high-power laser irradiation and it had become clear that the energy level calculations of SHM by Mayer's screening constants are not so good for partially ionized high-z ions. In 1982, R.M.More had improved the screening constant for calculating energy level of partially ionized high-z ions (4). In his article, Z_n is defined as

$$Z_n = Z - \sum_{m<n} \sigma_{nm} P_m - \frac{1}{2} \sigma_{nn} P_n,$$

and E_n is defined as

$$E_n = -\frac{Z_n^2}{2n^2} + \sum_{m>n} \sigma_{nm} \frac{Z_m}{m^2} P_m + \frac{1}{2} \sigma_{nn} \frac{Z_n}{n^2} P_n.$$

In his model, total energy is conserved in the case of ionization. But, the model cannot reproduce hydrogenic energy level. Marchand et al. (5) had improved the expression of Z_n to reproduce hydrogenic energy level as

$$Z_n = Z - \sum_{m<n} \sigma_{nm} P_m - \frac{1}{2} \sigma_{nn} \max(P_n - 1, 0).$$

There is different improvement by F.Perrot to include the energy level differences between different azimuthal quantum numbers (6).

But the above-mentioned models cannot express the differences between singlet and triplet of He-like ions. Energy level differences between He-like singlet and triplet are large enough and cannot be neglected for opacity calculations In the case of ions with a few electrons, the effect of spin-orbit interaction is not so important. For example, there are small energy differences among energy levels of He-like 3P's three states (J=0, 1, 2) and Li-like 2p's two states (J=1/2, 3/2). The energy differences between singlet and triplet states is from Pauli's exclusion principle. Therefore, if two different states (up or down) of same principal quantum number and azimuthal quantum number is added, Different states of singlet and triplet can be expressed in the framework of the SHM. From the physical point of view, this extension is corresponding with inner quantum number of quantum mechanics. In the model, total energy is given by

$$\langle H \rangle = \sum_k P_k \left(\frac{Z_k^2}{2n^2} - \frac{ZZ_k}{n^2} \right) + \frac{1}{2} \sum_{k \neq k'} P_k P_{k'} \frac{Z_k}{n^2} \sigma_{kk'} + \frac{1}{2} \sum_k \max(P_k - 1, 0) P_k \frac{Z_k}{n^2} \sigma_{kk}.$$

k indicate the states of n, l, $up/down$. First summation is the potential energy derived from hydrogen-like wave function of screened charge Z_k interacting the nuclear charge Z and kinetic energy in the coulomb field of Z_k. Second and third summation is from the inter-electronic interaction. Z_k is defined as

$$Z_k = Z - \sum_{n'l' < nl} \sum_{s'} P_{k'} \sigma_{kk'} - \frac{1}{2} P_{nls'} \sigma_{nlsnls'} - \frac{1}{2} \max(P_k - 1, 0) \sigma_{kk}$$

by the variational principle on P_n's of $<H>$. E_n is defined as

$$E_k = -\frac{Q_k^2}{2n^2} + \sum_{n'l' > nl} \sum_{s'} P_{k'} \frac{Q_{k'}}{n'^2} \sigma_{k'k} + \frac{1}{2} P_{nls'} \frac{Q_{nls'}}{n^2} \sigma_{nls'nls} + \frac{1}{2} \max(P_k - 1, 0) \frac{Q_k}{n^2} \sigma_{kk}.$$

In the model, total energy is conserved and hydrogenic energy level can be reproduced and express the different two states of the same principal quantum number and azimuthal quantum number, singlet and triplet. In the following section, applications to the He-like ions are shown.

APPLICATIONS TO HE-LIKE IONS

To determine screening constants, atomic data of energy level of high quality are needed. I chose Cowan's for He-like ground state (7), Accad et al's. for He atom (8), Martin et al's for Si (9) and P (10), and Corliss and Sugar's for Fe (11) ions.

Fig.1 shows the atomic number dependence of He-like ground state energy level. Horizontal axis is the atomic number to the power of three. From Fig.1, it is clear that the screening constant is proportional to the atomic number to the power of three. Therefore, it is natural that screening constant can be fitted for the equation

$$\sigma_{kk'} = aZ^3 + b$$

Table 1 shows the fitting parameters of the He-like n^1S, n^3S, n^1P, n^3P ($n \leq 5$).

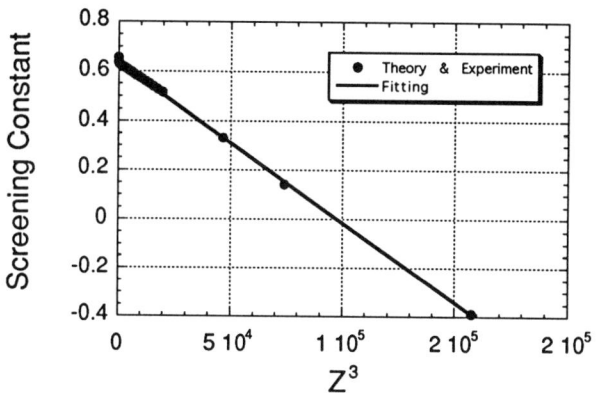

FIGURE 1. Screening constant of He-like ground states.

TABLE 1. Fitting parameter of screening constant

State	A (*10⁻⁶)	B
Ground	-6.5158	0.63603
$2\,^1S$	-7.1769	0.92989
$3\,^1S$	-5.6653	0.95023
$4\,^1S$	-4.6858	0.96273
$5\,^1S$	-3.9309	0.97043
$2\,^1P$	-2.3685	1.0331
$3\,^1P$	-2.4510	1.0188
$4\,^1P$	-2.1944	1.0137
$5\,^1P$	-1.9148	1.0111
$2\,^3S$	-8.1092	0.76297
$3\,^3S$	-6.3118	0.85196
$4\,^3S$	-5.1290	0.89170
$5\,^3S$	-4.2494	0.91456
$2\,^3P$	-3.8270	0.91495
$3\,^3P$	-3.5189	0.94632
$4\,^3P$	-3.0130	0.96135
$5\,^3P$	-2.5760	0.96967

TABLE 2. Energy level of He-like ground states by SHM

Element	$\sigma_{kk'}$	E (eV)	Detail	Error (%)
He	0.6360	25.31	24.59	+3
C	0.6346	391.7	392.1	-2
Al	0.6217	2085	2086	0.05
Fe	0.5215	8832	8828	0.05
U	-4.438	126528	?	?

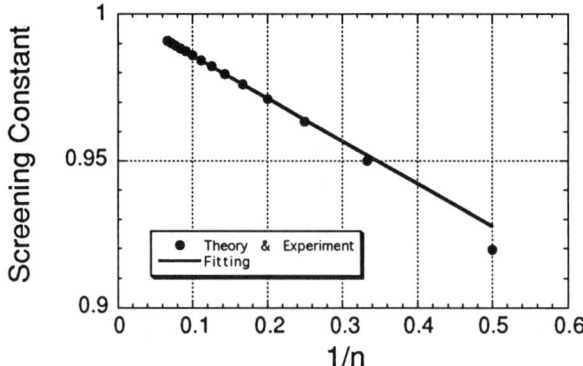

FIGURE 2. Principal quantum number dependence of He n^1S (n≥2).

Table 2 shows the calculated energy level in the case of He-like ground states by the SHM of the fitting parameter of Table 1. For low-z ions like He and C, calculated energy level by SHM have relatively large differences though small differences for high-z ions.

From Table 1, it seems that screening constants to highly excited states in same series have another dependence. Fig. 2 shows the dependence of the screening constants of He n^1S. Horizontal axis is the inverse of the principal quantum number n. It is clear that screening constant to highly excited states are proportional to the inverse of principal quantum number n. Therefore, screening constants can be fitted as

$$\sigma_{kk'} = b + \frac{c}{n}$$

From the two screening constants of He-like single and triplet, Li-like energy level can be calculated. The calculated energy level is not so bad: ~1% for Al ions.

SUMMARY

I have improved SHM for energy level calculation of high quality. By the model, energy level of He-like singlet and triplet can be expressed. The screening constant is proportional to the atomic number to the power of three and proportional to the inverse of principal quantum number n. For low-z ions, simple linear fitting of screening constant has less than 3% errors though less than 0.1% high-z ions. The simple linear fitting is not so good approximation and should be improved.

ACKNOWLEDGMENTS

The author would like to express his thanks to Profs. Y.Furutani and A.Fukuyama of Okayama University for their supports and encouragement.

REFERENCES

1. Slater,J.C., Phys.Rev. **36**, 57-64 (1930).
2. Mayer,H., "*Method of Opacity Calculations*" unpublished report LA-647, Los Alamos Scientific Laboratory, Los Alamos, New Mexico (1947).
3. Tsakiris,G.D., and Eidmann,K., J.Quant.Spectrosc.Radiat.Transfer **38**, 353-368 (1987).
4. More,R.M., J.Quant.Spectrosc.Radiat.Transfer **27**, 345-357 (1982).
5. Marchand,R., Caille,S., and Lee,Y.T., J.Quant.Spectrosc.Radiat.Transfer **43**, 149-154 (1990).
6. Perrot,F., Phys.Scripta **39**, 332-337 (1989).
7. Cowan,R.D., "T*he Theory of Atomic Structure and Spectra*", California, University of California Press, 1981, ch.1, pp.12-15.
8. Accad,Y., Pekeris,C.L., and Schiff,B., Phys.Rev.A **4**, 516-536 (1971).
9. Martin,W.C., and Zalubas,R., J.Phys.Chem.Ref.Data **12**, 323-375 (1983).
10. Martin,W.C., Zalubas,R., and Musgrove,A, J.Phys.Chem.Ref.Data **14**, 751-797 (1985).
11. Corliss,C., and Sugar,K., J.Phys.Chem.Ref.Data **11**, 135-241 (1982).

EXCITATION OF INTENSE SHOCK WAVES BY SOFT X-RADIATION FROM A Z-PINCH PLASMA

Fortov V.E., Dyabilin K., Lebedev M., Vorobiev O.,Yu.

HEDRC RAS, 127412 Moscow, Russia

Grabovskij E., Smirnov V.

TRINITI, 142092, Troizk,Russia

B.Goel

*Forschungszentrum Karlsruhe, INR,
Postfach 3640, D-76021, Karlsruhe, Germany*

Abstract. The paper presents the measurements of the shock waves intensities, generated by soft X-radiation in Al and Pb targets. The soft X-radiation was induced by the dynamic compression and heating of the plasma in the cylindrical Z-pinch geometry in the ANGARA-5-1 installation. 1D computer simulation of shock wave generation were performed with realistic EOS and energy transfer models.

INTRODUCTION.

A fundamental problem in the use of concentrated fluxes of charged particles and laser light in controlled [1] fusion and in the dynamic physics of high energy densities [2] is the substantial spatial nonuniformity of the power which is released. This nonuniformity disrupts the symmetry of the spherical compression of the fusion fuel and hinders the excitation of plane shock waves in the experiments on the behavior of matter under extreme conditions. One of the most effective ways to solve this problem is to use the x-ray emission from a plasma with an approximately thermal spectrum which arises when directed energy fluxes are applied to a target [3] or during the electrodynamic compression of cylindrical shells in a Z-pinch geometry [4]. The plane shock waves excited by this radiation, which is an extremely simple type of self-similar hydrodynamic flow, might be a more nature and highly rich source of experimental information on both the intensity of the incident x radiation and the physics of the interaction of this radiation with condensed targets.

Z-pinch plasma, produced in the big installations by the cylindrical liners electrodynamic compression seems to be one of the most favorable candidate for the source of the such X radiation.

The paper presents the measurements of the shock waves intensities, generated by soft X-radiation in Al and Pb targets. The scheme with the conversion of the laser to soft X-radiation, described in [5,6], is different from that used here: the soft X-radiation was induced by the dynamic compression and heating of the plasma in the cylindrical Z-pinch geometry in the ANGARA-5-1 installation[7,8]. As a result, the radiation pulse duration was about an order of magnitude more, with the power level being nearly the same as in [5,6].

EXPERIMENT AND THEORY

Eight modules of "ANGARA-5-1" pulse power generator allow store energy up to about 2 MJ and produce 1.2 MV, 150 ns halfsinusoidal pulse at 0.25 Ohm matched load. Disk shape magnetic selfinsulated current concentrator is used to collect current on the axis. Its efficiency is about 80%, inductance 25 nH. The regime with the current value I=3-4 MA and dI/dt=10 A/s is available with the plasma erosion opening switch.

For the plasma generation, the internal liner as a low density agar-agar cylinder with the implanted Mo was used (the density is less than 10, cylinder mass about 200 mg). The external liner diameter was 4 mm, wall thickness 0.2 mm and height 10 mm. As the external liner, the supersonic annular Xe jet with mass about 150 mg was used. The value of the pinch current was about 3.5 MA. Z-pinch plasma radiated soft X-r with the typical temperatures about 60-120 eV. This radiation was coming on the planar target which was positioned under the internal liner 1 mm part (Fig.1) [9]. The target was made as 16 mm Al and 80-200 mm Pb plates being connected together. The velocity of the shock wave was defined by the optical base method as the difference between the moments when the shock wave is coming from the inside on the step target free surface. For the guiding of the radiation, the optical fibers (quartz-polymer, length 80 m, 0.4 Db dumping, bandwidth 2 GGz) were used, providing the high enough noise defense of the experiment apparatus. The optical fibers (diameter 400 mm) were closely connected with the target: one with the Al, and the other with Pb surfaces. The distance between the centers of the fibers was less than 1mm and the effects of the radiative nonuniformity were nearly negligible. To eliminate the influence of the hard X-r, which can induce the light inside the fibers,they were positioned inside the steel tube. For the detecting of the optical radiation from the fibers, the silicon photodiodes with the time resolution less than 1ns were used. Under the experimental data processing, the time difference between the signals fronts was used, so the measurements time resolution less than 1.5 ns was provided.

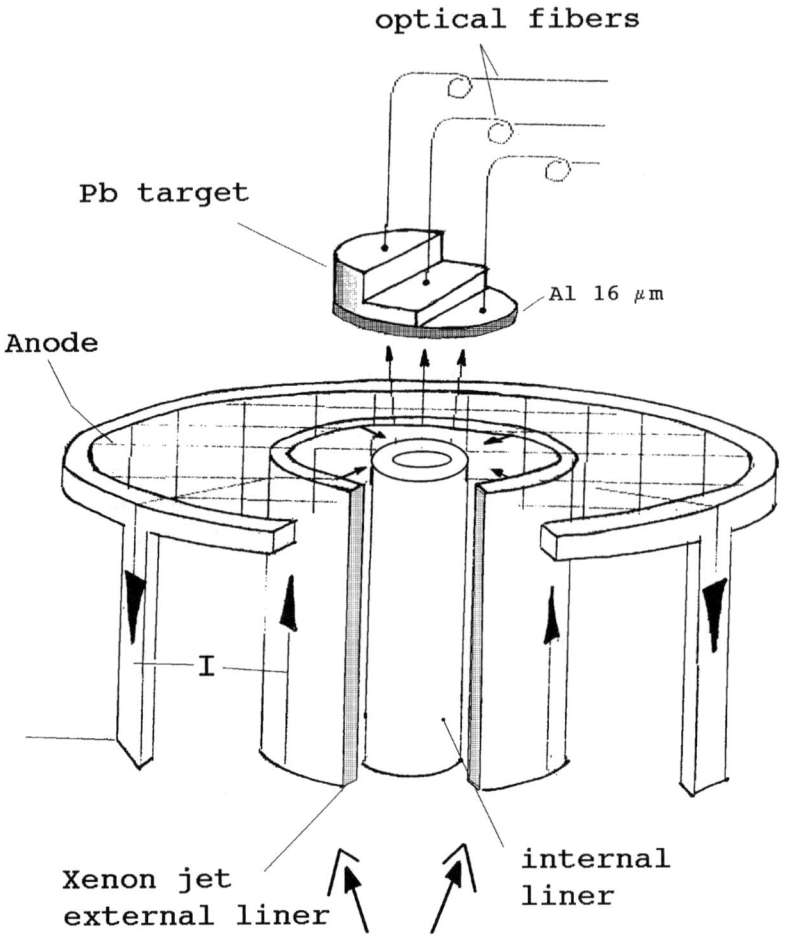

Figure.1 Experimental set-up.

To describe the dynamics of the target under X-ray action Euler equation closed by wide range semiempirical EOS [10,11] were used, taking into account melting, vaporization and ionization of matter. Calculations were performed also with EOS fittied to beam-target experiments at KALIF proton beam (Forschungszentrum Karlsruhe) [12] (see also the paper of Fortov V., Kanel G. et al).

Energy transfer by X-ray radiation was treated in multigroup diffusive approximation. Spectral opacities were used calculated in the frame of modified Hartri-Fock-Sletter model in wide range of temperatures and densities.

Euler equation were integrated by Godunov method. For calculation of energy transfer an implicit numerical scheme was employed. The results of the experiments and calculations are shown in Fig.2. The averaged (over the target volume) shock wave velocity is 7.3 0.6 km/s for 80 mm Pb thickness, and 4.6 0.3 km/s for 200 mm. In accordance with the Huguenot lead adiabate this means that the shock compression pressures are 3 Mbar and 0.9 Mbar correspondingly [13].

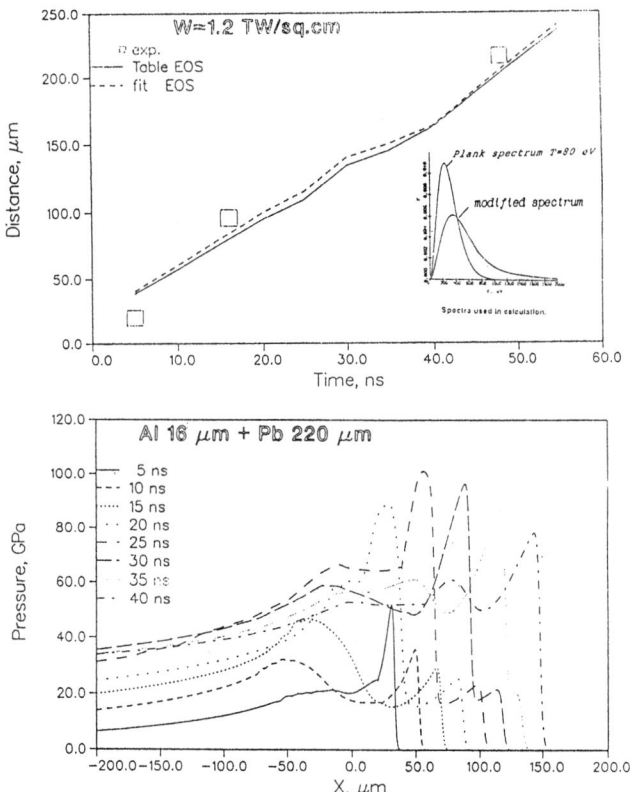

Figure.2 a) Shock front vs time position
b) Shock pressure evolution in the target

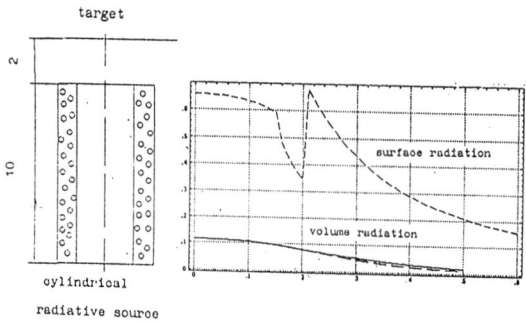

Figure.3 Radiated power distribution on the target surface

CONCLUSION

The results, presented above show the possibility of the intensive shock waves generation induced by the Z-pinch soft X-ray plasma radiation. At the flux power about TW/cm the shock compression value about some Mbar is achieved. The results of the theoretical modeling are seem to agree quite well with the experimental results.

The important question under the study was the space uniformity of irradiated power. Theoretical simulations predict the value of the nonuniformity about 5% within the liner size: Fig.3 shows the calculated space distribution of the radiative power on the target. Two curves correspond sources with the surface and volume radiative mechanisms. One can see that on the distance 1.5 mm (liner inner radius) the uniformity is good enough. One should note that the preliminary direct measurements of the shock wave arriving on free surface give some confirmations of this fact.

We wish to thank the ANGARA-5-1 staff. This work was supported by the Russian Fund for Fundamental Research (94-02-03430-a).

REFERENCES

1. J.J. Duderstadt, Inertial Confinement Fusion (New York, 1982)

2. S.I. Anisimov et. al., Sov. Phys. Usp. 27, 181 (1984)

3. D.L. Matthews et al., J. Appl. Phys. 54, 4260 (1983)

4. P.J. Turchi and W.L. Baker, J. Appl. Phys. 44, 4936 (1973)

5. T.Endo et al., Phys. Rev. Lett. 60, 1022 (1988)

6. Th.Lower, R.Sigel, Proceeding of 7 Int. Workshop of the Physics on Nonideal Plasmas. Markgrafenheide (1993)

7. A. Branitskij et al., Proc. of the 8-th Int. Conf. BEAMS-90, Novosibirsk,437 (1990)

8. Gasilov V. Zakharov S. Smirnov V. JETP Lett, (1991)

9. E. Grabovskij et al., JETP Lett., v.60, No. 1, 1, (1994)

10. A.Bushman, V.Fortov. Sov.Tech.Rev. B, Term. Phys., 1, 219,(1987)

11. W.Ebeling. Thermophysical Properties of Hot Dense Plasmas B.G.Teubner Verglasgeseltschaft, Stuttgart-Leipzig, (1991)

12. Vorobiev O.Yu., Goel B. Numerical simulation of foil acceleration experiments at KALIF, submitted for publication (1995)

13. A.V.Bushman, V.E.Fortov, Sov.Tech. Rev. B. Term. Phys.1,219 (1987)

Modelling of the Shock-induced Luminescence of free Metal Surface

Andrei Yu.Semenov

General Physics Institute
Russia Academy of Sciences,
Vavilov St. 38, 117942, Moscow, Russia x

Abstract. Our aims are the attempt to create a numerical instrument for adequate modelling of interest phenomena of the shock- induced luminosity of free metal surface. Numerical results in comparison with experiments are presented.

A set of experiments was devoted to the investigation of the shock-induced luminescence of free metal surface, [1]-[7]. These experiments can be described as follows (see. Fig.1): explosion- or power laser-generated shock wave propagates perpendicular to the free smooth surface of metal target. On the free surface a shock wave forms rarefaction wave and detector-pyrometer near the target begins to detect a hot metal luminosity. It permits to estimate an average heat temperature of the shock-induced luminescence in assumption that metal radiates as a black body. At the beginning of this process when shock wave emerges from free surface, the luminosity temperature (LT) is equal to the shock wave temperature. Further more the LT decreases because of the hot metal radiation absorption by cold metal in the rarefaction wave.

For the first time the temporal estimations of the shock-induced luminescence were presented in [8]. These estimations were based on the model metal equation of state and model opacity coefficients (the Kramers-Unzold formula). Therefore this approach does not permit to achieve an satisfactory agreement with the experiments.

© 1996 American Institute of Physics

The importance of theoretical description of the experiments under consideration is related, in particular, with the problem of possibility to perform the direct T-measurements in a shock wave.

This communication presents firsts the results of the adequate numerical modelling of LT. These numerical calculations are based on the approach of [8], but with the use of wide-range equation of states for metal [9],[10], wide-range semi- empirical model for the opacity coefficients and special numerical code for reliable numerical calculations of corresponding exponential integrals. The obtained results make it possible to achieve an agreement with the experimental data for Al ([1]-[2]), Pb and Bi ([5]-[7]) to verify some theoretical conceptions.

Results

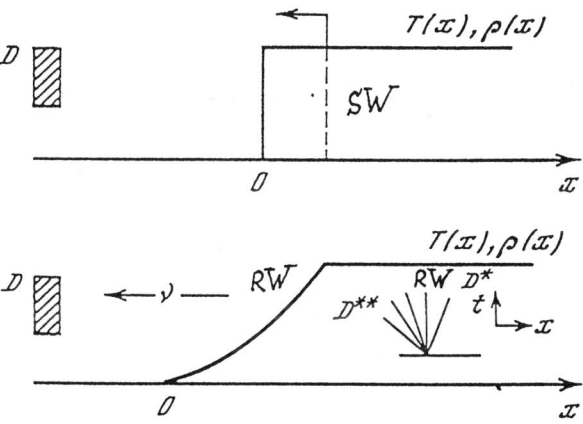

Fig. 1. Scheme of experiments. SW- Shock wave, D- detector, RW- rarefaction wave.

Fig.2-5 present results of investigation of the luminescence in lead for the following shock parameters: pressure- 1.66 Mb, shock wave velocity- 5.8 km/s.

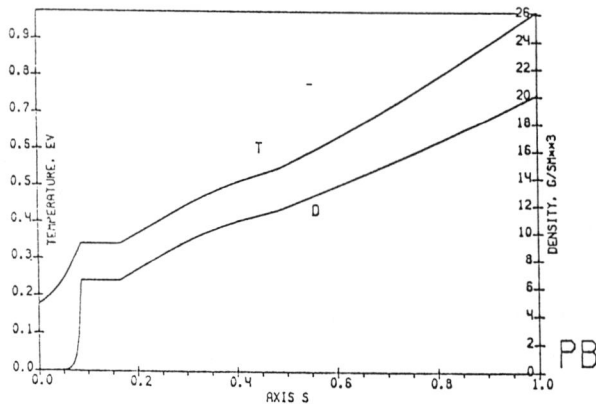

Fig.2: T- and ρ-profiles in rarefaction wave in self-similar variable s (s=0 is corresponding to vacuum, s=1 - to shock wave) along isentrope (with phase jump).

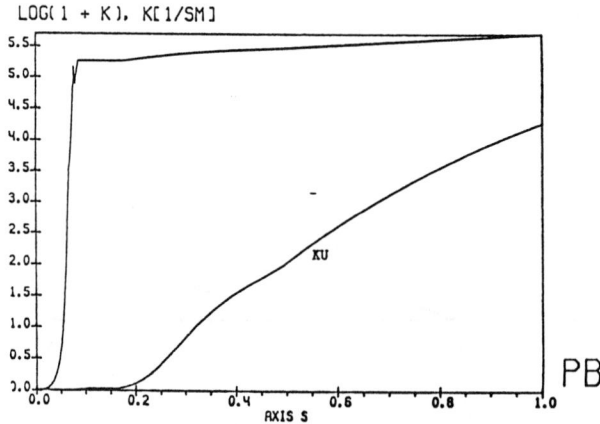

Fig.3: Opacity profiles in RW in accordance with complete model and Kramers-Unzold model [8] (KU).

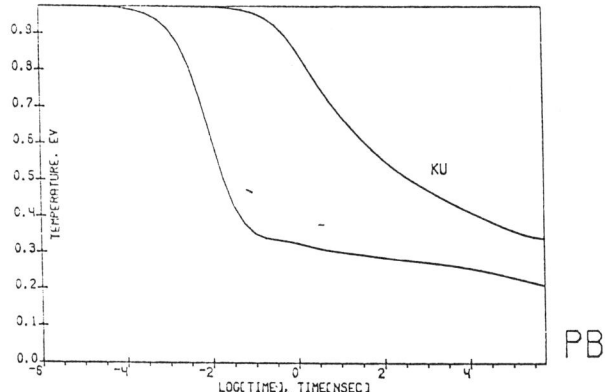

Fig.4: Luminosity temperature profile in time.

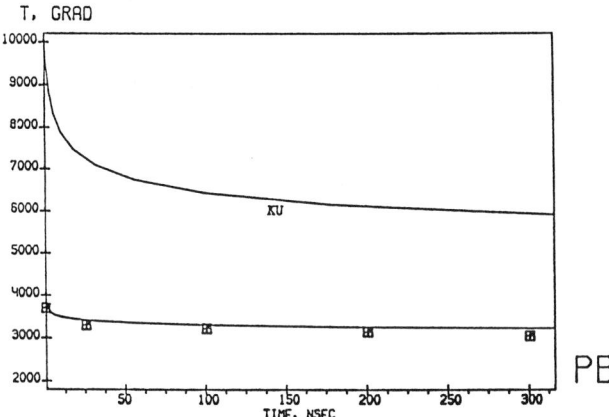

Fig.5: Luminosity temperature and comparison with experiments [6]-[7], λ=6000 A.

Fig.6-9 present results of investigation of the luminescence in bismuth with the following shock parameters: pressure- 4.2 Mb, shock wave velocity- 8.49 km/s.

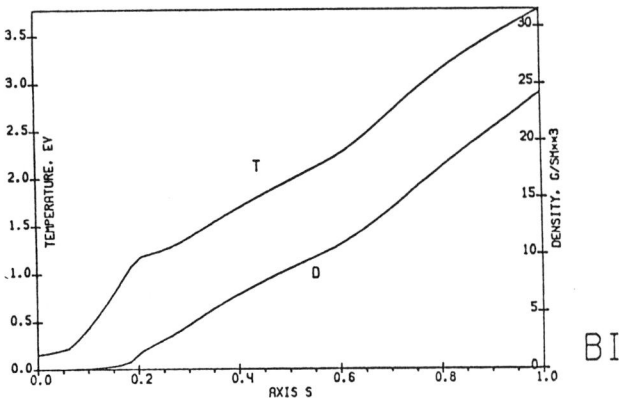

Fig.6: T- and ρ-profiles in rarefaction wave.

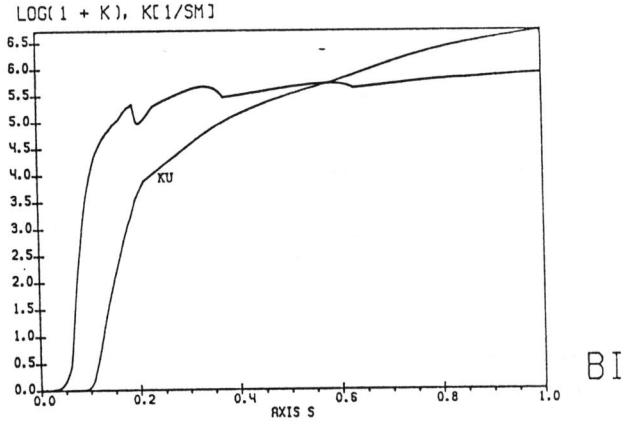

Fig.7: Opacity profiles in accordance with complete model and Kramers-Unzold model (KU).

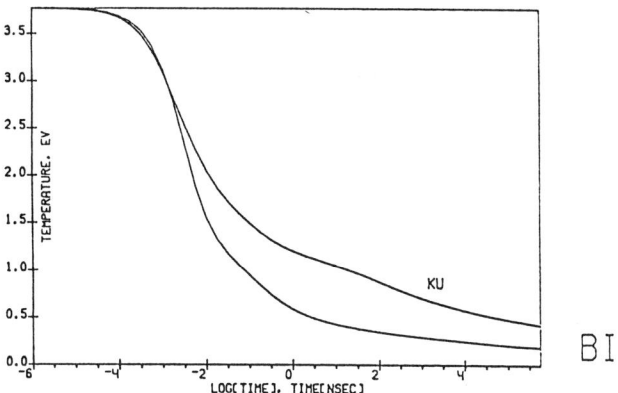

Fig.8: Luminosity temperature profile in time.

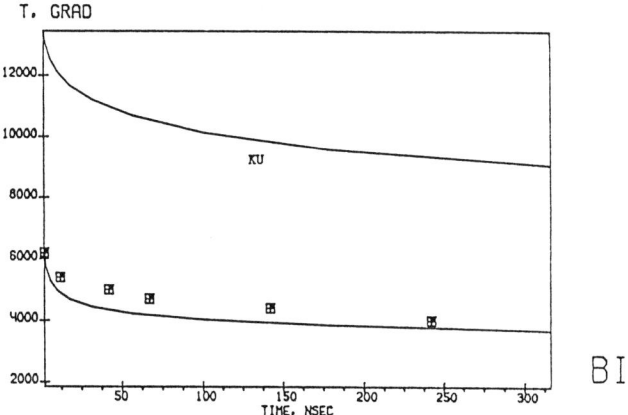

Fig.9: Luminosity temperature and comparison with experiments [5], λ=5650 A.

Fig.10-13 present results of investigation of the luminescence in aluminum with the following shock parameters: pressure- 7.89 Mb, shock wave velocity- 22 km/s.

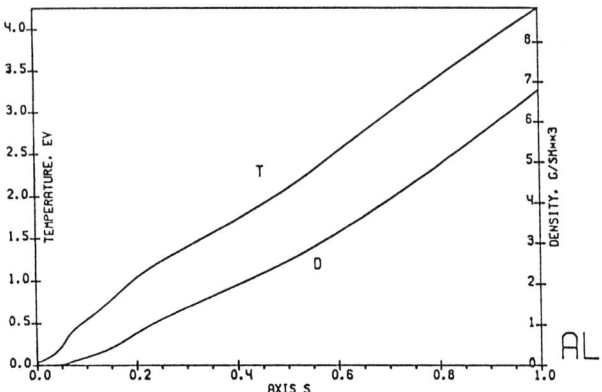

Fig.10: T- and ρ-profiles in rarefaction wave.

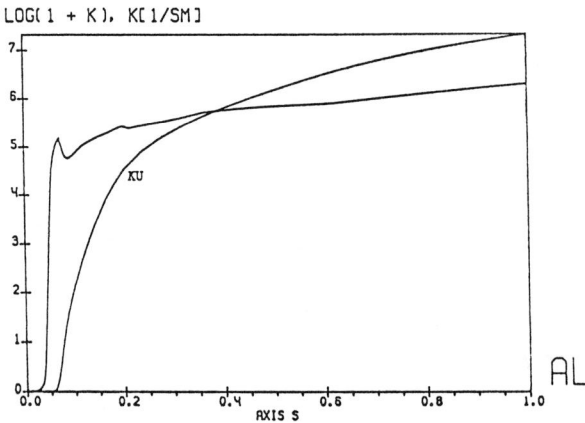

Fig.11: Opacity profiles in accordance with complete model and Kramers-Unzold model (KU).

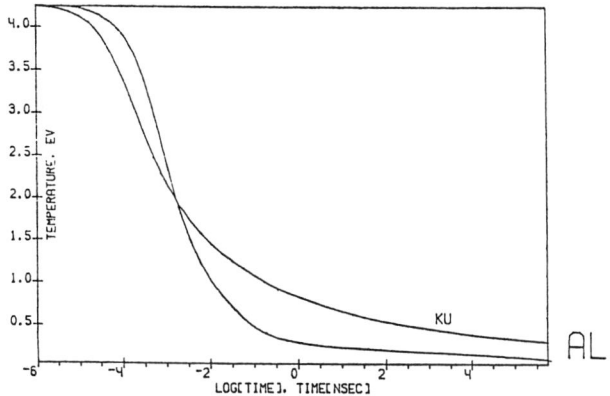

Fig. 12: Luminosity temperature profile in time.

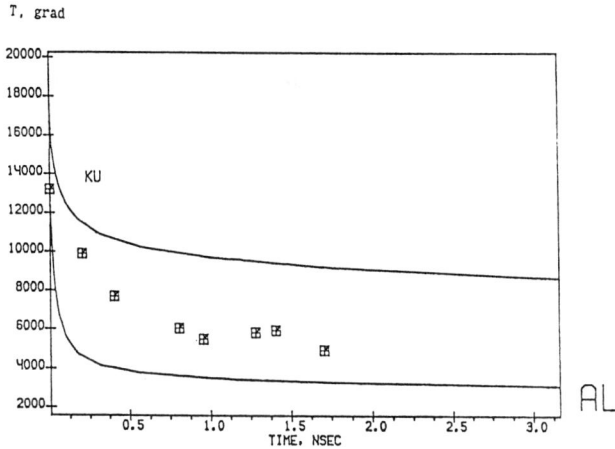

Fig. 13: Luminosity temperature and comparison with experiments [1], λ=6000 A.

This work is supported by RBRF (Russia), Grant 94-02-03413-a.

REFERENCES

1. Ng A., Parfeniuk D., Da Silva L. - *Phys.Rev.Let.* **54** 2604-2607 (1985).
2. Ng A., Parfeniuk D., Da Silva L. - *Opt.Commun.* **53** 389-393 (1985).
3. Ng A., Chin G., Da Silva L. et al- *Opt.Commun.* **72** 297-301 (1989).
4. Godwal B.K., Ng A., Da Silva L. et al- *Phys.Let.A.* **144** 26-30 (1990).
5. Kvitov S.V., Busman A.V. et all- *Zh.Eksp.Teor.Fiz.Let.* **53** 338-342 (1991).
6. Vovchenko V.I., Krasyuk I.K., Semenov A.Yu. "Ablative and dynamic characteristics of laser interaction with planar targets" in *Proceedings of General Physics Institute*. 1992 **36** pp.129-201.
7. Semenov A.Yu.,Polishuk Ya.B., Ternovoi V.Ya., Fortov V.E. Numerical modelling of the free metal surface shock induced luminescence. *Gen. Phys. Ins. Russ. Acad. Sci. Preprint* No.25. 1991.
8. Zel'dovich Ya.B, Raizer Yu.P., *Physics of shock waves and high temperature hydrodynamics phenomena.* New York: Academic Press, 1966.
9. Altshuller L.V., Busman A.V. et al - *Zh..Eksp.Teor.Fiz.* **78** 741-760 (1980).
10. Busman A.V., Kanel G.I., Ni A.L, Fortov V.E. *Dynamics of power interactions.* Chernogolovka, Chem.Physics Inst.USSR Acad.Sci.,1988.

Collisional High Temperature Or Solid State Plasma With Superconductivity Based On Laser-Plasma Interaction

Heinrich Hora

FB Elektrotechnik FH, Pf.120327, 93025 Regensburg, Germany and University of New South Wales, Sydney 2052, Australia

Abstract. The consequent magnetohydrodynamic theory of laser-plasma interaction arrived at several new nonlinear terms. A new effective collisional term in the generalized Ohm's law with a meaning to Langmuir solitons is used to derive conditions how in a real plasma with collisions, will lead to nonresistive (superconducting) electric currents at least into one spatial direction. This is predicted if special electron density gradients fit with the electron collisions and velocity profiles. Conditions for a realization for short times in laser produced plasmas are tools to study this mechanism. The stationary designs in semiconductor electron plasmas for semiconductive power lines at room (or other) temperature are outlined.

The desire to receive materials with an electric resistivity of the value zero is well known and has been realized for materials at very low temperatures as superconductors or at still low temperatures (100 K) as so called high temperature superconductors. The technical aim for transmission of electricity without losses of conductivity is well known. The following considerations are analyzing some new results of magnetohydrodynamic theory as known from high temperature plasmas can well be applied also to solid state plasmas in semiconductors or metals at room temperature.

Magnetohydrodynamics is the macroscopic description of plasmas as hydrodynamic fluids where, however, the properties of the magnetic response of plasmas has been included. The first use of a one-fluid model by Alfven (1942) led to the discovery of the magnetohydrodynamic (or Alfven-) waves [1]. A two-fluid model by combining (adding or subtracting) hydrodynamic Euler equations for electrons and ions of the plasma was derived by Schlüter [2]. The result was the generalized Ohm's law

$$\frac{m}{e^2 n_e}\left(\frac{d\mathbf{j}}{dt}+v\mathbf{j}\right)=\mathbf{E}+\frac{1}{c}\mathbf{v}\times\mathbf{H}+\frac{1}{en_e c}\mathbf{j}\times\mathbf{H}+\frac{c\nabla p}{en_e\left(1+\frac{1}{Z}\right)} \quad (1)$$

and the equation of motion

$$\mathbf{f} = -\nabla p + \frac{1}{c}\mathbf{j} \times \mathbf{H} - \frac{1}{4\pi}\frac{\omega_p^2}{\omega^2}\mathbf{E}\nabla\mathbf{E} \tag{2}$$

where (in Gaussian cgs units) m is the electron mass, e its electric charge and n_e the electron density, m_i the ion mass, Z its charge number and n_i its density, $\mathbf{E}$ the electric and $\mathbf{H}$ the magnetic field, $\mathbf{j}$ the current density $\mathbf{v}$ the net velocity of the plasma, ω_p the plasma frequency, ν the electron ion collision frequency, c the speed of light, and $\mathbf{p}$ the electron pressure. These equations very satisfactorily described the hydrodynamic state or motion of magnetically confined (mirrors and torous-like) plasma, of MHD generators using the second (Lorentz) and the third (Hall) term on the right hand side of (1) and other phenomena.

When laser-plasma interaction was treated, the historic ponderomotive force of Lord Rayleigh [3] and Helmholtz [4] had to be generalized to high frequencies, first achieved for low density plasma by Weibel [5]. The first treatment for general plasma density including dielectric response by Hora, Pfirsch and Schlüter [6] lead to a collisionless dielectric explosion of inhomogeneous plasma driven by the nonlinear electrodynamic forces. Subsequent treatment for obliquely incident laser radiation enforced a generalization of the equation of motion in order to provide the conservation of momentum such that instead of Eq. (2) we arrived at [7]

$$\mathbf{f} = -\nabla p + \frac{1}{c}\mathbf{j} \times \mathbf{H} + \frac{1}{4\pi}\mathbf{E}\nabla\cdot\mathbf{E} - \frac{1}{4\pi}\mathbf{E}\mathbf{E}\cdot\nabla\frac{\omega_p^2}{\omega^2+\nu^2}\left(1+i\frac{\nu}{\omega}\right)$$
$$-\frac{1}{4\pi}\frac{\omega_p^2}{\omega^2+\nu^2}\left(1+i\frac{\nu}{\omega}\right)\mathbf{E}\nabla\cdot\mathbf{E} - \frac{1}{4\pi}\frac{\omega_p^2}{\omega^2+\nu^2}\left(1+i\frac{\nu}{\omega}\right)\mathbf{E}\cdot\nabla\mathbf{E} \tag{3}$$

for quasi stationary laser fields (slow temporal variation of its intensity compared to the laser oscillation time $2\pi/\omega$ with the radian frequency ω) for the dispersive and dissipative plasma. For the general time dependence after a hefty controversy [8] about the terms involved, the most general formulation was reached [9] by adding further terms using the complex refractive index $\mathbf{n}$

$$\mathbf{f}_{NL} = \frac{1}{c}\mathbf{j}\times\mathbf{H} + \frac{1}{4\pi}\mathbf{E}\nabla\cdot\mathbf{E} + \frac{1}{4\pi}\left(1+\frac{1}{\omega}\frac{\partial}{\partial t}\right)\nabla\mathbf{E}\cdot\mathbf{E}(n^2-1) \tag{4}$$

where the electrodynamic part of this general nonlinear force contains ponderomotive and non-ponderomotive terms [10] [11]. The final generality of this formulation was proved by Lorentz and gauge invariance [12].

When deriving the two fluid equations from the Euler equations of electrons and ions for the final formulations (1) and (4) ([10], more detailed steps of the derivation were given in Ref. [11]) it was shown that the generalized Ohm's law appeared as

$$\frac{4\pi}{\omega^2}\left[\frac{d}{dt}\mathbf{j}+\mathbf{j}\left(v-\frac{1}{n_e}\frac{d}{dt}n_e\right)\right]=\mathbf{E}-\frac{1}{en_e c}\mathbf{j}\times\mathbf{H}+\mathbf{v}\times\frac{\mathbf{H}}{c}+\frac{c\nabla p_e}{en_e\left(1+\frac{1}{Z}\right)} \quad (5)$$

where the left hand side compared to Eq.(1) had a modified collision frequency

$$v_{mod}=v-\frac{1}{n_e}\frac{d}{dt}n_e \quad (6)$$

even with a further term due to the fact that the temporal differentiation of the current density in Eq. (5) is convective and not local which part we are neglecting in the following.

As we had discussed before [10], the additional term to the collision frequency can be related to Langmuir solitons at special conditions. The task of this paper is to find conditions where the modified collision frequency (6) is zero what would correspond to superconductivity. We are not discussing here the possibility of negative resistivity as known from Esaki diodes [13] or as laser mechanism needing an additional energy source.

We are aware that the derivations of the final and general two-fluid equations (4) and (5) contain several special assumptions of high frequency fields which limit to the dc-fields may need a further discussion. We further underline that the (usually because of space charge neutrality) neglected internal electric fields inside of an inhomogeneous plasma are included here as given by the second term on the right hand side in Eq. (4). This was necessarily enforced by the laser-plasma momentum balance [7]. The consequence of which is the addition of the Poisson equation (at least for one space dimension) [14] and in three dimensions, the spontaneous magnetic fields of laser plasma interaction are automatically appearing. These internal electric fields sometimes were put in question, e.g. as "Alfven's electric fields which are intuitively not clear" [15].

The condition that the modified collision frequency has to be zero requires (including the equation of continuity)

$$v=\frac{1}{n_e}\frac{d}{dt}n_e=\nabla\mathbf{v}_e\frac{1}{n_e}\nabla n_e \quad (7)$$

If we restrict ourself to a homogeneous electron velocity field $\mathbf{v}_e$, its divergence is then zero (to which point we shall return later). Expressing the collision frequency in a simplified way with a constant a which still is weakly dependent on the electron density and temperature by the Coulomb logarithm [10] and ignoring at this stage the quantum modification of the collision frequency as an expression of anomalous resistivity in quantitative agreement with experiments [10] we arrive at

$$\frac{a}{T}n_e=\frac{1}{n_e}\mathbf{v}_e\nabla n_e \quad (8)$$

If we ignore the ion current component - as possible in strongly n-doped semiconductor plasmas - and express $j = e n_e \mathbf{v}_e$, we arrive at

$$\frac{ae}{2T^{\frac{2}{3}}} = j\frac{1}{en_e^2}\nabla n_e = -2\mathbf{v}_e \nabla \frac{1}{n_e} \tag{9}$$

The task is then to generate an electron density profile such that its relation to the current density or the electron velocity fulfilles the relation (9). In the case of a divergent electron velocity field (what was excluded for deriving Eq.(9)) a further more generalized option of conditions is given for a vanishing modified collision frequency, i.e. an inhomogeneity produced compensation of the electron ion collision frequency.

For an experimental realization one may in the easiest way use a semiconductor electron plasma where the spatial inhomogeneity of the electron density can be designed appropriately and the zero-conductivity is then measured in a stationary way. A long zero-resistivity power line may then to be put together by adding up such elements with barrier free contacts in between.

Since laser produced plasmas are very flexible in generating density and temperature profiles one may well be able to generate similar resistivity-free (superconducting) high temperature plasmas for a short time to prove this special property.

When applying the same procedure of gradients in n_e and v for a zero resistivity MHD generator following Eq. (5) one has to take into account that the collision frequency v will be strongly dependent on the collisions with neutrals. This option should result in a non thermalizing 100% conversion of kinetic energy of plasma into electricity after the basic property of this mechanism may be confirmed by special studies in laser produced plasmas.

[1] Alfven, H. *Nature* **150**, 405 (1942)
[2] Schlüter, A. *Ztschr. Naturforsch.* **5a**, 72 (1950)
[3] Kelvin Lord Cambr. and Dublin Math. J., November (1845), see Electrodynamics and Magnetism p.32, Cambridge (1872)
[4] Helmholtz, H, see Pavlov,V.I. Sov. Phys. Usp. 21, 171 (1978)
[5] Weibel E. J. *Electronics and Control* **5**, 435 (1958)
[6] Hora, H., Pfirsch, D. and Schlüter, A., *Ztschr. Naturforsch.* **22A**, 278 (1967)
[7] Hora, H. *Phys. Fluids* **12**, 182 (1969)
[8] Zeidler, A., Schwab, H., and Mulser, P., *Phys. Fluids* **28**, 372 (1985)
[9] Hora, H. *Phys. Fluids* **28**, 3706 (1985)
[10] Hora, H. *Plasmas at high temperature and density*, Springer, Heidelberg (1991)
[11] Hora, H. *Elektrodynamik*, (S. Roderer, Regensburg 1994)
[12] Rowlands, T. *Plasma Physics and Controlled Fusion* **32**, 297 (1990)
[13] Esaki, L., *Rev. Mod. Physics* **46**, 237 (1974)
[14] Eliezer S. *et al Physics Reports* **172**, 339 (1989)
[15] Kulsrud, R., *Physics Today* **34**, No.4., 53 (3rd Col. 7th line) (1983)

N-body Lyapunov expansion rates in one component strongly coupled plasmas

Y.Ueshima, K.Nishihara
D.M.Barnett[A], T.Tajima[A], and H.Furukawa[B]

Institute of Laser Engineering, Osaka University, Suita ,Osaka 565 Japan.
[A]Institute for Fusion Studies, University of Texas at Austin, Texas 78712, USA.
[B]Institute for Laser Technology, Suita ,Osaka 565 Japan.

Abstract Phase space Lyapunov expansion rates are measured for the first time for Coulomb many body systems with the use of a 3-d particle code. The time averaged Lyapunov exponents, λ/ω_p, are found to be proportional to $\Gamma^{-2/5}$ and the cubic root of the diffusion coefficient in the range of $1<\Gamma<160$, where ω_p and Γ are plasma frequency and ion coupling constant, respectively. A large jump of the averaged Lyapunov exponent is observed near $\Gamma\sim170$, corresponding to the phase transition from liquid to solid. Instantaneous Lyapunov exponent has chaotic behavior and consists of three different spectra, flat, f^{-2} and f^{-1}.

INTRODUCTION

Recently there has been a grate deal of research efforts devoted to find relations between the Lyapunov exponents and macroscopic transport coefficients [1~6]. The Lyapunov expansion determines the rate at which initial state information is lost. Similarly transport coefficients generally involve the dissipation of knowledge of some microscopic quantity. For example, diffusion coefficient is determined from velocity auto correlation in time. We therefore conjecture that the Lyapunov expansion rates must be related to macroscopic transport coefficients. In this paper we present a relation between the phase space the Lyapunov expansion rates and the diffusion coefficients for ion one component strongly coupled plasmas with the use of a 3-d particle code.

The Lyapunov expansion rate is a mechanical characteristic quantity which can provide a useful window on microscopic evolution systems, because of its connection to phase mixing and loss of initial state information. The Lyapunov expansion rate may be definable even for stlongly non-equilibrium in which statistical quantities and transport coefficients are hard to calculate or have no meaning.

The separation between two adjacent trajectories in phase space expands exponentially when it is small enough. However, after a while the separation becomes small no longer and it will saturate. The saturation distances in momentum and position spaces are discussed. By imposing re-scaling method to prevent the saturation, the instantaneous Lyapunov expansion rates are observed for a long time. The instantaneous Lyapunov expansion rates may provide uniformity of the phase mixing and more precise information of microscopic process.

SIMULATION METHOD AND CONDITIONS

Ion one component plasmas are characterized by the ion coupling constant,

$$\Gamma \equiv \frac{e^2}{aT}, \quad \frac{4\pi}{3}a^3 \equiv \frac{1}{n} \equiv \frac{N}{V} \tag{1}$$

where e, T, a, n, N and V are, respectively, charge, temperature, ion sphere radius, number density, total number of ions, volume of a system. In this paper we consider plasmas of $1<\Gamma<160$. It is necessary for particle simulations of such dense plasmas to calculate precisely Coulomb interactions among distant and close particles. We perform the NVE (number of particles, volume and energy of a system) constant molecular dynamics simulation with the use of 3-d particle code SCOPE [7]. Particle-Particle Particle-Mesh method is used in the code to treat many particles, in which Coulomb forces among close particles are directly summed up and Particle-Mesh method is employed for the calculation of forces among distant particles. The system comprises 500 ions in periodic cubic box with uniform electron background. Newton's equation of motion is solved using a leap-frog technique which has the second order accuracy in time step. Unit time steps used are as follows; $\Delta\omega_p t = 0.006$ for $\Gamma=1$, $\Delta\omega_p t = 0.01$ for $\Gamma=2$, and $\Delta\omega_p t = 0.02$ for $\Gamma=5\sim200$, where ω_p is the plasma frequency.

To measure the Lyapunov expansion rates in 6N-dimensional phase space, we consider two systems, one is called as a reference system and the other as a displaced system. In the displaced system, initial position q_i and momentum p_i of the i-th particle (i=1~500) are slightly displaced from those of the corresponding particle in the reference system. The displacements are given by normal distribution with typical root mean square displacements of $5\ 10^{-3}$ mV_{th} in the momentum space, and $5\ 10^{-3}$ a in the position coordinates space, where m and V_{th} are ion mass and thermal speed ($mV_{th}^2=T$), respectively. We define 6N-phase space trajectories as $\Lambda(t) = (\ p_i(t),\ q_i(t)\ )$ in the reference system and $\Lambda'(t) = \Lambda(t)+\delta(t)$ in the displaced system. If the displaced trajectory diverges exponentially in time from the reference trajectory, the Lyapunov exponent is given by the time averaged phase space expansion rate. The instantaneous Lyapunov expansion rates are estimated by

$$\lambda(n\Delta t) = \frac{1}{\Delta t}\log\left(\frac{\|\delta(n\Delta t)\|}{\|\delta((n-1)\Delta t)\|}\right). \tag{1}$$

As will be shown later, after a while the separation between two trajectories is small no longer, and it will saturate. In order to observe the expansion rates for a long time, the displacement must be re-scaled [8~11]. The re-scaling factor is calculated by

$$f_r = \frac{\|\delta((n-1)\Delta t)\|}{\|\delta((n)\Delta t)\|}, \tag{2}$$

and then the new starting point of the re-scaled trajectory is given as follows,

$$\Lambda' = \Lambda(n\Delta t) + f_r\delta(n\Delta t). \tag{3}$$

RESULTS AND DISCUSSIONS

(1) Non Re-scaled Case

At first we examine without the re-scaling whether the displacement of the two trajectories diverges exponentially in time, for various ion coupling constants. Figures 1(a) and 1(b) show time variations of the displacements in momentum and position coordinates. In the figure, $\delta p \equiv \|\delta \mathbf{p}\|$, $\delta q \equiv \|\delta \mathbf{q}\|$ and λ_D is the Debye length. It is clearly seen that the displacements for all Γ diverge exponentially when the displacements are small enough. The exponential growth of the displacements show that the systems have more than one positive Lyapunov exponent. The expansion rates in the momentum and position coordinates spaces are the same values. The expansion rates decrease with the increase of the ion coupling constant. As the displacements reach certain values, the divergence speeds become slower. The deviation from the exponential growth occurs around the separation of $\delta p \sim mV_{th}$ and $\delta q \sim 3 \text{~} 4 \lambda_D$. The separation distance in the momentum space saturates at $\delta p \sim \sqrt{6}\, mV_{th}$, this saturation value can be easily understood as follows. From the definition of δp, it is given by

$$\overline{\delta p^2} \equiv \frac{1}{N}\sum_i^N (\mathbf{p}'_i - \mathbf{p}_i)^2 = \frac{1}{N}\sum_i^N (\mathbf{p}'^2_i - 2\mathbf{p}_i \cdot \mathbf{p}'_i + \mathbf{p}_i^2)$$
$$= 6(mV_{th})^2 - \frac{2}{N}\sum_i^N \mathbf{p}_i \cdot \mathbf{p}'_i \quad . \quad (4)$$

If the momenta of the reference and displaced systems, $\mathbf{p}_i$ and $\mathbf{p}'_i$, are not correlated with each other, the sum of their scalar product in Eq.(4) is zero. Once the momentum in the two systems is not correlated, the separation distance in the position coordinates diverges in proportional to $t^{1/2}$, since the particles diffuse in the real space due to random walk process.

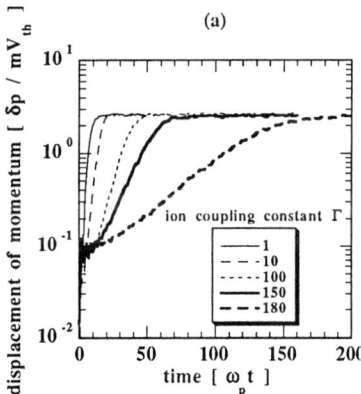
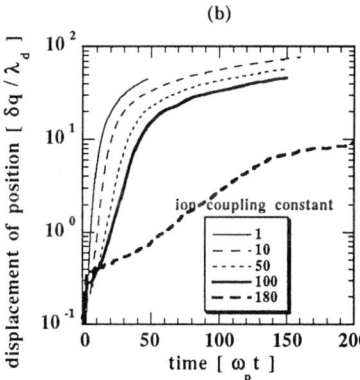

Figure 1 Time evolution of separation distances in momentum space (a) and real space for various ion coupling constants Γ=1,10.100,150,180.

(2) Re-scaled Case

As shown in the previous subsection, the expansion rates decrease with the increase of the ion coupling constant. To estimate the Lyapunov exponent more accurately, we calculate the instantaneous expansion rates with the re-scaled method for various ion coupling constants.

Figure 2 shows the instantaneous expansion rates for $\Gamma=1,10,50$ and 200. They start from small values and increase rapidly to large peak values, before declining slowly towards asymptotic values. This stage is named for Lyapunov transient time. In the transient time, an initial displacement seeks out the phase space eigen-direction with the largest positive eigenvalue of the Lyapunov exponents. As shown in Fig.2, the transient time increases as the increase of the ion coupling constant. The damping oscillations are observed for the large Γ of 50 and 200 in the transient time. It should be noted that any initial distribution of the displacement relaxes to a normal distribution within transient time.

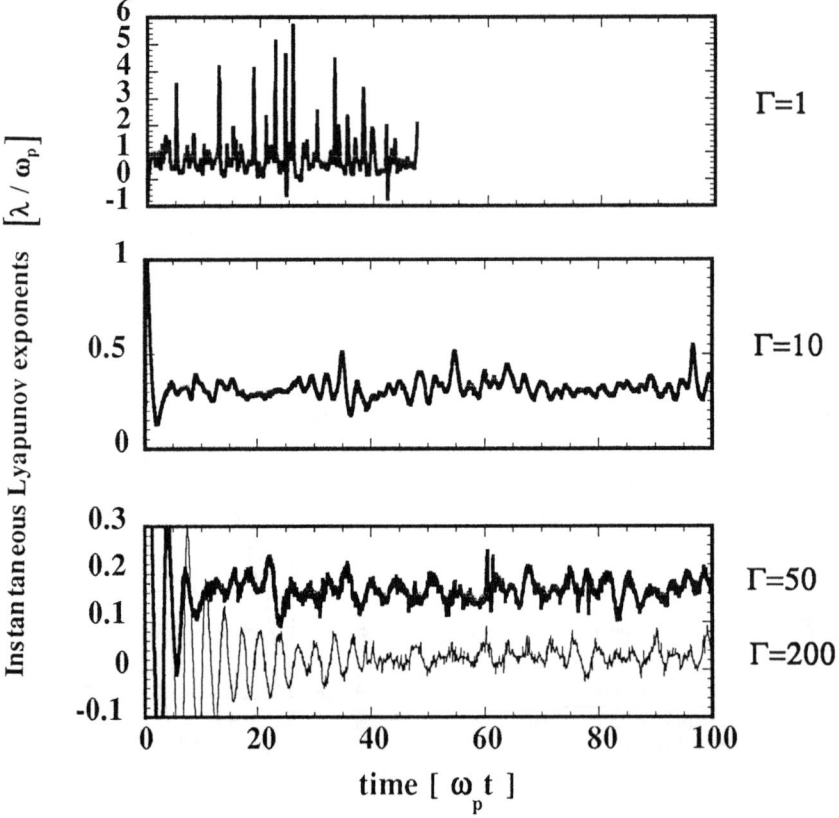

Figure 2   Instantaneous Lyapunov exponents for $\Gamma=1,10,50,200$.

All of the asymptotic expansion rates have positive values that may correspond to the maximum Lyapunov exponent, although their instantaneous values are chaotic in time. The chaotic behavior of the instantaneous Lyapunov exponent will be discussed later. Figure 3 shows the dependence of the time averaged Lyapunov exponents on the ion coupling constant. The averaged Lyapunov exponents are found to be proportional to $\Gamma^{-2/5}$ in the range of 1 ~ 160. The large jump of the Lyapunov exponent near Γ~170 corresponds to the phase transition from liquid to solid in one component plasma. The normalized diffusion coefficients, $D/a^2\omega_p$, are also found to be proportional to $\Gamma^{-6/5}$ in the range Γ<160. Those results indicate the averaged Lyapunov exponent to be proportional to the cubic root of the diffusion coefficient, $<\lambda>/\omega_p \sim (D/a^2\omega_p)^{1/3}$.

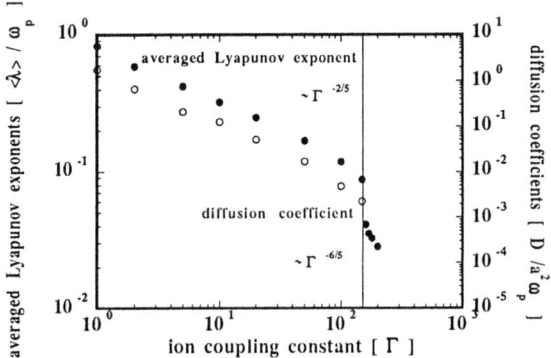

Figure 3 Dependence of time averaged Lyapunov exponents (closed circle) and diffusion coefficients (open circle) for ion coupling constant.

The chaotic behavior of the instantaneous Lyapunov exponents is investigated by taking their spectra. As shown in Fig.4, the instantaneous Lyapunov exponents consist of three different spectra. The first spectra around the plasma frequency may correspond to the collective mode in plasma. The others are f^{-2} and f^{-1} spectra in the high frequency range.

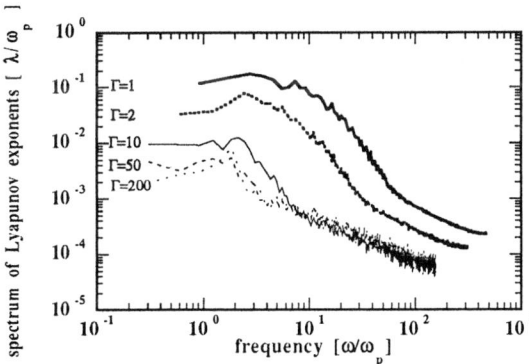

Figure 4 Spectrum of instantaneous Lyapunov exponents for Γ=1,10,50,200.

CONCLUSION

The Lyapunov expansion rates in 6N-dimensional phase space are investigated for the first time for Coulomb many body systems. The saturation of the separation in the momentum and position spaces are also discussed. The saturation in the momentum space occures, where the reference and displaced sysyems are not correlated each other. The time averaged Lyapunov exponents, λ/ω_p, are found to be proportional to $\Gamma^{-2/5}$ and the cubic root of the diffusion coefficient in the range of $1<\Gamma<160$, where ω_p and Γ are plasma frequency and ion coupling constant, respectively. A large jump of the averaged Lyapunov exponent is observed near $\Gamma\sim170$, corresponding to the phase transition from liquid to solid. Instantaneous Lyapunov exponent has chaotic behavior and consists of three different spectra, flat, f^{-2} and f^{-1}.

ACKNOWLEDGMENTS

One of the authors (Y.Ueshima) thanks Dr.M.Murakami, Dr.H.Sakagami, and Dr.S.Kato for useful comments. We also thank Professor K.Mima and Professor S.Nakai for their interest and encouragement. This work was partially supported by the Grant-in-Aid for Scientific Research from the Ministry of Education, Science, and Culture of Japan and by the Japan Society for the Promotion of Science for Japanese Junior Scientists.

REFERENCES

1. S.Chandhuri, G.Gangopadhyay and D.S.Ray,Phys.Rev.E. **47**,311(1993).
2. P.Gaspard and G.Nicolis,Phys.Rev.Lett. **65**,1693(1990).
3. D.M.Barnett, Lyapunov Exponents of Many Body Systems,Doctor Thesis,The University of Texas at Austin(1995).
4. C.Amitrano and R.S.Berry,Phys.Rev.Lett. **68**,729(1992).
5. D.J.Evans, E.D.J.Cohen and G.P.Morriss,Phys.Rev.A. **42**,5990(1990).
6. H.A.Posch and W.G.Hoover,Phys.Rev.A. **38**,473(1988).
7. K.Nishihara, Kakuyugo Kenkyu **66**,253(1991).
8. A.J.Lichtenberg and M.A.Lieberman, Regular and Stochastic Motion, Applied Mathematical Science v.38, Springer-Verlag, New York (1983).
9. N.G.van.Kampen, Stochastic Processes inPhysics and Chemistry, North-Holland Physics Publishing, Amsterdam (1983).
10. W.E.Wiesel,Phys.Rev.E. **47**,3686(1993).
11. D.J.Evans,Phys.Rev.Lett. **69**,395(1992).

Analyses of Stimulated Raman Scattering Light Spectrum and Plasma Temprature and Density for Cavity Targets *

Jia−Tai Zhang

Institute of Applied Physics and Computational Mathematics
P.O.Box 8009, Bejing 100088,China

ABSTRCT. An one and a half dimensional plasma Cloud−in−Cell(CIC) simulation code is used to study stimulated Raman scattering(SRS) in laser−plasma cavity targets from Shengguang−I Nd−glass laser facility. A feasible method for plasma density and temprature diagnosing is given and some results about plasma density profile are obtained by comparing theoretical and experimental spectrum of stimulated Raman scattering(SRS).

I. Introduction

Laser coronal plasma coupling is one of the important problems of laser inertial confinement fusion and X−ray laser in laboratory. In laser fusion, many process, such as laser absorbtion, laser−x−ray transform, the production of superthermal electron and the transmission of x−ray are clearly related to plasma condition. There are many difficulties in diagnosing plasma condition experimentally because of its small size($\sim 10^{-2}$ cm), high temperature($\sim 10^3$ ev), heavy density($\sim 10^{21} / cm^3$) and short duration($\sim 10^{-9}$ s) in laser fusion. Some variables of plasma condition, such as temperature, density, etc., which can not be easily measured directly in experiment so far, are important for theoretical study and experimental analysis. Generally speaking, there are two methods based on electromagnetic radiation in diagnosing plasma condition. First, observing and analysing plasma emission spectrum, one can obtain information of plasma condition[1−3]. Since the electron temperature is about $10^2 \sim 10^3$ kev, the plasma emission spectrum extends from ultra−violet ray to soft x−ray. Their energy of the ray is several orders higher than characteristic frequency of plasma. So it is not too difficult to analyse the

---

* Project supported by the National Hi−Technology Inertial ConfinementFusion Committee,China, the National Natural Science Foundtion of China, and the Science Foundation of China Academy of Engineering Physics.Wang Weixing ,Chang Tieqiang,Xu Linbao,Zhao Xuewei and Mei Qiyong et al.joined this work.

© 1996 American Institute of Physics

emission spectrum from the collective effect. Collective effect can influence some aspects of radiation, for example, the line-width and XUV spectrum with $h\nu \ll kT_e$ depend on binary collision and interaction of individual particle with radiation field. One can obtain the information of plasma density and temperature from XUV spectrum. Secondly, By measuring the scattered light of probe laser a plasma density can be obtained. The method is called the laser probe method[4]. The method is limited for high density plasma where laser can not arrive. A few variables measurable experimentally is related to plasma condition and the relation between them are well understood theoretically. It is expected to find some variables which can be measured easily and is sensitive to plasma condition. Comparing theoretical and experimental results, the useful information related to plasma condition can be obtained.

In this paper, numerical results of SRS using plasma particle simulation code are presented in Sec. II. Deducing plasma density profile and temperature from SRS spetrum are analyzed in Sec. III.

II. Plasma Particle Simulation

We designed $1\frac{1}{2}$-dimensional plasma cloud-in-cell(CIC) code. The SRS with SBS coexistence has been investigated by tracking 131072 simulated particles. The equations that are solved in the code are summarized as follows:

i) $1\frac{1}{2}$-dimensional Maxwell's equations

$$\frac{\partial E_x}{\partial x} = 4\pi\rho, \tag{1}$$

$$\frac{\partial B_z}{\partial x} = -\frac{1}{C}(\frac{\partial E_y}{\partial t} + 4\pi J), \tag{2}$$

$$\frac{\partial E_y}{\partial x} = -\frac{1}{C}\frac{\partial B_z}{\partial t} \tag{3}$$

ii) The velocities and positions of the particles are advanced in time according to the relativistic equations of motion:

$$\frac{dx_{\alpha i}}{dt} = P_{x\alpha i}/m_{\alpha i}, \tag{4}$$

$$\frac{dP_{x\alpha i}}{dt} = \mu_{\alpha i}[(E_x + E_x^{ext})_{\alpha i} + \frac{P_{y\alpha i}}{m_{\alpha i}C}(B_z + B_z^{ext})_{\alpha i}], \tag{5}$$

$$\frac{dP_{y\alpha i}}{dt} = \mu_{\alpha i}[(E_y + E_y^{ext})_{\alpha i} - \frac{P_{x\alpha i}}{m_{\alpha i}C}(B_z + B_z^{ext})_{\alpha i}], \tag{6}$$

$$m_{\alpha i} = [1 + (\frac{P_{\alpha i}}{C})^2]^{\frac{1}{2}}, \quad P_{\alpha i}^2 = P_{x\alpha i}^2 + P_{y\alpha i}^2 \tag{7}$$

$$P_{x\alpha i} = m_{\alpha i}u_{\alpha i}, \quad P_{y\alpha i} = m_{\alpha i}v_{\alpha i}, \tag{8}$$

$$m_{I_i} = m_{I_0}, \quad m_{e0} = 1,$$

where $\alpha = \begin{cases} I \\ e \end{cases}$ denote species of particles, i.e., electron and ion, i denotes the number of particles, μ and m denote charge and mass, respectively. $E_{x\alpha i}$ and $E_{y\alpha i}$ denote the electric field in x-direction and y-direction acting on the particles, $B_{z\alpha i}$ denote the magnetic field in z-direction acting on the particles, ext denote applied external field. Charge density ρ and current density J are computed by solving particle cloud equation[5], as well as the electric and magnetic field acting on the particles. Figure 1 gives the time evolution of the energy density (F_x, F_y, and F_z) of Langmuir wave and scattering electromagnetic wave under the condition of coupling of SRS and SBS.

Computed formulae of the field energy density are as follows:

$$F_x = \frac{1}{N} \sum_{l=0}^{N-1} E_{xl}^2, \tag{9}$$

$$F_y = \frac{1}{N} \sum_{l=0}^{N-1} E_{yl}^2, \tag{10}$$

$$F_z = \frac{1}{N} \sum_{l=0}^{N-1} E_{zl}^2, \tag{11}$$

where N is the number of spatial grid. Figure 2 shows the time history of electron velocity distribution function. The results in computer simulation in Figs. 1 and 2 show the evolution of SRS, at the first stage the energies of both waves grow exponentially in time (so called linear growth), after $\sim 120\omega_p^{-1}$, the saturation occurs, the electron distribution function broadens, its tail rises, its shape is a double-peak Maxwellian distribution. At an intensity of $I_0 = 8 \times 10^{14}$—$6 \times 10^{15} w/cm^2$, the energy fraction of hot electron is about 9%~20% of incident laser light energy. Temperature of hot electron $T_H \approx \frac{1}{2} m (\frac{\omega_E}{k_E})^2 \approx 40 \sim 60 Kev$.

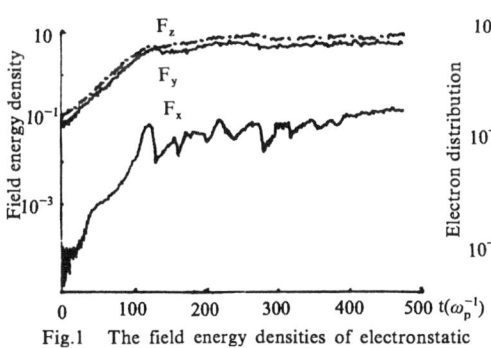

Fig.1 The field energy densities of electrostatic and scattering waves as a functions of time at $\lambda_0 = 1.053 \mu m$, $I_0 = 2 \times 10^{15} w/cm^2$.

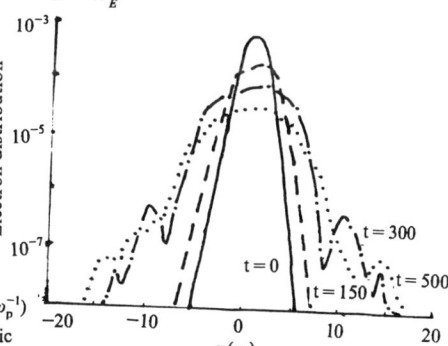

Fig.2 The evolution of electron distribution function at $\lambda_0 = 1.053 \mu m$, $I_0 = 2 \times 10^{15} w/cm^2$.

III Deducing Plasma Profile and Temperature from SRS Spetrum

The intensity $\Pi_\omega(T,I)$ of SRS scattered light involves information of plasma density profile. $\Pi_\omega(T,I)$ is the total spetrum measured experimentally.

First, introduce a concept——the density probability distribution of a plasma $P(n_e)$. Following consideration is made at $n_e < \frac{1}{4}n_c$. For one dimension plasma with length L, the portion of plasma whose density changes from n_e to $n_e + dn_e$ is given by

$$\frac{dx}{L} = \frac{1}{L}\frac{dx}{dn_e}dn_e \tag{12}$$

Define the density probability distribution[6]

$$P(n_e) = \frac{1}{L(dn_e/dx)} \tag{13}$$

Then the average electron density is

$$\bar{n}_e = \frac{1}{L}\int_0^L n_e(x)dx = \int_0^{\frac{1}{4}n_c} P(n_e)n_e dn_e \tag{14}$$

For the following two kinds of density profile

1). $n_e(x) = n_e(0)e^{-\frac{(L-x)}{L}}$, then

$$P(n_e) = \frac{1}{n_e}$$

2). $n_e(x) = const$, then

$$P(n_e) = \delta(n - n_0)$$

Following we consider the intensity of SRS spectrum. Plasma with different density has different contribution to total spectrum measured experimentally because of their different portion. Generally speaking, the contribution of plamsa with density n_e is proportional to its portion $\frac{\Delta x}{L}$, so we have

$$\Pi_\omega(T,I) = \sum_{\{n_e\}} I_\omega(n_e,I)\frac{\Delta x}{L} = \int_0^{\frac{1}{4}n_c} I_\omega(n_e,I)P(n_e)dn_e \tag{15}$$

where spectrum intensity $I_\omega(n_e,I)$ is obtained numerically. The density probability profile can be deduced inversely from the experimental and numerical spectrum. In order to obtain numerical spetrum we solve following three—wave coupling equations of SRS

$$(\frac{\partial}{\partial t} + V_0\frac{\partial}{\partial x} + \Gamma_0)E_0(x,t) = \gamma_0 E_1 E_2 exp(-i\int K(x')dx') \tag{16}$$

$$(\frac{\partial}{\partial t} + V_1\frac{\partial}{\partial x} + \Gamma_1)E_1(x,t) = \gamma_1 E_0 E_2 exp(i\int K(x')dx') \tag{17}$$

456

$$(\frac{\partial}{\partial t} + V_2 \frac{\partial}{\partial x} + \Gamma_2) E_2(x,t) = \gamma_2 E_0 E_1 exp(i\int K(x')dx') \tag{18}$$

which is induced from Maxwell's equations and fluid equations with WKB approximation under one-dimensional geometry, where E_i, V_i, Γ_i (i=0—laser wave, i=1—scattered light wave, i=2—Langmuir wave) are slow varying amplitude, group velocity and damping coefficient of the corresponging wave respectively, γ_i (i=1,2) is coupling coefficient, the phase mismatching $K = k_0 - k_1 - k_2$. It should be pointed out that our method discussed above involves implicitly the assumption that plasma density profile is nearly unchangable in the period of SRS. The experimental spetrum $\amalg_\omega (T,I)$ is an ensemble average effect and does not show the character of density profile at one time. Since SRS evolves much more rapidly than plasma density, the assumption is reasonable. In other words, the density profile deduced is the averaged one over a period of time.

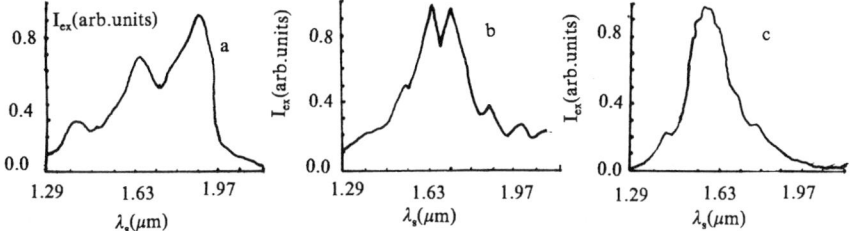

Fig.3   Experimental spectrum for three hohlraum targets.

Numerical solution of (16-18) shows SRS three wave grows with time. Figure 3 gives experimental spectrum for three hohlraum targets on Shenguang-I.Figure 4 gives theoretical spectrum of scattered light.Figure 5 shows the corresponding density profile roughly obtained by means of comparing theoretical and experimenal spectrum for three hohlraum targets plasma.The narrower SRS spectrum,the shorter the density scale lenghth.From spectrum analysis,the average density at $n_e < \frac{1}{4} n_c$ is 0.13-0.16n_c.The shape ofdensity distribution closely ralated to the target structure and laser intensity.

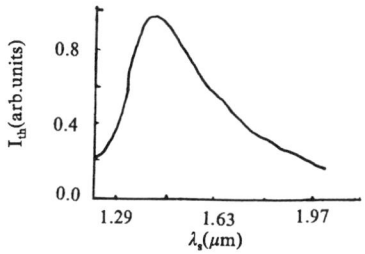

Fig.4   Theoretical spectrum of SRS.

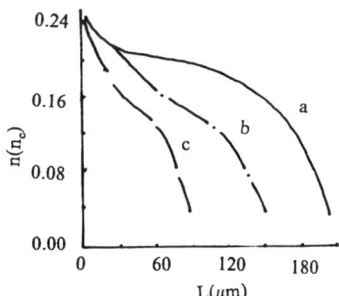

Fig.5   Density profile for three hohlraum targets plasma.

SRS light spectrum obtained from experiments cuts off steeply at the short wave. In a collisionless plasma this is because of Landau damping[7]. From the wave number and frequency match-conditions of SRS and the dispersion relations, the SRS light spectrum mesuured by the experiment using imitative OMA spectrograph is analyzed. For Langmuir cut-off wave number k_{EPW} to satisfing $k_{EPW} \cong 0.366$, we can obtain Landau damping $\gamma_L = -0.0674\, \omega_{pe}$, the thermal electron temperature of the corona region $T_e = 1.0 \sim 2.6$ KeV. For the majority of targets, T_e is about 1.5KeV.

ACKNOWLEDGMENT

We wish to thank professors Min Yu, Hansheng Pen and Xiantu He For their beneficial discussion and support, and the persons in target-making department and in Shenguang-I laser-facility department for their cooperation.

Reference

[1]. Peacock N.J. in Laser Plasma interactions, ed. Carins R.A. Sanderson J.J. pp. 711, Pub. Scottish University Summer School in Phys. Dept., Edinbergh university.

[2]. More R.M., in Applied Atomic Collision Phys. Vol. II, Pub. Academic Press, N.Y.1982.

[3]. Key, M.H. Hutcheon R.J., In Advances in Atomic and Molecular Phys., 16, (ed. Bates, Pub. Academics)1980,pp.201.

[4]. D.E. Evans and J. Katzenstein, "Laser Light Scattering in Laboratory Plasma" Rep. Prog. Phys. Vol.32, 207-271(1969).

[5]. Group CIC(Xu Linbao et al.)NuclearFusion and Plasma Physics,Vol.2,92(1982).(in Chinese);Zhang Jiatai et al.,Acta Phisica Sinica,Vol.40,1642(1991).

[6]. Zhang Jiatai,Wang Weixing,Chang Tieqiang,Chin. Phys.Lett.Vol.11,416(1994).

[7]. Zhang Jiatai,Acta Phisica Sinica,Vol.43,64(1994)(in Chinese)

Stimulated Raman scattering from two overlapped 527 nm laser beams

M. Tsukamoto,[a] K. A. Tanaka,[b] K. Mima, R. Kodama,
M. Kado, S. Nakaji, A. Nishiguchi,[c] M. Nakai, T. Norimatsu,
K. Nishihara, T. Yamanaka and S. Nakai

Institute of Laser Engineering, Osaka University,
Yamada-oka 2-6, Suita, Osaka 565 Japan
[a]*Present address : Research Center for Ultra High Energy Density Heat Source,*
Welding Research Institute, Osaka University,
11-1 Mihogaoka, Ibaraki, Osaka 567, Japan
[b]*Also the Department of Electromagnetic Energy Engineering,*
Osaka University, 2-1 Yamada-oka, Suita, Osaka 565 Japan
[c]*Osaka Institute of Technology, Asahiku, Osaka 535, Japan*

Abstract. We have observed stimulated Raman scattering (SRS) from two overlapped 527 nm laser beams. The short and long wavelength components were included in the SRS spectra. Plastic (CH) and aluminum (Al) planar targets were used, whose effective ion charges (Z) in underdense plasmas are 3.5 and 13, respectively. The intensity of the short wavelength component from Al (Z=13) target was stronger than that from CH target (Z=3.5). This Z dependence is different from the one using one laser beam irradiation in this study and the results shown by R. E. Turner et al. (1). We inferred the effect of two laser beam overlapping for the short wavelength component.

INTRODUCTION

Stimulated Raman scattering (SRS) is a three wave parametric instability in which an incident laser-light wave decays into a scattered-light wave and a electron plasma wave (2). The relations of the frequencies and wave vectors of these three waves are $\omega_0 = \omega_{SRS} + \omega_{EPW}$, $\mathbf{k}_0 = \mathbf{k}_{SRS} + \mathbf{k}_{EPW}$, where ω_0, ω_{SRS} and ω_{EPW} and $\mathbf{k}_0$, $\mathbf{k}_{SRS}$ and $\mathbf{k}_{EPW}$ denote the frequencies and the wave vectors of the incident electromagnetic light, scattered-light and electron-plasma waves.

In laser inertial confinement fusion (ICF) the scattered-light wave could reduce

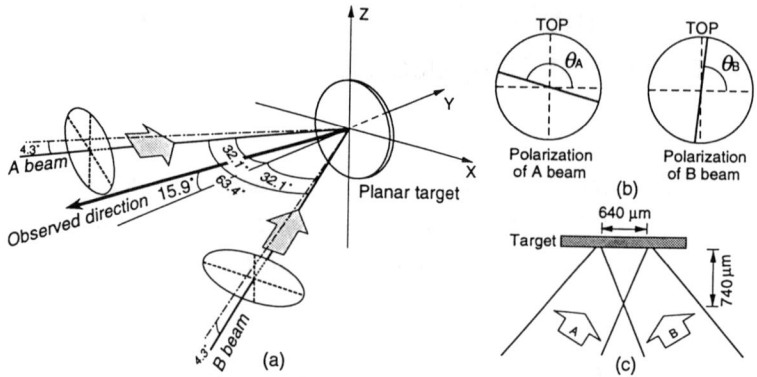

FIGURE 1. (a) This is the configuration of laser irradiation. (b) θ_A and θ_B are 162.5° and 85.5°, respectively. (c) The top view indicates lengths of overlapping.

the fusion efficiency by carrying the laser energy away from the fuel target, and the electron-plasma wave could preheat the fuel by hot-electrons through the wave damping (3). It is important for the achievement of high efficiency implosion to understand the detailed mechanisms of the SRS and to minimize or control it. In previous laser-plasma experiments thin foils or disk targets were irradiated with only one laser beam (one-beam experiments) (1, 4). However laser beams are overlapped on the fuel target in implosion experiment. In this paper, we report the first observation of SRS from two overlapped laser beams (two-beam experiments).

EXPERIMENTAL CONDITIONS

Experiments were performed with two beams of the GEKKO XII glass laser system at the Institute of Laser Engineering, Osaka University (5). The 527 nm light was obtained by frequency doubling of the Nd-doped glass laser light, 1053 nm, in a potassium dihydorogen phosplate (KDP) crystals. A random phased plate (RPP) was placed near the focusing lens (6). The 527 nm light was focused onto the target by a f/3 lens. Two beams (named A beam and B beam) were overlapped on the target as shown in Fig. 1(a) with a cartesian co-ordinate system with x, y and z axes. The shape of the laser pulse was approximately Gaussian, with a full width at half maximum (FWHM) of about 900 ps. Incident angle of each beam center was 32.1° and the angle between A beam and B beam was 63.4°. Polarizations of A and B beams were shown in Fig. 1(b), θ_A=162.5° and θ_B=85.5°. The top view of overlapped region was shown in Fig. 1 (c). Lengths of overlapping are 640 μm on x axis and 740 μm on y axis. The laser intensity in overlapped region was 3.5×10^{14} W/cm^2. We further performed one-beam

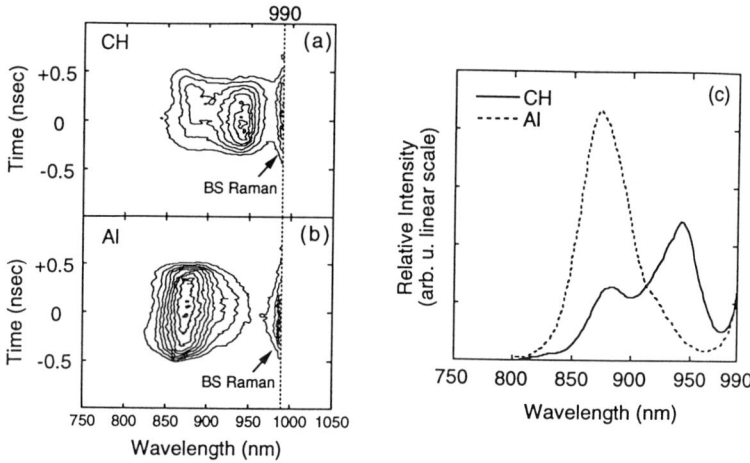

FIGURE 2. (a), (b) These are the streak images from CH and Al targets and (c) their time-integrated spectra in two-beam experiments.

experiments to compare the results of two-beam experiments. For one-beam experiments, the targets were irradiated with only A beam in Fig. 1 (a). The FWHM of the laser pulse and the intensity were 700 ps and 2.4×10^{14} W/cm^2. Plastic (CH : Z=3.5) and aluminum (Al : Z=13) disk targets were used in the two-beam and one-beam experiments. The thicknesses of plastic and aluminum targets were 14 µm and 5 µm, respectively and each diameter was 4 mm. These targets were chosen so that they would not burn through during the experiment. The angle between the observed direction and the target normal is 15.9°. The scattered light from the targets was collected by two concave aluminum mirrors and was guided to a 0.25 m grating spectrometer coupled to an optical streak camera (S-1 photo cathode).

EXPERIMENTAL RESULTS

Streak images from CH and Al targets in two-beam experiments are shown in Figs. 2 (a) and 2 (b), respectively. Each zero on vertical axes indicates the time of maximum laser intensity. The blue-shifted (BS) Raman light was also observed between 970 nm and 990 nm (7). To indicate the difference of these images between CH and Al target cases, the time-integrated spectra were shown in Fig. 2 (c). The solid and the dotted lines in Fig. 2 (c) shows a SRS spectrum from CH and Al targets in two-beam experiments, respectively. Figures 2 (a) and 2 (c) show two components in SRS spectra from CH target. The spectral peaks of the short and the long wavelength components are 880 nm and 940 nm, respectively.

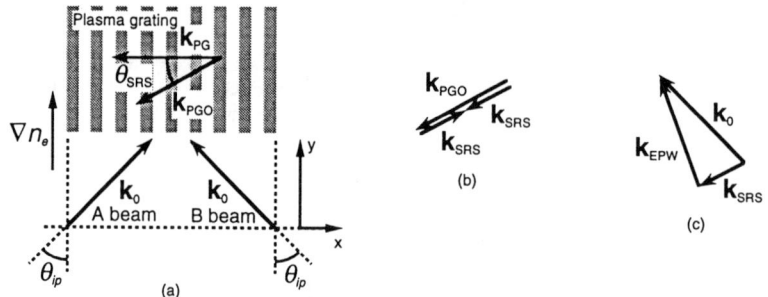

FIGURE 3. (a)The interference between A and B beams gives rise to the ponderomotive force. The modulation of density profile (plasma grating) is produced by the ponderomotive force. (b) This is a wave number condition for reflection of the scattered light due to the plasma grating. (c) This is a wave number condition for SRS.

For the long wavelength component from Al target the intensity is less than that for CH target. This tendency is same as the results in our one-beam experiments and the results by R. E. Turner et al. (1). The intensity of the short wavelength component increased when Al target were irradiated as shown in Fig. 2 (c). The increase was observed for the first time and these results didnot indicate the Z dependence of usually observed Raman light (1). Results in two-beam experiments indicate the short wavelength components were generated with some other mechanisms.

DISCUSSION AND SUMMARY

We discuss a generation mechanism of the short wavelength component and the Z dependence in the two-beam experiments. The interference between A and B beams gives rise to the ponderomotive force. The modulation of density profile (plasma grating) is produced by the ponderomotive force. The wave vector of the plasma grating has the relation of $\mathbf{k}_{PG}=2\mathbf{k}_0\sin\theta_{ip}$, where $\mathbf{k}_{PG}$, $\mathbf{k}_0$ and θ_{ip} denote the wave vectors of the plasma grating, the incident laser beams (A and B beams) and the incident angle of each laser beam in plasmas as shown in Fig. 3 (a). The direction of $\mathbf{k}_{PG}$ is perpendicular to the density gradient ∇n_e of expanding plasmas from the targets. $\mathbf{k}_{PGO}$ is a wave vector of the plasma grating on the oblique direction, $\mathbf{k}_{PGO}=\mathbf{k}_{PG}\cos\theta_{SRS}$, where θ_{SRS} is an angle between $\mathbf{k}_{PG}$ and $\mathbf{k}_{PGO}$ in Fig. 3 (a). When the condition of $\mathbf{k}_{PGO}=2\mathbf{k}_{SRS}$ is satisfied as shown in Fig. 3 (b), the scattered light can be reflected with the plasma grating. The reflected light and another reflected light propagated in the counter direction could make a standing wave of the scattered light with $\mathbf{k}_{SRS}$. When the wave number condition of SRS is satisfied as shown in

Fig. 3 (c), SRS can be grown with absolute mode since the scattered light with k_{SRS} is a standing wave. We infer that the standing wave is a noise source for the short wavelength component. The intensity of the standing wave increases as reflectivity due to the plasma grating increases. For steady state homogeneous plasmas, reflectivity of stimulated Brillouin scattering due to ion wave, R, is given by $R=\tanh^2\{\alpha(\delta n_i/n_i)L\}$, where α, δn_i, n_i and L are $(\pi/2)(n_e/n_{cr})\lambda_0(1-n_e/n_{cr})^{-1/2}$, the amplitude of ion wave, the back ground ion density and length of homogeneous plasmas (8). n_e, n_{cr} and λ_0 are the electron density, the critical density and the wavelength of incident laser. For inhomogeneous plasmas in two-beam experiment, the equation of reflectivity reduces to

$$R = \left(\frac{\pi}{2}\right)^2 \left(\frac{n_e}{n_{cr}}\right)^2 \left(1 - \frac{n_e}{n_{cr}}\right)^{-2} \left(\frac{\ell_{INT}}{\lambda_{SRS}}\right)^2 \left(\frac{\delta n_e}{n_e}\right)^2, \quad (1)$$

since exponent is small and ℓ_{INT}, λ_{SRS} and $\delta n_e/n_e$ are inserted instead of L, λ_0 and $\delta n_i/n_i$. Here ℓ_{INT}, λ_{SRS} and δn_e are the interaction length satisfied the SRS wave number condition, the wavelength of the scattered light and the amplitude of the electron density of the plasma grating, respectively. Equation (1) indicates the reflectivity of the scattered light with λ_{SRS} due to the plasma grating. ℓ_{INT} increases as the scale length L_p increases and as θ_{SRS} decreases in inhomogeneous plasmas.

When the plasma grating is produced, $\delta n_e/n_e$ could be estimated with the force equation $\nabla p_e = F_p$, where p_e is the electron pressure given by $p_e=(n_e+\delta n_e)T_e$ and F_p is the ponderomotive force given by $F_p=-(\omega_{pe}/\omega_0)^2(\nabla\langle E_{IF}^2\rangle/8\pi)$. T_e is the electron temperature. E_{IF} is the electric field of the interference fringes due to two beams and is given by $\mathbf{E}_{IF}=(\mathbf{E}_{Ax}\cos\theta_A \cdot \mathbf{E}_{Bx}\cos\theta_B) \cdot (\mathbf{E}_{Ax}\sin\theta_A \cdot \mathbf{E}_{Bx}\sin\theta_B)$, where E_{Ax} and E_{Bx} are the electric fields of A and B beams on the x axis, respectively. $\delta n_e/n_e$ is given by

$$\frac{\delta n_e}{n_e} = \frac{I_0}{cn_{cr}T_e}\left(\cos\theta_A\cos\theta_B\cos 2\theta_{ip} - \sin\theta_A\sin\theta_B\right), \quad (2)$$

where I_0 is the intensity of one incident laser beam.

T_e, L_p and laser absorption rate could be predicted with the one-dimensional (1-D) hydrodynamic code "HISHO" (9). T_e for Al target is higher than for CH target and L_p for Al target is shorter than for CH target. Then $\delta n_e/n_e$ for Al target is smaller than for CH target as indicated in Eq. (2) and ℓ_{INT} for Al target is shorter than for CH target since ℓ_{INT} decreases as L_p decreases. R for Al target is lower than for CH target and the intensity of the noise source due to the plasma grating for Al target is weaker than for CH target. However when the reflected light of the incident laser from the target surface influences the noise source, we cannot estimate the Z dependence of the short wavelength component with only Eq. (1). When the incident laser light is not reflected from the target surface, the plasma grating could be produced clearly. But in the experiments the incident laser light is reflected from the target surface. Another plasma gratings could be produced by the interference between two incident and two reflected laser beams. Then the noise source might be suppressed since another plasma gratings is overlapped on the main plasma grating for the noise source and $k_{PGO}=2k_{SRS}$ could not be satisfied. By estimation with 1-D hydrodynamic code " HISHO" laser absorption rate for Al target is higher than for CH target. The results indicate that the reflected light of incident laser from CH target surface suppress the noise source and generation of the short wavelength component more than for Al target.

In summary, we observed the SRS spectra composed with the short and the long wavelength components. The Z dependence of the short wavelength component was interpreted with the possible plasma gratings existing in plasmas produced with two overlapped laser beams.

ACKNOWLEDGMENTS

The authors wish to acknowledge the helpful comments by Dr. S. Kuruma, Dr. S. Kato and Mr. M. Honda. These experiments could not have been possible without considerable effort of the laser operation by the GOD group, and the target fabrication by the T group.

REFERENCES

1. R.E. Turner, Kent Estabrook, R. L. Kauffman, D. R. Bach, R. P. Drake, D. W. Phillion, B. F. Lasinski, E. M./ Campbell, W. L. Kruer, and E. A. Williams, Phys. Rev. Lett. **54**, 189 (1985).
2. W. L. Kruer, *The Physics of Laser Plasma Interactions* (Addison-Wealey Publishing Company, Inc., California 1987), p.73.
3. K. G. Estabrook, W. L. Kruer and B. F. Lasinski, Phys. Rev. Lett. **45**, 1399 (1980).
4. R. P. Drake, R. E. Turner, B. F. Lasinski, E. A. Williams, D. W. Phillion, K. G. Estabrook, W. L. Kruer, E. M. Campbell, K. R. Manes, and J. S. Hildum, Phys. Fluids **31**, 3130 (1988).
5. *World Survey of Activities in Controlled Fusion Research* [Nuclear Fusion special supplement 1991], (International Atomic Energy Agency, Vienna, 1991), p. 236.
6. Y. Kato, K. Mima, N. Miyanaga, S. Arinaga, Y. Kitagawa, M. Nakatsuka and C. Yamanaka, Phys. Rev. Lett. **53**, 1057 (1984).
7. M. Tsukamoto, K. A. Tanaka, K. Mima, M. Kado, S. Miyamoto, M. Nakai, T. Norimatsu, M. Takagi, K. Nishihara, T. Yamanaka, S. Nakai, and A. Nishiguchi, Phys. Plasmas **2**, 486 (1995).
8. W. L. Kruer, Phys. Fluids **23**, 1273 (1980).
9. See Ref. 5, p. 365.

The investigation of stimulated Raman scattering in laser produced plasma for heating wavelength 0.53µm.

E.A.Bolkhovitinov, V.Y.Bychenkov, M.O.Koshevoi, M.V.Osipov,
A.A.Rupasov, A.S.Shikanov, V.T.Tikhonchuk,
A.V.Kilpio[*], N.G.Kiselev[*], D.G.Kochiev[*], P.P.Pashinin[*],
E.V.Shashkov[*], Y.A.Suchkov[*].

Lebedev Physical Institute and [*]General Physics Institute, Moscow, Russia

Abstract. Experimental results of stimulated Raman scattering (SRS) investigation obtained on Nd-glass laser installation are explained on the basis of representations about possibility of plasma profile flat regions formation and their motion. Low and high electron density SRS spectrum cut-off are discussed as well as time-integrated and time-resolved SRS spectrum features.

Experimental results.

The stimulated Raman scattering in laser produced plasma was studied and a new scenario of time-integrated and time-resolved measurements was offered. Experimental results were obtained on a single-channel laser installation "Kamerton" [1]. The plasma was produced by 2.5ns and 30-80J (λ_0=527nm) laser pulse focused onto plane polyethylene targets of 5-20µm thicknesses. The focusing spot dia was ~100µm, the flux density of the heating radiation reached $3*10^{14}$ W/cm^2. The scheme of diagnostic set up is shown in Fig.1.

In experiments described in publications [2-4] the SRS was studied for massive targets, thin foils and the prepared plasma. An intermediate case of laser interaction with target, when a target thickness was greater than, but comparable with the depth of burning-through, was investigated in this report. That is the most interesting case for controlled fusion schemes because it corresponds to the largest coefficient of laser energy conversion into the target kinetic energy.

SRS was observed at different angles θ to the heating beam with different spectral resolution Δλ, time resolution Δt and spatial resolution Δx. As shown in Fig.1, the SRS spectra are registered by means of a many-channel optical analyser OVA-284 (θ~135°; $\Delta\lambda$~1nm; spectral region 580-900nm), spectrographs ISP-51 (θ~90°,180°;

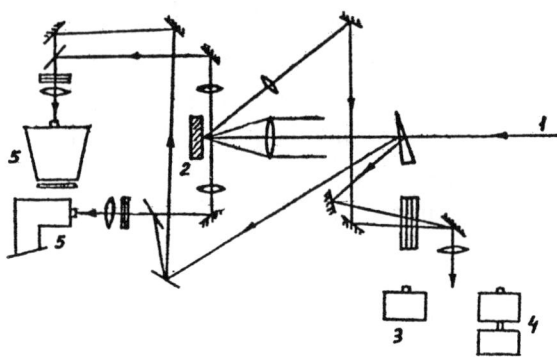

Figure 1. 1 - laser beam, 2 - target, 3 - OVA-284, 4 - spectrograph with streak-camera, 5 - spectrographs ISP-51, STE-1.

$\Delta\lambda$~0.5nm; Δx~20μm; spectral range 400-900nm), STE-1 (θ~90°-180°; $\Delta\lambda$~0.1nm; spectral range 450-900 nm), as well the spectrograph coupled with electron-optical chamber Hamamatsu-C979 (θ~135°; $\Delta\lambda$~1 nm; Δt~30 ps; spectral range 630-830 nm).

All the observed SRS spectra for angles observations θ=135° and 180° and 6 μm foil thickness were located in a wavelength range $\lambda_{min}<\lambda<\lambda_{max}$, where λ_{min}=610-650 nm and λ_{max}=690-750 nm. The typical densitometer tracing of time-integrated SRS spectrum is presented on Fig.2. In accordance with dispersion equations for Raman scattering each wavelength λ is connected with the definite scattering plasma density n:

$$n/n_c=(1-\lambda_0/\lambda)^2 \qquad (1)$$

where n_c - electron critical density, λ_0 – heating radiation wavelength. So, the measured values of λ_{min}= 640nm and λ_{max}=725nm for Fig.2 correspond to the density values of n_{min}= 0.030n_c and n_{max}=0.075n_c. Two scales of intensity modulations were observed with distance between peaks $\Delta\lambda$~0.5–1.0nm and 10-20nm. With increase of the foil thickness up to 20μm the SRS intensity decreases, the SRS spectrum is narrowed and shifted to longer wavelengths. At the angle 90° to the heating beam no SRS was recorded.

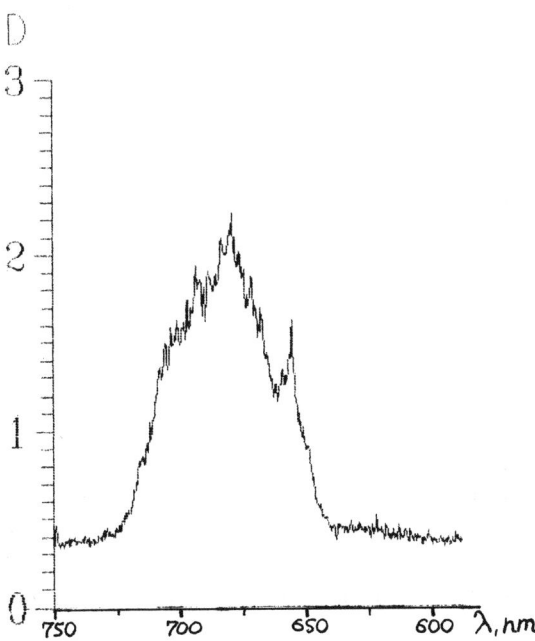

Figure 2. Typical densitogram of time-integrated SRS spectrum for 6μm thickness foil.

The time-resolved SRS spectra had a complex structure. Typical time-resolved spectrogram is presented on Fig.3a. In the initial part of the laser pulse the SRS radiation of long wavelengths arises and then during laser pulse duration the spectrum is shifted to the shorter wavelengths, i.e. the region of scattering is shifted to the more rare plasma. Therefore the SRS duration at separate wavelength is rather shot (~0.8ns) than the total duration (~2.5ns). There are two types of modulation - temporal (~0.1ns) and spectral (~50nm). Fig.3b-3d present intensity distribution obtained from time-resolved spectrum of Fig.3a for three time moments. For more thick foils the SRS dynamics is changed - the scattered radiation begins simultaneously almost in broad spectral range and there is no the strong dependence of radiation wavelength on time.

Discussions.

The fact of SRS observations suggests that the laser intensity is above the threshold level in the certain density region. Usually the low density (small wavelength) cut-off of the SRS spectrum is related to the rise of the SRS threshold due to the enhanced Landau damping in a low density plasma. Because of the exponential growth of the Landau damping with the wavenumber it is reasonable to consider the inequality

$$k_p r_{De} < 1/3 \qquad (2)$$

as the necessary condition for the SRS (where k_p - plasma wave wavenumber, r_{De} - electron Debay radius). For back scattering $k_p=2k_o$ (k_o - heating radiations wavenumber) and from (2) one can obtain the impression for electron temperature

$$T_e = (m_e c^2/36)*(n_{min}/n_c) \qquad (3)$$

From equations (1) and (3) for a minimum wavelength λ_{min}=630 nm the value of low density cut-off n_{min}=0.025n_c and the evaluation of electron temperature T_e=350eV can be obtained. We can accept this estimate as an overall plasma corona temperature which is valid for the higher densities too.

Now about high density cut-off n_{max}, which is equal to 0.09n_c from (1) for λ_{max}=750 nm. It is possible to suggest that this cut-off at electron density n_{max} is also related to the SRS threshold condition. The formula for SRS threshold:

$$v_{e,th}^2 / c^2 = v_{ei}^2 \omega_p / \omega_o^3 + \pi \lambda_o / L_n \qquad (4)$$

where v_e - electron oscillation velocity in laser field exiting SRS instability for $v_e>v_{e,th}$, v_{ei} - electron-ion collision frequency, L_n - electron density scale length.
The dissipative SRS threshold (the first item in eq.(4)) is below 10^{13} W/cm^2 for $T_e\sim$350eV. However the convective part of the threshold (the second item in (4)) is very large, so that we should assume the inhomogeneity scale length $L_n \sim$ 3cm in order to get the necessary threshold intensity about $3*10^{14}$ W/cm^2. This is an unrealistically large length. Meanwhile SRS simulations [5] in the model of homogeneous plasma slab show that the interaction length about 50μm is enough to amplify backscattering signal 400 times above the noise level. Therefore if we assume the existence of one or several local flat regions on the plasma density profile (similar to observed in [6]) it is possible to explain the exiting of SRS for used laser flux density. Such flat region can not stay too long, but the time scale about 10-20 ps is enough to excite SRS and get it into the saturation. Then the flat interaction region can by destroyed (due to the SRS itself or because of hydrodynamic motion of plasma) and appears later in the different place of plasma. This scenario suggests that the instantaneous SRS emission has a very narrow spectral line but it randomly changes its position during the pulse. This may by related to the peak structure of the observed SRS spectra. The motion of flat regions down along the density profile will give the decreasing the SRS radiation wavelength, illustrated by Fig.3b-3d. The high density SRS cut-off we can attribute to steeper gradients in the region n/n_c>0.1 where a laser beam can

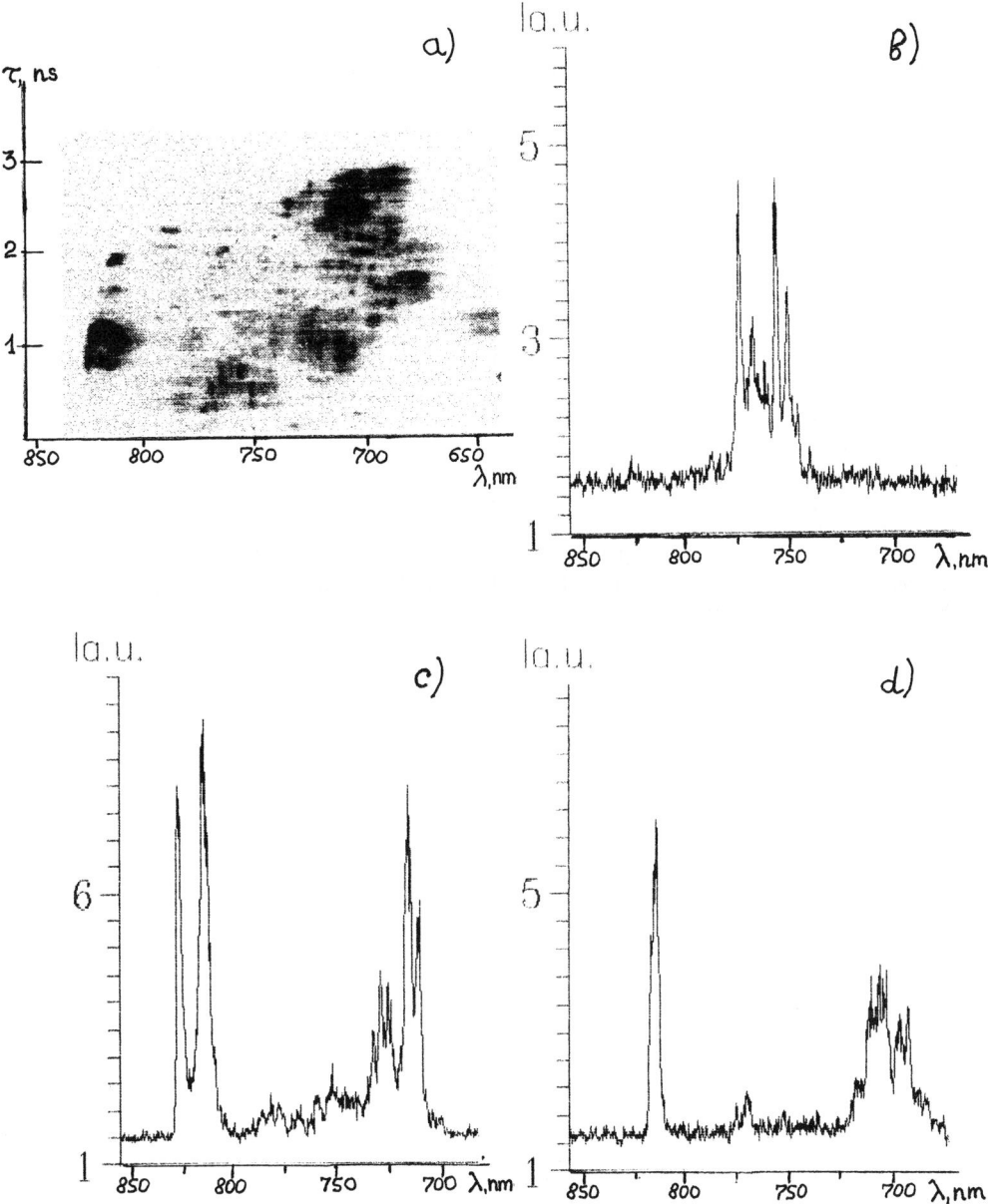

Figure 3. (a) - typical time-resolved SRS spectrum; (b),(c),(d) - SRS radiation intensity distributions for time moments 0.5ns, 0.9ns and 1.9ns, respectively.

not produce flat profiles of the necessary extent. Also the plasma from more thick foil targets has more steep density profile and therefore the possible flat regions should be shorter and generate less SRS that is in accordance with experimental measurements.

This work was carried out partially under support of International Science Foundation grant MM-1000 and Russian Foundation of Fundamental Researches grant 94-02-03864-a.

References

1. Kilpio A.V., Kiselev N.G., Kochiev D.G., Koshevoi M.O., Osipov M.V., Pashinin P.P., Rupasov A.A., Sarkisov G.S., Shashkov E.G., Shikanov A.S., Suchkov Yu.A. "Interaction of Nd-glass laser second harmonic radiation with plane targets on "Kamerton" installation," *in Proceedings of the 21st ECLIM*, 1991, pp.95-98.
2. Laser Programm Annual Report - 1986, LLNL UCRL, 50021-86, Livermore, 1986.
3. Walsh C.J. et al. *Phys.Rev.Letts*, **53**, 1445 (1984).
4. Bulatov E.D. et al. *in Proceeding 12st ECLIM*, 1979, p.13.
5. Kolber T., Rozmus W., Tikhonchuk V.T., *Phys. Plasmas*, v.**2(1)**, 256-273 (1995).
6. Zaharenkov Yu.A., Zorev N.N., Krohin O.N., Mikhailov Yu.A., Rupasov A.A., Sklizkov G.V., Shikanov A.S., Pisma v ZhETF, **21(9)**, 557-561 (1975).

Anomalous High Energy Electron Emission from Laser Plasma

Ivanov V.V., Knyazev A.K., Koutsenko A.V.,
Matzveiko A.A., Mikhailov Yu.A., Osetrov V.P.,
Popov A.I., Sklizkov G.V. and Starodub A.N.

*Laser Plasma Laboratory, P.N.Lebedev Physical Institute,
Leninsky Pr. 53, Moscow 117924, Russia*

ABSTRACT. An original method based on direct measurements of electron emission has been proposed. The energy of fast electrons experimentally has been observed up to the value of 200 keV with current flux density 10^5 A/cm^2. The dependencies of electron energy on radiation flux density as well as current density corresponding to cut off boundary of electron energy have been measured in the range of plasma temperatures from 70 to 200 eV. It is observed that a hydrodynamic instability leads to sharpening electron current pulse front at electron energy exceeding 100 eV. In the range of 1-10 keV the value of electron current more than 2 kA has been recorded. In this case, the current pulse duration is a few times more than the laser pulse, and the current pulse front is comparable with it.

The investigation of fast electrons with the energy rather higher than the energy of thermalized electrons generated from the laser plasma is important both for fundamental dense plasma physics and for some applications. In the light of an understanding of plasma phenomena which can be responsible for the generation of high energy electrons it is necessary to mention well known nonlinear phenomena as well. For example, the resonance absorption and the parametric instabilities in the vicinity of the critical point, and the two plasmon decay near one-quarter density, and SRS in the plasma corona (see for example [1]) can be favourable to an increase of the electron energy. Furthermore, the presence of fast electrons in the laser heated plasma can play a negative role for the direct compression of a laser fusion thermonuclear target. Even a small total energy of fast electrons as low as about one percent of the laser absorbed energy can lead to the preheating of the compressed core and it can result in catastrophyc decrease of the compression rate. As a practical matter, the possibility of high energy electron beam formation by a laser plasma opens up the way to the development of a new injector for the pulsed high current induction accelerator [2].

At the present time there are a set of investigations (for example [3,4,5]) dealt with the analysis of the super thermal electrons production in the range of energy from keV to MeV. There are also some experimental investigations related to the same problem at different experimental conditions (CO_2-laser, λ = 10.6 μm [6]; Nd-laser, λ =1.06 μm and its harmonics λ = 0.263 μm, λ = 0.526 μm) [7]. The maximum electron energy achieved is about 100 keV and it exceeds 1 MeV in some experiments [8,9]. The typical detectors used in the fast electron experiments are the electron spetrometer with photoemulsion and Si-diodes. These detectors are positioned at different distances from laser plasma electron source. All measurements have been done in a small solid angle. That is why, it was difficult to obtain experimental data on the integral electron current from the plasma. The same shortage is associated with another set of experiments when the indirect observation of the fast electrons is made by the processing of plasma X-ray radiation data [8].

In the present paper experimental data on investigation of the fast electrons generated from the laser plasma are given and the registration metodics for direct measurement of fast electrons energy and current is described. This methodics allows to record maximum energy of electrons simultaneously with the measurement of their current. Other words, it allows to measure the energy and the total number of the electrons in one shot. The experimental measurements have been carried out with the installation "PICO" which consists of two Nd-lasers and target chamber. The irradiation parameters are the following: λ = 1.06 μm, $\delta\lambda$ = 30 Å, laser pulse duration 2 ns, energy varies from 0.1 to 30 J and flux density from 10^{12} to $8 \cdot 10^{13}$ W/cm^2. Only nanosecond part of the installation has been used to irradiate the plane cylindrical target (12 mm length and 5 mm dia.; Cu and Al as a target material).

The schematic diagram is given in Fig.1. The cylindrical target is inserted into the coaxial capacitor which is connected with the

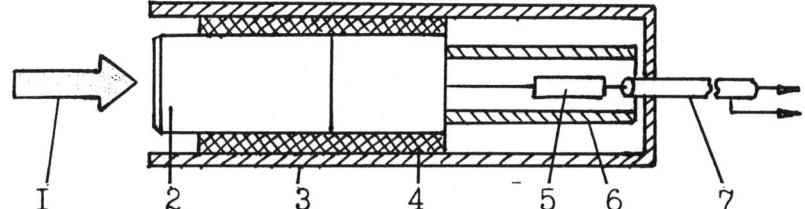

FIGURE 1. The schematic diagram of experimental setup : 1 − laser beam, 2 − target, 3 − coaxial capacitor, 4 − isolator, 5 − resistor, 6 − electromagnetic screen, 7 − transmitting line.

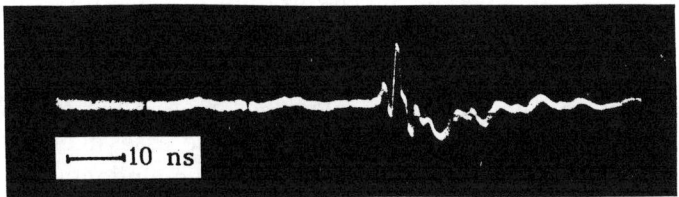

FIGURE 2. The time evolution of fast electrons current.

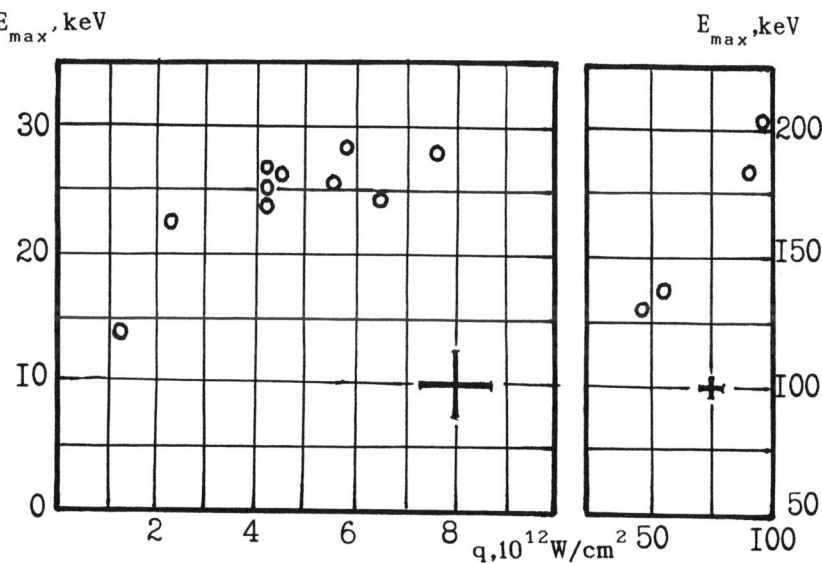

FIGURE 3. The dependency of maximum electron energy on laser flux density for AL target.

oscilloscope through the long transmitting line and the matched and calibrated adjustable attenuator. The temporal accuracy of the electron current pulse recording with respect to the beginning of the laser pulse is not worse than 0.5 ns. The time evolution of fast electrons current is shown in Fig.2. It is observed that the hydrodynamic instability leads to the sharpening of the current pulse front at the energy more than 100 keV.

The current pause has been observed during heating period in some shots. The value of the electron current more than 2 kA has been registered in the range of the electron energy from 1 to 10 keV. In this case, the current pulse duration is a few times more than the laser pulse and the current rise time is comparable with a half laser pulse duration.

The maximum energy of fast electrons more than 200 keV has been observed experimentally. In this case the assessment of current density give the value of 10^5 A/cm^2. The dependency of electron energy on laser radiation flux density at the focus has been measured in the range of plasma temperature from 70 to 200 eV which is evaluated by X-ray absorbing procedure. The current density from the target surface in the immediate vicinity has been estimated knowing focal spot diameter. Fig. 3 presents the dependency of maximum electron energy on laser flux density. The strong dependency (more than the second power of laser flux) has been observed at the 2 A level of the probe sensitivity.

Note that it is difficult to choose the only mechanism of electron acceleration in laser heated plasma. The theoretical evaluation of the maximum oscillatory energy determining the typical value of the fast electrons energy is given in [4]. With the supposition of resonance absorption and for the case of a linear plasma density profile the energy is determined as follows:

$$E_{max} \approx 4 e E_0 L \sqrt{fc/\pi \omega L} \approx 2 \cdot 10^{-6} \sqrt{f q_0 \lambda L} ,$$

where e - electron charge, E_0 - electric field strength of laser radiation, f - fraction of electromagnetic wave energy transformed to electrostatic field, L - typical plasma dimension in μm, ω and λ - frequency and wavelength (μm) of laser radiation, q_0 - laser radiation flux density in W/cm^2. The estimation for $f = 1$, $q_0 = 5 \cdot 10^{13}$ W/cm^2, $L = 50$ μm and $\lambda = 1.06$ μm gives $E_{max} = 100$ keV.

Approximately the same value can be obtained for the electron accelerated in the caviton field which is modelled in [3]. According to this model the superthermal peak appears on the electron energy distribution function due to high increase of the located field in a caviton. This peak is characterized by an effective "hot temperature" T_h, which can be approximated by the following expression:

$$T_h = 9{,}6 \frac{R_0^{0,2}}{A_b^{0,26}} \left(\frac{z+4}{14}\right)^{0,26} (q_0 \lambda)^{0,4} ,$$

where R_0 - evaporation wave front radius in cm, A_b - laser energy absorption coefficient in plasma, $[q_0 \lambda] = 10^{14}$ Wμm/cm^2, $[T_h] = $ keV. $T_h = 10$ keV for the typical experimental conditions. This value is in a good agreement with X-ray bremstrahlung measurements. The maximum energy of the fast electrons can be estimated as $E_{max} = 0{.}125\, D^2 T_h$, where $D = d/\lambda_D$ - halfwidth of field (caviton) near critical density expressed by Debye radius λ_D. For typical experimental conditions $D = 7$ and $E_{max} \approx 60$ keV.

The maximum electron energy estimated by these models are essentially less than the values obtained in the present work. The possible reason of the fast electron energy increase may be, for example, the direct acceleration of electrons in local plasma regions with small scale perturbations of the laser flux density connected with coherent effects at the target surface. The peak values of flux density fluctuations can be 10-100 times more than the average laser flux on the target surface.

Finally a few words about some interesting aspects related to fast electrons and X-ray generation in plasma created by ultrashort pulses ($\tau = 1$-10 ps) with flux density $q > 10^{19}$ W/cm^2. A small portion of plasma electrons can be accelerated by the laser field E_s up to the energy ($E_{max} = e E_s \lambda$) which is comparable with the rest mass $mc^2 = 0{.}51$ MeV. The maximum electron energy up to 0.5 - 1 MeV have been observed experimentally even at the moderate flux density typical for "Delfin" installation [8,9]. The possible generation of relativistic electrons in laser plasma allows to observe ultra hard X-rays (UHXR) with the photon energy about 1 MeV associated with the Compton scattering of the laser photons by the fast electrons.

The maximum energy of the scattered photons is $\hbar\omega' = 4 E_{max}^2 \hbar\omega /(mc^2)^2$ when $E_{max} > mc^2$. The photons are concentrated in a small angle $|\theta| \leq mc^2/E_{max}$ directed along the laser beam. If taking into account the electron energy obtained in our experiment 200 keV at $q = 8\cdot10^{13}$ W/cm^2 and the dependence $E_{max} \approx \sqrt{q}$ the value of scattered photon energy of $\approx 10^6$ eV is to be observed at flux densities of 10^{20} W/cm^2.

ACKNOWLEDGMENTS

This work was supported by International Science Foundation grant MN 5000 and by Russian Foundation for Basic Research grant 94-02-05794.

REFERENCES

1. Forslund, D.W., Kindel, J.M., Lee, K., Phys. Rev. Lett. **39**, 284-287 (1977).

2. Pavlovskii, A.I., Kuleshov, G.D., Sklizkov, G.V., Zysin, Yu.A., Gerasimov, A.I., Sov. Phys. - DAN **160**, 1, 68-70 (1965).

3. Rozanov, V.B., Shumskii, S.A., Theory of target compression by longwavelength laser radiation, Moscow: Nauka Press, 1990, pp. 8-33. (Proc. P.N.Lebedev Phys. Inst., Vol. 170).

4. Albritton, J., Koch, P., Phys. Fluids **18**, 1136-1139 (1975).

5. Kovrizhnih, L.M., Sakharov, A.S., Plasma Physics **5**, 840 (1979).

6. Ebrahim, N.A., Baldis, H.A., Joshi, C., Phys. Lett. **A84**, 253-254 (1981).

7. Rousseaux, C., Amiranoff, F., Labaune, C., Matthieussent, G., Phys. Fluids **B4** (8), 2589-2594 (1992).

8. Kalashnikov, M.P., Mikhailov, Yu.A., Lyapidevskii, V.K., Prorvich, V.A., Sklizkov, G.V., Sov. Phys.- Lebedev Institute Reports 8, 41-43 (1983).

9. Danilov, A.E., Mikhailov, Yu.A., Nikolaev, F.A., Sklizkov, G.V., Sov. Phys. - JETP Lett. **37**, 11, 522-525 (1983).

NONLINEAR INTERACTION OF HIGH POWER MICROWAVE WITH AN INHOMOGENEOUS PLASMA

X. Xu, H. Itoh, N. Yugami, B. Cros and Y. Nishida

Department of Electrical and Electronic Engineering,
Utsunomiya University, 2753 Ishii-Machi,
Utsunomiya, Tochigi 321, Japan
Laboratoire de Physique des Gaz et des Plasmas,
Bâtiment 212, Université Paris XI, 91405 Orsay, France.

Abstract

A frequency up-shift is observed in the interaction of a high power microwave with an unmagnetic inhomogeneous argon plasma. The induced frequency up-shift is considered to be related to a rapid expanding plasma which is created by the ponderomotive of a strong standing wave.

INTRODUCTION

Recently, the frequency shift has been studied in laser plasma interaction [1,2,3]. Theoretical analysis [2] shows that when a laser pulse propagates in the region of a ionization front in which the local density gradient is negative, it will be continuously up-shifted in frequency. This mechanism has been beautifully illustrated in a recent experiment by W. M. Wood et al. [3].

In our experiments, a high power electromagnetic wave ($f_0 = 9GHz$, $P_i \leq 250kW$, $\tau \simeq 1\mu s, \varepsilon_0 E_0^2/4n_{cr}KT_e \approx 0.1$) is launched into an unmagnetized argon plasma ($n_0 \leq 1 \times 10^{12} cm^{-3}$). In underdense region, frequency up-shift and some high frequency components have been observed. The dependence of the frequency up-shift and new frequency components on incident power and plasma density have also been investigated. We think this new phenomenon is related to a rapid expanding plasma which is created by the ponderomotive force arising from a standing wave. The incident microwave is partially reflected by a moving plasma and the frequency is up shofted due to the Doppler effect.

EXPERIMENTAL ARRANGEMENT

A cylindrical, azimuthally symmetric, unmagnetized argon plasma column was created in a vacuum chamber (32cm in diameter, 60cm in length). The outside surface of the vacuum chamber is covered with a large number of multi-dipole permanent magnets (cusp spacing $\simeq 4$ cm). The plasma is produced by a pulsed discharge between two directly heated LaB_6 cathodes and the chamber well (grounded). A typical plasma discharge pulse duration is $t_w = 2ms$ with a repetion rate of 10 Hz. The experimental parameters are: argon gas pressure $P_{Ar} = 2.0 \times 10^{-3}$ Torr, maximum plasma density $n_e \leq 1.0 \times 10^{12} cm^{-3}$, electron temperature $T_e \simeq 3 \sim 5eV$. The estimated electron-ion collision frequency is $\nu_{ei} \simeq 4.5 \times 10^6 sec^{-1}$, and ion neutral collision frequency is $\nu_{in} \simeq 10^5 sec^{-1}$. Within the experimental region of concern, the density gradient scale length $L = n_{cr}|dn_{cr}/dz|^{-1}$ is observed to be approximately linear along the chamber axis with $L \simeq 100cm \sim 120cm$, while in the radial direction there is a much weaker density gradient. Pulsed microwave radiation ($f_0 = 9GHz$) with a maximum rise-time $\tau_r \approx 150ns$ and a typical pulse width of $\tau \simeq 1\mu s$ is generated by a single magnetron tube with a repetion rate of 10Hz and launched by a rectangular horn antenna (aperture area= $13.5 \times 10.5 cm^2$) from the lower density side of the plasma along the chamber axis in the z direction. Because the pulse width of the microwave is short enough for the pulsed discharge, the effects of ionization and/or heating of background plasma electrons can safely be neglected. The plasma density perturbation and rf signal in the plasma are detected by a plane Langmuir probe and cylindrical probe, respectively. A block diagram of the experimental setup is shown in Fig.1.

EXPERIMENTAL RUSULTS

The frequency spectra of microwave pulses with and without plasma in the chamber at is shown in Fig.2. The solid and dashed lines represent the spectra in vacuum and in plasma, respectively. The frequency shift around 1MHz is clearly demonstrated and besides the frequency up-shift, some new frequency components have also been observed. for example, the highest frequency component is 9GHz+5.8MHz. The attenuation of the up-shifted rf signal is caused by the plasma.

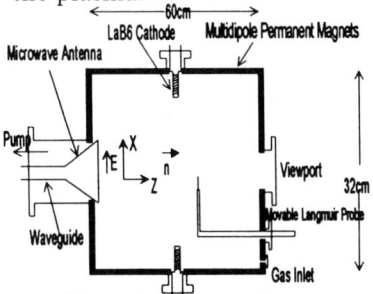

Fig.1 Experimental setup.

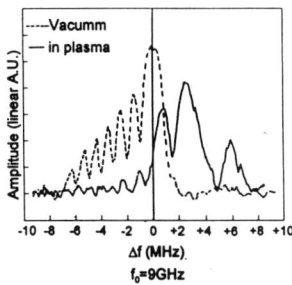

Fig.2 Frequency spectra of microwave without and with plasma.

Fig.3 shows that frequency up-shift and new frequency components for different plasma density. f_1 and f_m (m=2,3,4) represent the up-shift of center frequency (f_0= 9GHz) and the higher frequency components on the right side of f_1 in the spectra, respectively. It can be seen that at when the plasma density n_0 is increased from $0.03n_c$ ($n_c = 1 \times 10^{12} cm^{-3}$) to $0.76n_c$ (at Z=21cm, $n_0 \approx 0.06n_c \sim 0.9n_c$) the frequency up-shifts, which around 1MHz, are almost unchanged. But the frequencies of the higher frequency components in the spectra increase proportionally as the plasma density is increased. The frequency of f_2 shifts from 1.5MHz to 4.0MHz, and the frequency of f_3 shifts from 3.2MHz to 5.5MHz have been observed. So, it seems that the frequencies of the higher frequency components depend on plasma density more stronger than that of the frequency up-shift.

The relationship between the frequency up-shift, higher frequency components and the incident power is presented in Fig.4. We found that the frequency up-shift decreases from 1.5MHz to almost zero, and the frequencies of the higher frequency components also tend to become smaller as the incident power is changed from 250kW to 7.9kW. It is noted that when incident power $P_i \approx 7.9$kW the spectrum in the plasma is almost the same as that in vacuum. It implies that the threshold of incident power is around 7.9kW for the frequency up-shift observed in our experiments.

The velocity of plasma flow has also been investigated during the experiments. As shown in Fig.5, it is clear that the plasma flows propagate towards two directions from Z=11cm with a velocity $v \approx 1.0 \times 10^6 cm/s$. This velocity is almost the 5 times of c_s, the velocity of ion acoustic wave (In our case, $c_s \approx 2.5 \times 10^5 cm/s$).

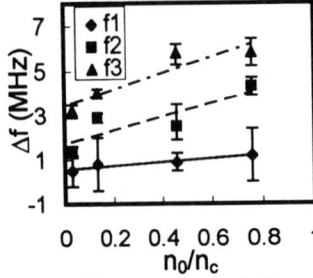

Fig.3 Frequency up-shift and higher frequency components vs. plasma density.

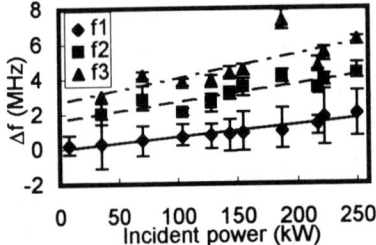

Fig.4 Frequency up-shift and higher frequency components vs. incident power.

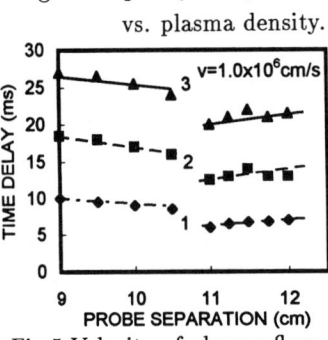

Fig.5 Velocity of plasma flow.

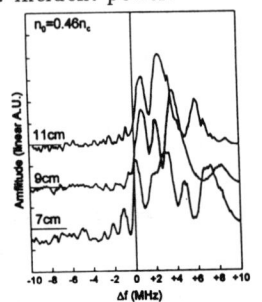

Fig.6 Spectra of microwave with plasma vs. position.

DISCUSSION

In the previous experiments of laser(microwave)-plasma interaction, only frequency red shift has been observed if there is no ionization front involved in the interaction. Such as in the experiments of stimulated Brillouin scattering (SBS) and stimulated Raman scattering (SRS). As shown above, it is obvious that in our experiments, the frequency up-shift and higher frequency components which observed in our experiments depend on the incident power of microwave and plasma density directly, and it also should be noted that the plasma flow propagates with a very high velocity. In our case because a very strong standing wave is built in the chamber, the plasma is expanded with a very high velocity by the pondermotive force of this standing wave. The incident wave is reflected partially by the moving plasma and the frequency is up shifted due to Doppler effect.

When the ponderomotive force acting on the plasma and pressure-gradient force in the plasma are taken into consideration, from the equation of motion the velocity of the expanding plasma is obtained as

$$v \approx \left[\frac{2P_i}{clA}\frac{\omega_p^2}{\omega^2} - K(T_i + T_e)\nabla n_0\right]/(m_i n_0) \quad , \tag{1}$$

where $\omega_p = (n_0 e^2/\epsilon_0 m_e)^{1/2}$, l are plasma frequency and gradient length of $\epsilon_0 \langle E^2 \rangle$, respectively. K is Boltzmann's constant. T_i and T_e are ion temperature and electron temperature, respectively. A is irradiated area of the microwave. If we substitute P_i= 250kW, f_0=9GHz, $T_e = 3$eV, A= $5 \times 10^{-3} m^2$, t=1 μs, $l = 7 \times 10^{-3}$ m, L=100cm, into Eq.(1) the velocity of the expanding plasma is $v = 3 \times 10^6$ cm/s.

When the incident wave encounters such a moving plasma, part of it is reflected by the plasma. In the frame of the moving plasma (the prime frame) the frequency of the incident wave is

$$\omega_0' = \gamma(1+\beta)\omega_0 \quad , \tag{2}$$

where $\gamma = (1-\beta^2)^{-1/2}$, $\beta = v/c$. Performing an inverse Lorentz transformation to get back to the lab frame, the wave frequency for the reflected wave is

$$\omega_r = \gamma^2(1+\beta)^2\omega_0^2 \quad . \tag{3}$$

Substituting $v = 1.0 \times 10^6$ cm/s into Eq.(3), the frequency up-shift due to the Doppler effect is

$$\nabla f = \frac{2\beta f_0}{1-\beta} = 1.8(MHz) \quad . \tag{4}$$

This result is almost in agreement with the experimental results.

The spectra as a function of position is shown in Fig.6. At different position, the frequency up-shift and higher frequency components are different. Because the intensity of the standing wave is changing with the position, the velocity of the expanding plasma also depends on position. When the incident wave travels along the Z axis, it is reflected by moving plasma with different velocity at the different positions. This is why we observed the obvious dependence of spectra on the position. In our results, it is noted that at some position the amplitude of f_1 is smaller than that of higher frequency components although the incident wave has not been reflected totally by the moving plasma. The signal which observed in our case is the superposition of the standing wave ($E_{Standing}=2E_0\cos kz\cos\omega t$, E_0 is the amplitude of the incident wave.) and up-shifted reflected wave ($E_r = RE_0 e^{i[(\omega+\Delta\omega)t-kz]}$, R is reflecting coefficient). Then the amplitude of the signal can be written as $|E| = |E_{Standing} + |E_r| = 2E_0\cos kz + RE_0$. The amplitude of the standing wave is changing with position z but the amplitude of the reflected wave is a constant if R is kept unchanged. So our results mentioned above are reasonable.

CONCLUSION

In summary, the frequency up-shift around 1MHz and some other higher frequency components in the range of 2MHz to 7MHz have been observed in our experiments. This is the first time the frequency up-shift(blue-shift) being observed in the experiment of microwave-plasma interaction when there is no ionization in the plasma. In our case, because a very strong standing wave is established in the chamber, the plasma is expanded rapidly ($v \approx 1 \times 10^6$ cm/s) by the collective ponderomotive force which arising from the electric field gradient of the standing wave. The incident wave is reflected partially by the moving plasma, and the frequency is up shifted due to the Doppler effect. In laser-plasma interaction, the frequency blue-shift for lower Z target has been observed by Phillion et al [4]. But the interpretation of the frequency shift maybe is not straightforward. Our results perhaps have some relationship with the explanation of the phenomenon in their experiments.

ACKNOWLEDGMENTS

This work is supported by a Grant-in-Aid for Science Research from Ministry of Education, Science and Culture of Japan.

REFERENCES

[1] S. C. Wilks, J. M. Dawson, and W. B. Mori, Phys. Rev. Lett. **61**, 337 (1988).
[2] P. Sprangle, E. Esarey, and A. Ting, Phys. Rev A **41**, 4463 (1990).
[3] W. M. Wood, C. W. Siders, M. C. Downer, Phys. Rev. Lett. **67**, 3523 (1991). W. M. Wood, C. W. Siders, M. C. Downer, IEEE Trans. Plasma Sic. **21**, 20 (1993).
[4] D. W. Phillion, W. L. Kruer and V. C. Rupert, Phys. Rev. Lett. **39**, 1405 (1977).

Interaction of 1.05µm and 0.53µm Lasers with Gold Disks

Shenye Liu   Yaonan Ding   Zhijian Zheng   and   Daoyuan Tang

Southwest Institute of Nuclear Physics and Chemistry of China
P. O. Box 525, Chengdu 610003, China

Abstract   Gold disks were irradiated with 1.05µm and 0.53µm lasers at pulse duration of ~ 0.8ns, intensity ranging from 5×10^{13} W/cm^2 to 4×10^{15} W/cm^2 on the SHEN GUANG I laser facility in China. The experimental results of laser absorption, scattering light, x-ray emission and plasma blow-off are presented in this paper. When the laser irradiated the gold disk obliquely, the angular distribution of scattered lights produced by 0.53µm lasers disagree with that predicted by the Brillouin scattering theory. The angular distribution is different from that reported previously by the others.

I INTRODUCTION

Interaction of high power laser with plasma is one of the basic topics which inertial confinement fusion(ICF) concerns. Laser can interact with plasma by many processes such as collision absorption, stimulating Brillouin scattering(SBS), stimulating Raman scattering(SRS) and so on. Higher absorption, less instability and fewer suprathermal electrons are what ICF persues. The present theory shows that short wavelength laser can suppress instability and enhance absorption of the laser. So hohlraum would be irradiated with shorter wavelength lasers in the future Chinese ICF program. It is neccessary to study the interaction of the short wavelength lasers with plasma experimentally in the Chinese ICF program.

The interaction of 1.05µm, 0.53µm lasers with gold disks under verious conditions has been studied in USA, Japan and France [1-7]. But it is the first time to study the interaction of 0.53µm lasers at energy upto 200J, intensity ranging from 5×10^{13} W/cm^2 to 4×10^{15} W/cm^2, pulse width of 0.8ns with gold disks on SHEN GUANG I in China. In the next section we will describe the experiments briefly. The experimental results and the discussion are reported in section III. The summary is given in section IV.

TABLE 1. The Experimental Conditions

λ_L	τ_L	Incident angle	F/Number	E_L
1.05µm	1.0ns to 0.8ns	10°, 30°, 45°	1.7	50J to 500J
0.53µm	~ 0.8ns	0°, 10°, 30°, 45°	1.7	20J to 200J

© 1996 American Institute of Physics

II EXPERIMENTS

The SHEN GUANG I laser facility which outputs laser energy about 1500J at 1.05μm wavelength, has two beams of laser light to enter the target chamber. In each experiment we used one beam of laser light to irradiate gold disks. In the experiments of 0.53μm laser, 1.05μm lasers were converted to 0.53μm laser with a KDP crystal. Tab.1 lists the experimental conditions. Gold disks, 15μm thick and 700μm in diameter, are supported by C_8H_8 film. Many of diagnostics have been developed in China in order to measure the absorption, scattered light and x-ray emission. An array of pyroelectric plasma calorimeters was used to measure the absorption and scattering of laser. The sub-keV x-ray energy is measured using an array of flat response XRD. The sub-keV x-ray spectra were measured by a multi-channel XRD spectrometer, a x-ray streak camera and a transmission grating spectrometer. A hard x-ray spectrometer was used for detecting hard x-ray from the plasma. We observed the spectra of scattering light including SRS and SBS using an OMA spectrometer.

III EXPERIMENTAL RESULTS AND DISCUSSION

A. Measured Absorption

Fig.1 shows the angular distribution of Au plasma energy from a gold disk irradiated with 0.53μm laser at normal incidence, having $I_L=5.2\times10^{14}$W/cm². θ in Fig.1 is the angle between the target normal and the detecting direction. D_{max} is the maxium of angular distribution. Note that the laser-produced plasma blow-off near target normal is the strongest. the similar structure angular distribution has been observed in the other experiments.

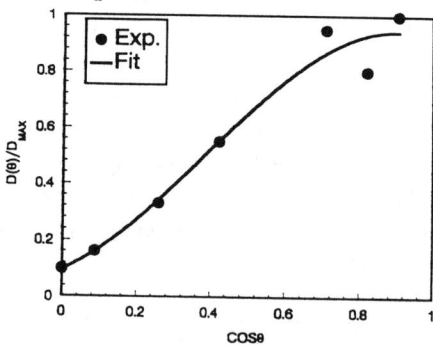

FIGURE 1. Angular distribution of plasma as a function of cos θ

The absorbed energy was obtained from the plasma energy distribution measured. Fig.2 (a) gives laser absorption of Au disks irradiated with 1.05μm and 0.53μm laser lights at incident angle of 45° in the present experiments, as a function of laser intensity. It can be observed that the absorption is lower at higher laser intensity because higher intensity of laser results in higher electron temperature and lower collision frequence of electrons with ions. The absorption as a function of incident angle is plotted in Fig.2(b). It shows that the absorption is lower at larger incident angle. This is because laser lights turn at turning point $\rho_t = \rho_c \cos^2\theta$. When laser lights enter plasma at larger angle, they can not enter the dense region and so the absorption is lower. Both Fig.2(a) and Fig.2(b) show that the absorption of short wavelength laser is higher because the short wavelength laser can enter the dense region($\rho_c \propto 1/\lambda^2_{\mu m}$). The curves in Fig.2(a) and Fig.2(b) are the numerical results.

We get a scaling law of absorption of gold disks irradiated with 1.05μm and 0.53 μm laser from the experimental data. The absorption of the Au disks is:

$$\eta_a = 1 - \exp[-1.7(I_{14}\lambda^2_{\mu m})^{-0.4}\cos^3\theta] \tag{1}$$

Where I_{14} is the intensity I_L, in units of 10^{14} W/cm^2, $\lambda_{\mu m}$ is the wavelength in microns, and θ is the incident angle of laser.

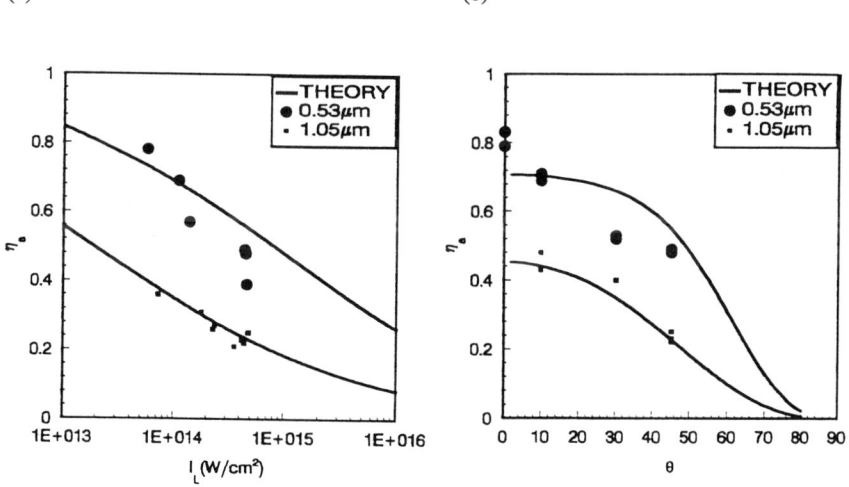

FIGURE 2. (a)The measured absorption as a function of intensity at 45° incident angle; (b) the measured absorption as a function of the incident angle, at intensity about 5×10^{14} W/cm^2.

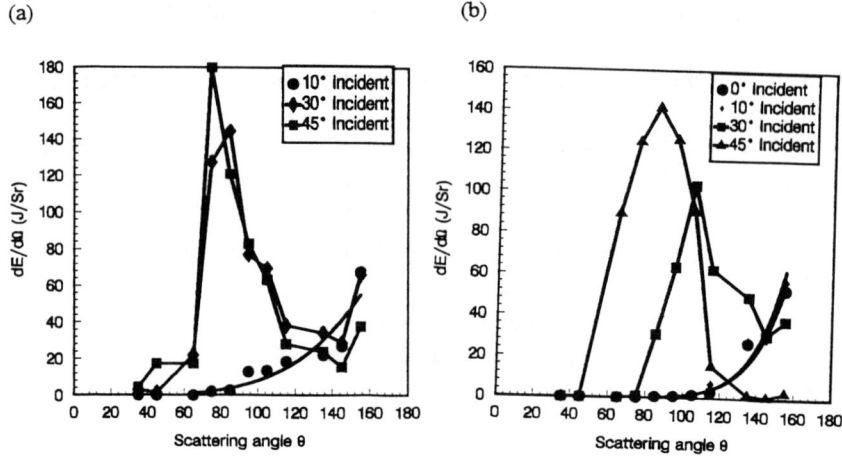

FIGURE 3. (a) The fluence of the scattered light from 1.05μm irradiation of Au disks as a function of scattering angle at 5×10^{14} W/cm^2; (b)the fluence of the scattered light from 0.53μm irradiation of Au disks as a function of scattering angle at $\sim 5\times10^{14}$ W/cm^2

B. Scattering of Laser Light

Fig.3(a) shows the measured angular distribution of scattered lights from Au disks irradiated with 1.05μm lasers. Fig.3(b) presents the results of the angular distribution of scattered lights from Au disks irradiated with 0.53μm lasers. Fig.3(a) and (b) all show that scattered lights at the specular direction are the strongest. So the scattered lights seem to come from the specular reflection and not from SBS as mentioned elsewhere[8].

Raman scattering into the lens is measured. The experimental results are listed in Tab.2. It can be seen that Raman scattering in 0.53μm laser experiments has been supressed because of enhancing collision absorption.

The experiments show that the Raman scattering lights from Au disks are much lower than those from the hohlraum. Raman spectra from Au disks irradiated with 1.05μm and 0.53μm laser are given in Fig.4(a) and Fig.4(b) respectively.

TABLE 2. Raman scattering light into the lens

$\lambda_{\mu m}$	E_{SRS}/E_L	I_{14}
1.05	1-2%	$\sim 5\times10^{14}$ W/cm^2
0.53	0.004%-0.001%	$\sim 5\times10^{14}$ W/cm^2

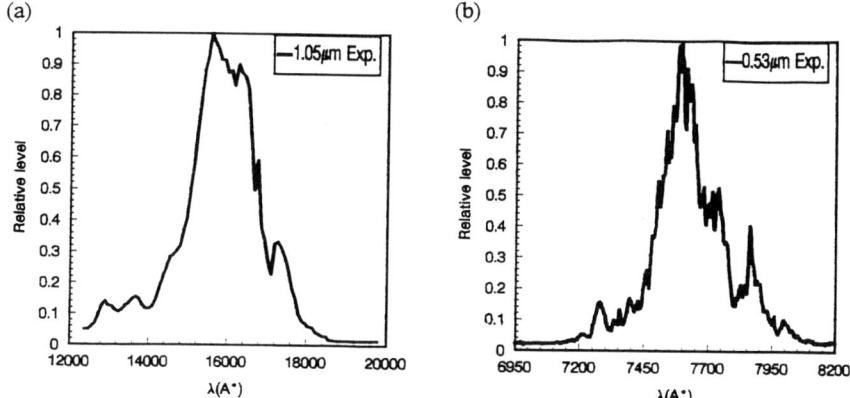

FIGURE 4. (a) Raman spectra in experiment of gold disks irradiated with 1.05μm laser, $I_L = 5 \times 10^{14}$ W/cm^2, at 45° incident angle; (b) Raman spectra in experiment of gold disks irradiated with 0.53μm laser, $I_L = 5 \times 10^{14}$ W/cm^2, at 45° incident angle.

C. X-ray Emission

Fig.5(a) gives the angular distribution of x-ray from 0.53μm laser-produced plasma. The x-ray conversion efficiency measured is about 0.30 to 0.35 for 1.05 μm laser experiments and about 0.37 to 0.48 for 0.53μm laser experiments. The hard x-ray spectra from Au disks irradiated with 1.05μm and o.53μm lasers are given in Fig.5(b). It shows that suprathermal electrons are supressed because of using 0.53μm laser to irradiate Au disks.

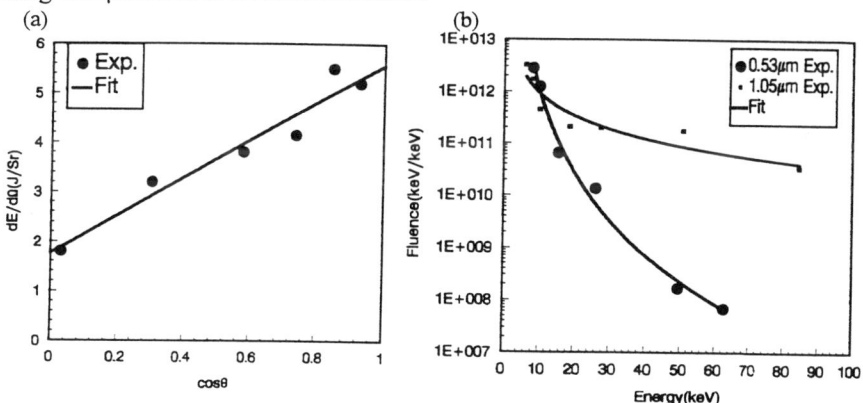

FIGURE 5. (a) The angular distribution of x-ray energy from 0.53μm laser-produced plasma; (b) suprathermal x-ray spectra from Au disks irradiated with 1.05μm and 0.53μm laser. $E_L \sim 200$J, $\tau \sim 0.8$ns, $I_L \sim 5 \times 10^{15}$ W/cm^2

IV SUMMARY

In this paper we have reported the experimental results about absorption, scattering, and x-ray emission of Au disks irradiated with 1.05μm and 0.53μm laser lights. The experiments have proved that the absorption of 0.53μm laser lights is higher and that suprathermal electrons and SRS from Au disks irradiated with 0.53μm laser are lower. The measured angular distribution of scattered lights has shown that a lot of scattered lights come from specular reflection and not from SBS. The absorption simulated numerically is comparable with the measured result. We have not reported the numerical results about x-ray emission and stimulating scattering. In future work it is required to calculate x-ray emission and stimulating scattering.

ACKNOWLEDGEMENTS

The authors have benefited by numerous useful discussion with Prof. Tie-qiang Chang and Prof. Jun Zhang at Institute of Applied Physics and Computational Mathematics, Beijing. These experiments could not have been finished without the dedicated efforts of the scientific and technical staffs of SHEN GUANG I laser facility. This work is supported by the national hi-tech plan.

REFERENCES

1. W. C. Mead et al Phys. Fluids 26, 2316(1983)
2. R. A. Hass et al Phys. Fluids 20, 322(1977)
3. R. P. Drake et al Phys. Fluids B3, 3477(1991)
4. William L. Kruer Phys. Fluids B3, 2356(1991)
5. E. M. Campbell Phys. Fluids B4, 3785(1992)
6. C. Garban-Labanne et al Phys. Fluids Vol. 28, 2580(1985)
7. J. S De Groot et al Phys. Fluids B4, 701(1992)
8. R. P. Drake et al Phys.Fluids B1, 1295(1989)

Instabilities of Nuclear Flames in Thermonuclear Supernovae

*Ken'ichi Nomoto, *Koichi Iwamoto, *Toshikazu Shigeyama
and †Hideaki Takabe

*Department of Astronomy and Research Center for the Early Universe, School of Science,
University of Tokyo, Tokyo 113, Japan
†Institute of Laser Engineering, Osaka University, Suita, Osaka 565, Japan

Abstract. In exploding white dwarfs (Type Ia supernovae), a nuclear flame propagates at a subsonic speed. The flame speed is a crucial model parameter and depends on the development of flame instabilities. We study microscopic instabilities of flames with a self-consistent linear stability analysis. In the dispersion relations thus obtained, the fastest branch can be interpreted as a thermal instability due to the strong temperature dependence of nuclear reaction rates. This instability grows faster than the hydrodynamic instabilities (e.g., Rayleigh-Taylor instabilities). The implications on the flame propagation are also discussed.

INTRODUCTION

A certain class of stars end their lives as supernova explosions to produce most of heavy elements in the Universe (1). The mechanisms of supernova explosions are:
i) gravitational collapse, bounce, and propagation of a shock wave in massive red (or blue) supergiant stars (Type II supernovae), and
ii) thermonuclear runaway and propagation of a nuclear flame in C+O white dwarfs (Type Ia supernovae).

Thorough understanding of these explosion mechanisms is necessary for the determination of the distance scale and age of the Universe as well as the origin and distribution of elements. Recent studies have revealed that instabilities and mixing associated with explosions are crucial in driving supernova explosions of both types. It has also been realized that the characteristics of the instabilities are rather common to supernova explosions and laser interactions. In particular, instabilities of a nuclear flame in the white dwarf are closely related to those of an ablation front.

For thermonuclear supernovae, carbon is ignited at the center of a C+O white dwarf when its mass increases to the Chandrasekhar mass due to mass accretion. Resulting carbon burning is so explosive as to incinerate the material into iron-peak elements. At such a high central density as $\sim 10^9$ g cm^{-3}, nuclear energy release is only ~ 20 % of the Fermi energy of degenerate electrons. Therefore, the resulting shock wave is not strong enough to directly form a detonation wave.

Instead a carbon deflagration wave propagates outward at a subsonic speed, i.e., on the time scale for heat transport (2). The outcome of deflagration strongly depends on the flame speed (2). To produce enough amount of radioactive ^{56}Ni ($\sim 0.6\ M_\odot$) to produce Type Ia supernovae in a deflagration wave, the flame speed must be accelerated to about one-fifth of the sound speed (3). If the flame speed is much slower, transition from deflagration to detonation (4) needs to be invoked.

The flame speed depends on the development of Rayleigh-Taylor instabilities at the flame front. Multi-dimensional hydrodynamical simulations of the flame propagation have been tried by several groups, though the results are still preliminary (5, 6, 7). The linear stability of the nuclear flame has been examined only with crude analytic approximations (8). Here we calculate an exact flame structure and perform linear stability analysis with a numerical approach taken for the study of an ablation front (9).

STRUCTURE OF STEADY DEFLAGRATION FRONT

The basic equations governing the background structure and the development of perturbations are those which describe the conservation of mass, momentum, and energy. In the co-moving frame moving at a constant velocity with the flame front, they are expressed as

$$\frac{\partial}{\partial t}\rho + \nabla \cdot (\rho \mathbf{v}) = 0, \qquad (2.1)$$

$$\rho\left(\frac{\partial}{\partial t}\mathbf{v} + \mathbf{v}\cdot\nabla\mathbf{v}\right) = -\nabla P - \rho\mathbf{g}, \qquad (2.2)$$

$$\rho\left(\frac{\partial}{\partial t}e + \mathbf{v}\cdot\nabla e\right) = -P\nabla\cdot\mathbf{v} - \nabla\cdot\mathbf{Q} + \rho\dot{S}, \qquad (2.3)$$

where P, ρ, T, e, Y_j, $\mathbf{g}$, and $\mathbf{v}$ denote the pressure, density, temperature, specific internal energy, chemical compositions, gravitational acceleration, and velocity, respectively, and $\dot{S}$ the energy generation rate by nuclear reactions.

The nuclear reaction rates F_j ($j = 1, \ldots, N$) of j-th element is related to the reaction equations for the mole fractions Y_j ($j = 1, \ldots, N$)

$$\frac{\partial}{\partial t}Y_j + \mathbf{v}\cdot\nabla Y_j = F_j(\rho, T, Y_1, \ldots, Y_N). \qquad (2.4)$$

The energy flux due to transport processes $\mathbf{Q}$ is given by Kirchihof's law as

$$\mathbf{Q} = -\sigma\nabla T, \qquad (2.5)$$

where σ is the conductivity due to radiation and thermal conduction. For degenerate white dwarf matter, the radiative energy transport is neglected, and we use the electron conductivity tabulated in (10).

Assuming a steady state and a one-dimensional planar geometry, i.e., $\frac{\partial}{\partial t} = 0$, $\nabla \to \frac{d}{dx}$, $\mathbf{v} \to u$, which is a good approximation since the width of the flame front is orders of magnitudes smaller than the pressure scale height, the basic equations described above are simplified into the system of first-order ordinary differential equations.

The system of these equations should be integrated with the initial values from far upstream point,

$$\rho = \rho_1, T = T_1, u = -v_{\text{def}}, Q = 0, Y_j = Y_{j,1} \qquad (x \to \infty). \qquad (2.6)$$

The flame velocity v_{def} is determined as a solution of eigenvalue problem, which satisfies the downstream boundary condition of vanishing the gradient of T,

$$Q = 0 \qquad (x \to -\infty) \qquad (2.7)$$

For $x \to -\infty$, $Q \to 0$, all the variables should approach their equilibrium values.

$$\rho \to \rho_2, T \to T_2, u \to u_2, Y_j \to Y_{j,2}. \qquad (2.8)$$

In order to solve the reaction chains, we adopted a nuclear reaction network including 14 species ^{4}He, ^{12}C, ^{16}O, ^{20}Ne, ^{24}Mg, ^{28}Si, ^{32}S, ^{36}Ar, ^{40}Ca, ^{44}Ti, ^{48}Cr, ^{52}Fe, ^{56}Ni, and ^{60}Zn with $(\alpha,p)(p,\gamma)$ channels being added to the (α,γ) reactions. Figure 1 shows a flame structure thus obtained for $\rho = 2 \times 10^9$ g cm^{-3} and 0.5 C + 0.5 O mixture. We should note that the width of the flame front is as thin as $\sim 10^{-4}$ cm.

LINEAR STABILITY ANALYSIS

The steady-state structures of the deflagration front obtained above are used for the background structures of the stability analysis. We simplify the nuclear reactions. Instead of fully solving the nuclear reaction network, we assume that carbon would be incinerated into ^{56}Ni with a nuclear energy release corresponding to the difference between their rest mass energies. Then equations (2.4) and (2.5) take the forms as

$$\dot{S} = qR, \qquad \frac{\partial}{\partial t}X_c + \mathbf{u} \cdot \nabla X_c = -R \qquad (3.1.1)$$

where X_c denotes the mass fraction of ^{12}C and q and R are given by

$$q = 7 \times 10^{17} \text{erg g}^{-1}\text{s}^{-1}, \qquad R = \frac{7}{36}\rho N_A \lambda X_c^2. \qquad (3.1.2)$$

The reaction rate is approximated by the ^{12}C+^{12}C reaction rate (11).

Perturbation Equations

The equations governing the perturbations are derived from the linearizationin of the basic equations (2.1) - (2.5) as follows:

$$\frac{\partial \rho_1}{\partial t} + \frac{\partial}{\partial x}(\rho_1 u_0 + \rho_0 u_1) + \rho_0 \frac{\partial v_1}{\partial y} = 0 \qquad (3.2.1)$$

$$\rho_0 \left(\frac{\partial u_1}{\partial t} + u_0 \frac{\partial u_1}{\partial x} + u_1 \frac{du_0}{dx} \right) + u_0 \frac{du_0}{dx} \rho_1 = -\frac{\partial p_1}{\partial x} - \rho_1 g \qquad (3.2.2)$$

$$\rho_0 \left(\frac{\partial v_1}{\partial t} + u_0 \frac{\partial v_1}{\partial x} \right) = -\frac{\partial p_1}{\partial y} \qquad (3.2.3)$$

$$\rho_0 \left(\frac{\partial e_1}{\partial t} + u_0 \frac{\partial e_1}{\partial x} + u_1 \frac{de_0}{dx} \right) + u_0 \frac{de_0}{dx} \rho_1$$
$$= -P_0 \left(\frac{\partial u_1}{\partial x} + \frac{\partial v_1}{\partial y} \right) - P_1 \frac{du_0}{dx} - \frac{\partial Q_{x1}}{\partial x} - \frac{\partial Q_{y1}}{\partial y} + (\rho q R)_1 \qquad (3.2.4)$$

$$Q_{x1} = -\sigma_0 \frac{\partial T_1}{\partial x} - \sigma_1 \frac{dT_0}{dx} \qquad (3.2.5)$$

$$Q_{y1} = -\sigma_0 \frac{\partial T_1}{\partial y} \qquad (3.2.6)$$

$$\frac{\partial X_{c,1}}{\partial t} + u_0 \frac{\partial X_{c,1}}{\partial x} + u_1 \frac{dX_{c,0}}{dx} = -R_1. \qquad (3.2.7)$$

These equations constitute the system for seven independent variables $\rho, T, u, v, Q_x, Q_y, X_c$. All other variables of first order should be expressed with these basic variables; P_1, e_1, σ_1, and R_1 are expanded in terms of ρ_1, T_1, and $X_{c,1}$.

Here we introduce another simplification. The reaction rate R strongly depends on the temperature, i.e., $R \propto \rho T^n X_c^2$ with $n = 20\text{-}30$, so that we neglect the derivative of R with respect to X_c. Since the dependence of P, e, σ on X_c is weak, we also neglect the derivatives of these variables with respect to X_c.

We assume the following form of solution,

$$f(x, y, t) = f(x) \exp(\gamma t + i K y). \qquad (3.2.8)$$

Substituting the form (3.2.8) into equations (3.2.1) - (3.2.7) and eliminating $Q_{y,1}$, these are reduced to the following system of the first order ordinary differential equations for five variables, $\rho_1(x), T_1(x), u_1(x), v'_1(x), Q_{x,1}(x)$, where $v'_1 \equiv i v_1$. We solve the following system of equations with appropriate boundary conditions:

$$u_0 \frac{d\rho_1}{dx} + \rho_0 \frac{du_1}{dx} = -\left(\gamma + \frac{du_0}{dx} \right) \rho_1 - \frac{d\rho_0}{dx} u_1 - K \rho_0 v'_1 \qquad (3.2.9)$$

$$\left(\frac{\partial P}{\partial \rho} \right) \frac{d\rho_1}{dx} + \left(\frac{\partial P}{\partial T} \right) \frac{dT_1}{dx} + M \rho_0 u_0 \frac{du_1}{dx} = -\left\{ \frac{d}{dx} \left(\frac{\partial P}{\partial \rho} \right) + M u_0 \frac{du_0}{dx} + M g \right\} \rho_1$$
$$- \frac{d}{dx} \left(\frac{\partial P}{\partial T} \right) T_1 - M \rho_0 \left(\gamma + \frac{du_0}{dx} \right) u_1 \qquad (3.2.10)$$

$$M \rho_0 u_0 \frac{dv'_1}{dx} = K \left(\frac{\partial P}{\partial \rho} \right) \rho_1 + K \left(\frac{\partial P}{\partial T} \right) T_1 - \gamma M \rho_0 v'_1 \qquad (3.2.11)$$

$$\rho_0 u_0 \left(\frac{\partial e}{\partial \rho}\right) \frac{d\rho_1}{dx} + \rho_0 u_0 \left(\frac{\partial e}{\partial T}\right) \frac{dT_1}{dx} + P_0 \frac{du_1}{dx} + \frac{dQ_{x,1}}{dx} =$$
$$\left\{ -\gamma \rho_0 \left(\frac{\partial e}{\partial \rho}\right) - u_0 \frac{de_0}{dx} - \rho_0 u_0 \frac{d}{dx}\left(\frac{\partial e}{\partial \rho}\right) + q \frac{\partial(\rho R)}{\partial \rho} - \left(\frac{\partial P}{\partial \rho}\right) \frac{du_0}{dx} \right\} \rho_1$$
$$+ \left\{ -\gamma \rho_0 \left(\frac{\partial e}{\partial T}\right) - \rho_0 u_0 \frac{d}{dx}\left(\frac{\partial e}{\partial T}\right) + \rho q \left(\frac{\partial R}{\partial T}\right) - \left(\frac{\partial P}{\partial T}\right) \frac{du_0}{dx} - K^2 \sigma_0 \right\} T_1$$
$$- \rho_0 \frac{de_0}{dx} u_1 - K P_0 v_1' \qquad (3.2.12)$$

$$\sigma_0 \frac{dT_1}{dx} = -\frac{dT_0}{dx}\left(\frac{\partial \sigma}{\partial \rho}\right)\rho_1 - \frac{dT_0}{dx}\left(\frac{\partial \sigma}{\partial T}\right)T_1 - Q_{x,1} \qquad (3.2.13)$$

All the variables are normalized to the downstream values of the several basic variables, i.e., $\rho/\rho_2 \to \rho, T/T_2 \to T, u/u_2 \to u, v/u_2 \to v, P/P_2 \to P, e/(P_2/\rho_2) \to e, Q_x/(u_2 P_2) \to Q_x, R/(u_2/d) \to R, q/(P_2/\rho_2) \to q, \sigma/(u_2 P_2 d/T_2) \to \sigma, g/(u_2^2/d) \to g$, where d is the width of the flame front. M is defined as $M = \rho_2 u_2^2/P_2$ and corresponds to the square of the ratio of deflagration velocity to sound velocity (v_{def}/C_S) as

$$M \equiv \frac{\rho_2 u_2^2}{P_2} \sim \frac{\rho v_{\text{def}}^2}{\rho C_S^2} \sim \left(\frac{v_{\text{def}}}{C_S}\right)^2. \qquad (3.2.14)$$

Numerical Method for the Eigenvalue Problem

The growth rate γ of the perturbations should be obtained by solving the eigenvalue problem, where, for a wavenumber of perturbations K, a special value of γ can give a physically meaningful solution. We adopt the same method to solve the problem as developed in (9). The equations (3.2.9) - (3.2.13) can be expressed in the matrix form,

$$\frac{d\phi}{dx} = A(K, \gamma, x)\phi \qquad (3.3.1)$$

where ϕ is a vector function and defined as

$$\phi(x) = \begin{pmatrix} \rho_1 \\ T_1 \\ u_1 \\ v_1' \\ Q_{x,1} \end{pmatrix}. \qquad (3.3.2)$$

As the densities, velocities, and temperatures are constant in the regions outside the flame front, the matrix A reduces to the constant matrix $A_1(K, \gamma)$ and $A_2(K, \gamma)$ in the upstream and downstream, respectively. The solutions outside the flame front should be constituted as a superposition of the eigenfunctions of A. As a result of actual calculations, we found out that for $\gamma \geq 0$, the eigenvalues of $A_j (j = 1, 2)$, $K_1^j, K_2^j, \cdots, K_5^j (j = 1, 2)$ are all real and distinct, and two of them are negative whereas three of them are positive ($K_1^j < K_2^j < 0 < K_3^j < K_4^j < K_5^j, j = 1, 2$). If we impose the physically meaningful boundary conditions that all the perturbations vanish far from the front, the asymptotic solutions in the upstream and downstream regions would be constructed as

$$\phi_{\text{up}}(x) = C_1 \phi_1^1(x) + C_2 \phi_2^1(x) \qquad (\text{upstream} : x \to \infty), \qquad (3.3.3)$$

$$\phi_{\text{down}}(x) = C_3 \phi_3^2(x) + C_4 \phi_4^2(x) + C_5 \phi_5^2(x) \qquad (\text{downstream} : x \to -\infty). \qquad (3.3.4)$$

We integrate the system of the equations taking these solutions as initial values from the points outside the front, and connected them at a certain point inside the front as

$$\phi_{\text{up}}(x_c) = \phi_{\text{down}}(x_c), \qquad (3.3.5)$$

which reduces to

$$\begin{pmatrix} \phi_1^1(x_c) & \phi_2^1(x_c) & \phi_3^2(x_c) & \phi_4^2(x_c) & \phi_5^2(x_c) \end{pmatrix} \begin{pmatrix} C_1 \\ C_2 \\ -C_3 \\ -C_4 \\ -C_5 \end{pmatrix} = 0 \qquad (3.3.6)$$

The condition of a nontrivial solution for (3.3.6) provides the dispersion relation.

Microscopic Instabilities of Deflagration Fronts

With the method described above, the stability of the steady deflagration front is studied. Figure 2 shows the dispersion relations for $\rho = 2 \times 10^9$ g cm^{-3} and 0.5 C + 0.5 O mixture. The wavenumber and the growth rate are normalized to d^{-1} and $v_{\text{def}} d^{-1}$, respectively, where d and v_{def} are the width and the velocity of the flame front ($d = 8.867 \times 10^{-6}$ cm and $v_{\text{def}} = 191.3$ km s^{-1} for the present case). Solid line shows the dispersion relation for no gravity ($g = 0$), while the dashed line corresponds to the case with $g = 100$. Here the gravitational acceleration g is normalized to v_{def}^2/d ($\sim 4.13 \times 10^{19}$ cm s^{-2} for the present case). This is much larger than the actual value of g in the interior of the white dwarf which is as small as $\sim 10^9$ cm s^{-2}. The mode profiles at $K = 1.0$ for no gravity are plotted in Figure 3.

DISCUSSION

Physical Interpretation of the Instability

The physical origin of the instability obtained in the previous section can be interpreted as follows:

The growth time scale is comparable to the time scale for the material to pass through the flame front ($1/\gamma \sim dv_{\text{def}}^{-1}$) and is also comparable to the time scale of nuclearenergy release ($e/\dot{S} = e/(Rq) \sim (P_2/\rho_2)(v_{\text{def}}/d)^{-1}(P_2/\rho_2)^{-1} = dv_{\text{def}}^{-1} \sim 1/\gamma$). The energy equations for the background structure and the perturbations take the following forms under isobaric condition, which is a good approximation here:

$$\rho u C_p \frac{dT}{dx} - \sigma \frac{d^2 T}{dx^2} = \rho q R \qquad (4.1)$$

$$C_p \left(\gamma \rho T_1 + j \frac{dT}{dx} \right) + \rho u C_p \frac{dT_1}{dx} - \sigma \frac{d^2 T_1}{dx^2} + K^2 \sigma T_1 = q(\rho R)_1 \qquad (4.2)$$

where j is the disturbance of mass flux, $j = (\rho u)_1 = \rho u_1 + \rho_1 u$. The conductivity σ and the specific heat C_p are assumed to be constant.

We make an order of magnitude estimate of the growth rate. Replacing the derivatives with respect to x with the divisions by flame width d,

$$\rho u C_p \frac{T}{d} - \sigma \frac{T}{d^2} = \rho q R, \qquad (4.3)$$

$$\gamma \rho C_p + \frac{\rho u C_p}{d} - \frac{\sigma}{d^2} + K^2 \sigma = n \frac{q \rho R}{T}, \qquad (4.4)$$

where the reaction rate is assumed to depend only on the temperature as $R \propto T^n$. And j is set to be zero since we are now interested in the thermal instability.

Eliminating the flame width d in (3.4.4) by (3.4.3), we obtain the following dispersion relation,

$$\gamma = \frac{1}{\rho C_p} \left(\frac{(n-1)\rho q R}{T} - \sigma K^2 \right). \qquad (4.5)$$

Substituting the typical value into each variable ($\rho \leftarrow \rho_2, T \leftarrow T_2, C_p \sim e/T \leftarrow P_2/\rho_2 T_2, q \leftarrow P_2/\rho_2, R \leftarrow v_{\text{def}}/d, \sigma \leftarrow v_{\text{def}} P_2 d/T_2$), we obtain the dispersion relation as

$$\gamma = \frac{v_{\text{def}}}{d} \left(n - 1 - (Kd)^2 \right). \qquad (4.6)$$

The stability problem of combustion in a solid propellant has been investigated Bychkov and Liberman (12). They took into account the front structure and obtained the dispersion relations analytically with some approximations and simple treatment of the burning front. Our estimate (4.6) is consistent with their results.

The Effect of Gravity

The gravity has an effect to make the front more unstable as seen in Figure 2. The increase in the growth rate is very small even if we take an artificially large strength of gravity. Therefore, the gravity is not important as long as the thermal instability is concerned.

CONCLUSIONS

Microscopic instabilities of steady deflagration fronts in white dwarfs are studied with a self-consistent linear stability analysis. In the dispersion relations, the fastest branch can be interpreted as the thermal instability due to the strong temperature dependence of reaction rates. This instability grows faster than the hydrodynamic instabilities (for example, Rayleigh-Taylor instabilities), and it could be a seed of the perturbation on scales comparable to and larger than the flame width. Study of the nonlinear behavior of this thermal instability is underway.

ACKNOWLEDGEMENTS

This work has been supported in part by the grant-in-Aid of the Ministry of Education, Science, and Culture in Japan for Scientific Research on Priority Areas (Astrophysics with Line X-rays and Gamma-rays: 05242102, 05242103, 06233101) and on the Center of Excellence (COE) program (Probing the Early Universe), and the Fellowships of the Japan Society for the Promotion of Science for Japanese Junior Scientists (4227).

REFERENCES

1 . Woosley S.E., and Weaver T.A. 1986, Ann. Rev. Astron. Ap. 24, 205
2 . Nomoto K., Sugimoto D., and Neo S. 1976, Ap& Space Sci. 39, L37
3 . Nomoto K., Thielemann F.-K., and Yokoi K. 1984, ApJ 286, 644
4 . Khokhlov A.M. 1991, A&A 245, 114
5 . Arnett W.D., and Livne E. 1994, ApJ 427, 314
6 . Khokhlov A.M. 1994, ApJ 424, L115
7 . Niemeyer J.C., and Hillebrandt W. 1995, ApJ, submitted
8 . Bychkov V.V., and Liberman M.A., 1995, A&A, in press
9 . Takabe H., Montierth L., and Morse R.L., 1983, Phys. Fluids 26, 2299
10 . Itoh N., Mitake S., Iyetomi H., and Ichimaru S. 1983, ApJ 273, 774
11 . Fowler W.A., Caughlan G.R., and Zimmerman B.A., 1975, ARAA 13, 69
12 . Bychkov V.V., and Liberman M.A., 1994, Phys. Rev. Lett. 73, 1998

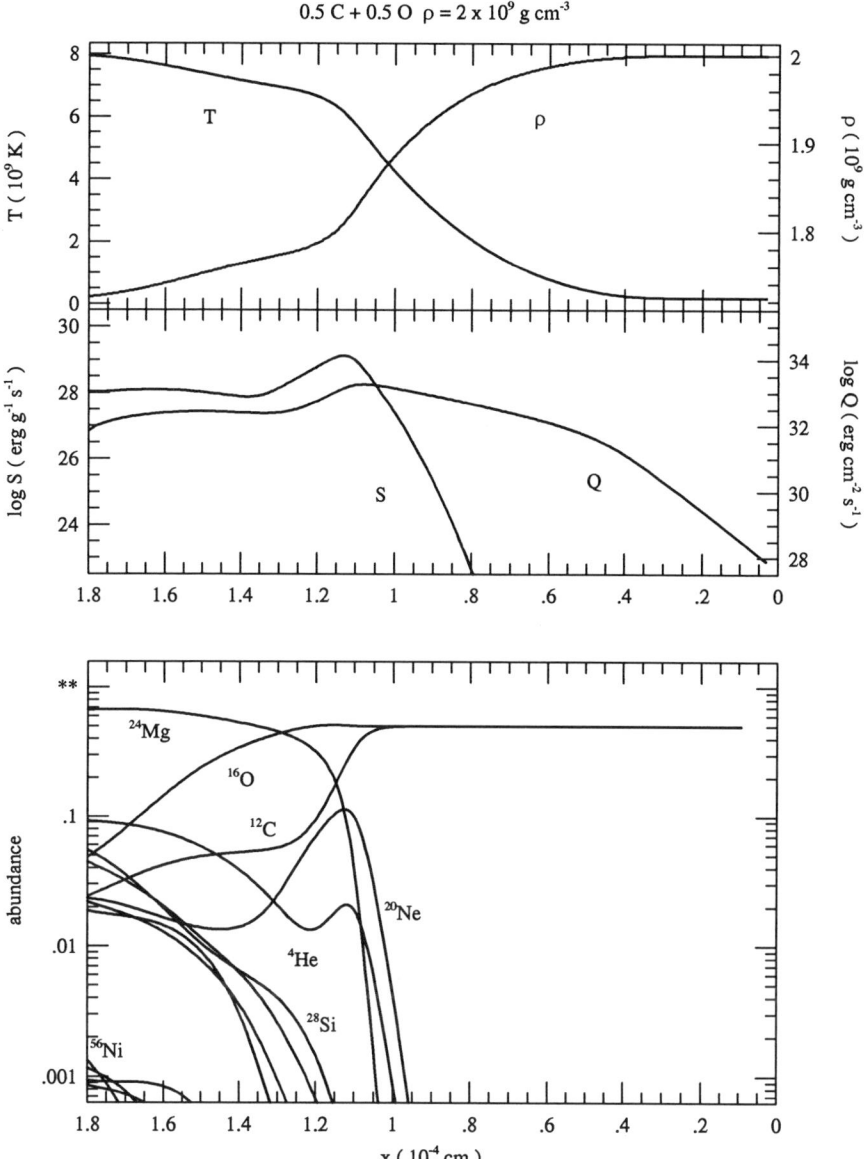

Figure 1: Structure of the deflagration front

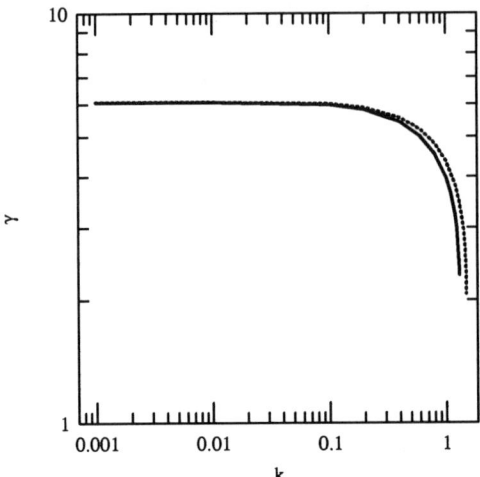

Figure 2: Dispersion relations

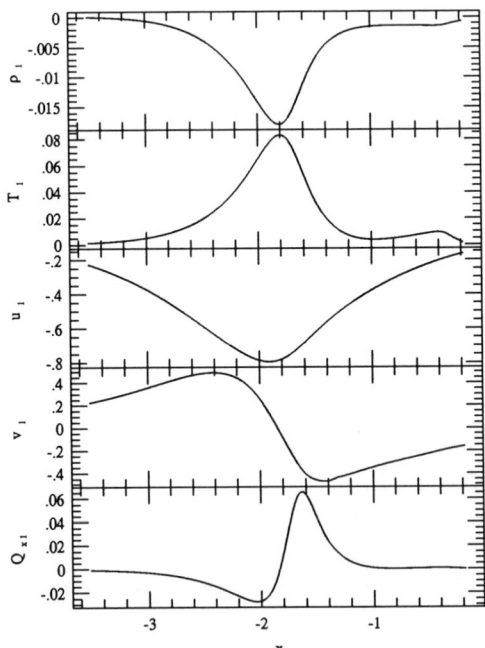

Figure 3: Mode profiles

Numerical Simulation of Jet-Like Structures in Laser Plasma

Andrei Yu. Semenov, Sergei F. Goncharov

General Physics Institute
Russia Academy of Sciences
Vavilov Street 38, Moscow 117942, Russia

Abstract. The aim of our investigation is verification of the hypothesis on gas dynamic mechanism of generation of long-life and large-scale jet-like structures in laser corona. Numerical results in comparison with experiments are presented.

A set of experimental investigations of laser plasma, produced near the solid surface, have revealed an interesting phenomenon "Jet-Like Structures" - JLS, for example, [1]-[6], review [7]. These structures are the plasma regions having abrupt and accurate density boundaries. These regions are clearly distinguishable at experimental shadowgrams as separate dark protuberances and are long-forming and long-living in comparison with laser pulse duration. Fig.1 presents the typical scheme of experiments of JLS generation. Here JLS are the protuberances of density outside focal spot. (H,R) are cylindrical coordinates.

The reasons or JLS generation are actively debating and discussing [3],[6], however a general point of view on this problem is absent.

Our numerical results of the large-scale JLS modelling and comparison with experimental data, for example, with reconstructing two-dimensional laser plasma density distributions ([5]) in some moments of time (General Physics Institute experiments on laser "UMI-35"), permit us in the first approximation to make a conclusion about gas-dynamic character of large-scale JLS. In particular, the numerical simulation demonstrates that JLS can appear as a result of gas dynamic interaction of the hot laser plasma region with cold background with non-uniform spatial density distribution . In some regions of initial data these numerical results are similar and stable: the JLS exist.

The computational modelling was based on a solution of two-dimensional and two-temperature gas dynamic systems. We used a reliable hybrid finite-difference method [7], which is essentially based on a characteristic relation of hyperbolic system.

© 1996 American Institute of Physics

Fig.2

Fig.3

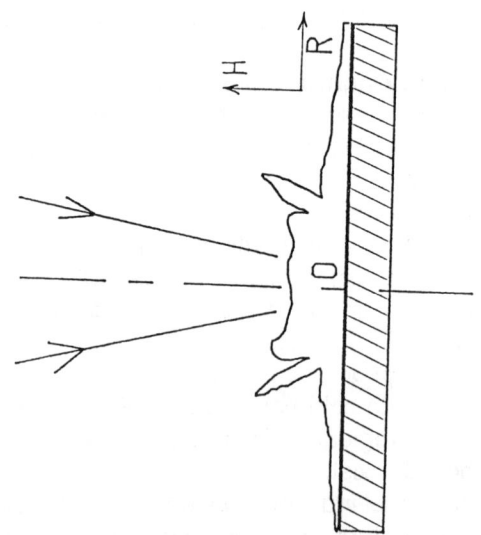

Fig.1

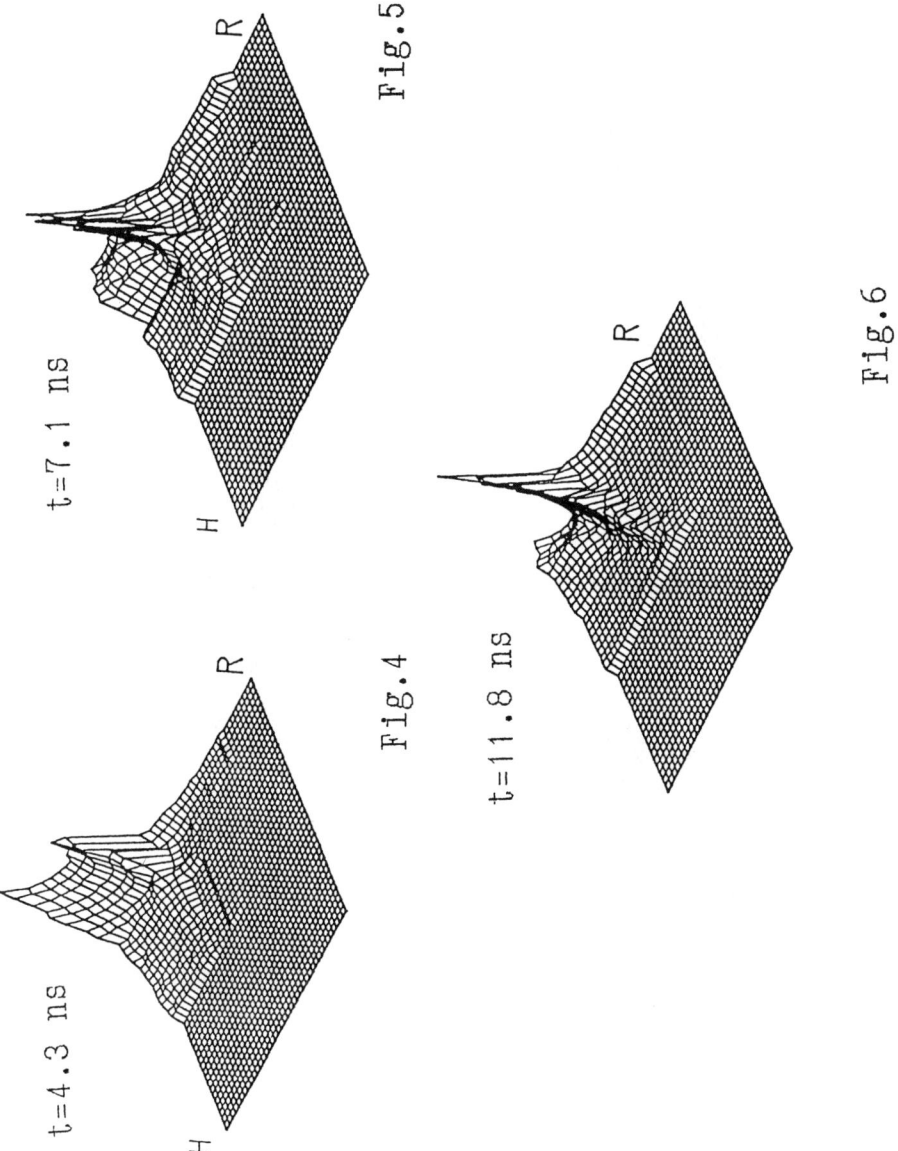

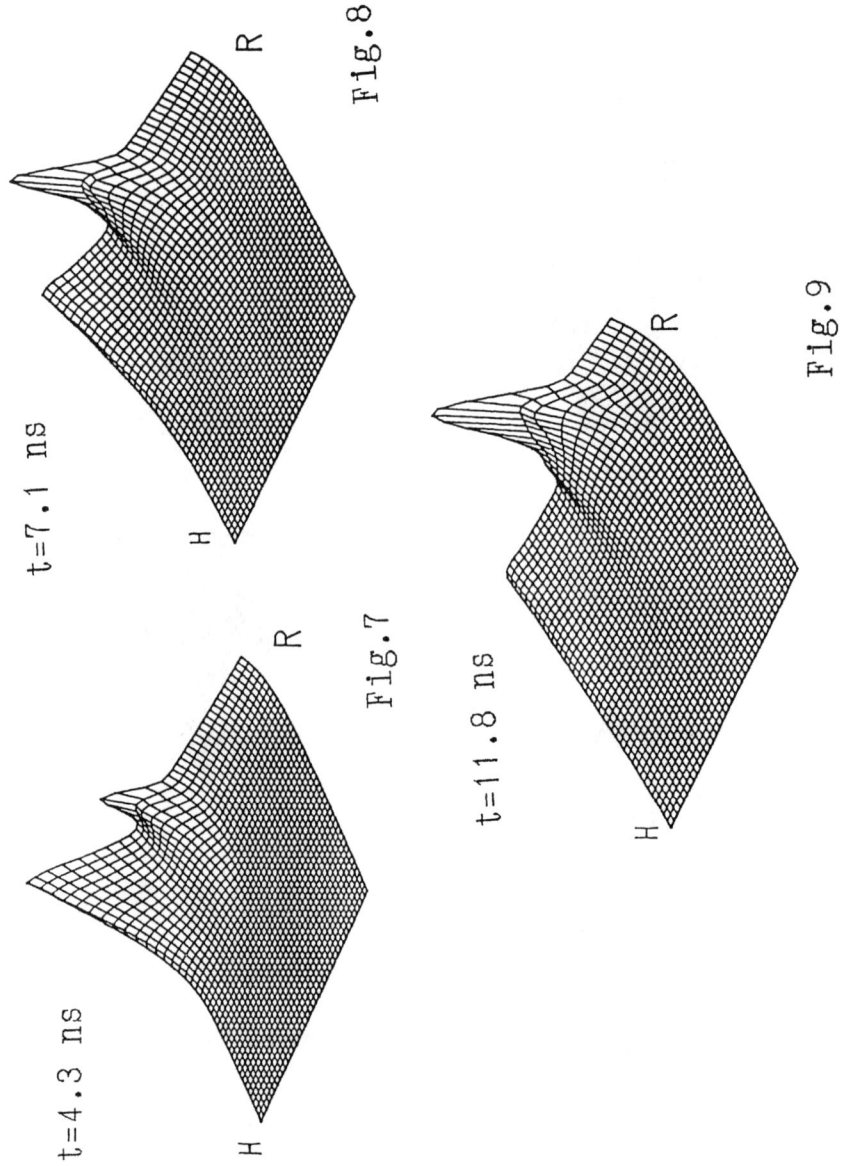

Fig.7
Fig.8
Fig.9

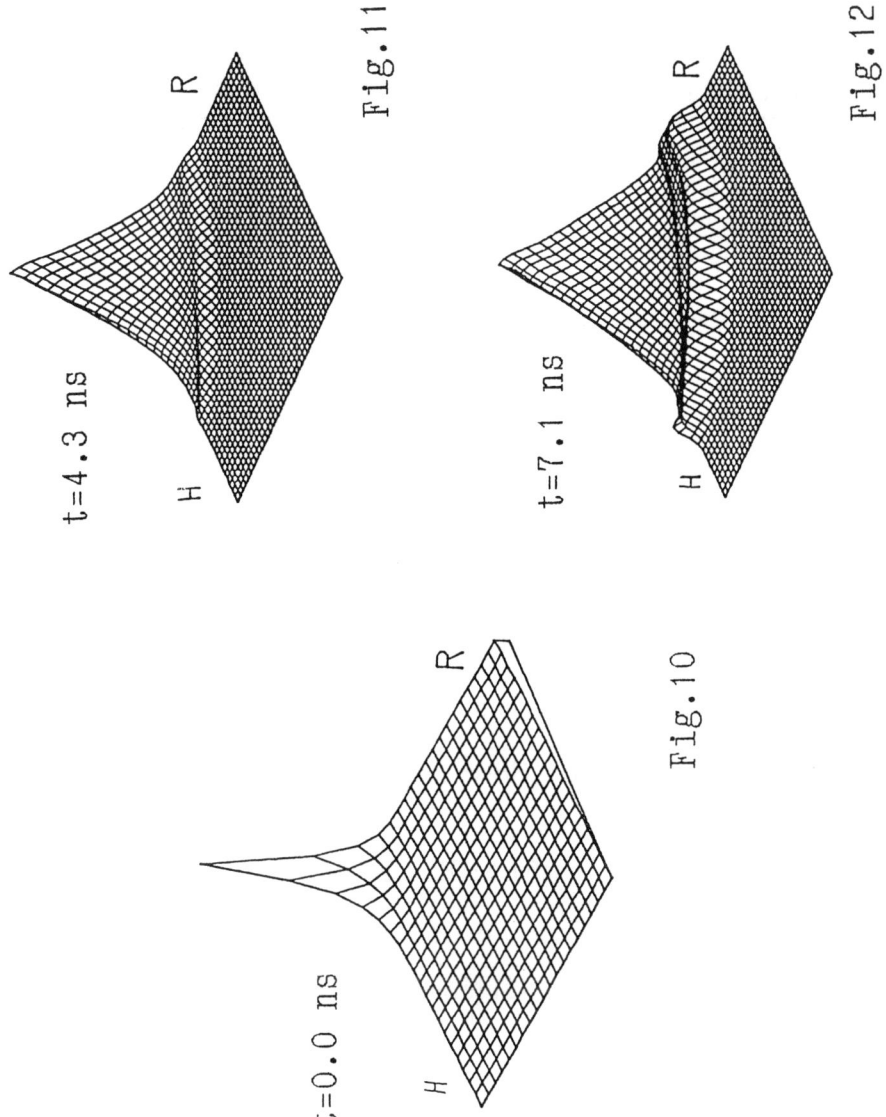

Fig.11 t=4.3 ns
Fig.12 t=7.1 ns
Fig.10 t=0.0 ns

Consider our results. Figs 2-3 present some typical experimental shadowgram (Fig.2) and interferogram (Fig.3) for mylar, $I=10^{13}$ W/cm^2, $\lambda=1.06$ μm, spot diameter = 480 μm, t=8 ns. Figs 4-6 are presented 2D experimental plasma density distribution reconstructing from interferograms by numerical solution of Abelian integral equations by modified Nestor-Olsen method (time in ns after laser pulse). Figs.7-9 are our numerical results corresponding to experimental results from Figs 4-6. The 2D initial density distribution for numerical results is showed on Fig.10:

$\rho(R,H)=\max\{\rho_1 \cdot \exp(-R/R_1-H/H_1), \rho_2 \cdot \exp(-R/R_2-H/H_2)\}$, $R \geq 0, H \geq 0$;

$H_1=30$ μm, $R_1=20$ μm, $H_2=40$ μm, $R_2=300$ μm, $\rho_1/\rho_2=30$. Index "1" marks he hot focal spot plasma (400 eV), "2"- to cold background plasma (0.5-5 eV). Velocity distribution is equal to zero. Figs 11-12 present for comparison the numerical computations results for plasma constant background. In this case we can see only standard shock wave structure and JLS is absent. Therefore the spatial background distribution is essential.

Our results. Numerical simulation demonstrates that the jet-like structures analogy can appear after laser pulse as a result of 2D gas dynamic interaction of the hot focal spot laser plasma region with cold plasma background with non-uniform spatial density distribution.

This work is supported by RBRF (Russia), Grant 94-02-03413-a.

REFERENCES

1. Willi O., Rumsby P.T., .Sartang S.-*IEEE J.*,**QE-17**,1909-1917 (1981).
2. Bondarenko Yu.A., Burdonski I.N. et al - *Soviet JETP*,**81**,170-179 (1981).
3. Thiell G.,Meyer B.- *Laser Part.Beams*,**3**,51-58 (1985).
4. Dhareswar L.J. et al - *J.Appl.Phys.*,**61**,4458.-4463 (1987).
5. Goncharov S.F.et at - *Sov.Phys.-Lebedev Inst.Reports*,**7**,24-26 (1987).
6. Xu Z.Z.et al - *Optics Comm.*,**69**,49-53 (1988).
7. Goncharov S.F. et al - *Gen.Phys.Institute Proceed.*,**36**, 228-246 (1992).
8. Semenov A.Yu.- *Soviet Math.Dokl.*, **30**, 602-605 (1984).

PIC Code Simulation and Theoretical Analysis on Generation of Weibel-type Instabilities

Kazuhito Satou and Toshio Okada

Faculty of Technology, Tokyo University of Agriculture and Technology,
2-24-16 Naka-machi, Koganei-shi, Tokyo 184.

Abstract. With the rapid progress of surroundings for ICF, especially laser technology, the occurence of self-generated strong magnetic fields and filamentary structures coupled with these fields have been observed in the ablation area of laser-irradiated targets by laser-plasma interaction experiments. According to these experiments and various computer simulations, these electromagnetic fields present exponential growth, with a growth rate of the order of $10^{10} \sim 10^{12} \text{s}^{-1}$, and convect towards the vacuum with group velocity of order of $10^4 \sim 10^6 \text{ms}^{-1}$. These phenomena may have serious consequences on target implosion. It has been suggested that Weibel-type electromagnetic instabilities have much to do with these observational results. In this paper, we'll report on the presence of Weibel-type instabilities in laser-produced plasmas with a theoretical analysis and will demonstrate with the aid of a computer simulation, PIC code.

INTRODUCTION

In the end of the 50's, E. S. Weibel theoretically estimated that transverse waves are self-generated in plasmas having electron velocity distributions which deviate sufficiently from the Gaussian (1).

After that, the dispersion relation equation and the growth rate of various Weibel-type instabilities have been studied by theoretical analyses and by some kinds of computer simulations, in both collisional and collisionless plasmas or in both linear and nonlinear stage (2–3). Moreover, it has been concluded that Weibel-like transverse waves reduce the heat flux in hot and dense plasmas, which play an important role in transport properties (4–5).

Recently, self-excited strong magnetic fields in the megagauss range have been observed in many laser-plasma interaction experiments. According to theoretical analyses and computer simulations, it seems that these magnetic fields have a great influence on the inhibition of heat flux, symmetrical target implosion, etc. (6–7). One possible mechanism for such an excitation of the critical electromagnetic modes is the above-mentioned Weibel-type instabilities.

The original Weibel instability grows as a standing wave, so it doesn't propagate. But, the observed electromagnetic waves convected with some group velocity. In this paper, we analyzed two types of Weibel instability and performed a PIC code simulation. At first, we consider the mechanism of the original Weibel instability, due to simple temperature anisotropy, and then we advance a different distribution model which has hot and cold electron parts and can explain the propagated mode.

Mechanism of Weibel Instability

We consider the single bi-Maxwellian which has a temperature anisotropy;

$$f(v_x, v_z) = \frac{n}{2\pi u_x u_z} \exp\left(-\frac{v_x^2}{2u_x^2} - \frac{v_z^2}{2u_z^2}\right), \qquad (1)$$

where n is the electron number density, u the thermal velocity and v the velocity. Subscripts x,z represent the x and the z direction, respectively. Now, we concider the Cartesian coordinates (x, y, z). It is thought that initial temperature anisotropy is formed by strong electric fields of the laser (8).

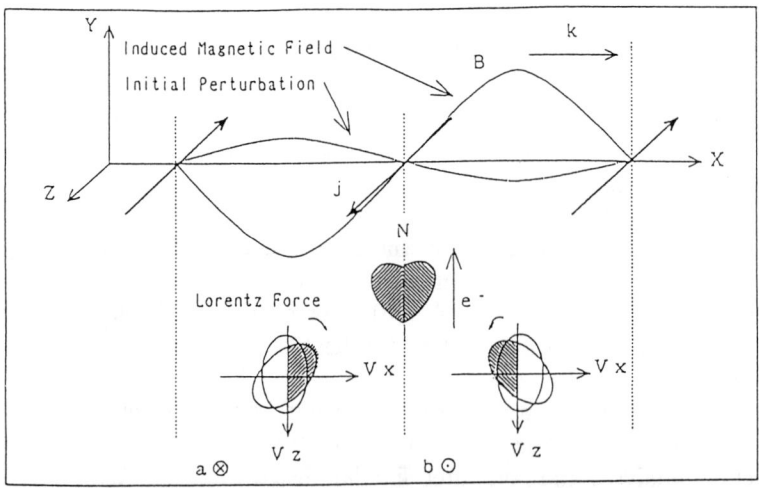

FIGURE 1. Mechanism of the self-generating magnetic fields.

Fig.1 shows the stage of the growth of a perturbation. We assume that electrons are under the distribution defined by Eq.1, and charge,current neutralization are satisfied completely. If we let thermal velocities $u_x < u_z$, electrons in region a and b are rotated to respective directions by the perturbed magnetic field in the y direction. Then it is thought that electrons near the node N belong to a "heart shaped" distribution composed by positive v_x in region a and negative v_x in region b. This electron distribution induced the current j in the z direction. At both sides of the node N, currents are induced in the opposite direction. All resultant currents enhance the initial perturbed magnetic field B.

The maximum growth rate of the Weibel instability for Eq.1 (from Ref.2) is obtained as follows;

$$\gamma = \left(\frac{8}{27\pi}\right)^{1/2} \omega_{pe} \frac{u_x}{c} \left(\frac{u_x}{u_z}\right)^2 \left(\frac{u_z^2}{u_x^2} - 1\right)^{3/2}, \qquad (2)$$

where ω_{pe} is the electron plasma frequancy and c the velocity of light.

We performed PIC code simulation on this phenomena. "PIC" means Particle-in-Cell, and it's a one of typical simulation schemes for performing microscopic instabilities affected by distribution function. Our PIC code is a two-dimensional in space (x, y) and a three-dimensional in velocity space (v_x, v_y, v_z). Ions are fixed in space as background and charge,current neutralizations are satisfied. We chose the initial thermal volocities; $u_x = 0.05c, u_y = 0.25c, u_z = 0.25c$.

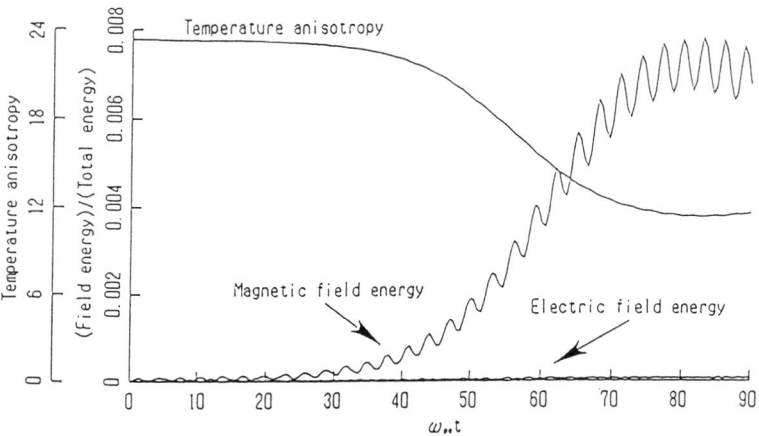

FIGURE 2. Profile of energy and temperature anisotropy vs. the time.

Fig.2 shows the profile of the magnetic field and the temperature anisotropy vs. the time normalized by ω_{pe}^{-1}. In case of $n = 10^{27} \mathrm{m}^{-3}$, the magnitude of maximum growth rate can be estimated to be about $8.9 \times 10^{13} \mathrm{s}^{-1}$, while $1.3 \times 10^{14} \mathrm{s}^{-1}$ is obtained from Eq.2. The intensity of resultant magnetic field is about 2.3MG, it stands comparison with data gatherd from experiments.

Propagation mode of Weibel-type Instabilities

It's well known that high energy electrons appear in the laser-plasma interaction. So we consider the following electron velocity distribution function to treat the ploblems of Weibel-type electromagnetic instabilities which propagate in the laser-irradiated target plasma.

$$f(v_x, v_z) = \frac{n\alpha}{2\pi u_x^h u_z^h} \exp\left[-\frac{(v_x - v_d^h)^2}{2u_x^{h2}} - \frac{v_z^2}{2u_z^{h2}}\right] \\ + \frac{n(1-\alpha)}{2\pi u_x^c u_z^c} \exp\left[-\frac{(v_x - v_d^c)^2}{2u_x^{c2}} - \frac{v_z^2}{2u_z^{c2}}\right], \quad (3)$$

where α is the ratio of hot electrons to all electrons, v_d the drift velocity and subscript h,c represent hot and cold electron part, respectively. This type of distribution have a temperature anisotropy, as well as an aforesaid single bi-Maxwellian, and may be more real than it.

In these systems, the dispersion relation of excited transverse waves is obtained as follows;

$$\omega^2 - (\omega_{pe}^2 + k^2 c^2) \\ + (1-\alpha)\omega_{pe}^2 \left(\frac{u_z^c}{u_x^c}\right)^2 W\left(\frac{\omega - kv_d^c}{ku_x^c}\right) \\ + \alpha \omega_{pe}^2 \left(\frac{u_z^h}{u_x^h}\right)^2 W\left(\frac{\omega - kv_d^h}{ku_x^h}\right) = 0, \quad (4)$$

where k is the wave number and W the W function connected with dielctric response. In the limit of $\left|\frac{\omega - kv_d^c}{ku_x^c}\right| \ll 1$ and $\left|\frac{\omega - kv_d^h}{ku_x^h}\right| \ll 1$, we obtain maximum growth rate;

$$\gamma = \left(\frac{8}{27\pi}\right)^{1/2} \frac{\omega_{pe}}{c} \left[\alpha \left(\frac{u_z^h}{u_x^h}\right)^2 + (1-\alpha)\left(\frac{u_z^c}{u_x^c}\right)^2 - 1\right]^{3/2} \\ \times \left[\alpha \frac{u_z^{h2}}{u_x^{h3}} + (1-\alpha)\frac{u_z^{c2}}{u_x^{c3}}\right]^{-1}. \quad (5)$$

Farthermore, we obtain the frequency of propagating waves;

$$\omega_r = \left(\frac{1}{3}\right)^{1/2} \frac{\omega_p}{c} \left[\alpha \left(\frac{u_z^h}{u_x^h}\right)^2 + (1-\alpha)\left(\frac{u_z^c}{u_x^c}\right)^2 - 1\right]^{1/2}$$
$$\times \left[\alpha v_d^h \frac{u_z^{h2}}{u_x^{h3}} + (1-\alpha)v_d^c \frac{u_z^{c2}}{u_x^{c3}}\right]\left[\alpha \frac{u_z^{h2}}{u_x^{h3}} + (1-\alpha)\frac{u_z^{c2}}{u_x^{c3}}\right]^{-1}. \quad (6)$$

Namely, our distribution function described by Eq.3 can have a propagating wave mode.

We performed PIC code simulation in order to compare with the original Weibel mode for initial conditions,

$$\begin{aligned}
&\alpha = 1/6 \\
&u_x^h = 0.10c, \quad u_y^h = 0.15c, \quad u_z^h = 0.15c, \quad v_d^h = 0.20c \\
&u_x^c = 0.03c, \quad u_y^c = 0.03c, \quad u_z^c = 0.03c, \quad v_d^c = -0.04c.
\end{aligned} \quad (7)$$

Using distribution function composed by these parameters, we obtained the profile of that fields vs. the time shown in Fig.3. As the result, we demonstrated the propagating mode of Weibel-type instability with the group velocity of about $1.1 \times 10^6 \text{ms}^{-1}$. These waves grew with the maximum growth rate of about $8.6 \times 10^{12} \text{s}^{-1}$. From Eq.4~6, the group velocity and the maximum growth rate of this propagation mode were calculated as $3.4 \times 10^6 \text{ms}^{-1}, 1.7 \times 10^{12} \text{s}^{-1}$, respectively.

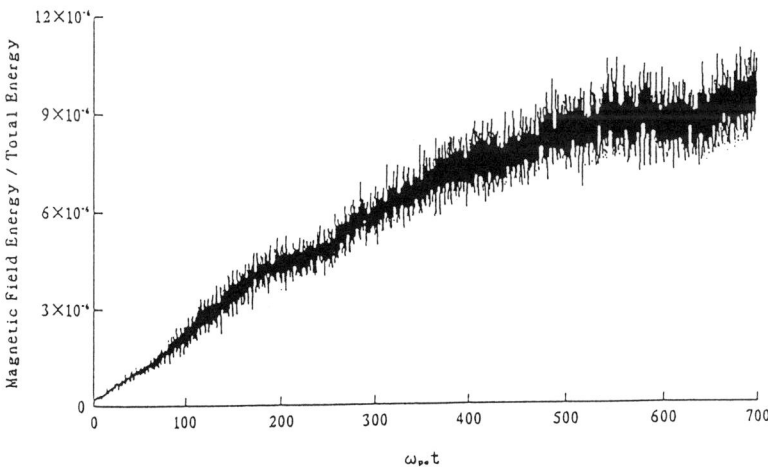

FIGURE 3. Profile of magnetic field energy vs. the time.

Conclusion. There is a sufficient possibility that some types of tempelature anisotropy will occur in the plasmas made from a strong-laser-irradiated targets (8). In these case, the Weibel-type electromagnetic instabilities will be able to self-excite. These modes convect in plasmas and have a serious influence on the inhibition of heat flux, deflagration wave, ablation pressure and symmetrical implosion, etc.. It is indispensable for the analysis of such instabilities to carry out many experiments with high power laser beam, and perform more detailed computer simlations which involve practical effects.

REFERENCES

1. E. S. Weibel, *Phys. Rev. Lett.* **2**, 83 (1959).
2. N. A. Krall and A. W. Trivelpiece, *Principles of Plasma Physics* (McGraw-Hill, New York, 1973), 490.
3. R. L. Morse and C. W. Nielson, *Phys. Fluids* **14**, 830 (1970).
4. A. Ramani, G. Laval, *Phys. Fluids* **21**, 980 (1978).
5. T. Okada, T. Yabe and K. Niu, *J. Plasma Phys.* **20**, 405 (1978).
6. T. Mochizuki, T. Yabe and K. Mima *et.al*, *Japanese J. Appl. Phys.* **19**, 645 (1980).
7. J. P. Matte, A. Bendib and J. F. Luciani, *Phys. Rev. Lett.* **58**, 2067 (1987).
8. J. C. Kieffer, J. P. Matte and H. Pepin *et.al*, *Phys. Rev. Lett.* **68**, 480 (1992).

Laser diagnostics of tokamak plasmas

T. Fukuda and †T. Matoba

Department of Fusion Plasma Research
†Department of ITER Project
Japan Atomic Energy Research Institute
Naka-machi, Ibaraki-ken 311-01, Japan

Abstract. Not only for the physics understanding of tokamak confinement, but also for the deliberate control of the plasma pressure and current density profiles, which is prerequisite for the improvement of fusion performances, tokamak diagnostics have demonstrated their effectiveness and capabilities to resolve various plasma parameters with detailed profile information. Above all, laser-aided diagnostics have pervasively implemented in the tokamak experiments to allow non-perturbative measurement of high-temperature plasmas, and they have hitherto made significant progress through large tokamak applications. This paper describes the recent development on the laser-aided diagnostics as well as the issues involved in the International Thermonuclear Reactor (ITER) application.

I. INTRODUCTION

Major experimental achievements in the recent tokamak research are high-lighted by the 10.7 MW fusion power production in the Tokamak Fusion Test Reactor (TFTR) D-T discharges and the fusion product of $1.2 \times 10^{21} m^{-3} \cdot s \cdot keV$ (equivalent Q_{DT}=0.46) at JAERI Tokamak 60 (JT-60), while equivalent Q_{DT} of 1.14 was achieved at Joint European Torus (JET) in 1991. Accordingly, substantial effort of the present tokamak research has been devoted to the attainment of steady-state improved confinement regimes, remote radiative cooling of divertor plasmas, and investigations on the tritium transport as well as the behavior of fusion products.[1-3] Not only for the physics understanding of tokamak confinement, but also for the deliberate control of the heat deposition and current density profiles, which is prerequisite for the suppression of magnetohudrodynamic (MHD) instability and sustainment of increased beta values,[4] tokamak diagnostics have demonstrated their effectiveness and capabilities to resolve various plasma parameters with detailed profile information.

Optically-pumped far-infrared laser interferometer, which reconstructs the electron density profile from the phase shift induced in the plasma, is one of the routine diagnostics. In addition, real-time feedback control of electron density is well established in large tokamaks. Faraday rotation of the polarization induced on the far-infrared laser has also been converted to the current density profile, which is intimately related to the MHD stability and the confinement properties. Stabilization of the laser frequency and the attainment of the single-mode oscillation were hereupon the crucial issues to retain the resolution against the long distance propagation in large tokamak applications.

Incoherent Thomson scattering spectrum of ruby laser radiation by the thermal motion of electrons provides detailed electron temperature profile, while total amount of the scattered intensity simultaneously yields the electron density profile. At JT-60, two sets of 10 J ruby lasers are independently fired every four seconds within the tokamak discharge period of 15 seconds. Scattered radiation is fed through bundled optical fibers to a 240 ch photo-diode array and 130 ch photo-multipliers. Detailed spatial resolution enables the estimate of local scale length of plasma parameters. 2 J YAG laser scattering system, which has repetitive firing capability at 10 Hz, has also been prepared at JT-60 to provide information on transient phenomena.

Termination of the steady-state high-performance discharge is often caused by the impurity influx from the divertor plates. Exploratory work on the laser-induced fluorescence diagnostic has been undertaken for the impurity and neutral particle flow measurements. In addition, turbulent transport is also intensively studied with either the far-forward or Bragg scattering of infrared lasers, in order to elucidate the ubiquitous L-mode confinement.

Laser-aided diagnostics for the profile measurement, namely interferometry and Thomson scattering are described respectively in section II and III. Physics-oriented diagnostics, mainly focused on the collective scattering, are discussed in section IV. Recent progress in the laser induced fluorescence diagnostic as well as the fusion product measurement are also mentioned respectively in section V and VI.

II. INTERFEROMETRY AND POLARIMETRY

A. Interferometry in large tokamaks

Recent thermonuclear fusion research is focused on the enlargement of plasma volume for the improvement of confinement properties, as represented by three large tokamak devices: JT-60, JET, and TFTR. For the understanding of plasma properties, such as the energy confinement and particle transport, interferometric measurements have been recognized to be an indispensable means of plasma diagnostics. Owing to the advent of FIR lasers in 1970,[5] interferometric plasma diagnostics has made a substantial progress by successfully adopting them as a scene beam for high density plasma applications. Although several endeavors, such as the development of the heterodyne detection technique with a twin laser system, have also been made for the improvement of fringe resolution during the mid-1970s, conventional interferometers reveal their intrinsic difficulties in interpreting the phase shift data with adequate resolution for large tokamak applications. Imposed requirements which previous interferometers envisage are mainly attributed to the long distance propagation of laser beams, and namely they are: (a) development of a single mode and frequency stabilized FIR laser, (b) investigations on the effect of enlarged beam diameter, especially concerning the distortion of the wave front due to the electron density gradient inside the propagating beam volume, (c) research on the stable and low-loss long distance propagation technique, (d) development of the optical components for the effective beam power distribution and polarization separation. Another aspect of limitation pertinent to large tokamak applications is the restricted diagnostic accessibility to obtain information over the full plasma volume. This condition would be forced more rigorously on the future large fusion devices, which are operated with D-T plasma components and equipped with a large blanket.

The FIR system at JT-60 employs 2 MHz beat-modulated twin CH_3OH lasers with Michelson configuration similar to that described by Wolfe et al.[5] The FIR lasers are mounted on a air-cushioned honeycomb bench for the mechanical stability, and they are located outside of the JT-60 torus hall to exclude the effect of stray magnetic field and external vibration. This layout leads to the total propagation distance of the probing beam over 130 m. The FIR beams, after being their propagation axes stabilized by the flexible light-guide and their optical parameters converted by the telescopic optics, are lead into the floor buried 40 m long tubes for the free-space propagation. Probing FIR laser beams are combined with 0.63 µm He-Ne laser beams for the vibration compensation. For the main plasma measurement corner-cube-reflectors (CCR) are installed inside the vacuum region and fixed below the diagnostic platform right above the JT-60 torus to dismiss the large structures above the floor level for reliability and compactness. The TFTR and JET interferometer system employs a massive frame (12 meters tall and 30 tons in weight for TFTR) which encircles the tokamak. The injection beam into the plasma and the returning beam trace a different beam path to suppress the optical feedback to FIR lasers as well as the cross talk inside the interferometer chamber, otherwise an optical isolator has to be developed, as seen in the TFTR interferometer.

For the application to large fusion devices, high frequency stability and mode purity of a source laser are required in order to obtain an adequate fringe resolution, since the difference of the optical path length between the probing and reference branch becomes inevitably large and the total propagation distance also becomes large. These factors can easily cause the deterioration of the fringe resolution and visibility due to the distortion of the wave front besides the loss of intensity. In the case of JT-60 interferometer, the total propagation distance exceeds 130 m, and the difference between the probing and reference branch is 40 m. The requirement imposed on the FIR lasers for the high resolution and totally adjustment-free JT-60 interferometer are: (1) the frequency and power stability control and (2) a mode purity, that is, the suppression of self-beat spectra due to the optical feedback of pump CO_2 laser beam or the multi-mode oscillation of FIR lasers. The achieved performances of the developed cw 118.8 μm CH_3OH waveguide laser, which employs a Ge etalon as a beam coupler and the cavity length modulation technique, are described in Ref. 6.

As for the frequency stability required, analytical estimate indicates that for the achievement of 1/100 fringe resolution requires the frequency stability of the probing laser less than 100 kHz ($\Delta f/f = 4 \times 10^{-8}$), and a stability of the offset laser frequency less than 1 MHz. Another factor that may deteriorate the fringe resolution for heterodyne detection is the power stability. A similar prediction as above shows that the amount of the amplitude modulation must not exceed 6%. Mode purity, especially the beam propagation with TEM_{oo} mode in a free-space which means the oscillation with EH_{11} waveguide mode inside the cavity is essential to achieve the optimum propagation for the total path length over 130 m and to retain the undistorted wave front inside a plasma, since the interferometer signal is produced by the averaged phase shift along the laser beam diameter.

A pump CO_2 laser has the output power is 77 W at 9.7 μm P(36) branch. The frequency stability of FIR lasers are mainly determined by the pump CO_2 laser frequency stability. The required frequency stability of the pump CO_2 laser to realize 1/100 fringe resolution can be estimated from the FM conversion coefficient. It is reported to be less than 0.1 to yield 1 MHz. In order to achieve this value, the pump CO_2 laser is (i) passively stabilized with Neoceram glass rods and a spacing, of which thickness is calculated so as to cancel out the thermal shrink of the cavity length. (ii) For the active stabilization, a reference CO_2 laser has been employed which is consisted of 1.3 m long discharge tube tuned to 9.7 μm P(36) branch with the same output coupler as the pump CO_2 laser. The passive stabilization technique employed on the pump CO_2 laser was also applied on the reference CO_2 laser. The output power is 7 W, and the center frequency f_o of the gain curve is locked in a way to keep the first derivative signal of the power tuning curve to zero. Absolute frequency stabilization scheme such as the application of opto-galvanic effect, Lamb dip, Stark effect, and an absorption cell (employed on the TFTR interferometer) for the achievement of f_o reproducibility is not required so far as to suppress the frequency fluctuation under 1 MHz. The frequency fluctuation of the reference CO_2 laser was measured to be less than 80 kHz_{p-p} by the tuning characteristics of the first derivative signal, and the offset frequency deviation was found to be less than 500 kHz_{p-p} by the F-V converted signal.

The output power of FIR laser is 34 mW. Mode purity is essentially governed by (i) the coupling device and (ii) the back talk of CO_2 laser radiation. (i) As to the coupling device, we have developed a 1048 mm thick Ge etalon coupler. It is placed at the Brewster angle for the p-polarized pump CO_2 laser beam and reflects the Q-branch s-polarized emission 95% by the resonant effect. The Ge etalon suppresses the possibility of other unnecessary oscillations at the same time. Reflectivity of the etalon for the p-polarized radiation is less than 2%. Therefore the simultaneous oscillation of 170.6 μm CH_3OH line is well suppressed. Since the pump CO_2 laser beam does not have to be collimated into a coupling-hole which is often seen in conventional FIR lasers, it is introduced into the FIR cavity without beam parameter modification with optical components. This is effective to realize the homogeneous pumping of laser media. Furthermore, the pump CO_2 laser beam, being unfocused, damages the ZnSe window very little. In comparison with conventional

FIR lasers with a hole-coupler, homogeneous pumping through the Ge etalon, which means that the pump CO_2 laser beam is not diffracted at the coupling-hole, yields a Gaussian-like transverse mode. (ii) Although the removal of optical feed back is one of the prime issues for the mode purity and stability of optically pumped FIR lasers, a few works have been reported on the suppression of self-beating oscillations. In order to restrain the optical feed back, a 10 mm diameter aperture has been placed in the CO_2 laser beam path; the reflected CO_2 laser beam at the concave mirror in the FIR cavity focuses right before the aperture and diverges so that only a fraction of the beam intensity goes back into the pump CO_2 laser cavity. The tunable output coupler is isolated from the pumping beam to assure the elimination of optical feed back even in a tuning operation of either pump CO_2 or FIR lasers.

The mesh output coupler, being off the pump CO_2 laser irradiation, provides an uniform output profile without any damage or deformation, and the effect of CH_3OH gas pressure ripple to the mesh output coupler was not observed. The smooth curvature and periodical single peaks of the tuning curve prove a single longitudinal mode oscillation. Frequency drift of the free-running probing laser and the stability of offset frequency for 10 seconds, which is the plasma pulse duration of the JT-60 tokamak, were less than 25 kHz and 2.5 kHz respectively, while the power stability for 10 seconds was 0.03 %. The far-field transverse mode pattern has been measured by scanning a pyro-electric detector to confirm the Gaussian shaped profile. The degree of linear polarization has also been measured with two grid polarizers one right next to another to find it is more than 99.98 %, which is adequate for the polarimetric measurement.

In order to reproduce the accomplished short term stability for 15 seconds for every tokamak plasma pulse, FIR lasers were also actively stabilized. The cavity length of the probing laser is AC modulated by mounting the total reflector on a PZT transducer in order to detect the first derivative signal with a lock-in amplifier. The cavity length is step-controlled with a stepping motor so as to bring the modulated output power signal to zero. The reference laser cavity is tuned so that the F-V converted beat signal frequency, produced by mixing the probing and reference laser beam, is locked to 2.0 MHz ± 100 kHz with a stepping motor. The stabilization control is sequentially actuated by a command signal sent from the JT-60 central control system. The probing laser cavity modulation and the reference laser cavity tuning with the stepping motor, which are carried out during tokamak plasma pulse intervals, are turned off 10 seconds before the plasma pulse to start free-running of the probe laser and DC offset frequency control with a PZT transducer for 30 s.

The developed FIR laser system has been in operation on JT-60 for two years yielding reliable data. It is totally adjustment and maintenance free for two weeks which is the period CO_2 gas mixture holds. 195 µm DCN laser system employed on the JET interferometer is

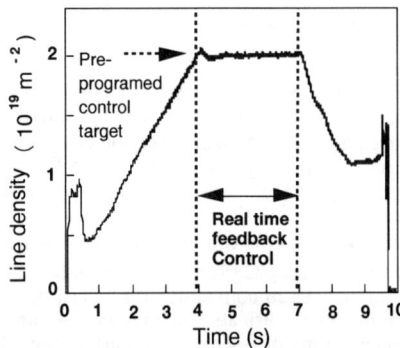

Fig. 1 Example of the real time density feedback control at JT-60.

not frequency stabilized and requires frequent maintenance mainly due to the vigorous carbonization and large amount of UV radiation on the laser components by the CD_4 (deutrated methane) gas discharge. Real-time feedback control of the electron density is routinely carried out at JT-60 as shown in Fig.1, which has enlarged the operation regime of JT-60. The signal to ratio, visibilities, and degrees of linear polarization at the injection side of vacuum windows for the FIR interferometer were 22-25 dB, 22-35%, and more than 99%, respectively. The whole interferometer system has been routinely in operation reliably for two years without any readjustment or attendant complications. The operation status is remotely controlled and monitored.

So far as the instrumentation of the future fusion devices is concerned, however, interferometric measurement presumably should undergo another stage of development. One of the requirements imposed on the application to the engineering fusion test devices is further research on the vibration compensation technique. Degradation of optical properties of molybdenum metallic mirrors inside the vacuum vessel and the deposition of the graphite granule on the vacuum windows have actually lead to the gradual reduction of the 0.63 μm He-Ne laser interferometer signal intensity. In order to retain the long-term reliability, the use of FIR sources both for the plasma phase-shift measurement and the vibration compensation is necessary. Several preliminary studies on the two-color FIR laser interferometry are being pursued in advance for JET and ITER. Another item of requirements is the development of solid-state FIR sources. The FIR laser system is still large and complicated, and damaged components due to the secular degradation have to be found and replaced in the years of operation. A solid-state FIR laser source can be installed close to the tokamak, according to the specific viewing chord, and it yields convenient replacement. FIR lasing based on hot hole transitions in Gallium doped germanium cell has been successfully achieved by Andoronov in 1984, and its lasing characteristics are intensively studied. The Gallium doped germanium hot carrier laser can potentially lase continuously in a wavelength range of 60 to 500 μm. The whole interferometer system could possibly be build in a disposable unit on the engineering fusion test devices, although it depends on the stability and gain performance of the germanium cell hot carrier laser.

B. Polarimetry

Further extension of the interferometric technique to the Faraday rotation measurement can potentially elucidate ubiquitous MHD activities and confinement mechanism through the evaluation of current density profiles, which is of basic and primary interest for a fusion plasma research. Although an application of Faraday rotation measurement to the plasma diagnostics was undertaken in the 1960s,[7] the first detection of the poloidal magnetic field in a tokamak plasma was performed by Kunz after a decade in 1978.[8] Stimulated by this experiment, intensive study for the improvement of resolution, such as the polarization modulation technique, have been pursued in the subsequent years. Nevertheless, the current density distribution has not yet been adequately diagnosed, suffering from the persisting experimental restrictions for the physics interpretation. In comparison with other approaches, such as the laser scattering technique and neutral lithium beam probe diagnostic, Faraday rotation measurement is one of the most prospective methods of diagnosing the current density profile, inasmuch as it provides continuous data throughout the entire plasma discharge. The primary reason that the wavelength of 118.8 μm radiation (the same as the TFTR system) was chosen is to suppress the elliptization of the linearly polarized probing beam (Cotton-Mouton effect), which is supposed to be significant in large tokamak plasmas, where high electron density and large toroidal induction along the long penetration distance are anticipated. The expected Faraday rotation angle and ellipticity are calculated to be 19 degrees and 0.015 (over 99.9% in terms of the degree of linear polarization) respectively. The rotation angle has been measured as an intensity ratio of the lock-in amplified 2 MHz heterodyne beat signals produced by mixing each orthogonal polarization components with the reference wave. The lock-in detected signal is low-pass filtered to yield the frequency response of 5 kHz. All the signal processing electronics have been designed so as to maintain linear responses in the range of interest, and calibration of the rotation angle with a half-wave etalon plate is carried out remotely for every plasma pulse with a motor-drive unit actuated by the timing signal sent from the JT-60 central control system. The resolution of Faraday rotation angle has been confirmed to be less than 0.25°.

Temporal evolution of the current density profile was evaluated by the Faraday rotation angle measurement combined with MHD equilibrium calculations, which provides continuous data throughout the entire plasma discharge. The measured rotation angle was converted to the plasma internal inductance

$$\ell_i \ [\ = \{2/a_p^2 B_{\theta(\rho=1)}^2\} \int_0^a B_\theta^2(\rho)\rho\,d\rho, \rho = r/a\ ].$$

Fig. 2 shows the evaluated ℓ_i for the case when the lower hybrid wave was applied to a 1 MA plasma to examine the efficiency of the current density profile.

III. THOMSON SCATTERING

Thomson scattering diagnostic is a standard tool in tokamaks for measuring simultaneously both electron temperature Te and electron density ne at the same spatial point with high time resolution of 0.1 to 50 nsec. Recent progress in the detector technology, such as the 2 dimensional photo diode array, has dramatically improved the spatial resolution of the incoherent Thomson scattering diagnostic, as represented by the TVTS system at TFTR. However, the requirement in the recent large tokamak experiment was to have the time evolution of the electron temperature and density profile, which is necessary to understand the physics of improved confinement and the cause of instability. Also for the transient transport analysis, time resolved profile measurement is of prime importance. As seemingly once established ECE diagnostic was found to be not always reliable in the recent tokamak experiment, due to the production of non-thermal electrons, significance of the Thomson scattering data has been stressed ever before. Therefore, various method of increasing the number of pulses were proposed.

Accordingly, a beam combine method of two ruby lasers has been developed for the JT-60 Thomson scattering system, that is mainly composed of a polarizer plate and a Faraday rotator. By this method provides the minimum time interval of 2 msec between the two laser pulses homogeneously combined. Also a high energy transmission of more than 95% has been attained. As a result, transient phenomena measurement can be available especially for pellet-injected plasmas by using the Thomson scattering system. The beam combine method applied to JT-60 is very attractive in terms of upgrading a Thomson scattering system to a high repetition diagnostic without any essential change of the existing equipments, and has been working with no serious trouble since 1989. The application of this method to a multi-laser system with three or more lasers is also discussed for the practical purpose of improving a repetition rate of high power pulsed laser.There are at least two approaches to realize the high repetition Thomson scattering. The first is to adopt higher repetition laser such as a Nd:YAG laser, Alexandrite laser (Cr : BeAl2O3) and Ti:sapphire laser. The Nd:YAG laser was firstly used in ASDEX with a repetition rate of 60 Hz. The recent progress of laser excitation method using such as a laser diode, which is most effective to a Nd:YAG laser, will make it possible to operate at higher repetition rate. A Nd:YAG laser with a laser diode excited amplifier is very attractive for a high repetition Thomson scattering to be achieved with high reliability due to no thermal problem of laser rod nor non-uniformity of beam profile. Unfortunately at present it costs too much to obtain the high laser energy and the uniformity, since low power level of a single laser diode (100 mJ/pulse) requires a large number of laser diodes for their purposes. Generally speaking, on the other hand, the Alexandrite laser and Ti:sapphire laser are under development from viewpoints of the reliability of laser system and the capability of high power. As a Q-switched laser with flash lamp pumping, they are available in commercial base with low power level of less than 1 J and with repetition frequency of less than 100 Hz only. Thus a high repetition laser of more than 100 Hz with high power of 5 to 10 J is not offered presently.

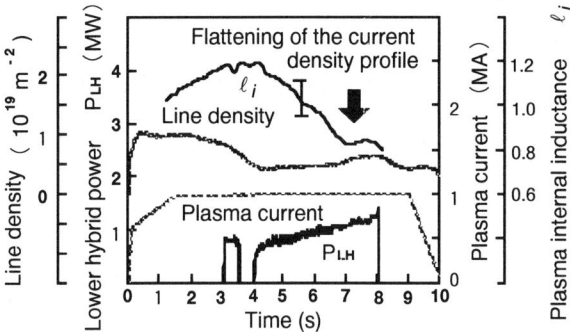

Fig. 2 Evolution of the current density profile during the lower hybrid current drive, evaluated from the Fadaray rotation diagnostic combined with the equilibrium code.

The second is to introduce a multi-laser system. There are two methods to construct a multi-laser system. One is a multi-beam path method as used in the DIII-D, and the other is a beam combine method presented here. These methods are characterized by use of two or more lasers, and by possessing capability of two kinds of laser operation: normal operation with a even time interval of exciting multiple lasers, and burst operation with a rapid succession of firing individual lasers at higher repetition rate of over several hundred Hz. In the multi-beam path method firing multiple lasers is not restricted by time. Therefore a simultaneous operation of multiple lasers is possible to increase the S/N ratio, and the minimum time interval of individual lasers in burst operation is limited by the data acquisition rate only. The width of a field of view may become inevitably wider, however, which results in a slight reduction in the signal to noise ratio. The DIII-D Thomson scattering system uses eight 20 Hz Nd:YAG lasers together with individually separated beam paths (a matrix of beams 2 wide by 4 high) in the near field and obtains a high repetition rate of 160 Hz in normal operation and of less than 10 kHz in burst operation. In the beam combine method, on the other hand, the simultaneous operation of multiple lasers is impossible, and the minimum time interval of individual lasers in burst operation is restricted by the driving time of a Faraday rotator as well as the data acquisition rate. The reduction of S/N ratio caused by widening a width of a view field can be avoided, however, since a single beam path can be used completely in the far field. Ruby laser has been used on a conventional Thomson scattering system, due to its high power and its wavelength suitable for highly sensitive detectors such as photomultiplier (PMT) and micro channel plate (MCP). But ruby laser has a low repetition rate of nominally 1 Hz (2.5 J) to 0.25 Hz (10 J) and a single pulse (20 J) due to its own feature of three-level laser. When the existing equipment of ruby laser, input optics, collection optics, spectrometers, detectors and so on are also used in an upgraded system, a multi-laser system is suitable for realizing a high repetition Thomson scattering. Otherwise most of equipment would be required to be replaced by other ones fit for another wavelength.

The Thomson scattering system for the JT-60 tokamak started its operation in 1986 with a single ruby laser. To make provision for the pellet experiment, another ruby laser was introduced together with the beam combiner in 1989. The original laser (PDS-4, JK Products Div. of Lumonics) and the additional laser (PDS-2s, JK Products Div. of Lumonics) have had capabilities of about 30 nsec pulse duration (FWHM), 20 mm beam diameter and less than 0.3 mrad beam divergence. The laser pulses were guided to the JT-60 device about 70 m away from the room where the lasers were located. On JT-60, beam combine method of two ruby lasers that mainly consists of a polarizer plate and a Faraday rotator was employed. Without any essential change of the existing equipment for the JT-60 Thomson scattering system, this beam combine method has improved twice the repetition rate compared with a single laser system in normal operation, and also has provided the utilization in transient phenomenon measurements with a minimum time interval of 2 msec with 10 J for each pulse in burst operation.

IV. COLLECTIVE SCATTERING

Coherent Thomson scattering, for which scattering wavelength is larger than the plasma debye length, observes the collective behavior of charged groups, while the incoherent scattering spectrum reflects the shape of electron velocity distribution. The significance of turbulent fluctuations on anomalous transport has been pervasively recognized in terms of the saturation in the ohmic confinement as well as H-mode transition. The reduction of turbulent fluctuations was actually detected in a limited edge plasma layer of the H-mode plasmas in most of the tokamaks either by reflectometry and collective scattering. Several works have been pursued for the determination of the precise thickness of this transport barrier with a reflectometric technique. Collective Thomson scattering diagnostic provides the crucial poloidal wavenumber which manifests the H-mode confinement properties.

For scattering studies, investigation of turbulent fluctuations have been attempted solely in terms of the parametric variations, such as the magnetic field or mass dependence of the fluctuation intensity and examination of the mixing length rule . However, the spectral structure of turbulent fluctuations will have to be numerically estimated using the ηi-mode and collisional trapped electron mode to compare with the locally measured spectrum. The crucial driving mechanism of turbulent fluctuations could thus be unveiled. Intrinsic difficulty of the reflectometric method in determining the suppression depth lies in the positional variation of the reflecting layer across the H-mode transition, although the fluctuation intensity has to be evaluated at a fixed position. For a fixed probing frequency, the ambient L-mode fluctuation intensity at the position where critical density layer is formed at the H-mode transition is larger than the value at the corresponding L-mode position. Therefore, the proposed method can possibly estimate the suppression depth to be less than the actual value.

In regard to the anomalous transport, a 212 GHz collective scattering diagnostics is being developed to identify the local spectral structure of microinstabilities at JT-60. The detection range of the normalized ky has been determined to be from 0.2 to 3.5, focusing on the ion temperature gradient mode and resistive MHD turbulence. Detailed density profile information can also enable the examination of the mixing length rule and the evaluation of the off-diagonal contribution in the coupled transport analysis. The scattering geometry is similar to ASDEX and DIII-D layout. Due to high toroidal magnetic field and dense plasma operation on JT-60, the available frequency region is rigorously restricted. 60 GHz region, which is utilized in TFTR with X-mode propagation, has limited availability, since this region is a narrow window for the probing wave to pass through, being right between the lower cutoff and upper hybrid resonance frequencies. The choice of the probing frequency in this region does not allow the magnetic field scanning. In addition, the availability of high power Watt level output millimeter-wave sources are limited to the frequency range of 220 GHz. The spectrum rapidly falls as around or , where the electron mode dominates. In order to fully diagnose the spectrum, high power transmitter is thus indispensable. Since X-mode wave is either significantly absorbed by the 2 nd ECE harmonic layer, of which frequency extends up to 350 GHz, or reflected below 190 GHz, 210 GHz O-mode wave was chosen. The ray and absorption of 210 GHz O-mode wave were calculated. The actual measurement region of the normalized wavenumber is 0.37 to 4.99 with the resolved spatial resolution of half the minor radius and 0.18 for the line integral measurement.

However , TFTR scattering diagnostic has prevailed the enhancement of fluctuations in the electron channel in the high density regime.[9] Together with the fact that numerous effort to resolve the turbulent transport is still carried on strongly, which means that collective scattering will continue to be an promising method of resolving the transport physics.

V. LASER INDUCED FLUORESCENCE

Neutrals and light impurities, which are released or emitted from the graphite limiter tiles employed in many tokamaks significantly influences the confinement and dilutes the deuterium fuel to reduce the neutron production rate. Recent results from the quasi-steady-state improved confinement experiment at JT-60 pointed out the termination of the high performance phase was triggered by the carbon influx from the divertor plate, which temperature exceeded 1200 °C. Therefore, the importance of understanding the behaviour of neutrals and impurity emission mechanism have been as of important issues as ever before, in the recent large tokamak experiment. Although the code estimate is possible using the DEGAS and divertor simulation model, combine with the D_α emission and Langmuir probe measurement, direct measurement is urged to examine the accuracy of code estimate. Since the D_α measurement has a number of drawbacks, such as (1) the signal is integrated along the line of sight, (2) applicability of low temperature divertor plasmas, (3) the evaluation of fluxes requires an estimate of local T_e and n_e information and the assumption of a complicated collisional-radiative excitation model.

Laser induced fluorescence diagnostic measures the emission from atoms that are excited by the laser to a specific energy state. Therefore, the laser wavelength is equal to the emitted light in the two-level system, while the former is shorter than the latter in the three-level system. Key issue of this diagnostic is availability of desirable laser sources. The wavelength region over 200 nm can be obtained either by a fundamental or harmonic frequency of the frequency tunable dye lasers. Recent progress in the lasers in the UV and VUV region have made this diagnostic as a efficient method of understanding the edge neutral and light impurity behaviour. This diagnostic is not only capable of measuring the atomic densities but also effective in evaluating the velocity distribution function from the spectral profiles. Much effort has been devoted for improving the signal to noise ratio against the background emission by e.g. modulating the laser power or seeking for the possibility of two-photon excitation. Also for the efficient spectral profile measurement, rapid frequency scanning method has also been proposed and demonstrated its feasibility.

For the detection of light impurities in the edge region of the plasma, high power VUV laser is necessary, of which wavelength is Raman shifted in H_2 or frequency tripled in a phase matched noble gas mixture. Bogen *et al.* produced L_α (121.6 nm, 150 W in 4 ns pulse width) by excimer laser pumped dye laser tripled in a gas mixture cell of krypton and argon, and measured the atomic density and velocity distribution in the boundary of the TEXTOR tokamak in 1989. The result indicated that most of the deuterium atoms have a velocity determined from the molecular dissociation process and not from the ion temperature of the edge plasma.

As the research effort being focused more in the divertor and edge physics in large tokamak experiment, together with the fact that its feasibility is intensively examined in medium-sized tokamaks, laser induced fluorescence may also be a potential candidate to be a routine diagnostic in large tokamaks.

VI. FUSION PRODUCT DIAGNOSTIC

Documentation of the ion term in the spectral density function was intensively carried out in the 1960 s, mostly in high-density low-temperature pinch plasmas. However, demonstration of the ion temperature measurement in a tokamak plasma was first carried out by Woskoboinikow *et al.* on ALCATOR-C in 1983, using a 385 µm D_2O laser. Behn *et al.* made refined measurement with more reasonable signal to noise ratio on the TCA tokamak in 1988. However, this method turned out to be too much complicated and delicate to be routinely operated in a tokamak experiment, and much simpler method, as represented by the charge exchange recombination was employed instead. Similar situation has happened also for the Rutherford scattering diagnostic.

The major cause of difficulty was the development of high power sources to enable the adequate scattering intensity against the significantly reduced cross section, in comparison with the non-thermal fluctuation measurement. However, as this technique was claimed to be one of the most promising candidate for the measurement of the density and birth velocity distribution function of alpha-particles, many of the laboratories re-examined its feasibility. JET and TFTR, which both have the D-T campaign programme, started the design studies in 1988. JET employed 140 GHz gyrotron, which can produce the power of up to 1 MW on a CW basis. The frequency was chosen to be right in the dip of the ECE spectrum to eliminate the background emission. The scattering angle was chosen to be between 20 to 40 degrees, for which Salpeter parameter is well above 1.

As was the case for the demonstration experiment, production of the high power source has been difficult to result in the delay of the actual measurement. The ITER R & D requires either 200 MW with the pulse length of 100 ns and the repetition frequency of 20 kHz or 1 GW with 100 ns and 1 kHz at 1.5 THz.

VII. SUMMARY AND CONCLUSION

Laser-aided diagnostics have hitherto made significant progress through large tokamak applications. However, another great stride has to be made toward the ITER application. Besides further integration of the present technology is required, e.g. stability over thousands of seconds, inhospitable environment induce additional stringent demands. Nonetheless, it is inconceivable to proceed with ITER without the laser-aided diagnostics. FELs are proposed for the measurement of density and velocity distribution function of alpha-particles on ITER. Development of high-power repetitive lasers are also precious for the LIDAR diagnostic which is a laser radar technique.

REFERENCES

[1] T. Nishitani, S. Ishida, M. Kikuchi, A. Azumi, M. Yamagiwa, T. Fujita, Y. Kamada, Y. Kawano, Y. Koide, T. Hatae, M. Mori, and S. Tsuji, Nucl. Fusion **34**, 1069 (1994).
[2] O. Naito and the JT-60 Team, Phys. and Controlled Fusion **35**, B215 (1993).
[3] M. Mori and the JT-60 Team, Plasma Phys. Controlled Fusion **36**, B181 (1994).
[4] T. Ozeki, M. Azumi, S. Tokuda, and S. Ishida, Nucl. Fusion **33**, 1025 (1993).
[5] S. M. Wolfe, K. J. Button, J. Waldman, and D. R. Cohn, Appl. Optics **15**, 2645 (1976).
[6] T. Fukuda and A. Nagashima, Rev. Sci. Instrum. **60**, 1080 (1989).
[7] I. S. falconer and S. A. Ramsden, J. Appl. Phys. **39**, 3449 (1968).
[8] W. Kunz and G. Dodel, IR Phys. **18**, 773 (1978).
[9] N. Bretz, P. Efthimion, and J. Doane, Rev. Sci. Instrum **59**, 1538 (1988).

Properties of high-Z laser-produced plasma determined by means of ion diagnostics

J. Wołowski, P. Parys, E. Woryna, L. Láska*, K. Mašek*, K. Rohlena*, W. Mróz**, and J. Farny***

Institute of Plasma Physics and Laser Microfusion, 00-908 Warsaw, P.O. Box 49, Poland,
**Institute of Physics, Acad. Sci. Czech Rep., 180 40 Prague 8, Na Slovance 2, Czech Republic,*
***Institute of Optoelectronics, Military University of Technology, 01-489 Warsaw 49, Poland,*
****95445 Bayreuth, Meranierring 28/7, Germany*

Abstract. The Nd:glass laser (E_L<20J, $\tau_L \approx$1ns, λ=1.06μm, $I_L \leq 10^{14}$Wcm^{-2}) and the iodine laser PERUN (E_L<50J, $\tau_L \approx$350 ps, λ=1.315μm, $I_L \leq 10^{15}$Wcm^{-2}) were used for production of medium- and high-Z laser-produced plasmas. Properties of highly charged ion streams emitted from laser-produced plasma were studied by ion diagnostic methods: ion collectors and an electrostatic ion analyzer. The results obtained demonstrate the possibility of getting high charge states and high maximum energies of ions (for Ta: z_{max}=53, E_{max}=4.8MeV). Results of calculations taking into account the two plasma temperatures are consistent with experimental results at more detailed analysis of the influence of recombination on the expanding plasma parameters. The nonthermal phenomena generated hot electrons and fast ions as well as decreased the absorption efficiency observed in experiments.

INTRODUCTION

During expansion of a hot laser-produced plasma, most of the thermal energy of the plasma is transformed into the kinetic energy of ions. But in the process of interaction of the laser beams with high-Z targets, more than 50% of the absorbed laser energy may be re-emitted by the hot plasma as X-rays. A part of the X-rays is once again absorbed in the target outside the laser focus where it creates a dense and colder plasma. Generally, as a result of interaction of the laser radiation with medium- and high-Z targets, two groups of ions (thermal and "slow") are emitted. In the case of long-wavelength high-power (λ_L > 0.5 μm, I_L > 10^{13} Wcm^{-2}) laser-plasma interactions, non-thermal processes in the plasma can generate a group of "fast" ions. Therefore, the measurement of the ion spectra and ion charge state are required for providing information on the absorption mechanisms of the incident laser light, plasma parameters, as well as ionization and recombination processes occurring in the plasma.

© 1996 American Institute of Physics

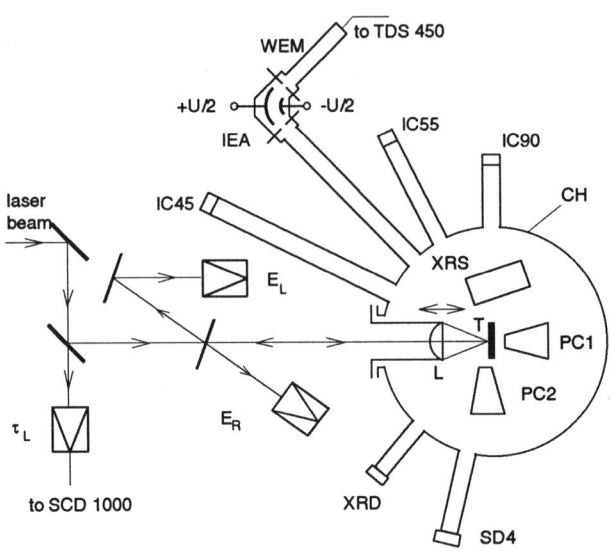

FIGURE 1. Experimental arrangement used at the IPPLM, Warsaw.

Studies of high-Z plasma produced with high-power laser are directed towards determination of physical processes in such plasma (e.g., energy transformation and transport, ionization and recombination) as well as optimization of the indirect-drive laser fusion, X-ray laser-plasma sources, and sources of multicharged ions for heavy ion accelerators (1,2). The paper presents results of investigations of physical phenomena occurring in a plasma of medium and high atomic number produced with lasers of about 1 µm in wavelength at power densities of $10^{13} \div 10^{15}$ Wcm^{-2}. Experimental investigations of the plasma were performed mainly by means of corpuscular diagnostics. They are completed with results of numerical computations performed with the use of models simulating the phenomena under investigation.

EXPERIMENTAL SETUPS

The investigations were performed at the Institute of Plasma Physics and Laser Microfusion, Warsaw, with the use of a Nd laser system of parameters: $\lambda_L = 1.06$ µm, $E_L < 20$ J, $\tau_L \approx 1$ ns (3), and at the Institute of Physics, AS CR in Prague with the use of an iodine laser system PERUN of parameters: $\lambda_L = 1.315$ µm, $E_L \approx 30$ J, and $\tau_L \approx 350$ ps (4). Thick Al, Cu, Mo, Ta, Pt and Pb slabs were used as targets. The ions from laser-produced plasmas were investigated using mainly the time-of-flight technique. Both a cylindrical electrostatic ion-energy analyzer (IEA) and ion collectors (IC) were used to measure ion fluxes emitted from plasma (5). Ion collector pulses from the IC and ion spectra from the IEA were registered with

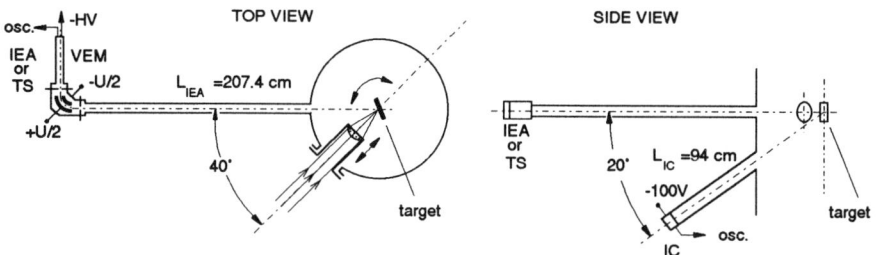

FIGURE 2. Experimental arrangement used at the IP ASCR, Prague.

digitizing oscilloscopes.

Schematic diagram of the experimental setup used at the IPPLM is shown in Figure 1. The laser beam was focused with an aspherical lens with the accuracy of 20 µm on the target surface. The electrostatic ion-energy analyzer was located at the angle of 45° at the distance of 230.2 cm from the target. Three ICs were located at the 45°, 55°, and 90° angles at the distances of 77.9, 75.3, and 46.1 cm, respectively.

The diagram of experimental system used at the Institute of Physics in Prague is shown in Figure 2. The PERUN laser beam was focused with an $f/2$ optics on the target (tilt angle 30°). The accuracy of the target position settlement was 50 µm. The IEA was located 207.4 cm away from the plasma in the plane of the laser beam axis and the target normal, 10° to the target normal in opposite direction to the laser beam. The IC was located below the laser-target plane underneath the IEA. The flight path of the IC was 94 cm.

RESULTS

By means of the simplest apparatus for corpuscular diagnostics of plasma, planar ion collectors, we obtained important data about laser-produced plasma and about plasma behavior. The ion velocity distribution, the charge and energy carried by ions reaching collector were obtained from the ion collector signals. In the case of medium- and high-Z targets, ion collector oscillograms, generally, consist of three ion groups: fast, thermal and slow. However, in some laser shots, we observed two groups of fast ions. In Figure 3a, an example of ion collector signal for Al plasma is shown. The ion velocity distribution, $d(N\bar{z})/dv = f(v)$, for this oscillogram is presented in Figure 3b. Fast ions registered in our experiments had velocities higher than 5×10^7 cm/s, thermal ions - in the range of $(2 \div 5) \times 10^7$ cm/s, and slow ions - below 2×10^7 cm/s.

The IEA was used for determination of energy distribution of different ion species, the number of particular ion species, the total number of ions and the total energy carried by them, the average charge state of the plasma, and the abundance of ion species measured at a long distance from the target. In order to get the time,

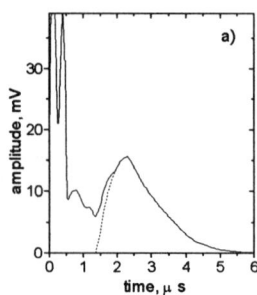
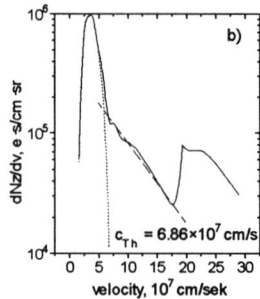

FIGURE 3. Ion collector signal with two groups of fast ions (a) and corresponding ion velocity distribution (b).

velocity or energy distribution of ions, a large number of ion spectra have to be measured at given working conditions of the laser system changing the deflection potential from one laser shot to another. Figure 4 shows an example of the energy distribution of Ta ions measured by means of the IEA. The average charge state as a function of the ion energy and the percentage abundance of Ta plasma are determined (6) on the basis of Figure 4. The highest recognized charge state of Ta ions was 53+ and the maximum energy of Ta ions was about 4.8 MeV. The peak ion current density was about 10 mA/cm^2.

DISCUSSION AND CONCLUSIONS

The mechanism of production of the ions slower than thermal ions was studied in (7). The number of slow ions and energy carried by them increase quickly with an increase in the Z-number of the target material. These dependences indicate that the slow ions are generated mainly outside the laser focus as a result of heating by X-rays emitted from the region of laser-beam absorption. The thermal ion group consists of low charge state ions. The faster ion group consists of low charge state ions. The faster ion group in the collector signal (Fig. 3), if analysed by the IEA, is found to be composed of the high charge state of ions and of light contamination ions (H, C, and O). Since the velocity distribution of the fast ions can be accurately approximated by an exponential relation (8), the slope of ion velocity distribution plot determines the ion sound velocity from which the hot electron temperature can be estimated.

If it is assumed (Fig. 3) that the slower group is composed of Al ions, a fitting procedure gives T_{eh} = 11.4 keV. The mean Al ion energy is $\overline{E}_{if}$ = 150 keV. The faster ion group is to be most likely interpretable as a fast proton signal since all the other impurity ions such as C, N, and O ions, give a signal largely overlapping with the fast Al ion peak. For the fast protons, the two-temperature plasma models render $T_{eh} \approx 5$ keV. The mean energy of the proton group, $\overline{E}_{if}$, is 27 keV (4).

The mean energy of fast ions can be also calculated on the basis of the model

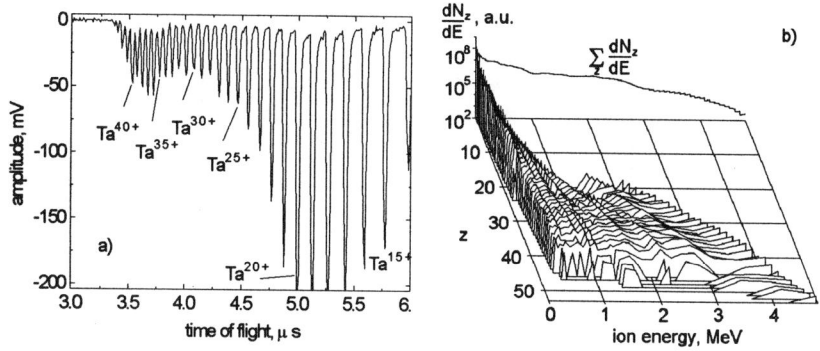

FIGURE 4. Ta ions spectrum from the IEA (a) and energy distributions of Ta ions (b).

given by Morse and Nielsen (9): $\overline{E}_{if} = 1.25 \overline{z}_0 T_{eh}$. With $T_{eh} = 11.4$ keV, is $\overline{E}_{if} = 165$ keV obtained from this relation for fast Al ions, and $\overline{E}_{if} = 6$ keV for fast H ions. This compaobtares well with the experimental value of energy determined for Al, but the discrepancy for the protons is fairly large ($\overline{E}_{if} = 27$ keV).

The most likely mechanism accelerating the electrons in our case are parametric instabilities, the presence of which is indirectly confirmed by the broad spectra of plasma emission, with the use of the second harmonic of the laser frequency, found in the related series of experiments (10).

Using the modified Shearer-Barnes formula (11) for $\overline{z}_0 = f(T_e)$ and the relation $\overline{E}_i = C(\overline{z}_0 + \alpha) T_e$, where $C = 4$ and $\alpha = T_i/T_e = 1$ for isothermal plasma expansion (8,11), the electron temperature and $\overline{z}_0$ can be determined. The best power fit to the experimental dependence $T_e \propto f(I_L \lambda_L^2)$ can be expressed as $T_e \propto (I_L \lambda_L^2)^{0.1}$. However, simple analytical models (12,13) render somewhat steeper dependence of T_e on I_L. The confrontation of our measurements with these predictions points out towards a diminishing absorption efficiency of the incident laser power with the growing intensity in the focus. There are several reasons why the absorption efficiency is be reduced, i.e., filamentation due to the hot spots in the beam profile, density profile steeping by the ponderomotive force, and other phenomena, which likely occur when working with a short, intense pulse of long-wavelength laser.

In parallel with the experimental work, a series of computational modeling of the far expansion zone combined with ionization and recombination kinetics were performed. According to the single-temperature, model the high charge states should persist due to the falling temperature inside the recombination zone extending over the distance of about 1 cm away from the target (14). A way of explaining the occurrence of the high charge state in the far expansion zone is opened by the existence of fast ion group in the collector signals (Fig. 3), which invariably contains the high charge states. The presence of the fast group means that the plasma time of evolution follows the mechanism of two-temperature isothermal expansion. A series of self-similar model calculations were performed

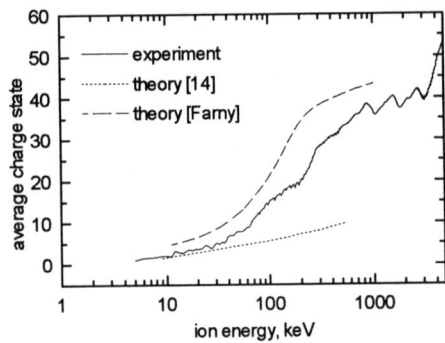

FIGURE 5. Average charge state of Ta plasma as a function of ion energy.

assuming that the energy deposition process renders a two-temperature electron energy distribution, which in turn leads to a two-temperature exponential density profile. Starting from this solution, the plasma expansion can be followed by applying the same time to a simple ion balancing scheme using the "*Trip*" code. The recombination heating is also included. The charge distribution at 2 m from the target, calculated on the basis of this simplified model, is seen in Figure 5. For comparison, the experimental data as well as the result of calculations performed using the model (14) are shown in this figure.

REFERENCES

1. Haseroth, H., and Hora, H., *New Methods and Technologies,* Chapter IV Advances of Accelerators Physics and Technologies, Singapore: World Scientific Publishing Company (ed. Schopper, H.), 1993.
2. Mróz, W. et al., *Rev. Sci. Instrum.* **65**, 1272-1274.
3. Badziak, J. et al., "Nd:glass lasers for laser-matter interaction studies at the IPPLM", in *Proceedings of the 21st European Conference on Laser Interaction with Matter,* Warsaw, Poland, October 21-25, 1991, pp. 131-133.
4. Chvojka, M. et al., *Czech. J. Phys.* **44**, 851- 864 (1994).
5. Denus, S. et al., *J. Tech. Phys.* **2**, 121-141 (1977).
6. Woryna, E. et al., *Laser and Particle Beams,* submitted for publication.
7. Mróz, W. et al., *Laser and Particle Beams* **10**, 689-696 (1992).
8. Wickens, L.M. et al., *Phys. Fluids* **24**, 1894 (1981).
9. Morse, R.L., and Nielsen, C.W., *Phys. Fluids* **16**, 909 (1973).
10. Mašek, K. et al., "Generation of second harmonics in an iodine-laser-produced plasma", in *Proceedings of the SPIE,* 1992, Vol.1980, pp. 109-116.
11. Farny, J., Ph.D. Thesis, Warsaw, 1985.
12. Fabbro, R., et al., *Phys. Fluids* **28**, 1463 (1985).
13. De Groot, J.S. et al., *Phys. Fluids B* **4**, 701 (1992).
14. Latyshev, S.V., and Roudskoy, I., *Fizika Plazmy* **11**, 1175 (1985).

ICF Burn-History Measurements Using 17-MeV Fusion Gamma Rays

R. A. Lerche*, M. D. Cable* and P. G. Dendooven[†]

Lawrence Livermore National Laboratory, P.O. Box 5508, L-473, Livermore, CA 94550
[†]University of Jyvaskyla, Accelerator Laboratory, P.O. Box 35, FIN-40351 Jyvaskyla, Finland

Abstract. Fusion reaction rate for inertial-confinement fusion (ICF) experiments at the Nova Laser Facility is measured with 30-ps resolution using a high-speed neutron detector. We are investigating a measurement technique based on the 16.7-MeV gamma rays that are released in deuterium-tritium fusion. Our concept is to convert gamma-ray energy into a fast burst of Cerenkov light that can be recorded with a high-speed optical detector. We have detected fusion gamma rays in preliminary experiments conducted at Nova where we used a tungsten/aerogel converter to generate Cerenkov light and an optical streak camera to record the signal.

INTRODUCTION

The fusion reaction-rate (or burn history) for an inertial-confinement fusion (ICF) target is a valuable piece of information for researchers studying laser-target interactions. It provides a sensitive measure of their ability to accurately model the interaction process. The centroid of the burn history, which is often referred to as the "bang time," depends on the coupling between the incident laser energy and the target capsule, and on the hydrodynamics of the capsule implosion. The details of the burn history, which are often characterized by a burn width, are related to plasma conditions at the time of peak target compression.

At the Nova Laser Facility, we use fusion neutrons from the D(d,n)^{3}He and T(d,n)^{4}He reactions to directly measure the thermonuclear burn rate for ICF targets (1). Temporal resolutions of 30 ps are achieved for deuterium-tritium (DT) neutron yields as low as 5×10^8. Since fusion neutrons are nearly monoenergetic and most of them escape from a small (< 200 μm diam) emission region (2) without collision, they preserve burn history information as they travel radially outward towards a fast detector some distance away.

There are, however, two fundamental limits to the temporal resolution that can be achieved by direct neutron detection. Doppler broadening of the neutron energy spectra causes a spread in the arrival times of the neutrons at a detector, and detector thickness produces a time spread caused by detection point uncertainty. For 14-

MeV DT neutrons, the time spread caused by the thermal motion of the plasma ions is given by $\Delta t = 122\sqrt{T_i} \times d$, where the time spread Δt is in ps, the plasma ion temperature T_i is in keV, and the target-to-detector distance d is in meters. Detection point uncertainty for DT neutrons is nearly 20 ps per mm. At Nova we typically make reaction rate measurements with a 1-mm thick detector placed between 2 and 20 cm from a target.

At the future National Ignition Facility (NIF), it is unlikely that reaction-rate measurements with 30-ps resolution can be made directly with fusion neutrons. At this megajoule laser facility we expect an "exclusion zone" to exist in the center of the target chamber. Its radius will be determined by experiment energy and will probably range from 50 cm to 5 meters. To achieve 30-ps resolution, neutron based measurements at NIF will require either a small neutron-to-radiation (gamma ray or light) converter close to the target or a neutron energy selection mechanism.

We are studying the possible use of fusion gamma rays as an alternative to fusion neutrons for making burn history measurements (3). A gamma-ray based measurement is attractive because gamma rays are virtually unaffected by the plasma temperature, have a large interaction cross section in many materials, and produce no time dispersion traveling to a distant detector. The T(d,γ)^{5}He reaction produces gamma rays with energies up to 16.7 MeV. The major disadvantage of this reaction is its low branching ratio of 5×10^{-5}. In this paper we describe the first observation of fusion gamma rays in an ICF experiment and discuss their possible use for future burn history measurements.

GAMMA-RAY DETECTOR CONCEPT

Figure 1 shows the basic concept for a gamma-ray based ICF burn history measurement. A target filled with a deuterium-tritium fuel mixture is heated and compressed with intense laser or x-ray radiation. The burning fuel isotropically emits fusion neutrons and gamma rays which travel radially outward from the compressed target core. Some of the gamma rays interact with a two-stage converter to produce Cerenkov light. In the first converter stage, gamma rays produce forward directed, relativistic electrons and positrons by Compton scattering and pair production interactions. The charged particles move into the second converter stage where they produce Cerenkov light. An optical system collects the Cerenkov light and relays it to a fast optical detector such as a high-speed streak camera for recording.

The converter offers the designer a number of options. Final detector design will focus on providing an instrument with good time resolution ($\Delta t < 30$ ps) and sensitivity. The choice of low-Z or high-Z material for the first-stage gamma-ray converter determines whether the primary interaction is Compton scattering or pair production. For 16.7 MeV gamma rays, the interaction cross-section for high-Z

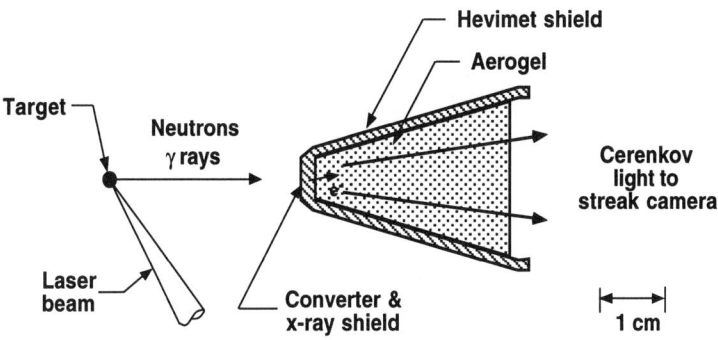

Figure 1. Hevimet nose cone acts as a γ ray to charged particle converter as well as a radiation and light shield.

materials is approximately 100 times greater than it is for low-Z materials. The range of the resulting electrons and positrons, however, is much greater in low-Z materials. We estimate that about 15% of the 16.7 MeV gamma rays incident on a high-Z converter that is several millimeters thick produce electron-positron pairs that enter an adjacent Cerenkov converter.

The second converter stage determines the characteristics of the Cerenkov light. The threshold energy for production of Cerenkov radiation, the cone angle for its emission, and the number of photons emitted per centimeter of track length depend on the index of refraction n of the converter material (4). Cerenkov light is produced in a material only when the speed of a charged particle exceeds that of light in that media. Thus, Cerenkov light is produced only when β the velocity of a particle relative to that of light in a vacuum and the index of refraction are such that $n\beta > 1$. Cerenkov radiation is emitted into a cone whose half angle θ relative to the direction of the charged particle motion is given by $\cos\theta = (1/n\beta)$. The photon production rate into the visible spectrum (400 - 700 nm) is given by

$$\frac{dN}{dx} = 490 \left(1 - \frac{1}{n^2\beta^2}\right) \text{ photons/cm.}$$

Detector geometry can be selected to enhance collection of the Cerenkov light (5).

FUSION GAMMA RAYS OBSERVED

We recently conducted a set of direct-drive target experiments at the Nova Laser Facility to access our ability to detect fusion gamma rays. Yields from 1-mm diam deuterium-tritium filled glass ball targets irradiated with 1-ns square pulses ranged from 10^{12} to 2×10^{13} DT neutrons. For these experiments, we adapted equipment normally used for our neutron burn history measurements (1). The standard Hevimet (90% tungsten) nose cone (Fig. 1) acted as the first converter stage in which gamma rays interact primarily by pair production to produce electron-positron pairs. A 0.241 g/cm^3 silica aerogel filled the interior of the nose cone replacing the standard 1-mm thick fast plastic scintillator. The aerogel acted as the second converter stage, converting relativistic particle energy into Cerenkov light. With an index of refraction of 1.06, the threshold energy for an electron to produce Cerenkov light in the aerogel is 1.03 MeV and the maximum cone angle for the emitted radiation is 19°. We estimate photon production at the rate of 53 photons per centimeter of track length. An f/2 optical telescope relayed light from the aerogel to a streak camera. System temporal response is better than 15 ps (FWHM). No modifications were made to the equipment to optimize Cerenkov light collection. Indeed, the telescope optics were designed to pass wavelengths between 350 and 450 nm.

Figure 2 shows streak camera signals recorded for target-to-aerogel distances of 2, 3, and 4 cm. Each signal has a large 550 ps wide pulse produced by 14-MeV target neutrons interacting with the silica aerogel. The burn duration for this type target measured in similar experiments is nominally 200 ps FWHM. The 550 ps corresponds to the neutron transit time across the 3.4 cm thick piece of aerogel. In Fig. 2 the neutron-induced signals are normalized, temporally aligned, and overlaid. This allows us to easily observe a small pulse at the foot of the neutron peak. Its time relative to the neutron pulse changes with target-to-aerogel distance in a manner consistent with fusion gamma rays. A 1 cm increase in target-to-aerogel distance produces a 162 ps increase in time between gamma ray and neutron induced signals.

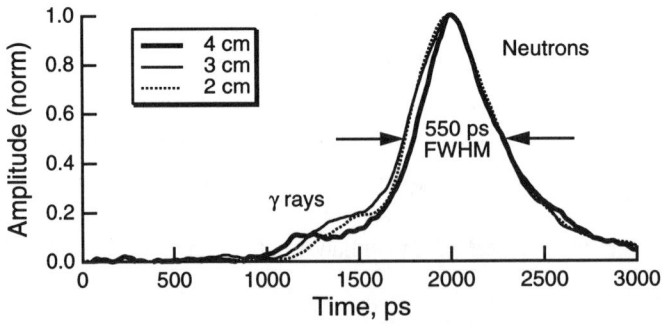

Figure 2. Aerogel signal intensity versus time for several target-to-aerogel distances. Neutron-induced signals are normalized, temporally aligned and overlaid. Presented in this fashion, change in gamma-gay signal timing relative to the neutron signal is easily observed. Time base is relative to incident laser power for 4 cm gamma ray data.

The streak camera simultaneously records a fiducial signal along with the aerogel signal. This allows us to determine the time of a gamma ray signal relative to the laser power incident on a target. The time of the gamma ray signal corresponds to the neutron emission time that we measure with a separate bang time detector (6). For these targets, the nominal neutron emission time is at 1 ns.

An additional experiment was performed with an aluminum nose cone replacing the Hevimet nose cone. In this configuration, the pair production cross section for 16.7-MeV gamma rays is reduced by a factor of 25 and the primary interaction mechanism changes to Compton scattering. The results (see Fig. 3) are consistent with the small peak being caused by pair production in the high-Z nose cone. The gamma ray signal observed with the Hevimet nose cone is not observed with the aerogel inside the aluminum nose cone. Also of note, with the substantially reduced shielding of the aluminum nose cone, there is no significant x-ray signal observed between the start of target irradiation and bang time.

DISCUSSION

These experiments demonstrate our ability to detect fusion gamma rays emitted by the T(d,γ)^{5}He reaction for Nova targets yielding 10^{12} to 10^{13} neutrons. Several pieces of experimental evidence are consistent with this conclusion. First the change in time-of-flight separation between gamma ray and neutron pulses is consistent with the target-to-aerogel distances used. Second, the gamma ray signals occur at the bang time of the target as determined with a separate neutron detector. Finally, the gamma ray signal is strongly dependent on the change of first stage converter material from tungsten to aluminum in a manner consistent with pair production.

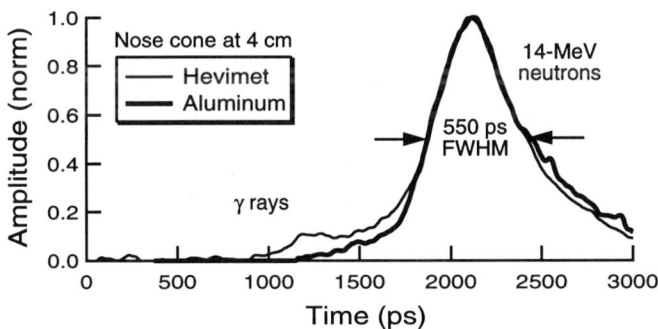

Figure 3. Aerogel signal intensity versus time for Hevimet and aluminum nose cones. Gamma-ray signal depends on high-Z converter material.

Additionally, the reduced shielding of the aluminum nose cone resulted in no noticeable increase in detected x rays, suggesting that we are indeed seeing fusion gamma rays and not target x rays.

Future effort will be directed at developing a sensitive burn history detector based on fusion gamma rays. To retain temporal response < 30 ps, we imagine the detector will remain a streak camera. There are several areas where significant improvements to sensitivity can be made. Improved optical transmission of the second converter stage will certainly enhance the signal. Our aerogel sample has relatively poor (< 20% through 7 mm) transmission for wavelengths < 500 nm. Similarly, optical transmission of the relay optics between the converter and streak camera can be greatly improved. Current optics were designed to transmit wavelengths between 350 and 450 nm. The streak camera has an S20 photocathode with a UV-grade sapphire window and good sensitivity for wavelengths between 250 and 700 nm. Additional gains in sensitivity can also surely be made with careful selection of converter materials and converter geometry.

ACKNOWLEDGMENTS

We gratefully acknowledge G. Mant for his continued assistance with the experimental equipment. We also thank the Nova Operations staff for their excellent support during this work. This work was performed under the auspices of the U. S. Department of Energy by the Lawrence Livermore National Laboratory under Contract No. W-7405-ENG-48.

REFERENCES

1. Lerche, R. A., Phillion, D. W., Tietbohl, G. L., *Rev. Sci. Instrum.* **66** 933-935 (1995).
2. Ress, D., Lerche, R. A., Ellis, R. J., Lane, S. M., Nugent, K. A., *Science* **241**, 956-958, (19 August 1988).
3. Lerche, R. A., Cable, M. D., Phillion, D. W. *Rev. Sci. Instrum.* **61**, 3187-3189 (1990).
4. Arther H. Snell, Editor, *Nuclear Instruments and their Uses, Volume 1* , New York, John Wiley & Sons, Inc., 1962, pp. 166-171.
5. Lewis, K., Moran, M. J., Hall, J., Graser, M., *Rev. Sci. Instrum.* **63**, 1988-1990 (1992).
6. Lerche, R. A., Kania, D. R., Lane, S. M., "Neutron Emission Time Measurements at Nova," in Laser Interaction and Related Plasma Phenomena, Vol. 8, edited by Heinrich Hora and George H. Miley, Plenum Press, NY, NY, 1988, pp 541-547.

Target laser plasma scale effect in electron temperature measurements by electric potential probe technique

I.A.Bufetov, G.A.Bufetova, V.B.Fedorov and S.B.Kravtsov.

General Physics Institute of the Russian Academy of Sciences
Vavilov St., 38, Moscow, 117942, Russia,

Abstract. Target laser plasma electron temperature dependence $T_e(t,I)$ on time ($t\sim10$ns) and on laser flux intensity ($I\sim10^9 \div 10^{13}$ W/cm^2) by electric potential probe technique was measured. We have shown for the first time that probe signal corresponds to the plasma electron temperature T_e only if the plasma size $l \sim d$ is small in comparison with the distance L of the probe from the plane target, $\eta = (d/L) < 1$. In the opposite case, when the scale parameter $\eta > 1$, the probe signal changes considerably, it can even change its polarity. It is experimentally shown that in these conditions the probe signal is formed mainly as the induction signal from the target plasma toroidal current system and corresponding circular magnetic fields. It was shown that the plasma temperature does not depend on the wavelength of laser radiation in the range $(1.06 \div 0.265)\mu$ and its dependence on the laser radiation intensity may be approximated as $T_e \sim I^\alpha$, where $\alpha = 0.33 \pm 0.02$, in the whole investigated range of experimental conditions. Maximal plasma temperature $T_e \sim 300$ eV has been obtained.

EXPERIMENTAL SETUP

The experiments on the plasma heating at the plane Al and Cu targets under the nanosecond pulses of the first, second and fourth harmonics of the Nd-glass laser were carried out. As the source of radiation we used a Nd-laser, which (Fig.1) includes master oscillator (MO) and multistage amplifier. MO, based on the Ga-Sc-Gd garnet crystal, operated on one transverse and one longitudinal mode. The pulse duration was 10 ns. Laser pulse had an asymmetric time envelope. The sharp front of the pulse (duration ~ 1 ns) was formed by an electro-optical shutter. The rear side of the pulse was smoother. Its shape was determined by the MO generation process. After amplification in silicate glass stages and spatial filtering the laser pulse energy reached 20 J, divergence being $8\cdot 10^{-5}$ rad (wavelength 1060 nm). After the frequency conversion of the radiation by means of the KDP crystals the 2d and 4th harmonics radiation appeared with the pulse energy up to 10 J (in 4th harmonic). The irradiance of the target surface was $I = (10^9 \div 10^{13})$ W/cm^2 the focusing spot diameter changing in the range $d = (0.08 \div 3.1)$ mm

© 1996 American Institute of Physics

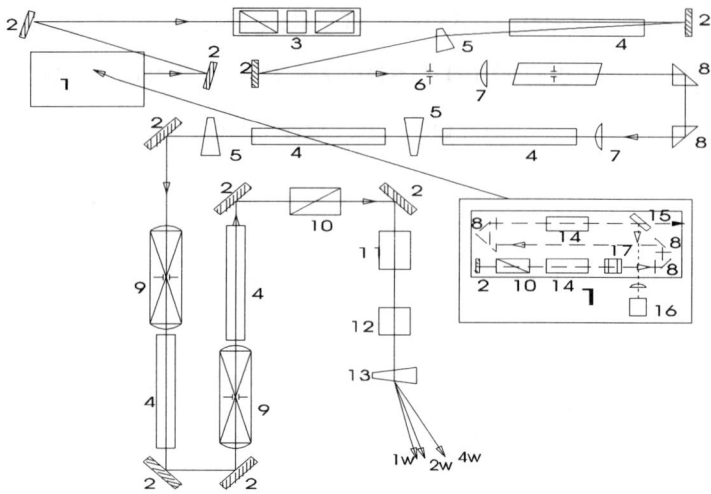

FIGURE 1. Laser installation. 1-MO unit; 2-mirrors; 3-electro-optical shutter; 4-Nd-glass rod; 5-glass prism; 6-diaphragm; 7-lens of the first vacuum space filter; 8-turning prism; 9-vacuum space filter; 10-polarizator; 11-KDP converter $1\omega \rightarrow 2\omega$; 12-KDP converter $2\omega \rightarrow 4\omega$; 13-quarz dispersing prism; 14-GSG-garnet rod; 15-glass plate; 16-optically driven spark gap; 17-mode selector unit.

Applying the electric potential probe technique for measuring the plasma parameters we have discovered the early unknown scale effect, which, in case of large focusing spot diameters, may distort the probe signal form, up to changing its polarity.

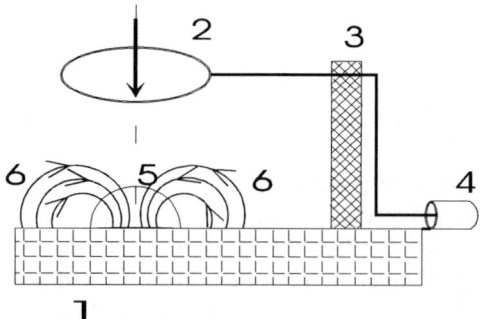

FIGURE 2. Scheme of experiment: 1-metal target, 2-electrical probe, 3-isolator, 4-coaxial 75 Ω cable, 5-laser plasma, 6-fountain-like currents. Laser beam direction is shown by the arrow.

THE PROBE SIGNAL ANALYSIS

We have investigated the e.m.f. of the double charge boundary layer of the laser metal target plasma The measurements of the e.m.f. were carried out by electric potential probe technique (Fig.2).

It was shown that the probe signal corresponds to the double charge layer potential at the plasma boundary only if plasma size $l \sim d$ is small in comparison

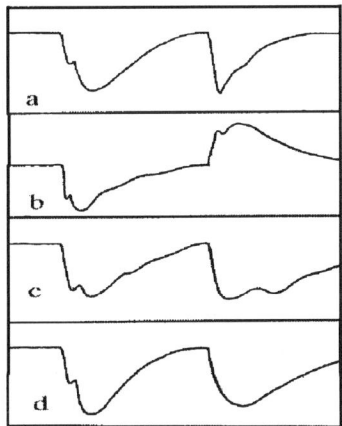

FIGURE 3. Some results of electric potential probe diagnostics of nanosecond optical target discharge. The first pulse at each trace is the laser one, the second pulse is the electric probe signal. Time delay between pulses is equal 22ns. Parameters of experiments are:

a) **D**= 40mm, **d**=0.08mm, **I** = 8.7·10^{11} W/cm^2, **U**= -552 V

b) **D**= 40mm, **d**=3.1mm, **I**=6.5·10^9 W/cm^2, **U**= +32 V

c) **D**= 40mm, **d**=3.1mm, **I**=10^9 W/cm^2, **U**= -40 V

Laser plasma is surrounded by dielectric cylinder of internal diameter 10mm and of height 2.5mm

d) **D**= 4mm, **d**=3.1mm, **I**=10^{10} W/cm^2, **U**= -113 V

D- target size, d-focal spot diameter, **I**-intensity, **U**-maximum amplitude of electric probe signal.

with the distance $L \approx 5$ mm of the probe from the target, $(d/L)=\eta <1$ (Fig.3a).

If the scale parameter η increases and becomes greater than 1, the probe signal changes considerably, it can even change its polarity (Fig.3b). It was shown experimentally (Fig.3b,c,d) that in these conditions (Fig.3b) the probe signal was formed mainly as the induction signal from the target plasma toroidal current system and corresponding circular magnetic fields. This effect reveals itself most distinctly in experiments with the forth harmonics of Nd-laser.

The results of measurements of electric probe potential U with $\eta < 1$ may be used to investigate the electron temperature of the laser driven plasma according to $T_e \sim U$ (1). In the opposite case ($\eta > 1$) the experimental results may give information about the fountain-like system of electric currents and magnetic fields in the region of the target.

ELECTRON TEMPERATURE MEASUREMENTS

We have used the first possibility to determine the plasma electron temperature T_e. The method (1) is based on the fact, that the double charge layer potential U_d at the plasma boundary corresponds to the plasma temperature according to relation $eU_d \sim kT_e$, where e is the charge of electron, k is the Boltzmann's constant, and $U \approx U_d$. The scheme of the experiment is shown on fig.4.

In these experiments the condition $\eta < 1$ is fulfilled if the focusing spot changes in the range $d = (0.08 \div 1)$ mm. In the case $d = 3.1$ mm the system of toroidal currents was suppressed by means of target diameter limiting to the size of the focusing spot, that corresponds to the fig.3d case.

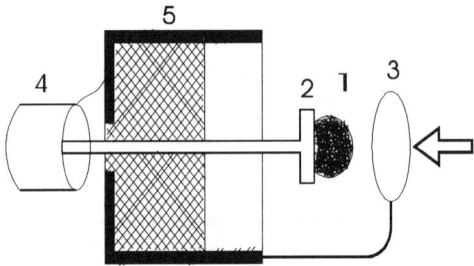

FIGURE 4. The scheme of the target construction. The arrow shows the propagation direction of the laser radiation. 1 - laser plasma; 2 - metal target; 3 - probe ring; 4 - radio-frequency cable; 5 - coaxial connector.

The asymmetry of laser pulse enable us to show experimentally, that in these experiments' conditions the plasma temperature maximum (during the laser pulse) T_e is reached in the region of one-dimensional movement of the plasma near the target, which at values of light intensity $I > 10^{11}$ W/cm^2 occurs much earlier then maximum of the laser pulse amplitude is reached.

The potential difference between the probe and the metal target was measured by the high-speed oscilloscope with time-resolution better then 1 ns.

The results of the measurements of the maximum plasma electron temperature in all experiments are presented on fig.5, as a dependence on radiation intensity I, for the first(o), second($\square$) and forth(∇) harmonics of laser radiation. In spite of the considerable dispersion of experimental data, it is possible to infer, that T_e does not depend, accuracy being 30%, on the radiation frequency. As to dependence $T_e(I)$, it may be approximated for all radiation wavelengths as $T_e(I) \sim I^\alpha$,

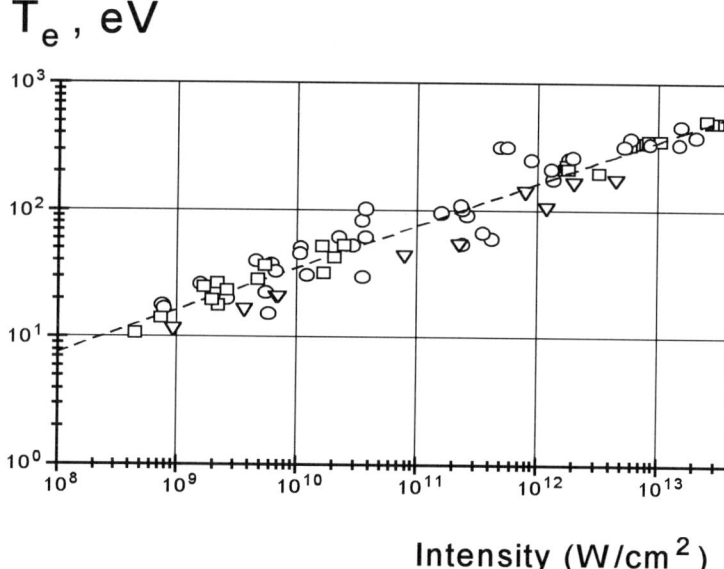

FIGURE 5. The dependence of the electron plasma temperature T_e on the intensity of laser radiation. The dotted line shows the approximate function $T_e(I) \sim I^\alpha$.

$\alpha = 0.33 \pm 0.02$.

At intensities $I \sim 10^{13}$ W/cm^2 the measurements of T_e, based on the X-ray diagnostic of the plasma radiation, were carried out. The measured absolute values of T_e on the lower and upper boundaries of the interval of the laser radiation intensities are (16 ± 2) eV and (350 ± 50) eV, correspondingly.

This study was supported by RFFI grant (N 94-02-05301) and by ISF grant (N M6J000).

REFERENCES

1. Motilev S.L., Pashinin P.P., *JTP* (Russia), **48**, 742-744 (1978).

X-ray Emission Computed Tomography with Attenuation Correction for ICF Research

Yen-Wei Chen

*Department of Electrical and Electronic Engineering, Faculty of Engineering
University of the Ryukyus, 1 Senbaru, Nishihara, Okinawa 903-01, Japan
Tel: +81-98-895-2221 ext.2778, Fax: +81-98-895-4377
E-mail: chen@tec.u-ryukyu.ac.jp*

Abstract. An attenuation correction method was proposed for laser-produced plasma emission computed tomography (ECT), which is based on relation of attenuation coefficient and emission coefficient in plasma. Simulation results show that the reconstructed images are dramatically improved in comparison to reconstructions without attenuation correction.

INTRODUCTION

In inertial confinement fusion (ICF) research, the implosion symmetry is one of the most important issues to achieve high density compression. The x-ray images obtained by pinhole camera and coded aperture imaging have provided direct information on uniformity of the compressed core. However, because these images are two-dimensional projections of the three-dimensional spherical implosion targets, these may not be enough to evaluate the implosion symmetry. If attenuation of x-ray within plasma is neglected, the intensity of the obtained two-dimensional x-ray image is approximately proportional to a line integral of the three-dimensional x-ray distribution emitted from imploded target. The reconstruction of three-dimensional x-ray distribution from its projections is just a linear inversion problem. In order to obtain tomographic pictures of the imploded target, we have successfully developed an x-ray emission computed tomographic (ECT) technique to reconstruct three-dimensional compressed core from pinhole camera images [1] or uniformly redundant arrays (URA) coded aperture images [2].

On the other hand, in recent laser fusion experiments, high density compressions have been achieved [3]. Thus, neglecting attenuation of x-ray within the plasma is not a valid approximation. It is necessary to develop a new ECT technique with attenuation correction. In recent medical ECT, there have been several methods [4], [5], [6] proposed for attenuation correction. Most correction or compensation methods proposed have assumed, for simplicity, a uniform attenuation coefficient distribution. In laser plasma ECT, the attenuation coefficient distribution is dependent on plasma density and plasma temperature, which are unknown parameters and non-uniform. In this paper we will present a new attenuation correction for laser plasma ECT based on the relation between the attenuation coefficient and emission coefficient in plasma.

ATTENUATED PROJECTION

For simplicity, we consider a two-dimensional case here. The projection geometry is shown in Fig.1. Let $f(x,y)$ and $\mu(x,y)$ represent a two-dimensional activity distribution and the attenuation coefficient distribution of the plasma, respectively. The attenuated projection $P_\phi(r)$ with the projection angle ϕ is given as

$$P_\phi(r) = \int_{-\infty}^{\infty} f(r,s) \cdot \exp\left[-\int_{-G_\phi(r)}^{s} \mu(r,s') \, ds'\right] ds, \qquad (1)$$

where

$$r = x \cdot \cos\phi + y \cdot \sin\phi \qquad (2a)$$
$$s = -x \cdot \sin\phi + y \cdot \cos\phi \qquad (2b)$$

and $G_\phi(r)$ is the distance from the center of rotation to body contour. Computed tomography (CT) is one of the inversion techniques to estimate the activity distribution $f(x,y)$ from its several projections $P_\phi(r)$. In conventional ECT, the attenuation is neglected ($\mu(x,Y)=0$). The activity distribution $f(x,y)$ can be easily obtained by Radon transform or other methods [7]. In recent medical ECT with attenuation correction, the attenuation distribution was assumed to be known. Thus the unknown parameter in Eq.(1) is only $f(x,y)$ and it is possible to obtain $f(x,y)$ from the projection data. If $\mu(x,y)$ is unknown, there are two unknown parameters in Eq.(1); it is impossible to obtain the solution $f(x,y)$ from only the projection data. Hence, it is necessary to know another information on $f(x,y)$ or $\mu(x,y)$ in order to solve for activity distribution $f(x,y)$.

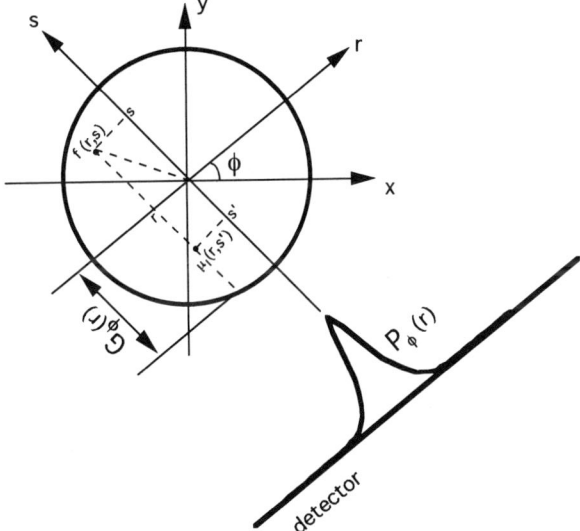

FIGURE 1. Projection geometry.

In laser plasma, when the x-ray photon energy $h\upsilon$ in the observed region is higher than the ionization energy of the plasma, the dominant radiation attenuation (absorption) process is inverse-bremsstrahlung. The attenuation (absorption) coefficient $\mu(x,y)$ [8] is given

as

$$\mu(x,y) \propto \rho(x,y)^2 \cdot T(x,y)^{-1/2} \cdot (h\upsilon)^{-3}, \tag{3}$$

where $\rho(x,y)$ and $T(x,y)$ are plasma density and temperature distribution, respectively, which are unknown and non-uniform. The dependances of attenuation coefficient (μ) on plasma density (ρ) and temperature (T) in a CD plasma are shown in Fig.2(a) and the attenuation of x-ray is shown in Fig.2(b) as a fuction of area density ρR.

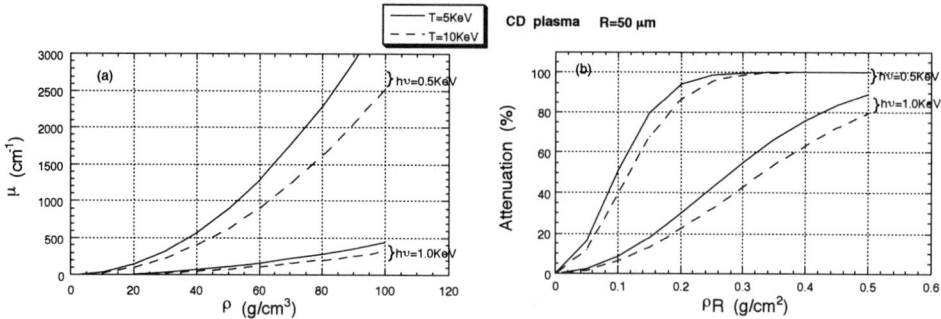

FIGURE 2. The attenuation coefficient (a) and the attenuation (b) in a CD plasma.

In such plasma, the dominant of radiation process is bremsstrahlung. By assuming a Maxwellian electron-energy distribution, the radiation emission intensity $f(x,y)$ [8] is given as

$$f(x,y) \propto \rho(x,y)^2 \cdot T(x,y)^{-1/2} \cdot \exp\left[-\frac{h\upsilon}{kT(x,y)}\right], \tag{4}$$

where k is the Boltzmann constant. From Eqs.(3) and (4), it is easy to get the relation between $\mu(x,y)$ and $f(x,y)$ as

$$\mu(x,y) \propto f(x,y) \cdot \exp\left[-\frac{h\upsilon}{kT(x,y)}\right]. \tag{5}$$

Assuming the temperature T is much larger than x-ray photon energy $h\upsilon$ (typical values of T and $h\upsilon$ are 10KeV and 1KeV), $\exp[-h\upsilon / kT(x,y)]$ in Eq.(5) is approximately a uniform constant of 1. Thus $\mu(x,y)$ is directly proportional to $f(x,y)$ in the space, which can be written as

$$\mu(x,y) = \beta \cdot f(x,y), \tag{6}$$

where β is a constant determined by x-ray photon energy, atomic number Z of plasma and detection efficiency which are known parameters. Thus the projection $P_\phi(r)$ of Eq.(1) can be rewritten as

$$P_\phi(r) = \int_{-\infty}^{\infty} f(r,s) \cdot \exp\left[-\beta \int_{-G_\phi(r)}^{s} f(r,s') ds'\right] ds. \tag{7}$$

As shown in Eq.(7), the unknown parameter in projection is just f. It is now possible to obtain the real solution f from the projections.

SIMUNATION RESULTS

We carried out the computer simulations to demonstrate the capability of this method. A typical iterative algorithm known as the algebraic reconstruction technique (ART) [1], [7] was used here for reconstruction. The algorithm is shown as follows:

$$f^{k+1}(x,y) = f^k(x,y) \cdot P_\phi(r) / R^k_\phi(r), \tag{8}$$

where $f^k(x,y)$ is the reconstruction obtained after k th iteration, and $R^k_\phi(r)$ is its attenuated projection. Figure 3(a) shows the phantom used in the simulation. It consists of 51x51 pixels, which represents a typical implosion target. There is a hot core at the center surrounded by cold plasma. Figure 3(b) shows a reconstruction result of low-density plasma (attenuation=0) after 10 iterations from 10 one-dimensional projections. Figure 3(c) and 3(d) show the reconstruction result of high-density plasma (attenuation=23%) without attenuation correction and with attenuation correction, respectively. In order to make a quantitative comparison, we show the profiles of 3(b), 3(c) and 3(d), which are taken across the center in horizontal directions indicated by arrows, in Figs. 3(b'), 3(c') and 3(d'), respectively. Dashed lines are phantoms which are taken from Fig.3(a) and solid lines are reconstructions.

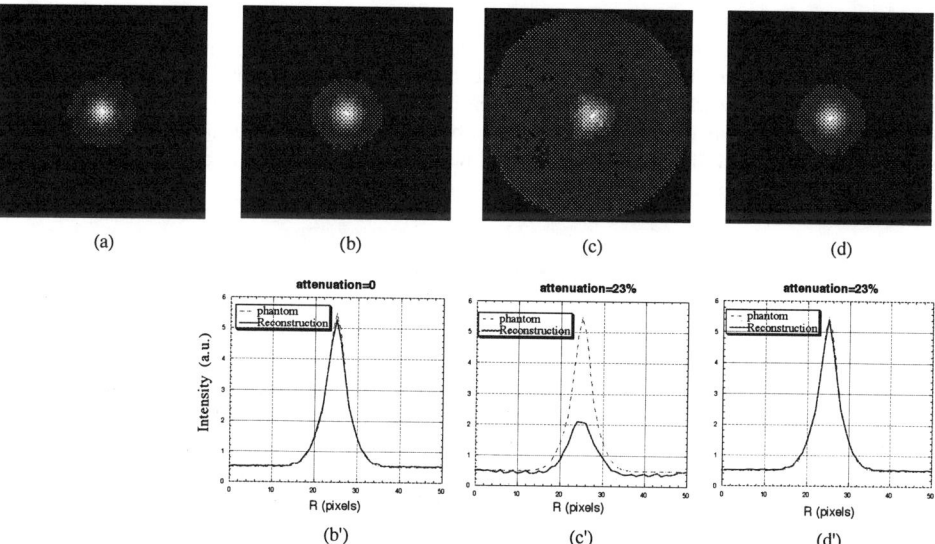

FIGURE 3. Comparison of the reconstructed images and phantom. (a) 51x51 pixels phantom. (b) and (b') Reconstruction of low-density plasma (attenuation=0). (c) and (c') Reconstruction of high-density plasma (attenuation=23%) without attenuation correction. (d) and (d') Reconstruction of high-density plasma (attenuation=23%) with attenuation correction.

Furthermore, we show a normalized rms error ζ (rms of the difference projection/ rms of the true projection) in Fig.4 as a function of attenuation.

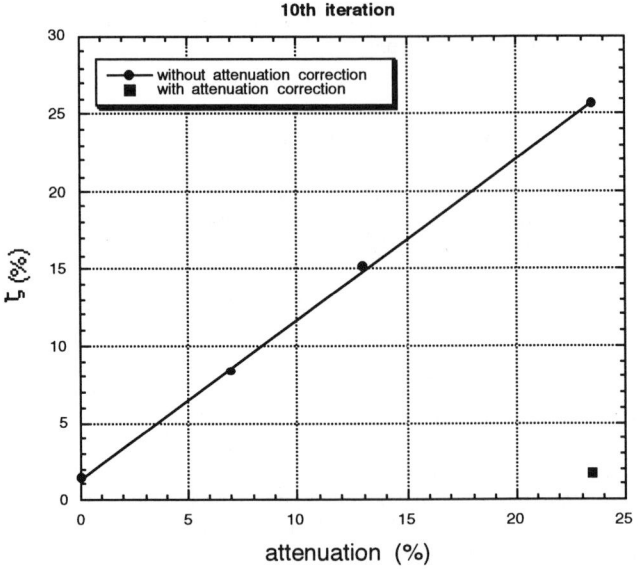

FIGURE 4. Normalized rms errors ζ of reconstruction as a function of attenuation.

As shown in Figs. 3(b) and 4, for low-density plasma (attenuation=0), ART can provide a good reconstruction with an rms error ζ of 1.5%, which is the accuracy of the conventional ART algorithm. For high-density plasma (with attenuation), if we do not correct or compensate the attenuation, the reconstruction is very poor and the rms error of reconstruction will linearly increase with increasing attenuation, while by using the proposed attenuation correction method it is possible to obtain a good reconstruction with a small rms error which is almost the same as that of low-density plasma (no attenuation) even for higher density plasma.

CONCLUSION

We have developed a new attenuation correction method for laser plasma ECT, which is based on relation of attenuation coefficient and emission coefficient in plasma. Simulation results show that the reconstructed images are dramatically improved in comparison to reconstructions without attenuation correction and it is possible to obtain an equally good reconstruction as that of low-density plasma (no attenuation) even for higher density plasma by using the proposed attenuation correction method. This method is expected to be an important diagnostic tool in future high density laser fusion experiment.

RERFERENCES

1. Chen, Y.-W., Miyanaga, N., Yamanaka, M., Nakai, M., Tanaka, K.A., Nishihara, K., Yamanaka, T. and Nakai, S., "Three-dimensional imaging of laser imploded targets," *J.*

Appl. Phys., vol. 68, no. 4, pp.1483-1488, 1990.
2. Chen, Y.-W., Yamanaka, M., Miyanaga, N., Yamanaka, T., Nakai, S., Yamanaka, C. and Tamura, S., "Three-dimensional reconstruction of laser-irradiated targets using URA coded aperture cameras," *Opt. Commun.*, vol. 71, no. 5, pp.249-255, 1989.
3. Nakai, S., Mima, K., Yamanaka, M., Azechi, H., Miyanaga, N., Nishiguchi, A., Nakaishi, H., Chen, Y.-W., Setsuhara, Y., Norreys, P.A., Yamanaka, T., Nishihara, K., Tanaka, K., Nakai, M., Kodama, R., Katayama, M., Kato, Y., Sasaki, H., Jitsuno, T., Yoshida, T., Kanabe, T., Yokotani, A., Norimatsu, T., Takagi, M., Katayama, H., Izawa, Y. and Yamanaka, C., "High density compression of hollow-shell target by Gekko-XII and laser fusion research at ILE, Osaka University," *Laser Interaction and Related Plasma Phenomena*, vol.9, pp.25-67, 1989.
4. Tanaka, E., "Quantitative image reconstruction weighted backprojection for single photon emission computed tomography," *J. Comp. Asst. Tomogr.*, vol.7, pp.692-700, 1981.
5. Chang, L. T., "A method for attenuation correction in radionuclide computed tomography," *IEEE Trans. Nucl. Sci.*, vol. NS-25, pp.638-643, 1987.
6. Inoue, T., Kose, K. and Hasekawa, A., "Image reconstruction algorithm for single-photon-emission computed tomography with uniform attenuation," *Phys. Med. Biol.*, vol.34, pp.299-304, 1989.
7. Herman, G. T., *Image Reconstruction from Projections*, Academic, New York, 1980.
8. Zel'dovich, Ya. B. and Raizer, Yu. P., *Physics and Shock Waves and High-Temperature Hydrodynamic Phenomena*, Academic, New York, 1966.

Two dimensionally space-resolved electron temperature measurement of fusion plasma by x-ray monochromatic imaging method

K.Fujita, H.Nishimura, I.Uschmann[*], E.Förster[*], H.Takabe, Y.Kato,
and S.Nakai

Institute of Laser Engineering, Osaka University,
2-6 Yamadaoka, Suita, Osaka 565 Japan

Max-Planck-Arbeitsgruppe Röntgenoptik an der Friedrich-Shiller-Universität Jena[*]
Max-Wien-Platz 1, 07743 Jena, Germany

ABSTRACT Electron temperature distributions of laser created fusion plasma were measured by using toroidally bent Bragg crystals. A tiny amount of argon was seeded in deuterium fuel gas and monochromatic images of Ar^{+17} (1s-3p) Lyβ and Ar^{+16} ($1s^2$-1s3p) Heβ lines were taken to provide temperature distribution of the compressed core from their intensity ratios. A fusion core created by laser-generated x rays in a micro-cavity showed the temperature structure corresponding to the illumination asymmetry caused by the cavity irradiation geometry. The experimental distribution of the line-ratio of Lyβ to Heβ was compared with the postprocessed outputs from one dimensional simulation, assuming perfect spherical implosion, to discuss degradation of pellet implosion.

Introduction

ICF researches have been extensively made toward achieving laboratory demonstration of ignition and burn of fusion pellets driven with an extremely uniform heating source. It is of great importance to understand the influence of fluid instability occurring during spherical convergence of the fusion pellet which may result in discrepancies in neutron yields between the experiments and model predictions [1]. The implosion performance has been measured by using conventional spectroscopic measurements and neutron yields [2]. It was deduced from this study that the implosion proceeds safely until the beginning of the deceleration phase but further compression at the stagnation phase collapses due to fluid mixing occurring at the pusher / fuel contact surface. Further study, therefore, has been needed to know the detailed structure of the compressed core. For this purpose, we developed an x-ray monochromatic camera (XMC) using toroidally bent crystals to provide two-dimensionally (2-D) space-resolved electron temperature distributions of compressed core plasmas as a novel diagnostic [3]. Observation was made for the core plasma created with laser-generated x rays.

Experiments and Analysis

Implosion experiments were performed using the GEKKO XII Nd:glass laser. A laser pulse of 5 kJ energy and 351nm wavelength was focused with the two-bundle illumination system. The pulse waveform was a Gaussian with a full width at half maximum (FWHM) duration of 0.75 nsec. Figure 1 shows the configuration of the x-ray-driven fusion target and laser beams. Target irradiation condition was chosen to minimize implosion asymmetries with the 3-D illumination calculation code[4].

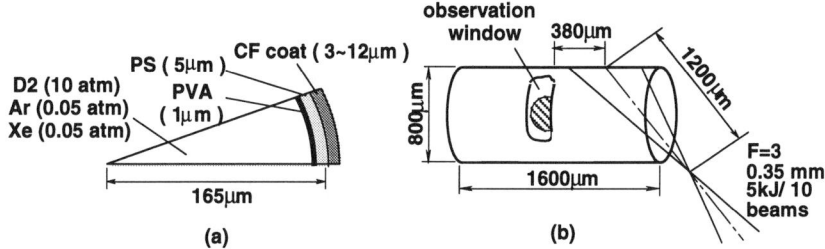

FIGURE 1. The configuration of the x-ray-driven fusion target and laser beams. (a)CF coated PS PVA shell, (b) cylindrical cavity made of gold.

Cylindrical Cannonball targets were used. The target consisted of a cylindrical cavity made of gold and a fuel capsule. The inner diameter and the length of the cylinder were 800 and 1600 μm respectively, and the wall thickness of the cylinder was 10 μm. The cylinder had an observation window on the side wall. The window was covered with polystyrene film to keep gold plasmas away from line of the observation. The fuel capsule was set at the center of the cylinder cavity. Plastic shell was used as the fuel capsule. A typical capsule shell had a 160 μm inner diameter, 9 to 18 μm thick wall consisted of three layers. The inner layer was 1 μm thick polyvinylalcohl (PVA) to seal a mixture gas of deuterium fuel (10 atm), argon (0.05 atm) and xenon (0.05 atm). The middle layer was 6 μm thick polystyrene(PS). The most outer layer was 3 to 12 μm thick polytetrafluoroethylene (so called Teflon $CF_{1.3}$) as an x-ray absorber and ablator. The variation of the CF coat thickness controls implosion behavior from the stagnation-free mode to the high compression mode with high radial convergence at the stagnation phase.

Imploded core plasmas were diagnosed by x-ray spectroscopic method: The XMC provided 2-D space-resolved monochromatic images of Ar Heβ and Lyβ lines. An x-ray flat-crystal spectrometer (XCS) provided spatially integrated Ar spectra. These spectra were recorded in a time-integrated manner with absolutely calibrated Fuji MI-FX film. The XMC consisted of toroidally bent Bragg crystals: Ge(311) for Ar Heβ line and Si(311) for Lyβ line. The spectral window was about 300. The spatial resolution was less than 10 μm. The XCS used a pentaerythritol (PET(002)) flat crystal. The spectral resolution ε/Δε at Lyβ line was about 3000 mainly determined by finite source size. Typical Ar spectra are shown in Fig.2. The images are already reduced with respect to x-ray film response, filter transmission, aperture size of XMC, reflectivities of

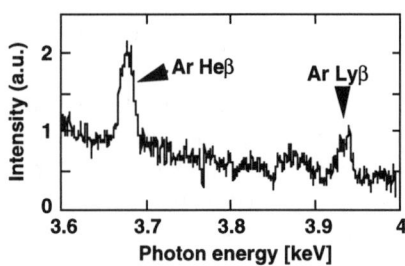

FIGURE 2. Typical space-averaged Ar spectra from the compressed core

the crystals, and the line width of each spectrum [1]. The calibrated monochromatic images, however, still contain the continuum component as a background. This component was subtracted, assuming that the ratio of line to continuum is constant

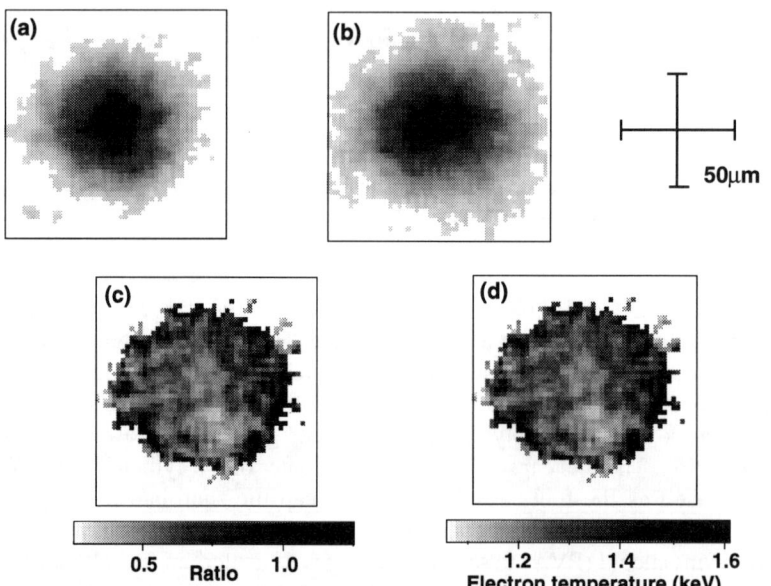

FIGURE 3. Typical experimental results: (a) monochromatic images of Ar Heβ line, (b) Lyβ line, (c) intensity ratio of Lyβ/ Heβ and (d) electron temperature.

at all spatial positions, by referring experimental data from XCS. The typical monochromatic images after this correction are shown in Fig.3 (a) for Heβ line and (b) for Lyβ line. Spatial distributions of the intensity ratio of Lyβ to Heβ lines, which reflect the spatial distributions of electron temperature, were obtained from the both images. The result is shown in Fig.3 (c). It was confirmed that space-averaged line intensity ratio obtained with XMC and that with XCS for the same laser shot are consistent each other. The electron temperature and density of necessary, were extracted by the help of spectrum analysis code RATION [5]. Space-averaged electron temperature for the typical shot was about 1.3 keV. The electron density, 3×10^{23}/cc, used in this evaluation was determined by the Ar Heβ line width from XCS. Using this density, a temperature map (Fig. 3(d)) of core plasma was obtained from the ratio distribution (Fig.3(c)). It is seen in Fig.3 (d) that relatively cooler region near the center of the core is surrounded by somewhat hotter regions. This trend can be seen in the four shots for different ablator thicknesses. In Fig.3 (d), furthermore, there are seen four cooler spots looking like a four-leaf clover of nearly 25 μm diameter near the center of the core. Such a mode number of 4 asymmetry structure may be resulted from the illumination asymmetry caused from the present capsule irradiation geometry.

Comparison with simulation and discussion

Pellet implosion was simulated with 1-D hydrodynamic code ILESTA-1D. Details of the simulation code and calculation modeling are described elsewhere [6,7]. The simulation results were postprocessed by the help of RATION, assuming optically thin plasma, to compare with the experimental results. First, data tables of Ar Heβ and Lyβ line intensities were prepared as a function of electron temperature and density. Next, temporal and spatial evolution of the Ar line emission was obtained, using outputs from the ILESTA simulation. Electron temperature and density at each time for each mass coordinate was correlated to the Ar line emission. Finally, ray-tracing and time-integration were made to match with the experimental conditions. The ray-trace was carried out by simply accumulating the emission intensities along the line of observation. Then, expected monochromatic images were replicated by time-integrating image after the ray-trace.

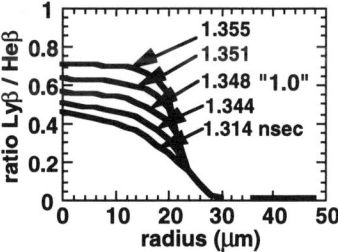

FIGURE 4. One of the postprocessed simulation results. Intensity ratio is rapidly changing at around the time marked "1.0" corresponding to the begining of the deceleration phase.

It is seen that intensities of the calculated monochromatic images monotonically increase until the maximum compression, at 1.390 nsec in this case, because the intensity of Ar emission is a sensitive function of electron temperature. Hence the experimental images may represent the maximum electron temperature achieved in a compression process although the images were obtained in a time-integrated manner. This hypothesis is validated in detail in Ref.[2]. Furthermore, temporal evolution of the ratio of Lyβ to Heβ is rapidly changing at around the final phase of implosion (see Fig.4). Thus the ratio will be useful for evaluating the implosion performance and degrees attained.

Figure 5 shows the radial distributions of the ratio averaged over azimuthal direction for four different ablator thicknesses. The experimental results for the four shots show a trend that the ratio is nearly flat at a region close to the core center but becomes higher in its surrounding region. For comparison, simulation results are shown for three different times for each ablator thickness : one profile is marked "0.5" which represents the time when the converging shock wave first reaches the pellet center. In the same manner, the mark "1.0" represents when the reflected shock wave from the center encounters the pusher fuel contact surface. The "max" represents the ratio profile at the maximum compression. It is obvious that the experimental results correspond to the profile marked "1.0" except for the outer-half of the profile. As discussed above, if the implosion process is terminated at a moment, this result supports the conclusion of the study of Ref.2 that the fuel is stably compressed until the beginning of pusher deceleration marked "1.0".But, in

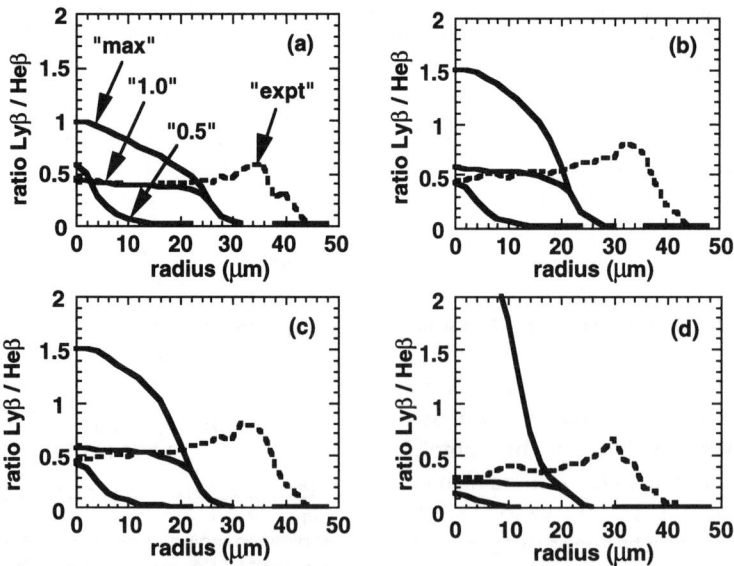

FIGURE 5. Ratio distributions experiments (dashed line) vs simulations (solid lines). (a) Initial areal density of the shell is 1.37 mg/cm^2, (b) 1.94, (c)2.45 and (d) 3.40.

the outer region, the experimental profiles are very different from the simulated one. Regardless the accurate value of temperature, the larger ratio results in higher temperature for the temperature around 1 keV. So, the hump profile in the outer-region suggests higher temperature than the core center. One possible explanation for this is that the reflected shock wave from the core center propagates into the pusher-fuel contact surface which is perturbed by fluid instability and the temperature rises due to numerous local reflections and propagations of the scattered shock waves. It should be, however, also noted that the experimental results are sensitive to the subtraction process of the continuum x-ray component. Therefore intensity distribution of the continuum should be measured. For this purpose, the background channel of XMC tuned at a slightly different wavelength are being prepared.

ACKNOWLEDGEMENTS

The authors acknowledge R.W.Lee for offering the RATION code. The authors also thank the experimental crew of the GEKKO XII facility.

REFERENCES

1. For example, H.Nishimura et al. *Proceedings of 14th International Conference Plasma Physics and Controlled Fusion Research*, Würzburg, (IAEA, Vienna, 1993) Vol.**3**, 97
2. H.Nishimura et al. *Phys.Plasmas*, June 1995 issue
3. I.Uschmann, E.Förster et al. *Rev.Sci.Instrum.* **66**, 734 (1995)
4. M.Nakamura et al. *Laser Part. Beams* **10**, 421 (1992)
5. R.W.Lee et al. *J.Q.R.S.T.* **32**, 91(1984)
6. H.Takabe et al. *Phys.Fluids* **31**, 379 (1988)
7. H.Takabe and T.Nishikawa, *J.Q.R.S.T.* **51**, 379 (1994)

Development of Multi Channel Neutron Spectrometer at GEKKO XII Laser Fusion Facility

N. Izumi, T. Yamagajo, T. Nakano, T. Kasai, H. Azechi, Y. Kato and S. Nakai
Institute of Laser Engineering, Osaka University 2-6 Yamada-oka, Suita, Osaka 565, JAPAN

T. Iida,
Department of Nuclear Engineering, Osaka University, 2-1 Yamada-oka Suita, Osaka 565, JAPAN

Abstract. A high efficiency time-of-flight neutron spectrometer is under construction at the GEKKO XII laser fusion facility. Neutron spectra measured with this system is used for diagnosing fuel areal density and ion temperature of inertial-confinement-fusion targets. This system consists of 960 plastic scintillation detectors and a data acquisition electronics. Fuel areal density is deduced from DT neutron spectra produced in initially pure deuterium fuel. Ion temperature is also deduced from primary DD neutron energy spread. The spectrometer was designed to measure relatively rare secondary neutrons from neutron yield as low as 7×10^5 (100 detector hits). It is also capable to measure 2.45 MeV neutron spectra in lower neutron yield experiment as low as 5×10^5 neutron. We have constructed a 1/10 module for testing performance of the system, established calibration technique and used it in implosion experiment at the GEKKO XII laser fusion facility to demonstrate ion temperature measurement.

INTRODUCTION

Fuel areal density can be deduced using secondary nuclear reactions in an initially pure deuterium fuel. This technique has been proposed[1,2] and demonstrated experimentally[3]. Primary reactions in pure deuterium fuel are

$$D + D \rightarrow T(1.01 \text{MeV}) + p(3.02 \text{MeV}) \quad (1)$$
$$D + D \rightarrow {}^3He(0.82 \text{MeV}) + n(2.45 \text{MeV}).$$

Products of primary reaction cause secondary reactions:

$$T (< 1.01 \text{ MeV}) + D \rightarrow {}^4He + n\ (11.8 - 17.1 \text{ MeV}) \quad (2)$$
$${}^3He (< 0.82 \text{ MeV}) + D \rightarrow {}^4He + p\ (12.5 - 17.4 \text{ MeV}).$$

In the case of small areal density or high electron temperature, where energy loss of the test particles, ${}^3H$ and T, are negligible, the reaction number ratios of primaries to secondaries are directory proportional to the fuel areal density. If primary products are assumed to be generated in entire region of a spherically uniform fuel, the relations are given by

$$\frac{Y_{2n}}{Y_{1p}} = 0.09 \langle \rho R \rangle \quad , \quad \frac{Y_{2p}}{Y_{1n}} = 0.14 \langle \rho R \rangle, \tag{3}$$

where Y_{1p}, (Y_{2p}) is primary (secondary) proton yield, Y_{1n}, (Y_{2n}) is primary (secondary) neutron yield and $\langle \rho R \rangle$ is the averaged fuel areal density. For higher areal density region, the relations of the reaction number ratio and the areal density become more complex because the cross sections of secondary reactions vary along slow down of the test particle. The measurement of the fuel areal density for the significant energy loss case can be made by measuring secondary DT neutron yields and secondary D^3He proton yields simultaneously[3]. As proposed[4], it is also possible to determine energy loss of triton from kinetic broadening of secondary neutron energy spectra.

In this paper, we report the present status of the neutron spectrometer at the GEEKO laser system for secondary neutron measurement. The spectrometer consists of an array of scintillation detectors, each of which acts as a single hit detector.

DETECTOR DESCRIPTION

The detector array is located at 13.4 m from the target chamber center. The detectors are shielded by 4-cm lead and a 2.2-cm iron to reject γ-ray and x-ray background pulses. Each detector consists of a plastic scintillator (BC408, 6-cm dia × 6-cm thick), photomultiplier tubes (PMT, Thorn EMI 9902KSB95), a high voltage bleeding base. All scintillator surfaces except for a window to be coupled to the PMT are painted using BC-620 reflector paint in order to obtain high light output and better pulse height response. The PMT is 10 stage liner focus type and has a 1.5" photocathode window. The photocathode is optically coupled to the scintillator surface using silicone optical coupling compound (OKEN6262A).

FIGURE 1 shows high voltage bleeding circuit. Each bleeding circuit has a 20 turn cermet trimmer (R17, 2MΩ) to adjust PMT gain discrepancies. Once each

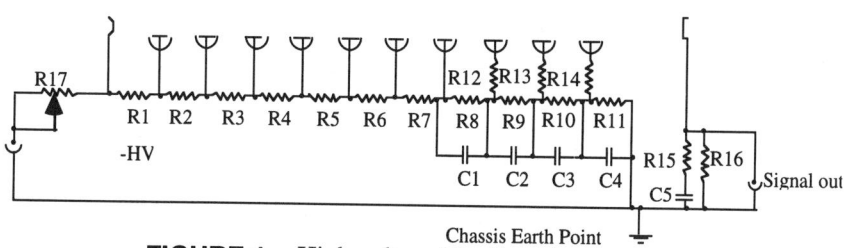

FIGURE 1. High voltage bleeding circuit

PMT gain is adjusted, all channels are treated as it has the same gain. Since uniform high voltage was supplied parallel to all detectors, only 1 high voltage power supply is needed for 960 channel detectors. A reflection terminator (R15, 50 Ω) is implemented to reject multiple-pulsing due to reflections.

In order to compensate timing walk caused by varying amplitudes of signal, PMT signals are led to constant fraction discriminators (LeCroy 3420). Timing signal is picked off by time to digital converters (LeCroy 1877) and recorded in 500 ps time bin width. The data of detection timings are accumulated in a computer.

CALIBRATION

Gain adjustment was made by using ^{60}Co γ rays. Detector gains were adjusted so that the Compton edge appears at the same pulse height. In order to confirm that this simplified gain adjustment is reliable, the detectors, which were already gain adjusted, were tested by using ion accelerator type neutron generator OKTAVIAN at Osaka University. The time width of the deuteron beam was about 3 ns. DD protons were counted *in situ* using a silicon surface barrier detector for yield monitoring. Activation measurement was also carried out for cross calibration. FIGURE 2 shows the TOF spectra of DD neutrons. The time origin corresponds to the timing of deuterium beam just bombarding the target. Thin line shows the TOF spectra with a 0.5-cm lead γ-ray shield in front of the detector, whereas the thick line shows that with a 5-cm lead. The first peak at ~4 ns is due to prompt x rays and γ rays emitted from the structure around the target. This peak can be eliminated with a 5-cm lead. The second peak shows the neutron signals. The signal pulses of shaded region on FIGURE 2 were used to

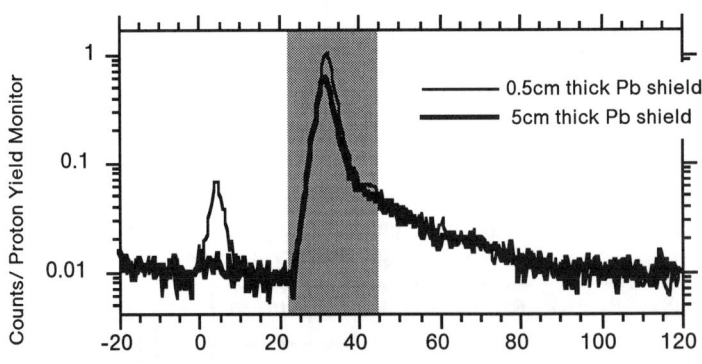

FIGURE 2. Time-of-flight spectra of DD neutrons from an accelerator type neutron generator The time origin corresponds to the time of deuteron beam bonbardment on the target.

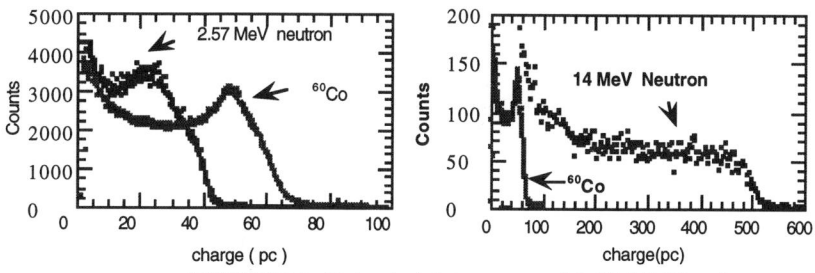

FIGURE 3. Pulse height spectra of 2.57-MeV and 14.1-MeV neutrons along with that for ^{60}Co gamma ray.

make pulse height spectra. The same selection was also carried out in DT neutron experiment. FIGURE 3 shows pulse height spectra of 2.57-MeV and 14.1-MeV neutrons as well as that of ^{60}Co γ rays. The detector-to-detector fluctuation of the full energy edge was about 3 % for DD neutron and 7 % for DT neutron at standard deviation. This result assures reliability of PMT gain calibration using ^{60}Co γ rays.

TEST MODULE DEMONSTRATION

We have constructed a 96 channel test module and used it for ion temperature measurement. The module was located at 11.24 m from the target chamber center. In this experiment we used leading edge discriminator (LeCroy 4413) in burst guard mode to record the width of signal pulse. The time to digital converter records the leading edge and the trailing edge of the signal pulses. FIGURE 4

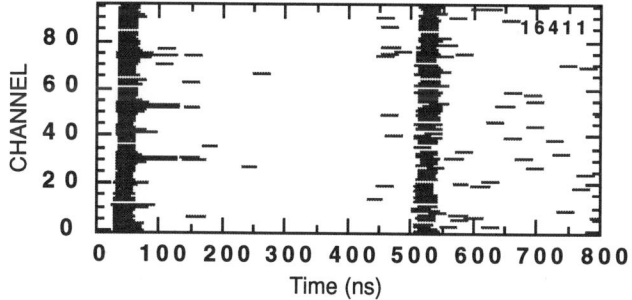

FIGURE 4. Raw timing data map obtaind by detector module. The first group of pulses is prompt gamma rays and the second group is primary neutron signal.

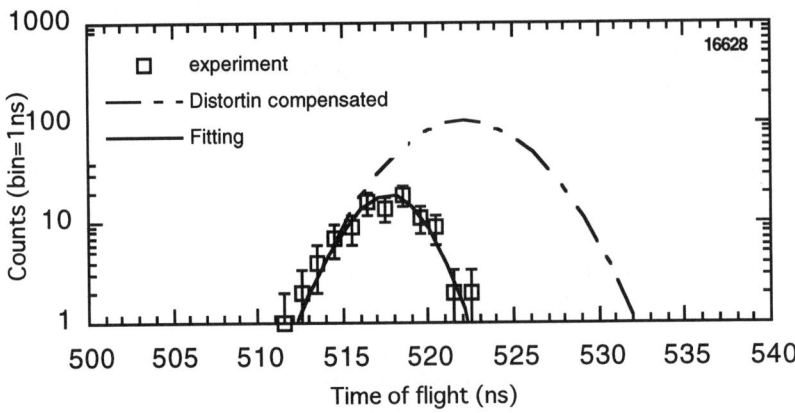

FIGURE 5. Primary neutron time spectra
Spectra distortion caused by detector satulation was compensated using "first hit" method. (Gaussian spectra was assumed)

shows the raw data of a pure DD fuel implosion experiment. Each horizontal bar shows the time above the discriminator threshold. Length of the bar shows the pulse width and left side of the bar shows arriving time of signal. The first and the second groups of the bars correspond to prompt x or γ rays and primary neutrons respectively. Prompt x rays are used as a time fiducial to compensate the discrepancy of each PMT's intrinsic time delay. FIGURE 5 shows the TOF spectrum of primary neutrons at 2.5×10^7 neutron yield. Because this system was not optimized to measure such high yield, the spectrum was distorted by detector saturation. Assuming a Gaussian distribution of neutron energy spectrum, it is possible to deduce original energy spread (first hit mode operation[5]). The dashed line of FIGURE 6 shows compensated spectra. Ion temperature deduced from this spectrum was 1.09 ± 0.16 keV.

SUMMARY

A high efficiency neutron spectrometer is under construction. This system is optimized to detect relatively low yield secondary neutrons from initially pure deuterium filled fuel for measuring fuel areal density. It is also capable to measure primary neutron spectra for ion temperature measurements. A 1/10 scale module was made for testing performance of this system. All detector gains are adjusted using ^{60}Co γ-ray source. Accuracy of gain adjustment was tested using 2.57-MeV and 14.1-MeV neutron source. Ion temperature measurement in implosion experiment was demonstrated at GEKKOXII laser fusion facility.

REFERENCES

[1] E. G. Gamalii, S.Yu, Gus'kov, O. N. Krokhin, and V. B. Ruzanov, JETP Lett. **21**, 70 (1975)
[2] T. E. Blue and D. B. Harris, Nucl. Sci. Eng. **77**, 463 (1981)
[3] H. Azechi, N. Miyanaga, R. O. Stapf, K. Itoga, H. Nakaishi, M. Yamanaka, H. Shiraga, S. Ido, K. Nishihara, Y. Izawa, T. Yamanaka and C. Yamanaka, Appl. Phys. Lett. **49**, 555 (1986).
[4] M. D. Cable, S. P. Hatchett, J. Appl. Phys. **62**, 2233 (1987).
[5] R. A. Lerche, S. P. Hatchet, M. D. Cable, M. B. Nelson, Rev. Sci. Instrum. **63** 4877 (1992).

4. High Intensities, Short Pulse Interactions

Generation of Coherent XUV Radiation With Sub-Picosecond KrF Lasers

A.A. Offenberger[1], S.G. Preston[2], M. Zepf[2], C.G. Smith[2], W.J. Blyth[2], M.H. Key[2,3], J.S. Wark[2], A. Djaoui[3] and D. Neely[3]

[1] Department of Electrical Engineering, University of Alberta, Edmonton, Alberta, Canada T6G 2G7
[2] Clarendon Laboratory, Department of Physics, University of Oxford, Oxford OX1 3PU, United Kingdom
[3] Rutherford Appleton Laboratory, Chilton, Didcot, Oxon, OX11 0QX, United Kingdom

Abstract We report on recent experiments utilizing 350-400 femtosecond pulses of KrF laser radiation to: i) generate high order odd-integer harmonics in helium (37th, 67 Å) and neon (35th, 71 Å) and; ii) demonstrate nonlinear enhancement of xuv line emission in optically ionized nitrogen. The high harmonics are shown to arise from He^+, Ne^+ and Ne^{2+} ions which permit higher conversion efficiency in the xuv than is possible from neutrals. The observed density dependence of the xuv emission may imply a low electron temperature in the optically ionized, recombining plasma.

INTRODUCTION

The short wavelength and broad bandwidth of KrF lasers make them well suited for generating coherent extreme ultraviolet (xuv) radiation via recombination of highly stripped ions or by harmonic generation. We review the plasma physics associated with optical field ionization of atoms at high laser intensity (I_L); discuss experimental results on electron heating by intense KrF laser pulses of duration τ_L to identify the parameter space (I_L, n_e) for producing low temperature plasmas at high electron density (n_e) appropriate to recombination xuv lasers and; motivate the use of short wavelength lasers (KrF) for harmonic generation from ions (compared to neutrals) as an alternative approach to generating coherent xuv radiation. We report nonlinear enhancement of xuv emission in several Li-like nitrogen transitions (e.g., 3d-2p [247.6 Å], 4d-2p [186.1 Å]) and what is believed to be the shortest wavelength achieved to date by harmonic generation ($\lambda = 67$ Å in He^+).

High field optical ionization and recombination cascade may provide a feasible scheme for pumping xuv lasers (1,2). To be practicable, however, a low electron temperature T_e at high electron density n_e is required to minimize collisional excitation and maximize recombination. Since high laser intensities ($I_L \sim 10^{16}$-10^{19} W/cm^2) are required to produce highly stripped ions, there will be accompanying electron heating - the amount depending on laser (I_L, τ_L) and plasma (n_e, T_e) parameters that are ultimately determined by the need to produce a significant population of Li-like or H-like excited states. Moreover, the intensity must be sufficient to empty the lower states and collisional excitation small enough to achieve reasonable population inversion and lasing efficiency. This places severe constraints on the allowable heating.

The electron heating mechanisms include: above threshold ionization (ATI) heating ($\sim I\lambda^2$), important at low n_e; nonlinear inverse bremsstrahlung (IB) absorption ($\sim n_e^2/I^{3/2}\lambda_\mu$), important with increasing n_e; and stimulated Raman scattering (SRS), with temporal growth rate $\sim (I\lambda^2)^{1/2}$, the most important at high n_e. Since the number of e-foldings of the SRS instability and the IB heating both increase with pulse duration, for a given intensity, short pulse and short wavelength lasers are essential to minimize heating.

We have previously measured the electron heating in helium and neon gas targets irradiated by 350-400 fs pulses of KrF laser radiation at a focused intensity of 10^{18} W/cm^2 (3). From those measurements, it can be concluded that heating from ATI and IB is modest for $n_e \cdot I \leq 3\times 10^{37}$ W/cm^5 and pulse durations $\tau_L \approx 380$ fs. Similarly, from the higher density results, one can conclude that heating from SRS is acceptable for intensities $I_L < 6\times 10^{17}$ W/cm^2 at density $n_e < 5\times 10^{20}$ cm^{-3} and pulse duration $\tau_L \approx 380$ fs, as well as for higher I_L and shorter τ_L. Thus, for nitrogen density $\approx 10^{19}$ cm^{-3} and $I_L \leq 3\times 10^{16}$ W/cm^2 in the present experiment, moderate heating is anticipated.

In the case of harmonic generation, the maximum harmonic photon energy $E_{max} \approx I_p + 3U_P$ (4), where I_p=ionization energy and U_P=ponderomotive energy $\sim I_L\lambda^2$. This would suggest that higher ionization potential I_P, laser intensity I_L and laser wavelength λ are all desirable for maximizing the photon energy and hence minimizing wavelength. However, there is a trade-off between cut-off energy and generation efficiency. The laser intensity I_L is limited by saturation of ionization (i.e., fully stripped state) and the harmonic conversion efficiency is a function of the laser wavelength. Compared to neutrals, ions have higher ionization potential I_p and saturation intensity I_L. In addition, harmonic emission from ions is predicted to be higher for short wavelength laser drivers (4) as a consequence of reduced time for diffusion of the electron wavefunction during electron tunneling through the Coulomb barrier and re-colliding with the ion core. Thus, the possibility of using KrF lasers to extend harmonic generation to shorter xuv wavelengths at greater efficiency was another objective of the present study.

EXPERIMENTAL

The experiments were conducted using the SPRITE laser facilities at Rutherford. For the present experiment, the pertinent laser parameters are: pulse energy (on target) ≤ 250 mJ; pulse length $\approx 350\text{-}400$fps; focused intensity $\leq 2\times 10^{17}$ W/cm^2. A 100 cm off-axis parabolic focusing mirror produced a 6x diffraction limited focal spot of ≈ 20 μm and a focal depth exceeding 700 μm. For most of the results to be discussed here, the working laser intensity was kept to $\approx 3\times 10^{16}$ W/cm^2, determined by the conflicting requirements of optically ionizing but not excessively heating electrons in the nitrogen recombination experiment, and avoiding ionization saturation in the harmonic generation experiments. Problems of incoherence, laser beam propagation and phase matching were partially mitigated in the present study by working with a KrF laser, moderate density gas targets and long focal length mirror.

A pulsed gas jet target (0.5 mm x 2 mm aperture) was produced by a high pressure pulsed solenoid valve, enabling neutral densities n_o of $10^{16}\text{-}10^{20}$ cm^{-3}. Nitrogen ($n_o \sim 10^{18}\text{-}10^{19}$ cm^{-3}) was chosen for the recombination measurements, since the required parameters (I_L, n_e, T_e) to achieve gain could be relaxed. Helium and neon ($n_o \sim 10^{17}\text{-}10^{18}$ cm^{-3}) were used for the harmonic generation measurements, helium for large I_p and neon for multi-electron capability. To ensure the focal depth exceeded the plasma length, the gas target was oriented to intersect the laser beam along the smaller dimension, i.e., interaction length of 500 μm.

Harmonics and xuv emission were collected by a gold coated grazing incidence cylindrical mirror, dispersed by a flat-field grazing incidence xuv spectrometer (1200 g/mm Hitachi grating, $\vartheta = 3.77°$) and detected by a Galileo double microchannel plate fibre-optically coupled to an intensified CCD camera with 4x reduction. The effective solid angle was 1.5×10^{-5} steradians, the useful spectral range ~ 200 Å and the resolution ~ 1 Å. The MCP/CCD camera response was calibrated separately. Carbon or aluminum filters of varying thickness were used to block the incipient KrF laser beam and attenuate low order harmonics or Li-like nitrogen emission.

RESULTS AND DISCUSSION - HARMONIC GENERATION

A typical harmonic spectrum for He ($n_o \sim 5\times 10^{17}$ cm^{-3}), uncorrected for filter or instrument response, is shown in Figure 1(a). Harmonics up to the 37th can be clearly seen in the 1st order spectrum (the high level of 2nd order is due to the blaze angle being optimized for $\lambda < 100$ Å in 2nd order). The harmonic conversion efficiency - corrected for MCP response, grating and focusing mirror reflectivity, collection solid angle of the spectrometer and focusing

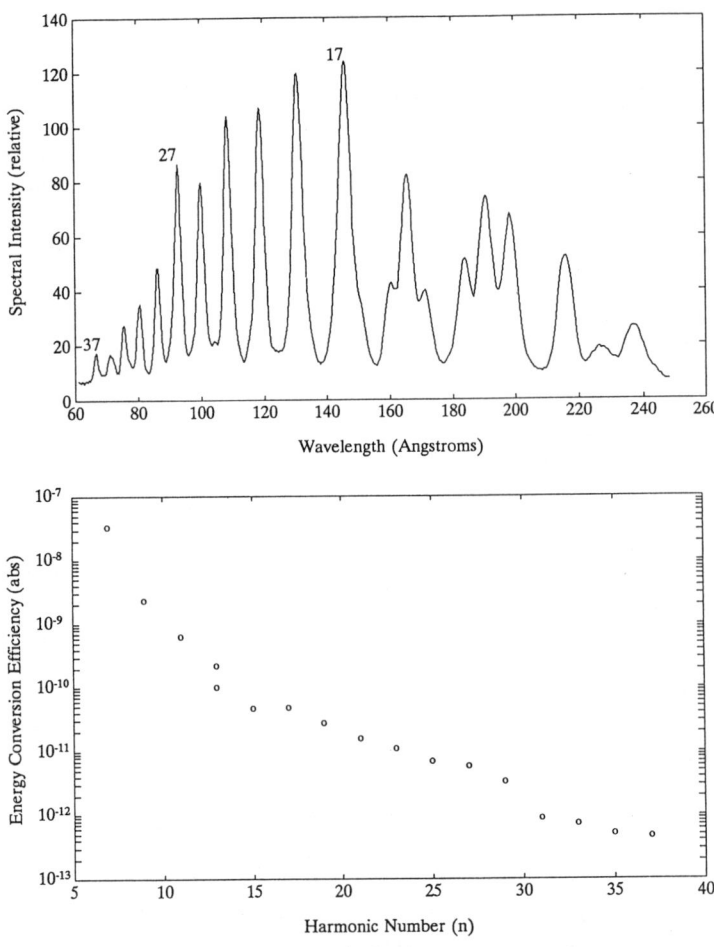

FIGURE 1. Harmonic spectrum (a) and energy conversion efficiency for helium (b).

mirror and filter transmission - is shown in Figure 1(b), assuming harmonic emission covered the full solid angle of the laser beam.

Similar results are found for Ne. In contrast to He, only harmonics up to n=35 were observed for all focusing and gas target conditions. While the conversion efficiency is comparable for both species at higher harmonics, helium is approximately an order of magnitude larger than neon for n=7,9,11.

Exact comparison of theory and experiment is difficult because the net generation depends on time and volume integration of unknown experimental conditions (phase variations, spatial variations in intensity and propagation effects). Nonetheless, calculations (in progress) based on the single active electron approximation, show that He$^+$ contributions dominate over neutral He for n>15 and that the spectrum is due entirely to He$^+$ for n>25. Likewise, for Ne it is found that Ne$^+$ becomes the principal source of harmonics for n>9 and Ne^{++} is responsible for harmonics above 21. The extension of the plateau to n=37 at measurable levels confirms the improvement of conversion efficiency in harmonic generation of high energy photons from He$^+$, Ne$^+$ and Ne^{2+} ions.

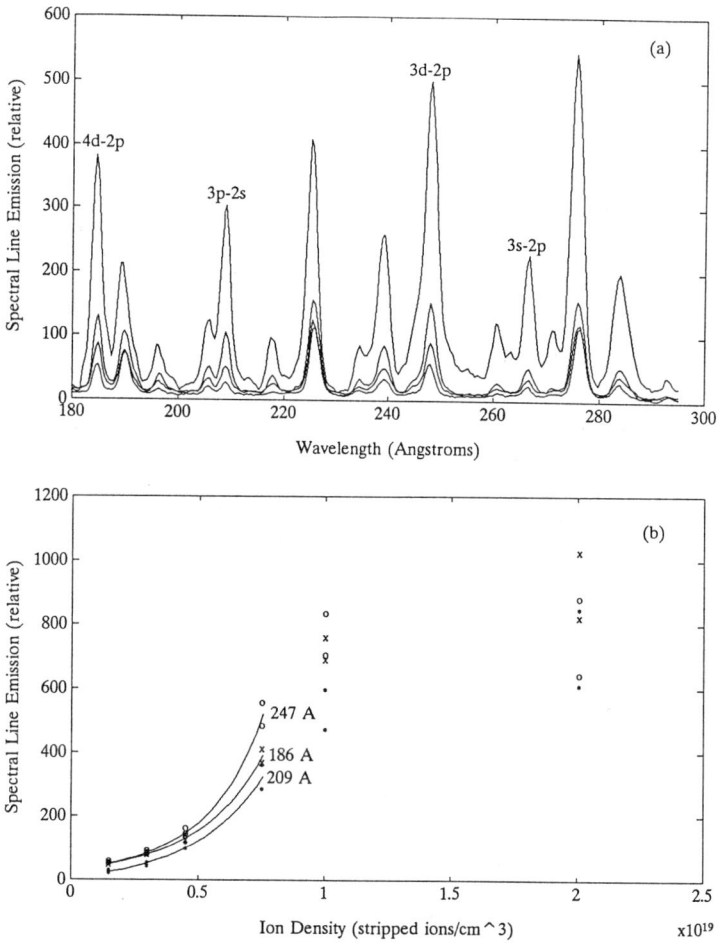

FIGURE 2. Dependence of xuv spectra (a) and time integrated spectral intensities on ion density (b) for Li-like (N^{4+}) nitrogen emission. The spectra in (a) are averages of three shots for ion densities of (1.5, 3, 4.5, 7.5)x10^{18} cm^{-3} (weakest to strongest emission). Harmonic generation gives rise to the spectral peaks at 191Å, 226Å, 276Å.

RESULTS & DISCUSSION - XUV EMISSION IN NITROGEN

For approximately constant laser irradiance, the line emission from Li-like nitrogen ions in the recombining plasma was measured as a function of gas density. The wavelength coverage of the spectrometer allowed observation of the important Li-like transitions (spectral lines), i.e., 3d-2p (247.6 Å); 4d-2p (186.1 Å); 3p-2s (209.3 Å) and 3s-2p (266.3 Å). The observed spectra and variation of (time integrated) spectral intensities with ion density are shown in Figure 2.

Figure 2(b) summarizes the variation with density for $N_{ion} \leq 2 \times 10^{19}$ cm^{-2}. For $N_{ion} > 10^{19}$ cm^{-3}, the apparent "saturation" is likely due to a combination of plasma effects. For $N_{ion} = 7.5 \times 10^{18}$ cm^{-3} and Z=5 (ionization to He-like), the associated electron density $N_e = 3.8 \times 10^{19}$ cm^{-3}. Laser beam refraction, electron heating and collisionality become more important beyond such densities, even for KrF lasers.

The xuv emission is seen to increase nonlinearly with density for $N_{ion} \leq 7.5 \times 10^{18}$ cm^{-3}. The circles, crosses and stars represent the estimated errors in the data for the three xuv transitions identified in Figure 2(b) and the solid curves are best fits to the average data in the lower density region. Detailed analysis is in progress to determine if the nonlinear enhancement is related to population inversion generated by recombination. This would require and may imply a low temperature plasma ($T_e \approx 14$ eV).

ACKNOWLEDGEMENTS

We wish to thank A. Sanpera and K. Burnett at Oxford University for calculations and comments on the harmonic generation.

REFERENCES

1. Burnett, N.H. and P.B. Corkum, J. Opt. Soc. Am. B **6**, 1195 (1989); Burnett, N.H. and Enright, G.D., IEEE J. Quantum. Electron. **26**, 1797 (1990).
2. Eder, D.C., Amendt, P. and Wilks, S.C., Phys.Rev. A **45**, 6761 (1992).
3. Blyth, W.J. et al, Phys. Rev. Lett. **74**, 554 (1994).
4. Krause, J.L., Schafer, K.J. and Kulander, K.C., Phys. Rev. Lett. **68**, 3535 (1992); Corkum, P.B., Phys. Rev. Lett. **71**, 1994 (1993); Lewenstein, M. et al, Phys.Rev. A **49**, 2117 (1994).

Scenarios of plasma formation with intense fs laser pulses

P. Mulser, A. Al-Khateeb, D. Bauer, A. Saemann, and R. Schneider

Theoretical Quantum Electronics, Technische Hochschule, Hochschulstr. 4A, D-64289 Darmstadt, FRG

Abstract. A non-standard fluid model is developed to simulate superintense pulse-solid interaction on the fastest time scale in the collisional regime. Ionization by field emission is the shortest process, occuring during fractions of a laser cycle and leading to ionization dephasing. At later stages ionization by electron impact may become significant in higher-Z materials. As another significant result collisional absorption is revealed to occur to appreciable degrees.

1. Introduction

A new class of plasmas in a parameter range not accessible before is created by ultrashort superintense laser pulses interacting with solids. Laser systems delivering such pulses are now available in numerous laboratories. In the present contribution the early stage, ionization and inverse bremsstrahlung absorption, is studied at irradiances ranging from 10^{16} to 10^{19} Wcm$^{-2}\mu$m^2 and typical pulse lengths of 100 fs. When matter is exposed to such intensities it is rapidly ionized to a high degree, and a hot plasma is formed. In the case of a high-Z solid target free electron densities of at least one order of magnitude higher than in a metallic conductor are achieved. This means that plasma and solid state physics can be extended into a new parameter region not accessible before.

At the irradiances under consideration there is almost no difference in the optical behaviour between dielectric and conducting targets. Under the action of the strong laser field of magnitude

$$\hat{E}\,[\text{Vcm}^{-1}] = 27.5 \times \{I\,[\text{Wcm}^{-2}]\}^{1/2} \tag{1}$$

(see Table 1, field amplitude $\hat{E}$) the first electrons are set free in a fraction of a cycle by field ionization. As their concentration increases by this process, ionization by electron impact may become dominant. The target becomes highly conducting and reflecting. The penetrating light is absorbed in a layer of a fraction of a wavelength from which most of the energy is transported into deeper layers of the target by the electrons (heat conduction). At the end of the light pulse the plasma slab reaches a typical thickness from 0.1 to 0.3 μm. Detailed calculations show that in the initial stage of plasma formation and heating the electron-ion collision frequency ν_{ei} assumes values as high as $\nu_{ei} \simeq 10^{16}$ s^{-1} which is well above the laser frequency ω. In such a situation the electron quiver motion is determined by the friction force and collisional absorption is appreciable. However, as the plasma is heated up to higher electron temperatures T_e collisionality reduces strongly ($\nu_{ei}^2 \ll \omega^2$) and the quiver motion is governed by inertia. As becomes clear from the values of quiver energy W in Table 1 and as will be seen later, in fast rising pulses collisional absorption reduces almost to zero at a rather early stage of the laser-matter interaction. At 10^{18} Wcm^{-2}

© 1996 American Institute of Physics

Table 1. Electric field amplitude $\hat{E}$ in Vcm^{-1}, normalized maximum oscillation speed $\beta_0 = v_{os}/c$, oscillation amplitude $\hat{\delta}_y$ in Å, the ratio $\hat{\delta}_x/\hat{\delta}_y$ and averaged oscillation energy W in keV are given as a function of Nd and KrF laser intensities.

I	$\hat{E}$	$\lambda = 1060$ nm				$\lambda = 248$ nm			
[Wcm^{-2}]	[Vcm^{-1}]	β_{os}	$\hat{\delta}_y$[Å]	$\hat{\delta}_x/\hat{\delta}_y$	W[keV]	β_{os}	$\hat{\delta}_y$[Å]	$\hat{\delta}_x/\hat{\delta}_y$	W[keV]
10^{16}	3×10^9	0.09	140	1.1×10^{-2}	1.06	0.02	7.7	2.6×10^{-3}	0.06
10^{17}	9×10^9	0.275	440	3.6×10^{-2}	10.4	0.07	24	8.3×10^{-3}	0.58
10^{18}	3×10^{10}	0.68	1280	9.5×10^{-2}	97.3	0.21	77	2.6×10^{-2}	5.8
10^{19}	9×10^{10}	0.95	2120	0.16	651	0.56	238	7.5×10^{-2}	55
10^{20}	3×10^{11}	0.995	2355	0.175	2830	0.91	463	0.15	415

Nd laser intensity the cycle-averaged oscillation energy amounts to 100 keV with an oscillation amplitude $\hat{\delta}$ of 1300 Å. This means that during one cycle an electron sweeps over 350 atoms in the solid. At shorter wavelengths λ the excursion and the oscillation energy both reduce as $(\lambda/\lambda_{ND})^2$. Nevertheless, in all gases and in condensed matter of light atoms the collision frequency is rather determined by the electronic quiver speed than by the thermal velocity. The evaluation of the plasma formation and heating process is modelled by calculating the ionization rates and incorporating them in a fluid description of the induced currents and the internal energy increase. The standard fluid description is modified by additional terms originating from fast ionization (e.g. ionization dephasing[1]). The electromagnetic field distribution is determined from the wave equation, solved on the fastest time scale. Multiphoton ionization is studied by classical and quantum models. It is found that tunneling plays a minor role in comparison to the free streaming over the potential barrier. For answering the question how efficient collisional absorption is, the initial electron energy distribution just after field ionization has to be known. The collision frequency in a superstrong laser field is calculated for $T_e \ll W$ from an oscillator model. For its instantaneous value an approximate expression is assumed whereas its cycle-averaged value is calculatd exactly. Owing to extremely rapid ionization and the strong dependence of the electron-ion collision frequency on its quiver motion the refractive index becomes nonlinear and changes on a short time scale. As a consequence, frequency upshifts and harmonic generation are observed. With the increase of the electron energy, thermal or nonthermal, the collision frequency decreases and the gain in oscillation energy over a mean free path starts exceeding the amount of dissipated energy. At this instant a runaway phenomenon sets in which is in perfect analogy to the dc runaway effect in ohmically heated toroidal plasma devices, and absorption reduces to purely collective character.

2. The fluid model

The laser interacts with the electrons of density $n_e(x,t)$, velocity $v(x,t)$, ideal gas pressure $p(x,t)$ and electron temperature $T(x,t)$. The ions are assumed to form the neutralizing cold background for the electronic fluid. Owing to charge quasineutrality their flow velocity is identical with the cycle-averaged electron speed,

$v_i(x,t) = \overline{v(x,t)}$. The conservation equations for the electrons (mass, momentum, energy) assume a non-standard form which allows for the following effects: fast ionization, and absence of above-threshold ionization (ATI) energy just after the single ionization process has occured. Owing to fast ionization frequency spreading occurs and the fully time-dependent wave equation for the laser field,

$$\nabla^2 E - \frac{1}{c^2}\frac{\partial^2}{\partial t^2} E = \frac{1}{\varepsilon_0 c^2}\frac{\partial j}{\partial t}, \qquad (2)$$

has to be solved ($j = -e n_e v$). Since, in this context here we are not interested in resonance absorption or other collective plasma oscillations, $\nabla E = 0$ is set. The non-relativistic Boltzmann equation for the electron distribution function $f(x, u, t)$ reads

$$\frac{\partial f}{\partial t} + u \nabla f - \frac{e}{m_e}(E + u \times B)\nabla_u f = g_I + g_{ee} + g_{ei} + g_{en}. \qquad (3)$$

The collision terms on the RHS stand for the ionization $g_I(x, u, t)$ by electron collisions and field emission, elastic electron-electron, electron-ion and electron-neutral collisions $g_{ee}(x, u, t)$, $g_{ei}(x, u, t)$ and $g_{en}(x, u, t)$. For times $t \leq 100$ fs viscosity and collisional heat transfer to the ions are not significant. g_{ee} is important for thermalization. We assume that isotropization among the electrons always occurs during the time intervals considered and suppress g_{ee} in our further considerations. The electron-ion interaction term g_{ei} at position x and time t is given by

$$g_{ei}(u) = \int \{f(u')f_i(u'_i) - f(u)f_i(u_i)\} \sigma_\Omega |u - u_i| d\Omega du'_i = n_i \{f(u') - f(u)\} \sigma_\Omega |u| d\Omega \qquad (4)$$

owing to $|u_i| \ll |u|$ in the ion distribution function $f_i(x, u_i, t)$. Replacing the correctly screened Coulomb cross section σ_Ω and f_i in (4) by $\sigma_\Omega^{(n)}$ and f_n for electrons interacting with neutrals g_{en} is obtained. Integrating (3) over u yields the electron balance equation

$$\lambda_I = \partial_t n_e + \nabla(n_e v) = dn_e/dt + n_e \nabla v,$$

$$\lambda_I = \int g_I(u) du, \quad \int g_{ei}(u) du = \int g_{en}(u) du = 0. \qquad (5)$$

All kinds of recombination processes can be neglected. Multiplying (3) by $m_e u$ and integrating over u leads to the momentum balance

$$\frac{d}{dt}(m_e n_e v) + m_e n_e v(\nabla v) + \nabla P = -e n_e (E + v \times B) - (\nu_I + \nu_{ei} + \nu_{en}) m_e n_e v, \qquad (6)$$

with $\int g_I |u| du = 0$ and the electron-ion collision frequency

$$\nu_{ei} = -\frac{v n_i}{n_e |v|^2} \int \{f(u') - f(u)\} \sigma_\Omega |u| u du d\Omega = \frac{v n_i}{n_e |v|^2} \int f(u) \sigma_\Omega |u| (u - u') du d\Omega \qquad (7)$$

and ν_{en} accordingly. P is the electron pressure tensor defined in the standard way. In the following we set for it $P_{ij} = p \delta_{ij}$, $p = n_e k_B T$. Momentum loss due to field ionization is described by $m_e n_e v \nu_I$. Collisional ionization contributes to this term to the degree to which v is involved in the ionization process. The remaining fraction is due to the thermal motion and does not lead to a change in momentum density. Multiplying (6) by v yields, with the kinetic energy $\mathcal{E} = m_e v^2/2$,

$$n_e \frac{d}{dt}\mathcal{E} + 2\lambda_I \mathcal{E} + v\nabla P + 2(\nu_{ei} + \nu_{en})n_e \mathcal{E} = jE - 2\nu_I n_e \mathcal{E}. \tag{8}$$

It shows that $2\nu_I n_e \mathcal{E} = \lambda_I E_I$ has to be set; hence

$$\nu_I = \frac{\lambda_I}{n_e}\frac{E_I}{2\mathcal{E}} \tag{9}$$

is the friction coefficient for momentum density loss by ionization. It originates from the force term in (3). In collisional ionization the eventual kinetic energy given to the ejected electron in excess to E_I does influence the electron distribution function f, but does not lead to changes in momentum and energy balances and is therefore automatically included. In field ionization, it leads to an effective increase of internal energy which, however, is much lower than the ATI energy. Rapid ionization causes a phase shift of v relative to E (ionization dephasing[1]).

Energy conservation is deduced from the moment of $m_e u^2/2$. If the small energy transfer from electrons to the ions is neglected the energy balance reads

$$\frac{\partial}{\partial t} n_e \left(\frac{m_e}{2}v^2 + \varepsilon\right) + \nabla n_e v \left\{\frac{m_e}{2}v^2 + \varepsilon + \frac{P}{n_e}\right\} + \nabla q_e = jE - \lambda_I E_I \tag{10}$$

Thereby ε is the internal energy per electron which we identify with $\varepsilon = 3k_B T/2$; $q_e = (m_e/2)\int (u-v)^2(u-v)f du$ is the electron heat flow density. With the help of (5), (6), (9) and the vector identity $\nabla v^2/2 = (v\nabla)v + v \times \nabla \times v$ (10) reduces to the energy balance for the thermal energy ε,

$$\frac{d}{dt}n_e \varepsilon + n_e \varepsilon \nabla v = -(P\nabla)v - \nabla q_e + \frac{m_e}{2}\lambda_I v^2 + (\nu_{ei} + \nu_{en})m_e n_e v^2. \tag{11}$$

It has an immediate physical interpretation: Energy must be supplied to heat by mechanical work $-P\nabla)v$ (i.e. $-pdV$ work) and friction (term containing $\nu_{ei} + \nu_{en}$) and to compensate energy losses from divergent flows with $\nabla v \neq 0$ and from electron heat conduction. The non-standard terms are $dn_e\varepsilon/dt$ instead of $n_e d\varepsilon/dt$ and $\lambda_I m_e v^2/2$.

Collisional ionization is sufficiently well described here by the Lotz formula[2]

$$\sigma_I^Z = 2.76\pi a_0^2 \left(\frac{E_I^H}{E_I}\right)^2 \frac{\ln(\mathcal{E}/E_I)}{\mathcal{E}/E_I[\text{eV}]} \; [\text{cm}^2]; \tag{12}$$

a_0 Bohr radius, Z charge of the ion after ionization, $\mathcal{E}$ electron energy. The rate of change of n_e is

$$\frac{dn_e}{dt} = n_e |v| \sum_{Z\geq 1} \sigma_I^Z n_i^Z = \lambda_I - n_e \nabla v. \tag{13}$$

3. Field ionization

Multiphoton ionization occurs to an appreciable degree at $I \simeq 10^{12}$ Wcm^{-2} or at an order of magnitude higher value. The ionization cross section shows a very strong dependence on the laser frequency ω when an integer number of photons becomes resonant with intermediate transitions. In the absence of such resonances the probability P for ionization by N photons is proportional to I^N, I laser intensity, and as a rule multiple ionization occurs sequentially. Exceptions from this rule, however, seem also to exist with their enhancement at increased intensities. With increasing laser energy above-threshold ionization (ATI), level shift due to the dynamical Stark effect and higher level excitations become more and more important. Various approaches have therefore been used at moderately high intensities ($I \lesssim 10^{16}$ Wcm^{-2})[3]. Above $I\lambda^2 \simeq 10^{16}$ Wcm$^{-2}\mu$m multiphoton ionization is more adequately described by tunneling through the distorted atomic potential and by field emission. By choosing a Coulomb potential and equating its maximum U_m in the laser field ("critical field" strength E_c) to the atomic energy levels ionization stages as high as indicated in Table 2 are possible at laser intensities of $I = 10^{16}, 10^{18}, 10^{20}$ and 10^{22} Wcm^{-2}.

Table 2. Classical ionization degrees at various laser intensities.

I[Wcm^{-2}]	C	Al	Cu	Ag	Au
10^{16}	+ 4	+ 3	+ 4	+ 6	+ 8
10^{18}	+ 5	+ 10	+ 17	+ 19	+ 23
10^{20}	+ 6	+ 11	+ 26	+ 37	+ 51
10^{22}	+ 6	+ 13	+ 27	+ 45	+ 69

Deviations from these values are possible because of tunneling and Stark level shifts. To test the adequacy of a classical ionization model for laser fields $E \simeq E_c$ the ionization dynamics is solved quantum mechanically in a one-dimensional soft Coulomb potential,

$$i\hbar\frac{\partial}{\partial t}|\psi\rangle = \left\{\frac{1}{2m_e}\left(p + \Re e\hat{A}e^{-i\omega t}\right)^2 - e\phi(x)\right\}|\psi\rangle,$$

$$\phi(x) = \frac{\kappa}{(0.1 + x^2)^{1/2}}, \quad \kappa = \frac{Ze}{4\pi\varepsilon_0}. \tag{14}$$

The classical calculation was started with an ensemble of 50 noninteracting electrons, all of energy $\mathcal{E}$, but phases randomly distributed. In Fig. 1a,b,d $|\psi|(x,t)|^2$ is compared with the classical trajectory density (d) for a sinusoidal laser field of $\hat{E} = 1.55\ E_c$. Excellent agreement is found. For a direct comparison with d the 3D plot a is projected into the (x,t)-plane in b. Even fine-structures are reproduced: The shadow around $x = 0$ and times $t = 80 - 150$ a.u. in b has its classical correspondence in the particles in d which oscillate in the same time interval around $x = 0$. Discrepancies only occur for times $t < 70$ a.u. with $E(t) < E_c$ due to tunneling which is absent in d. In Fig. 1c the corresponding phase space plot shows the adiabatic deformation of the classical orbits under the influence of the laser field of frequency 0.183 a.u. (KrF). It also reveals that the ionization from a 1D potential occurs at velocities which are much reduced in comparison to those in the undisturbed potential. In Fig. 2a a comparison is made between the quantum mechanical and classical ionization degrees for an instantaneously switched on static field of strength $\hat{E} = 1.13$ a.u. and a KrF laser field $\hat{E}\sin\omega t$. Apart from an insignificant time shift of several atomic units (1

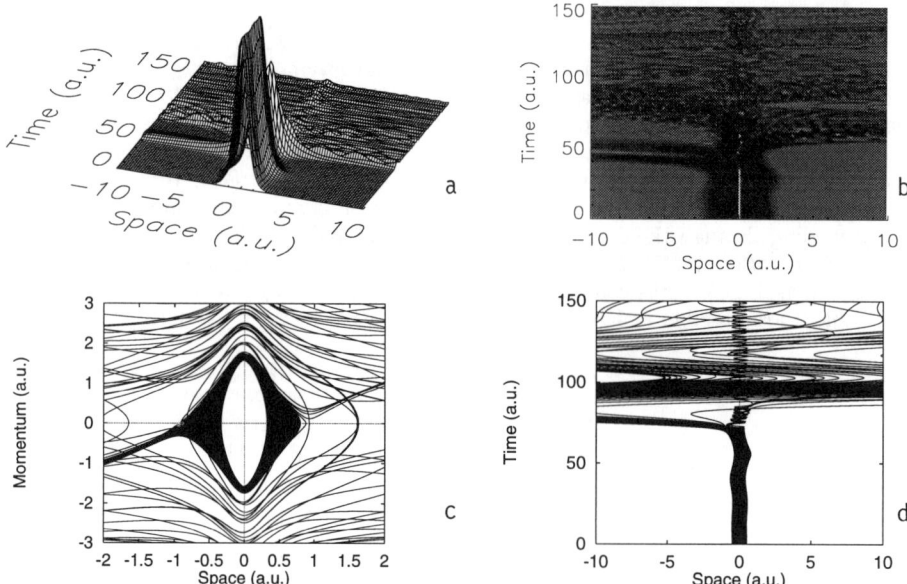

Figure 1. a: Electron probability distribution $|\psi(x,t)|^2$ under the action of a sinusoidal field, $\hat{E} = 1.13$ a.u.; b: contour plot of a; c: classical orbits of 50 electrons with random phases in phase space; d: classical orbits $x(t)$ of c; to be compared with b.

a.u. = 0.024 fs) both pairs of curves show identical slopes, i.e. ionization rates of $\Gamma = 4.5 \times 10^{16}$ s^{-1} from the ground state with $E_I = 45.6$ eV in $V(x) = -e\phi(x)$ from (14). In Fig. 2b the energies are shown as functions of time for the situation in a. Again, there is surprising agreement between quantum and classical behaviour as soon as $\hat{E}(t)$ exceeds the critical field E_c for a given state by typically 15 %. Just after ionization the electron energies are low (1 a.u. = 27.21 eV). From Figs. 1,2 we conclude that well above E_c there is good agreement between quantum and classical 1D field ionization; tunneling is not significant.

In order to model field ionization in 3D a Coulomb potential with the tunneling region around its maximum is fitted by a parabolic barrier. The transmission factor $T(\vartheta)$ under the angle ϑ to the electric field direction is given by

$$T(\vartheta) = \frac{1}{1+e^{-\varepsilon(\vartheta)}}, \quad \varepsilon(\vartheta) = -\frac{2\pi}{\hbar}\left(\frac{m_e}{2}\right)^{1/2}\left(\frac{Z}{4\pi\varepsilon_0 e}\right)^{1/4}\frac{E_I - e|U_m|\cos\vartheta}{|E(t)\cos\vartheta|^{3/4}}. \quad (15)$$

For a Coulomb potential the maximum is $|U_m| = (eZ/\pi\varepsilon_0)^{1/2}|E(t)|^{1/2}$. The transmission $T(\vartheta)$ has to be averaged over half the solid angle. To accomplish this, ε is brought into the form

$$\varepsilon = C\frac{1-\alpha\cos\vartheta}{\cos^{3/4}(\vartheta)}, \quad \alpha = \left|\frac{eU_m}{E_I}\right|, \quad C = -\frac{2\pi}{\hbar}\left(\frac{m_e}{2}\right)^{1/2}\left(\frac{Z}{4\pi\varepsilon_0 e}\right)^{1/4}\frac{E_I}{|E(t)|^{3/4}},$$

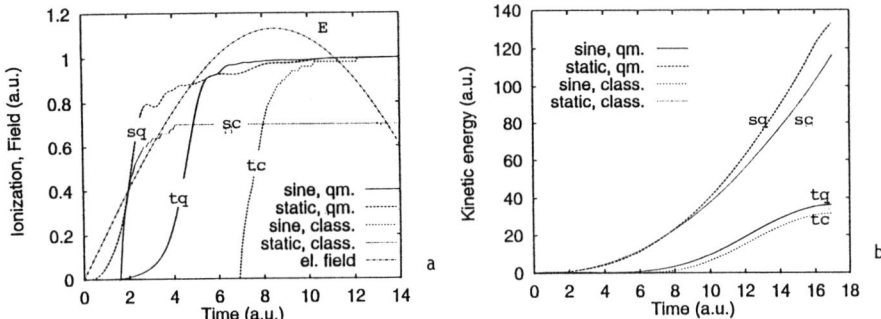

Figure 2. a: Classical and quantum mechanical ionization degrees as functions of time in the field of Fig. 1, static and sinusoidal; b: corresponding energies of the free escaping electrons as functions of time. sc, sq: static class., quant.; tc, tq: time-dependent class., quant.

and expanded in $y = \cos\vartheta$ according to $\varepsilon \simeq C[7 - 3\alpha - (3 + \alpha)y]/4$ to yield

$$T(y) \simeq \frac{e^{-\beta y}}{A + e^{-\beta y}}; \quad A = \exp\left(-\frac{7 - 3\alpha}{4}C\right), \quad \beta = \frac{3 + \alpha}{4}C.$$

From averaging over ϑ

$$T = \langle T(\vartheta) \rangle = \frac{1}{|\beta|} \ln \frac{A + e^{|\beta|}}{A + 1} \qquad (16)$$

results. The most crucial part now is to determine the relevant tunneling current. We do it by estimating the transit time τ of the electron in the potential. From the uncertainty relation one obtains

$$\Delta p = (\langle p^2 \rangle - \langle p \rangle^2)^{1/2} = \langle p^2 \rangle^{1/2} = m_e \bar{v} = \frac{\hbar}{2r_0} \Rightarrow \frac{1}{2} m_e \bar{v}^2 \frac{4r_0}{\bar{v}} = E_I \tau = \hbar,$$

and hence the recurrence time $\tau = \hbar/E_I$. This leads to the ionization rate G used in numerical modelling of field ionization,

$$G = \frac{E_I}{\hbar} \frac{1}{|\beta|} \ln \frac{A + e^{|\beta|}}{A + 1}; \quad \Rightarrow \frac{dn_e}{dt} = G(Zn_0 - n_e). \qquad (17)$$

4. Collisional absorption

A typical situation in a strong laser field and low-Z target is characterized by $\hat{v}_{os} \gg v_{th}$, $v_{th} = (k_B T/m_e)^{1/2}$, when the pulse is in its maximum. Then the motion of an electron in the neighbourhood of an ion of charge Ze is governed by the oscillator model for the deviation δ from the purely sinusoidal orbit[4],

$$\ddot{\delta} + \omega_p^2 \delta = -\frac{\kappa e}{m_e} \frac{\boldsymbol{r}(t)}{|\boldsymbol{r}(t)|^3}. \qquad (18)$$

The electron oscillates with amplitude p around its center with coordinates (a, b) according to
$$r(t) = b e_\perp + (a + p\sin\omega t)e_\parallel$$
in field direction $e_\parallel$. Collisional absorption is obtained by calculating the oscillation energy $E(t) = m_e(\dot{\delta}^2 + \omega_p^2 \delta^2)/2$ for a fixed pair of parameters (a,b) and then by averaging this expression over one oscillation period and all positions (a,b). We were able to do this analytically in terms of Anger functions $\mathbf{J}_\sigma$ and modified Bessel functions K_0 and K_1. With the collision parameter $b_0 = \max(b_\perp, \lambda_B)$, $b_\perp$ for perpendicular deflection, $\lambda_B = \hbar/m_e \hat{v}_{os}$ reduced de Broglie wavelength, one arrives in this way at the expression for the cycle-averaged $\overline{\nu}_{ei}$ [5],

$$\overline{\nu}_{ei} = 8\pi \frac{\kappa^2 e^2}{m_e^2} \frac{n_i}{\hat{v}_{os}^3} \int_{-\infty}^{+\infty} \mathbf{J}_\sigma^2(x) \frac{b_0}{p}|x| K_0\left(\frac{b_0}{p}|x|\right) K_1\left(\frac{b_0}{p}|x|\right) dx. \qquad (19)$$

For integer index $\sigma = \omega_p/\omega$ the Anger function $\mathbf{J}_\sigma$ coincides with the Bessel function J_σ. For such cases an asymptotic expansion is possible in the form

$$\overline{\nu}_{ei} = 8\frac{\kappa^2 e^2}{m_e^2}\frac{n_i}{\hat{v}_{os}^3}\left\{\ln^2\left(\frac{p}{b_0}\right) + (8\ln 2 - 4\lambda)\ln\frac{p}{b_0} + 16\ln^2 2 - 16\ln 2\lambda + 4\lambda^2 - \frac{\pi^2}{12}\right\},$$

$$\lambda = \lambda(0) = 0, \quad \lambda = \lambda(\sigma) = \sum_{k=1}^{\sigma}\frac{1}{2k-1}. \qquad (20)$$

which is an excellent approximation of (19) for $p/b_0 > 20$ and $\sigma \leq 10$. This is a remarkable result under several aspects. Although v_{os} becomes zero at maximum excursion no singularity appears in $\overline{\nu}_{ei}$ (i). Almost all analytical expressions derived so far contain correction terms of the form $\ln(\hat{v}_{os}/v_{th})$ which can be of limited validity only; they diverge for $v_{th} \to 0$ (ii). The familiar Coulomb logarithm known from the literature is replaced by a more sophisticated function of b_0/p (iii). The integral I in (19) is shown in Fig. 3 as a function of $\sigma = \omega_p/\omega$. There are weak resonances

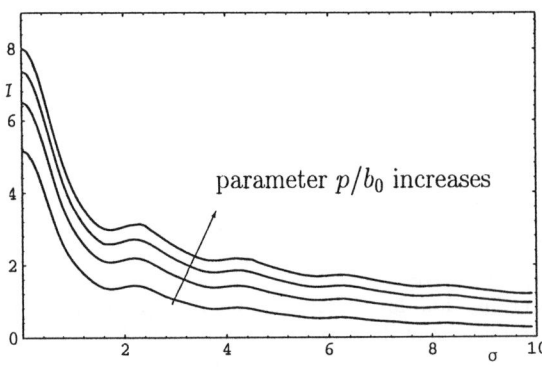

Figure 3. Integral I from (19) as a function of $\sigma = \omega_p/\omega$ and with $p/b_0 = 20, 40, 60, 80$.

at even integers of σ; in addition, the figure shows that in the underdense plasma screening becomes less and less significant as ω_p reduces to zero (iv).

On the fast time scale the instantaneous value of ν_{ei} is needed. Unfortunately for this no accurate analytical formula exists until now. However, it can be shown that in a wide range of the parameter p/b_0 a good approximation results when the amplitude $\hat{v}_{os}$ in (19) and (20) is replaced by $\alpha(v_{os}(t) + v_0)$ where v_0 provides for a thermal or Fermi velocity cutoff and α is chosen such that on the cycle average (19) or (20) results. In the following section a simpler, Spitzer-oriented ν_{ei} is used in which the number 8 of (20) is replaced by $4(2\pi)^{1/2}/3$, for the bracket the ordinary Coulomb logarithm is taken and $\hat{v}_{os}^3$ is substituted by $(v_{os}^2(t) + v_{th}^2)^{3/2}$. As expected there are always stages in the pulse-target interaction during which absorption and temperature depend rather sensitively from variations of ν_{ei}.

5. Interaction scenarios

The interaction of laser pulses in the intensity range $I = 10^{16} - 10^{18}$ Wcm^{-2} at Nd frequency with solid hydrogen, helium, aluminum and copper is studied numerically under perpendicular incidence. The formulae used are those developed in the foregoing sections. The wave equation (2) is solved by the method of characteristics. Although the laser cycle is subdivided into several 10^3 sometimes, expecially owing to very fast field ionization ($\Gamma = 10^{16}$ s^{-1} and higher), time resolution was a problem. On the other hand the simple model is so rich of phenomena that numerous additional runs will be needed to develope a general interaction scenario.

One of the most prominent features is the strong deviation of v and j from the sinusoidal shape of the driving field, and their phase shifts. From (17) $\lambda_I = G$ follows for a typical situation during field ionization. According to Sec. 3 G easily assumes values as high as 5×10^{16} s^{-1} for $I \geq 10^{16}$ Wcm^{-2}. As a consequence, from (6) $j = \sigma E$, $\sigma = \varepsilon_0 \omega_p^2 E/\nu_I$ and $\mathcal{E}_{os} = 2 W (\omega/\nu_{\text{eff}})^2$, result with $\nu_{\text{eff}} = \nu_I + \lambda_I/n_e$. The vacuum oscillation energy W is thus strongly reduced, typically to $\mathcal{E}_{os} \simeq E_I$ and hence $\sigma = 2 \varepsilon_0 \omega_p^2/G$. For $n_e = 2 \times 10^{29}$ m^{-3} this means $\sigma = 2.5 \times 10^{16} \varepsilon_0$ [mks] at $\omega = \omega_{Nd}$ and a refractive index squared $\eta^2 = 1 + i \varepsilon_0\sigma/\omega = 1 + 14$ i follows. This means that absorption A by ionization dephasing alone is as high as $A = 4\eta_r/[(1 + \eta_r)^2 + \eta_i^2] = 0.55$. High values of the velocity dependent collision frequency ν_{ei} and especially of ν_I cause considerable deviations from a harmonic motion, as illustrated by Fig. 4.

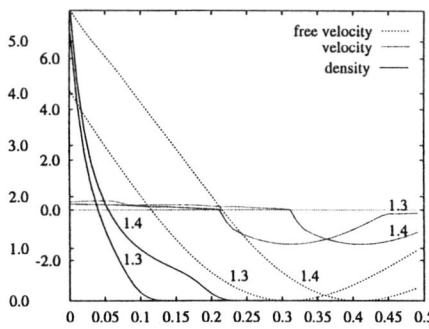

Figure 4. Ionization dephasing. During ionization the oscillation velocity is heavily damped and deformed in comparison to the free oscillation of an electron (dotted lines). Abscissa in μm, electron density n_e in units of 10^{22} cm^{-3}, velocity v in 10^8 cms^{-1}, times in fs.

Here, $v_{os}(x,t)$ is shown and compared with its vacuum value at two different times (1.3 and 1.4 fs) for a Nd laser pulse of $I = 10^{17}$ Wcm^{-2} impinging onto carbon C^{4+}. The laser pulse is switched on with a Gaussian time profile and is held constant after 3 oscillations. As a consequence of the anharmonicities in j higher harmonics are produced. So in one computer run in helium with $I = 10^{17}$ Wcm^{-2} the 5.6

harmonic component was strong, before complete ionization, in forward direction and only weak in backward direction. The shift of 0.6 ω was a consequence of fast ionization induced Doppler effect. The reflection coefficient R and the maximum electron temperature T as functions of time for the foregoing run with C^{4+} are shown in Fig. 5. Electronic heat conduction with Spitzer's conductivity coefficient was used (setting $q_e = 0$ leads to a maximum T which is increased by 30%). As soon as the total collision frequency $\nu = \nu_I + \nu_{ei} + \nu_{en}$ falls below ω strong reflection sets

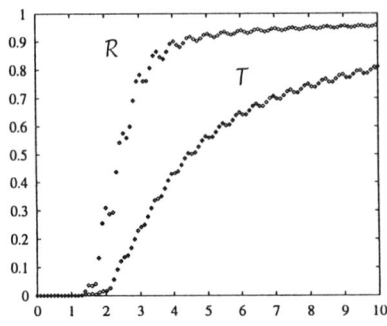

Figure 5. Reflectivity R and electron temperature T in C^{4+} at $I = 10^{17}$ Wcm^{-2} (Nd) as functions of laser cycles. Unity for T corresponds to 1 keV electron temperature.

in. For $\nu = \omega$ a fractional absorption $A = 0.96\, \omega/\omega_p$ is obtained (corresponding to $A = 7.3\%$ in Fig. 5), provided $\omega_p/\omega \gg 1$ holds. The evolution of the electron density n_e and the total electric field E in space and time are shown in Fig. 6.

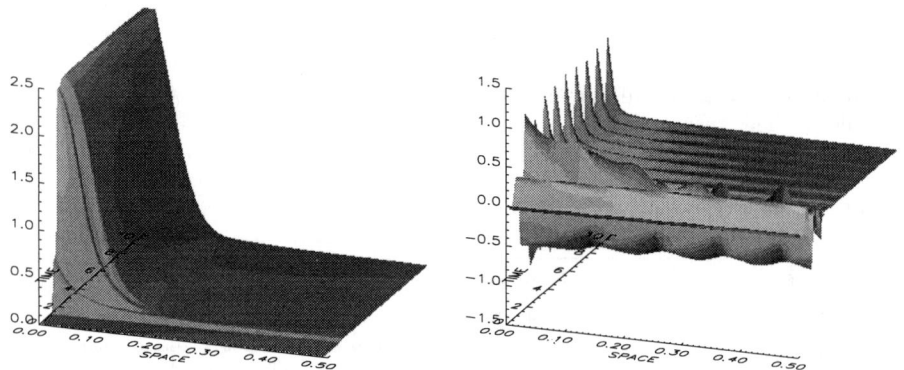

Figure 6. Temporal evolution of n_e and E for $I = 10^{17}$ Wcm^{-2} onto C^{4+} in units of laser cycles. Space coordinate in μm; n_e is given in units of 10^{23} cm^{-3}, E in units of 10^9 Vcm^{-1}.

References

1. F. Cornolti, P. Mulser, and M. Hahn, Laser & Particle Beams **9**, 465 (1991).
2. W. Lotz, Z. Phys. **232**, 101 (1970).
3. W. Becker et al., J. Opt. Soc. Am. B **4**, 743 (1987).
   H.R. Reiss, Phys. Rev. A **22**, 1786 (1980); Prog. Quant. Electr. **16**, 1 (1992).
4. P. Mulser, K. Niu, and R.C. Arnold, Nucl. Inst. Meth. Phys. Res. A **278**, 89 (1989).
5. A. Saemann, P. Mulser, *Collision Frequency in Intense Laser Fields*, subm. f. publ.

Vlasov Simulation of Superintense Laser-Solid Interaction

H. Ruhl

Theoretische Quantenelektronik,
TH-Darmstadt, Hochschulstrasse 4A, 64289 Darmstadt, Germany

Abstract The energy deposition by a single, obliquely incident p-polarized laser beam of ultrashort pulse duration is studied by relativistic Vlasov simulations. It is demonstrated that the magnitude of fractional absorption is determined by the slow time scale evolution of the electron distribution function. In particular, secular (dc) electric and magnetic fields control the degree of absorption. These fields are generated at the vacuum-plasma interface. Their magnitude increases with growing irradiation. It will be shown that the dc magnetic field provides a collective energy gain-limiter.

INTRODUCTION

Terrawatt laser systems have created novel experimental conditions for intense laser-plasma interation studies. It is now possible to produce near solid density plasmas with scale lengths of a fraction of the vacuum wavelength. Plasmas of this kind have a variety of interesting properties. For instance, experiments performed with s- and p-polarized light demonstrate that absorption is enhanced for p-polarised light, thus indicating that a considerable fraction of the absorption is purely collective (1,2).

Since there is evidence for strong collective energy deposition, a detailed knowledge of the relevant absorption mechanisms is of both theoretical and practical concern. It is well known that collective absorption of ultrashort laser pulses leads to the generation of energetic particles. For instance, there is experimental evidence for MeV electrons generated by ultrashort laser pulses (3,4,5,6). For normal incidence Kruer *et. al.* (7) proposed a mechanism called $\mathbf{j} \times \mathbf{B}$-heating. Brunel (8) tried to explain absorption and fast particle generation for p-polarized light in steep gradients but this model is purely electrostatic. Thus, it cannot be applied to a single laser beam geometry without modification, as Brunel already pointed out himself.

The analysis of energy deposition in this paper is based on Vlasov calculations for p-polarized light. Use is made of the boosting technique which

© 1996 American Institute of Physics

was first introduced by Bourdier (9,10,11). The Vlasov simulations reveal that secular time scales are important for a detailed picture of collective ultrashort laser pulse absorption. Since ultrashort laser pulses produce highly overdense plasmas with steep density gradients strong, oscillating electric and magnetic fields E_x and B_z will be present in the skin layer which exert the force $F_x = -e(E_x + v_y B_z)$ onto the electrons. The transverse velocity is given by $v_y = cp_y/(m^2c^2 + p_x^2 + p_y^2)^{1/2}$. Due to the conservation of the transverse canonical momentum we conclude that p_y consists of two contributions, the first of which stems from the oscillation velocity and the second from the boost velocity needed to perform the transformation into the frame of normal incidence. Hence, it is given by $p_y = p_{os} - \bar{\beta}\bar{\gamma}mc$ where $\bar{\beta} = \sin\theta$, $\bar{\gamma} = 1/\cos\theta$ and $p_{os} = eA_y$ holds. For $p_{os} \ll \bar{\beta}\bar{\gamma}mc$ we obtain $v_y \approx -c\bar{\beta} + (p_y + \bar{\beta}\bar{\gamma}mc)/m\bar{\gamma}^3$.

Depending on the angle of incidence and the magnitude of the oscillation velocity, three different mechanisms of particle acceleration exist. For small angles of incidence the force on the electrons is of second order in the fields. It oscillates with a frequency twice the oscillation frequency and is always positive. Thus, we observe electron jets which are generated twice a cycle. This mechanism is known as **j** × **B**-heating (7). It is a push-push mechanism. For oblique incidence and $p_{os} \ll \bar{\beta}\bar{\gamma}mc$ we expect to find electron jets of two different energies since the force on the electrons changes magnitude whenever it changes sign (see the above expansion of v_y and Fig. 3). The force on the electrons is basically first order in the fields. This mechanism is the well known push-pull mechanism. It is this mechanism which will prove to be sensitive to the dc magnetic field. Ultimately, for $p_{os} \gg \bar{\beta}\bar{\gamma}mc$ the angle of incidence can be neglected. The absorption is then equal to the case of s-polarisation for the same high irradiances, since the force on the electrons is of the form $F_x \approx -e(e_x + c|B_z|)$ which is basically of first order for arbitary angle of incidence. Hence, fractional absorption should be best for normal incidence. We will concentrate on the push-pull mechanism in the course of this paper. We believe that nonlocal first order transport is closest to a non-electrostatic Brunel mechanism. It should be pointed out that all the mechanisms, in particular the Brunel mechanism, work for step-like density profiles as well. The electrons then never enter the vacuum.

MODEL

At intensities above $I\lambda^2 = 10^{16}$ Wcm$^{-2}\mu$m^2 matter is highly ionized in a few cycles of the light wave by field ionization and by collisions. As the electron quiver energy considerably exceeds the thermal velocity, collisions are negligible and efficient light absorption is mainly due to collective effects. For a pulse lengths of a few laser cycles ($\tau < 50$ fs) as considered here no substantial ion motion occurs. The model used to calculate collective absorption in condensed targets assumes that the target is fully ionized, the ions forming an immobile, highly overdense background separated from the vacuum by a finite transition layer. The electrons are free to move in the plane of light beam incidence. The equations are given in covariant form to ease their transformation to different systems of reference, in particular to the system of reference for normal incidence. The Vlasov equations for electrons and ions are given by

$$\left(\frac{dx^\mu}{d\tau}\frac{\partial}{\partial x^\mu} + \frac{dp^\mu}{d\tau}\frac{\partial}{\partial p^\mu}\right) f^{(i/e)} = 0. \quad (1)$$

The equations of motion are

$$\frac{dx^\mu}{d\tau} = \frac{p^\mu}{m_{i/e}}, \quad \frac{dp^\mu}{d\tau} = \pm\frac{eZ_{i/e}}{m_{i/e}c} F^{\mu\nu} p_\nu. \quad (2)$$

Maxwells's equations are given by

$$\frac{\partial}{\partial x^\mu} F^{\mu\nu} = \frac{J^\nu}{c\epsilon_0}, \quad \frac{\partial}{\partial x^\mu} \tilde{F}^{\mu\nu} = 0. \quad (3)$$

The correlation between the current densities and the distribution functions are

$$J^{(i/e)}_\nu = \pm e Z^{(i/e)} \int d^4p \, c \, p_\nu \, 2 \, \theta(p_0) \, \delta\left(p^\mu p_\mu - m_{i/e}^2 c^2\right) f^{(i/e)}, \quad (4)$$

and ultimately

$$J_\nu = J_\nu^e + J_\nu^i. \quad (5)$$

The quantities $Z^{(i/e)}$ indicate the charge states of electrons and ions while $F^{\mu\nu} = \partial^\mu A^\nu - \partial^\nu A^\mu$ is the well-known Maxwell tensor and $\tilde{F}^{\mu\nu}$ its dual. The positive signs refer to the ions. Equations (1) to (5) are used to describe the interaction of a p-polarized plane wave with an ideal plasma. The ions are kept immobile. The x-axis points along the direction of the density gradient. The numerical integration of Eq. (1) is performed by a splitting scheme, which treats the relativistic Lorentz force correctly up to second order in Δt where Δt is the time step. The size of the simulation box is 6 λ_{vac} in space and 60 mv_{th} for each axis in momentum space. The spatial resolution was taken as small as a thermal Debye length. Thus, up to 1000 gridpoints in space and 200 gridpoints for each momentum direction where taken. The boundary conditions for the electron distribution function were chosen such that thermal electrons were allowed to enter the simulation box in the overdense region. The fast electrons generated at the vacuum-plasma interface which ultimately reached the end of the simulation box were allowed to escape freely. Energy and particle conservation proved to be excellent during the whole simulation time. The initial temperature for the electron distribution was chosen to be 10keV which is too high for $I\lambda^2 \approx 10^{18}$ Wcm$^{-2}\mu$m^2 but which might become more reasonable for irradiances of $I\lambda^2 \approx 10^{19}$ Wcm$^{-2}\mu$m^2 as planned for future experiments.

RESULTS

Of particular interest is the overall cycle-averaged fractional absorption coefficient A vs angle of incidence for different parameters (see Fig. 1). It shows aspects important for a detailed picture of nonlinear energy deposition: (i) for short scale lengths maximum absorption occurs around $\theta = 75°$; (ii) the magnitude of the dc electric and magnetic fields generated on the target surface correlate with the magnitude of absorption (see Fig. 1); (iii) for $I\lambda^2 = 10^{19}$ Wcm$^{-2}\mu$m^2 fractional absorption is very small and remains almost constant vs angle of incidence. Figure 1 shows the correlation between the magnitude of the dc fields and fractional absorption. It gives the angular scaling of the peak dc electric and magnetic fields for $I\lambda^2 = 10^{17}$ Wcm$^{-2}\mu$m^2 and $I\lambda^2 = 10^{18}$ Wcm$^{-2}\mu$m^2. We observe that for $I\lambda^2 = 10^{17}$ Wcm$^{-2}\mu$m^2 the dc fields E_x and B_z are small (E_x and $cB_z \leq 4.0 \cdot 10^5$V$/\lambda$ for $\theta < 70°$). The corresponding absorption profile is of simple shape with a maximum value of $A \simeq 56\%$. For $I\lambda^2 = 10^{18}$ Wcm$^{-2}\mu$m^2 the dc fields E_x and cB_z grow by a factor of six (E_x and $cB_z \leq 2.4 \cdot 10^6$V$/\lambda$ for $\theta < 70°$). The associated peak absorption decreases by almost a factor of two compared with the previous case. For a qualitative explanation of the origin of the dc fields the concept of light pressure is helpful even though the simple minded application of this concept does not yield the observed angular scaling of the dc electric and magnetic fields. Light pressure ponderomotively steepens the density profile and hence in the largely overdense plasma ($n_e = 25n_c$) a charge double layer is generated, positive ($n_e < n_i$) at the vacuum boundary and negative ($n_e < n_i$) inside. This leads to the dc electric field. In the system of normal incidence the charge imbalance gives rise to a narrow current double layer which, according to Ampere's law, leads to a well localized dc magnetic field. A detailed analysis of the secular fields has to take nonlocal first order electron transport into account. This approach will yield the desired angular scaling of the dc fields.

To prove the correlation between the dc fields and reduced fractional absorption it is helpful to analyze energy deposition vs space. Figure 2 shows the fractional energy deposition vs space for $I\lambda^2 = 10^{18}$ Wcm$^{-2}\mu$m^2. We define the fractional energy gain as the area under the positive part of the time averaged quantity $\overline{j_y E_y}$ while the fractional energy loss corresponds to the negative part. It should be noted that in the frame of normal incidence the quantity $\overline{j_x E_x}$ does not contribute to time averaged energy deposition. The reason for the disappearence of $\overline{j_x E_x}$ is the equation $\partial_t E_x = -j_x/\epsilon_0$ which holds in this frame. For steady state solutions $\overline{j_x E_x}$ must disappear. Figure 1 shows the fractional energy gain and loss vs angle of incidence. We observe that the fractional energy gain is drastically reduced with increasing irradiation while the ratio of energy loss to gain remains almost constant. Thus, a mechanism must exist which inhibits the energy gain of the electrons with increasing irradiation. We know that in the frame of normal incidence the electrons are accelerated by the electric and magnetic fields. For large angles of incidence and $p_{os} \ll \bar{\beta}\bar{\gamma}mc$ we obtain $F_x \approx -e(E_x - c\bar{\beta}B_z)$ where $E_x = E_{x0} + E_{x1}$ and $B_z = B_{z0} + B_{z1}$ holds. The field E_{x0} denotes the dc field and E_{x1} the fields on the fast time scales. The same applies for the magnetic fields. The first two plots of Fig. 2 give the spatial distribution of the dc electric and magnetic fields and the corresponding time-averaged squared amplitudes vs space. They allow to estimate the amplitudes of the fields on the fast time scales. For $\theta = 70°$ and $I\lambda^2 = 10^{17}$ Wcm$^{-2}\mu$m^2 (see second plot of Fig. 1) we obtain $E_{x0}/E_{x1} = B_{z0}/B_{z1} \approx 0.22$ while for the same angle and $I\lambda^2 = 10^{18}$ Wcm$^{-2}\mu$m^2 the ratio is $E_{x0}/E_{x1} = B_{z0}/B_{z1} \approx 0.45$. For small angles the ratios B_{z0}/B_{z1} vanish while the ratios E_{x0}/E_{x1} remain finite. From the first and second plots of Fig. 2 we observe that the dc parts of the electric and magnetic fields are always positive precisely where the fields on the fast time scales drive the electrons. The electromagnetic fields can efficiently deposit energy into those electrons which have been pulled into the underdense plasma. For these electrons the pushing part of $F_x = -e(E_x - c\bar{\beta}B_z)$ is very effective. The same pushing part is very inefficient when depositing energy into electrons in the overdense plasma since these electrons rapidly escape the driving fields. It is the pulling part of the force $F_x = -e(E_x - c\bar{\beta}B_z)$ which determines the number of electrons in the underdense plasma that can be effectively pushed back into the target. Consequently, the efficiency of the pulling force determines the magnitude of fractional absorption. Since the the dc force $F_x = -e(E_x - c\bar{\beta}B_z)$ is always positive the pulling force on the electrons is inhibited. This inhibition is small for $I\lambda^2 = 10^{17}$ Wcm$^{-2}\mu$m^2. But for $I\lambda^2 = 10^{18}$ Wcm$^{-2}\mu$m^2 it is considerable. Consequently fractional absorption drops. We would like to call this mechanism collective gain inhibition.

Figure 3 shows electron jets for the case $I\lambda^2 = 10^{18}$ Wcm$^{-2}\mu$m^2 for different angles of incidence. It decently supports the assumption that the simple force term for the force on the electrons given in the introduction to this article is enough to analyze the different regimes of collective energy deposition, e.g.

that for $p_{os} \gg \bar{\beta}\bar{\gamma}mc$ the angle of incidence is irrelevant (see Fig. 1).

SUMMARY

A definition of the Brunel mechanism for a single laser beam has been given. It has been shown that Brunel deposition relies on first order nonlocal electron transport. It has been demonstrated that first order nonlocal electron transport is extremely sensitive to the slow time scale evolution of the electron distribution function. In particular, dc electric and magnetic fields provide efficient energy gain limiters. These fields cannot be avoided and grow with increasing irradiation. For extremely high irradiances ($p_{os} \gg \bar{\beta}\bar{\gamma}mc$) all the energy deposition is due to first order electron transport, e.g. non-electrostatic Brunel deposition from our definition. For this case s- and p-deposition cannot be distinguished any more. The interaction has the tendency to become universal with respect to s- and p-polarisation.

REFERENCES

1. H. M. Milchberg and R. R. Freeman, *Phys. Fluids* **2**, 1395 (1990).

2. U. Teubner, J. Bergmann, B. van Wonterghem, F. P. Schäfer, and R. Sauerbrey, *Phys. Rev. Lett.* **70**, 794 (1993).

3. J. P. Matte, J. C. Kieffer, S. Ethier, and M. Chaker, *Phys. Rev. Lett.* **72**, 1208 (1994).

4. H. Chen, B. Soom, B. Yaakobi, S. Uchida, and D. D. Meyerhofer, *Phys. Rev. Lett.* **70**, 3431 (1993).

5. C. Rousseaux, F. Amiranoff, C. Labaune, and G. Matthieussent, *Phys. Fluids B* **8**, 2589 (1992).

6. H. Hamster, A. Sullivan, S. Gordon, and R. W. Falcone, *Phys. Rev. Lett.* **49**, 671 (1993).

7. W. L. Kruer and Kent Estabrook, *Phys. Fluids* **28**, 431 (1985).

8. F. Brunel, Phys. Rev. Lett. **59**, 52, (1987); *Phys. Fluids* **31**, 2714 (1988).

9. A. Bourdier, *Phys. Fluids* **26**, 1804 (1983).

10. P. Gibbon and A. R. Bell, *Phys. Rev. Lett.* **68**, 1535 (1992).

11. P. Gibbon, *Phys. Rev. Lett.* **73**, 664 (1994).

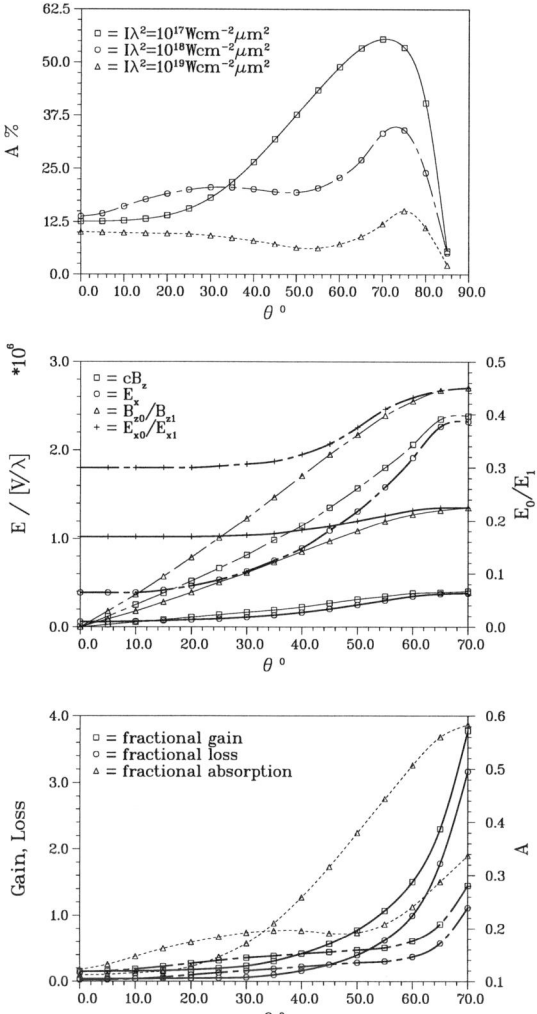

Figure 1. Fractional absorption, dc electric and magnetic fields and fractional energy gain vs angle of incidence. The chained-dashed curves belong to $I\lambda^2 = 10^{18}$ Wcm$^{-2}\mu$m^2. The parameters common to all curves are $T_e = 10$ keV, $n/n_c = 25$ and $L/\lambda = 0.023$. The fractional absorption is given by the difference between fractional energy gain and loss. The marked points have been calculated.

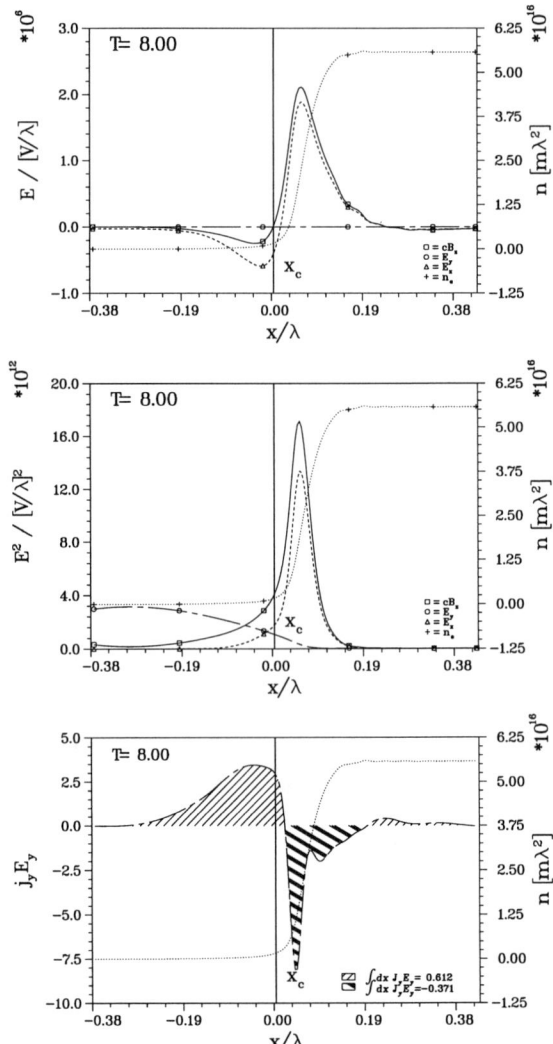

Figure 2. Secular electric and magnetic fields, time-averaged squared field amplitudes and time-averaged energy deposition by $\overline{j_y E_y}$ in arbitrary units vs space for $I\lambda^2 = 10^{18}$ Wcm$^{-2}\mu$m^2, $T_e = 10$ keV, $n/n_c = 25$, $L/\lambda = 0.023$ and $\theta = 60°$. The values of the integrals give the fractional energy gain and loss. The time is given in full cycles.

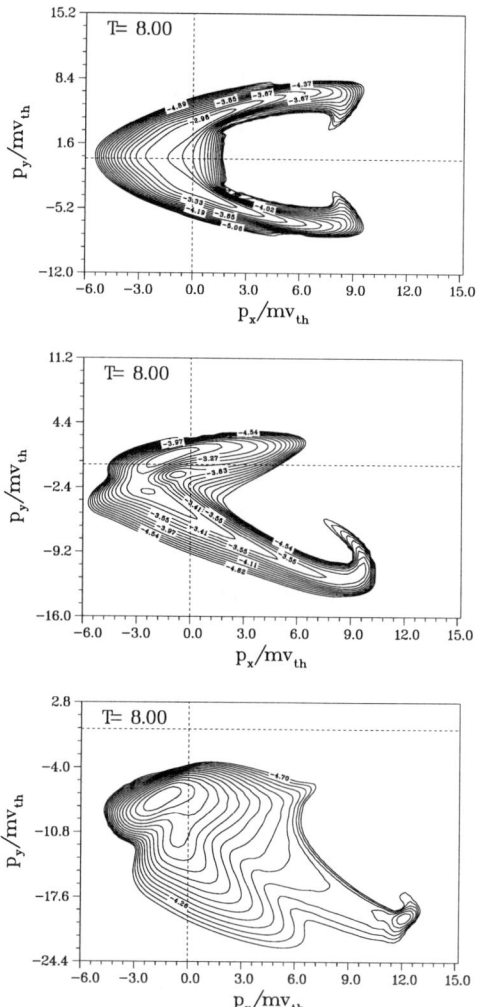

Figure 3. Contour line plot of the time-averaged electron distribution function $f_e(p_x, p_y, t)$ at the critical density in logarithmic scale for $\theta = 0°$ (top), $\theta = 30°$ (middle) and $\theta = 60°$ (bottom). The parameters are $I\lambda^2 = 10^{18}$ Wcm$^{-2}\mu$m^2, $n/n_c = 25$, $T_e = 10$ keV and $L/\lambda = 0.023$. The time is given in full cycles.

Radiative pumping generated by a hot electron Kα x-ray source in femtosecond laser-produced plasma experiments

A. Rousse*[†], J.P. Geindre*, P. Audebert*, F. Fallies* & J.C. Gauthier*
A. Dos Santos°, G.Grillon°, A. Mysyrowicz° & A. Antonetti°

*Laboratoire pour l'Utilisation des Lasers Intenses, Ecole Polytechnique, 91128 Palaiseau, France.
°Laboratoire d'Optique Appliquée, Ecole Nationale des Sciences et Techniques Avancee, 91120 Palaiseau, France.
[†]Institute for Physical and Chemical Research (RIKEN), Hirosawa 2-1, Wako-shi, Saitama 351-01 Japan.

Abstract. We demonstrate the first observation of X-ray fluorescence resulting from inner-shell photoionization by a subpicosecond laser-produced plasma x-ray source. 8 keV electrons produced by a 100 fs laser at $3 \; 10^{16}$ Wcm^{-2} intensity create K-shell vacancies in a Ti layer. This generates a 4.51 keV Kα x-ray pulse which is used to photoionize a Ca sample. Fluorescence of Ca at 3.69 keV (Kα) and 4.01 keV (Kβ) is observed as a result of Ti Kα photoionization.

INTRODUCTION

The development of bright and ultra-fast x-ray sources from picosecond and subpicosecond laser-produced plasmas has shown a very rapid progress in the last years (1-4). To be useful for applications, such x-ray sources require high brightness, quasi-monochromaticity, short time duration, high repetition rate and good shot to shot reproducibility. Up to now, most experiments have dealt with the thermal x-ray emission resulting from the hot ($T_e > 100$ eV) laser-generated plasma (2, 5-7). For this type of emission, the factors controlling the x-ray pulse duration and intensity have been established (8). Short duration and high brightness can be obtained when laser energy deposition occurs at near solid density (9). This is difficult to achieve because of the limited contrast ratio of short pulse lasers (3).

It has been shown recently (2,4,10,11) that non thermal emission related to the energetic electrons created during the non linear interaction of the laser field with the plasma can be a very promising alternative for short pulse x-ray generation. Such energetic electrons produce Kα fluorescence line radiation due to relaxation of the outer-shells electrons following a single K-shell ionization process. Quasi-monochromaticity is achieved because the temperature increase due to the electron energy deposition (which would produce a weak Kα wavelength shift) is small (4). Being produced by highly non linear effects, the fast electrons are thought to last no longer than the duration of the laser pulse. The duration of the Kα emission itself is thus limited by the electron thermalization time.

The present work describes the first attempt to use this bright and ultra-fast monochromatic x-ray source as a pump pulse to radiatively perturb a sample set beyond the edge of the penetration depth of the hot electrons. Nanosecond laser experiments have already investigated radiative pumping to study ionization and excitation dynamics of radiative transitions (12,13) and to control the population inversion in x-ray laser schemes (14,15). These studies involved x-ray sources generated from thermal plasmas and required very accurate time dependent simulations to optimize the electronic density and temperature of the source and photo-pumped plasmas. In our case, the experiment involves one hot electron plasma source but the pump radiation generator and the sample are both solid materials which remain at low temperatures. Our demonstration of fluorescence is a very incentive result for the study of solid-phase systems in basic atomic physics (16), molecular photoionization dynamics (17,18) and photoionization x-ray laser schemes in the femtosecond regime (19,20).

NUMERICAL SIMULATIONS OF THE RADIATIVE PUMPING

The electron and x-ray photon sources produced by the short pulse laser have been fully characterized in our previous experiments (4). To summarize, we have found that, at laser intensities around 10^{16} Wcm^{-2}, 12% of the laser energy was converted into a quasi-maxwellian hot electron distribution function with an average temperature of 8 keV. In the following, the Kα radiation generated by these electrons will be called the pump source. The thickness of the radiative pump layer has to be much larger than the maximum penetration depth of the hot electrons to avoid direct Kα excitation by electrons in the photo-pumped layer. The fluorescence intensity in the photo-pumped layer is proportional to the photoelectric absorption cross section and to its thickness. For low-Z materials, most of the fluorescence radiation comes from Kα lines because the K-shell photoelectric cross section is ten times larger than the one for the other shells (21).

The optimization of the fluorescence signal has led us to choose the layer materials and thicknesses by taking into account the constraints imposed by the detector efficiency and the x-ray optical properties of the materials (22). We have used our Monte Carlo model of electron energy deposition both for the design of the target assembly and to analyze the data (4). Results show that high fluorescence efficiency is obtained for (i) a high intensity of hot electron Kα pump source, (ii) a high absorption rate of the pump source into the photo-pumped layer, (iii) a small Auger effect (non radiative energy transfer which can compete which Kα radiative decay), (iv) the smallest pump layer thickness to reduce the auto-absorption of the pump Kα line from its point of emission down to the photo-pumped layer, (v) a sufficiently large photo-pumped layer to absorb most of the pumping radiation, but sufficiently small to let the auto-absorption of the fluorescence Kα radiation to be negligible.

Requirement (ii) involves the use of two materials with close values of Z so that the absorbing K-edge of the photo-pumped element coincides with the energy of the pumping Kα radiation. If requirements (i) and (ii) are taken into account the highest fluorescence intensity is produced for medium Z (Z≈13) materials. At lower Z, the Auger effect is too large and at higher Z, the hot electron Kα pump source (with our 8 keV electron distribution) is too weak despite the great increase of the Kα radiative transition probability. Taking into account the experimental constraints imposed by the detector efficiency and target fabrication, the experimental demonstration of radiative pumping has been studied with the couple Ti (Z=22) and Ca (Z=20). Requirements (iv) and (v) have been optimized with Monte Carlo simulations shown in figure 1 which gives the Ti pump and Ca fluorescence Kα line intensities as a function of Calcium thickness. We assume that the detector is set at 30° from the backside target normal. Intensities are calculated for a titanium thickness of 25 μm, which is sufficiently large compared to the hot electrons penetration depth in titanium (12 μm). The Ti pump intensity decreases at large values of Ca thickness due to opacity effects. The highest fluorescence is achieved for a calcium thickness of 10 μm for which opacity effects on the Kα line balance the fluorescence generated by the Ti Kα pump source. The highest intensity of the resulting fluorescent signal is 3 10^6 photons/str for a calcium thickness of 10 μm.

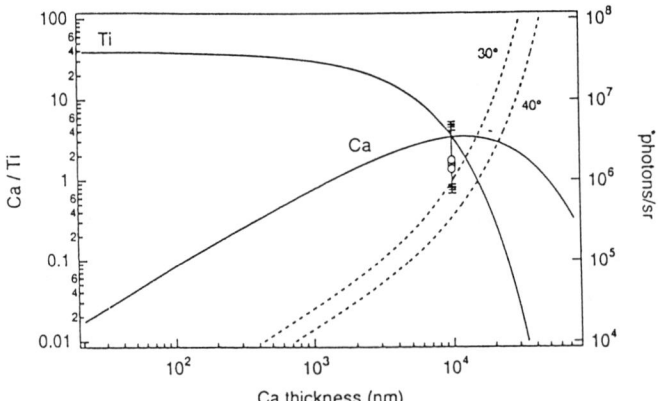

FIGURE1. Number of photons per steradian for the pump (Ti) and the fluorescence (Ca) as a function of calcium thickness (right-hand scale). Titanium thickness is 25 μm. Experimental (circles) and calculated (broken curves) Ca to Ti Kα line ratios (left scales).

EXPERIMENTAL SETUP

The experimental system has already been presented elsewhere (17). Briefly, a 100 fs (FWHM), 10 Hz, 1.5 mJ dye laser pulse at 620 nm was focused using a f/8 lens on targets at intensities up to 3×10^{16} Wcm^{-2}. A controlled amplified spontaneous emission prepulse (23) was used to enhance the hot electron conversion efficiency of the plasma source. Tri-layered targets were made of a titanium foil over which was deposited an aluminum film of varying thickness on the laser side and a calcium film on the opposite side (see the inset in figure 2). As discussed above, the titanium and calcium layer thicknesses were 25 μm and 10 μm, respectively. Time integrated Kα emission from the calcium and titanium layers was measured with a cooled (-40°C) x-ray sensitive charge-coupled-device (CCD) camera operating in the single photon counting mode. It was set at an angle of 30° from the target normal. The use of a CCD was dictated by the small number of photons (≈ 300) that have to be detected at each shot. The CCD was absolutely calibrated by measuring its energy response with the synchrotron source at the Laboratoire pour l'Utilisation du Rayonnement Electromagnetique (LURE) facility. At incident photon energies around 4 keV, this detector achieved a spectral resolution of the order of 150 eV that was sufficient to isolate the Kα lines of the two materials.

EXPERIMENTAL RESULTS AND COMPARISON WITH THE SIMULATIONS

The histogram of a 10 laser shots CCD recording, obtained from a target with an aluminum film thickness of 300 Å, is shown in figure 2(a). Kα lines from Ti pump source (4.51 keV) and from calcium (3.69 keV) are clearly observed. The Kβ lines of Ca and Ti are also weakly visible around 4 and 5 keV, respectively. X-rays with higher energies are also detected as a signature of the bremsstrahlung radiation from hot electrons. Several experimental time-integrated Ca to Ti Kα line ratios are shown in figure 1. Results take into account the specific response of the CCD at the two different Kα energies. Also plotted in figure 1 is the line intensity ratio estimated from numerical simulations. The detector angle has been set at 30° and 40° from the target normal to show the sensitivity of the calculations to the variations of the fluorescence optical pathlength in the Ca layer. Despite the large variation of the predicted ratio as a function of calcium thickness, an outstanding agreement is obtained (within a factor of 2) between experiment and simulations. The error bars show the uncertainty in detector calibration; the reproducibility of the data is about 20%.

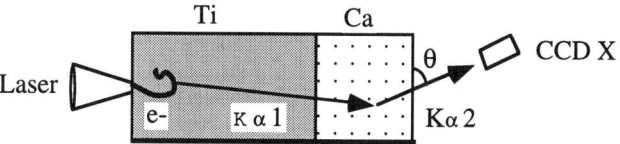

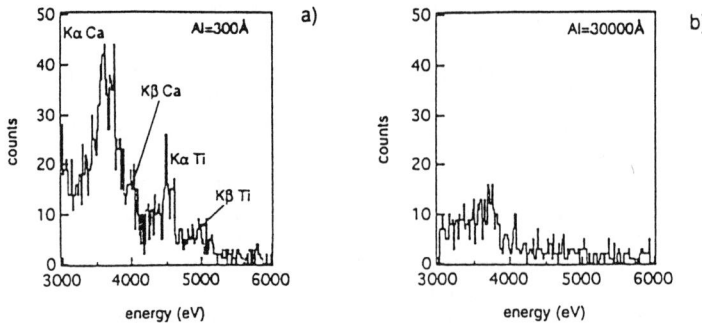

FIGURE 2. Histograms of x-ray CCD data (10 laser shots) obtained with 25 μm Ti and 10 μm Ca targets. (a) Al layer is 300 Å, (b) Al layer is 30 000 Å. The inset shows the target configuration.

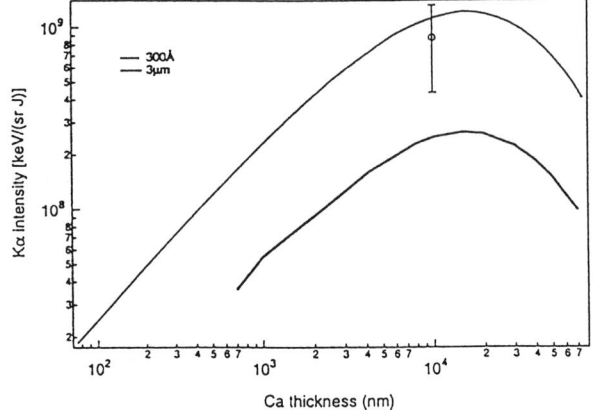

FIGURE 3. Calculated Kα fluorescence emission of calcium thickness of the two different aluminum layers used in the experiments. The absolute fluorescent intensity corresponding to the measurement of figure 2(a) is shown for comparison.

Without changing the experimental conditions, we have increased the aluminum thickness to 3 μm. Results are shown in figure 2(b). The Ca Kα line intensity is reduced by a factor 3 and exhibits a poor contrast ratio to the bremsstrahlung emission background. Since most of the electrons with energies below 20 keV were stopped by the 3 μm Al layer, this experiment indicates that a significant fraction of the observed fluorescence is produced by electrons in the distribution with energies above 20 keV. The calculated Kα fluorescence emission for the two aluminum thicknesses is shown in figure 3 together with the absolute fluorescence intensity measurement of figure 2(a) for 300 Å Al layer. For 10 μm thick calcium, the fluorescence is reduced by a factor 4.5 in reasonable agreement with the experimental results.

CONCLUSION

In conclusion, we have demonstrated for the first time the generation of x-ray fluorescence induced by K-shell photoionization using a short pulse x-ray source from a femtosecond laser-produced plasma. Hot electrons generated during the laser-solid interaction at the target surface produce a quasi-monochromatic Kα x-ray pump source which is used to photoionized a sample layer. This paper demonstrates that the hot electron Kα source can be used efficiently as a bright x-ray pump/probe pulse. With lasers of increasingly larger focused intensities, our work opens the way to material and plasma studies with short pulse x-rays. Targets consisting of an electron converter layer (such as aluminum) deposited on a Kα generator could be used to backlight plasmas for absorption spectroscopy or even to heat a sample in a very short time to allow equation of state and transport coefficients studies of matter at solid density and temperatures in the 1-10 eV range.

REFERENCES

1. Murnane M.M., Kapteyn H.C., Rosen M.D. and Falcone R.W., *Science* **251**, 531 (1991)
2. Audebert P., Geindre J.P., Gauthier J.C., Mysyrowicz A., Chambaret J.P. and Antonetti A., *Europhys. Lett.* **19**, 189 (1992)
3. Kieffer J.C., Chaker M., Cote Y, Beaudoin Y, Pepin H, Chien C.Y., Coe S. and Mourou G., *Phys. Fluids B* **5**, 2330 (1993)
4. Rousse A., Audebert P., Geindre J.P., F. Fallies & J.C. Gauthier A. Dos Santos, G.Grillon, A. Mysyrowicz & A. Antonetti, *Phys. Rev. E* **50**, 2200 (1994)
5. Cobble J.A., Schappert J.T., Jones L.A., Taylor A.J., Kyrala J.A. and Fulton R.D., *J. Appl. Phys.* **69**, 3369 (1991)
6. Murnane M.M., Kapteyn H.C., Gordon S.P., Bokor J., Glytsis E.N. and Falcone R.W., *Appl. Phys. Lett.* **62**, 1068 (1993)
7. Gordon S.P., Donnelly T., Sullivan A., Hamster H. and Falcone R.W., *Opt. Lett.* **19**, 7 (1994)
8. Milchberg H.M., Lyubomirsky I. and Durfy C.G., *Phys. Rev. Lett.* **67**, 2654 (1991)
9. Rosen M., *in proceedings of the conference SPIE* **1229**, 160 (1990)
10. Chen H., Soon B., Yaakobi B., Uchida S. and Meyerhofer D., *Phys. Rev. Lett.* **70**, 3431 (1993)
11. Meyerhofer D.D., Chen H., Delettrez J.A., Soom B., Uchida S. and Yaakobi B., *Phys. Fluids B* **5**, 2584 (1993)
12. Monnier P., Chenais-Popowics C., Geindre J.P. and Gauthier J.C., *Phys.Rev. A* **38**, 2508 (1988)
13. Back C.A., Lee R.W. and Chenais-Popowics C., *Phys. Rev. Lett.* **63**, 1471 (1989)
14. Goodwin D.G. and Fill E.E., *J. Appl. Phys.* **64**, 1005 (1988)
15. Elton R.C., *X-ray lasers*, San Diego Ca: Academic
16. Filipponi A., Tyson T.A., Hodgson K.O. and Mobilio S., *Phys. Rev. A* **48**, 1328 (1993)
17. Rosker M.J., Wise F.W. and Tang C.L., *Phys. Rev. Lett.* **57**, 321 (1986)
18. Teng T., Huang H.W. and Olah G.A., *Biochem.* **26**, 8066 (1987)
19. Murnane M., Kapteyn H. and Falcone R., *IEEE J. Quantum Electron.* **25**, 2417 (1989)
20. Kapteyn H.C., *Appl. Opt.* **31**, 4391 (1992)
21. Krause M.O., *J. Phys. Chem. Data* **8**. 307 (1979)
22. Henke B.L., Lee P., Tanaka T., Simabukuro M.L. and Fujikawa B.K., *At. Data Nucl. Data Tables* **27**, 1 (1982)
23. Gauthier J.C., *Nato Advanced Studies on Laser Interaction with Atoms, Solids and Plasmas*, ed More R.M., New York, Plenum Press

Ultra-Intense, Short Pulse Laser-Plasma Interactions with Applications to the Fast Ignitor

S. C. Wilks, W. L. Kruer, P. E. Young, J. Hammer, and M. Tabak

X-Division and Dept. of Physics
Lawrence Livermore National Laboratory, Livermore, CA 94550

Abstract. Due to the advent of chirped pulse amplification (CPA) as an efficient means of creating ultra-high intensity laser light ($I > 5 \times 10^{17}$ W/cm^2) in pulses less than a few picoseconds, new ideas for achieving ignition and gain in DT targets with less than 1 megajoule of input energy are currently being pursued. Two types of powerful lasers are employed in this scheme: (1) channeling beams and (2) ignition beams. The current state of laser-plasma interactions relating to this fusion scheme will be discussed. In particular, plasma physics issues in the ultra-intense regime are crucial to the success of this scheme. We compare simulation and experimental results in this highly nonlinear regime.

INTRODUCTION

The fast ignitor (FI) fusion concept[1] is a promising new method of potentially obtaining dramatically higher gain than conventional inertial confinement fusion (ICF), for a given amount of input energy. More importantly, this high gain may be achieved at significantly lower driver energy (~100 kJ) than is theoretically possible with conventional isobaric models of ICF [2]. The recent availability of ultra-intense, short pulse lasers allows for the delivery of a considerable amount of energy to a compressed core in a short time, (~ 10 psec) thus allowing for higher gains at lower input energies. A detailed outline of the FI is given in reference [1]. In this paper, we concentrate on the plasma physics issues that arise in this scheme.

A brief description of the fast ignitor will now be given , First, a standard (directly driven) implosion is created, in the usual way. This is envisioned to be a high density compression, such as that obtained at Osaka[3], in which they achieved 600g/cc. It is important to note that since the FI does not rely on shock timing for ignition, mix and associated Rayleigh-Taylor instabilities are not a concern. Next, a 100-psec, 10^{18} W/cm^2 laser is directed at this imploded pellet, creating a channel evacuated of plasma through the underdense portions of the corona. This laser also pushes on the critical surface. In fact, although the critical surface starts out at about 200 μm from the center of the compressed core, it is thought that this laser can push the critical surface to about 150 μm from the center of the core. The third step in the process is the firing of a second, more intense (~10^{20} W/cm^2), shorter (~10 psec) laser which is sent down the channel created by the first laser pulse. When this intense laser interacts with the dense plasma at

© 1996 American Institute of Physics

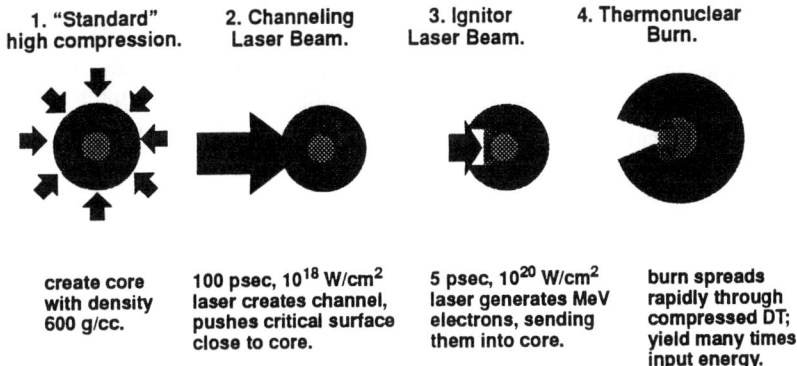

FIGURE 1. Schematic of the four steps involved in the fast ignitor fusion concept.

the end of the channel, a substantial number of MeV electrons are then generated. These electrons transport to the core, and initiate a burn in the compressed DT, thus yielding many times the total input energy. The entire process is summarized in Fig. 1. As seen in the above description, this scheme relies heavily on laser-plasma interaction in a relatively unexplored regime of parameter space; namely ultra-intense, ultra-short laser plasma interactions with both underdense and overdense plasmas.

CHANNELING LASER ISSUES

We begin with the channeling laser issues. The first item is the competition between filamentation and self-focusing of the laser beam in this underdense portion of the corona. Clearly, filamentation will be detrimental to channel creation, whereas self-focusing will help to form a channel. This has been studied both theoretically and experimentally. Early experiments by P. Young[4] using a 100 psec, 10^{15} W/cm^2 pulse shot into a 1/10 - 1/4 n_{cr} plasma with parabolic profile created by pre-exploding a CH foil, showed that a channel forms early, but does not exist past the peak of the density profile. Subsequent simulations, such as that shown in Fig. 2a, showed that the laser was actually filamenting, then "spraying" apart. This spreading of the laser beam was later confirmed by experiments by P. Young [5]; the results are shown in Fig 2b. Additional simulations and experiments predicted that if the pulse was longer, or equivalently, if the intensity was higher, that channel formation would be possible. This has since been observed, and details regarding the exact relation between various laser and plasma parameters that allow channel formation can be found in Ref. 6.

Once the channel has formed, it is important that the laser push the critical surface closer to the core. Hole boring is a key element of the FI, and an important question is how deep the laser can "drill" or push into the overdense plasma. Hole boring has been observed in simulations with laser intensities of 10^{19} W/cm^2, pushing a density step of 4 n_{cr} approximately 3 μm[7].

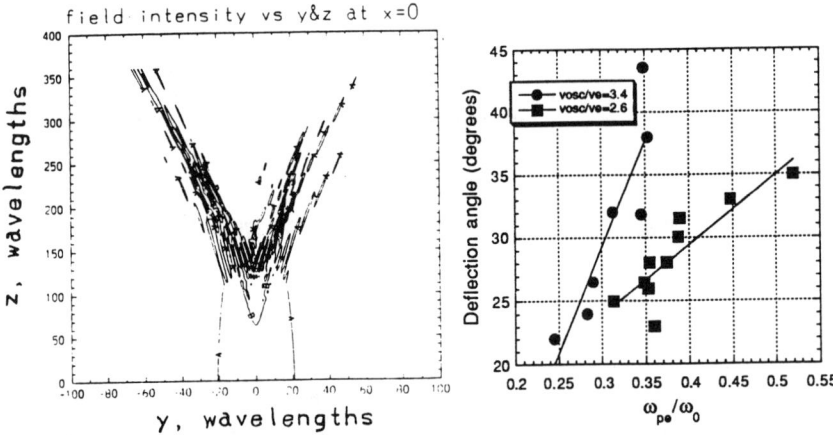

Figure 2. (a) Field intensity from a 2-D hydro simulation at t = 50 psec. The laser propagates from bottom to top. Note channeling of laser in lower portion of plot. Subsequent filamentation results in laser beam spray at an angle. (b) Experimentally measured ngular beam deflection as a function plasma density for two intensities.

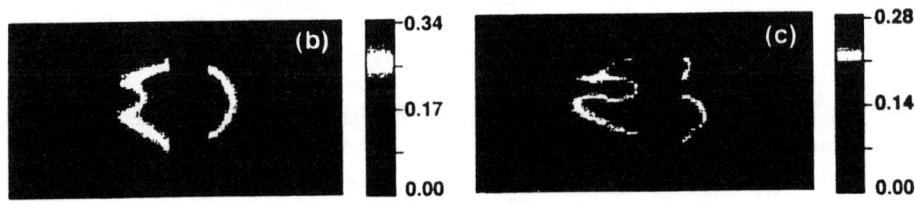

Figure 3. Experimental data showing channel formation at (a) 50 psec and (b) 400 psec, for an $I\lambda^2 = 10^{16}$ W/cm^2 laser pulse traversing a parabolic density profile plasma with peak density n = 1/4 n_{cr}. The laser propagates from left to right in these enhanced density interferograms.

By balancing the momentum flux of the ion mass flow with the light pressure, it is found that the critical surface recedes at a velocity given by.[8]

$$\frac{v_f}{c} = \sqrt{\frac{n_{cr}}{n_p} \frac{Zm}{M} \frac{I\lambda^2}{2.74 \times 10^{18}}}$$

where n_p/n_{cr} is the density at the end of the channel M is ion mass, m is the electron mass, Z is the charge state of the ions, I is the intensity and λ is the wavelength of the incident laser. Clearly, the reflected light will be red-shifted, since the surface it is reflecting off of is receding. This has been experimentally observed by Kalashnikov, Nickles et. al.[9] Thus, the ability of the laser to push the critical surface at least in one regime has been demonstrated.

IGNITOR BEAM ISSUES

We now discuss a few of the laser plasma physics issues concerned with the second laser pulse, the so-called ignitor pulse. The first issue is maintaining the integrity of the ignitor pulse as it propagates down the channel. Even with channel creation, there will most likely be a small amount of residual plasma in the channel, as shown in Fig. 4. Even plasma densities of 10^{-3} n_{cr} in the channel can cause detrimental effects to the ignitor pulse, since the intensity is of order 10^{20} W/cm^2 and the growth rate of instabilities such as relativistic filamentation and Stimulated Raman Scattering (both back and forward) depend on the density and intensity. For example since the power threshold relativistic filamentation[10] is given by $P_{cr} \sim 16\ (n/n_{cr})$ GW, a 10 pSec pulse with a 10 µm FWHM width in a $10^{-3}\ n_{cr}$ plasma is well above threshold. This may ultimately cause the beam to "spray" similar to what happens in ponderomotive filamentation as discussed above.

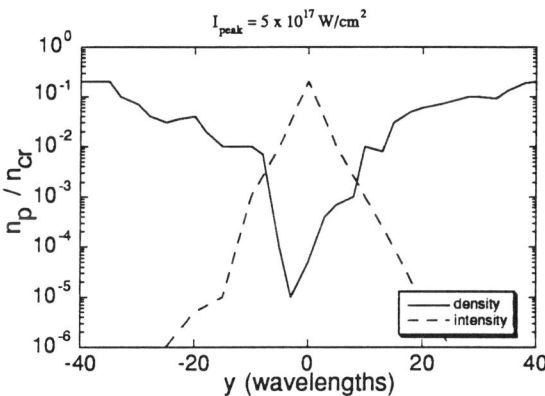

Figure 4. Transverse cut near peak of density profile from a hydro simulation, showing deep channel formation (solid line) due to intense laser (dotted line), with peak intensity $I \lambda^2 = 10^{17}$ W/cm^2.

In addition, SRS forward scatter[11] has been shown to cause transverse beam break-up in ultra-short (100 femtoseconds) intense laser pulses. Let us briefly consider the effects of SRS backscatter on the ignitor pulse in detail. Previous studies of 1pSec pulses with intensities >10^{17} W/cm^2 have shown that considerable SRS backscatter is possible.[12] This is an important concern for the ignitor pulse, in that a substantial fraction of the ignitor energy (~1/2) could easily be scattered back out of the hole. Many recent experiments with picosecond, intense lasers have noticed a broadening of the backscattered spectrum.[13,14] Through PIC simulation studies, we have attributed this to the following mechanism. Previously[12], we have shown that when the product of the growth rates and the pulses length (τ_p) is such that

$$\gamma \tau_p \cong \frac{1}{2} \frac{v_{osc}}{c} \sqrt{\frac{\omega_p}{\omega_0}} \tau_p \sim 15,$$

substantial SRS occurs. This results not only in backscattered light, but also in a large fraction (~50%) of the plasma electrons heated to an effective temperature of ~ 18 keV. The back half of the pulse can not only scatter off the cooler electrons near the initial temperature of the plasma (at ~ 5keV), but also Compton Scatter off of the high temperature component of the plasma. This two temperature plasma distribution has been seen in simulations, such as that shown in Fig. 5.

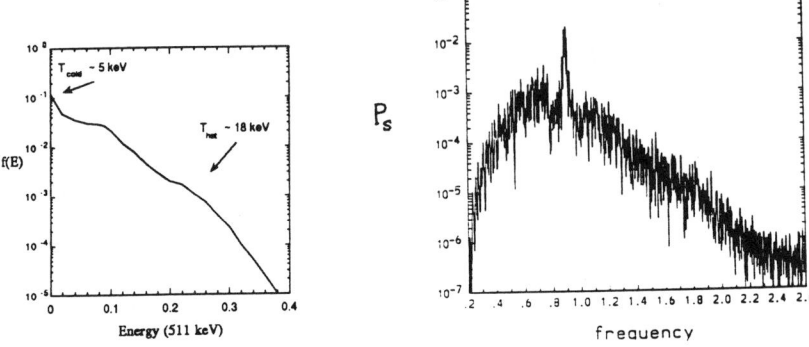

Figure 5. (a.) electron distribution function of plasma electrons, after intense laser (I = 10^{18} W/cm^2) has passed by, showing cold component (Te ~ 5 keV) and hot component (Th ~ 18 keV). (c) Resulting backscatter spectra, showing broad feature associated with Compton Scattering, and sharp peak associated with SRS off cold component.

This is a good example of how unexpected phenomena can occur, and add difficulties otherwise not foreseen in this new regime of ultra-intense laser plasma interactions. It is currently thought that although these parametric instabilities may play a role early on, in the 10 psec pulse, that they will in fact be transitory, thus allowing for propagation of the ignitor pulse down the channel.

Assuming that the ignitor pulse has propagated to the steepened critical surface, we now turn our attention to the generation of hot electrons that will ultimately be deposited into the compressed core. Two things are important in this step; 1.) significant absorption into the hot electrons must be achieved, and 2.) these electrons must have the correct energy (~1 MeV) to have a range in the compressed core corresponding to a 3.5 MeV alpha particle. Theory and simulation [7,8] have indicated that for $I\lambda^2 > 5 \times 10^{17}$ W μm/cm^2, the energy of electrons produced at a steepened critical surface is roughly twice the ponderomotive potential associated with the laser. This fact, coupled with the fact that 1-2 MeV electrons have the same range as 3.5 MeV alphas, dictates that $\sim 10^{19} - 10^{20}$ Wμm/cm^2 should be used for the ignitor pulse. The second point is that PIC simulations see extremely good absorption into hot electrons, even at very high (~ 100 n_{cr}) densities, of 2-dimensional effects are taken into account.[15] Experiments are only now beginning to access this regime, but initial results look promising.[16] Theoretical work by Yang, et al[17], that discusses various absorption mechanisms, is beginning to unravel the complex mechanisms governing absorption of ultra-intense pulses on critical surfaces. Another issue has recently been brought up, that may have consequences for the ignitor pulse. Self-induced transparency[18] may allow the ignitor pulse to propagate past the critical surface, and thus generate energetic electrons even closer to the core.

The final step, propagation of hot electrons to the compressed core has been studied by Glinsky, Mason, Hammer and Tabak,[19] and the reader is referred to this paper for details. One interesting note is that the B-fields seen in the overdense plasma published in previous work[7,8] have also been seen in Ref. 19, and may help guide the hot electrons to the core.

CONCLUSIONS

In conclusion, we have reviewed some of the important plasma physics issues relevant to the Fast Ignitor fusion concept. In particular, channeling beam issues such as channel creation and hole-boring have been shown to be feasible both through simulation and experimentally. Secondly, ignitor beam issues, such as the efficient generation of MeV electrons at a steepened critical surface. It is encouraging that many of the important aspects of the fast ignitor have been observed in experiment. However, these are individual experiments which test only one specific aspect of the FI at a time. As the expertise in laser building progresses, a system capable of doing an integrated experiment will emerge, thus allowing for a true test of this novel fusion concept. *This work was performed under the auspices of the U.S. Department of Energy by Lawrence Livermore National Laboratory under contract No. W-7405-Eng-48.

REFERENCES

1. M. Tabak, J. Hammer, M. Glinsky, W. L. Kruer, S. C. Wilks, J. Woodworth, E. M. Campbell, and M. Perry, Phys. Plasma, **1** 1626 (1994).

2. J. Meyer-er-Vehn, Nucl. Fusion **22**, 561 (1982).

3. H. Azechi, T. Jitsano, T. Kanabe, M. Katayons, K. Mima, N. Miyanaga, M. Naki, S. Naki, H. Nakaishi, M. Nakatsuka, A. Nishiguchi, P.A. Norrays, Y. Setsuhara, M. Takagi, M. Yamanaka, and C. Yamanaka, Laser Part. Beams, **9**, 2 (1991).

4. P. E. Young, Phys. Fluids B 3, 2331 (1991).

5. S. C. Wilks, P. E. Young, J. Hammer, M. Tabak, and W. L. Kruer, Phys. Rev. Lett. **73**, 2994 (1994).

6. P. E. Young, J. Hammer, W. L. Kruer and S. C. Wilks, Phys. Rev. Letter, to be published (1995).

7. S. C. Wilks, W. L. Kruer, M. Tabak, and A. B. Langdon, Phys. Rev. Lett, **69**, 1383 (1993).

8. S. C. Wilks, Phys. Fluids B, **5**, 2603 (1993).

9. Kalashnikov, P. E. Nickles, Th. Schlegel, M. Schnuere, F. Billardt, I. Will, W. Sandner, and N.N. Demchenko, Phys. Rev. Lett., **73**, 260 (1994).

10. C. E. Max, J. Arons, and A. B. Langdon, Phys. Rev. Lett, **33**, 2091 (1974); W. B. Mori, C. Joshi, J. M. Dawson, D. W. Forslund and J. M. Kindel, Phys. Rev Lett. **60**, 1298 (1988); A. Borisov, et.al. Phys. Rev. A, **45**, 5830 (1992).

11. T. Antonsen and P. Mora, Phys. Rev. Lett, **69**, 2204 (1992); W. B. Mori, C. Decker, D. E. Hinkel, T. Katsouleas, Phys. Rev. Lett., **72**, 1482 (1994).

12. S. C. Wilks, W. L. Kruer, E. A. Williams, P. Amendt, D. Eder, Phys. Plasmas, **2** 274 (1995).

13. C. Coverdale, Phd Thesis (1995).

14. A. Ting, K. M. Krushelnick, A. Fisher, C. Manka, H. R. Burris, NRL Report No. 6790-95-7667 (1995).

15. W. L. Kruer and S. C. Wilks, LIRPP proceedings, Monterey Meeting, 105, editor G. Miley, (1994).

16. C. Darrow, D. Klem, private communications; D. Price, private communication.

17. T. -Y. B. Yang, W. L. Kruer, R. M. More, and A. B. Langdon, to be published Phys. Plasmas (1995).

18. G. Bonnaud, to be published in Phys. Rev. Letts. (1995).

19. M. E. Glinsky, R. J. Mason, and M. Tabak, Bull. Am. Phys. Soc. **38**, 2080 (1993).

Large Amplitude Wakefield Excitation and Particle Acceleration in High Density Plasma for Plasma based Accelerator

Yasushi Nishida

Department of Electrical and Electronic Engineering, Utsunomiya University

Utsunomiya, Tochigi 321, Japan

Abstract. Recent progress on plasma based accelerators and related plasma phenomena are reviewed on both laser based accelerators and the results from the microwave plsasma interaction.

INTRODUCTION

High energy particle accelerators demand new scheme for further high energy accelerators with much higher acceleration gradient and thus compact facilities. In the conventional schemes the acceleration gradient could be limited by breakdown within the wave guide facilities which must be employed for adjusting the velocity of the wave to that of the particles. The maximum field strength in this case could be achieved to about 100 MV/m. If we employ the plasma, we don't worry about the breakdown problem and the maximum field strength of the plasma wave is expected theoretically $E_{max} \approx \delta\sqrt{n}$ [V/cm]. Here n is the plasma density in cm^{-3} and δ is the amplitude of the excited wave. For example, if the plasma density n is 10^{18} cm^{-3}, $E_{max} \approx 10^9$ [V/cm] which is about 200-1000 times of the maximum field strength expected on the conventional system.

PLASMA BASED ACCELERATORS

Proposed realistic plasma based accelerators are, typically,

[1] Beatwave Accelerator (BWA), [2] Wakefield Accelerator including (a) laser wakefield accelerator (LWA) and (b) plasma wakefield accelerator (PWA) and [3] Crossfield Accelerator (VpxB accelerator or Surfatron).

Concepts [1] and [2] are the potential acceleration, i.e. the potential difference in the wave trough limits the particle kinetic energy. The VpxB accelerator or Surfatron has a static magnetic field applied perpendicular to the wave propagation direction and the particles are accelerated in parallel to the wave front. In this scheme the self-stabilisation mechanism of the trapped particles in the wave trough exists, and this is an advanced accelerator concept with respect to those of [1] and [2]. Acceleration principle is illustrated in Fig.1.

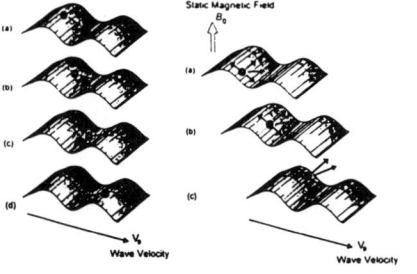

Fig.1. Acceleration principle. (a)Potential acceleration and (b) cross field acceleration.

[1] Beatwave Accelerator

The most mature accelerator scheme developed so far is the BWA. This utilises two electromagnetic wave beams with slightly different frequencies injected into the plasma (1). The two waves satisfy the Manley-Rowe relations, and the lowest frequency excites the plasma wave resonantly, if the frequency matches the intrinsic plasma frequency of the background plasma. Since the velocity of the plasma wave is nearly the speed of light, the plasma wave can be used as a driver wave for the accelerator. The maximum energy gain is limited by the dephasing between a trapped particles and the wave, which occurs over a length L given by

$$L \cong \Delta W / eE = 2\gamma_{ph}^3 / k_0$$

where ΔW is the maximum energy of the particles obtained by the potential well of the wave and $\gamma_{ph} \approx \omega_0 - \omega_p$ is the Lorentz factor measured with wave phase velocity v_p.

The experimental works have been conducted in number of laboratories, including laser-plasma and microwave-plasma interactions. Recent measurements done by UCLA group (2) show that the electrons injected at 2.0 MeV are accelerated to 30 MeV in the acceleration length of 1 cm, indi-

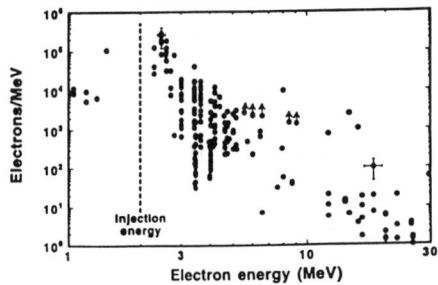

Fig.2. Electron energy spectrum observed in BWA experiment at UCLA (Ref.2).

cating the acceleration gradient of almost 3 GeV/m (Fig.2). In this experiment the wave amplitude of 30% is observed. This shows the promising feasibility of the beatwave based linear accelerator on table-top.

[2] Wakefield Accelerator

Laser Wakefield Accelerator

The beatwave scheme requires two laser beams and a plasma frequency must precisely be tuned to the difference frequency between them. This imposes a uniformity constraint on the plasma density in the acceleration area, or within an acceleration length L. An alternative idea is the wakefield accelerator originally presented by Tajima and Dawson (1). An intense laser pulse with short pulse width in the order of the plasma wave period is injected into a plasma, and leaves a strong wakefield behind it. In order to excite a sufficiently strong wakefield, the laser pulse must be intense enough and be focused into a small region, the length of which is limited by the Rayleigh length. The theoretical maximum energy gain in a plasma accelerator can be achieved if the plasma wave exists over a distance much greater than the Rayleigh length.

Recent preliminary experimental works done by the team of KEK, Osaka Univ., Utsunomiya Univ. and Nagoya Univ. have observed an electron acceleration of about 18 MeV in the length of 0.6 mm, resulting in the acceleration gradient of 30 GeV/m. The acceleration length might be limited by the Rayleigth length.

We also observed wave excitation in the ion wave range by irradiating short pulse microwave (2.86 GHz, pulse width $\approx$ 1-2/fpi) (3). When the short microwave is launched into the nonuniform plasma, the packet of electron plasma wave is excited at the resonance layer, which propagated down with the group velocity of the plasma wave. After shut-off the pulse, large amplitude waves ($\geq$ 30%) are excited to run down with a velocity of about 10 times of the acoustic speed. From the dispersion relation, these waves are interpreted a pseudo wave excited by high energy ion component (i.e. the wakefield excited by ion bunch) and a wakefield excited by a plasma wave packet.

Optical Guiding

Two concepts for overcoming the diffraction of laser light have been proposed: relativistic guiding and plasma channel optical guiding. In relativistic guiding, the relativistic electrons oscillate back and forth with relativistic velocity and the increase of mass reduces the plasma frequency. Since a plasma has a dielectric constant less than unity, the decrease of plasma frequency increase it at the area where the wave intensity is stronger. This results in a system similar to the optical fiber. A total power for effective relativistic optical guiding has a threshold power of

$16(\omega/\omega_p)^2$ GW (4). The change in refractive index depends on the changes both in the electron mass and in the electron density.

If the plasma radial shape is controlled somehow to make minimum density along the axis, the laser light could be trapped over the Rayleigh length because larger refractive index on the axis area. This is again a similar structure of the optical fiber. C.Durfee et al (5) have recently demonstrated such a plasma waveguide by using two laser pulses. The first pulse form a channel for the second. A Nd:YAG laser, $\lambda = 1.064 \mu$ m with final total pulse energy 250mJ, is focused to a spot size 10μ m with 4×10^{14} W/cm^2. The length of the plasma formed by the initial pulse determines the extent to which a second pulse can be guided. An example is shown in Fig.3 (5).

We used an intense microwave to observe the wave guiding. When the plasma density is minimised along the axis by inserting thin polyimide film (1 cm width x 0.1 mm thick x 25 cm length), the microwave makes a hole to go into overdense side and is trapped in the radial direction (see Fig.4).

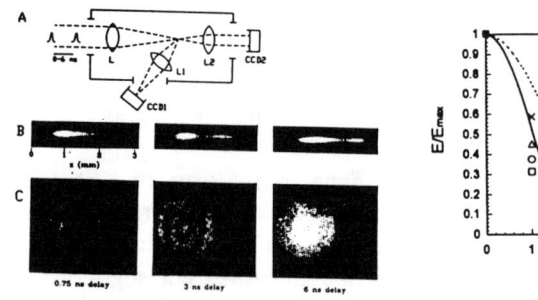

Fig.3. (a)Optical arrangement. (b)Fluorescence images (CCD$_1$) for various delays of pulse 2. (c)Probe patterns (CCD$_2$) for the same delay as (b) (Ref.5).

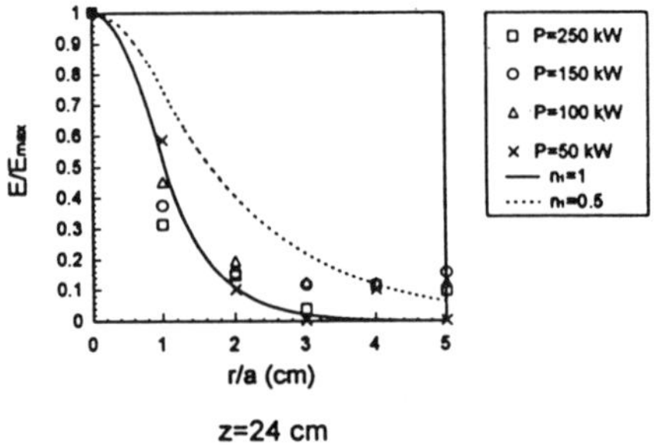

Fig.4. Electric field frofile in radial direction with theoretical results of two cases of plasma density.

Frequency Up-Shifting and Photon Acceleration

It was proposed that a short, intense laser pulse from a plasma by multiphoton ionisation resulting in time-dependent plasma expansion to produce spectral change. This process is called photon acceleration or up-shifting of the incident light spectrum. During the intensity rise of the laser pulse, a plasma begins to form, lowering the refractive index and the blueshift of the incident light results. Such an up-shifting are observed by a laser of 620 nm wave length with 100 fs pulse width in a gas cell (6), and the ionization-induced blueshift of 30 nm is resulted (Fig.5). There is another result of using the microwave associated with ionisation by picosecond lasers (7).

We used an intense microwave (≤ 250 kW) of 9 GHz irradiated in underdense plasma confined in rather small chamber. The microwave makes strong standing wave which expand the plasma very quickly by Ponderomotive force and the blueshift of 1-10 MHz is observed, but no redshift at all (Fig.6).

PROBE SPECTRA:

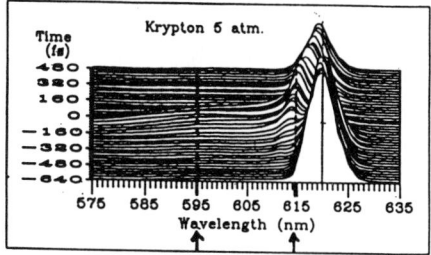
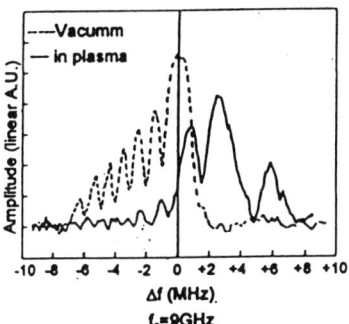

Fig.5. Time-resolved blueshifts measured temporal structure of ionization fronts (Ref.6).

Fig.6. Frequency spectra of microwave with and without plasma. Center frequency is 9.0GHz.

Plasma Wakefield Accelerator

Electron bunch can also be used to excite wakefield (8-10). The excitation mechanism by a bunch of electron is essentially the same as a laser wakefield, instead using electron bunch. The advantage may be a stiffness of the driver bunch, no erosion of the bunch is expected, but that two high energy accelerators are required is disadvantage. The transformer ration, R, is defined as the ratio on an intense bunch at lower energy decelerated while accelerating a dilute higher energy particles. R should be as large as possible and is limited to $2\pi N$ for a triangular pulse, if the rise time is over N-plasma period. When we use the ion bunch instead of electrons, the wakefield is excited in the ion wave range (11) which is useful for understanding the precise physics such as the bunch shape, transformer ratio and so forth.

[3] Crossfield Accelerator

The original acceleration mechanism is proposed by Sugihara and Mizuno (12) and experimentally confirmed by Nishida et al. (13). The accelerator concept in relativistic regime is proposed by Katsouleas and Dawson (14). The trapped particles are stable enough in the wave trough and the radiationloss is almost the same order of magnitude with the convensional linac. In this aspect, this accelertor is advanced concept with respect ot other plasma based accelerators. So far no experimental works in the relativistic regime are reported.

CONCLUSION

Plasma based acceleraors are progressing steadily to realize the GeV-order, table-top accelerators. The high energy physics requires not only the desired particle energy but also a appropriate luminosity. We have a table-top, tera-watt with fs order lasers and the experimental works could be accelerated quickly to show the realistic accelerators. The high power, high frequency (50-100GHz) microwaves also could realize the plasma based accelertors because of much higher efficiency, if the high power, short enogh microwave sources are available.

ACKNOWLEDGMENTS

A part of the present works is supported by the Grant-in-Aid for Scientific Research and for Co-operative Research (A) from The Ministry of Education, Science and Culture, Japan.

REFERENCES

1. Tajima, T. and Dawson, J.M., *Phys. Rev. Lett.* **43**, 267-270 (1979).
2. Everett, M. et al., *Nature* **368**, 527-529 (1994).
3. Nishida,Y., Kusaka,S. and Yugami,N., *Physica Scrip.* **T52**, 65-68 (1994).
4. Sprangle, P. et al., *Appl. Phys. Lett.* **53**, 2146 (1988), *Phys. Rev. Lett.* **64**, 2011-2014 (1990), *Phys. Rev. Lett.* **67**, 2021-2024 (1991).
5. Durfee III, C.G. and Milchberg, H.M., *Phys, Rev. Lett.* **71**, 2409-2412 (1993).
6. Wood, W.M., Siders, C.W. and Downer, M.C., *Phys. Rev. Lett.* **67**, 3523-3526 (1991).
7. Savage Jr., R., Brogle, R., Mori, W. and Joshi, C., *Phys. Rev. Lett.* **68**, 946-949 (1992).
8. Chen,C., Dawson, J.M., Huff, R.W. and Katsouleas,T., *Phys. Rev. Lett.* **55**, 693-696 (1985).
9. Rosenzweig, J.B. et al., *Phys. Rev. Lett.* **61**, 98-101 (1988).
10. Nakanishi, H. et al. *Part. Accl.*, **32**, 203-208 (1990).
11. Nishida,Y. et al., *Phys. Rev. Lett.* **66**, 2328-2331 (1991).
12. Sugihara, R. and Midzuno, Y., *J. Phys. Soc. Jap.* **47**, 1290 (1979).
13. Nishida, Y. et al., *Phys. Lett.* **105A**, 300-303 (1984).
14. Katsouleas, T. and Dawson, J.M., *Phys. Rev. Lett.* **51**, 392-395 (1983).

Self-Modulation of Short Intense Laser Pulse: Opportunities For Wake-Field Accelerator

N.E.Andreev*, L.M.Gorbunov** and V.I. Kirsanov*

*High Energy Density Research Center,
Institute for High Temperatures of Russian Academy of Sciences,
Izhorskaya 13/19, Moscow 127412, Russia*

**Lebedev Physical Institute, Russian Academy of Sciences
Leninskii prospect 53, Moscow 117924, Russia*

Abstract. Comparative analytical and numerical study of the laser pulse self-modulation and the plasma wake-field excitation is introduced for the cases when the self-modulation is triggered by the relativistic self-focusing and when it is initiated by the second frequency-shifted weak-intensity laser pulse. Both cases of a homogeneous plasma and a preformed plasma channel are considered. Basing on the obtained results, the new scheme of stimulation (by a second weak-intensity laser pulse) and maintaining (using a preformed plasma channel) of the strong wake-field excitation is proposed. This scheme gives an opportunity to provide enhanced acceleration at the distances of tens of the Reyleigh lengths and to obtain gigaelectronvolt energy of the accelerated electrons using the present-day laser technology.

1. INTRODUCTION

The concept of laser accelerator (LA) of charged particles in a plasma [1] offers a possibility to design a compact table-top accelerator with multi-GeV accelerating gradient. At present, there are three commonly accepted approaches to excitation of electron plasma waves (EPW) in the LA. These approaches differ mainly in the pulse length L_{pul} (compared to $\lambda_p = 2\pi/k_p = 2\pi\omega_p/v_g \approx 2\pi\omega_p/c$, where ω_p is the electron plasma frequency) and shape of a laser pulse and, consequently, in nonresonant (Laser Wake-Field Accelerator or LWFA, $L_{pul} \ll \lambda_p$) [2], self-resonant (Self-Modulated LWFA or SM-LWFA, $L_{pul} \gg \lambda_p$, initially smooth pulse) [3], or resonant (Plasma Beat Wave Accelerator or PBWA, $L_{pul} \gg \lambda_p$, initially modulated pulse) [1] excitation of a EPW.

Already the first proof-of-principle experiments have shown that all the basic schemes of the LA can provide the rate as high as GeV/m (or even several tens of GeV/m [4]). To obtain not only a high acceleration rate but also to reach multi-

GeV energy of the accelerated particles, one needs to have a long-distance transportation (over many the Rayleigh length, $Z_R = \frac{1}{2}(k_0 L_{10}^2)$) of laser pulses in plasmas without the defraction spreading.

From the standpoint of practical realization, the most simple opportunity to prevent the diffraction spreading of a pulse is parabolic channels of a plasma density, which have been experimentally realized and studied by Dufree and Milchberg [5]. Such channels were first suggested for guiding of short pulse (LWFA) by Esarey and collaborators [6] and for guiding of a longer pulse (SM-LWFA, where the pulse modulation was stimulated by a second, weak-intensity, frequency-shifted laser pulse) by Andreev and collaborators [7].

Here we report the results of the pulse self-modulation in a homogeneous plasma and plasma channels. Peculiarities and advantages of this approach to EPW excitation for particle acceleration are discussed. A comparative study of the basic accelerating schemes for the channel-guided pulse propagation is carried out.

2. BASIC EQUATIONS

Our study is based on the set of equation for an envelope of an axisymmetric laser pulse ($a = (eE/m\omega_0 c)$) and the hydrodynamic equation for the normalized electron density perturbations ($n = \delta n / N_0$) in a rarefied ($\lambda_0 \ll \lambda_p$) plasma, which transverse density profile is either a parabolic $N = N_0[1 + (r/R)^2]$ (R is the channel radius) or a homogeneous ($R \to \infty$)[3, 7]

$$(2i\omega \frac{\partial a}{\partial t} + 2c\Delta_\perp a + \frac{\omega_p^2}{4}|a|^2 a = \omega_p^2 a[n + (r/R)^2] \tag{1}$$

$$v_g^2 \frac{\partial^2 n}{\partial \xi^2} + \omega_p^2 n = \frac{c^2}{4}\Delta|a|^2 \tag{2}$$

Here, it is also assumed that $R\omega_p / c \gg 1$ and $|a| < 1$. Below, we consider the pulse with initial Gaussian shape in both longitudinal and transverse directions: $a = a_0 \exp(-\xi^2 / L^2 - r^2 / L_{10}^2)$.

For our analysis we also use equations (1) and (2) reduced (in the para-axial approximation) to the equation for a normalized pulse spot size $f = L_\perp(\xi,t)/L_{10}$ [9,10].

3. SM-LWFA IN A HOMOGENEOUS PLASMA

In the standard SM-LWFA, a laser pulse, which is not modulated initially, penetrates into a homogeneous underdense ($\omega_p \ll \omega_0$) plasma and, because of the strong resonance modulation instability (that can be either 1D [8] or 3D [9] dimensional in nature), a modulation of the pulse (with wavelengh λ_p) develops. As a result, the pulse generates a EPW resonantly (during the time that is sufficient for particle acceleration, $t > T_a = \lambda_0 \gamma^3 / 2$) in the same manner as a two-frequency pulse in the PBWA scheme. Seed perturbations needed for instability onset can be provided either by a sharp leading edge of the pulse or can be simulated by the second weak-intensity, frequency-shifted pulse, or even can arise due to steepening of the pulse longitudinal profile during its relativistic self-focusing (SF). In a homogeneous plasma, for laser pulses with duration of several plasma periods self-modulation (SM) develops for

$$P_0 \gtrsim P_c \approx 16.2(\lambda_p/\lambda_0)^2 \, 10^9 W = 16.2\gamma^2 10^9 W \qquad (3)$$

In the SF stage before the pulse SM occurs, the solution of the equation for f (ignoring the ponderomotive nonlinearity) predicts the growth in the longitudinal gradients of the on-axis field of the pulse [10]. As soon as the typical scale of the on-axis variations of the pulse-amplitude becomes comparable with λ_p, the excitation of the EPW occurs that acts as a seed for SM of the pulse. The time t_m needed for triggering of the SM is of order Z_R/c (see Fig. 1). The character of the approximate analytical dependencies for this time [10] (for $P_{0m} \approx P_c$, $t_m \propto (Z_R/c)(k_p L)$ and, for $P_{0m} > P_c$, $t_m \propto (P_{om}/P_c - 1)^{-1/2}$), correlates with the results of the numerical calculations [using (1) and (2)] shown in Fig.1.

Our results also indicates that ponderomotive nonlinearity can also be important for this process because during the relativistic SF the size of the focal sport can become comparable with λ_p. Figure 2 shows the evolution of peak-pulse and EPW amplitudes for parameters close to those of simulations [4]. Solid curves corresponds to both relativistic and ponderomotive nonlinearities incorporated. Dots show the same, but when only 1D EPW excitation taken into account [this corresponds to neglecting of the transverse derivatives in the Laplas operator on the right-hand side of (2)]. Even for a sufficiently wide (here, $k_p L_{10} = 20$) laser pulse, when the relativistic SF trigger SM, the ponderomotive nonlinearity is also important determining duration of the SM stage and efficiency of EPW excitation.

For the standard SM-LFWA, condition (3) can be treated as a limitation on γ (and, consequently, on the energy gain of accelerated particles). From diagram in Fig.3 we can see that, for $P = 10TW$, the SM is limited by $\gamma < 25$, i.e., the corresponding possible particle energy gain is less than 1 GeV. In the same figure, for comparison, dashed curves show the regions where the diffraction limitation of the on plasma wave excitation region is important (this is introduced here as $P_0 \gtrsim 5.6 a_0^2 \gamma^3 10^9 W$). Physically, only the parts of dashed curves below the solid give the correct prediction because the expression for their above parts should be corrected with taking into account the relativistical guiding.

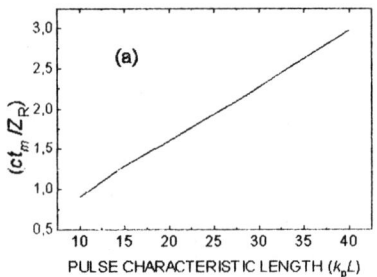 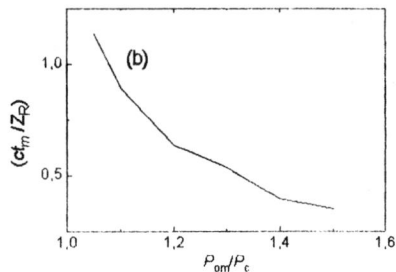

FIGURE 1. Numerical results for time t_m needed for the SF to trigger excitation of EPW (up to $n = 0.025$) for $\gamma = 22.5$ and $a_0 = 0.27$. (a) shows t_m for fixed $P_{0m} \approx P_c$ and $k_p L_1 = 20$, versus $k_p L_1$; (b) shows t_m versus P_{0m}/P_c when initial $k_p L = kL_1 = 20$ are fixed.

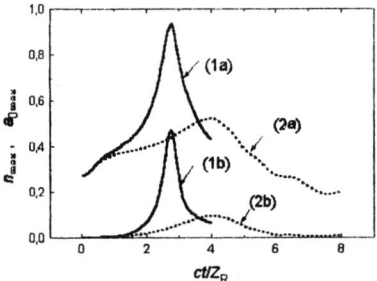 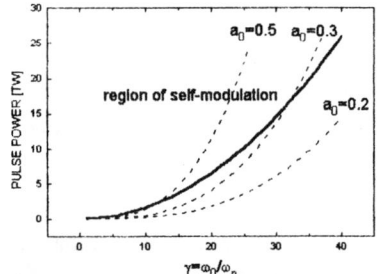

FIGURE 2. For $P \approx P_c$, solid curves show (1a) the maximum of the laser field and (1b) amplitude of excited EPW as functions of time. Dotted curves show the same when ponderomotive nonlinearity is neglected. ($\gamma=22.5$, $k_p L = k_p L_1 = 20$, $a_0=0.27$)

FIGURE 3. In the plane (P,γ), the region corresponding to SF and SM of a smooth-pulse is above the solid curve and those corresponding to the linear diffraction limitation, $L_a < 2Z_R$, are below dashed curves (for $a_0 = 0.2$, 0.3, and 0.5).

4. PLASMA CHANNELS AND SM-LWFA

Sufficiently long and smooth pulse can propagate in a parabolic channel with no modulation over many Z_R. Because of this, to stimulate the modulation of the pulse (with $P_{0m} < P_c$), we suggest to use a small-amplitude second pulse with frequency shifted to ω_p with respect to ω_0 [7].

We studied the excitation of EPW and its dependence on the amplitude of the second pulse using equations (1) and (2) [7, 10, 11]: the SM behaves typically for an instability and (i) the duration of the stage of EPW excitation is sufficiently long (Fig. 4) compared to T_a; (ii) t_m and the peak EPW amplitude depends weakly on the amplitude of the second pulse (Fig 4); (iii) excitation of EPW is insensitive to

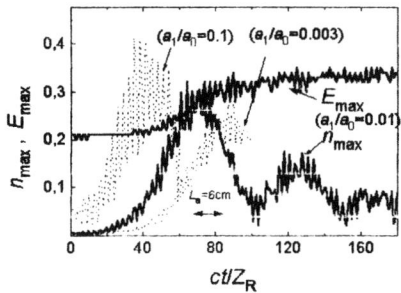 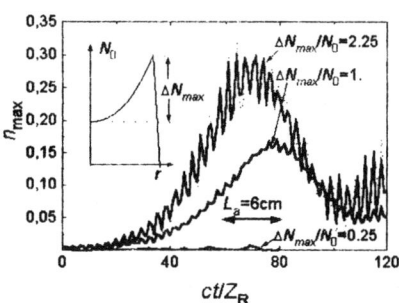

FIGURE 4 Evolution of E_{max} and n_{max} for SM of a laser pulse in a channel initiated by a frequency-shifted second laser pulse with amplitude $a_1 = 0.01 a_0$ (solid curves) and n_{max} for $a_1 / a_0 = 0.1$ and 0.003 (dots).

FIGURE 5. EPW excitation in a leaky channel. Solid curves are for depth of the channel $\Delta N_{max} / N_0 = 0.25, 1, 2.25$. Dots corresponds to the parabolic channel. Other parameters are the same as in Fig.4.

the sign of the frequency shift and only weakly depends on the detuning of this shift from the value of ω_p [11].

Our simulation have shown that the EPW excited in the channel pulsates (see Fig. 4). These pulsations are caused by the pulsations of the laser field, which cannot be eliminated (because of the variation in the laser field along the pulse) by an adjustment of the pulse-channel parameters.

Using equation for f [9,10], we can obtain expressions [10] for pulsations of f in the channel $f^2(\xi,\eta) = 1 + \left[\Omega^{-2}\left(1 - P_0(\xi)/P_c\right) - 1\right]\sin^2(\Omega\eta)$, the period of these pulsations of $T_p = \pi\Omega^{-1}\left(\frac{1}{2}\left(k_0 L_{\perp 0}^2\right)\right) = \pi\left(\lambda_p / \lambda_0\right) R / c$, and the channel radius corresponding to the minimal pulsation: $R = \frac{1}{2}\left(k_p L_{\perp 0}^2\right)\left(1 - P_{0m}/P_c\right)$.

Deviation of the channel shape from a parabolic, when a parabolic growth of the plasma density with distance from the axis changes to a rapid decrease [5], can be of importance for realization of the channel guiding because the part of laser radiation can escape through the side boundary (leaky channel). For this case, the results in Fig.5 show that, for a channel-density depth larger than the value of the on-axis density, this effect is not pronounced during t_m and T_a.

5. COMPARISON OF LASER ACCELERATING SCHEMES IN CONFIGURATION WITH A PLASMA CHANNEL

The study of EPW excitation [11] shows that it is virtually possible, for the basic LA schemes, to obtain multi-GeV energies of the accelerated electrons using the present-day terawatt lasers and channel-guided pulse propagation.

However, because the production of the plasma channel with a given parameters is a technically difficult problem, the effect of variations of the channel and plasma parameters on the excitation of the plasma wave is to be studied.

We studied analytically and numerically [11] the effect of mismatch between the focal spot of the laser pulse ($2L_{10}$) and the channel width ($2R$) on the pulse pulsations and the pulsations of the wake EPW and found that it can have strong effect on the resonant excitation (either for the PBWA or SM-LWFA). However, it turns out that, for the laser peak power $P_{0m} \ll P_c$, in underdense plasmas, the pulsations in a fairly narrow channel in a low-density plasma have the period T_p sufficiently less than the acceleration time T_a and only the field averaged over pulsations is important for the particle acceleration. Because of this, for all basic accelerating schemes, it is not necessary to minimize the pulsations in the channel by a precise adjustment of the laser — channel parameters.

Our simulations for PBWA in a channel, revealed a typical resonance curve (for 1J energy in two-frequency, 1 ps laser pulse, i.e. for the same parameters as in simulations [7]) that allowed less than ten-percent detuning of the on-axis plasma density. In contrast to this, for the SM-LWFA, even forty percent density variations have not changed substantially the wake amplitude (which was approximately twice less than that maximum in PBWA case for a pulse of the same energy), though the delay in reaching the maximum of wake-field amplitude grows as the detuning increases (for the 20 and 40 percent detunings, the delay was 10 for 35 percent, correspondingly). The plasma wave excitation in the LWFA also was weakly affected by the on-axis density detunings, but to have the same amplitude of the plasma wave, as in the case of the SM-LWFA, almost an order increase in the laser power was necessary.

Our results indicate that, for $\gamma < 20$, SM-LWFA in homogeneous plasma (in our example, for 1.25J in 1ps pulse, $\Delta W_a = 0.2\text{GeV}$ on 0.6cm), for greater γ, SM-LWFA in a channel (in our example, for 1,2 J in 1.2ps pulse and channel with the radius 60μm and depth greater than on-axis density $n_0 = 4 \; 10^{17}$ cm^{-3}, $\Delta W_a = 1\text{GeV}$ on 6cm), require a moderate pulse power and can provide an enhanced acceleration, i.e., can be considered as the best candidates for the design of the LA.

REFERENCES

1. T.Tajima and J.M.Dawson, Phys. Rev. Lett. **43**, 267, (1979).
2. L.M.Gorbunov and V.I.Kirsanov, Sov. Phys. JETP, **93**, 260 (1987)]. P.Sprangle, E.Esarey, A. T Ting, and G.Joyce. Appl. Phys. Lett., **53**, 2146, (1988).
3. N.E.Andreev, L.M.Gorbunov, V.I.Kirsanov, et al., JETP Lett., **55**, (10), 571 (1992).
4. K.Nakajima, D.Fisher, T.Kawakubo, et al., (submitted to Phys. Rev. Lett., 1995)
5. G.G. Dufree III and H.M.Milchberg, Phys. Rev. Lett. **71**, 2142 (1993).
6. E.Esarey, P.Sprangle, J.Krall, et al., Phys. Fluids **B5**(7), 2690 (1993).
7. N.E.Andreev, L.M.Gorbunov, V.I.Kirsanov, et al., JETP Letters, **60**(10), 713 (1994).
8. T.M.Antonsen, Jr. and P.Mora, Phys. Rev. Lett. **69**, 2204, (1992). N.E.Andreev, et al., Physica Scripta **49**, 101 (1994). W.B.Mori, et al., Phys. Rev. Lett. **72**, 1482 (1994). A.S.Sakharov and V.I.Kirsanov, Phys..Rev. E **49**, 3274 (1994)
9. E.Esarey, J.Krall, and P.Sprangle, Phys. Rev. Lett. **72**, 2887 (1994)
10. N.E.Andreev, L.M.Gorbunov, and V.I.Kirsanov, Phys. Plasmas **2**(6), (1995) (to appear).
11. N.E.Andreev, L.M.Gorbunov, and V.I.Kirsanov, Fiz. Plazmy **21**, (1995) (to appear).

Plasma-Core Induced Self-Guiding of an Ionizing Ultrashort Pulse in Gases

A.V.Kim*, D.Anderson+, M.Lisak+, V.A.Mironov*, A.M.Sergeev*, L.Stenflo++

*Institute of Applied Physics of the Russian Academy of Sciences
46 Ulyanov St., Nizhny Novgorod 603600, Russia
+Chalmers University of Technology, S-41296 Gothenburg, Sweden
++Umea University, S-90187 Umea, Sweden

Abstract. The propagation of high-power laser radiation producing rapid ionization in dense gases is analyzed by using a nonlinear electrodynamical model. It is shown that the interplay between the Kerr-type and the defocusing ionization nonlinearities may lead to the plasma-core induced self-guiding effect of ultrashort laser pulses. Steady-state self-consistent laser-plasma structures are considered.

INTRODUCTION

The creation of long plasma channels by means of optically field-induced ionization and the propagation of short, high-intensity laser pulses within these plasmas is a problem that has important implications for XUV laser development and laser-driven particle acceleration (1). It may also play a fundamental role in the process of significant frequency spectrum transformation of intense laser pulses that are focused into dense gases (1). In this paper, we examine the effect of laser channeling due to the creation of self-sustained plasma-field structures in the presence of strong ionization of a dense gas with a Kerr-type atomic nonlinearity. It seems to be quite paradoxical that the ionization which usually has a strong defocusing effect on the radiation in the created plasma region (2,3) may be the origin of waveguiding. Nevertheless, recent experimental results (4,5), where short-pulse self-channeling has been observed over distances of several Rayleigh lengths, showed that the presence of a strong ionization does not destroy the guiding effect. The aim of this paper is to demonstrate that the interplay between the Kerr-type focusing nonlinearity and the threshold-type ionization nonlinearity may result in the formation of steady-state plasma-field structures where the guiding along the created plasma core is attained owing to the gas ionization. The combined effect of focusing and defocusing nonlinearities (heating and ionization, Kerr-effect and ionization) has been studied theoretically in (6,7). However, the attention in Refs.

(6,7) has been paid to the competition between these nonlinearities rather than to their coexistence. Here, we shell analyze the opportunity to create a long quasi-steady-state waveguide, aiming at an increase of the laser-plasma interaction length for the above mentioned applications.

BASIC EQUATIONS

We start the analysis by first considering ionization processes which occur in dense gases in the presence of laser radiation. The evolution of the plasma electron density, $N(r,t)$, is described by the ionization balance equation

$$\frac{\partial N}{\partial t} = \gamma(|E|)N_m \qquad (1)$$

where the ionization rate γ is a function of amplitude of the electromagnetic field, $|E|$, N_m is the neutral gas density. We have omitted the dependence of γ on the phase of the field which may be essential for energy balance in the case of linearly polarized radiation (8,9,) but which is not important for the spatial field structure. Our analysis neglects other mechanisms of electron density change, e.g. ponderomotive and thermal effects which are usually investigated in the case of already created fully ionized plasmas. During a short laser pulses (<1ps), these nonlinearities can not significantly affect the refractive index in comparison with the rapid ionization that enhances the electron density by many orders of magnitude. The ionization rate function is characterized by a sharp dependence on the field intensity, which is experimentally treated as the existence of an ionization threshold. Theoretically, the rate of tunneling ionization γ is usually approximated by an exponentially growing function of $|E|$, (10). In order to be able to describe the plasma structure analytically we will adopt the following simplified model for the function $\gamma(|E|)$:

$$\gamma(|E|) = \begin{cases} \infty, & |E| > E_{th} \\ 0, & |E| \leq E_{th} \end{cases} \qquad (2)$$

where E_{th} is the field threshold value for ionization which depends on the sort of gas and on the laser radiation parameters such as carrier frequency and field polarization. The infinite value of the ionization rate in Eq.(2) means that, for a field exceeding locally the threshold value, the electron density immediately becomes high enough (with respect to the pulse duration) to stop a further field growth due to the refraction. We assume also that the gas pressure is sufficient to provide the required number of free electrons. Thus, in the model of instantaneous threshold-type ionization the electric field turns out to be self-limited at the threshold value inside the plasma region. Computer simulations with more sophisticated ionization models (2,3) have demonstrated similar effects of self-limitation. Formally, Eqs.(1),(2) imply that, in the presence of the laser pulse and under the condition of self-limitation, a quasi-steady-state of the plasma can be described by

$$N = \begin{cases} N(r), & |E| = E_{th} \\ 0, & |E| < E_{th} \end{cases} \quad (3)$$

This expression is obviously valid for the bulk of the pulse and does not hold near the front where the plasma density remains non-zero in the region of evanescent field.

The neutral component of the gas is responsible for the Kerr effect which arises due to the nonlinear motion of bound electrons in the upper atomic shells. The bound electron nonlinear response is local in time (at least for the available laser pulse duration > 10fs) and follows the intensity of the optical field. An atom subjected to ionization usually does not contribute to the Kerr effect because of the essential difference between the atomic and ionic nonlinear polarization. We consider rather dense gases where a small ionization degree is sufficient to prevent the field growth due to the Kerr-effect self-focusing. Because the bulk of the atoms remains neutral, the characteristic Kerr nonlinearity field can be taken as being independent of the free electron density. At the same time, the contribution of the ionization nonlinearity to the change of the gas refractive index is significant even at a small ionization degree due to a strong polarization response of detached electrons as compare to the bound ones. Hence, we have two nonlinearities of different signs and of different dependencies on the laser intensity.

Let us now proceed by deducing the spatial structure of the electromagnetic field in the emerging plasma. Our aim is to demonstrate the guiding effect keeping in mind that in the reference frame connected with the propagating laser pulse the field distribution is quasi-stationary and localized in the transverse direction. Note that the very leading low-intensity part of the pulse, that obeys the linear diffraction law, can not be guided and therefore it is excluded from the consideration. The conditions which facilitate the implementation of the guiding effect are a weak radiation focusing and a large beam aperture geometry. Considering the propagation of an s-polarized electromagnetic wave, we use the nonlinear Helmholtz equation to describe the evolution of the slowly varying amplitude of the electric field $E(r) exp(i\omega t)$. Thus

$$\nabla^2 E + k^2 \left(\varepsilon + \alpha |E|^2 \right) E = 0 \quad (4)$$

where $k = \omega/c$ is the laser wave number in vacuum, α is the coefficient of the Kerr nonlinearity, $\varepsilon = 1 - (1 + i\nu/\omega) N/N_c$ is the plasma permittivity, ν is the collision frequency, $N_c = m(\omega^2 + \nu^2)/4\pi e^2$ is the critical electron density.

The equation system (3),(4) describes consistently the following electromagnetic problem: with a given source of electromagnetic radiation we have to find an electron density distribution such that inside the plasma region, where $N(r) \neq 0$, the electric field amplitude is equal to the threshold value for ionization; outside the plasma, where $N(r)=0$, the field amplitude is below the threshold.

We express the electric field as $E = A\,exp(i\varphi)$ and separate the real and imaginary parts of Eq.(4) to obtain

$$\nabla\left(|A|^2 \nabla\varphi\right) = \delta n |A|^2 \tag{5}$$

$$\nabla^2 A - (\nabla\varphi)^2 A + \left(1 - n + \alpha|A|^2\right)A = 0 \tag{6}$$

where we have introduced the dimensionless variables $kr \to r$, $N/N_c = n$, and $\nu/\omega = \delta$. Substituting Eq.(3) into Eqs.(5), (6) yields the following relations:

(i) Inside the plasma region, where A is equal to the constant E_{th}, we have

$$\nabla^2 \varphi = \delta n \tag{7}$$

$$(\nabla\varphi)^2 = 1 - n + \alpha E_{th}^2 \tag{8}$$

(ii) Outside the plasma region, where $n=0$ and $|A| < E_{th}$, one obtains

$$\nabla\left(|A|^2 \nabla\varphi\right) = 0 \tag{9}$$

$$\nabla^2 A + \left(1 - (\nabla\varphi)^2 + \alpha|A|^2\right)A = 0 \tag{10}$$

At the plasma boundary the continuity conditions for the field amplitude and its derivative must be satisfied.

STEADY-STATE STRUCTURES

The equation system (7)-(10) together with the boundary conditions describes all stationary plasma structures which can be produced by powerful laser radiation in a dense gas. However, to obtain analytical results which clearly demonstrate the physical significance of the problem we will restrict the analysis by considering most simple field configuration. Focusing our attention on a two dimensional model in which the light field propagates along the z-axis, we look for solutions of Eqs.(7)-(10) in the form $A = y^o A(x)$, $\varphi = -hz + \phi(x)$ and $n=n(x)$, where h is the wave number along the direction of propagation.

Inside the plasma region Eqs.(7) and (8) then reduce to

$$\frac{d}{dx}\sqrt{n_m - n} = \delta n \qquad (11)$$

where $n_m = 1 + \alpha E_{th}^2 - h^2$. Integrating (11), we obtain

$$n(x) = \begin{cases} n_m \sec h^2(\delta n^{1/2} x), & |x| \leq d \\ 0, & |x| > d \end{cases} \qquad (12)$$

where d is the plasma transverse dimension, normalized by k. Outside the plasma Eq.(9) yields $A^2 (d\phi/dx) = J = const$, i.e. the transverse power flux density is non zero due to the plasma absorption. Substituting this into Eq.(10), we find

$$\frac{d^2 A}{dx^2} - \frac{J^2}{A^3} + (1 - h^2 + \alpha A^2) A = 0 \qquad (13)$$

Let us first analyze a simplified but instructive case in which absorption effects can be neglected, i.e. $\delta = \nu / \omega = 0$. Then Eq.(12) yields $n(x) = n_m = const$ for $|x| \leq d$; which describes a homogeneous plasma channel. In the region outside the plasma, it follows from Eqs.(9), (10) that J=0 and the localized solution of Eq.(13) is given by

$$A(x) = \sqrt{\frac{2(h^2 - 1)}{\alpha}} \sec h\left(\sqrt{(h^2 - 1)}(x - d)\right), \quad |x| \geq d \qquad (14)$$

where $h > 1$. Applying the continuity conditions at $x = d$, we find that $A(x = d) = E_{th}$ which gives $h = \sqrt{1 + \alpha E_{th}^2/2}$. Thus, the self-sustained filament has the following structures:

$$n(x) = \begin{cases} \alpha E_{th}^2 / 2, & |x| \leq d \\ 0, & |x| > d \end{cases} \qquad (15)$$

$$A(x) = \begin{cases} E_{th}, & |x| \leq d \\ E_{th} \sec h\left(\sqrt{\alpha E_{th}^2/2}(x - d)\right), & |x| > d \end{cases} \qquad (16)$$

The plasma dimension, d, is determined by the total incident beam power, $P = \int_{-\infty}^{\infty} A^2 dx$. It follows from (16) that $d = (P - P_{th})/2E_{th}^2$, where $P_{th} = 2\sqrt{2E_{th}^2/\alpha}$. Thus, to create a plasma filament defined by (15),(16) it is of course necessary to exceed the threshold field for ionization by focusing. In addition the beam power has to be larger than the threshold power value P_{th}.

In the presence of absorption effects ($\delta \neq 0$) a longitudinally unlimited plasma filament supported by a localized wave field cannot exist. Due to absorption of laser radiation, the waveguide plasma has a limited longitudinal size, and the supporting field structure will give rise to a transverse energy flux across the plasma boundary. Since $J \neq 0$, Eq.(13) has only periodic solutions. Then, depending on the normalized wave number value, h, two cases can be distinguished. If $h<1$, the value of h can be fixed and defined by the external conditions of the problem (unlike the above analysis where h has been determined by the continuity of the field). A similar problem but without local nonlinearity ($\alpha = 0$) has been solved in Ref.(11). It can been shown that the effect of $\alpha \neq 0$ does not change the solution qualitatively. If $h>1$, the local nonlinearity plays an important role as it is responsible for the existence of localized plasma structures. Note that, in experiments, the energy flow in the transverse cross-section supporting the dissipative filament can be produced by using the axicon lens focusing.

ACKNOWLEDGMENTS

This work was supported in part by the International Science Foundation and the Russian Basic Science Foundation.

REFERENCES

1. See, for example, the special issue of *IEEE Trans. on Plasma Sci.* **21**, *No*1 (1993).
2. Rankin, R., Capjack, C.E., Burnett, N.M., and Corcum, P.B., *Opt.Lett.* **16**, 835-837 (1991)
3. Leemans, W.P., Clayton, C.E., Mori, W.B. et al., *Phys.Rev.* A **46**, 1091-1105 (1992).
4. Sullivan, A., Hamster, H., Gordon, S.P. et al., *Opt.Lett.* **19**, 1544-1546 (1994).
5. Braun, A., Korn, G., Liu, X. et al., *Opt.Lett.* **20**, 73-75 (1995).
6. Litvak, A.G., Mironov, V.A., Fraiman, G.M., and Yunakovskii, A.D., *Sov.J.Plasma Phys.* **1**, 31- 40 (1975).
7. Liu, X., and Umstadter, D., in *Short Wavelength V: Physics with Intense Laser Pulses*, Vol.17, 1993 OSA Technical Digest Series (Optical Society of America, Washington D.C.), pp.45-47.
8. Corcum, P.B., Burnett, N.M., and Brunel, F., *Phys.Rev.Lett.* **62**, 125-128 (1989).
9. Gildenburg, V.B., Kim, A.V., and Sergeev, A.M., *JETP Lett.* **51**, 104-107 (1990).
10. Keldysh, L.V., *Sov.Phys.JETP* **20**, 1307-1312 (1965).
11. Gildenburg, V.B., and Golubev, S.V., *Sov.Phys.JETP* **40**, 46-49 (1974).

Anomalous Attenuation of Laser Light due to Ion Acoustic Decay Instability (IADI) in Large Scale Plasma relevant to Laser Fusion

Katsu Mizuno, P. E. Young*, and K. Estabrook*

Univ. of Calif., Livermore-Davis & PPRI
L-418, Lawrence Livermore National Laboratory, PO Box 808, Livermore, CA 94551 USA

*Univ.. of Calif., Lawrence Livermore National Laboratory, PO Box 808, Livermore CA 94551 USA

The IADI causes anomalous absorption of laser light with a moderate intensity laser in a large scale plasma. Because of the weak density modification due to the moderate intensity laser, the instability width can be large. The anomalous attenuation length of the laser can be much smaller than the instability width along the density gradient, indicating that most of laser energy may be anomalously absorbed by the IADI. An instability width of only a few laser wavelengths is sufficient to absorb most of the laser energy. Experimental results are consistent with the large scale plasma and the strong absorption.

INTRODUCTION

The Ion Acoustic Parametric Decay Instability (IADI)[1], in which an electromagnetic wave decays into an electron plasma wave (epw) and an ion acoustic wave (iaw) near the critical density n_c (where the electromagnetic wave frequency equals the plasma frequency), is a fundamentally important subject in plasma physics. It has been studied by numerous authors in laser plasma interactions[2-3], microwave experiments[4], and ionospheric studies. One of the main issues of the IADI in laser produced plasmas is to understand whether or not it is important in the large scale plasmas relevant to laser fusion. The IADI is important because significant hot and/or warm electron heating can occur even when it is relatively weak, if the unstable volume is large enough[3]. If it is excited, it has many important applications such as hot and warm electron heating, anomalous enhancement of lateral heat transport, and anomalous DC resistivity.

When electromagnetic wave excites an epw, the energy is deposited in hot/warm electrons. Previously, hot electron heating was calculated[5] in a plasma irradiated by a high intensity laser ($I\lambda^2 \sim 1 \times 10^{15} \sim 10^{18}$ W-$\mu m^2/cm^2$). The IADI was excited at a steep plasma. The plasma wave amplitude was large, and the instability width was small. The self-consistent plasma density was steep, and the instability was localized in a small region. Because of the strong excitation of the epw, a significant amount of hot electrons was heated even in the small heating region. It was thought relatively easy to avoid the hot electron heating since a high intensity laser were needed to heat them. In contrast, we find that the IADI threshold is quite low and reaches homogeneous plasma collisional values in a laser produced large scale plasma.[3] Because of the weak laser intensity, the epw is excited only moderately. However, the density profile modification is also weak, and the plasma density gradient is gentle near the instability region, implying that the instability width can be large. The electron heating is mostly determined by the product of the plasma wave amplitude, and the instability width (of the interaction time of the electron with the instability). For high intensity laser, we have a large amplitude epw, and a small instability width. On the other hand, for moderate intensity laser, we have a moderate amplitude epw, and a large instability width. An important point is that a relatively weak instability in a large region can heat (hot and warm) electrons as much as a strong instability heats them in a small region. When the IADI is excited on a shallow long scale length plasma, relatively low intensity laser can anomalously heat electrons.

EXPERIMENTAL ARRANGEMENTS

The experiments were performed using the Janus Laser facility at the Lawrence Livermore National Laboratory, and GDL at the Laboratory for Laser Energetics, University of Rochester. Laser wavelength λ_L = 1.06 µm and 0.53 µm, the laser pulse length τ_L = 1.0 nsec, and the maximum laser energy 200 J. The laser intensity I_L was varied from 10^{12}-3×10^{15} W/cm^2 by controlling the laser energy, and the spot size independently. The laser normally irradiates a planar CH target. The laser light was focused through an f/2 (or f/3) lens onto the target. The target was thick enough (50µm) that no burn through was observed. We measured the emission spectrum near the second harmonic ($2\omega_L$) of the incident laser, which was collected at 135° and 180° from the axis of the incident laser, and in the plane of the laser electric field.

EXPERIMENTS
Second Harmonic Spectrum

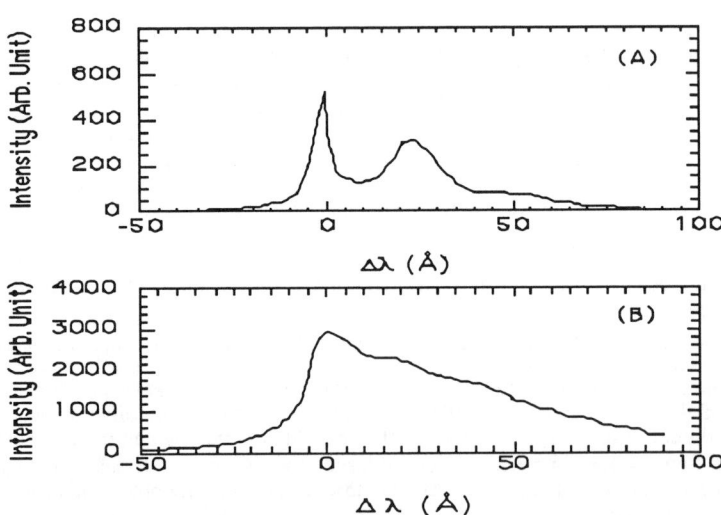

FIGURE 1. Second harmonic spectrum: (A) I_L = 3 x 10^{13} W/cm^2, and (B) 2×10^{14} W/cm^2.

We have measured emission spectrum near the second harmonic of laser light. The IADI excites the epw and the iaw, which satisfy the relation $\omega_L = \omega_{epw} + \Omega_{iaw}$, where ω_L, ω_{epw}, and Ω_{iaw} are the frequencies of the laser light, the epw and the iaw. Two electron plasma waves ($\omega_{epw} = \omega_L - \Omega_{iaw}$ and $\omega_{epw'} = \omega_L - \Omega_{iaw'}$) coupled to produce an electromagnetic wave emission. The frequency is approximately $2\omega_{epw} \approx 2\omega_L - 2\Omega_{iaw}$. In our previous paper, we reported the IADI near the threshold, when it was excited weakly. When the IADI is excited weakly near the threshold, we saw well defined Stokes peaks. Figure 1 shows the second harmonic spectra for two laser intensities; a weak intensity slightly above the IADI threshold, and a moderately high intensity. The right-hand signal is a Stokes signal emitted from the electron plasma waves excited by the IADI. A sharp well-defined Stokes mode is excited with the weak intensity laser irradiation (Fig. 1 (A)). The peak appears near the Landau damping cutoff of the epw at $k\lambda_{De} \sim 0.23$. When the incident laser intensity increases, the spectral shape is quite different from one shown in (A). The spectrum becomes broad (Fig. 1 (B)), and the original Stokes peak is now barely distinguishable. The spectral broadening of the Stokes mode is of interest. It may be relevant to the degree of turbulence of the epw.

Second Harmonic Emission versus Laser Spot Size

As shown in the previous paper[3], the IADI shifted from convective-loss regime to uniform plasma regime when laser spot size increased. The IADI can be excited in a large volume. Hence, we expect to see the increase of the IADI emissions. We have seen that the Stokes intensity increases with the laser spot size. For the small spot of the diameter D=100 μm, no Stokes spectrum was seen because the laser intensity is lower than the IADI threshold. When the laser spot size is large enough that the laser intensity is above the IADI threshold values, the Stokes signal increased with D. In order to keep the laser intensity constant, we increased the total incident laser energy with increasing D. The normalized Stokes signal increased with the laser spot diameter D.

ANOMALOUS ABSORPTION OF LASER LIGHT

Anomalous Collision Frequency Estimated from One-Dimensional Particle in Cell (PIC) Computer Simulation

An important parameter to characterize anomalous absorption of laser light is anomalous collision frequency υ^*. The anomalous collision frequency is the heating rate of electrons by electron plasma wave. We have the definition of the anomalous collision frequency.

$$\upsilon^* = (dT/dt)(\omega_0^2/\omega_{pe}^2)(8\pi/E_L^2) \qquad (1)$$

We estimated the υ^* by measuring the temporal increase of the total plasma energy density, dT/dt, using one dimensional electrostatic PIC computer simulation code, where E_L is laser electric field. Figure 2 shows the υ^* vs the local laser intensity I (laser intensity at $0.9\, n_c$) for a fixed plasma density, $n/n_c \sim 0.9$, which is the mid point between $0.8\, n_c$ and n_c, and it is slightly above the $0.86\, n_c$, where the IADI is most unstable. We plotted the υ^* versus the normalized laser intensity $I\lambda_L^2/T_e$. The swelling effect of the laser light was included. The υ^* increases with $I\lambda_L^2/T_e$ until the shifting point intensity of 3×10^{14} W-μm²/cm²keV, and above the intensity ν^* increases slowly. In the moderate (laser) intensity regime below the shifting point, the anomalous collision frequency scales as

$$\frac{\nu^*}{\omega_L} \approx 4.1 \times 10^{-10} \times \left(\frac{I\lambda_L^2}{T_e}\right)^{0.56} \tag{2}$$

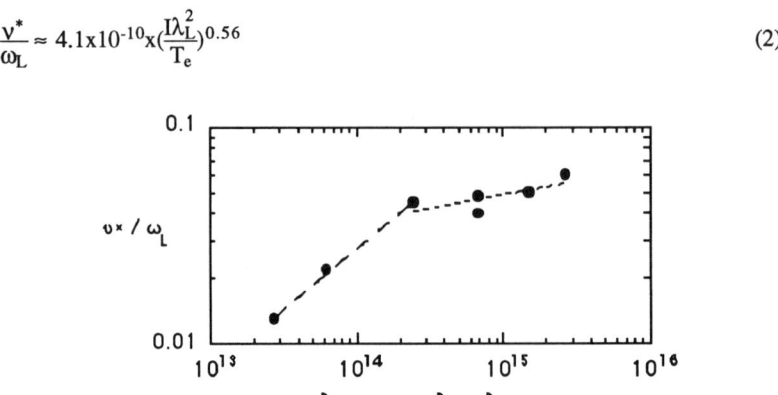

FIGURE 2. Anomalous collision frequency vs. the normalized laser intensity $I\lambda_L^2/T_e$ (W-μm²/cm²keV) obtained from one dimensional particle simulations.

Anomalous attenuation of the laser light

The transfer of energy from laser to electrons via the IADI (the rate of energy loss from laser light) is given by $\nu^* E_L^2/8\pi$. When we consider the spatial problem, the (e-folding) attenuation length of laser energy δ is given by v_g/ν^* in term of the group velocity v_g of laser light in plasma. Therefore we have

$$\frac{\delta}{\lambda_L} \approx 6 \times 10^{-2} \frac{1}{\left(\frac{\nu^*}{\omega_L}\right)} \tag{3}$$

, which depends strongly on the anomalous collision frequency ν^*. In deriving the equation (3), the group velocity of the laser light is assumed to be $0.37c$ (c is the speed of light), which is estimated at a plasma density $n/n_c = 0.86$, where the IADI linear growth rate has a peak value.

By combining Eqs. (2), and (3), we can estimate the laser attenuation length due to the IADI as

$$\frac{\delta}{\lambda_L} \approx 1.5 \times 10^8 \times \left(\frac{I\lambda_L^2}{T_e}\right)^{-0.56} \tag{4}.$$

For a simple estimate, we ignore the plasma density dependence of ν^*. This simple approximation will be justified, since $\delta \ll L_{IADI}$ as is discussed in the following section. The accurate value depends on the detail of the density profile, and local value of $\upsilon*$. The important point is that our simple estimates indicate that the attenuation length of the laser can be much smaller than the instability width as shown in the following section.

Plasma Density Profile

We have made computer calculations of the plasma density profile using the 2-dimensional LASNEX computer code[6]. The calculations were made using our experimental parameters: the 1.06 μm laser, with 1 nsec Gaussian pulse, was focused onto a 50 μm thick planar CH target using f/2 lens (the laser spot size was 500 μm). The laser intensity was 3×10^{13}, and 10^{14} W/cm^2, and the flux limiter was f = 0.1.

The plasma scale length was long near the instability region. The electron temperature was about 0.7 keV. It is well known that the IADI can be excited[3] at the densities $0.8 < n/n_c < 1$. The lower limit density is determined by the Landau damping cut off of the epw near the $k_{epw}\lambda_{De} \approx 0.3$ (k_{epw}, and λ_{De} are wave number of the epw, and Debye length). Let's define the length between n_c and $0.8\, n_c$ as the instability width, L_{IADI}. For the above parameters, the length L_{IADI} was quite large, about 30 μm for the both laser intensities of 3×10^{13}, and 10^{14} W/cm^2.

Laser light attenuation length, and the instability width

We can now compare the laser attenuation length and the instability width. For laser intensity 3×10^{13} W/cm^2 (10^{14}), the electron temperature is about 0.7 keV (1), so we have $\delta/\lambda_L \sim 3$ (2) using Eq. (4). On the other hand, $L_{IADI}/\lambda_L \sim 30$ for the both laser intensities. The attenuation length is much smaller than the instability width. Therefore, we predict without the accurate spatial profile of $\upsilon*$ that the most laser energy which reaches the instability region is absorbed by the IADI. The absorbed energy is deposited to hot and warm electrons.

Two-Dimensional Electromagnetic Computer Simulation Results

These results are consistent with the calculations using the two-dimensional relativistic particle and electromagnetic field simulation code ZOHAR[7]. Figure 2 shows laser absorption versus L_{IADI}. It shows that the instability width L_{IADI} of only a few laser wavelengths may be sufficient to absorb most of laser energy. The ZOHAR simulations were made with high laser intensities to minimize the numerical noise. However, $\upsilon*$ saturates at high laser intensity as shown in Fig. 2, so that the results should be insensitive to the laser intensity. In fact, when we increased the laser intensity from 5×10^{15} to 1×10^{16} W/cm^2, no significant change of the absorption was seen. For comparison, we plotted the theoretical values calculated using the $\upsilon*$ given in Fig. 2. The simple estimates give reasonably good agreements with ZOHAR results. The important point is that there was no drastic change of the anomalous absorption as laser intensity increased from 10^{14} to 5×10^{15} W/cm^2.

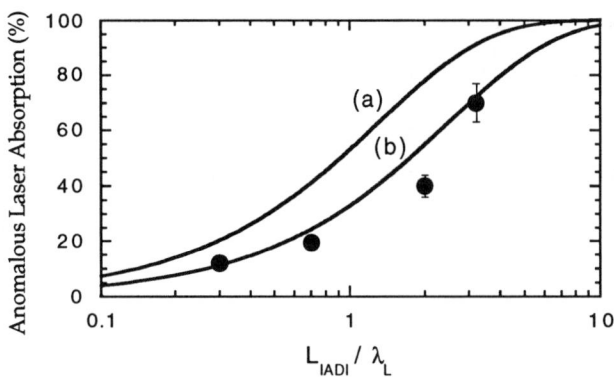

FIGURE 3. The solid circles are fraction of laser absorption calculated by ZOHAR simulations. The curves are the theoretical values for (a) : 5×10^{15} W/cm^2 and 4 keV, and (b): $I = 10^{14}$ W/cm^2, and Te = 1 keV.

SUMMARY

In summary, anomalous absorption in a large scale plasma is quite different from those of small scale plasmas. The Ion Acoustic Decay Instability (IADI) may cause anomalous laser absorption with a relatively weak laser intensity in a large scale plasma. The anomalous attenuation length of the laser can be only a few laser wavelengths in width. These are consistent with the 2-dimensional electromagnetic field computer simulation results. The experiments indicate that the threshold of the IADI is low, so that the IADI is excited on a shallow, long scale length plasma. The measured results of the second harmonic signals are consistent with a strong anomalous absorption by the IADI in a large scale plasma.

ACKNOWLEDGMENTS

We acknowledge the encouragement of J. Knauer, J. S. DeGroot, W. L. Kruer, E. M. Campbell. This work was partially supported by the Lawrence Livermore National Laboratory Laser Fusion program, and was partially supported by the National Laser Users Facility at Univ. of Rochester, with financial support from the U. S. Department of Energy.

REFERENCES

1. W. L. Kruer, The Physics of Laser Plasma Interactions (Addition-Wesley, Reading, MA, 1988); K. Nishikawa, J Phys. Soc. Jpn. **24**, 916, and 1152 (1968).
2. C. Yamanaka et al, Phys. Rev. Lett. **30**, 594 (1973); K. Tanaka et al, Phys. Fluids **27**, 2187 (1984); F. Dahmani et al, Phys. Fluids **B3**, 2558 (1991).
3. K. Mizuno et al, Phys. Rev. Lett. **65**, 428 (1990); K. Mizuno et al, Phys. Fluids B **3**, 1983 (1991); K. Mizuno et al, Phys. Rev. Lett. 73, 2704 (1994).
4. K. Mizuno et al, Phys. Rev. Lett. **52**, 271 (1984); K. Mizuno et al, Phys. Rev. Lett. **56**, 2184 (1986).
5. K. Estabrook, and W. L. Kruer, Phys. Fluids **26**, 1888 (1983).
6. G. B. Zimmerman and W. L. Kruer, Comments Plasma Phys. Controlled Fusion **2**, 51 (1975).
7. A. B. Langdon, and B. F. Lasinski, In Methods in Computational Physics, edited by J. Killeen (Academic, New York, 1976), Vol. 16, p327.

Evolution of fs High-Power Laser-Produced Plasma

K.Shimizu[1], M.Kuwabara[2], K.Go[1], H.Takakuwa[2], H.Kitazawa[1], and S.Karashima[3]

1. Interdisciplinary Graduate School of Science and Engineering, Tokyo Institute of Technology, 4259 Nagatsuta-cho, Midori-ku, Yokohama 227, Japan
2. Faculty of Engineering, Tokyo Institute of Technology, 2-12-1 O-okayama, Meguro-ku, Tokyo 152, Japan
3. Faculty of Engineering, Science University of Tokyo, 1-3 Kagurazaka, Shinjuku-ku, Tokyo 162, Japan

Abstract. Spatial and temporal behavior of a fs high-power laser-produced plasma is investigated for carbon and magnesium targets irradiated at 10^{16} Wcm^{-2}, in an evolution time of 500 fs in which plasma expansion would be negligible. The plasma is taken to be a Maxwellian collisional plasma with an ion density of 4.0×10^{22} cm^{-3}, the wave-wave coupling between the plasma and laser field being neglected. As a result, we found that a completely ionized carbon plasma with an electron temperature of 600 eV is realized, and for magnesium about 40% of the total plasma ions are completely ionized with an electron temperature of 900 eV.

INTRODUCTION

The interaction of very short laser pulses with solid has become an important field of study with the recent development of intense picosecond and subpicosecond lasers, in view of their applications to soft X-ray lasers.

A highly ionized plasma with high density is produced on a solid surface irradiated by a short-pulse laser with high power. In the initial stage of the plasma production, the laser power is supposed to be absorbed by nonlinear multiphoton processes such as inverse bremsstrahlung and tunnel ionization. However, it seems to be improbable that helium-like and hydrogen-like ions are produced only by the tunnel ionization process in the short-pulse duration. Electron-ion collisional ionization would also be effective with the evolution of the plasma. Probably, those ions produced by the tunnel ionization shift to higher charge states by the elctron-ion collisional process.

On the perspective mentioned above, we devote to investigate the heating and moderation of plasma electrons and the ionization in the strong electromagnetic field.

MODEL CALCULATIONS

In the present calculation we assume a cold, uniform plasma with an ion density of 4.0×10^{22} cm^{-3} and an electron temperature of 10 eV at the initial stage of evolution of a laser-produced plasma. The pumping laser pulse ($\lambda = 250$ nm) with the Gaussian shape of 400-fs FWHM is normally incident onto a target surface with a peak intensity of 10^{16} Wcm^{-2}. The strength E of the electric field in the plasma is determined by solving a linearized Maxwell equation Eq.(1) with one dimension (1):

$$\nabla^2 E - \nabla \cdot (\nabla \cdot E) + \frac{\omega_{pe}^2}{c^2} \varepsilon E = 0 \ , \tag{1}$$

$$\varepsilon = 1 - \frac{\omega_{pe}^2}{\omega(\omega + i\nu_{ei})} \ . \tag{2}$$

The dielectric function ε of the plasma is assumed to change continuously at the boundary between plasma and vacuum.

In the initial plasma with high density, the electron-ion collision frequency ν_{ei} is comparable to the frequency ω of the oscillating laser field, so that the density of the plasma current generated by the electromagnetic field is considerably reduced as compared with that in a collisionless plasma. Therefore, the laser field is able to penetrate into an overdense plasma. Actually, this field is very effective in heating the initial plasma.

The electron plasma is heated by inverse bremsstrahlung, tunnel ionization and three-body recombination. And, the electron cooling is made by the collisional ionization. Taking into account those processes, we find Eq.(3) for electron energy conservation:

$$\frac{d}{dt}(n_e \frac{3}{2} k_B T_e) = \frac{1}{2} n_e m \upsilon_{os}^2 \nu_{ei} + \sum_{q,n} \frac{1}{2} N_{q,n} m \overline{\upsilon_{os}^2 R_{q,n}^t} + \sum_{q,n} E_{q,n} N_{q,n} R_{q,n}^{cr}$$

$$- \sum_{q,n} E_{q,n} N_{q,n} S_{q,n} + \nabla \cdot (\kappa \nabla k_B T_e) \ , \tag{3}$$

where n_e is the electron density, υ_{os} is the electron quiver velocity along the oscillating electric field, S is the rate coefficient for collisional ionization, R^{cr} is the rate coefficient for three-body recombination, R^t is the rate coefficient for tunnel ionization (2), $N_{q,n}$ is the population density of a q-charged ion at the state n, $E_{q,n}$ is the ionization energy, and κ is the thermal conductivity. Each term of the right hand side of Eq.(3) shows the contribution of classical inverse bremsstrahlung, tunnel ionization, three-body recombination, collisional ionization and electron thermal conduction. Here, since the diffusive heat flow is lager than the free-streeming in a strong plasma temperature gradient, we use a flux limit of 0.03 to avoid the breakdown of the thermal diffusive term.

The ion fraction is calculated using a rate equation which includes collisional excitation and deexcitation, radiative transition, radiative recombination, three-body recombination, and tunnel ionization. Also, the electron velocity distribution is assumed to be Maxwellian. In the present study, the electron thermal velocity v_e is much larger than the electron quiver velocity v_{os}, and the laser frequency ω is smaller than electron-electron and electron-ion collision frequencies. Therefore, it is conceivable that the electron velocity distribution turns quickly into the Maxwellian (3).

The calculation is performed for the plasma medium of carbon and magnesium. The energy levels from the ground state to the excited state with a principal quantum number of 10 are included in the rate equation. The plasma expansion is negligible, because the initial plasma is cold and the duration of the incident pulse is extremely short. Moreover, the plasma produced by the short pulse on the solid surface has a steep density gradient and the laser wave length is short, so that the wave-wave coupling between the plasma and laser field such as resonance absorption and Raman scattering could be neglected.

RESULTS AND DISCUSSION

Figures 1 and 2 show the temporal and spatial behaviors of the electric field, average ion charge, fractional ion abundance, electron temperature, ratio of the electron quiver velocity v_{os} along the oscillating field to the electron thermal velocity v_e, and electron-ion collision frequency in carbon and magnesium plasmas.

As seen from Figs.1(a) and 2(a), the field strength decreases with the laser wave propagation into the plasma due to the absorption and reflection. Also, it decays as time progresses, because an increase of electron temperature causes a decrease of the electron-ion collision frequency, which enhances the plasma current density along the laser electric field and consequently the electromagnetic wave reflection (see Figs. 1(f) and 2(f)).

Figures 1(b) and 2(b) show the average ion charge. Although the average ion charge states $\langle z \rangle \leq 3$ are reached by the tunnel ionization in the extremely short time, a completely ionized plasma is mainly generated by collisional cascade excitation and by successive collisional and tunnel ionizations from excited states. Such a carbon plasma is produced on the plasma surface after 400 fsec (see Figs.1(b) and 1(c)). For magnesium, the fraction of completely ionized ions on the plasma surface increases rapidly during 400 fs and 600 fs, after the peaking of the incident laser pulse, and amounts to about 40% of the total ions at 600 fs (see

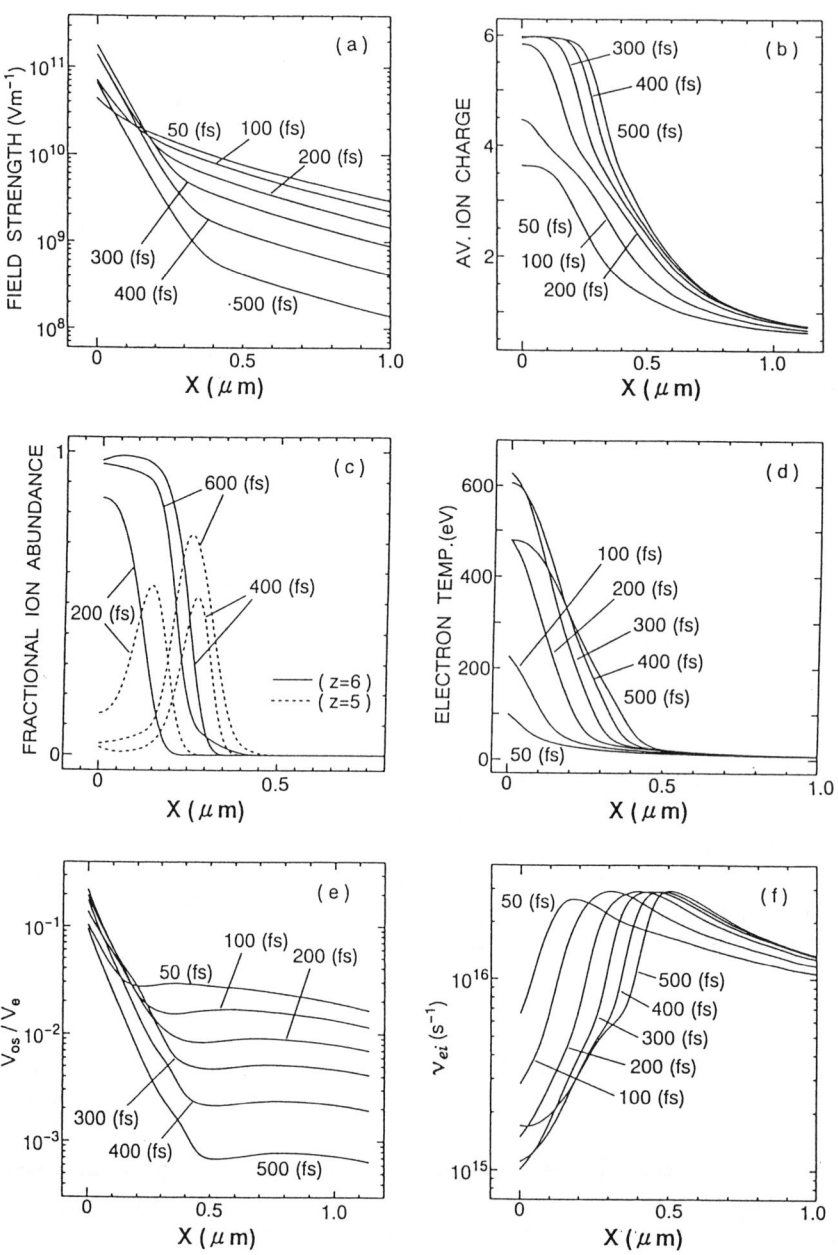

FIGURE 1. Temporal and spatial behavior of the electric field, average ion charge, fractional ion abundance, electron temperature, ratio of v_{os} to v_e, and electron-ion collision frequency in carbon plasma.

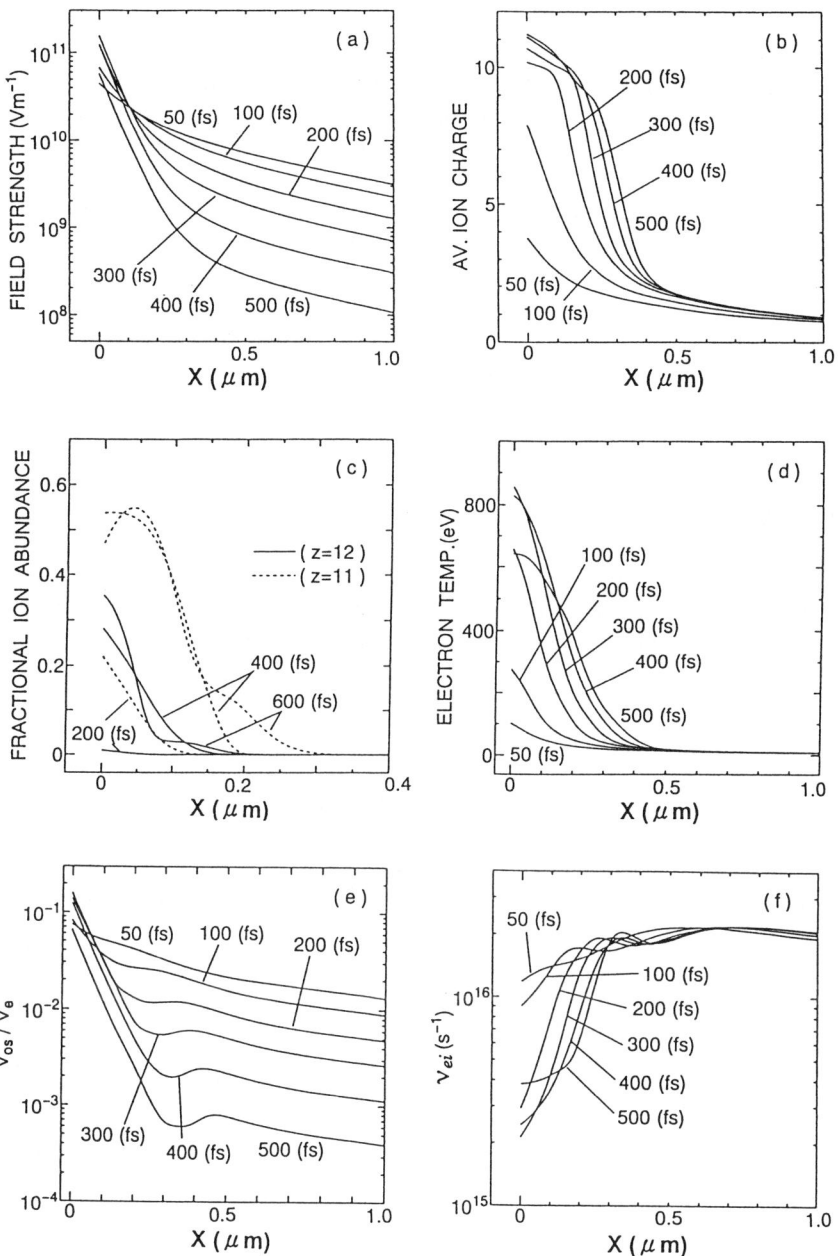

FIGURE 2. Temporal and spatial behavior of the electric field, average ion charge, fractional ion abundance, electron temperature, ratio of v_{os} to v_e, and electron-ion collision frequency in magnesium plasma.

Fig.2(c)). The electron temperature on the plasma surface is then about 900eV (see Fig.2(d)). From those results, we find that the magnesium plasma is rapidly approaching to the ionization balance at this electron temperature.

Figures 1(d) and 2(d) show the electron temperature. The electron plasma heating is due to the inverse bremsstrahlung and tunnel ionization processes. In the present calculation, however, the electron quiver velocity v_{os} is much smaller than the electron thermal velocity, as seen from Figs.1(e) and 2(e), and therefore the plasma heating due to inverse bremsstrahlung is dominant.

SUMMARY

We found that the completely ionized plasma is produced on the solid carbon target irradiated by a 400-fs laser pulse (λ =250 nm) at an intensity of 10^{16} Wcm^{-2}, and for magnesium about 40% of the total plasma ions on the plasma surface are completely ionized. However, the electron velocity distribution was taken to be Maxwellian. In future the heating and ionization in a non-Maxwellian plasma should be investigated by solving a Fokker-Planck equation. Also, the wave-wave coupling between the plasma and electromagnetic wave field should be take into account for a low density plasma near the boundary between plasma and vacuum.

ACKNOWLEDGMENTS

This work has been performed as one of the research projects of Tokyo Electric Power Company's Endorsed Chair of Highly Sophisticated Energy System at the Department of Electrical and Electronic Engineering in the Tokyo Institute of Technology. The authors gratefully acknowledge a financial support from Tokyo Electric Power Company.

REFERENCES

1. Kruer, W. L., *The Physics of Laser Plasma Interaction*, CALIFORNIA: ADDISON-WESLEY, 1988, ch.5, pp.45-52.
2. Ammosov, M.V., Delone, N.B., and Krainov, V.P., *Sov. Phys. JETP* **64**, 1191-1194 (1987)
3. Langdon, A. B., *Phys. Rev. Lett.* **44**, 575-579 (1980).

Theoretical Analysis of the Multi-Species Ion Plasma Interacting with an Ultra Intense Laser

S. Miyamoto[*], S. Kato[†] and K. Mima[*]

[*]Institute of Laser Engeneering, Osaka University, 2 - 6, Yamadaoka, Suita, 565, Japan
[†]Institute for Laser Science, University of Electro-Communications, Chofu, Tokyo, 182, Japan

Abstract. The recent development of laser technology, the irradiation above 10^{18} W/cm² have been possible. Then fast ignitor concept has proposed. It requires a ultra-intense short pulse laser. The simulation of ultra intense laser interaction with solid density plasma is presented in this paper, and fast ion and electron emission, neutron yeild due to D-D reaction and absorption of lasr energy are discussed

INTRODUCTION

The development of short-pulse high intensity lasers has opened new regime in the study of laser produced plasmas. At intensities higher than 10^{18} W/cm², the electron quiver velocity for 1 μm radiation become relativistic and the radiation pressure reaches more than 300 Mbar. In this regime, generation of fast particle (1,2) and MeV hard X-ray (3) have been predicted.

The fast ignitor concept proposed recently by Tabak et al. (4) could drastically change the ignition requirement for inertial confinement fusion program. The fast ignitor scheme is that short intense laser is injected at the time around the maximum compression and additional heating on the compressed core by the energetic particles follows.

Recently, it was reported that the fast ion creation from picosecond laser produeced plasma had been studied for fast ignitor (5). Then, fusion yields and neutron emissions were mesured from deuterated polystyrene targets irradiated with focused intensity of above 10^{18} W/cm^2. The fusion yeild was 4.4×10^9.

PARTICLE SIMULATION

We performed a particle simulation using a one dimensional relativistic full electro-magnetic code. In this code, a particle has one space compornent and three velocity compornents, and all fields have whole compornents described in one dimensional space. Three kind of particle's motions(electron, deuteron and carbon) are computed to take account of a real plasma.

Our simulation system is as follows; plasma thickness is 6 μm and 22 μm vacuum lie in the each side of plasma to avoid the disturbance of boundary. Simulation runs up to about 0.16 psec, so that any particle does not reach to system boundary. Laser light is injected from left side of the system at a constant intensity through a run. Other simulation parameters are as follows; laser wavelength is 1.06 μm. The ratio of electron density with cretical density is 300. Initial temperature is 3 keV. there are 8049 grid and 3×15000 super particles in the system.

ELECTRON AND ION ACCELERATION

When the laser is injected, by its photon pressure, plasma is pushed and it forms a collisionless shock. As the shock propagats, ions are accelerated at the shock front. Those ions have very high energy. As the result of simulation, we get energy distribution of electron and ion as fig.1. From the ion distribution, we can find that ions are splited two part; the high energy component and the low energy one. The high energy compornent in the distribution is the ion that is accelerated at a shock front. And its energy increases avobe 1 MeV as laser intensity increase, as indicated in fig.2. From the figure we can find that the peak energy of high energy component in ion distribution has a clear correlation with laser intensity, so that it is identified the ion accelerated by laser.

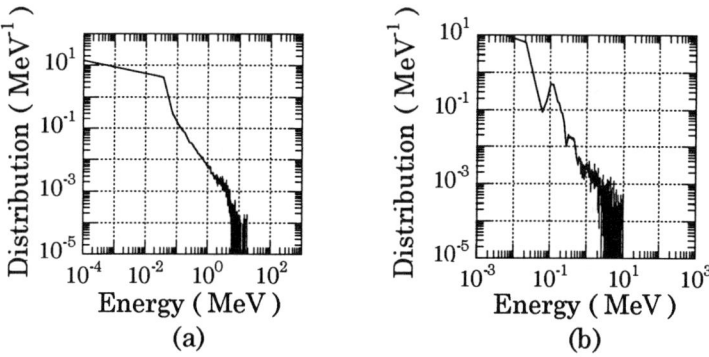

FIGURE 1. Energy distribution function of electron (a) and deuteron (b). Laser intensity is 1×10^{19} W/cm² and pulse width is 0.16 psec.

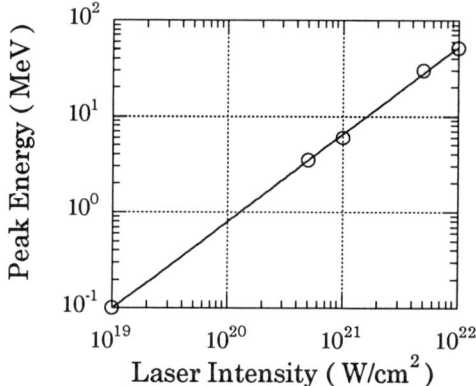

FIGURE 2. Intensity dependence of peak energy in deuteron distribution function shown in fig.1.

This hot ion has a strong directivity in its anglar velocity distribution. It is presented in fig. 3. High energy ion is almost distributed in forward direction, but low energy component widly spread all over angle. It is considered that this directivity results from our 1-D calculation. We think the directivity is not so strong in the case of 2-D or 3-D calculation.

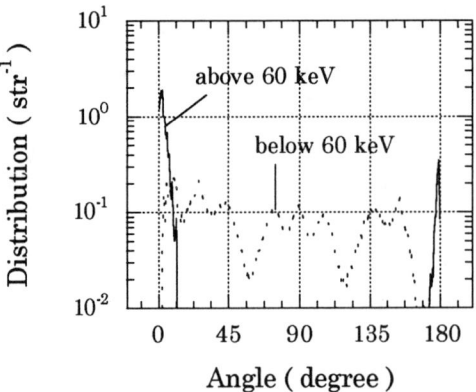

FIGURE 3. Anglar distribution of deuteron. Laser propagation direction is 0 angle.

NEUTRON YIELD

When an ultra intense laser is irradiated into a plasma, hot directive deuterium ions are produced. These deutrons travel into a plasma and decelrate due to collision with electrons. Then it reacts with another deuteron in some probability. We consider the total probability that one deuteron reacts with another untill stop. This probability represented as

$$P(E_o) = n_D \int_{E_o}^{0} v_i \sigma(E_i) \frac{dt}{dE_i} dE_i . \qquad (1)$$

where E_o is energy of accelerated deuteron, $dE_i/dt = -n_{ei} E_i$, n_{ei} is electron-deuteron collision frequency. To get neutron yeild, ion distribution function determined from simulation is multiplied to P, and integrated over energy.

The result is shown in fig.4. Fig4 (a) shows a laser intensity dependence. As the laser intensity increases, neutron yield increases as was expected. Each simulation run has been performed up to 0.16 psec. Fig.4 (b) shows a dependense of neutron yield with laser pulse width. From this figure we can expect that the neutron yield will become the order of 10^{17} m^{-2} in the case of 10^{19} W/m^2, 1 psec irradiation. Assuming that laser focul area is 10^{-8} m^2, we get the neutron yield $\approx 10^9$.

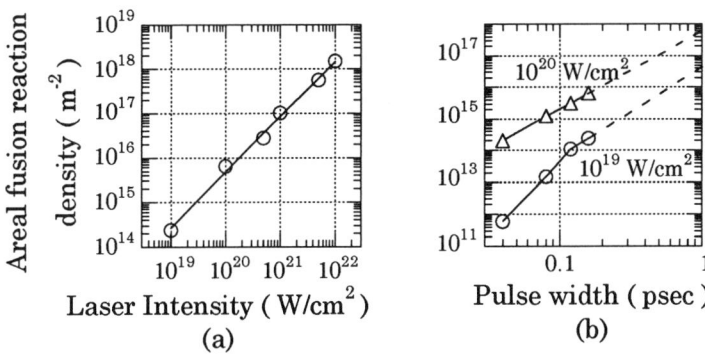

FIGURE 4. Dependence of areal fusion reaction density with laser intensity (a) and pulse width (b).

LASER ENERGY ABSORPTION

Laser energy conversion ratio to particle kinetic energy is plotted in fig.5 at the intensity from 10^{19} W/cm² to 10^{21}. The virtical axis is the ratio of total increas of electron's or ion's kinetic energy to laser energy at 0.16 psec. When the intensity is low, laser energy is converted mainly to electron. But, according to the increase in intensity, the ratio of ion increases. And the total conversion ratio is the order of 10 % all over intensities.

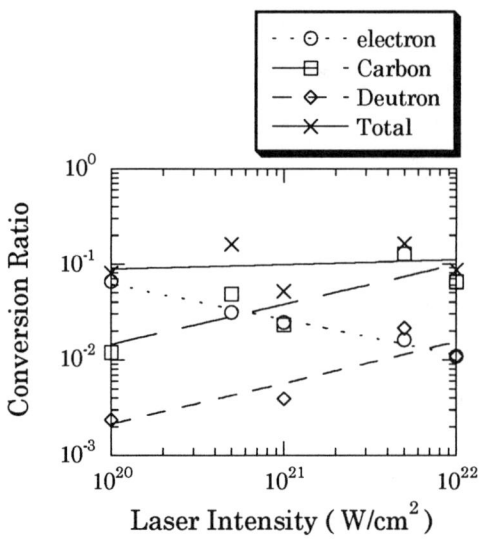

FIGURE 5. Energy conversion ratio of each species.

SUMMARY

We performed a series of computer simulation. As the result of simulation, fast ion and electron acceleration was measured. The fusion yeild is explained by the reaction between fast deuteron accelerated by an intense laser and low energy bulk deuteron. Finally, laser energy conversion to plasma was evaluated, and its ratio was the order of 10 % all over intensities.

REFERENCES

1. Brunel, F., Phys. Rev. Lett. **59**, 52 (1987).
2. Denavit, J., Phys. Rev. Lett. **69**, 3052 (1992).
3. Kmetec, J. D., et al., Phys. Rev. Lett. **68**, 1527 (1992).
4. Tabak, M., et al., Physics of Plasma **1**, 1626 (1994).
5. Few, A. P., et al., Phys. Rev. Lett. **73**, 1801 (1994).

Higher harmonics generation in dense plasmas with an intense ultra-short pulse laser

S. Kato, A. Nishiguchi[1], S. Miyamoto[2] and K. Mima[2]

Institute for Laser Science, Univ. of Electro-Communications, Chofu, Tokyo, 182, Japan
[1] Osaka Institute of Technology, Asahi-ku, Osaka, 535, Japan
[2] Institute of Laser Engineering, Osaka University, Suita, Osaka, 565, Japan

Abstract. In plasmas produced by irradiating an intense ultra-short pulse laser on a solid target, the intense laser radiation directly interacts with a dense plasma. In such a case, the plasma screening effects are important for the response to the laser field. For such plasmas, the dynamical form factor of plasma electron and ion is derived. By using the dynamical form factor, the electron density fluctuation around an ion and the vector potential are obtained. As the results, the high-order harmonic generation in dense plasmas irradiated by an intense laser field is investigated.

INTRODUCTION

Recently, the nonlinear dynamics in an intense laser field have become an interesting area. In particular, plasmas produced by the ultra-short pulse laser have attracted much interest as novel sources of radiation. The emission includes coherent radiation at harmonics of the laser frequency, desecrate x-ray related with the bound-bound transition [1], and hard x-ray bremsstrahlung radiation due to high energy electron [2,3]. The higher harmonics generation due to anharmonic motion of the bound electrons [4-7] and response of underdense [8-10] and overdense plasmas [11-13] have been recently investigated.

The ultra-short pulse laser can heat a solid surface before the surface plasma expands. In such a case, the plasma density scale length is less than the laser skin depth. Consequently, the ultra-short pulse laser directly interacts with the solid-density plasma. The large space charge fluctuations are induced to generate strong electrostatic waves [14]. As a result, strongly anharmonic motion of electrons would develop, and generate higher harmonics. So far, the plasma screening effects on the high-order harmonic generation in dense plasmas have not been investigated, although the effects have been discussed on the inverse bremsstrahlung absorption [15-17] and induced electrostatic fields [14]. Therefore, this paper discusses the induced field in the relation with the plasma polarization.

The electron density fluctuations and the vector potential are derived by using the linear response theory. Moreover, the ion-ion correlation and the electron screening are in particular important in the strongly coupled plasmas. The electromagnetic radiation is due to the electron current-density fluctuation around an ion. The plasma screening effects are estimated by using the hypernetted-chain equation and the linear response theory.

DERIVATION OF ELECTROMAGNETIC RADIATION

A electromagnetic radiation due to the current-density fluctuations in a dense plasma irradiated by an intense laser field is derived. We assumed that the laser light is linearly polarized and the laser intensity is constant, namely $E_L(t) = E_0 \sin(\omega_0 t)$, where E_0 and ω_0 are the laser electric field and the laser frequency, respectively. When the electron quiver speed is comparable to the speed of light, the relativistic effects are important for the plasma dynamics. The electron motion in the perpendicular excursion length with the laser electric field is neglected in this model. Namely, the mass correction for the electron excursion length is only included. The relativistic effects are discussed in detail in the summary. The formulation is analogous to our previous paper for the induced electrostatic fields [14]. In the following analysis, the effects of bound electron is neglected, because the number of free electrons contributing to generating the electron density fluctuation is much greater than the number of bound electron in the highly ionized plasmas.

The electron density fluctuation in the laboratory frame is obtained as

$$\delta n_e^{lab}(\mathbf{r},t) = \delta n_e^{osc}(\mathbf{r} - \mathbf{r}_0(t), t) \quad , \tag{1}$$

where δn_e^{osc} represent the electron density fluctuation in the frame that oscillates with an electron excursion length $\mathbf{r}_0(t)$ as follows [14]:

$$\delta n_e^{osc}(\mathbf{r},t) = \sum_{n=-\infty}^{\infty} \exp(in\omega_0 t)$$
$$\times \int \frac{d\mathbf{k}}{(2\pi)^3} Z^* \left[\frac{1-\varepsilon(\mathbf{k},-n\omega_0)}{\varepsilon(\mathbf{k},-n\omega_0)} S(\mathbf{k}) \right] J_n(\mathbf{k}\cdot\mathbf{r}_0) \exp(i\mathbf{k}\cdot\mathbf{r}) \quad , \tag{2}$$

where $S(\mathbf{k})$, $\varepsilon(\mathbf{k},\omega)$ and $J_n(x)$ are the ion form factor, the dielectric response function and a Bessel function of the first kind, respectively. The electron excursion length $r_0 = eE_0/m_e\omega_0^2 [1+(eE_0/m_e\omega_0 c)^2/2]^{-1/2}$, where e, m_e and c are the electron charge, mass and the speed of light, respectively.

The detailed formulae of the electron dielectric response function and the ion static form factor were described in the previous paper [14]. This oscillation frame was introduced by analyzing the inverse bremsstrahlung in the plasma [15-17]. Substituting Eq.(2) into Eq.(1), we have

$$\delta n_e^{lab}(\mathbf{r},t) = \sum_{n=-\infty}^{\infty} \exp(in\omega_0 t) \int \frac{d\mathbf{k}}{(2\pi)^3} \delta n_e^n(\mathbf{k}) \exp(i\mathbf{k}\cdot\mathbf{r}) \quad , \tag{3}$$

where,

$$\delta n_e^n(\mathbf{k}) = \sum_{m=-\infty}^{\infty} F^m(\mathbf{k}) B_n^m(\mathbf{k}\cdot\mathbf{r}_0) \quad , \tag{4}$$

$$F^m(\mathbf{k}) = Z^* \left[\frac{1-\varepsilon(\mathbf{k},-m\omega_0)}{\varepsilon(\mathbf{k},-m\omega_0)} S(\mathbf{k}) \right] \quad , \tag{5}$$

$$B_n^m(\mathbf{k}\cdot\mathbf{r}_0) = J_m(\mathbf{k}\cdot\mathbf{r}_0) J_{m-n}(\mathbf{k}\cdot\mathbf{r}_0) \quad . \tag{6}$$

$F^m(x)$ and $B_n^m(x)$ represent the plasma screening factor and nonlinearity of the electron excursion length, namely the laser intensity, respectively.

The current density induced in the plasma through coupling between laser and the density fluctuations in the plasma is obtained [18]. The Fourier components of the current-density fluctuations is

$$\mathbf{J}(\mathbf{k},\omega) = \frac{e^2 n_e \mathbf{E}_0}{2m_e \omega} [\delta n_e(\mathbf{k},\omega-\omega_0) - \delta n_e(\mathbf{k},\omega+\omega_0)] \quad , \tag{7}$$

where n_e is the electron density and $\delta n_e(\mathbf{k},\omega)$ is the Fourier components of Eq.(3) as follows:

$$\delta n_e(\mathbf{k},\omega) = \sum_{n,m=-\infty}^{\infty} F^m(\mathbf{k}) B_n^m(\mathbf{k}\cdot\mathbf{r}_0) 2\pi\delta(\omega+n\omega_0) \quad . \tag{8}$$

The Fourier component of the vector potential $\mathbf{A}_\omega(\mathbf{r})$ far away from the emission region are determined as follows.

$$\mathbf{A}_\omega(\mathbf{r}) = \frac{\exp(i\mathbf{k}_{em}\cdot\mathbf{r}')}{cr} \int_V \mathbf{J}_\omega(\mathbf{r}') \exp(-i\mathbf{k}_{em}\cdot\mathbf{r}') d\mathbf{r}' \quad , \tag{9}$$

where $\mathbf{k}_{em}$ and V are the wave number of the electromagnetic wave and the volume where a intense laser interacts with overdense plasma, respectively, and $\mathbf{J}_\omega(\mathbf{r})$ is the temporal Fourier component of the current.

Substituting Eq.(7) and Eq.(8) into Eq.(9), we obtain the vector potential as follows:

$$\mathbf{A}_{N\omega_0}(\mathbf{r}) = \frac{e^2 n_e V \mathbf{E}_0}{2m_e N\omega_0} \frac{\exp(i\mathbf{k}_{em}^N \cdot \mathbf{r}')}{cr} \alpha(\mathbf{k}_{em}^N, N\omega_0) \quad , \tag{10}$$

where

$$\alpha(\mathbf{k}, N\omega_0) = \sum_{m=-\infty}^{\infty} F^m(\mathbf{k}) J_m(\mathbf{k}\cdot\mathbf{r}_0) [J_{m+N-1}(\mathbf{k}\cdot\mathbf{r}_0) - J_{m+N+1}(\mathbf{k}\cdot\mathbf{r}_0)] \quad . \tag{11}$$

and k^N_{em} is the wave number of the N-th order harmonic emission.

The Fourier component of time-averaged power radiation per unit solid angle by Eq.(10) is

$$\frac{dP_{N\omega_0}}{d\Omega} = \frac{1}{4} \frac{cE_0^2}{8\pi} \left(\frac{k_{em}^N c}{N\omega_0}\right)^2 r_e^2 \left|\alpha(k_{em}^N, N\omega_0)\right|^2 \sin^2\theta \, (n_e V)^2 \quad , \tag{13}$$

where $r_e = e^2/m_e c^2$ is the classical electron radius and θ is the angle between $\mathbf{E}_0$ and $\mathbf{k}^N_{em}$.

NUMERICAL CALCULATIONS

In this section, the total power spectra in the overdense plasmas are calculated by integrating Eq.(13) over all solid angle for the laser wavelength is 0.53 μm, in this case, the electron cutoff density n_c is 4.0×10^{21} cm^{-3}. The ion temperatures are assumed to be the electron temperatures. The temperature is 1 keV. For a fully ionized aluminum, a solid density n_s is 8.6×10^{23} cm^{-3}, namely, $n_s/n_c = 215$. We assumed that the volume V does not depend on the plasma density. However, in the overdense plasma, the volume depend on the skin depth c/ω_p, where ω_p is the plasma frequency. The plasma screening factor F^m has a peak at $|m| \approx \omega_0/\omega_p = (n_e/n_c)^{1/2}$, namely, the frequency is resonance near the plasma frequency, when $k^N_{em} \ll k_D$.

Figures 1 (a) and 1 (b) show the electron density dependence of the total power spectrum for the laser intensities are 1×10^{18} W/cm^2 and 2×10^{18} W/cm^2, respectively. The peaks of spectrum are at $N \approx 2\,(n_e/n_c)^{1/2}$. The power spectra corresponding to $n_e = 100n_c$ and n_s appear in $I_L = 2\times10^{18}$ W/cm^2. The high-order harmonics are from 16th to 30th for $n_e = 100n_c$ and from 28th to 43th $n_e = n_s$. For $n_e = n_s$ in Fig. 1(b), there is a peak at $N = 41$, because the dispersion relation is satisfied at the vicinity of $N \approx 41$ and $m = 15$, namely, the wave number of electromagnetic wave corresponds with the wave number of the plasma wave and the imaginary part of the dielectric function is negligible small. In such a case, the electron-ion collision frequency became important for the damping of the wave. However, this model is not included the collisional process. Therefore, this strong peak is actually reduced.

Figures 2 (a) and 2 (b) show the laser intensity dependence of the power spectrum for the cutoff and solid densities, respectively. For the cutoff density in Fig. 2(a), the power of second and fifth order harmonics in $I_L = 1\times10^{18}$ W/cm^2 are 10^2 and 10^6 times greater than the harmonics in $I_L = 1\times10^{17}$ W/cm^2, respectively. The generation of the high-order harmonics is strongly depend on the laser intensity, For the solid-density plasma in Fig. 2(b), the peak power of the high-order harmonics in $I_L = 2\times10^{18}$ W/cm^2 is 10^8 times greater than the harmonics in $I_L = 1\times10^{18}$ W/cm^2.

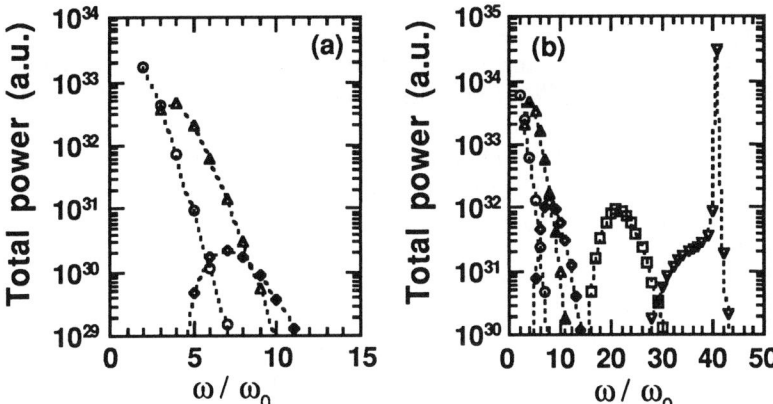

FIGURE 1. Total power spectrum of electromagnetic radiation as a function of the laser intensity. The circle, triangle, diamond, square and inverse triangle show $n_e = n_c$, $4n_c$, $10n_c$, $100n_c$ and n_s, respectively. Figures (a) and (b) are for $I_L = 1 \times 10^{18}$ W/cm^2 and 2×10^{18} W/cm^2, respectively.

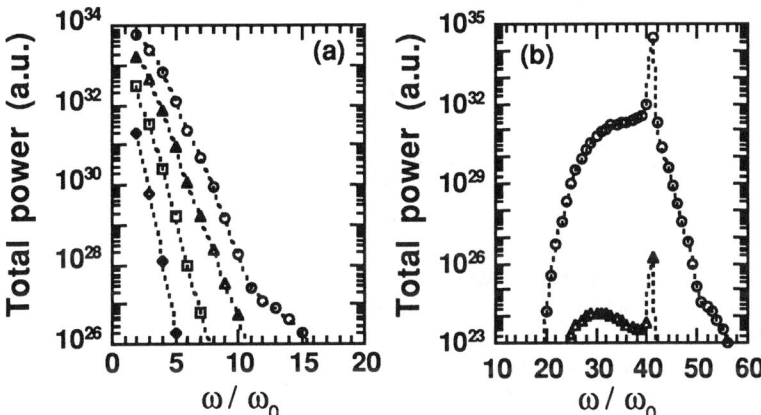

FIGURE 2. Total power spectrum of electromagnetic radiation as a function of the electron density. The diamond, square, triangle, and circle show $I_L = 1 \times 10^{17}$ W/cm^2, 4×10^{17} W/cm^2, 1×10^{18} W/cm^2 and 2×10^{18} W/cm^2, respectively. Figures (a) and (b) are for the cutoff and solid densities, respectively.

SUMMARY

We derive the electron density fluctuation around an ion in a dense plasma irradiated by an intense laser field. The total power spectra of the electromagnetic radiation due to the electron current-density fluctuation are evaluated. There are both odd and even harmonics due to the nonlinear response of electron current-density fluctuation. As the results, the harmonics above 30th appear in the solid density interacting with $I_L = 2 \times 10^{18}$ W/cm^2. The harmonics under 10th are generated at the vicinity of the cutoff density.

The free electron in a plane monochromatic electromagnetic wave performs the well-known figure 8 orbit. The electron quiver speed is comparable to the speed of light, when the laser intensity $I_L \lambda^2$ is higher than 10^{18} Wμm^2/cm^2, where I_L and λ are the laser intensity and wavelength, respectively. The ratios of the parallel excursion length with the laser electric field to the perpendicular one is 17.7% in the extreme relativistic case. The ratios are 3.6 and 9.5%, for the laser intensities $I_L \lambda^2$ are 10^{17} and 10^{18} Wμm^2/cm^2, respectively. Therefore, the total power spectra of the electromagnetic radiation are maybe modified by this motion. For example, the magnetic field of the laser radiation will play very important roles in the direction of the emission, which has been neglected in the analysis of this paper.

When the excursion length is grater than the average ion radius, the electron collides a lot of ions. In this case, not only the current-density fluctuation but also the bremsstrahlung emission is important. If the ions are strongly coupled in dense plasmas, this bremsstrahlung emission will be coherent radiation [19].

REFERENCES

[1] Chaker, M., Kieffer, J. C., Matte, J. P., Pepin, H., Audebert, P., Maine, P., Strickland, D., Bado, P., and Mourou, G., Phys. Fluids **B3**, 167 (1991).
[2] Kmetec, J. D., Gordon, III C. L., Macklin, J. J., Lemoff, B. E., Brown, G. S., and Harris, S. E., Phys. Rev. Lett. **68**, 1527 (1992).
[3] Chichkov, B. N., Kato, Y., and Murakami, M., Phys. Rev. **A46**, 4512 (1992).
[4] Kulander, K. C. and Shore, B. W., Phys. Rev. Lett. **62**, 524 (1989).
[5] Eberly, J. H., Su Q., and Javanainan, J., J. Opt. Soc. Am. **B6**, 1289 (1989).
[6] Krause, J. L., Schafer, K. J., and. Kulander, K. C, Phys. Rev. Lett. **68**, 3535 (1992).
[7] Reed, V. C. and Burnett, K., Phys. Rev. **A46**, 424 (1992).
[8] Brunel, F., J. Opt. Soc. Am. **B7**, 521 (1990).
[9] Sprangle, P., Esarey, E., and Ting, A., Phys. Rev. Lett. **46**, 2011 (1990), Sprangle, P., Esarey, E., and Ting, A., Phys. Rev. **A41**, (1990).
[10] Rax, J. M. and Fisch, N. J., Phys. Rev. Lett. **69**, 772 (1992).
[11] Carman, R. L., Forslund, D. W., and Kindel, J. M., Phys. Rev. Lett. **46**, 29 (1981).
[12] Bezzerides, B., Jones, R. D. and Forslund, D. W., Phys. Rev. Lett. **49**, 202 (1982).
[13] Grebogi, C., Tripithi, V. K., and Chen, H. H., Phys. Fluids **26**, 1904 (1983).
[14] Kato, S., Nishiguchi, A., and Mima, K., Phys. Rev. **E50**, 2193 (1994).
[15] Kato, S., Kawakami, R., and Mima, K., Phys. Rev. **A43**, 5560 (1991).
[16] Dawson, J. and Oberman, C., Phys. Fluids **5**, 517 (1962).
[17] Jones, R. D. and Lee, K., Phys. Fluids **25**, 2307 (1982).
[18] Ichimaru, S., Basic Principles of Plasma Physics (Benjamin, Reading, Mass.,1973).
[19] Hüller, S., and Meyer-ter-Vehn, J. Meyer-ter-Vehn, Phys. Rev. **A48**, 3906 (1993).

Super-Strong Laser Field Generation and Their Interaction with Solid Target in Vacuum

A.A.Andreev, V.I.Bayanov, A.B.Vankov, A.A.Kozlov, V.A.Komarov, I.V.Kurnin, N.A.Solovyev, S.A.Chizhov, V.E.Yashin

Institute for Laser Physics, SC "Vavilov State Optical Institute"
2, Birzhevaya line, 199034, SSt.Petersburg, Russia

Abstracts. We consider interaction of superstrong laser fields with condenced targets in vacuum. Our experiments were made on our laser system for the picosecond pulse generation, its recompression with chirp introduction and regeneration amplification. The results of laser radiation absorption and conversion efficiency into fast particles and X-ray are presented for a broad range of parameters.

1. INTRODUCTION

Ultrashort (< 1 ps) laser pulse interaction in vacuum with the hard target attracts significant interest during last years. Such an interest is caused by the possibility to realize the source of fast particles, appropriate for various applications, including even the scheme of "fast ignition" laser inertial confinement fusion (1-3). Superdense hot plasma, produced by such an interaction, is a new object. Special features of such an object are now intensely studied (4,5). In (6-8) the laser radiation with various state of polarization absorption and scattering in such a plasma was studied; in (9-12) numerical modeling of such an interaction was carried out. It was found out, that under the power fluency densities of $\sim 10^{17}$ W/cm^2, the basic mechanisms, leading to radiation absorption and particles acceleration, are the well known resonant absorption and ambipolar ions acceleration (13,14), which reveal themselves in the case of long pulse interaction with the matter as well. However, in the case of supershort pulse, when the plasma gradients are significant, these processes development is different from that in the case of plasma with low rate of density nonuniformity. In this work we have studied the picosecond laser pulse (energy <0.5 J and contrast < 10^7) interaction with the Al target in vacuum. It was found out that p-polarized radiation absorption exceeds significantly that of s-polarized. Significant part of the energy absorbed in both cases is converted into

© 1996 American Institute of Physics

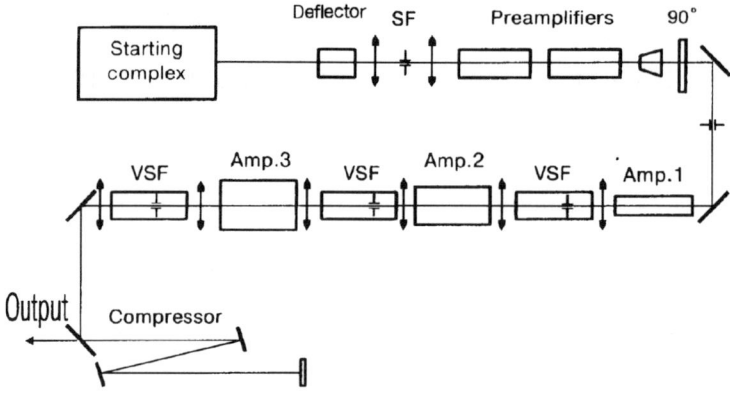

FIGURE 1. Laser setup schematics.

heating and acceleration of the fast particle, emitted in the direction, mirror-like to the laser radiation beam incidence direction with respect to the flat Al target.

2. EXPERIMENTAL CONDITIONS

Laser setup, used in these experiment, uses the scheme of the chirped pulse amplification with the consequent its compression in diffraction gratings compressor (see Fig.1) (15). Primary pulse with the duration 1 ps was generated in the neodymium glass master oscillator with the self mode locking and negative feedback loop. This pulse was then stretched up to duration of $\sim$ 450 ps and at the same time chirped in the decompressor, created on the base of diffraction gratings with the telescope between them. Preliminary amplification of this pulse up to the energy of $\sim$ 3 mJ was made in the neodymium glass regenerative amplifier. The scheme, based on the electrooptical deflector, mounted at the regenerative amplifier output, improved the amplified pulse contrast with respect to the preceding prepulses up to 10^6. This chirped pulse was then amplified in the main amplifier up to the energy of 1-2 J and compressed in the double-pass diffraction gratings compressor down to the duration of 1.5 ps. Maximal energy of the compressed pulse was limited at the level of $\sim$ 0.5 J due to the optical damage threshold and grating's size (110 x 110 mm). Radiation divergence was 10^{-4}, providing the intensity on the target $\sim 10^{17}$ W/cm^2.

The flat Al target was mounted in the center of vacuum chamber, evacuated down to 10^{-4} torr. The target mounting made it possible to vary the target angle with respect to the laser beam and to its polarization plane (see Fig.2). Laser radiation was focused onto the target by the aspherical lens (diameter 50 mm, focal length 120 mm, lens thickness 12 mm, excluding thus the nonlinear distortions action). Hence, the laser focal spot with the diameter 12 μm contained 70% of the total energy E_L.

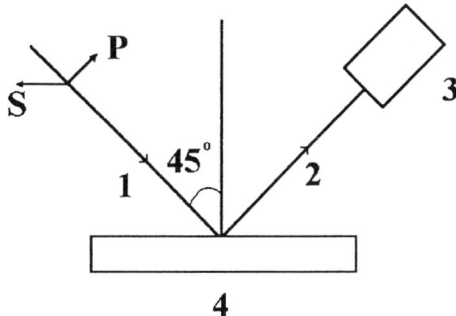

FIGURE 2. Target unit schematics. (1) incident radiation; (2) reflected radiation; (3) calorimeter or ion collector; (4) Al target.

Reflected light energy was measured in the mirror - like direction, using the lens *D:F=1* mounted on the distance 40 mm off the target, and the calorimeter IMO-2M. Spatial distribution of scattered light was estimated using the photo-sensitive paper, covering the cylinder with radius 30 mm, mounted along the target (see Fig.3).

The ion velocity distribution was measured by time-of-flight technique using charge collector (16). The collector was equipped with the receiving element as cilinder cavity form and biased mesh (75 V) to reduce the emission of secondary electrons from the detector surface. The charge collector was aligned at 45° with respect to the target normal. The distance from the target to the detector and solid angle of detector were fixed to be 25 cm and 1.25×10^{-3} sr, respectively, throughout the experiments.

3. NUMERICAL MODEL

3.1. Absorption

Under the to-date experimental conditions, when the laser intensity exceeds 10^{16} W/cm^2, and pulse duration is less than 1 ps, plasma nonuniformity characteristic scale is usually lower than the radiation, penetrating down to the skin-layer thickness, wavelength λ. Temporal scale of such a hydrodynamical characteristics as electron temperature T_e and concentration N_e applicability exceeds that of electrodynamical characteristics. Hence radiation absorption may be calculated using the steady-state wave equations. In the case of s-polarized wave we can use the equation for the local dielectric permeability with the given $T_e(x)$ and $N_e(x)$. Uni-dimensional approximation is appropriate in the case, because the length of the plasma produced is lower, than the laser spot size. Plasma dissipation properties are characterized by the Landay damping and by the electrons collisions with the ions.

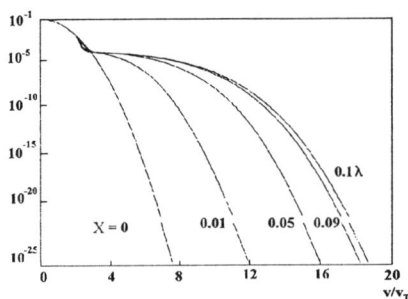

FIGURE 3. Electron velocity distribution for $L=0.2\lambda$, $T_e=1$ keV, $\theta=45°$, $I\lambda^2=10^{16}$ W $\mu m^2/cm^2$.

This system of equations makes it possible to determine the absorption of the p-polarized wave in the nonuniform plasma and modification of the electron distribution function (in the starting moment it is Maxwell one) due to the Landay damping.

In the Fig.3 is shown the calculated distribution function. One can see formation of the quasi-linear plato and of the tail of fast electrons, generated due to the resonant absorption.

3.2. Hydrodynamics with the account of ponderomotive forces and ionization kinetics

The hydrodynamical parameters of the plasma, produced by the laser pulse action onto the hard target in vacuum, were calculated with the help of Lagrange code "Ion" (19). This code uses the model of dual temperature hydrodynamics with the account of non-local electron thermal conductivity (appropriate, when the scale of temperature nonuniformity is lower than the length of electron free pass) and of the ion viscosity. This approach is calculation also the ion composition in the kinetic approximation. In the subnanosecond temporal range the calculations, using this code, were in good agreement with the experiment in calculations of the X-ray output under the given absorption.

On the each temporal stage of the hydrodynamical code the numerical model, presented above, was used for light absorption calculation, basing on the hydrodynamical parameters, obtained on the previous stage. Both bremsstrahlung and resonant absorptions were taken into account in the case of p-polarized radiation. As to the resonant mechanism, the part of the energy, possessed by the fast electrons, was evaluated from the distribution function. After it some part of this energy was expended onto the fast ions acceleration, while the other part - onto the large target heating. Ponderomotive pressure was taken into account in the movement equation similarly to

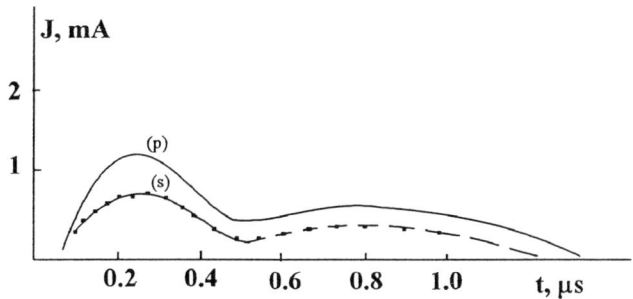

FIGURE 4. Oscillogram of the ion current, registered under the angle of θ=135°.

(20). Noteworthy, that in our case of large density gradient and the comparatively low laser intensity the influence of the ponderomotive pressure is not large during the first half of the pulse, while during the second the gradient varies significantly.

4. EXPERIMENTAL RESULTS AND DISCUSSION

The results of the laser radiation scattering diagram by the flat target (incidence angle 45°) investigation show some broadening of the mirror-like reflected radiation component. Possible reason is the reflecting plasma surface distortion due to the intense laser radiation ponderomotive pressure.

We did not measure the plasma scattered energy along the overall corporal angle. However, possible estimation of such an energy shows that its value is in the order of magnitude lower then that for the mirror-like (specular) reflection.

Dependence of the reflectivity coefficient versus incident laser radiation energy for s and p-polarizations shows the energy absorption growth with the incident energy growth.for both polarizations. However, for s-polarized radiation reflectivity is much higher than for the p-polarized radiation. The reason of growth in the case of p-polarization is that the ponderomotive pressure sharpens the plasma density profile from $L_N=0.4$ μm to $L_N=0.2$ μm, increasing thus the resonant absorption (19).

Calculation for the s-polarized radiation, which is absorbed only via bremsstrahlung mechanism in the simple model, also gives the value much lower than the experimentally measured. However, under our experimental conditions the ponderomotive pressure can result in the nonuniform distribution across the laser spot. Hence, sufficient p-polarized component may be produced by the primarily s-polarized radiation.

In the Fig.4 is shown the oscillogram of the ion current (registered under the angle of 135°) and ion velocities spectrum for the Al plasma (laser radiation intensity 10^{16} W/cm^2, incidence angle 45°).

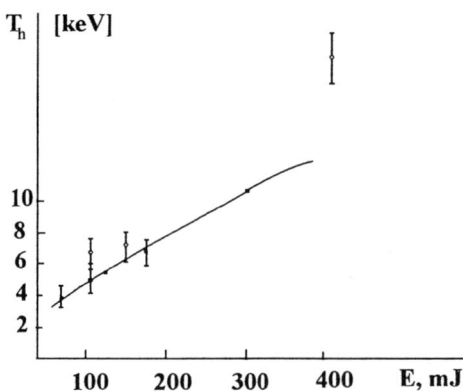

FIGURE 5. Fast electrons temperature dependence vs. incident laser energy.

Experimentally measured ion spectrums may be to a good extent explained in the frame of simple dual-temperature model (23). Evaluation of T_e from the curve *(dN/dV)* slope from the set of calculated curves, we choose the most close to the experimental data.

In the Fig.5 is shown the fast electrons temperature dependence vs. incident laser energy. In the figure the experimental data is accompanied by the calculation results for the realized experimental conditions. One can see, that laser energy increase results in the hot electrons temperature growth. Empiric formula $T_h = E^{0.44}$ is valid for this theoretical and experimental curves.

Creation of the fast particles group, flying under the angle $\theta = 45°$, possessing the sufficient part of the energy under the comparatively low laser radiation densities $\sim 10^{16}$ W/cm^2, hardly may be interpreted in the frame of resonant absorption mechanism only.

One of another mechanisms may be the surface waves generation by laser radiation on the plasma with the sharp gradient (21). Such a surface wave can accelerate the particles, accelerated according to the resonant absorption mechanism, along the target surface, providing thus the necessary component of their velocity.

This research was in part made possible by the Grant no. RP2000 from the International Science Foundadtion.

REFERENCES

1. Luter-Davis B. et al. Laser Phys. 1991,v.1,p.325
2. Rosen M.D. Proc.SPIE, 1991,v.1229,p.160

3. Perry M.D. et al.1994 In Proceedings of Conference on High Field Interactions and Short Wavelength Generation, St.Malo,ThB2.
4. Murnane M.,Kaypten H.C.,Rosen M.D.,Falcone R.W. Science,251,531 (1991)
5. Milchberg H.,Freemon R.R.,Davey S.C.,More R.M. Phys.Rev.Lett. 1988,v.61.p.2364; Achmanov S.A. et al., Quantum Electr.(Sov.) 1991, v.18, p.278 (in Russian)
6. Teubner U.,Bergmann J., van Wonterghem B. et al. Phys.Rev.Lett. 1993,v.70,p.794
7. Meyerhofer D.D.,Chen H.,Delettrez J.A. et al. Phys.Fluids.1993. v.B5,p.2584.
8. Sauerbrey R. ey al. 1992 In Proceedings of the 3rd International Colloquium on X-Ray Lasers,Schliersee,paper O-15.
9. Kieffer J.C.,et al. IEEE J.Quant.Electr.1989,25,2640.
10. Fedosejevs R., et al. Appl.Phys.B. 1990,50,79.
11. Gibbon P.,Bell A.R.,Phys.Rev.Lett. 1992,68,1535.
12. Drska L.,et al. Laser and Particle Beams,1992,10,461.
13. Joshi C., Mori W.B.,Katsouleas T. et al. 1984,Nature,311,525.; Silin V.P., Sov.Uspekhi Fiz.Nauk, 1985, v.145, p.225 (in Russian)
14. Taijma T.,Dawson J.M. 1979,Phys.Rev.Lett.43,267.; Gurevitc A.V., Mesherkin A.P., Sov.JETP, v.80, n.5, p.1810, 1981 (in Russian)
15. Vankov A.V. et al. Proceedings of SPIE, 1994, vol.2095, p.87
16. V.M.Komarov, Pribory i technica eksperimenta N1, 167, (1987) (in Russian)
17. Radiotechnique (Sov.), v.17, Moscow, 1978, p.166 (in Russian)
18. Andreev N.E., Sc.D.Dissertation Autosummary., Moscow, 1987 (in Russian)
19. Andreev A.A. et al. Proceedings of SPIE, vol.2097 Laser Applications (1993) p.409
20. Audebert P. et al. Europhys.Lett.,19(3),p. 189 (1992)
21. Gamaly E.G. Phys.Rev. A,1994,v.135,p.1032
22. Zozula A.A., Silin V.P., Tikhochuk V.T. Sov.J.of Plasma Phys., 1987, v.35, p.130 (in Russian)
23. Wickens R.,Allen J.,Rumsby P. Phys.Rev. Lett. v. 41,243,1978

Simulation Study for Interaction between Ultra Power Laser and Dense Plasma Slab

Hitoshi Sakagami* and Kunioki Mima†

*Computer Engineering, Himeji Institute of Technology
2167 Shosha, Himeji, Hyogo 671-22, Japan
†Institute of Laser Engineering, Osaka University
2-6 Yamada-oka, Suita, Osaka 565, Japan

Abstract. The interaction between ultra intensity laser and dense plasma is investigated with the use of a 1-1/2 dimensional electromagnetic, relativistic both electron and ion, particle-in-cell code, EMPAC. When laser intensity is increased, the laser front can propagate into the plasma slab even if its density is greater than the critical density and the plasma is classically opaque. The propagation velocity of the laser front is compared with the analytical model. The intensity threshold is characterized as a function of plasma density and ion-electron mass ratio.

INTRODUCTION

Recent progressing laser technology has made it possible to generate ultra intensity subpicosecond pulses, and experiments are now being carried out to explore new regimes of laser-matter interaction (1-3). When the plasma is irradiated by such laser with intensities up to $I_L\lambda_L^2$ ~ 10^{20} W/cm^2-µm^2, electrons oscillating in the field of the laser wave are strongly relativistic. The physics of interaction of such ultra intensity laser with the plasma substantially differs from that of the lower intensity case. It was predicted that the ultra power laser would be able to propagate into a classically opaque overdense plasma by increasing the inertial electron mass and hence decreasing the effective electron plasma frequency (4-6). One of the most interest applications envisaged for those laser systems are the fast ignitor (7). The key concept of the fast ignitor is to ignite the compressed hot core by energetic ions and suprathermal electrons generated by an ultra intensity pulse, which is irradiated after a main laser pulse causes the implosion.

We have been investigating the interaction between ultra intensity laser and dense plasma with the use of a 1-1/2 dimensional electromagnetic, relativistic both electron and ion, particle-in-cell code, EMPAC. A dense plasma slab, which is thick enough not to fade away by expansion to vacuum during simulations, is introduced in the simulation system. Long regions of vacuum are appended both sides of the plasma slab, and no artificial boundary conditions, such as the reemission of escaping particles, are needed.

ANOMALOUS PENETRATION

When the effective electron plasma frequency is reduced below the laser frequency by the relativistic effect, the laser can penetrate into the plasma which is classically overdense but effectively underdense (8). The ion density and the laser field profiles obtained from a typical simulation result are shown in Fig. 1. The anomalous penetration of the laser occurs and the laser front propagates deeper and deeper into the overdense plasma with a steady velocity. The propagation velocities are measured from the trajectories of the laser front and are shown in Fig. 2 as a function of the laser intensity. The analytic

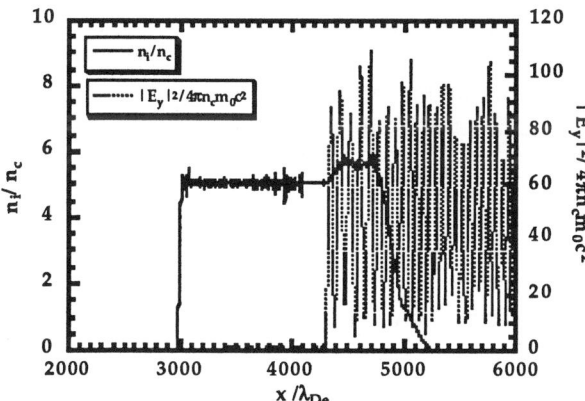

FIGURE 1. The ion density and the laser field profiles obtained from a typical simulation result for $I_L \lambda_L^2 = 10^{20}$ W/cm^2-μm^2, $m_i/m_e = 1836$, and $n_0/n_c = 5$.

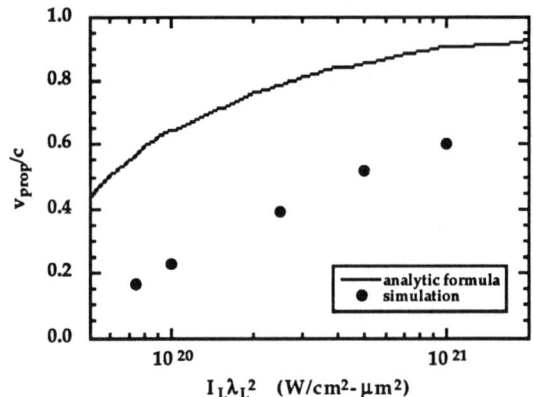

FIGURE 2. The propagation velocity for $m_i/m_e = 1836$ and $n_0/n_c = 5$, where the solid circles and the solid line represent the simulation results and the analytic formula, respectively.

formula, which was simply derived from the dispersion relation as the group velocity of the laser, is described as follows:

$$\frac{v_{prop}}{c} = \sqrt{1 - \frac{n}{n_c \gamma}} \quad (1)$$

where v_{prop} is the propagation velocity, c is the light velocity in vacuum, n is the plasma density, n_c is the critical density, and γ is the Lorentz factor. Ignoring the oscillation in the longitudinal field and considering the quivering velocity in the transverse field, the simple estimation for γ is obtained as follows:

$$\gamma = \sqrt{1 + \frac{I_L \lambda_L^2}{1.37 \times 10^{18}}}. \quad (2)$$

The propagation velocity increases with the laser intensity, and there is much discrepancy between the simulation results and the analytic formula. Namely the velocity obtained by simulations is much lower than the analytical prediction.

Another simulations with immobile ions, in which the ion inertia has no effect, are performed to estimate the threshold of the laser intensity for the anomalous penetration with different plasma

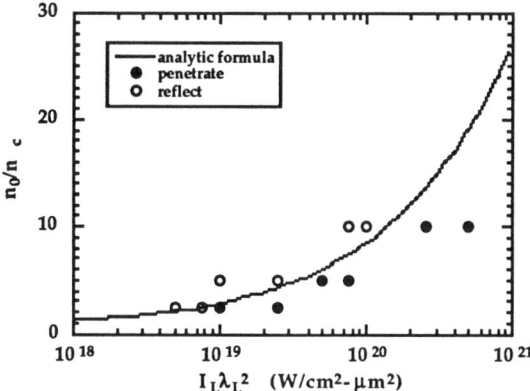

FIGURE 3. The transition diagram between the anomalous penetration and the reflection for immobile ions, where the solid, hollow circles and solid line indicate the penetration, the reflection and the analytic formula, respectively.

densities. Figure 3 shows the transition between the anomalous penetration regime and the reflection regime, and the analytic formula is obtained from Eq. (1) with $v_{prop} = 0$, where the lower part of the analytic curve indicates the anomalous penetration regime. There is remarkable agreement between them.

ION INERTIA EFFECT

We have also performed the simulations with different ion masses to investigate the effect of the ion inertia to the anomalous penetration. Figure 4 also shows the transition between the anomalous penetration regime and the reflection regime, and the solid line indicates the analytic threshold, where the right part of the solid line is the anomalous penetration regime. It is clearly seen that light ions prevent the laser from invading into the overdense plasma even though its intensity is greater than the threshold and higher the laser intensity is, lighter ion is needed for the infiltration.

In the reflection regime the ultra intensity laser strike the plasma slab, the collisionless shock is launched by the extremely high pressure of the laser, compressing the plasma twice the initial density just as the strong shock limitation. The light pressure also pushes electrons, but

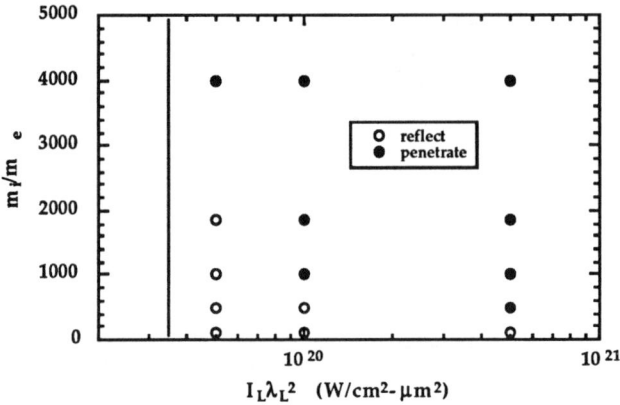

FIGURE 4. The transition diagram between the anomalous penetration and the reflection for $n_0/n_c = 5$, where the solid, hollow circles and the solid line indicate the penetration, the reflection and the analytic formula, respectively.

ions still remain because of their large mass. The large electrostatic field is generated by the strong charge separation and produces very energetic ion beam inward (9,10). This ion beam, which is propagating ahead of the shock, and the unperturbed ion bring about the two-stream instability and subsequent oscillatory profile of density. The plasma edge is also traveling inward with a steady velocity and a redshifting in frequency of the reflected light is observed. We measure the traveling velocity and the backscattered frequency, and find a good agreement with the usual Doppler formula.

In the anomalous penetration regime, quivering electrons increase their inertial mass and hence decreasing the effective electron plasma frequency below the laser frequency. Thus the laser does not reflect at the plasma edge and the laser pressure is not high enough to launch the collisionless shock. The plasma front moves with a velocity, which can be estimated by balancing the momentum flux of the mass flow with the laser pressure. If the velocity is so fast that electrons can not have enough time to get energy from the laser, the inertial electron mass does not increase so much and the laser is reflected by the plasma. This mechanism raises the threshold intensity of the anomalous penetration for a fixed plasma density. Lighter ions are, shorter is the interaction time and higher laser intensity is needed for the penetration. The dependency of the threshold intensity on the ion mass in Fig. 4 can qualitatively explained by this mechanism. It is noted that the laser is not reflected due to the higher density which is achieved by the compression with the collisionless shock. It is decided for the current various parameters that the anomalous penetration should occur at first. If they are the reflection condition, and then the laser is reflected, followed by the collisionless shock which is launched by the extremely high pressure of the laser.

SUMMARY

We have investigated, through a series of simulations with the 1-1/2 dimensional electromagnetic, relativistic particle-in-cell code EMPAC, the anomalous penetration, namely the laser could propagate into a classically opaque overdense plasma by increasing the inertial electron mass and hence decreasing the effective electron plasma frequency. If ions have infinite mass and the ion inertia has no effect, we have found that the analytical threshold of the anomalous penetration were in good agreement with simulation results. As a finite ion mass, light ions were found to raise the threshold intensity of the anomalous penetration for a fixed plasma density.

ACKNOWLEDGMENT

We would like to thank Dr. S. Kato and S. Miyamoto for fruitful discussions.

REFERENCES

1. Main, P., and Mourou, G., *Opt. Lett.* **13**, 467 (1988).
2. Perry, M. D., Patterson, F. G., and Weston, J., *Opt. Lett.* **15**, 381 (1990).
3. Patterson,F. G., Gonzales, R., and Perry,M. D., *Opt. Lett.* **16**, 1107 (1991).
4. Akhiezer, A. I., and Polovin, P. V., *Sov. Phys. JETP* **3**, 696 (1956).
5. Kaw, P. and Dawson, J. , *Phys. Fluids* **13**, 472 (1970).
6. Steiger, A. D., and Woods, C. H., *Phys. Rev. A* **5**, 1467 (1972).
7. Tabak, M., Hammer, J. H., Kruer, W. L., Wilks, S. C., Glinsky, M. E., Woodworth, J. G., and Mason, R. L., "The Fast Ignitor", presented at 23rd ECLIM, Oxford, England, September 19-23, 1994.
8. Lefebvre, E., and Bonnaud, G., *Phys. Rev. Lett.* **74**, 2002 (1995).
9. Wilks, S. C., Kruer, W. L., Tabak, M., and Langdon, A. B., *Phys. Rev. Lett.* **69**, 1383 (1992).
10. Denavit, J., *Phys. Rev. Lett.* **69**, 3052 (1992).

Study of Supra-thermal Electrons and K-α X-rays from High Intensity 500 fs Laser-Produced Plasmas

J. Dunn, B.K.F. Young, A.K. Hankla, A.D. Conder, W.E. White and R.E. Stewart

Lawrence Livermore National Laboratory, P.O. Box 808, Livermore, CA 94551 USA

Abstract. We describe recent laser-solid interaction experiments using the 500 fs Janus Nd:glass (1053 nm) laser presently at 1.5 TW power level. The laser beam path is enclosed in vacuum from the compressor to the target and is focused using an off-axis paraboloid. Optical diagnostics monitor the near field pattern, focal spot, spectrum, temporal shape and pre-pulse level. A 12 µm diameter (FWHM) focal spot is achieved (2.5× diffraction limit) corresponding to a peak irradiance of 8×10^{17} W cm^{-2} on target. A suite of x-ray diagnostics characterize the x-ray emission from the plasma. We present results for normal incidence irradiation of high-Z (Zn, Ge, Mo, Sn) solid targets. The supra-thermal electrons produced in the short scale length plasma have temperature T_H >100 keV and can efficiently fluoresce the cold K-α lines in the 8 - 30 keV energy range.

1. INTRODUCTION

The development of compact high power subpicosecond lasers based on chirped pulse amplification (CPA) has introduced a new area of high intensity, ultrashort plasma physics in the last decade (1 - 4). Plasmas of high density and temperature >100 eV are produced when intense ultrashort laser pulses are focused on solid targets (5 - 7). These can be utilized as sources of bright, short lived x-rays from the thermal plasma (7, 8). Experimental work has characterized the absorption (9 - 12) and the production of supra-thermal electrons (12, 13) in the intensity regime of $I \lambda^2 = 10^{15} - 10^{18}$ W cm^{-2} µm^2. Very recently fast electrons from subpicosecond plasmas have been identified as a mechanism for producing short bursts of x-ray continuum and line emission (12, 14 - 16).

In this paper, we describe recent laser-solid interaction experiments with the 500 fs Janus laser at peak irradiances near 10^{18} W cm^{-2} with known laser characteristics. We present results for normal incidence irradiation of high-Z (Zn, Ge, Mo, Sn) targets. The supra-thermal electrons produced in the short scale length plasma have a temperature T_H >100 keV and can efficiently fluoresce cold K-α lines in the hard x-ray 8 - 30 keV energy range.

2. JANUS-1ps LASER DESCRIPTION

The Janus-1ps laser has a number of key features: (a) the beam path is enclosed in vacuum from compressor to target; (b) the laser is well characterized for every shot; (c) the laser is integrated into the Janus system allowing the combination of three beams of 500 fs, 100 ps, 1 ns duration for channeling and long scale length pre-formed plasma experiments; (d) an upgrade path to shorter pulses ~250 fs by using mixed silicate/phosphate glass and higher energy ~10 J on target corresponding to $>10^{19}$ W cm^{-2} is available.

The Janus-1ps laser is a hybrid CPA system based on a Ti:Sapphire oscillator and regenerative amplifier front end tuned to 1053 nm wavelength with Nd:phosphate glass power amplifiers. The 100 fs, 1 nJ oscillator output pulse train is stretched to 1.5 ns (FWHM) in a single diffraction grating stretcher: gain narrowing during amplification reduces this to 400 ps. Using polarization rotation, a single chirped pulse is selected and injected into the regenerative amplifier. After multiple passes the pulse is switched out by means of a Pockels cell. It is accompanied by pre-pulses separated by 9 ns corresponding to the cavity's round trip time. A pair of Pockels cells is used to minimize the intensity of these pre-pulses to better than 9 orders of magnitude down from the main pulse. The 1 mJ pulse is amplified using 7 mm, 16 mm and 25 mm diameter rods to ~2 J energy. After amplification, the beam is enlarged to 8 cm diameter and is recompressed in a four grating vacuum compressor to 500 fs. The short pulse beam is relayed under vacuum to the target chamber where it is focused to a 12 μm (FWHM) spot size (2.5× diffraction limit) by a 10 cm diameter off-axis paraboloid with 30 cm focal length. A 5 μm thick nitrocellulose debris shield, protecting the paraboloid, is the only post compression transmissive optic, thus minimizing self-focusing effects. The throughput loss due to the compressor and optics after the final

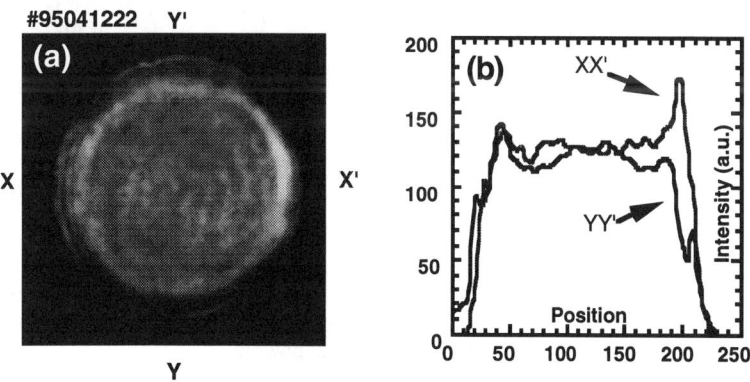

FIGURE 1. (a) Near field beam image at full power is measured by CCD camera. (b) Lineout showing good flat top profile across the beam diameter.

amplification stage results in a maximum of 700 mJ energy on target. With the temporal and focus shape described below, this laser energy corresponds to a peak irradiance of 8×10^{17} W cm^{-2}. The laser can be fired at a repetition rate of 1 shot every 3 minutes and the energy is repeatable to within ± 5%.

Laser parameters including energy, temporal shape, pre-pulse, near field image, focal spot and spectrum are monitored on every shot. The near field beam pattern is recorded by a charge-coupled device (CCD) camera system. Figure 1 shows a typical flat top profile measured at full power after the 25 mm rod. After amplification and appropriate attenuation, the laser focus from the paraboloid is imaged with an f/2 microscope objective with 2μm resolution onto an IR calibrated CCD camera. Figure 2 (a) and (b) show the image and lineout of the 12.6×13.0 μm^2 (FWHM) focus spot measured at full power.

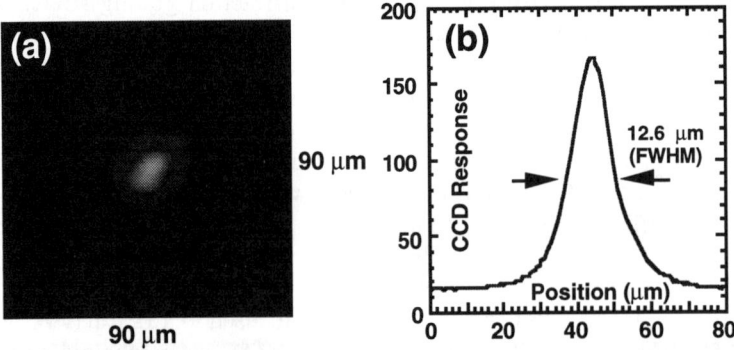

FIGURE 2. (a) Image and (b) lineout of the laser focus spot (12.6×13.0 μm^2 (FWHM)) measured at full power.

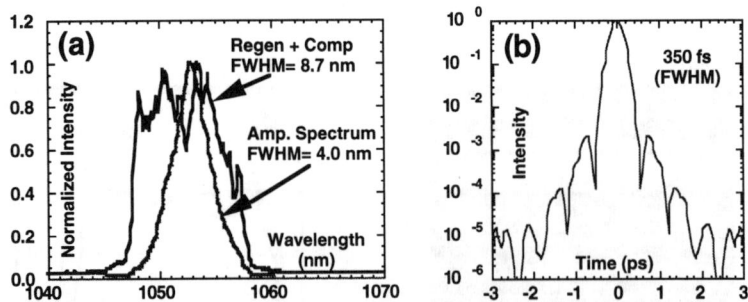

FIGURE 3. (a) Spectrum of the regenerative amplifier beam after the compressor and at full power. (b) Semi-logarithmic plot showing expected pulse shape obtained from the Fourier transform of the laser spectrum taken at full power.

The laser spectrum is monitored at the stretcher and compressor. Figure 3 (a), shows the spectrum at 1053 nm for the regenerative amplifier beam after the compressor with a measured bandwidth of 8.7 nm (FWHM). For full system power shots, 4.0 nm (FWHM) is measured after the compressor: this is sufficient bandwidth to support 400 - 500 fs (FWHM) pulse widths. Figure 3. (b) is a semi-logarithmic plot showing the expected laser pulse shape obtained by taking the Fourier transform of the spectrum after the compressor at full power. The central feature best fits a *sech²* function with a 350 fs (FWHM). The finite size of the gratings results in spectral clipping of the diffracted beam during compression: this introduces a *pedestal* on the main pulse at a level 2×10^{-3}, 10^{-4}, 10^{-5} of the peak intensity for -0.7 ps, -1.4 ps, -2.2 ps, respectively.

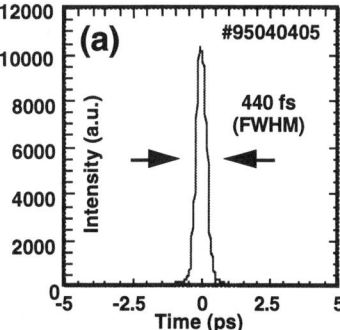 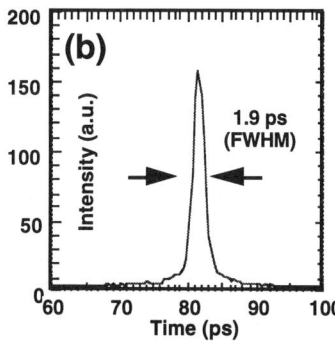

FIGURE 4. (a) Pulse width of 440 fs (FWHM) is measured using a second order single shot autocorrelator, and (b) 1.9 ps instrument limited width from optical streak camera.

The temporal shape and *pre-pulse* are studied as follows. A calibrated second order single shot autocorrelator with 10^4 dynamic range in a 6 ps time window measures the pulse after compression; Figure 4 (a) shows a 440 fs (FWHM) laser pulse at full power. A third order scanning autocorrelator with more dynamic range has measured the laser pulse and confirmed the pedestal of Figure 3 (b). A Hamamatsu C1587 optical streak camera is used to study the 2 ps to 1 ns time frame. The intensifier readout is lens-coupled to a 16-bit CCD camera (17) giving $10^2 - 10^3$ dynamic range relative to the intensifier noise floor. Figure 4 (b) shows a measured width of 1.9 ps (FWHM) from the streaked image: although limited by the streak camera instrumental resolution (> 1.3 ps) this is in agreement with Figure 4 (a). A fast calibrated photo-diode monitors back reflected laser energy from the target and can detect the presence of a pre-pulse 9 ns before the main pulse due to leakage from the the regenerative amplifier. At full power, this pre-pulse level is measured to be $< 3 \times 10^{-10}$ of peak intensity.

3. EXPERIMENTAL DESCRIPTION AND RESULTS

A suite of x-ray instruments have characterized the spatial, spectral and temporal characteristics of the x-ray emission from the high intensity laser-produced plasma. These include x-ray sensitive CCD cameras, a 7-channel filtered NaI scintillator array, a pin diode array, a modified Kentech x-ray streak camera with 2 ps time resolution and various crystal spectrometers. We describe initial results for high-Z slabs where the target is irradiated at *normal incidence* to the laser axis. The 15× magnification x-ray sensitive CCD pinhole camera, described in (18), has 5 - 15 µm diameter apertures positioned ~15 mm from the plasma and filtered with 25 µm Be. The CCD readout allows immediate access to time integrated soft 0.8 - 8 keV x-ray images of the laser spot. Optimum focusing of the laser can be be found in a few shots. It also provides spatial information of the laser intensity for accurate determination of the irradiance parameter. Figure 5 (a) is a profile plot of the soft x-ray image emitted from a Ge slab irradiated with 370 mJ energy at near best focus at 2×10^{17} W cm^{-2}. The measured 14.4 µm (FWHM) from the x-ray spot size is in close agreement with the laser optical focus images recorded at high power, see Figure 2. Spectra recorded simultaneously show that the 8 - 10 Å Ne- and Na-like n = 3 - 2 Ge lines are the brightest x-ray features. High Z targets e.g. Ge ionized to the L-shell are observed to generally radiate stronger 1 - 3 keV x-rays than K-shell ionized Al. Detectable x-ray images can be recorded for >50 mJ laser energy on a Ge slab.

The CCD has excellent linearity between x-ray energy and electron-hole pair production, requiring 3.65 eV x-ray energy/ electron-hole pair (19, 20). When low noise CCDs are cooled, this inherent energy dispersive characteristic combined with high quantum detection efficiency (>50% at 4 keV) and good 2-dimensional spatial resolution (<25 µm) allows the detection and discrimination of single x-ray photons (0.5 - 20 keV). For these reasons, the CCD detector has been utilized extensively by the astrophysical community. It has recently been applied to 3 mJ, 100 fs laser-plasmas (15, 16). We use a 1024 × 1024 pixels CCD camera cooled to -35°C (17) with low read noise 8 electrons rms. and dark current measured at 0.5 electron pixel^{-1} s^{-1}. With a pixel full well capacity of 250 k electrons this gives excellent 15-bit dynamic range. The pulse height spectrum of known K-α lines self-calibrates the detector energy response and agrees with the optical calibration of linearity to better than 0.7%. An algorithm is used to look for charge in isolated pixels produced by x-rays absorbed in the depletion region of the detector.

The CCD detector was placed 93 cm from the plasma at 15° from the target normal. A filter array of 50 µm Cu, 50 µm Y, 50 µm Nb, 50 µm Ta was mounted in a nose cone ~12 cm from the CCD detection plane to provide different energy channels. No fluorescence of the array was observed during the experiment. The photon flux, mainly continuum emission, in the x-ray channels below 8 keV was significantly high to make detection of single photon events difficult for laser energies >100 mJ. Therefore, target materials of Zn with 8.63 keV K-α and higher were studied. At ~10^{18} W cm^{-2}, energetic events identified as 50 - 200 keV continuum x-rays could be detected uniformly across the CCD. The NaI scintillator array measured supra-thermal electron temperatures $T_H > 100$ keV.

Figure 5 (b) indicates the observed x-ray pulse height spectrum for single pixel events from a 25 μm thick Mo foil irradiated at full power. The Mo K-α and K-β fluorescence lines at 17.5, 19.6 keV, respectively, are labelled. The background between 5 - 20 keV is mainly due to continuum x-rays but may include partial events from the Mo K-α line. The peak at 1.5 keV is a combination of thermal noise and partial events. The CCD energy resolution of 210 eV (FWHM) at ~17.5 keV, determined from (19), gives E/ΔE ~ 80 for the detector.

The focus spot was made slightly larger by moving the target towards the paraboloid by 200 μm into the converging laser beam. The laser energy was varied from 20 to 600 mJ corresponding to $5 \times 10^{15} - 2 \times 10^{17}$ W cm^{-2} while the focus remained constant. Figure 5 (c) shows the dependence of the 17.5 keV Mo K-α fluorescence as a function of laser energy. There are a number of interesting features. The K-α fluorescence is observed to scale linearly with the laser energy and the conversion efficiency remains constant over nearly 2 orders of magnitude laser intensity. Secondly, at low intensity the K-α fluorescence is still generated efficiently which indicates a hot electron temperature T_H ~ 20 keV.

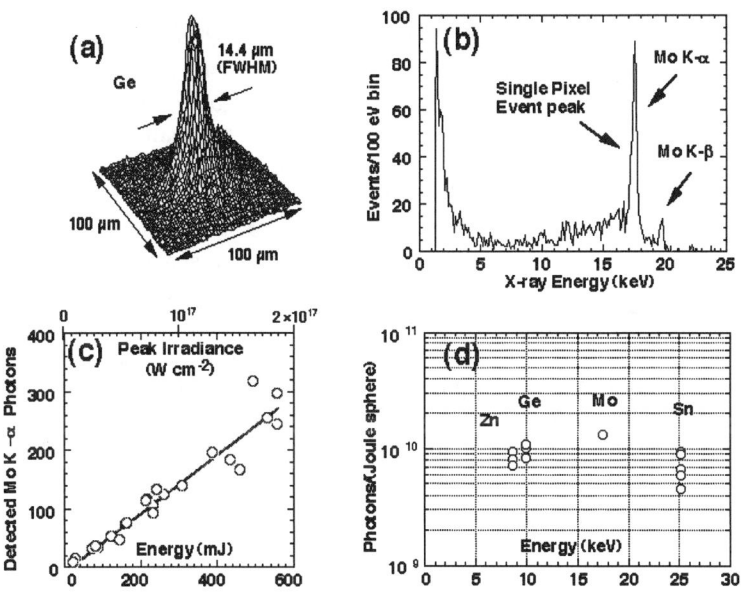

FIGURE 5. (a) Soft x-ray image of the plasma produced by irradiating an optically polished Ge slab with 370 mJ energy at best focus. The focal spot size is 14.4μm (FWHM). (b) Pulse height spectrum showing intense Mo K-α and K-β fluorescence lines at 17.5, 19.6 keV, respectively from laser heated Mo target. (c) Detected Mo K-α photons as a function of laser energy from 20 mJ to 600 mJ. (d) K-α yield measured at high intensity 5×10^{17} W cm^{-2}. (The two lowest Sn data points are at 2×10^{17} W cm^{-2}).

Figure 5 (d) shows the measured K-α yield for different targets at best focus ~5 × 10^{17} W cm^{-2} and normal incidence. The measured K-α yield takes account of the CCD quantum detection efficiency for the depletion region thickness, the filter response and the solid angle, assuming isotropic x-ray emission. Yields of >10^{10} photons /(Joule sphere) are measured for all target materials.

4. DISCUSSION AND CONCLUSIONS

The role of the amplified spontaneous emission (ASE) pedestal of the laser pulse has been discussed for 100 fs experiments (14 - 16). An ASE pedestal 2 ns before the main pulse and at 10^{-6} lower intensity has been reported as necessary for the efficient production of energetic x-rays >20 keV for ~10^{18} W cm^{-2} intensities (14). Similarly, it has been noted at 3 × 10^{16} W cm^{-2} that a controlled ASE pedestal enhances the hot electron conversion efficiency and 1.5 - 6.4 keV K-α fluorescence production (15, 16). The intensity of the pedestal is close to the damage threshold of the target material (15) and rippling of the target surface by the ASE is a possible explanation for this phenomena; higher thermal x-ray emission has also been seen for 100 fs irradiation of textured surfaces (8).

In comparison, this experiment at 500 fs, 1053 nm at near 10^{18} W cm^{-2}, has a pre-pulse level ~10^9 W cm^{-2} which is close to or below the expected target damage. If a *pre-pulse* arriving nanoseconds before the main pulse has an effect on hot electron generation and x-ray production, then some change in the K-α emission would be noted as the laser intensity is reduced. This is not observed in Figure 5 (c) which exhibits a linear scaling of the K-α yield with laser energy down to 5 × 10^{15} W cm^{-2}. Differences in the temporal shapes of the pulses may be part of the explanation. The 500 fs laser pulse is 4× longer and has a significant *pedestal* intensity 3 - 5 ps before the peak of the pulse, see Figure 3 (b). This would pre-form a short scalelength plasma and establish a critical density surface before the peak of the pulse. Significant absorption into hot electrons is observed as inferred by the K-α emission for the 500 fs pulses. The details of the laser absorption mechanism at normal incidence is beyond the scope of this discussion.

Table 1, on next page, shows the absolute 8 - 25 keV K-α conversion efficiency for the studied targets: Mo and Sn have a maximum of ~4 × 10^{-5} of the incident laser energy converted into 17.5 and 25.2 keV monochromatic x-rays. Zn and Ge have a 50% lower value. Studies of high intensity 100 fs experiments (16) observe hot electron pumped 1.5 - 6.4 keV K-α fluorescence for Al, Ca, Fe targets with 12% absorption into 8 keV hot electrons. Quoted conversion efficiencies of ~8 × 10^{-5} for 6.4 keV x-rays are higher than reported here. This slightly higher conversion at lower x-ray energies may be explained by additional pumping from the tail of the thermal electron distribution.

In conclusion, we have shown that high intensity 500 fs laser plasmas can be used as an intense source of hard, 8 - 30 keV, monochromatic x-rays.

Table 1. K-α X-ray Conversion Efficiency

Z	Element	E (keV)	Conversion efficiency into K-α
30	Zinc	8.63	1.3×10^{-5}
32	Germanium	9.88	1.7×10^{-5}
42	Molybdenum	17.5	3.7×10^{-5}
50	Tin	25.2	3.7×10^{-5}

ACKNOWLEDGMENTS

We thank D. Price and G. Guethlein for assistance with the hard x-ray scintillator and P. Young for help with the optical streak camera measurements. The excellent technical support of S. Shiromizu and J. Hunter is much appreciated. Thanks to Mark Eckart for continuing support of this research. This work was performed under the auspices of the U.S. Department of Energy by the Lawrence Livermore National Laboratory under Contract No. W-7405-Eng-48.

REFERENCES

1. Strickland, D., and Mourou, G., *Opt. Commun.* **56**, 219 (1985).
2. Milchberg, H. M., et al, *Phys. Rev. Lett.* **61**, 2364 (1988).
3. Kieffer, J. C., et al, *Phys. Rev. Lett.* **62**, 760 (1989).
4. Murnane, M. M., Kapleyn, H. C., and Falcone, R. W., *Phys. Rev. Lett.* **62**, 155 (1989).
5. Cobble, J. A., et al, *Phys. Rev. A.* **39**, 454 (1989).
6. Audebert, P., et al, *Europhys. Lett.* **19**, 189 (1992).
7. Kieffer, J. C., et al, *Phys. Fluids B* **5**, 2330 (1993).
8. Gordon, S. P., et al, *Opt. Lett.* **19**, 7 (1994).
9. Fedosejevs, R., et al, *Appl. Phys. B* **50**, 79 (1990).
10. Klem, D. E., Darrow, C., Lane, S., and Perry, M. D., in *Proceedings of Short-Pulse High-Intensity Lasers and Applications II SPIE Vol.* **1860**, 1993, pp. 98.
11. More, R. M., et al, these proceedings (1995).
12. Nickles, P. V., et al, these proceedings (1995).
13. Darrow, C., Lane, S., Klem, D. E., and Perry, M. D., in *Proceedings of Short-Pulse High-Intensity Lasers and Applications II SPIE Vol.* **1860**, 1993, pp. 46.
14. Kmetec, J. D., et al, *Phys. Rev. Lett.* **68**, 1527 (1992).
15. Rousse, A., et al, *J. Phys. B: At. Mol. Opt. Phys.* **27**, L697 (1994).
16. Rousse, A., et al, *Phys. Rev. E* **50**, 2200 (1994).
17. Conder, A. D., Dunn, J., and Young, B. K. F., *Rev. Sci. Instrum.* **66** (1) 709 (1995).
18. Dunn, J., Young, B. K. F, and Shiromizu, S. J., *Rev. Sci. Instrum.* **66** (1) 706 (1995).
19. Lumb, D. H., and Hopkinson, G. R., *Nucl. Instrum. Methods* **216** 431 (1983).
20. Lumb, D. H., and Holland, A. D., *Nucl. Instrum. Methods* A**273** 696 (1988).

Radiation Properties from Ultra Short Pulse Laser Produced Plasma

N.Hasegawa, H.Nakagawa, H.Yoneda, K.Ueda, and H.Takuma
Institute for Laser Science, University of Electro-communications

A ps UV laser pulse was used to irradiate solid slab targets at intensity $I=10^{14}-10^{16} W/cm^2$. Anisotropy of the electron distribution function was measured by x-ray polarization spectroscopy. Even with time-integrated measurements, the strong polarization of F $1s^2-1s2p(^1S)$ line was observed at lower intensity condition. It denoted that the beam-like electrons generated at the interaction region penetrated into the over-density critical region.

Introduction

Plasma generated by the ultra-short-pulse laser has very attractive features for many applications and atomic physics. Specially, the high density and high charged but low-temperature plasma produced by the short-wavelength laser have been expected for achieve the compact x-ray laser medium. In such plasmas, the electron mean-free-path becomes comparable to the plasma scale length and the electron distribution function is apart from the Maxwellian distribution. Therefore, normal x-ray spectroscopy assuming the thermal equilibrium condition do not adaptive to accurate diagnostics. It is necessary to measure the electron distribution function directly. Though Thomson scattering method are used to measure it in the lower density plasma, there is no available laser for probing high density plasma like UV irradiated solid targets. In addition, since the relaxation of the high velocity electron is key point to determine these interaction physics, only the difference from the equilibrium such as anisotropy has to be measured. By these reasons, the x-ray polarization spectroscopy[1] was used to diagnose high density plasma in this paper. As there is many literature for polarization in the plasma[2], the detail explanation of the mechanism of the polarization of atomic line is not mentioned here. A simple explanation is below; when the atom excited by the electron collision anisotropically, they keep this information as the nonuniform excitation of the magnetic sub levels. If there is no change through the hyper-fine interaction, the atoms emit the radiation which is polarized in the direction of anisotropy of electrons. The origin of anisotropy is considered to be generated by the quiver motion of the electrons with electric field of the laser light, high energy beam-like electrons generated in the interaction region penetrate into the higher density region, and so on. since it is easy to image the larger anisotropy will be occurred in the lower temperature plasma, this diagnostic will be suitable for the plasma irradiated by the ultra-short-pulse and short wavelength lasers.

Experimental set up

KrF laser system

In the interaraction with high intensity lasers, the contrast of pre-pulse and main pulse is critical point for the generation of high density plasmas. In our system, saturable absorber for UV and visible light can control the pre-pulse energy. The picosecond pulse was generated by a short cavity dye laser (Lamda-Physik FL4000T modified) at 497nm. The second-harmonic pulse converted by a BBO crystal was amplified by a double-pass discharge amplifier. The saturable absorbers were inserted between the amplifiers to keep the prepulse energy to be the oscillator level. After pre-amplification, the pulse passed through the saturable absorber again and was amplified up to 2J level by a large aperture e-beam pumped KrF amplifier. The amplified laser pulse was focused on a CF_2 target with F/3 aspherical lens. Irradiation intensity was chosen for $1 \times 10^{14} - 5 \times 10^{16} W/cm^2$ by changing the focusing condition. Minimum focal spot was $20 \mu m$, and power contrast ratio of the pre-pulse to the main pulse was 10^7 at the target surface. It was confirmed that these small prepulse energy could not affect main plasma parameters in this experiment. Instead of good contrast prepulse, the post-pulse could not be controlled completely because UV saturable absorber, acridine, has slow recover time component. (the absorption recorded to 95% in 0.4ns, but the subsequently recover time is 115ns) The contrast ratio of the post pulse achieved in this experiments was $10^4 \sim 10^5$.

Spectrometer

The polarization states of the emitted x ray lines were measured by the polarized spectrometer. This was consisted with a pair of flat crystals which are made by cleaving a large crystal to obtain the same surface condition of Bragg reflection. Each crystal surface was aligned as the reflection angles because orthogonal direction each other. The estimated reflectivity contrast of s- and p-

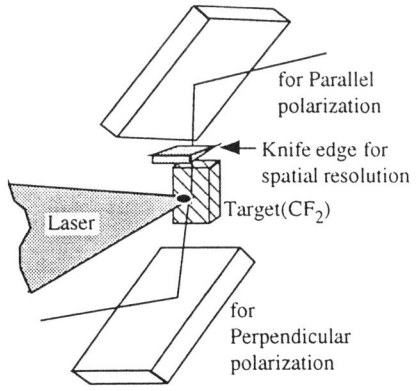

Fig. 1 Polarization spectrometer
  A pair of flat crystals modify cleaving one large crystal were aligned as the Bragg reflection angle(~39°). Knife edge was settled near (~1.0mm) the target plasma to measure the spatial profile of radiation.

polarization was more than 10 at the incident angle of 45±5 degree. The x-ray from the plasma passed through the visible and UV cut filters and after Bragg reflection they were recorded on the Kodak DEF film.

To measure the beam-like electron from the interaction region to the over-dense plasma, one surface of the crystals was aligned to be perpendicular to the target surface. The schematic drawing of this spectrometer and the target is shown in Fig.1. The detected line and the crystals were chosen to be F $1s^2$-$1s2p$ (^1S) and KAP because the respectively. The energy affecting the polarization was 1~3keV and it was suitable for detecting the distortion of the electron distribution function in the relative low temperature plasma.

A knife-edge was put on near the target surface for measure the spatial profile of the line intensity. The magnification of this configuration was an hundred and the spatial resolution of 15μm was achieved in the parallel polarization spectrometer.

Experimental result and discussion

Plasma scale length

Before the polarization spectroscopic measurements, the absorption properties of the CF_2 target was measured at $I=1\times10^{14}W/cm^2$ and pulse length 2ps. Figure 2 shows that the fraction of scattered energy as a function of the incident angle. It denoted that the highest absorption was occurred at $\theta=55°$. From the theory of the resonant absorption, the plasma scale length was estimated to be 0.16λ. If this expansion was assumed to be the thermal motion, the electron temperature had to be the small value (about 8eV).

This indicated that there is the other force like the ponderomotive one to push the plasma toward the high density region. At the polarization spectroscopic measurements, the incident angle was fixed to be 40° because the electron temperature might increase ten times larger at $10^{16}W/cm^2$ irradiance.

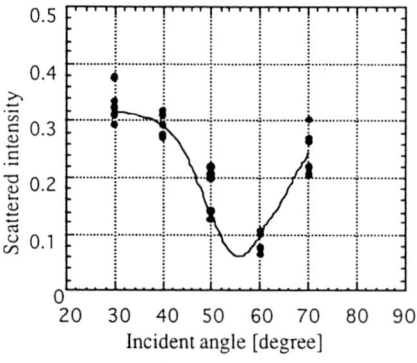

Fig. 2 The absorption of 2ps KrF laser was measured as a function of incident angle. The maximum absorption was obtained θ~55°. The plasma scale length deduced from the theory of resonant absorption was 0.16λ.

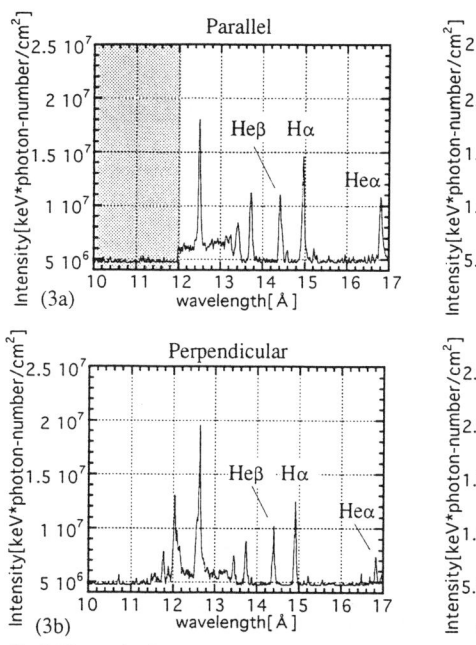
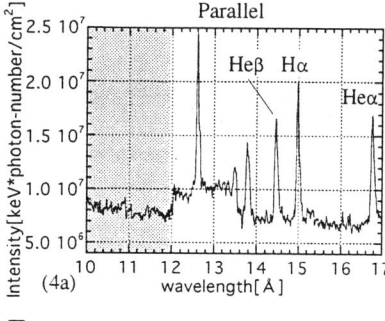
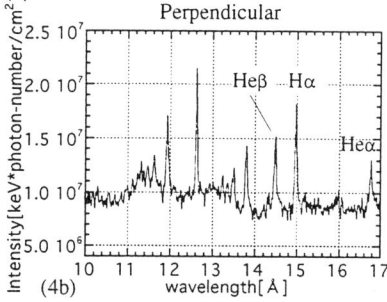

Fig. 3 The result of the polarization spectrometer. The incident angle of laser light was 40° and the irradiance was 5×10^{14} W/cm^2. 3a is a spectrum of parallel component and 3b is perpendicular one. The polarization of He$_\alpha$ line was P=0.41. The hatched region on parallel spectrum means outside of the view in the spectrum.

Fig. 4 The result of the polarization spectrometer. The irradiance was 2×10^{16} W/cm^2. The other notion were the same as Fig. 3. The polarization of He$_\alpha$ line was P=0.34.

Polarization spectroscopy

The emission line of He and H like F ion were observed by the polarization spectrometer at irradiance I=5×10^{14}W/cm^2 and 2×10^{16}W/cm^2. The results of the time integrated and spatial integrated spectrum were shown in Fig.3 and Fig.4, respectively.

In both spectrums of the parallel and the perpendicular spectrometer, almost the same intensity of H like line was obtained that denoted the reflectivity of both crystal was almost equal. To obtain the polarization state with higher accuracy, intensity of 1s-2p line was used as a reference line because it is known that this line is not polarized[3]. The polarization state was defined by this formula,

$$P = \frac{\left(I_{1s^2-1s2p}/I_{1s-2p}\right)_{parallel} - \left(I_{1s^2-1s2p}/I_{1s-2p}\right)_{perpendicular}}{\left(I_{1s^2-1s2p}/I_{1s-2p}\right)_{parallel} + \left(I_{1s^2-1s2p}/I_{1s-2p}\right)_{perpendicular}}$$

In both irradiance, the large polarization of $1s^2$-$1s2p$ (^1S) line was observed even with time and spatial integrated measurements. (P=0.41 for I=5×10^{14}W/cm^2 and P=0.34 for I=2×10^{16}W/cm^2.) The positive sign of polarization denoted beam-like distribution of the electron. The electron density estimated from Stark broadening of the $1s^2$-$1s3p$ line were 1×10^{22}cm^{-3} for I=2×10^{16}W/cm^2. This density was as large as the density of laser turning point ($n_e=n_c\cos^2\theta$). Those results mean that the fast electron generated at interaction region penetrated into higher density region and created the large anisotropy of the electron distribution function. In the previous experiment[1], in which 1μm and 1ps laser light was used for irradiating Al targets at normal incident, the smaller and negative polarization was obtained.(P=-0.25) There are two reason for explanation of this difference. At first, as the KrF laser was used in this paper, the electron temperature of the background was smaller than those of 1μm case and the larger anisotropy was performed in the plasma. The second, obliquely incident laser light generated the larger number of the super thermal electron by the resonant absorption process and these electron were transported to the higher density region. These physical figures also explained the dependence of irradiance on the polarization. At higher intensity, the mean electron velocity of the electron increased and decreased the anisotropy at around 1keV energy of the electron. Two wavelength polarization spectroscopy which will detect the different energy regions of the electrons may be clarified the shift of the mean velocity of the electron at higher irradiance.

Spatial profiles of atomic lines

Fig. 5 shows that the results of the spatial profile measurements of $1s^2$-$1s2p$ and $1s$-$2p$ line intensity as a function of the distance from target surface. The emission of the $1s^2$-$1s2p$ line was occurred at the target surface because the spatial spread of this line was almost equal to the spatial resolution of this diagnostics. That was consisted with the Stark broadening density measurements

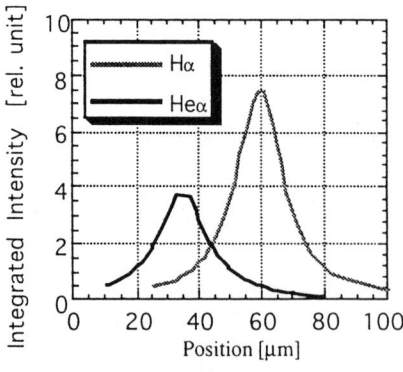

Fig. 5. F-H$_\alpha$ and He$_\alpha$ spatial profile at I=5×10^{14}W/cm^2. The spatial resolution of this measurements was 15μm. H$_\alpha$ emission was occurred 25μm far from the position of the He$_\alpha$ line emission.

in which He-line. The separation of emission position of The reason of $1s^2$-$1s2p$ and $1s$-$2p$ lines were observed. The reason of this was considered that the ionization occurred at the later time of the main pulse because of the slow relaxation of the electron energy to the ion system or heating by the post-pulse energy. The further experiments with more clean laser beam are planed with the Pockels cell switching in this institute.

Conclusion

The anisotropy of electron distribution function of KrF irradiating plasma was measured by the x-ray polarization spectroscopy. The measured polarization denoted that the hot electron generated at the interaction region penetrated into higher density region and the beam-like distribution function was performed. The measured anisotropy was larger than the previous experiments of 1μm laser. The reason of this large polarization was considered to be the lower background temperature in UV laser interaction and the hot electron generation in the obliquely incident. These mean the polarization spectroscopic method was powerful and only tool for diagnose the high density plasma produced by the ultra-short pulse UV laser light.

Reference

[1] J. C. Kieffer, J. P. Matte, M. Chaker, Y. Beaudoin, C. Y. Chien, S. Coe, G. Mourou, J. Dubau and M. K. Inal, Phys. Rev. E 48, 4648 (1993).
[2] J. C. Kieffer, J. P. Matte, H. Pepin, M. Chaker, Y. Beaudoin, T. W. Jhonston, C. Y. Chien, S. Coe, G. Mourou and J. Dubau, Phys. Rev. Lett. 68, 480 (1992).
[3] M. K. Inal and J. Dubau, J. Phys. B 22, 3329 (1989).

Dependence of x-ray yield from aluminum plasma produced by a pair of femtosecond Ti: sapphire laser pulses on pulse time separation

Hidetoshi Nakano, Tadashi Nishikawa, Hyeyoung Ahn, and Naoshi Uesugi

NTT Basic Research Laboratories
3-1, Morinosato Wakamiya, Atsugi-shi, Kanagawa Pref. 243-01, Japan

Abstract The influence of a prepulse on soft x-ray emission in the range of 50 ~ 200 Å from Al plasma produced by 130-fs Ti:sapphire laser pulses was studied at an intensity of 10^{14} W/cm^2 at normal incidence. A prepulse with an intensity of 10^{13} W/cm^2 caused enhancement of soft x-ray emission with a large time separation between prepulse and an intense main pulse. However at 800 nm, a prepulse resulted in reduction with small pulse time separation, contrary to previous reports. This observed trends were qualitatively explained in terms of the absorption dependence on scale length.

With the recent development of high-power ultrashort pulse lasers, laser-produced plasmas have become more attractive for their potential use as efficient and ultrafast x-ray sources (1, 2). One possible way to enhance x-ray yield is through use of a prepulse that plays an important role by forming gaseous plasma before the incidence of an intense main pulse (3). The importance of the additional prepulse for pumping x-ray lasers has been shown both theoretically (4) and experimentally (5). In earlier experiments with short main pulse, the ASE pedestal was regarded as the prepulse (6, 7), and the ASE-prepulse was found to play an enhancement or reduction role depending on the experimental conditions (6). The enhancement properties of short prepulses have been reported (3, 8-10) using UV to visible pulses. However detailed and systematic study on the influence of subpicosecond prepulse, including spectral information and dependence of prepulse parameters, has been reported only with UV pulses (9).

In this paper, we describe the experimental results of investigation into the enhancement of soft x-rays emitted from Al plasma produced by femtosecond Ti:sapphire laser pulses at 800 nm. Soft x-ray spectra were measured by changing pulse time separation between prepulse and main pulse with various combinations of prepulse and main pulse intensities. At a large pulse time separation, the soft x-ray yield was drastically enhanced. However, in our experiments, the slight reduction of soft x-ray yield was observed in all cases at a small pulse time separation in the range of 10 ~ 100 ps. This has not yet been reported with short prepulses. The experimental results were compared with calculations based on the collisional absorption in plasma to support a qualitative understanding of the observed results.

The laser system, which utilized chirped pulse amplification, consisted of a passively mode-locked Ti:sapphire oscillator, a pulse stretcher, a regenerative amplifier, a two-stage linear amplifier chain, and a pulse compressor. It operated at 10 Hz, and provided 130-fs, 50-mJ optical pulses at a wavelength of 800 nm. The laser pulse was split into two parts with a variable time delay τ_D: a prepulse and a main pulse. Both pulses were focused onto the same spot on a flat target at normal incidence with a 500-mm focal length lens. The spatial overlap of the two pulses was monitored by a Si CCD after magnifying optics. A flat aluminum-deposited glass was used as a target to avoid the effects due to surface irregularities (11). The thickness of the Al layer was about 3000 Å, which was thicker than the penetration depth (6, 12). The target was mounted on a rotating stage to expose a fresh surface at each laser shot. A flat-field grazing-incidence spectrograph was mounted at 45° to the target normal to measure the time-integrated soft x-ray spectra in the range from 50 to 200 Å. A ruled unequal groove spacing grating with a nominal groove number of 1200 mm^{-1} was installed in the spectrograph. A microchannel plate combined with a linear CCD array was used as a detector.

In Fig. 1, the total yield of soft x-ray is plotted as a function of laser intensity. In this figure, the results of a solid Al target are also shown for comparison. From

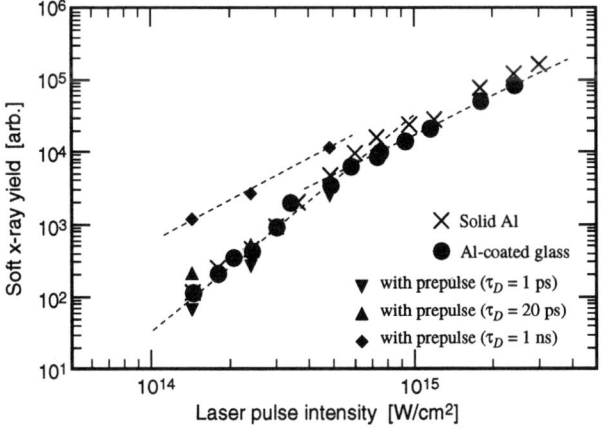

Figure 1. Total soft x-ray yield as a function of laser pulse intensity.

this figure, it is clear that the 3000-Å thickness of the aluminum layer was enough for the subsequent experiments with a laser intensity below 10^{15} W/cm². This figure shows that the total soft x-ray yield in the range from 50 to 200 Å is a good fit to a power law with an exponent of 3.0 for the intensity below 5×10^{14} W/cm² and 1.8 for higher intensity where there is no prepulse. In this figure, the soft x-ray yield is plotted as a function of the main pulse intensity preceded with a prepulse of 2.0×10^{13}-W/cm² intensity for various pulse time separations. With a prepulse, the yield is fit to a power law with an exponent of about 3, which is close to the value for the single pulse case, at a small separation and 1.8 at a 1.0-ns separation. This means that the prepulse is more effective for lower intensities at a large pulse time separation.

Figure 2 shows the dependence of the enhancement in total soft x-ray yield due to the prepulse on the pulse time separation between two pulses for prepulse intensities of 5.9×10^{12}, 2.0×10^{13}, and 5.9×10^{13} W/cm² while keeping the main pulse intensity at 2.4×10^{14} W/cm². At a pulse time separation larger than 100 ps, the soft x-ray yield increased with pulse time separation as expected. The maximum enhancement we observed was over 30. However at a small pulse time separation, a slight decrease of the soft x-ray yield was observed. The reduction in soft x-ray yield was not reported in a similar work with KrF laser pulses at 248 nm (9). Figure 2 also reveals that the region of soft x-ray yield reduction shifts toward the smaller pulse time separation with increasing prepulse intensity. This tendency can be explained by the expansion properties of the preformed plasma that is formed by

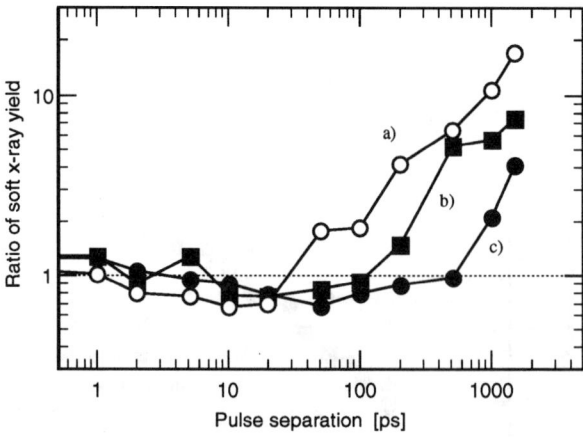

Figure 2. Ratio of total soft x-ray yield between with and without prepulse as a function of pulse time separation for various prepulse intensities. a) $I_{pre} = 5.9 \times 10^{13}$ W/cm², b) $I_{pre} = 2.0 \times 10^{13}$ W/cm², c) $I_{pre} = 5.9 \times 10^{12}$ W/cm². The main pulse intensity I_{main} was 2.4×10^{14} W/cm² in all cases.

the prepulse irradiance. The expansion velocity of the preformed plasma v_{exp} is related to plasma temperature T_e as $v_{exp} = \sqrt{Zk_BT_e/m_i}$, where Z and m_i are average ionization degree and ion mass, respectively. With higher prepulse intensities, the initial temperature and the ionization degree of preformed plasma increases. Therefore, soft x-ray reduction was observed at shorter pulse time separations with higher prepulse intensity.

For normal incidence, collisional absorption is the dominant absorption process for the laser light. As a first approximation to estimate the prepulse influence, we calculated the light absorption of the preformed plasma. For this calculation, we numerically solved the Helmholtz wave equations (13) with a density gradient profile given as the Riemann solution of the hydrodynamic equations (14): $n(x,t) = n_0[3/4 - x/(4v_{exp}t)]^3$. The solid density n_0 and the electron-ion collisional frequency v_{ei} were assumed to be 1.6×10^{23} cm^{-3} and 6×10^{15} s^{-1} (2.55ω). We chose values that were listed in a paper by Wang and Downer (15) since the prepulse intensity was similar to their experimental conditions. In Fig. 3, calculated absorptions are shown as a function of scale length L ($= v_{exp}t$) for collisional frequencies v_{ei} of 4.0ω, 2.55ω, and 1.5ω, where ω represents the angular frequency of the laser light. In the figure, scale length was normalized by the wavelength of the laser light. According to these results, absorption decreases initially to a minimum value near $L \sim 0.3\lambda$, and then increases with scale length for all cases. This trend of collisional absorption qualitatively agrees with the depen-

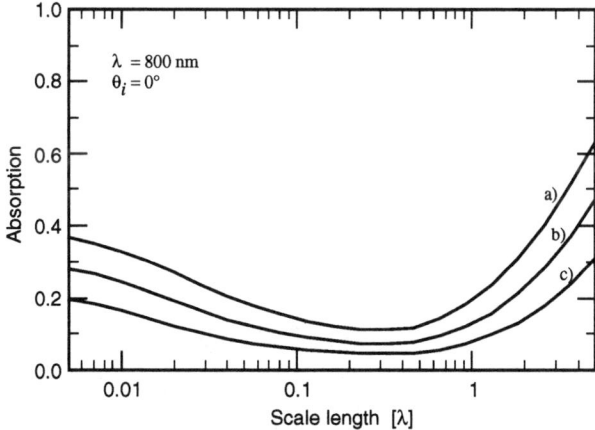

Figure 3. Calculated dependence of light absorption at 800 nm in plasma on scale length. Scale length was normalized by the wavelength of light. Calculations were carried out in case of s-polarized light with an incidence angle of 0°. a) v_{ei} = 4.0ω, b) v_{ei} = 2.55ω, c) v_{ei} = 1.5ω.

dence of x-ray yield on pulse time separation shown in Fig. 2. By assuming electron temperature T_e and ionization degree Z of the preformed plasm as 10 eV and 3, respectively (15), and neglecting their time evolution during plasma expansion, the scale length of 250 nm corresponds to the pulse time separation τ_D around 25 ps. Figure 2 shows that the soft x-ray yield reached a minimum at the pulse time separation around 20 ps when the prepulse intensity was 2.0×10^{13} W/cm². This is denoted as b) in Fig. 2. Since electron temperature T_e depends on laser intensity I as, $T_e \propto I^{4/9}$ (16), and plasma expansion velocity v_{exp} depends on T_e as, $v_{exp} \propto T_e^{1/2}$, the pulse time separations where the soft x-ray yield reaches a minimum were estimated to be 19 and 33 ps in case of a) and c) in Fig. 2, respectively. On the other hand, Fig. 3 shows the soft x-ray yield reached minimum value at pulse time separations around 10 ~ 20 and 50 ps in case of a) and c), respectively.

Figure 4 shows observation of soft x-ray spectra at various pulse time separations. Intensities of the main pulse and the prepulse were 2.4×10^{14} and 2.0×10^{13} W/cm², respectively. These figures reveal that the amount of soft x-ray emission changed almost uniformly over the whole spectral range at small separation. For small pulse time separation, clear peaks were not observed in a soft x-ray spectrum. At a large pulse time separations, some peaks appeared in a soft x-ray spectrum. The peaks appear in the curve, around 68, 78, and 90 Å generally correspond to from AlVIII, AlVII and AlVI ions emissions, respectively. This result shows that the degree of ionization after the main pulse irradiation increased at a large pulse time separation

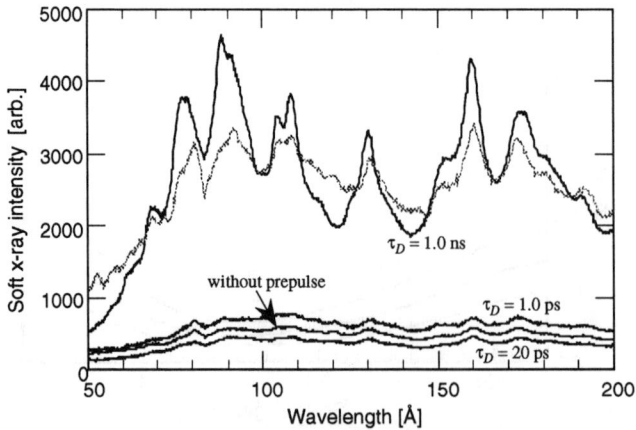

Figure 4. Spectra of soft x-ray emission with various pulse time separations. The intensities of the main pulse and the prepulse were 2.4×10¹⁴ W/cm² and 2.0×10¹³ W/cm², respectively. The dotted curve is a spectrum with a single pulse having an intensity of 4×10¹⁴ W/cm².

compared with that at a small separation. However the drastic change in spectrum, which was reported by Teubner et al. (9), was not observed. As shown in Fig. 1, a single pulse with a 4×10^{14}-W/cm^2 intensity could emit almost the same amount of soft x-rays as that emitted when the pulse time separation was 1 ns. However, in the single pulse case, the continuum on the short wavelength side ($\lambda < 70$ Å) was much higher than that of the double pulse cases although there was no obvious peak.

In summary, we have investigated the influence of prepulse in soft x-ray emissions from aluminum plasma produced by femtosecond Ti:sapphire laser pulses at normal incidence. We observed that soft x-ray emission was enhanced by a prepulse, except when the pulse time separation was small. Small separation reduced the soft x-ray yield. These trends were qualitatively explained by the collisional absorption properties of laser light in preformed plasma.

REFERENCES

1 Murnane, M. M., Kapteyn, H. C., Gordon, S. P., and Falcone, R. W., Appl. Phys. B **58**, 261 (1994).
2 Kieffer, J. C., Chaker, M., Matte, J. P., Pépin, H., Côté, C. Y., Beaudoin, Y., Johnston, T. W., Chien, C. Y., Coe, S., Mourou, G. , and Peyrusse, O., Phys. Fluids B **5**, 2676 (1993).
3 Kühlke, D., Herpers, U., and von der Linde, D., Appl. Phys. Lett. **50**, 1785 (1987).
4 Burnett, N. H. and Enright, G. D., IEEE J. Quantum Electron. QE-**26**, 1797 (1990).
5 Sher, M. H. and Benerofe, S. J., J. Opt. Soc. Am. B **8**, 2437 (1991).
6 Murnane, M. M., Kapteyn, H. C., and Falcone, R. W., Phys. Rev. Lett. **62**, 155 (1989).
7 Nam, C. H., Tighe, W., Valeo, E., and Suckewer, S., Appl. Phys. B **50**, 275 (1990).
8 Tom, H. W. K. and Wood, II, O. R., Appl. Phys. Lett. **54**, 517 (1989).
9 Teubner, U., Kühnle, G., and Schäfer, F. P., Appl. Phys. B **54**, 493 (1992).
10 Kieffer, J. C., Chaker, M., Matte, J. P., Côté, C. Y., Beaudoin, Y., Jiang, Z., Chien, C. Y., Coe, S., Mourou, G., Peyrusse, O., and Gilles, D., in *Short-Pulse High-Intensity Lasers and Applications II*, Edited by Baldis, H. A., Proc. SPIE **1860**, 127 (1993).
11 Kapteyn, H. C., Murnane, M. M., Szoke, A., Hawryluk, A., and Falcone, R. W., in *Ultrafast Phenomena VII*, edited by Harris, C. B., Ippen, E. P., Mourou, G. A., and Zewail, A. H. (Springer-Verlag Berlin, Heidelberg, 1990), p. 122.
12 Zigler, A., Burkhalter, P. G., Nagel, D. J., Rosen, M. D., Boyer, K., Gibson, G., Luk, T. S., McPherson, A., and Rhodes, C. K., Appl. Phys. Lett. **59**, 534 (1991).
13 Milchberg, H. M. and Freeman, R. R., J. Opt. Soc. Am. B **6**, 1351 (1989).
14 More, R. M., Warren, K. H., Young, D. A., and Zimmerman, G. B., Phys. Fluids **31**, 3059 (1988).
15 Wang, X. Y. and Downer, M. C., Opt. Lett. **17**, 1450 (1992).
16 Fedosejevs, R., Ottmann, R., Sigel, R., Kühnle, G., Szatmári, S., and Schäfer, F. P., Appl. Phys. B **50**, 79 (1990).

Applications of a 30-fs Multiterawatt Laser (A): Generation and Time-gated Imaging of Laser-Produced X-rays for Medical Applications

C. P. J. Barty

University of California, San Diego
Urey Hall, MC 0339
La Jolla, CA 92093-0339

C. L. Gordon III, B. E. Lemoff and G. Y. Yin

Stanford University
Edward L. Ginzton Laboratory
Stanford, CA 94305

P. M. Bell

Lawrence Livermore National Laboratory
L-484, P.O. Box 808
Livermore, CA 94550

Abstract: 30-fs, multiterawatt laser pulses are focused to intensities of >10^{18} W/cm^2 onto a solid Ta target to generate x-rays (10-30 keV) for diagnostic imaging. Time gated detection is demonstrated as a technique for removal of scattered radiation and for the improvement of image contrast by a factor of nearly 5.

INTRODUCTION

In 1992 it was reported that hard x-rays with energies up to 2 MeV could be produced from an ultrashort pulse driven laser produced plasma (1). In that experiment the output of a 0.5-TW laser system capable of producing 125-fs, 60-mJ pulses was focused to an intensity of > 10^{18} W/cm^2 onto a solid Ta target. Preceding this pulse was a long duration, low intensity prepulse which created a low density plasma in front of the target. The main laser pulse accelerated electrons in this plasma to MeV energies. These electrons then collided with the target material and produced Bremsstrahlung x-rays. This process is very similar to that of a conventional x-ray tube in which electrons are accelerated from a cathode and create x-rays at an anode. Indeed it was determined in the 1992 experiment that the conversion efficiency of laser light to x-rays scaled as the peak electric field strength of the laser pulse much the same way as the efficiency in a conventional rotating anode device scales as the anode/cathode fall voltage. Although x-rays with energies up to 2 MeV were measured, the majority of the flux was estimated to be in the 20 to 150 keV range or the "diagnostic" x-ray range commonly used for medical radiographic imaging (2). This fact suggests that ultrashort-pulse-pumped, LPP x-ray sources may have uses in medical imaging applications. In particular, there are two distinct differences between the LPP x-ray source and conventional x-ray sources, namely source duration and source size. The duration of the LPP source is believed to be <1 ps or nearly 6 orders of magnitude shorter than conventional x-ray devices and allows different modalities of imaging to be employed. The source size of the LPP x-ray source may be as much as 100 times smaller than conventional devices and thus may allow imaging of much smaller features than previously possible. A comparison of the two sources is given in Table 1.

Measurements of the flux from the 1992 experiment suggests that in order to be competitive with conventional devices, higher repetition rate sources must be developed. From the experimentally determined scaling with electric field, it is evident that shorter duration pulses will require less energy to produce the same x-ray yield and thus may allow higher repetition systems. In the present experiments a new laser system was utilized which was capable of producing 30-fs, 4-TW, nearly diffraction limited pulses at a repetition rate of 10 Hz (3). X-rays generated with this device were used in a novel imaging arrangement which capable of producing significantly higher contrast images.

TIME GATED DETECTION

In traditional x-ray imaging, three things may happen to the x-ray photons as they pass through a sample: they may be absorbed by

	Pulsed X-Ray Tube	LPP Source (past)	LPP Source (future)
Shots/Sec	30	5	(1000)
Deposited J	2.75 J	40 mJ	(10 mJ)
Efficiency	1%	0.3%	(1%)
Average Power	825 mW	0.6 mW	(100 mW)
X-Ray Δt	5 ms	<1 ps	(~100 fs)
Source Size	1 mm X 1 mm	100 μm X 100 μm	(100 μm X 10 μm)

TABLE 1. Comparison of Rotating Anode and Laser Generated X-Ray Sources
TIME GATED DETECTION

In traditional x-ray imaging, three things may happen to the x-ray photons as they pass through a sample: they may be absorbed by dense material, they may undergo a scattering event, or they may pass through the sample unaffected. It is the difference in flux between the unaffected or ballistic photons and those that are absorbed that is responsible for the formation of an image at the detector. Scattered photons that also reach the detector will decrease the contrast ratio and increase the noise of the image. In mammalian samples, this can be a significant problem. Typically, for every one ballistic photon passing through 20 cm of mammalian tissue with 2 cm of bone density material in the path, there will be ~ 0.1 absorption events and ~ 7 scattering events.

To illustrate this problem, we consider the hypothetical case of a material that has only absorption and no scattering. In this case the contrast ratio, CR, of the image that is formed is

$$CR = \Delta N/N_B \qquad \text{Eq. (1)}$$

where N_B is the total number of x-ray photons incident on the sample and ΔN is the number difference of the ballistic and absorbed photons. The image will have a signal-to-noise ratio that is determined by the statistics of the x-rays. At low photon numbers, the x-rays may be modeled by a Poisson distribution. Thus, the signal-to-noise ratio, SNR, is given by

$$SNR = \Delta N/\sqrt{N_B} = CR\sqrt{N_B} \qquad \text{Eq. (2)}$$

Now if we consider the same material but add N_s scattered photons to the image then CR and SNR become

$$CR = \Delta N/(N_B + N_S) \qquad \text{Eq. (3)}$$

$$SNR = \Delta N/\sqrt{N_B + N_S} \qquad \text{Eq. (4)}$$

Clearly, CR and SNR are decreased by the presence of scattering. It is possible to increase CR and SNR in either case by increasing the x-ray flux. However, this is not practical when working with live samples. In this situation, the minimum acceptable SNR of the image determines the flux that is used.

Because the x-rays from a laser-driven source have a duration that is short (< 1 ps) with respect to the transit time through the sample (~ 1 ns), it may be possible to eliminate the effects of scattered photons by temporal discrimination. When compared with ballistic photons, scattered photons travel a longer path and thus experience a longer delay in reaching the same location at the detector. If one uses a detector which may be turned off rapidly after the arrival of the ballistic photons, then the scattered photons may be removed from the image. For the same SNR, it is then possible to construct an image with much less flux. For mammalian objects, this can be seen by setting the right-hand sides of Eqs. (2) and (4) equal and making the substitution $N_s = 7N_B$. In this case, the number of photons needed to form the image is eight times less than that needed with time-integrated detection. An eight times reduction in dosage would greatly benefit many medical procedures. For instance, in x-ray coronary angiography, a patient may receive a full year's dose in a single 30-minute session.

The degree of image improvement or dose reduction will be limited by the speed with which the detector may be gated. Relatively simple microchannel plate (MCP) detectors with gate times as short as 50 ps have been developed for time-resolved studies of x-ray emission in inertial confinement fusion experiments (4). To estimate the degree of x-ray reduction that may be possible with such detectors, a quasi-3D numerical simulation was performed. In this simulation, a temporal delta function of x-ray photons was inputted uniformly over one side of a 20 cm x 20 cm 2D area. Each x-ray had a probability of 1 to scatter within a 5 cm mean path. The x-rays had an isotropic angular scatter pattern which is characteristic of Compton scattering in the medical x-ray regime. The path lengths of the x-rays that exited the lower side of the square (the location of the detector) were computed and transformed into a traversal time. The traversal time of the ballistic photon was subtracted from the traversal time of the scattered photon and then a histogram of the photon number versus time was computed. The final output of the calculations was the temporal behavior of the scattered photons. The results of this model for a sample which is 4 times thicker than the mean scattering distance are shown in Figure 1. In this case only 3.3% of the scattered photons arrive at the detector during the first

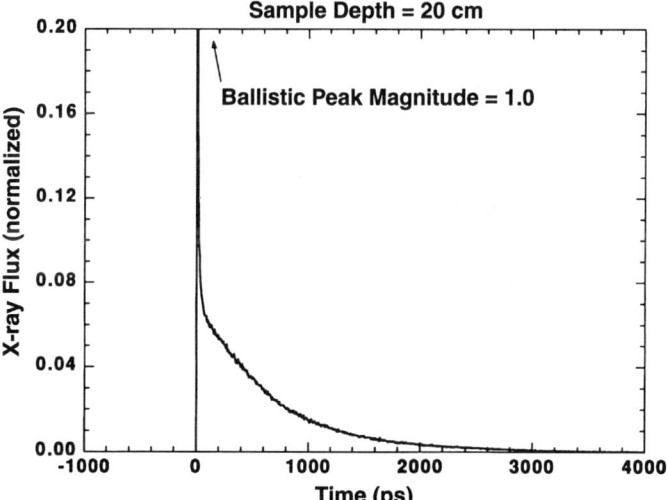

FIGURE 1. Predicted Temporal History of the X-rays Through a Sample Whose Thickness is 4 Times the Mean Scattering Distance

50 ps. Thus one expects to be able to achieve nearly the ideal limit for reduction in dose to the patient with this detector.

EXPERIMENT

Experimental verification (5) of the utility of time gated detection was performed with the arrangement depicted schematically in Figure 2. The response of a human sample was approximated by a phantom consisting of a metallic (predominantly Fe) object suspended vertically in a plastic container filled with water. The metallic object was 1.5 mm wide by 5 cm long with a hexagonal cross section, and the container was 17 cm thick in the direction of the x-ray beam), 30 cm wide and 40 cm high. The phantom was placed between the gated detector input window and laser plasma source at a location which resulted in an image magnification of approximately 2. Figure 3 shows the resulting shadow graphs for three separate tests. In (a) the water was removed from the container and thus no scattering occurred. Approximately 40 laser shots were used. In (b) water was placed in the container and a dc potential of 900 V. In (c) the MCP was pulses with a ~900 V, ~100 ps pulse. Because of the attenuation of the water approximately 2400 shots were required to obtain images (b) and (c). The x-ray exposure rate from the laser produced plasma source was measured to be ~833 mR/pulse at 46 cm from the laser plasma. Total exposures in case (b) and (c) were equal and were monitored by a separate 1 mm thick NaI scintillation detector. From the horizontal lineouts of each image it is

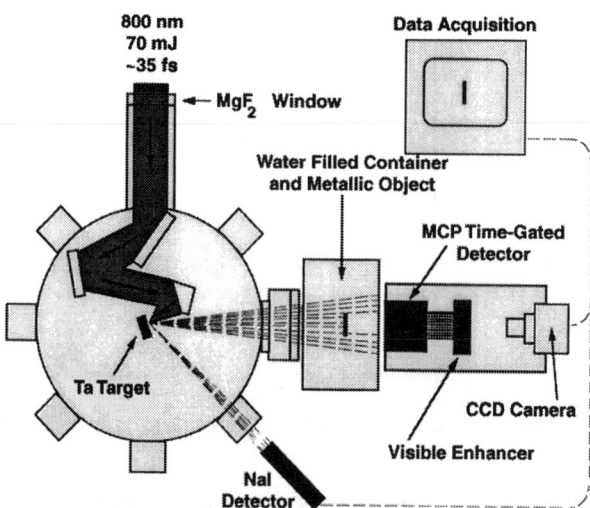

FIGURE 2. Schematic Diagram of the Time-gated X-Ray Imaging Aparatus

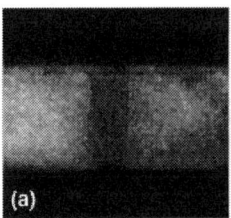 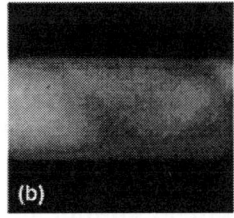 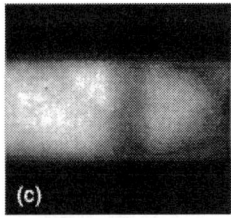

FIGURE 3. Recorded CCD Images of a 1.5 mm Wide Metal Object

possible to estimate the contrast in the three cases. The results are 0.34, 0.06 and 0.27 for (a), (b), and (c) respectively. The factor of ~5 improvement in contrast between dc imaging and gated imaging is significant. Traditional angular discrimination-based methods of scatter reduction typically result in a contrast improvements of the order of 3 (6).

It should be noted that in the above tests no effort was made to optimize the x-ray generation, employ digital image enhancement techniques or to optimize the detector for operation in the 10 to 30 keV range of our source. Such optimizations and tests of methods to reduce jitter between the gating high voltage and the arrival of ballistic x-rays at the detector are the subject of ongoing investigations.

ACKNOWLEDGMENTS

The authors would like to thank Stephen E. Harris for inspiration in this work, Leonard Trulson for assistance with the high voltage pulser, and Don Busick for guidance with the x-ray beam measurements. This work was supported by the U.S. Air Force Office of Scientific Research, by the U.S. Army Research Office, and by generous equipment loans from the Lawrence Livermore National Laboratory.

REFERENCES

1. J. D. Kmetec, C. L. Gordon III, J. J. Macklin, B. E. Lemoff, G. S. Brown and S. E. Harris, *Phys. Rev. Lett.*, **68**, 1527-1530, 1992.
2. K. Herrlin, G. Svahn, C. Olsson, H. Petterson, C. Tillman, A. Persson, C. Wahlstrom and S. Svanberg, *Radiology*, **189**, 65, 1993.
3. C. P. J. Barty, C. L. Gordon III and B. E. Lemoff, *Opt. Lett.*, **19**, 1442-1444, 1994.
4. P. M. Bell, J. D. Kilkenny, O. L. Landon, R. L. Hanks and D. K. Bradley, *Rev. Sci. Instrum.*, **63**, 5072, 1992.
5. C. L. Gordon III, G. Y. Yin, B. E. Lemoff, P. M. Bell and C. P. J. Barty, *Opt. Lett.*, **20**, 1056-1058, 1995.
6. T. S. Curry III, J. E. Dowdy and R. C. Murry Jr., Philadelphia, Pensylvania: Lea & Febiger, 1988, p. 88.

Electron Acceleration by Transverse Electromagnetic Wave Supplemented with Crossed Static Magnetic Field

Noboru Yugami, Syuichi Sanjou, and Yasushi Nishida

Department of Electrical and Electronic Engineering,
Utsunomiya University,
2753 Ishii-machi, Utsunomiya, Tochigi 321, Japan

abstract. Electron linear accelerator, using slow tranverse electromagnetic wave (TE wave) supplemented with crossed static magnetic field, has been demonstrated. An energy gain of 12.0 keV for electrons is observed from an incident energy of 62 keV in a 0.5 m accelerator, when an external magnetic field of 2.0 G is applied. The results show the feasibility of higher gradient and compact accelerator using an intense laser.

INTRODUCTION

Recent interests in charged particle accelerators are focused on generating much higher acceleration gradient than the one possible in the present conventional accelerators, which have a typical acceleration gradient of about 10 - 20 MeV/m. In the last decade some new acceleration schemes, e.g., using a plasma, including a plasma wake field accelerator (PWFA), a plasma beat wave accelerator (PBWA) and a $v_p \times B_0$ accelerator (or a surfatron), have been proposed, [1-6] where v_p is the wave phase velocity and B_0 is the applied static magnetic field. In addition, the possibility of charged particle acceleration using a transverse electric field supplemented with a weak external static magnetic field in parallel with the wave magnetic field has also been discussed.[7] There were no experimental evidences so far, to our best knowledge, and using TE mode for driver wave gives the feasibility of higher gradient and compact accelerators, because the slow wave structure or mode converter is not necessary for high energy accelerators.

THEORY

It is convenient at first to review the acceleration mechanism briefly (Fig. 1). Suppose that the transverse electromagnetic wave which has the maximum electric field E_x and the maximum magnetic field B_y in the x and y direction, respectively, propagates with the phase velocity v_p in the z direction. When an external static magnetic field B_0 less than B_y is applied in the y direction, there exists two magnetic neutral points (A and B) where the electric field is non-zero. Around point B, the Lorentz force acts on the particles which are accelerated by the electric field in the z direction toward the magnetic neutral point. Therefore the particles bunch there, the trapping point, and are accelerated by the wave electric field in the x direction, continuously.

The trapping condition is obtained by the balance of the force in the z direction, $|B_y| > \gamma_p^2 B_0$. The inequality represents the unlimited acceleration condition. In other words, if this condition is satisfied, the electron can be accelerated continuously without detrapping from the wave trough.

EXPERIMENTS

Precise experimental investigations have been performed in order to confirm the theoretical predictions. A schematic view of the experimental set up is shown in Fig. 2(a), and the slow-wave structure of the TE mode as shown in Fig. 2(b). The dimensions of the slow wave structure are listed in Table I. The wave guide wall is made of the copper.

Table. Dimensions of the Dielectric Wave Guide

Dielectric materials	Macorl and Folsteright
Accelerator length	48 cm
Thickness Δd	50 mm
Width $2h$	50 mm
Separation $2d$	13～33 mm
Hight $\ell = 2d + 2\Delta d$	113～133 mm

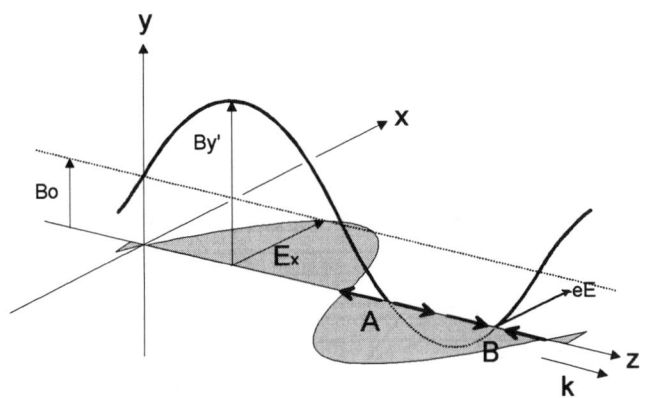

Fig. 1 Schematic View of TE Acceleration Mechanism

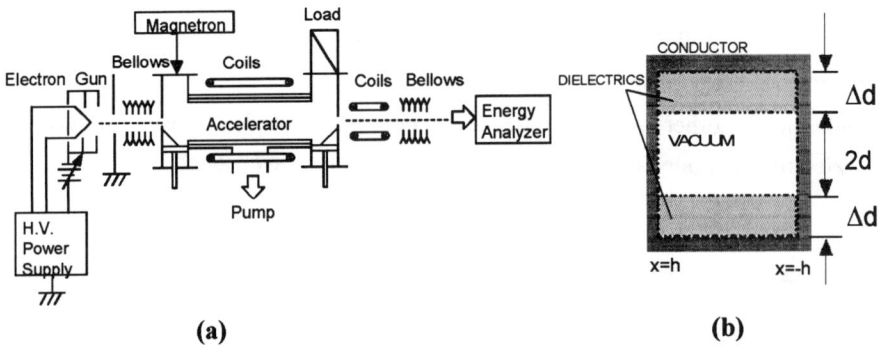

Fig. 2 (a) Experimental Setup (b) Cross View of Slow Wave Structure

This structure consists of 20 dielectric blocks and 48 cm long in the wave propagation direction. The blocks are installed upper and lower side of the wave guide with the separation for the electron path. The injected electrona are initially accelerated by a high-voltage DC power supply, with maximum energy up to 100 keV and a current of 1 mA. The electrons accelerated through the accelerator are analyzed by a magnetic field bending type electron energy analyzer.. The maximum resolution of the energy analyzer is less than 0.1 keV. This instrument is calibrated with the above-mentioned electron beam source. The calibration is carried out by changing the acceleration voltages without RF power before and after the main experiments.

The pulsed electromagnetic wave of a maximum power of 10 kW is generated by a magnetron with a typical pulse width of 5 µs with repetition of 10

Hz.

The static, vertical magnetic field for a $v_p \times B_0$ acceleration is generated by a pair of saddle-shaped external coils. A maximum field strength of 10 G is measured at the center of the accelerator. This value should be strong enough to demonstrate the $v_p \times B_0$ principle with TE mode under the present experimental parameters.

The relative dielectric constant can be estimated to be 6.3 calculated after adjusting the measured phase velocity with the calculated one from the dispersion relation of the electromagnetic wave in the wave guide. Using this value, we have designed the phase velocity of 0.46c (corresponding energy of 65 keV) with $\Delta d / d = 7.7$.

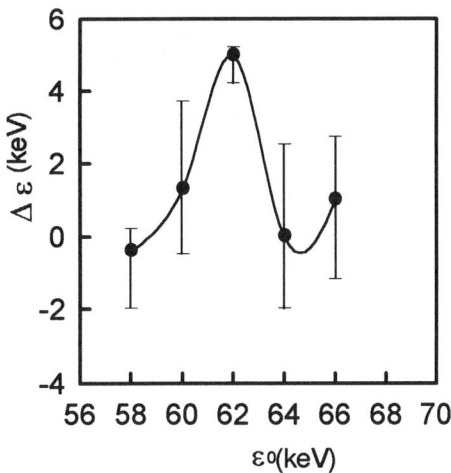

Fig. 3 Electron Energy Increment without External Magnetic Field

Figure 3 shows the electron energy increment without an external applied magnetic field. The maximum electron energy increment, $\Delta \varepsilon = 5.0$ keV is observed when an incident electron energy ε_0 is 62 keV. This implies that this slow wave srtucture with the RF frequency of 2.45 GHz resonates the electrons with the energy of 62 keV. The vertical error bars indicate the variation in electron energy oriented from the unstableness of beam transparecy through the wave guide, because the electrons experience the force from the induced charge on the dielectric surface. The resonance at the electron energy of 62 keV is different from the designed value of 65 keV. This disagreement may be due to the home-made structure. These results indicate that the dielectric slow wave structure can work as the electron accelerator driven by a pure transverse electromagnetic wave propagating with nearly equal to the electron beam velocity. Therefore the relativistic electron beam could be accelerated by an intense transverse electromagnetic waves, for example, high power lasers, in vacuum without using any metalic cavities. However unlimited acceleration would not be expected because there is no electron trapping mechanism in the wave potential when no external magnitic field is applied.

Typical examples of observed energy spectrum of electron are shown in Fig. 4, when an incident electron energy is 62 keV. Arrows indicate the peak electron fluxes where energy is measured. When the static magnetic field is increased, the peak electron energy increases until $B_0 = 2.0$ G. For further increase of the magnetic field ($B_0 > 2$ G), no acceleration occurs any more as the trapping condition for the particles, (Eq. (4)), is violated.

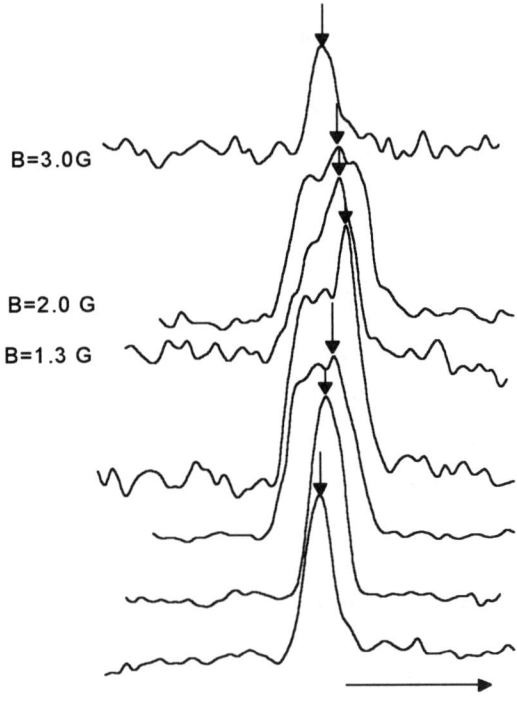

Fig. 4 Typical Electron Energy Spectrum

An example of the experimental results is shown in Fig. 5. Here, the horizontal and vertical axes indicate the strength of the applied external magnetic field and the energy increment of electrons which is defined by the differences between the accelerated energy ε and the energy without the static magenitc field, i.e. $\varepsilon - \varepsilon$. Symbols indicated by solid circles, triangles and squares stand for the value of the incident electron energy of ε = 60, 62, and 66 keV, respectively. The maximum energy increment of 7.0 keV is observed at the incident energy of 62 keV with the external magnetic field of 2.0 G. Note that the resonant interaction between the electromagnetic wave and the electrons occurs, such as shown in Fig. 3, and even at $B = 0$, the electron energy is increased.

Up to 2 G, the energy increment increases as the applied magnetic field, while over 2 G, it saturates or even decreases in all incident energy. These results imply that the electron detrapping occurs from the wave potential.

In our experiments, the trapping condition is not necessary satisfied because the magnetic field B of the wave can be estimated to be of the order of 1.3 G. The inequality requires the complete equilibrium in the phase space. In the present experiments, however, it takes about t>20 ns to reach the steady state, while the accerelation ceases t<3.6 ns in the present experiments as is estimated from the electron travelling time throughout the wave guide when the incident energy of the electron is 62 keV.

The electromagnetic wave with a different mode, for example, a slow longitudinal wave (TM mode), could propagate in the wave guide, so that such a wave can accelerate the electron. Althogh we have not directly investigated the existence of the TM mode in our wave guide, we have calculated the propagation of the TM mode with the same manner as the analysis for the TE mode.

CONCLUSION

In conclusion, the electron acceleration using slow transverse electromagnetic wave (TE wave) supplemented with the static magnetic filed has been demonstrated. The energy gain of 12.0 keV for electrons is observed from an incident energy of 62 keV in a 0.5 m accelerator, when an external magnetic field of 2.0 G is applied. The results show the feasibility of the higher gradient and compact accelerator using an intense laser.

REFERENCES

1. T. Tajima and J. M. Dawson, Phys. Rev. Lett. **43**, 267 (1979).
2. J. M. Dawson, R. W. Huff and T. Katsouleas, Phys. Rev. Lett. **54**, 693 (1985).
3. Nishida, M. Yoshizumi and R. Sugihara, Phys. Fluids **28**, 1574 (1985).
4. Katsouleas and J. M. Dawson, Phys. Rev. Lett. **51**, 392 (1983).
5. Nishida and T. Shinozaki, Phys. Rev. Lett. **65**, 2386 (1990).
6. Nishida, N. Yugami, H. Onihashi, T. Taura and K. Otsuka, Phys. Rev. Lett. **66**, 1854 (1991); N. Yugami, H. Onihashi, K. Otsuka T. Taura and Y. Nishida, Jap. J. Appl. Phys. **32**, 5703 (1993).
7. Takeuchi, K. Sakai, M. Matsumoto and R. Sugihara, Phys. Lett. **A 122**, 257 (1987); S. Takeuchi, K. Sakai, M. Matsumoto and R. Sugihara, IEEE Trans. on Plasma Sci. **PS-15**, 251 (1987).

Large Amplitude Wakefield in an Ion Wave Regime Excited by Short Microwave Pulse

Noboru Yugami and Yasushi Nishida

Department of Electrical and Electronic Engineering
Utsunomiya University
2753 Ishii-machi, Utsunomiya, Tochigi 321, Japan

Brigitte Cros and Gille Matthieussent

Laboratoire de Physique des Gaz et des Plasma
Batiment 212, Universite Paris-Sud
91405 Orsay, Cedex, France

abstract. When a short microwave pulse of the order of an ion wave period is irradiated in a nonuniform unmagnetized plasma, non-linear large amplitude waves ($\delta n / n_0 \approx 40 \%$) in an ion wave regime have been observed. The observed large amplitude waves can be interpreted by the model based on the Mach-cone structure induced by the supersonic ion bunch accelerated by a wave field at the resonance layer. Before growing the large amplitude ion wave, hot ion beam with the energy (≈ 25 eV) was observed by a tiny Faraday cup. The energy of ion beam is good agreement of predicted by the Mach-cone theory.

INTRODUCTION

To generate a large amplitude electron plasma wave and ion wave is important for the study of wakefield acceleration experiments. So far, large amplitude ion wave($\delta n / n_0 \approx 20 \%$) have been observed in a double plasma device[1,2] by injecting ion bunch in a target plasma. There are two kinds of ion wakefield have been observed; the one is in a single ion species[1], while the other is in three components plasma with negative ions.[2]

In our present experiments, a single plasma and a short microwave pulse are employed to excite the ion wakefield wave. Although the exciter of the wakefield, the microwave pulse does not propagate, i.e. remaining in the resonance region, the excited hot ion beam propagates after shutting off the rf field with the velocity much larger than that of an ion acoustic speed. In these, that the pulse width of the microwave should have a period of the ion oscillation ($\tau_{pi} = 2\pi / \omega_{pi}$) is the key natures for exciting wakefields. The excited wakefield has maximum amplitude of about $\delta n / n_0 \approx 40 \%$.

EXPERIMENTAL SETUP

The experiments are carried out in cylindrical, nonuniform unmagnetized argon plasma produced by a pulse discharge. Typical plasma parameters of the plasma are as follows; density $n_0 \approx 2 \times 10^{11}$ cm^{-3}, electron temperature $T_e \approx 3-5$ eV, ion temperature $T_i \approx T_e/10$. The typical density gradient scale length in the axial direction is $L_z \approx 100-150$ cm. All the plasma parameters including fluctuating wave forms are diagnosed by tiny probes. When density fluctuations are measured, the probe is biased into the electron saturation current region (≈ 40 eV). For measuring the ion energy, a small Faraday cup is used, which can detect ions flowing down towards the lower density side.

The p-polarized microwave pulse with frequency $f = 2.86$ GHz and a maximum power of 10 kW is irradiated from a horn antenna located at lower end of plasma density. The pulse width ranges from 60 ns to 1 μs in full width at the half maximum (FWHM) with a repetition of 10 Hz, and typical fall-off time, τ_{fall}, ranged from 80 to 200 ns, where τ_{fall} is defined as the time difference between 10 % and 90 % of the maximum value of the microwave power.

EXPERIMENTAL RESULTS

When a narrow microwave pulse is irradiated, a large amplitude electron plasma wave can be excited at the critical layer where $\omega = \omega_{pe}$. Figure 1 shows typical density fluctuations, n, detected by the probe located at a resonance point ($n_c \approx 2 \times 10^{11}$ cm^{-3}) as a parameter of different microwave pulse width (shown in bottom traces). The incident microwave denoted by (a), (b) and (c) have a pulse width of 60, 250 and 400 ns with fall-off time of 80, 170 and 180 ns, respectively, at the maximum power of 4.5 kW. Numbers from 1 to 5 denoted in the trace (a) indicate indexes for later convenience. The maximum amplitude, $\delta n / n_0$, corresponding to the peak "5" in (a), (b) and (c) are 0.38, 0.28 and 0.23, respectively.

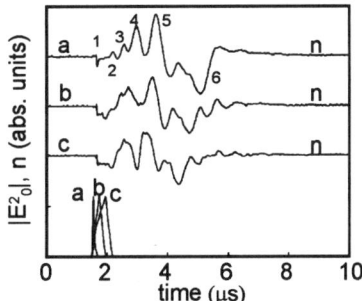

Fig. 1 Temporal Evolution of Ion Waves with Different RF Pulse Width

All the wave forms indicate that large amplitude density fluctuations are excited after shutting the microwave off.

Figure 2 shows a time-space distribution of the fluctuations obtained by scanning the probe in the z direction. The waves were excited at the resonance point ($z \approx 22$ cm) and propagated down the lower density region. The first wave, which is denoted 1 in Fig. 1, is excited at almost the same time of shutting the microwave off and propagates faster than other waves (2 - 5). The velocity of the first wave is close to the electron thermal velocity, 1×10^8 cm/s. After the electron plasma wave propagated out, another waves, marked 2-5 in Fig. 1 which have frequencies in the ion wave regime, are excited. These waves have the phase velocity of 5×10^6 cm/s and propagate with decreasing their velocities.

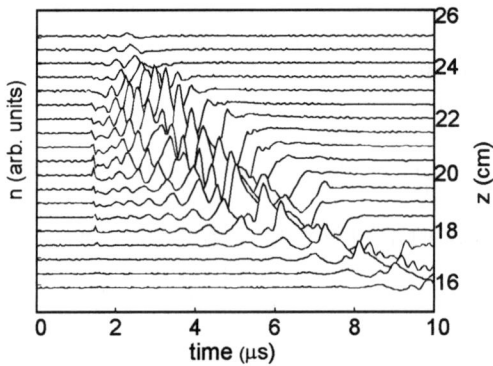

Fig.2 Spatial-Temporal Evolution Ion wave

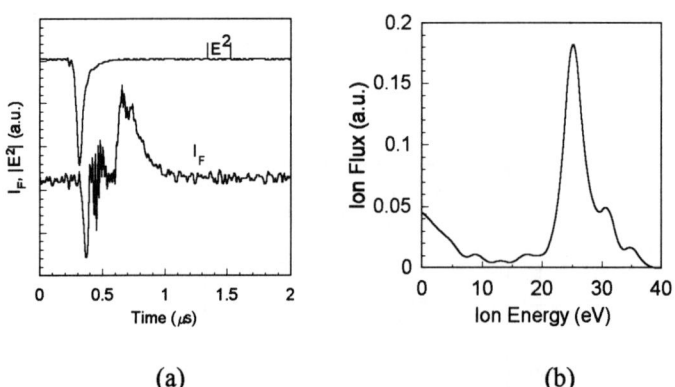

(a)   (b)

Fig.3 (a) Example of temporal variation of the collector current (b) Distribution Function

An example of the Faraday cup wave form is shown in Fig. 3(a). Spike in the wave form around τ=0.4 μs is due to a high energy electron signal whose energy is > 180 eV. The Faraday cup was applied only -180 V as a bias voltage to repel the hot electron, because the break down occurred in the instrument. The spike corresponds with the RF signal, therefore the electrons generate around resonance absorption layer. A positive signal of ions are found around 0.5 μs after turn-on of the microwave. The energy distribution is also shown in Fig. 3(b). Strong high energy ions are shot out with typical energy of 25 eV. These ions travel down toward the lower density side. The bunch of high energy ions can pull-in electrons into the ion bunch area by the Coulomb force from the surrounding plasma. Just after the ion bunch running out, the ambipolar potential originated from the negative space charge can drive ion waves, resulting in excitation of the wakefield in the ion wave regime. The wave velocity should be very close to the traveling velocity of the ion bunch with supersonic speed. This situation is even similar to the previous results performed in the double plasma device,[1,2] in which a bunch of ions was shot out by applying the pulse between two plasmas to excite the ion wave wakefield.

DISCUSSION

The mechanism of exciting the present wave was considered. At the critical layer, the electron plasma wave can be excited resonantly to propagate down the density gradient as a wave packet (width : τ_w) with a group velocity v_g. This velocity was typically $v_g \approx 4.3 \times 10^7$ cm/sec. As the wave packet has large enough amplitude, the ponderomotive force of this packet can drive the ion wave resonantly because of $\tau_w \approx \tau_{pi}$. Because the velocity of the wave packet is much larger than Cs, the speed of the ion wave, the wave exciter looks like to exist in "all the layer" at the same time where the wave packet went trough, just like the one observed experimentally (see Fig. 3(a)). At the same time, a bunch of high energy electrons and ions were shot out as a result of strong acceleration by the excited waves.

Here, the packet of electron plasma wave propagates with the velocity v_g, and the Gaussian shape of the wave packet is assumed with the characteristic width ζ_0 in the z direction. The ion bunch can excite the lower frequency cavity like component. We assume that the ion bunch has a mean velocity v, which is greater or of the order of Cs and moves down through the plasma in lower density area with finite extent. Here in this case, we consider the phenomena in 2-dimensional space for simplicity. The density of the bunch, therefore, is assumed to be a function of $z - vt$ and y. The density perturbation excited by the external perturbation, $\delta n_{ext.}(y, z - vt)$, is given by

$$\frac{d^2\tilde{n}}{dt^2} - C_s^2 \nabla^2 \tilde{n} = -C_s^2 \nabla^2 \delta n_{ext.}(y, z - vt).$$

Here, the supersonic motion of the ion bunch will result in the excitation of ion acoustic Mach cone structures as it can be shown by taking Fourier transform in space and Laplace transform in time of Eq.(2). If the bunch is located near $y \approx 0$ and $z \approx 0$ at $t \approx 0$, one finally obtain the density perturbation

$$\tilde{n}(y,z,t) = \frac{j}{2} \frac{1}{C_s v \left(1 - \frac{C_s^2}{v^2}\right)^{1/2}} \int_{-\infty}^{+\infty} d\omega \omega \delta n_0 \left(\frac{\omega}{v}, \frac{\omega}{C_s}\left(1 - \frac{C_s^2}{v^2}\right)^{1/2}\right)$$

$$\times \exp(j\omega t) \cdot \exp\left(-\frac{\omega}{v} z\right)$$

$$\times \left[\exp\left(-j\frac{\omega}{C_s}\left(1 - \frac{C_s^2}{v^2}\right)^{1/2} y Y(y)\right) + \exp\left(j\frac{\omega}{C_s}\left(1 - \frac{C_s^2}{v^2}\right)^{1/2} y Y(-y)\right)\right]$$

where $n_{ext.}(t=0) = \delta n_0$ and $Y(\theta)$ is the step function defined as $Y(\theta) = 0$ for $\theta < 0$, and $Y(\theta) = 1$ for $\theta > 0$.

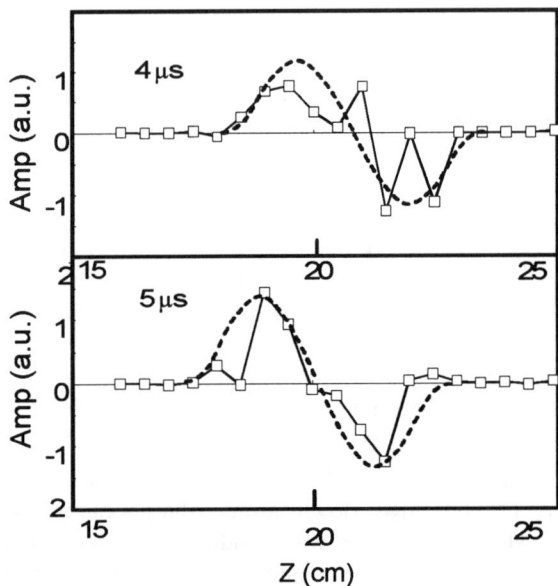

Fig. 4 Snapshot of Density Perturbation Amplitude. Dashied Lines Indicate Calculated Results.

When a Gaussian shape is assumed for the initial bunch of ions with characteristic width δy_0, and δz_0, one can obtain the condition for reaching the maximum amplitude at which $\Psi^2 = (\Delta\Psi)^2$ is held, where

$$\Psi^{\pm} = t - \frac{z}{t} \pm \frac{-y}{v\tan\alpha}, \quad \Delta\Psi = \frac{1}{v^2}\left(\delta z_0^2 + \frac{\delta y_0^2}{\tan^2\alpha}\right)$$

The behavior of the function $\Psi\exp(-\Psi^2/2(\Delta\Psi)^2)$ corresponds to the density modulation, and the numerical results can be compared with experiments such as shown in Fig. 4 by dotted lines. Here, the amplitude of the numerical calculation was fit to the experiments at $t=5$ μs and $\Delta y_0 = \Delta z_0 = 2$ mm is assumed. The characteristic features observed in the experiments are in reasonable coincidence with the present model.

CONCLUSION

In conclusion, we have observed the large amplitude ion wave regime by irradiating narrow width microwave pulse with ion oscillation range. The maximum amplitude is about $\delta n/n_0 = 40\%$. While the lower frequency cavity like ion wave can be interpreted by the Mach cone structure associated with the bunch of ions with supersonic velocity, which is accelerated by the plasma wave. The numerical results show fairly good agreement with the experiments.

Acknowledgments

A part of the present work was supported by the Grant-in-Aid from the Ministry of Education, Science and Culture, Japan. Two of authors (B. C. and G. M.) appreciate JSPS and CNRS for the help of France-Japan collaboration.

REFERENCES

1. Nishida, T. Okazaki, N. Yugami, and T. Nagasawa, Phys. Rev. Lett. **66**, 2328 (1991).
2. W. Aossey, J. E. Wlliams, H. S. Kim, J. Cooney, Y. C. Hsu and K. E. Lonngren, Phys. Rev. E .**47**, 2759 (1993).

Thomson Scattering Measurement of the Beat-Wave Excited Relativistic Plasma Waves

Yoneyoshi Kitagawa, Shin Watanabe, Katsuyuki Kawase,
Kiyonobu Sawai, and Sadao Nakai

Institute of Laser Engineering, Osaka University
Yamadaoka 2-6, Suita, 565, Japan

Abstract. double-line CO_2 laser, of 10.59-μm and 9.57-μm lines or 10.3-μm, illuminated a hydrogen gas. The beat of two lines or a forward Raman scattering can excite a relativistic plasma wave, accelerating hot electrons in the plasma to high energy of above 20 MeV. On the other hand, a backward Raman scattering occurs in a single line laser illuminated plasma, exciting a slow plasma wave and accelerating plasma electrons to a few hundred keV. A second harmonic YAG laser is used for Thomson scattering measurements of these plasma waves.

BEAT-WAVE ACCELERATION OF ELECTRONS IN A PLASMA

Since Tajima and Dawson have proposed the laser beat-wave accelerator as a promising collective acceleration scheme[1], there have been many theoretical and experimental efforts. We, for the first time, demonstrated that a laser beat wave can accelerate high energy electrons in a plasma[2].

A double-line CO_2 laser, coming from the LEKKO VIII electron-beam controlled laser system, has a 0.4-ns rise and 1-ns full width at half maximum pulse containing 150 J each in the 10.59-μm and 9.57-μm lines. The laser is focused into 1 to 0.3-mm spot size. At the laser rise, the intense laser light fully ionizes a puffed hydrogen gas by tunnel ionization or multi-ionization processes. Thus produced plasma has the resonant density so as the beat-wave frequency equals the plasma frequency corresponding to $1.1 \times 10^{17}/cm^3$. The beat of the two frequencies drives an electron plasma wave propagating through the plasma. The electron plasma wave has a phase velocity equal to the group velocity of the beating laser lights. Since the laser frequencies are both much greater than the plasma frequency, the group velocity is close to the speed of light. An electron injected with a velocity nearly equal to the speed of light, can be trapped in the thus driven plasma wave and can be accelerated to high energies. The plasma wave of the amplitude of ~6% is observed at the resonant density[2]. The field is ~1.5 GV/m or more. We observed, only at this resonant point, that hot electrons in the plasma with the energies of less than 6 MeV are accelerated upto ~ 21 MeV, shown in Fig. 1(a). The number of forward-accelerated electrons (10~21MeV) attains more than 30,000 in the laser cone. Single-line irradiation at resonance does not show any acceleration, as in Fig. 1(b). For the experimental parameters, we simulated how the beat-wave excited plasma wave trap and accelerate injected 0.6 MeV electrons, by treating plasmas in 1-d fluid and injected electrons in particle code. The result shown in Fig. 2 is at the 2 mm down from the focal point . The laser power is 1.2×10^{17} W/cm^2 ($v_{osc} = 0.1c$). The right upper figure shows time-integrated electron spectra, suggesting that the electrons are accelerated upto 28 MeV.

Now we proceed to inject pulse-powered 1MeV electrons into this resonant plasma and to accelerate them. The observed energetic electrons support the possibility that a beat-wave accelerator mechanism may have been operative.

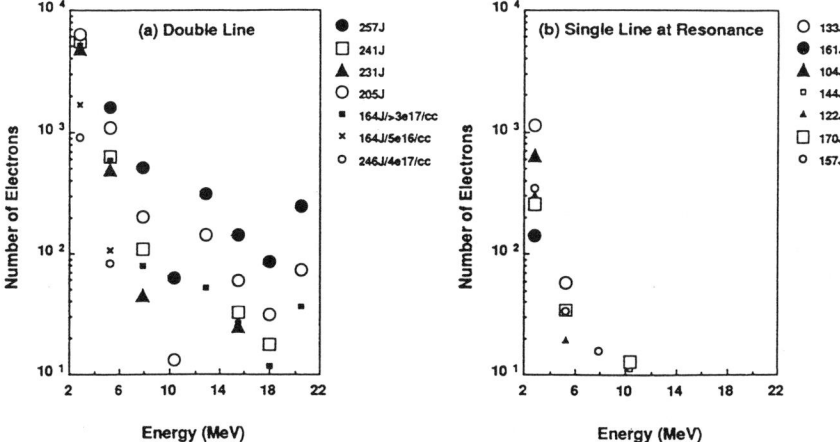

FIGURE. 1. Energy spectra of the electrons forward emitted from the hydrogen plasma. Number of Electrons per channel (a) with the double line both at and out of resonance and (b) with the single line irradiation at the resonance($1-2 \times 10^{17}$/cm^3). Same symbols are from the same shot, of which the parameters are as laser energies / n_0. Vertical error is one electron (detection limit). Horizontal bar is channel width.

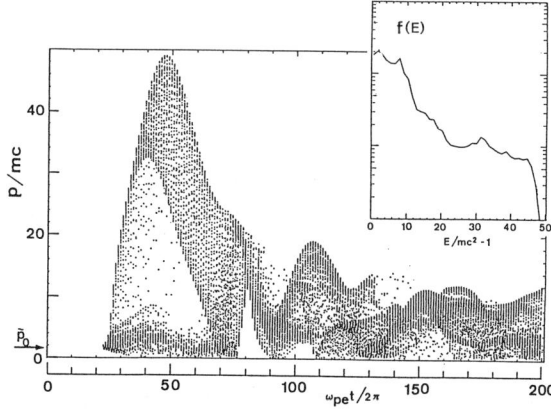

FIGURE. 2. Simulated time variation of accelerated electrons and time integrated energy spectra.

THOMSON SCATTERING MEASUREMENT OF PLASMA WAVES

A plasma wave of a phase velocity $\beta_\phi = (1-\gamma_\phi^{-2})^{1/2}$ and an amplitude $\varepsilon = \delta n/n_0$ can accelerate injected electrons up to $\sim 2\gamma_\phi \varepsilon mc^2$, where $\gamma_\phi = \omega_0/\omega_p$ and ω_0/ω_p are laser and plasma frequencies, respectively. The incident laser decays into a forward slow plasma wave and a backward reflected light due to a backward Raman scattering above a relatively low threshold laser intensity of order of $\sim 10^{-12}$ W/cm^2. For the intensity of $\sim 10^{-14}$ W/cm^2, a forward Raman scattering decays into a forward relativistic fast plasma wave and a forward scattered light. The two frequency laser forces to excite a fast plasma wave more easily than the forward Raman scattering. Thomson scattering measurement is now believed to be the best way to determine the amplitude ε. Since the plasma wave works as a moving grating against the incident laser light, we can simply describe the scattered power P_s by a Bragg scattering formula as:

$$\frac{P_s}{P_0} = \left[\frac{\pi}{2}\varepsilon\frac{n_0}{n_c}\frac{L_z}{\lambda_{YAG}}\right]^2, \qquad (1)$$

where P_0 is the peak power of the incident 2nd harmonic YAG light of $\lambda_{YAG} = 532$nm. n_0 and n_c are plasma density and critical density for 532nm, respectively. The YAG laser is injected perpendicular to the CO$_2$ laser axis, so that L_z is minimum between its spot size and the plasma wave length. Figure 3 is the Thomson scattering diagram. The YAG laser 2nd harmonic output is 400mJ in 8 ns pulse width. The CO$_2$ laser pulse slicer (Pockels cell) output of 10kV in 100ps rise triggers the YAG oscillator Q-switch, so that the synchronization between the CO$_2$ laser and the YAG laser is less than 1 ns. The scattering angle θ is 1.49 mrad for the forward Raman scattering and 106 mrad for the backward Raman, respectively.

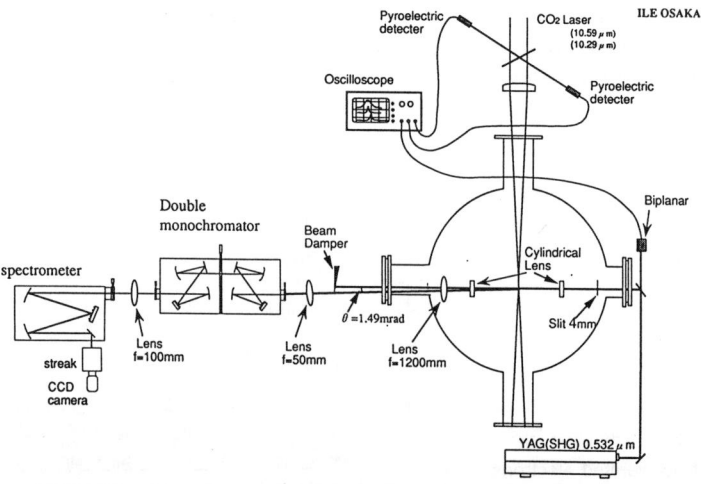

FIGURE 3. Diagram of YAG laser Thomson scattering measurement set up. The figure is for the Forward Raman or the beat-wave excitation. For the backward Raman scattering, θ = 106 mrad.

The scattered signals are detected above the laser energy of 250 J. Figure 4 shows a typical streak image of the slow plasma wave due to the backward Raman scattering. The maximum P_s/P_0 was 10^{-6}. At the left hand is a stray light from the incident 532-nm YAG laser. First, we can know the electron density of the the CO_2 laser ionized plasma from the spectral separation at the timing when the CO_2 laser excites the slow plasma wave. The plasma wave is excited during 300ps around the CO_2 laser peak. Second, the detected power P_s yields the amplitude ε by using Eq. (1). Thus ε is plotted as a function of n_0 in Fig. 5. $\varepsilon = 1.5$ % for $n_0 = 5 \sim 8\times10^{16}$ cm^{-3}. The phase velocity is 1.1×10^9 cm/s, suggesting that ε of 1.5 % accelerates hot electrons to 600 keV, which the beat-wave-excited fast plasma wave can trap and accelerate.

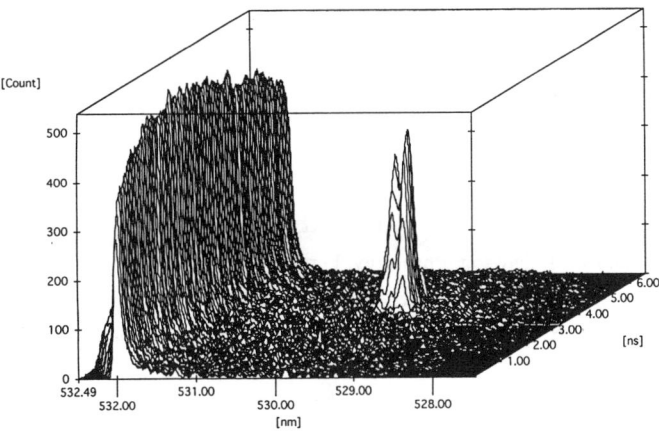

FIGURE 4. Typical streak image of a slow plasma wave due to the backward Raman scattering. At the left hand is a stray light from the incident 532-nm YAG laser, of which the pulse width is ~8ns.

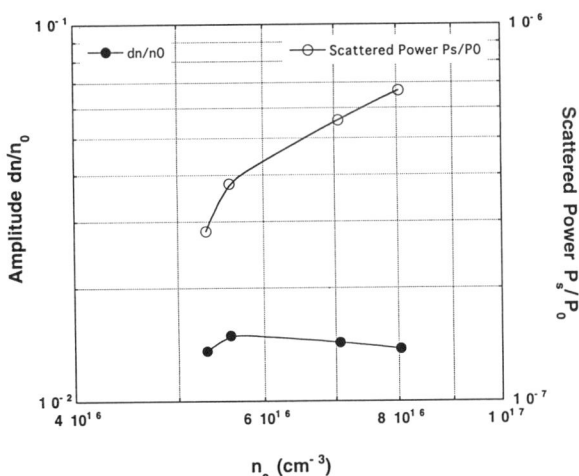

FIGURE 5. Plasma wave amplitude ε due to backward Raman scattering versus electron density n_0.

SUMMARY AND NEXT ISSUE

By using the double-line CO_2 laser beat, we have accelerated hot electrons of 0.6~1 MeV probably due to backward or forward Raman scattering to more than 20 MeV. Number of the accelerated electrons were so far $10^3 \sim 10^4$. Thomson scattering measurement is now in proceed to determine the beat-wave excited fast plasma wave amplitude.

The next issue of the beat-wave acceleration is to increase the number and quality of the accelerated particles by phase-matching between the particles and the plasma wave potential and by short-bunching the particles.

Subnanosecond intense laser system is now in construction, which is used to demonstrate the laser wakefield acceleration.

ACKNOWLEDGEMENTS

We acknowledge Prof. H. Takabe for his cooperation on simulation and Prof. K. Mima and Prof. K. Nishihara for their encouragements.

REFERENCES

[1] T. Tajima and J. M. Dawson, Phys. Rev. Letters **56**, 267 (1979).
[2] Y. Kitagawa et al., Phys. Rev. Letters **68**, 48 (1992).
[3] L. D. Landau and E. M. Lifshitz, *"Electrodynamics in continuous media"*, Chap. 15.

Electron Acceleration by Longitudinal Electric Field Generated by Colinearly Overlapped Two Laser Beams

Satoshi Takeuchi

Electronic Engineering and Computer Science
Yamanashi University, Takeda 4-3-11, Kofu, Japan

Abstract. A localized beam like Gaussian beam in vacuum has an axial longitudinal electric field because of a finite radius caused by focusing. By the use of this longitudinal field, a new idea to efficiently accelerate electrons is presented. Two Gaussian beams having the fundamental mode TEM_{00} are directly applied to the electron acceleration, because an optimal longitudinal electric field can be created by overlapping the beams each other. In the overlapped region, when the phase difference between the two travelling waves is just a half wavelength, the optimal field has a significant peak of an axial field and its intensity pattern is similar to that of a TEM_{10} mode usually observed in the Gaussian beam. While, the transverse electric and magnetic fields cancel each other in this region. Hence, electrons that resonantly interact with the beams can be stably accelerated.

INTRODUCTION

Energy density of a laser beam is usually much greater than that of microwaves, and hence many ideas of using powerful lasers to accelerate electrons have been proposed.[1,2] Since, however, the laser beam is essentially a transverse electromagnetic wave, directions of acceleration in those schemes are transverse and this gives rise to complexity of configuration. By the use of the longitudinal electric field of a single beam, novel ideas of electron acceleration have been reported,[3-6] in which the longitudinal field is formally described by $E_z = -\int \nabla_t \cdot \mathbf{E}_t dz$, where subscript t means the transverse component. However, it is known that the excitation of TM modes of laser beams is a difficult task because of its short wavelength.

We report here another new idea to accelerate electrons by the axial longitudinal field as shown in the above abstract. In this method, no structure is needed to convert beam mode from TE to TM, hence the optimal longitudinal field can be generated without loss of energy caused by mode conversion.

GAUSSIAN LASER BEAM

To analytically derive the optimal field in the electron acceleration, all components of beam field must be explicitly described. Then, we firstly show that all components are given by vectorial analysis.[3,6] Suppose that the electric field components are

$$(E_x, E_y, E_z) = (f,\ 0,\ g)\exp(ikz - i\omega t), \tag{1}$$

where the functions $f(x,y,z)$ and $g(x,y,z)$ stand for beam envelopes, and $\omega/k = c$ with the wave number k and the frequency ω of travelling wave. Substituting E_x into the wave equation, we obtain Helmholtz type equation in the form

$$\frac{\partial^2 f}{\partial x^2} + \frac{\partial^2 f}{\partial y^2} + 2ik\frac{\partial f}{\partial z} = 0, \tag{2}$$

under the paraxial approximation $|\partial^2 f/\partial z^2| \ll |2ik\partial f/\partial z|$. This equation has a complex solution with Gaussian type:

$$f_{n,m} = \frac{A}{F^{(n+m+2)/2}} H_n(\mu) H_m(\nu) \exp\left(-\frac{r^2}{F}\right), \tag{3}$$

here $\mu = x/\sqrt{F}$, $\nu = y/\sqrt{F}$, $r^2 = x^2 + y^2$ and $F = -2z/ik + w_0^2$ is a complex value with a beam waist w_0, further H is the Hermite polynomial. Here if we use the fundamental mode $(n,m) = (0,0)$, the functions f and g are given by

$$f = \frac{A}{F}\exp\left(-\frac{r^2}{F}\right),\quad g = \frac{2x}{ikF}f, \tag{4}$$

where $\nabla \cdot \mathbf{E} = 0$ and another paraxial approximation $|\partial g/\partial z| \ll |ikg|$ are used. Consequently, concrete forms of electromagnetic fields can be written by

$$E_x = \frac{A}{F}\exp\left(-\frac{r^2}{F} + ikz - i\omega t\right),\quad E_y = 0,\quad E_z = \frac{2x}{ikF}E_x, \tag{5}$$

$$B_x = \frac{4xy}{ikF}E_x,\quad B_y = \left(1 + 2\frac{y^2 - x^2}{k^2 F^2}\right)E_x,\quad B_z = \frac{2y}{ikF}E_x. \tag{6}$$

SETTING OF TWO BEAMS

Now, taking advantage of the axial symmetric feature of the Gaussian beam derived above, we symmetrically arrange two beams with respect to the xz plane and to the yz plane. We set the beam axis of one beam to $(x,y) = (-r_0, 0)$ and that of the other to $(x,y) = (r_0, 0)$. We call, for simplicity, the former Beam1 and the latter Beam2. According to eqs. (5)

and (6), all field components of the two beams are described by replacing the variable x by $x + r_0$ for Beam1 and by replacing x by $x - r_0$ for Beam2, respectively. Therefore, these beams colinearly overlap each other and the gap distance between axes is $2r_0$. It is important that the phase difference between the two travelling waves is $i\pi$ and it corresponds to just a half wavelength $\lambda/2 = \pi/k$. As a result, the electron that interacts with the two beams in vacuum experiences the overlapped fields given by

$$h_{j1} + h_{j2} = h_j(x + r_0, y, z, t) + h_j(x - r_0, y, z, t) \exp(i\pi), \tag{7}$$

where h_j stands for the field of j-component, i.e., E_j or B_j, $(j = x, y, z)$, and subscripts 1 and 2 denote Beam1 and Beam2, respectively.

ENERGY GAIN

To investigate the electron acceleration, we must consider the relation between a moving velocity of electron and the phase velocity of travelling wave. The reason for this is that the phase velocity of the Gaussian beam is usually greater than the speed of light c, then the energy gain of electron strongly depends on the phase relation. Suppose that a test electron having a high value of Lorentz factor γ, its velocity is nearly equal to c because of $\gamma \gg 1$. Such an electron starts from $z = -\ell$ at $t = 0$ and passes $z = \ell$ along the z axis through the overlapped region. By the use of a transit time given as $t_p \equiv 2\ell/c$, the phase relation between the electron and the travelling wave can be approximated as

$$kz - \omega t \approx -k\ell - \omega t_p(1 - v_z/c) \approx -k\ell, \tag{8}$$

where $z \approx v_z t$ and $(1 - v_z/c) \approx 1/2\gamma^2$ are used. This implies that the electron located at the initial phase $\psi_0 \equiv -k\ell$ moves together with the travelling wave, and hence it experiences stationary fields even if time elapses. In particular, near the z axis, i.e., $(x, y) = (0, 0)$, the electron mainly experiences transverse field $E_{x1} + E_{x2} \approx 0$ and $B_{y1} + B_{y2} \approx 0$, because these fields cancel each other in the overlapped region. While, it feels the axial longitudinal electric field. Hence, the electron that resonantly moves together with the travelling wave is efficiently accelerated, so that its energy gain is derived by

$$mc^2(\gamma - \gamma_0) = q \int E_{z1} + E_{z2} \exp(i\pi) \, dz, \tag{9}$$

$$= \exp(i\psi_0) \frac{4qAr_0}{ik} \int_{-\ell}^{\ell} \frac{1}{F^2} \exp\left(-\frac{r_0^2}{F}\right) dz, \tag{10}$$

where m and q stand for the mass and the charge, and γ is the Lorentz factor defined by $\gamma = (1 - v_z^2/c^2)^{-1/2}$. The real part of energy gain, i.e.,

$G \equiv \Re\{mc^2(\gamma - \gamma_0)\}$, can be written by

$$G = \frac{4qA}{w_0 \xi} \exp\left(-\frac{\xi^2}{\eta^2+1}\right) \sin\left(\frac{\xi^2 \eta}{\eta^2+1}\right) \sin\psi_0, \qquad (11)$$

here we used $\xi \equiv r_0/w_0$, $\eta \equiv \ell/\ell_0$ and the Rayleigh range $\ell_0 \equiv kw_0^2/2$.

If the gain G has a maximum value, as indicated in the phase relation, the electron will be fixed at an optimal initial position, i.e., $\sin\psi_0 = 1$. Using this estimation, we derive an optimal energy gain of an accelerated electron. Partial derivatives $\partial G/\partial \xi = 0$ and $\partial G/\partial \eta = 0$ lead the relation between ξ and η:

$$2\xi^2 = \eta^2 - 1, \quad \tan\left(\frac{\xi^2 \eta}{\eta^2+1}\right) = \frac{\eta^2 - 1}{2\eta}. \qquad (12)$$

By numerical calculation, we obtain a pair of solutions $\xi = 1.369$ and $\eta = 2.179$. While, a peak intensity of a single laser beam is given by $E_0 \equiv A/w_0^2$ at beam waist. Substituting these values into eq.(11), we finally obtain a handy-type formula of the maximum energy gain:

$$\begin{aligned}G_{\max} &= 1.376 w_0(\text{cm}) q E_0(\text{V/cm}), \\ &= 0.267 \times 10^2 \sqrt{P(\text{W/cm}^2) w_0^2(\text{cm})} \ \text{eV}.\end{aligned} \qquad (13)$$

This is similar to that given by the TM-mode acceleration.[3,6]

Figure 1 shows the optimal beam profiles on the xz plane at $t=0$, where two beams are colinearly overlapped each other. Wave train of the electric field is depicted and the regions of strong field are shown by dark streaks. The longitudinal electric field is shown in Fig.1(a). The overlapped region of two beams in the center of figure is mainly occupied by the strong electric field. While, in the case of the transverse electric field depicted in Fig.1(b), field intensity in the overlapped region is so week. Hence, the electron resonantly interacting with the longitudinal field can be stably accelerated.

SUMMARY

We showed that the optimal longitudinal electric field is created and the transverse fields cancel each other by the colinear overlapping of two beams. In the case of the optimal acceleration presented here, the gap distance between two beams is $2r_0 = 2.728 w_0$ and the acceleration length is $R_a = 2\eta = 4.358\ell_0$. As denoted by the optimal energy gain, $P \times \pi w_0^2$ is constant provided the total out put power of laser beam is fixed, while the acceleration length $R_a = 2.179 kw_0^2$ is reduced if we have stronger focusing.

We present examples of the optimal energy gain and the acceleration length for some lasers in TABLE 1. We will obtain 1TeV electron in an acceleration device of below several kilometers long which is composed by a few ten thousands units of acceleration elements. This is still within the present state of laser technology.

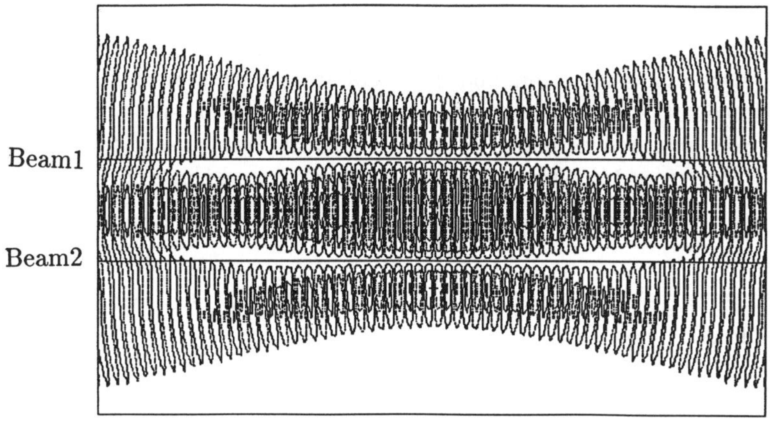

Beam1

Beam2

(a) Longitudinal electric field $E_{z1} + E_{z2}\exp(i\pi)$

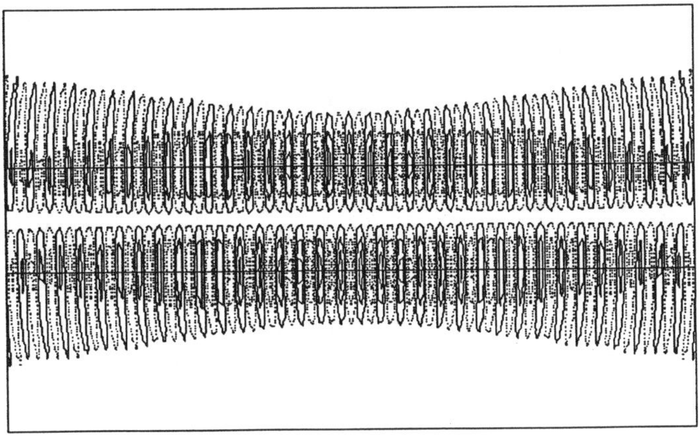

(b) Transverse electric field $E_{x1} + E_{x2}\exp(i\pi)$

FIGURE 1. Optimal beam profiles of colinearly overlapped two beams on the xz plane.

TABLE 1 The optimal energy gain and the acceleration length for some lasers.

laser	$\lambda(\mu m)$	P(W/cm^2)	w_0(cm)	G_{max}(eV)	R_a(cm)
CO_2	10	10^{15}	10^{-2}	8.4×10^6	1.37
Nd:glass	1	10^{17}	10^{-3}	8.5×10^6	1.3×10^{-1}
Ti:sapphire	1	10^{18}	10^{-3}	2.7×10^7	1.3×10^{-1}

ACKNOWLEDGEMENTS

This work was carried out under the collaboration research program of National Institute for Fusion Science and of Institute of Laser Engineering of Osaka University. It was partly supported by the Grant-in-Aid for General Science Research of Ministry of Education, Science and Culture.

REFERENCES

1. ed. Mills,F.F.,*Advanced Accelerator Concepts*, AIP Conf. Proc. No.156, (AIP, New York, 1987).

2. ed. Joshi,C.,*Advanced Accelerator Concepts*, AIP Conf. Proc. No.193, (AIP, New York, 1989).

3. Takeuchi,S., Sugihara,R., and Shimoda,K., "Electron Acceleration by Longitudinal Electric field of a Gaussian Laser Beam", J. Phys. Soc. Jpn., **63**, 1186-1193 (1994).

4. Scully,M.O., "A Simple Laser Linac", Appl. Phys. **B 51**, 238-241 (1990).

5. Scully,M.O. and Zubairy, M.S., "Simple laser accelerator: Optics and particle dynamics", Phys. Rev. A, **44**, 2656-2663 (1991).

6. ed. Hora,H., and Miley,G.H., *Laser Interaction and Related Plasma Phenomena, Vol.9,* (Plenum Press, New York, 1990), Caspers,E., and Jensen,E., "Particle acceleration with the axial electric field of a TEM10 mode laser beam", pp.459-466.

7. Shimoda,K.,"Vectorial Analysis of the Gaussian Beams of Light", J. Phys. Soc. Jpn. **60**, 141-145 (1991); ibid, 1432.

Properties of spectra of the reflected and transmitted radiation during propagation of relativistically strong laser pulses in underdense plasmas

S. V. Bulanov[1], T. Zh. Esirkepov[2] and N. M. Naumova[1]

[1] *General Physics Institute of the Russian Academy of Sciences, Vavilov str. 38, 117942 Moscow, Russia*
[2] *Moscow Institute of Physics and Technology, Dolgoprudny, 141730 Moscow Region, Russia*

Abstract. Particle-in-cell simulation has been performed to study the spatial-temporal evolution of the pulse propagating in an underdense plasma. The spectra both of the reflected and transmitted radiation are investigated. The spectrum structure of the reflected radiation is due to the backward stimulated Raman scattering meanwhile the transmitted radiation structure is mainly due to the nonlinear self-phase-modulation. The influence of the pulse shape on the transmitted radiation spectrum is revealed. The dependence of the main features of the spectrum and the self-consistent pulse distortion is found. The pulse distortion is accompanied by the relativistic electrons generation.

Recent experiments on the interaction of high-intensity ultra-short laser pulses with underdense plasmas [1] have stimulated the work the results of which we present below. A detailed analysis is needed to investigate carefully a quite complicated dependence of the backscattered and transmitted radiation on the parameters of a plasma and laser pulse. Here we present the results of the numerical simulation with 1(2/2)D relativistic code of the laser pulse propagation in a plasma. The code used has been described in Ref. [2, 3]. The PIC simulation helps us to investigate carefully the interaction of laser pulses with plasmas since it provides in citu information about the pulse and electron distribution evolution with time. The comparison of this information with that obtained from the reflected and transmitted radiation and fast electron distribution should elucidate the physical mechanism that leads to the observed pattern of the frequency spectra. We consider the evolution of fairly smooth circular polarized laser pulses with the length equal to 100 oscillating periods per pulse. The pulse intensity value is expressed in term of the dimensionless amplitude $a = eE/m\omega_0 c$, which equals to v/c in the non-relativistic limit. The pulse is incident on an uniform plasma with density $n = 0.01 n_c$, where $n_c = m\omega_0^2/4\pi e^2$ is the critical density value. The ion motion is unimportant for these parameters and they are assumed to be at rest. At the initial time the laser pulse is chosen to be smooth and long enough to do not excite the Langmuir wave. Characteristic times of the backward and forward stimulated Raman scattering (SRS) growth are correspondingly [2, 4, 5] $t_{BSRS} \approx 10\omega_0^{-1}$ and $t_{FSRS} \approx 500\omega_0^{-1}$ for our range of parameters.

© 1996 American Institute of Physics

To have a sufficient time for the FSRS and BSRS to develop the pulse evolution has been investigated during the time of the pulse passage through the plasma with the length of a few pulse lengths. The length of the plasma slab is equal to $10000c/\omega_0$. In Fig.1 the frequency spectra of the reflected (left) and transmitted (middle) radiation and the transverse component of the electric field (right) are shown as functions of the pulse intensity. The dimensionless amplitude a varies from 0.4 (a) up to 2 (d). The spectra and electric field profile are shown for $t = 5000\omega_0^{-1}$ and $t = 10000\omega_0^{-1}$. On the left row the reflected wave frequency spectrum temporal development is shown. We see that with the increasing of the wave amplitude the spectrum is broadening. At that time the pulse modulation at its rare edge appears (right row), it is stronger for greater pulse amplitude (d). The transmitted radiation spectrum is presented in the middle row. We see broadening of the spectrum with time. Stronger modulation of the reflected radiation spectrum than of the transmitted one is due to the reflected radiation passes the way twice: first time in forward and second time in backward direction.

In Fig.2 the transmitted radiation spectrum for $t = 6600\omega_0^{-1}$ is shown for $a = 0.8$ (a) and $a = 2$ (b), respectively. We see the typical features of the spectrum as it follows. The spectrum has a minimum near the carrier frequency ω_0 for $a < 1$ and a maximum near ω_0 for $a > 1$. This dependence can be related to the pulse self-phase-modulation. According to [6] the pulse spectrum can be expressed in the form

$$|E(\omega)|^2 = \left| \int_0^\infty \mathcal{E}(t) e^{-i\omega_0 t + i\Delta\phi(t)} dt \right|, \quad (1)$$

with the self-frequency-modulation being

$$\Delta\omega(t) = -\partial(\Delta\phi)/\partial t. \quad (2)$$

The self-phase-modulation when the relativistically strong laser pulse propagates in a plasma is

$$\Delta\phi(t) \approx \frac{t\omega_p^2}{2\omega_0} \left(1 - \frac{1}{\sqrt{1 + a^2(t)}} \right) \quad (3)$$

To obtain this expression we describe the electromagnetic wave packet in the envelope quasistatic approximation. Eq. (3) describes relativistic increase of the electron mass and change of the electron density inside the wave packet.

For a symmetrical pulse the self-frequency-modulation, $\Delta\omega$, has a minimum in the leading part of the pulse and it has a maximum at the rare part of the pulse. In Fig.3(a) the pulse amplitude and the difference between ω_0 and the local frequency value are presented for $a = 1$ and $t = 6000\omega_0^{-1}$. This dependence is obtained as a result of PIC simulation. In Fig.3(b) the pulse spectrum in comparison with the initial one is shown.

In Fig.4 the frequency spectra calculated with the model dependence of $\Delta\phi$ given by Eq.(3), are shown for $a = 0.4$ (a) and $a = 2$ (b) and for different

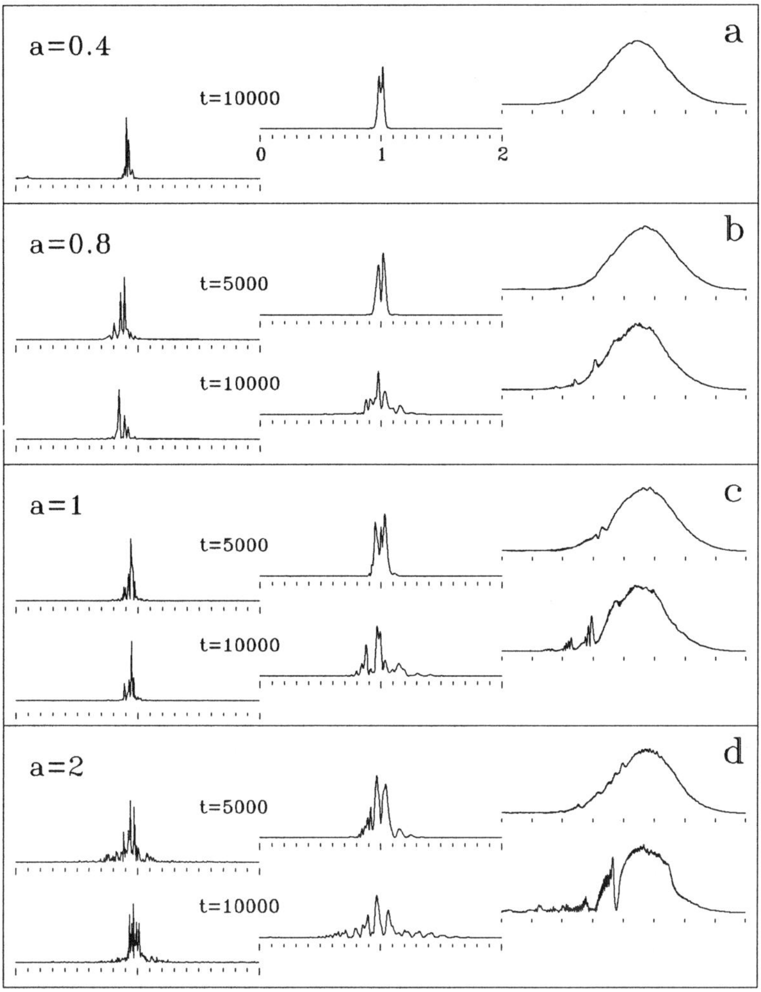

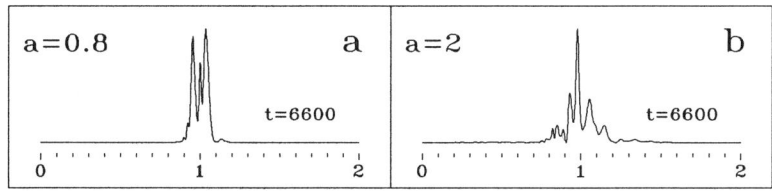

FIGURE 1. Frequency spectra of the reflected (left) and transmitted radiation and the pulse envelope for the amplitudes (a) $a = 0.4$, (b) $a = 0.8$, (c) $a = 1$ and (d) $a = 2$ at the time $t = 5000\omega_0^{-1}$ and $t = 10000\omega_0^{-1}$.

FIGURE 2. Comparison of the frequency spectra of the transmitted radiation for the amplitudes (a) $a = 0.8$ and (b) $a = 2$ at the time $t = 6600\omega_0^{-1}$.

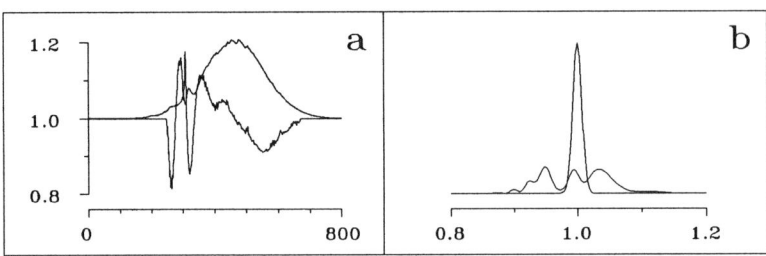

FIGURE 3. Transverse electric field and carrier frequency along the pulse (a) and frequency spectrum (b) at $t = 6000\omega_0^{-1}$.

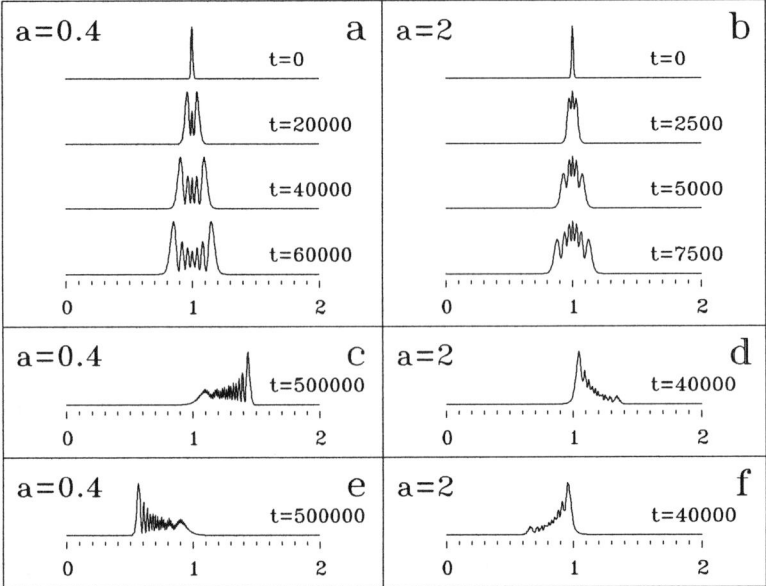

FIGURE 4. Evolution of frequency spectra of smooth symmetrical laser pulses in underdense plasmas ($n = 0.01 n_c$) for the amplitudes $a = 0.4$ (a) and $a = 2$ (b) due to the self-phase-modulation. Frequency spectra of the pulses with sharp front edge influenced by the self-phase-modulation for $a = 0.4$ (c) and $a = 2$ (d); the same for sharp rare edge: $a = 0.4$ (e) and $a = 2$ (f).

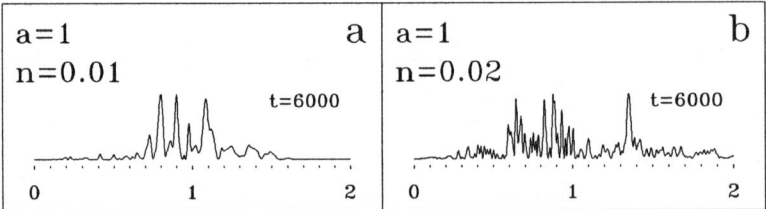

FIGURE 5. Frequency spectra of the transmitted radiation for the pulse intensity $a = 1$ and different density values: $n = 0.01 n_c$ (a) and $n = 0.02 n_c$ (b) for the time $t = 6000\omega_0^{-1}$.

t when the pulse has symmetrical profile. One can see that in the nonrelativistic limit, $a < 1$, the self-modulated spectrum has a minimum in the central part, while in the ultrarelativistic limit, when $a > 1$, the spectrum has a maximum (Fig.4(b)). The influence of the pulse non-symmetry results in the frequency spectrum non-symmetry as is shown in Fig.4(c-f) for a sharp front edge (c,e) and for a sharp rare edge (d,f), for the pulse amplitude magnitudes being equal to $a = 0.4$ (c,d) and $a = 2$ (e,f).

The influence of the plasma density on the pulse evolution is illustrated in Fig.5, where the spectra of the pulses with $a = 1$ propagating in the plasma with the density value equal to $0.01n_c$ and $0.02n_c$, respectively, are shown for $t = 6000\omega_0^{-1}$.

The process of SRS possesses a quite complicated temporal development, as it has been emphasized in papers [2, 3, 7]. For the amplitude $a = 2$ and plasma density $n = 0.04n_c$ and the same pulse length the characteristic times of backward and forward SRS development are $t_{BSRS} = 6\omega_0^{-1}$ and $t_{FSRS} = 125\omega_0^{-1}$, respectively. The evolution of this pulse in plasmas is demonstrated in Fig.6, where the transverse electric field is shown for $t = 0 \div 9000\omega_0^{-1}$ in the (x-ct,t) coordinate system. The backward SRS development appears from the very beginning of the pulse entrance into the plasma. Afterwards the modulations with the length of the order of the Langmuir wave length λ_p appear on the pulse envelope those are related to the forward SRS. This process results in strong plasma wave excitation and the effective acceleration of the plasma electrons up to the energy of order of $100mc^2$ in the forward direction. The most effective wake wave excitation (and the pulse destruction) and electron acceleration takes place at $t \approx 2000 \div 5000\omega_0^{-1}$. In Fig.7 electron phase space is presented for this time. In addition, we observed that the electrons are accelerating in the backward direction with the energy gain being of order of $10mc^2$. For the time $t = 9000\omega_0^{-1}$ the pulse has lost the main part of its energy and comprises a few separate pieces with the length value approximately equal to λ_p.

The presented results demonstrate novel features of the interaction of high-intensity sub-picosecond pulses with underdense plasmas. There is apparent dependence of the spectra of the reflected and transmitted radiation on the pulse amplitude, length and profile and plasma density. This could be used to identify the pulse shape and amplitude changes in a plasma.

REFERENCES

1. Darrow, C. B., Coverdale, C. A., Crane, J. K., Perry, M. D., Mori, W. B., Decker, C., Joshi, C., and Clayton, C., *Bulletin of the American Physical Society* **39**, 1519 (1994).
2. Bulanov, S. V., Inovenkov I. N., Kirsanov V. I., Naumova N. M., and Sakharov A. S., *Phys. Fluids B* **4**, 1935-1942 (1992).
3. Bulanov, S. V., Esirkepov, T. Zh., Kamenets, F. F., Naumova, N. M., *Plasma Physics Reports* **21** (1995).
4. Mori, W. B., Decker, C. D., Hinkel, D. E., and Katsouleas, T., *Phys. Rev. Lett.* **72**, 1482-1485 (1994).
5. Sakharov, A. S., and Kirsanov, V. I, *Phys. Rev. E* **4**, 3274-3282 (1994).
6. Shen, Y. R. The principles of nonlinear optics. New York: John Wiley and Sons, 1984, ch.17, pp.310-316.
7. Decker, C., Mori, W. B. and Katsouleas, T., *Phys. Rev. E* **50**, 3338 (1994).

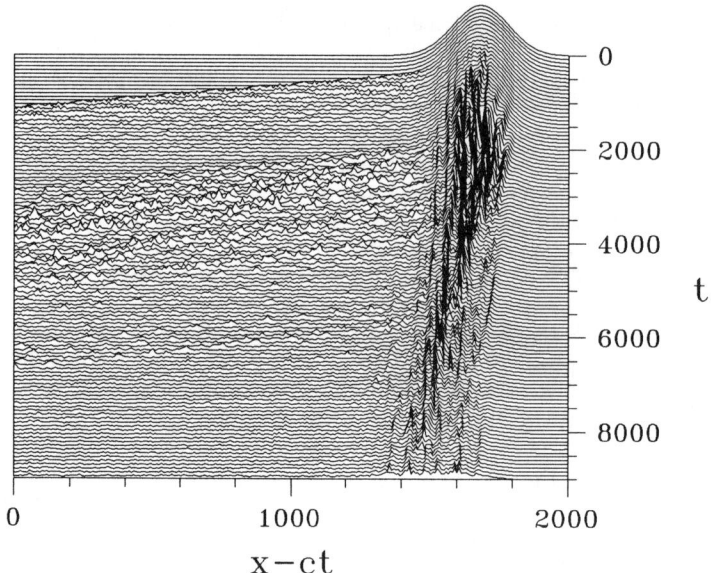

FIGURE 6. The evolution of the pulse with the amplitude $a = 2$ in a plasma with density $n = 0.04 n_c$ in the $(x - ct, t)$ coordinates.

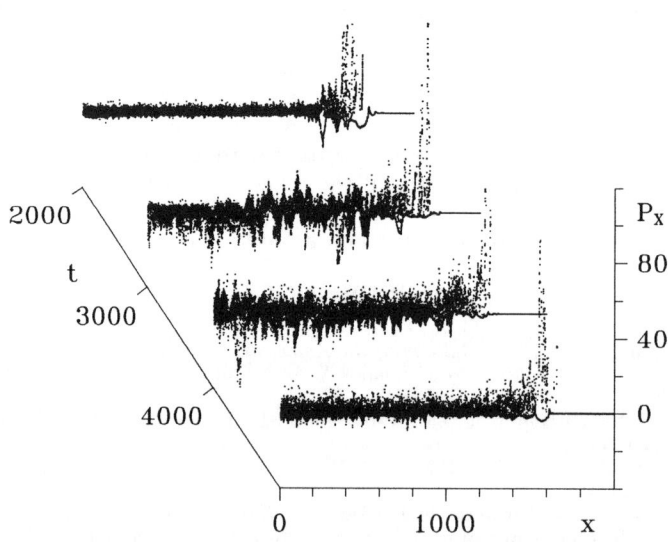

FIGURE 7. Electron phase space (P_x, x) corresponding to Fig.6 for $t = 2000 \div 5000 \omega_0^{-1}$.

Stimulated Raman Scattering and Short Laser Pulse Evolution

N.E.Andreev [a], L.M.Gorbunov [b]

[a] *High Energy Density Research Center of Russian Academy of Sciences, Izhorskaya 13/19, Moscow 127412, Russia*
[b] *P.N.Lebedev Physical Institute of Russian Academy of Sciences, Leninskii pr.,53, Moscow 117924, Russia*

Abstract. The basic equations are formulated and used to study the self-consistent space-time evolution of a intense ultra-short laser pulse and near-backward stimulated Raman scattering (SRS). The spectral intensity of detected in laboratory frame Raman scarrered radiation is determined through the thermal density fluctuations ahead of the pulse by solving the boundary problem in the pulse frame. The results of numerical simulations of self-consistent nonlinear pulse propagation and SRS are presented that confirm our analitical predictions.

INTRODUCTION

One of the physical effects that are expected to be important for the nonlinear dynamics of intense laser radiation in plasmas is the stimulated Raman backscattering. Recently, the SRS attrected attention in relation to subpicosecond intense laser pulses [1-6]. In underdense plasmas the SRS was considered mainly as a possible channel of the pulse energy losses. However, the low intensive SRS can be used as a diagnostic of the fast laser pulse evolution.

The SRS of ultrashort laser pulses is quite different from that in the case of long pulses. The scattered radiation appearing at the leading front of the pulse increases bing convected towards its trailing edge. As a result, the SRS of short pulses is a convective instability (in the pulse frame) and the spectrum and intensity of backscattered radiation depend essentially on the intensity distribution in the laser pulse [2,4,6]. Moreover, being sufficiently intensive, the SRS may significantly affect the pulse propagation.

In this report we formulate the basic equations to describe self consistently both pulse propagation and backscattered radiation. The analytic theory of SRS is developed for the conditions when SRS action on the laser pulse evolution can be ignored. A few numerical results are presented to demonstrate the temporal evolution of the laser pulse and SRS during the intensive wake-field generation.

BASIC EQUATIONS

To describe the electron motion, we use the set of the hydrodynamic equations that takes into account the second and third order terms with respect to the electric field and that has the form:

$$\frac{\partial \mathbf{v}}{\partial t} + (\mathbf{v}\nabla)\mathbf{v} = \frac{e}{m}\left\{\left[1 - \frac{v^2}{c^2}\right]\mathbf{E} - \frac{\mathbf{v}}{c^2}(\mathbf{E}\mathbf{v}) + \frac{1}{c}[\mathbf{v}\mathbf{B}]\right\} \quad (1)$$

$$\frac{\partial n}{\partial t} + \text{div}(n\mathbf{v}) = 0, \quad (2)$$

where n and $\mathbf{v}$ are electron density and velocity, respectively; $\mathbf{B}$ is the magnetic induction vector related to the electric field $\mathbf{E}$ by the Maxwell equation.

To describe a space-time evolution of an intense short laser pulse and backscattered radiation in plasmas, we introduce high frequency (HF) electric field in the form

$$\mathbf{E}(\mathbf{r}, t) = \frac{1}{2}\sum_{\sigma=\pm 1}[\mathbf{E}_\sigma(\mathbf{r}, t)\exp(-i\omega_0 t + i\sigma k_0 z) + \text{c.c.}], \quad (3)$$

where $\sigma=1$ corresponds to the radiation associated with laser pulse and $\sigma = -1$ with the radiation scattered bakward. The amplitudes $\mathbf{E}_\sigma$ are assumed to be slowly varying functions of time and space compared to the scales ω_0^{-1} and k_0^{-1}. We denote here the carrier frequency and wave number as ω_0 and k_0, respectively, and assume that the pulse propagates along the 0Z-axis.

Using representation (3), we can obtain, from Eq.(1) in the second order approximation with respect to electric field, the slowly varying in time ponderomotive force. This force produces two kinds of electron density perturbations in a plasma. The first type corresponds to slowly varying in space perturbations and is responsible for such effects as self-focusing, filamentation and self-modulation of radiation. In the case of a plasma approximation with immobile ions, the equation for this type of electron density perturbations δn_0 has the form

$$\left[\frac{\partial^2}{\partial t^2} + \omega_p^2\right]\frac{\delta n_0}{n_0} = \frac{e^2}{4m^2\omega_0^2}\Delta\sum_{\sigma=\pm 1}|\mathbf{E}_\sigma|^2 \quad (4)$$

where n_0 is the background electron density, $\omega_p = \sqrt{4\pi e^2 n_0/m}$ is the plasma frequency, and Δ is the Laplace operator.

The second kind of the electron density perturbations $\delta\tilde{n}$ corresponds to the small scale space variations that are responsible for the stimulated Raman scattering. In accordance with the space dependence of the ponderomotive force the density perturbation $\delta\tilde{n}$ can be represented in the form

$\delta \tilde{n} = \frac{1}{2}[\delta n_2 \exp(2ik_0 z) + \text{c.c}]$, where the envelope amplitude δn_2 varies slowly in space. The value δn_2 is determined by the following equation

$$\left[\frac{\partial^2}{\partial t^2} + \omega_p^2\right]\frac{\delta n_2}{n_0} = \frac{2e^2 k_0^2}{m^2\omega_0^2}\left[1 - \frac{i}{k_0}\frac{\partial}{\partial z} - \frac{\Delta_\perp}{4k_0^2}\right]\mathbf{E}_1 \mathbf{E}_{-1}^*, \qquad (5)$$

where $\Delta_\perp$ is the transverse part of the Laplace operator.

To derive the equations determining the variation of the amplitude $\mathbf{E}_\sigma$ it is necessary to calculate the density of HF current in a plasma $\mathbf{j} = \mathbf{j}_0 + \mathbf{j}_{n1}^{(1)} + \mathbf{j}_{n2}^{(2)}$, where $\mathbf{j}_0 = en_0\mathbf{v}_1$ is a linear, with respect to the electric field, part of the current, determined by the linear solution of Eq.(1) for $\mathbf{v}_1$. The nonlinear current $\mathbf{j}_{n1}^{(1)}$ is proportional to the third power of the electric field and is related to the incorporation of the relativistic effects (so called the relativistic nonlinearity). The nonlinear current $\mathbf{j}_{n1}^{(2)} = e\mathbf{v}_1(\delta n_0 + \delta \tilde{n})$ arises as the result of the electron density perturbations, which are discussed above. As a result, the equations for the amplitudes $\mathbf{E}_\sigma$ (as it follows from the Maxwell equations) take the form [4]

$$\left\{2i\sigma k_0 \frac{\partial}{\partial z} + \frac{2i\omega_0}{c^2}\frac{\partial}{\partial t} + \Delta_\perp + \frac{\omega_p^2}{c^2}\left[\frac{e^2}{4m^2\omega_0^2 c^2}\sum_{\sigma=\pm 1}|\mathbf{E}_\sigma|^2 - \frac{\delta n_0}{n_0}\right]\right\}\mathbf{E}_{\sigma\perp} =$$

$$= \frac{\omega_p^2}{2c^2}\left\{\mathbf{E}_{-\sigma\perp}\frac{\delta n_{\sigma 2}}{n_0} + \frac{c^2}{\omega_c^2}\left[2\nabla_\perp\left(\frac{\delta n_{\sigma 2}}{n_0}\nabla_\perp \mathbf{E}_{-\sigma\perp}\right) + \nabla_\perp\left(\mathbf{E}_{-\sigma\perp}\nabla_\perp \frac{\delta n_{\sigma 2}}{n_0}\right)\right]\right\}$$
(6)

where for $\sigma = -1$ the notation $\delta n_{\sigma 2} = \delta n_2^*$ is used; $\mathbf{E}_{\sigma\perp}$ is the transverse component of the amplitude $\mathbf{E}_\sigma$ and the transverse part of the operator ∇ is denoted as $\nabla_\perp$.

The basic set of equations (4) - (6) can be used to study the self-consistent evolution of a laser pulse incorporating the near backward scattering of the radiation. It should be noted that in Eq. (5) only the direct excitation of scattering plasma waves by the HF electric fields of the pulse and scattered radiation is taken into account, whereas the influence of the density perturbations δn_0 (coused mainly by the pulse) is ignored.

If the intensity of scattered radiation is fairly low, then considering the pulse evolution, we can ignore the effect of the backscattering and the basic equations can be separated into two subsets. One subset determines the pulse propagation and has been used in Refs. [7,8]. The other subset describes the SRS of a laser pulse with a shape that is either given or is determined by the mentioned above subset of Eqs.(4) and (6) (for $\sigma=+1$) where the influence of scattered field is ignored [4,6].

The equations for the backward scattering of the radiation can be simplified in the conditions when the transverse space dependence of δn_2^* is connected only with the transverse space variation in the pulse field $\mathbf{E}_{1\perp} = \mathbf{E}_0$. In this case, two last terms in the right hand side of Eqs.(6) are proportional to the small factor $(k_0 r_0)^{-2}$ and can be omitted. In addition, it is possible to neglect the terms containing operator $\Delta_\perp$ both in Eq.(6) and in Eq.(5). As a result we have

$$i\left[\frac{\partial}{\partial t} - v_g \frac{\partial}{\partial z}\right] \mathbf{E}' = \frac{\omega_p^2}{4\omega_0} \nu_2 \mathbf{E}_0(\mathbf{r}_\perp) \tag{7}$$

$$\left[\frac{\partial^2}{\partial t^2} + \omega_p^2\right] \nu_2 = -\frac{2e^2 k_0^2}{m^2 \omega_p^2} \mathbf{E}_0^*(\mathbf{r}_\perp) \mathbf{E}', \tag{8}$$

where $v_g = c^2 k_0/\omega_0$ is the group velocity and we introduced the notations $\nu_2 = \delta n_2^*/\delta n_0$ and $\mathbf{E}' = \mathbf{E}_{-1\perp}$ and neglected the small term of the order of $(k_0 L)^{-1}$.

NEAR-BACKWARD SCATTERING. LINEAR APPROXIMATION

In this section the form of a laser pulse is assumed to be given and Eqs.(6) (with $\sigma = -1$) and (5) are studied by analytical methods. To investigate the transverse effects in the general case, it is convenient to introduce the Fourier transform with respect to variable $\mathbf{r}_\perp$ and also to use variables $\xi = z - \mathbf{v}_g t$ and $\eta = t$ corresponding to the frame moving with the pulse group velocity $\mathbf{v}_g$. We assume the simplest shape of the pulse with length L: the value of $\mathbf{E}_0$ is constant for $|\mathbf{r}_\perp| \leq r_0$ and is zero for $|\mathbf{r}_\perp| > r_0$. Introducing the Fourier transforms with respect to variable η we obtain from Eqs. (5) and (6) for $|\mathbf{k}_\perp| > r_0^{-1}$ [6]:

$$\left[\left(\omega - 2i\mathbf{v}_g \frac{\partial}{\partial \xi}\right) - \frac{c^2 \mathbf{k}_\perp^2}{2\omega_0}\right] \left(\mathbf{E}_0^* \mathbf{E}'(\mathbf{k}_\perp)\right) = \frac{\omega_p^2}{4\omega_0} \nu_2(\mathbf{k}_\perp) \left(|\mathbf{E}_0|^2 - \frac{c^2}{\omega_0^2} |\mathbf{E}_0 \mathbf{k}_\perp|^2\right) \tag{9}$$

$$\left[\left(i\omega + \mathbf{v}_g \frac{\partial}{\partial \xi}\right)^2 + \omega_p^2\right] \nu_2(\mathbf{k}_\perp) = \frac{2e^2 k_0^2}{m^2 \omega_0^2} \left(1 - \frac{i}{k_0} \frac{\partial}{\partial \xi} + \frac{\mathbf{k}_\perp^2}{4k_0^2}\right) \left(\mathbf{E}_0^* \mathbf{E}'(\mathbf{k}_\perp)\right)$$

The boundary conditions for this set correspond to the following assumptions: in the region ahead of the pulse ($\xi \geq L$), the electron density perturbations are given (and supposed to be continuous at the leading edge of the pulse ($\xi = L$)) and the intensity of the spontaneous electromagnetic radiation with frequencies close to ω_0 is negligible. The spectrum of the density perturbations corresponds to a thermal equilibrium plasma with a temperature T. It should be noted that the detected frequency in the laboratory frame ω_d is connected with that in the pulse frame ω (in Eqs.(9)) as follows:

$$\omega_d = \omega_0 + (\omega + k_\perp^2 v_g/2k_0)/2 \tag{10}$$

The spectrum of the scattered electric field can be obtained by using the solution of the system (9) as follows:

$$P(\omega_d, \mathbf{k}_\perp) = \int d\mathbf{k} \int d\mathbf{k}' d\omega' < \mathbf{E}'(\omega_d, \mathbf{k}') \mathbf{E}'^*(\omega_d', \mathbf{k}') > =$$

$$= 2|\mathbf{E}_0|^2 \left(\frac{\omega_p^2}{8\omega_0}\right)^2 \frac{k_0^2 T}{v_g^3 e^2 n_0^2} \sum_{s=\pm 1} \{[\frac{\text{sh}(\kappa L)}{\kappa}(1 - (k_1' - k_3')D_1^{(S)}) + \quad (11)$$

$$+ D_1^{(S)} \sin((k_1' - k_3')L)]^2 + \left(D_1^{(S)}[\text{ch}(\kappa L) - \cos((k_1' - k_3')L)]\right)^2\}$$

where $D^{(S)} = \frac{1}{\Delta}[2k_1 + (3\omega + 2s\omega_p)/2v_g - k_1^2/4k_0]; s = \pm 1; \Delta = \kappa^2 + (k_1 - k_3)^2$; k_1 and κ are the real and imaginary parts of the two complex conjugate roots, and k_3 is the third real root of the cubic characteristic equation of the system (9) for ω determined by Eq.(10).

For example, in the weak coupling limit $\left(\kappa \ll \omega_p/c \text{ or } v_e/c \ll (\omega_p/\omega_0)^{1/2}\right)$, for a linearly polarized radiation of the pulse, we obtain using Eqs.(11) and (10) the following expression for the spectral density in the laboratory frame:

$$P(\omega_d, \mathbf{k}_\perp) = 16\pi T \frac{\omega_p}{\omega_0 c} \exp\left\{G_0\left[\left(1 - (\frac{\mathbf{k}_\perp}{2k_0})^2\right)\left(1 - \frac{(\mathbf{k}_\perp \mathbf{e})^2}{k_0^2}\right) - \frac{\Delta\omega_d^2}{\Delta\omega_0^2}\right]^{1/2}\right\}$$

where $\Delta\omega_0 = (v_E/c)(\omega_0/2\omega_p)^{1/2}, G_0 = L\Delta\omega_0/c$ is the maximum value of the enhancement factor corresponding to the backward SRS, $\Delta\omega_d = \omega_d - (\omega_0 - \omega_p)$. This formula shows that density of scattered radiation energy decrease as both the deviations of frequency (from the resonance value $\omega_0 - \omega_p$) and scattering angle (from that corresponding to direct backscattering) increase. By integrating $P(\omega_d, \mathbf{k}_\perp)$ over the frequencies and $\mathbf{k}_\perp$ we can evaluate the total energy density of the scattered radiation and estimate the conditions when SRS has a weak effect on the laser-pulse propagation [4,6].

BACKWARD SCATTERING. NUMERICAL RESULTS

To take into account the nonlinear pulse dynamics in the consideration of SRS, we incorporated the back action of SRS on the pulse dynamics and carried out the numerical calculations of the self-consistent evolution of the pulse and backscattered radiation using Eqs.(6) (with $\sigma=+1$), (4) and (7)-(8).

It was assumed that the phase front of the pulse is initially plane and that the amplitude has a Gaussian shape in both directions (ξ, r). In the case of neodymium-glass laser, the parameter values used in the calculation correspond to a pulse duration of 175 fs, to an energy of 3.5 J, to a spot size $r_0=75$ mkm and to a maximum initial intensity $\simeq 10^{17}$ W/cm^2. The background plasma density was $2 \cdot 10^{18}$ cm-3. Fig.1 shows the on-axis dynamics of the laser

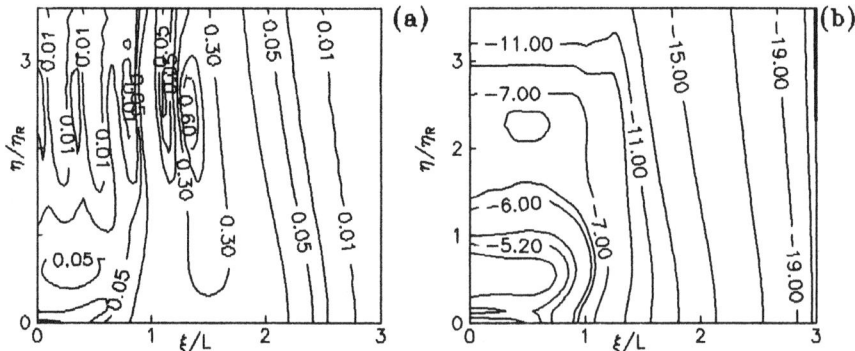

Figure 1: Contour lines for on-axsis ($r = 0$) time history of the laser pulse amplitude $a = v_E/c = |e\mathbf{E}_0/m\omega_0 c|$ (plot (a)) and that for $\ln\left(a'\right) = \ln\left|e\mathbf{E}'/m\omega_0 c\right|$ of the backscsttered radiation (plot (b)).

pulse amplitude during the self-modulation [6-8] (plot (a)) and also that for the field of the backscattered radiation (plot(b)). The time is mesured in the units of the Raylleigh time $\eta_R = k_0 r_0^2/2c$. Because the amplitude of the scattered field is relatively small, the corresponding pulse depletion is weak and can be traced only in the evolution of the pulse trailing part. In the case under consideration, in accordance with the mentioned above estimations of the total reflected energy [4,6], the feedback action of SRS on both the nonlinear pulse dynamics and plasma wake-field excitation is rather small. Consequently, for chosen parameters, during pulse propagation, SRS results in energy losses that do not attain 2%.

In conclusion, the numerical calculations confirm our analitical predictions of the conditions when the energy losses due to SRS are small.

REFERENCES.

[1] V.K.Tripathi, C.S.Liu, Phys. Fluids B3, 468 (1991).
[2] C.B.Darrow, C.Coverdale, M.D.Perry, W.B.Mori,
 C.Clayton, K.Marsh, and C.Joshi, Phys.Rev. Lett. 69(3), 442 (1992).
[3] S.V.Bulanov, I.N.Inovenkov, V.I.Kirsanov, N.M.Naumova, and
 A.S.Sakharov, Phys. Fluids B. 4(7), 1935 (1992).
[4] N.E.Andreev and L.M.Gorbunov, Proc. of 7th Int.Conf. "Laser Optics",
 1993, St-Petersburg, June 21-25, 1993, Part 2, p. 652; Proc. SPIE,
 2097, 437 (1994).
[5] A.S.Sakharov and V.I.Kirsanov, Phys..Rev..E 49, 3274, (1994).
[6] N.E.Andreev, V.I.Kirsanov, L.M.Gorbunov, Phys. Plasmas **2** (6) (1995)
[7] N.E.Andreev, L.M.Gorbunov, V.I.Kirsanov et al,
 Sov. JETP Lett. 55,(10), 571 (1992); Physica Scripta 49, 101 (1994).
[8] T.M.Antonsen, Jr. and P.Mora, Phys. Rev. Lett. 69, 2204, (1992)

Stimulated Scattering of Radiation at Interaction of Short Laser Pulse with Dense Plasma

A.A.Andreev, A.N.Sutyagin

Institute for Laser Physics, SC "Vavilov State Optical Institute",
12, Birzhevaya line, 199034, St.Petersburg, Russia

Abstracts. We have find out that in the case of the laser pulse with pico- or subpicosecond duration its energy part, scattered via SBS in the strongly inhomogeneous short plasma, reaches the value of fraction of percent for the intensity of $\leq 10^{17}$ W/cm^2, if there is no prepulse, and for the intensity of $\leq 10^{15}$ W/cm^2, if such a prepulse is presented. This fact may be of use in the experimental evaluation of the pulse contrast value and influence.

In this paper we discuss the stimulated Brillouin scattering (SBS) of the pulses with pico- or subpicosecond duration, which occurs in the strongly inhomogeneous, completely ionized, short laser plasma, containing the boundaries of radiation reflection. With the purpose of two dimensional description of SBS process we use the following set of equations (see (1)):

$$\left(2i\frac{c_s}{c}\frac{\partial}{\partial \tilde{t}} + \frac{\partial^2}{\partial \tilde{x}^2} + k_{xn}^2\right)\tilde{E}_n = (1-i\nu)\sum_{m \neq 0}\tilde{E}_{n-m}\delta\tilde{N}_m$$

$$\left(\frac{\partial^2}{\partial \tilde{t}^2} + 2\Gamma_s^{(n)}\frac{\partial}{\partial \tilde{t}} - \frac{\partial^2}{\partial \tilde{x}^2} + (nk_{II})^2\right)\delta\tilde{N}_n = \frac{1}{4}N_0\left[\frac{\partial^2}{\partial \tilde{x}^2} - (nk_{II})^2\right](|\tilde{E}|^2)_n$$

Here $\tilde{E}(\tilde{x},\tilde{y},\tilde{t}) = \sum \tilde{E}(\tilde{x},\tilde{t})\exp(i n k_{II}\tilde{y})$ is the dimensionless complex amplitude of the s-polarized wave,"

$$\tilde{E} = \frac{v_E}{v_{T_e}} = \frac{E(x,y,t)}{\{4\pi n_c(\bar{z}T_e + T_i)\}^{1/2}},$$

v_E - the velocity of electron oscillations in the wave field, v_{TE} - the thermal velocity of electrons, T_e, and T_i - the temperature of electrons and ions correspondingly, $\bar{z}$ - the average charge of electrons, $n_c = n_{ec}/\bar{z}$ - the critical density of ions, $n_{ec} = m_e\omega_0^2/4\pi e^2$, $\delta\tilde{N}(\tilde{x},\tilde{y},\tilde{t}) = \tilde{N}(\tilde{x},\tilde{y},\tilde{t}) - \tilde{N}_0(\tilde{x}) = \sum_n \delta\tilde{N}_n(\tilde{x},\tilde{t})\cdot\exp(i n k_{II}\tilde{y})$, - the plasma density perturbations, caused by the ponderomotive pressure, $(\tilde{x},\tilde{y}) = k_0(x,y)$ - the dimensionless coordinates, $\tilde{x}$ - the coordinate, orthogonal to the layer, $\tilde{x}$ varies in the range $[0, D_x]$, $\tilde{x} = 0$ corresponds to the starting boundary of the layer, $\tilde{x} = D_x$ corresponds to the final boundary of the layer, where the plasma equals the critical value, $\tilde{y}$ - the coordinate, parallel to the layer, $\tilde{y}$ varies in the range $[0, D_y]$, $\tilde{t} = \omega_s(k_0) t$, $k_0 = \omega_0/c$, ω_0 - the incident radiation frequency, $\omega_s = c_s k_0$, c_s - sound velocity, $k_{II} = 2\pi/D_y$, $k_{xn}^2 = \cos^2\theta_n - (1 - iv)(\tilde{N}_0 + \delta\tilde{N}_0)$; $\cos^2\theta_n = 1 - (n k_{II})^2$; $v = v_{ei}/\omega_0$; v_{ei} - the frequency of electron - ion collisions. We shall consider that the harmonics n = +1 corresponds to the pumping wave incidence direction with the angle of $\theta_1 \approx 30^o$ to the normal to the layer, n = -1 corresponds to the backward scattering, n = +1 - to the mirror-like scattering and reflection from the dense plasma boundary correspondingly. The following approximation was chosen for the ion sound damping decrement:

$$\Gamma_s^{(n)} = \Gamma_{s0} \cdot \left[n k_{II} - \frac{1}{2 k_{x1}} \frac{\partial^2}{\partial \tilde{x}^2} \right] + \Gamma_{s1}; \qquad v_{ii} < \omega_s$$

$$\Gamma_s^{(n)} = \left[\frac{\gamma_{se}}{\omega_s(k_0)} \right] \cdot \left[n k_{II} - \frac{1}{2 k_{x1}} \frac{\partial^2}{\partial \tilde{x}^2} \right] + \frac{0.8 T_i}{z T_e} \cdot \frac{\omega_s(k_0)}{v_{ii}} \cdot \left[n^2 k_{II}^2 - \frac{\partial^2}{\partial \tilde{x}^2} \right]; \qquad v_{ii} > \omega_s$$

Here

$$\Gamma_{s0} = \frac{\gamma_{se} + \gamma_{si}}{\omega_s(k_0)}; \quad \Gamma_{s1} = \frac{\gamma_{sii}}{\omega_s(k_0)}; \quad k_{x1} = (\cos^2\theta_1 - \tilde{N}_0)^{1/2} \text{ for } 4 k_{x1}^2 > k_{xc}^2 \text{ and } k_{x1} = \frac{k_{xc}}{2}$$

for $4 k_{x1}^2 < k_{xc}^2$; $k_{xc} = 2 L_c^{-1/3}$; $L_c = c_s t_p$, - the scale of density inhomogeneity in the critical point. We assume that the plasma density profile is the exponential one with the inhomogeneity scale $L_c = c_s t_p$, t_p - pulse duration.

I. It is well known (see(2)), that in general case the plasma density can be disturbed via following four processes:

a) due to the plasma displacement by the light pressure: $\delta\tilde{N}_0(\tilde{x}) \approx \frac{1}{4} \sum_{\sigma = \pm 1} \left| \tilde{E}_\sigma^{(0)} \right|^2$, $E_{+1}^{(0)}$, $E_{-1}^{(0)}$ - the amplitudes of the incident pumping wave and of the wave, reflected by the layer;

b) due to the density modulation by the pumping wave beats:

$$\delta\tilde{N}_0 = \sum_{\sigma = \pm 1} (\tilde{E}_{0\sigma} \tilde{E}_{0-\sigma}^*) \cdot e^{2 i \sigma k_x^{(0)} \cdot x}$$

here $k_x^{(0)} = k_0 \cdot \sqrt{\cos^2\theta - \tilde{N}_0}$. This case corresponds to the SBS in the direction of the mirror-like reflection with the excitation of two Stokes satellites with $\Delta\omega \sim \omega_s$;

c) due to the following process:

$$\delta \tilde{N}_2 \sim e^{-i\omega t} \sum_{\sigma=\pm 1} b^*(\omega) e^{-2i\sigma k_x^{(0)} x} (\tilde{E}_{0-\sigma} \cdot \tilde{E}_\sigma^*)$$

here $b(\omega) = \omega_{sb}^2 \left[2\omega \gamma_s(\omega) - i(\omega_{sb}^2 - \omega^2) \right]^{-1}$; $\omega_{sb} = 2k_0 c_s \sqrt{1 - N_0(\tilde{x})}$; $\tilde{E}_\sigma$ - the amplitudes of the scattered waves. This equation corresponds to to the excitation of acoustic waves, collinear to the pumping radiation waves and describes thus the backward SBS;

d) due to the following process:

$$\delta \tilde{N}_2 \sim e^{-i\omega t} d^*(\omega) \sum_{\sigma=\pm 1} \tilde{E}_{0\sigma} \tilde{E}_\sigma^*$$

here $d(\omega) = \omega_{sd}^2 \left[2\omega \gamma_s(\omega) - i(\omega_{sd}^2 - \omega^2) \right]^{-1}$; $\omega_{sd} = 2k_0 c_s \sin\theta_1$.

This equation corresponds to the excitation of the acoustic waves, propagating strictly along the plasma layer. It corresponds to the case of double SBS; it is well known (see (3)), that this process corresponds to the excitation of the absolute instability, the threshold of which is usually lower than that of backward SBS.

II. On the first stage we have studied the SBS of sub-ps pulses for the case when the plasma is generated due to the very scattering pulse heating action; this pulse has sufficiently high contrast, giving the possibility to neglect the prepulse action. For the pulse duration of several ps or less (down to the values of 1 ps fraction) the relation $L_c/\lambda_0 < 1$ is valid (λ_0 - wavelength of the incident radiation). Numerical simulation by the code "DSMBS" (see (1)) has shown very low level of backward SBS scattering for the range of pumping intensities I = 0.1..10 PW/cm^2 and pumping pulse durations t_p = 0.5..3 ps. In the Fig.1 is shown the dependence of the backward reflection coefficient R vs. steady-state SBS increment $\kappa_0 = \frac{1}{8} \frac{\omega_s}{\gamma_s} |\tilde{E}_0|^2 \cdot 2\pi \frac{L_c}{\lambda_0}$. One can see that the value of R

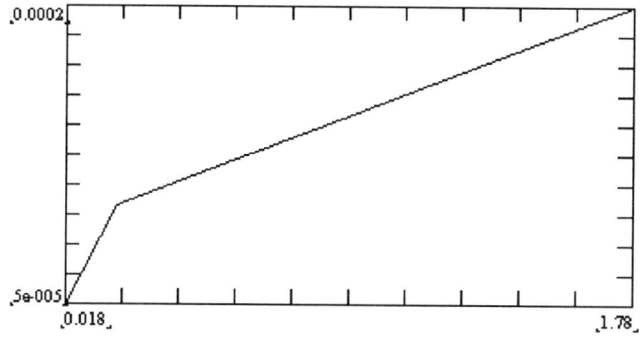

FIGURE 1. Dependence of the backward SBS reflectivity vs. sound amplification gain.

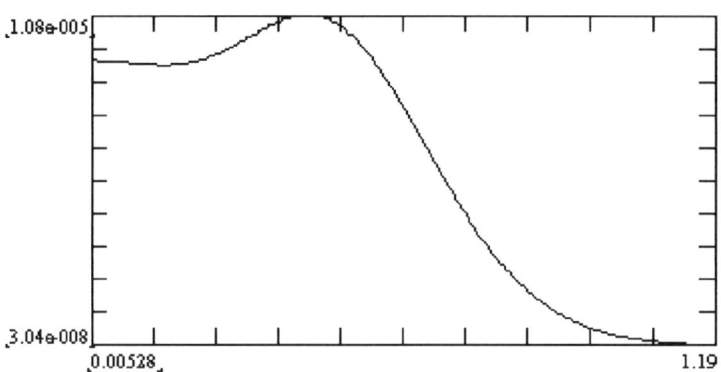

FIGURE 2. Distribution of the backward scattered radiation field inside the plasma layer.

is $\leq 10^{-4}$. In these calculations we took the value of electron temperature T_e in the range 300..1000 eV; such an evaluation results from the uni-dimensional hydrodynamics calculation with the account of backward breamstrehlung absorption. For such a super-short laser pulse the ion temperature T_i meets the relation $\dfrac{T_i}{T_e} \leq 10^{-2}$; the sound damping equals $\Gamma_{s0} \approx 0.1$.

We have supposed to obtain the much higher value of R for the case $L_c < \lambda_0$: seemingly, in this case there is the possibility of excitation of the absolute backward instability - the case of double SBS, when the sound is running along the plasma layer (case **Id**). However, our calculations have revealed the situation, which is quite different from the double SBS case.

According to the calculation results, for the case of $L_c < \lambda_0$ and under the sufficiently high pumping intensities, the scattering is completely determined by the process of plasma displacement by the intense light field, i.e., by the radiation scattering on the perturbations $\delta\widetilde{N}_0(\tilde{x})$ (case **Ia**). In the Fig.2 is shown the distribution of the field harmonics $n = -1$ inside the plasma layer, corresponding to this situation.

III. Quite another situation is realized in the case of scattering of the high-contrast pulses with the duration, exceeding somewhat 5-10 ps, when $L_c/\lambda_0 > 1$ is valid. One can see (Fig.3) that the scattering reflectivity $R \leq 10^{-2}$ and is much higher than in the case of $L_c/\lambda_0 < 1$. Here the electron temperature value was the same ($T_e = 300..1000$ eV) and pump intensities $I=10^{14}..10^{16}$ W/cm². Quite different SBS processes are observed for the intensities of $I \approx 10^{14}$ W/cm² and $\approx 10^{15}$ W/cm². In the case of $I \approx 10^{14}$ W/cm² one can observe the acoustic oscillations modulation δN_2 in the region of the comparatively disperse plasma; to our opinion, this situation corresponds to the

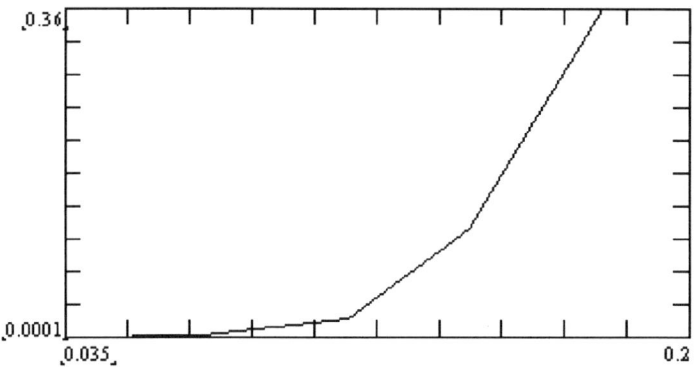

FIGURE 3. Dependence of the backward SBS reflectivity vs. sound amplification gain.

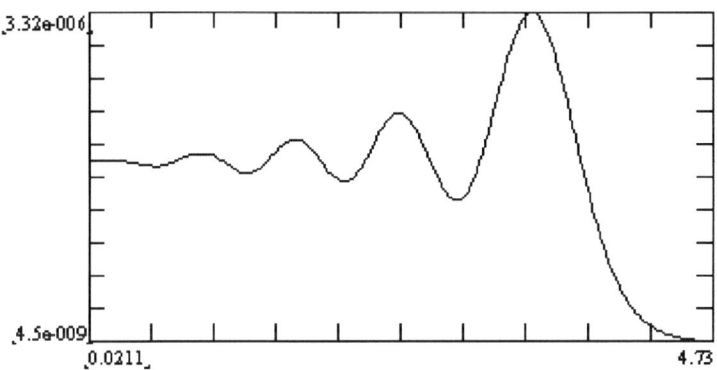

FIGURE 4. Distribution of the backward scattered radiation field inside the plasma layer.

development of backward SBS instability. In the Fig.4 is shown the distribution of the field harmonics ($n=-1$) in this case. As to the intensity range $I \approx 10^{15}$ W/cm^2, here the acoustic perturbations in the disperse plasma are dumped and the acoustic harmonics $\delta \tilde{N}_2$ is maximal in the region of dense plasma nearby the critical surface. I.e., in this case the sound waves run along the layer and the reflectivity can be much higher $R \sim 10^{-2}$. It can even reach the yet higher value of $R \sim 10^{-1}$ for the intensities $I = 10^{15}..10^{16}$ W/cm^2. To our opinion, in this case the scattering is completely determined by the process of double backward SBS.

IV. Let us, at last, discuss the case of the low-contrast sub-ps pulse scattering, when there is no possibility to neglect the prepulse action onto the plasma. Let it be the pulse duration t_p, prepulse duration t_{pr} and the energies - W_p and W_{pr} correspondingly. Hence

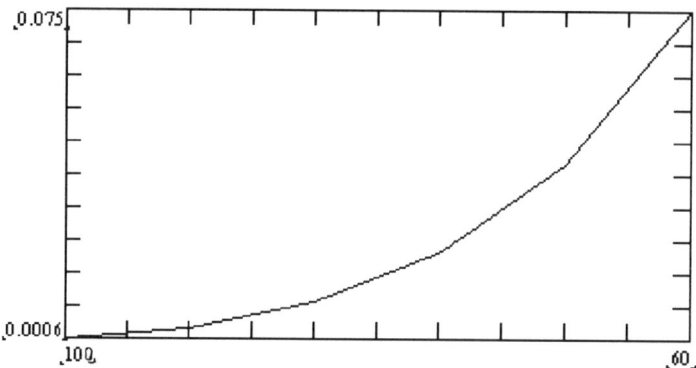

FIGURE 5. Dependence of the bacward scattering reflectivity vs. incident radiation contrast

the intensity contrast is $K_I=K_W K_t$; here $K_W=W_p/W_{pr}$, $K_t=t_{pr}/t_p$. We shall discuss the case of super-short pulse (t_p~1 ps) and contrast K_I=10..100; prepulse intensity I_{pr}~10^{13} W/cm². In this case the prepulse is scattered on the plasma with the inhomogeneity scale L_c, which is determined only by the prepulse: $L_c=c_s t_{pr}$; the sound velocity is determined by the plasma temperature, established by the prepulse action. In this case the backward scattering level will be determined by the main pulse scattering on the long inhomogeneous plasma with $L_c=c_s t_{pr}$. One can see (Fig.5) that in this case the rate of backward scattering can be rather large $(R\sim 10^{-1})$; the scattering nature is determined first of all by the absolute instability development (double SBS), if the threshold value of

$$\kappa_0^{th} = \frac{1}{2\cos\theta_1} \ln^{-1}\left(\frac{\omega_0 L_c}{c}\cos\theta_1\right)^{2/3}$$

is overcame (see (1-3)). Note, that even in the case of rather modest pump intensities result in significant variation of the super-short pulse scattering nature and in significant growth of the backward scattered intensity in comparison with the case of plasma, created by the very super-short pulse (see **II**). That is why these processes, realized while super-short laser pulse interacts with the laser-produced plasma, can be used as the criterion of the contrast influence rate and possible presence of prepulse.

REFERENCES

1. Andreev A.A., Andreev N.E., Sutyagin A.N., Tikchonchuk V.T. Sov.J.Plasma Phys., **15**, 944-950, (1989).
2. Andreev N.E., Chegotov M.V., Tikchonchuk V.T., Preprint FIAN, n.169 (1988).
3. Andreev N.E., Silin V.P., Tikchonchuk V.T., Preprint FIAN, n.19 (1987).
4. Hinkel D.E., Williams E.A., Berger R.L., forthcoming in Physics of Plasma, October (1994).

Problem of Plasma Wakefields with Low Phase Velocities and Relativistic Selffocusing of Short Laser Pulses

L.N.Tsintsadze*

Department of Materials Science, Faculty of Science,
Hiroshima University, Higashi-Hiroshima 739, Japan

Abstract

A generalized set of equations for nonstationary three dimensional plasma wakefield and selffocusing of relativistic intense electromagnetic (e.m.) wave is obtained. The generation of larg-amplitude plasma wakefields with low phase velocities by an intense short laser pulses is studied and the dependence of the maximum amplitude of the excited plasma wakefields on the group velocity of laser pulses is determined.

---

*Permanent Address: Plasma Physics Department, Institute of Physics, Georgian Academy of Sciences, Tbilisi 77, Republic of Georgia

Currently, one of the most important problems of plasma physics is the study of interaction of relativistic intense e.m. waves with plasma. Such interest to this phenomenon is caused by the possibility of plasma heating and realization of laser ICF [1-3] as well as excitation of wakefields. A special interest to the latter is connected with the design of new plasma methods of acceleration of charged particles [3-6]. As it was shown, the increase of radiation intensity changes essentially oscillating properties of plasma [7], the structure of soliton [7], leads to parametric resonance in the electron plasma [8], affects the propagation of shock waves [9], relativistic modulational and filamentational [10,11,12] instabilities. In [13] was shown that relativistic effects in the problem of selffocusing of radiation leads to plasma compression in the range of moving focus, that is a new phenomenon in isotropic plasma.

Due to complexity of the set of equations describing the processes of selffocusing or excitation of the three dimensional wave fields, these processes in the above papers were investigated under numerous assumptions. Obviously the most important is the problem of derivation of more realistic system.

Let us consider propagation of relativistically strong circularly polarised high frequency e.m. wave along axis z in the collisionless plasma. We use the set of Maxwell's equations and relativistic hydrodynamics equations of electrons neglecting thermal motion. Plasma ions are assumed to be immobile.

$$\Delta A_\perp - \frac{1}{c^2}\frac{\partial^2 A_\perp}{\partial t^2} = \frac{4\pi e^2}{m_o c^2}\frac{n A_\perp}{\gamma} . \tag{1}$$

$$\frac{d\vec{p_e}}{dt} = -e\left[\vec{E} + \frac{1}{c}[\vec{v_e} \times \vec{B}]\right] . \tag{2}$$

$$\frac{\partial n_e}{\partial t} + div\, n_e \vec{v_e} = 0 . \tag{3}$$

$$div\, \vec{E} = 4\pi e(n_o - n_e) \tag{4}$$

where

$$A_\perp = A_x + iA_y, \quad \vec{E} = -\frac{1}{c}\frac{\partial \vec{A}}{\partial t} - \nabla\varphi,$$

$$\vec{p_e} = \frac{m_o \vec{v_e}}{(1 - v_e^2/c^2)^{1/2}}, \quad \gamma = \left(1 + \frac{p^2}{m_o^2 c^2}\right)^{1/2} .$$

Using the traditional approach, the vector potential is represented as

$$A_\perp = (\hat{e}_x + i\hat{e}_y)A_o(t, z, r_\perp)exp\, i(k_o z - \omega_o t) . \tag{5}$$

then eq. (1) for slowly changing in time and coordinate of the complex amplitude are rewritten as wellknown form

$$\left(\Delta_\perp + \frac{\partial^2}{\partial z^2} - \frac{1}{c^2}\frac{\partial^2}{\partial t^2}\right) + \frac{2i\omega_\circ}{c^2}\left(\frac{\partial}{\partial t} + \frac{k_\circ c^2}{\omega_\circ}\frac{\partial}{\partial z}\right)A_\circ + \frac{\omega_\circ^2}{c^2}\left(1 - \frac{k_\circ^2 c^2}{\omega_\circ^2}\right)A_\circ = \frac{\omega_p^2}{c^2}\frac{1+\delta n/n_\circ}{\gamma}A_\circ . \quad (6)$$

where

$$\delta n = n_e - n_\circ, \quad \omega_p^2 = \frac{4\pi e^2 n_\circ}{m_\circ} .$$

The usual expression before the first derivative along z in the second bracket of the left part of eq. (6) is taken as group velocity ($v_g = k_\circ c^2/\omega_\circ$). Actually this equation agrees with the group velocity in the nonrelativistic case. As concerns relativistically intense waves, when oscillating velocities of the electrons become of the order of speed of light, in this case it is impossible to introduce the motion of the group velocity. There is no equation of dispersion. Only in the approximation of the geometric optics one can obtain equation the form of which agrees with the dispersion one

$$\frac{k_\circ^2 c^2}{\omega_\circ^2} = 1 - \frac{\omega_p^2(n)}{\omega_\circ^2 \gamma} . \quad (7)$$

Since in eq. (7) the density and relativistic factor change in the space and time , the latter is the equation of the state relativistically strong e.m. wave. In this connection definition of $k_\circ$ and $\omega_\circ$ in the plasma becomes problematic. Indeed, we could have introduced the vacuum values of $k_\circ$ and $\omega_\circ$, then $v_g = c$ and any change of the value of the wave vector has to be included into the change of amplitude $A_\circ$. Otherwise, one may assume that the amplitude in certain point is constant, then ex. (7) becomes dispersion equation with constant values $n_\circ$ and $\gamma_{\circ\perp} = (1 + (eA_{\circ\perp}/m_\circ c^2)^2)^{1/2}$.

It follows from eq. (7) that if $\omega_\circ \gg \omega_p/\gamma^{1/2}$ then $v_g \simeq c$ (almost vacuum situation) and $\partial_z^2 A_\circ \simeq (1/c^2)\partial_t^2 A_\circ$ is in the eq. (6). Further, the structure of eq. (6) indicates at the character of dependence $A_\circ$ upon coordinates: 1.If nonstationary process of propagation of a strong wave is considered,

$$A_\circ = A_\circ(\vec{r}_\perp, z - v_g t, t) \quad (8)$$

with account of $v_g = k_\circ c^2/\omega_\circ \sim c$ eq. (6) will be of the form

$$\Delta_\perp A_\circ + \frac{2i\omega_\circ}{c^2}\frac{\partial A_\circ}{\partial t} + \frac{2v_g}{c^2}\frac{\partial^2 A_\circ}{\partial \xi \partial t} + \frac{\omega_\circ^2}{c^2}\left(1 - \frac{k_\circ^2 c^2}{\omega_\circ^2}\right)A_\circ = \frac{\omega_p^2}{c^2}\frac{1+\delta n/n_\circ}{\gamma}A_\circ . \quad (9)$$

In the case of nonstationary selffocusing

$$A_\circ = A_\circ(\vec{r}_\perp, z - v_g t, z) \quad (10)$$

$$\Delta_\perp A_\circ + 2ik_\circ \frac{\partial A_\circ}{\partial z} + 2\frac{\partial^2 A_\circ}{\partial z \partial \xi} + \frac{\omega_\circ^2}{c^2}\left(1 - \frac{k_\circ^2 c^2}{\omega_\circ^2}\right)A_\circ = \frac{\omega_p^2}{c^2}\frac{1+\delta n/n_\circ}{\gamma}A_\circ .\tag{11}$$

where $\xi = z - v_g t$.

Note that in some papers relationship $\omega_\circ^2 = \omega_p^2 + k_\circ^2 c^2$ was used incorrectly, which in relativistic plasma is of no physical sence that leads to the erroneous Schrödinger equation.

Up to-date there is no precise equations for the three dimensional relativistic intensive e.m. wave, describing nonlinear dynamics. As a matter of fact, in previous papers the expression contained only transverse momentum, caused by the field of pumping. As it was shown in [5, 6, 14] concerning the laser accelerator on wakefield, the longitudinal momentum $<p_\parallel>$ (caused by ponderomotive force) becomes of the same order that $|\tilde{p}_\perp|$, but in ultrarelativistic case is much more, meaning that γ should be represented as $(1+(\tilde{p}_\perp^2 + <p_\parallel>^2)/m_\circ^2 c^2)^{1/2}$. All this concerned the one dimensional plasma motion in the wave field. The problems becomes complicable when the wave amplitude depends upon three coordinates. As a matter of fact, in this case a ponderomotive force appears, acting in transverse direction relative to propagation of the pumping wave. It results in formation of the mean (over the period of pumping field $\tau = 2\pi/\omega_\circ$) transverse momentum, i.e. $\vec{p}_\perp = <\vec{p}_\perp> + \tilde{\vec{p}}_\perp$. Presence of $<\vec{p}_\perp>$ in equations of electron motion and Maxwell's makes the problem practically unsolvable.

The present paper gives relationship which allows to simplify the study of propagation of relativistic intensive wave in three dimensional case, not loosing the generality.

In ordar to show the above said, let us write the equation of electron motion without account of gas dynamic pressure in the form

$$\frac{\partial \vec{p}}{\partial t} = -e\vec{E} - m_\circ c^2 \nabla \gamma .\tag{12}$$

where $\vec{p}$ - total momentume. Let us represent $\vec{p}$ as

$$\vec{p}_\perp = <\vec{p}_\perp> + \tilde{\vec{p}}_\perp ,$$

$$p_\parallel = <p_\parallel> .\tag{13}$$

Substituting expression for momentume $\vec{p}$ (13) into (12) and averaging over period $\tau = 2\pi/\omega_\circ$,

$$\partial_t <\vec{p}_\perp> = \nabla_\perp (e\varphi - m_\circ c^2 <\gamma>) .\tag{14}$$

$$\partial_t <p_\parallel> = \nabla_\parallel (e\varphi - m_\circ c^2 <\gamma>) .\tag{15}$$

Here it is accounted for that $<\vec{E}> = -\nabla\varphi$

$$<\gamma> = \frac{\omega_\circ}{2\pi}\int_0^{2\pi/\omega_\circ} dt\left(1 + \frac{(<\vec{p}_\perp> + \tilde{\vec{p}}_\perp)^2 + <p_\parallel>^2}{m_\circ^2 c^2}\right)^{1/2} .\tag{16}$$

From (14) and (15) we can easily get

$$\nabla_\| <\vec{p}_\perp> = \nabla_\perp <p_\|> \tag{17}$$

Relationship (17) allows to compare $<p_\perp>$ with $<p_\|>$. If the averaged physical values along propagation of the pumping change faster than in the transverse direction, it folows from (17) that $<p_\perp> \ll <p_\|>$ since

$$\frac{<p_\perp>}{<p_\|>} \simeq \frac{\ell_\|}{\ell_\perp} \ll 1. \tag{18}$$

In this case $<\vec{p}_\perp>^2$ can be neglected in (16) as compared with $<p_\|>^2$ and $<\gamma>$ is written

$$\gamma = \left(1 + \frac{\tilde{p}_\perp^2 + <p_\|>^2}{m_o^2 c^2}\right)^{1/2}. \tag{19}$$

Condition ($\ell_\| < \ell_\perp$) for such problems as selffocusing or selfchannelling or generation of wakefield is fulfilled well. In order to close the set of equations it is necessary to obtain the equation for the potential φ. One should find nonlinear perturbation of the electron density $\delta n = n_e - n_o$. Due to strong relativism it is impossible to express in reasonable functions or to write the equation, which could be solved in three dimensional case. But such possibility exists if one uses the above condition and $\partial_\xi f \gg \nabla_\perp f \gg \partial_z f$. Then the Poisson eq. (4) and the continuity equation will be of the form

$$\partial_\xi^2 \varphi = -4\pi e \delta n. \tag{20}$$

$$\frac{\delta n}{n_o} = \frac{v_\|}{v_g - v_\|} \tag{21}$$

where $v_\| = <p_\|>/m_o\gamma$.

It follows from (15) that

$$-v_g <p_\|> = e\varphi - m_o c^2 \gamma + C \tag{22}$$

where C - the integration constant. If on the infinity minus $<p_\|> = \varphi = A_\perp = 0$, then $C = m_o c^2$ and if on the infinity $<p_\|> = \varphi = 0$, $A_\perp = A_{o\perp}$, then $C = m_o c^2 \sqrt{1 + u_{o\perp}^2}$ and in dimensionless limits (22) is written as

$$\beta u_\| = \gamma - Y. \tag{23}$$

where $\beta = v_g/c$, $u_\| = <p_\|>/m_o c$, $Y = \phi + 1$, or $Y = \phi + \sqrt{1 + u_{o\perp}^2}$ $\phi = e\varphi/m_o c^2$ and $u_{o\perp} = eA_{o\perp}/m_o c^2$.

In the transparent plasma or in the case of strong relativism, when $\omega_o^2 \gg \omega_p^2/\gamma$, one can assume $\beta \simeq 1$ in (9) and (11). Then from (4) and (23), (21), (9) and (11) we obtain the set of two equations in dimensionless units

$$\frac{\partial^2 \phi}{\partial \rho_\xi^2} = \frac{1}{2}\left(\frac{1+|u_\perp|^2}{Y} - 1\right). \tag{24}$$

$$2i\frac{\omega_o}{\omega_p}\frac{\partial u_\perp}{\partial \tau} + \Delta_{\rho_\perp} u_\perp + 2\frac{\partial^2 u_\perp}{\partial \rho_\xi \partial \tau} + \frac{\omega_o^2}{\omega_p^2}\left(1 - \frac{k_o^2 c^2}{\omega_o^2}\right)u_\perp - \frac{u_\perp}{Y} = 0. \tag{25}$$

where $\rho_\xi = k_p \xi$, $\rho_\perp = k_p r_\perp$, $\tau = \omega_p t$, $k_p = \omega_p/c$.

In the case when $\phi(\rho_\xi, \rho_\perp, z)$, eq. (11) is as follows

$$2i\frac{k_o c}{\omega_p}\frac{\partial u_\perp}{\partial \rho_z} + \Delta_{\rho_\perp} u_\perp + 2\frac{\partial^2 u_\perp}{\partial \rho_\xi \partial \rho_z} + \frac{\omega_o^2}{\omega_p^2}\left(1 - \frac{k_o^2 c^2}{\omega_o^2}\right)u_\perp - \frac{u_\perp}{Y} = 0. \tag{26}$$

In the general case when β is arbitrary value, equation for ϕ becomes

$$\frac{\partial^2 \phi}{\partial \rho_\xi^2} = \gamma_p^2 \left\{ \left(1 - \frac{1}{\gamma_p^2}\right)^{1/2} \frac{Y}{(Y^2 - \gamma_\perp^2/\gamma_p^2)^{1/2}} - 1 \right\}. \tag{27}$$

where

$$\gamma_p = \frac{1}{\sqrt{1-\beta^2}}.$$

Obviously, at $\gamma_p \to \infty$, eq. (27) becomes (24).

The authors of [7, 14] in order to obtain maximum electron energy, assumed that $\beta \sim 1$. But in this case increases hard the length on which electrons should accelerate, since $\ell \simeq (\omega_o^2/\omega_p^2)(c/\omega_p)$. This would be insufficient for the new type of laser accelerators. At $\beta \sim 1$ they could not answer the question: How does the maximum amplitude of driven waves depend on the group velocity of the laser (or on the phase velocity of excited wakefield)? Besides, at $\beta \sim 1$ the connection of the passed transverse wave of e.m. field with plasma is very weak. Therefore more interesting is the study of the case $\beta < 1$.

Now we can write the coupled set of field equations which describes the 1D nonlinear laser-plasma interaction within the quasistatic approximation for laser pulses of arbitrary polarization and arbitrary intensities:

$$\left(\frac{1}{\gamma_p^2}\frac{\partial^2}{\partial \rho_\xi^2} + 2v_p\frac{\partial^2}{\partial \rho_\xi \partial \tau} - \frac{\partial^2}{\partial \tau^2}\right)u_\perp = \frac{v_p u_\perp}{\sqrt{(1+\phi)^2 - (1+u_\perp^2)/\gamma_p^2}}. \tag{28}$$

$$\frac{\partial^2 \phi}{\partial \rho_\xi^2} = \gamma_p^2\left\{v_p \frac{1+\phi}{\sqrt{(1+\phi)^2 - (1+u_\perp^2)/\gamma_p^2}} - 1\right\}. \tag{29}$$

Here v_p ($\to v_p/c$) is the dimensionless phase velocity which equals the group velocity of the laser pulse.

Let us first examine the case of a square-shaped laser radiation pulse having the circular polarization which propagates in the z direction with unchanged shape and the group velocity $v_g = v_p$.

Integrating eq. (29) we obtain

$$\left(\frac{d\phi}{d\rho_\xi}\right)^2 + 2\gamma_p^2\left\{1+\phi - v_p\left[(1+\phi)^2 - \frac{\gamma_{o\perp}^2}{\gamma_p^2}\right]^{1/2}\right\} = 2\gamma_p^2\left\{1 - v_p\left[1 - \frac{\gamma_{o\perp}^2}{\gamma_p^2}\right]^{1/2}\right\} \quad (30)$$

where $\gamma_{o\perp}^2 = 1 + u_{o\perp}^2$.

Analysing eq. (30) we find the following expression for the maximum amplitude of the electric field ($E = -d\phi/d\rho_\xi$) as a function of the pump amplitude and the phase velocity:

$$E_{max} = \sqrt{2}\left(\gamma_p^2 - \sqrt{(\gamma_p^2-1)(\gamma_p^2-\gamma_{o\perp}^2)} - \gamma_{o\perp}\right)^{1/2}. \quad (31)$$

With the help of eq. (31) we can write:

$$E_{max}^2(\gamma_{o\perp}, v_{p1}) - E_{max}^2(\gamma_{o\perp}, v_{p2}) = \left(\sqrt{\gamma_{p1}^2-1} - \sqrt{\gamma_{p1}^2-\gamma_{o\perp}^2}\right)^2 - \\ \left(\sqrt{\gamma_{p2}^2-1} - \sqrt{\gamma_{p2}^2-\gamma_{o\perp}^2}\right)^2 \quad (32)$$

where $E_{max}(\gamma_{o\perp}, v_{p1})$ and $E_{max}(\gamma_{o\perp}, v_{p2})$ denote the electric field maximum amplitude generated by the given pump $-\gamma_{o\perp}$, at the different phase velocities v_{p1} and v_{p2}, respectively.

The numerical analysis of eq. (29) in the case of the Gaussian pulses confirm that the lower is the phase velocity of the wakefield, the greater is its maximum amplitude. The Fig. 1 shows the dependence of the wakefield amplitude on the phase velocity for Gaussian 1 psec pulses ($\tau_{FWHM} = 1$ psec, τ_{FWHM} — full width at half maximum) of various amplitudes.

In Fig. 2 the numerical solutions of wakefields produced by two Gaussian pulses with $A_{\perp max} = 1.4 m_o c^2/e$ (solid line) and $A_{\perp max} = 2.8 m_o c^2/e$ (dashed line) are presented.

Note that the both pulses have identical duration, $\tau_{FWHM} = 0.1$ psec, but in the first case (solid line) $v_{p1} = 0.95c$ ($\gamma_{p1} = 3.16$), whereas in the second one (dashed line) $v_{p2} = 0.99c$ ($\gamma_{p2} = 10$).

Thus, we can conclude that Gaussian pulses with lower amplitudes and lower group velocities are able to generate the wakefields having greater amplitudes than wakefields produced by Gaussian pulses with higher amplitudes but phase velocities close to the speed of light.

Acknowledgments

The author wishes to thank Prof. K.Nishikawa for support and hospitality.

We acknowledge the support of Japan Society for the Promotion of Science through the JSPS Postdoctoral Fellowship Program.

References

1. C.E.Max, in Laser Plasma Interaction, ed. by R.Balian and J.C.Adam, (North-Holland Publishing Co., Amsterdam, 1982).

2. J.L.Bobin, Phys. Rep. **122**, 173 (1985).

3. T.Tajima and J.M.Dawson, Phys. Rev. Lett. **43**, 267 (1979).

4. T.Katsouleas et al. In New Developments in Particle Acceleration Technique, ed. by S.Turner, ICEA-CAS Workshop Proc. v. II, 401 (1987).

5. V.I.Berezhiani and I.G.Murusidze, Phys. Lett. **148**, 2146 (1990).

6. S.V.Bulanov, V.I.Kirsanov, A.C.Sakharov, Sov. J.Plasma Physics **16**, 935 (1990).

7. V.I.Berezhiani, N.L.Tsintsadze, D.D.Tskhakaya, J.Plasma Phys. **24**, 15 (1980).

8. N.L.Tsintsadze, Zh. Eksp. Teor. Fiz. **59**, 1250 (1970).

9. J.F.Drake, Y.C.Lee, K.Nishikawa, N.L.Tsintsadze, Phys. Rev. Lett. **36**, 196 (1976).

10. P.K.Shukla, R.Bharuthram, N.L.Tsintsadze, Phys. Rev. A **35**, 4889 (1987).

11. P.K.Shukla, R.Bharuthram, N.L.Tsintsadze, Physica Scripta **38**, 578 (1988).

12. L.N.Tsintsadze, Sov. J.Plasma Phys. **17**, 872 (1991).

13. D.P.Garuchava, N.L.Tsintsadze, D.D.Tskhakaya, Zh. Eksp. Teor. Fiz. **98**, 1558 (1990).

14. T.Katsouleas, C.Joshi, J.Dawson, F.F.Chen, W.Mori, C.Darrow and D.Omstadter, in Laser Acceleration of Particles, AIP Conf. Proc. N. 130, ed. by C.Joshi and T.Katsouleas, (New Yor, 1985).

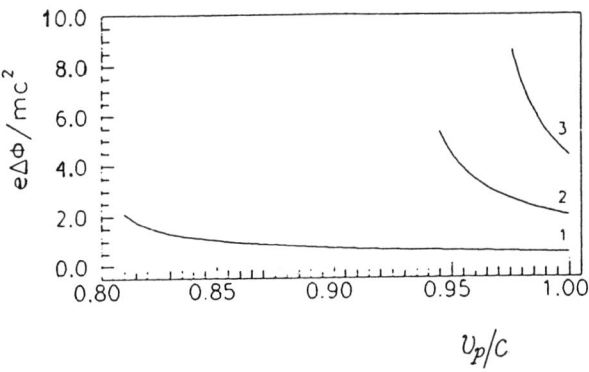

Fig.1 Dependence of the wakefield amplitude on the phase velocity for Gaussian pulses τ_{FWHM} = 1 psec) of various amplitudes. Curve 1 corresponds to $A_{\perp max} = 1.4\ mc^2/e$; curve 2 – $A_{\perp max} = 2.8\ mc^2/e$; curve 3 – $A_{\perp max} = 4\ mc^2/e$.

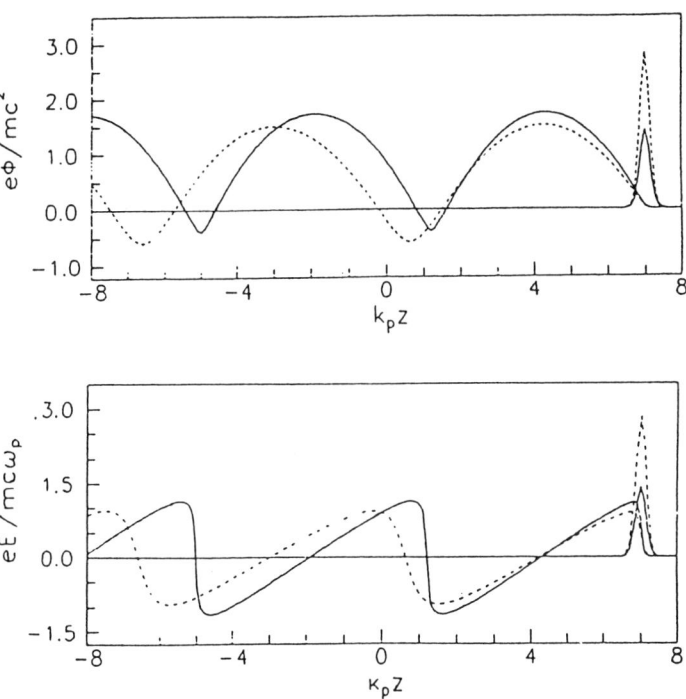

Fig.2 The electric fields and electrostatic potentials of the wakefields excited by two Gaussian pulses with identical duration. $\tau_{FWHM} = 0.1$ psec, and different amplitudes. The solid curve corresponds to the case $A_{\perp max} = 1.4\ mc^2/e$ and $v_p = 0.95\ c$. The dashed curve corresponds to $A_{\perp max} = 2.8\ mc^2/e$ and $v_p = 0.99\ c$.

The inertia of tunneling ionization and high-order harmonic shifting in the nonlinear single-atom response

A.V. Kim[1], E.V. Vanin[1], A.M. Sergeev[1],
D. Farina[2], M. Lontano[2],
M.C. Downer[3]

[1]Institute of Applied Physics, Russian Academy of Sciences,
603600 Nizhny Novgorod, Russia
[2]Istituto di Fisica del Plasma, Consiglio Nazionale delle Ricerche,
EURATOM-ENEA-CNR Association, 20133 Milano, Italy
[3]University of Texas at Austin,
Austin, Texas 78712, USA

Abstract. We demonstrate that the ionization-induced high-order harmonic emission mechanism can be affected by a new frequency-shifting effect which is present directly in the single-atom response at fixed laser frequency and monotonically varying laser intensity. By analyzing the motion of newly-born electrons we found that the phases of their collisions with "parent" ions are influenced by the dynamics of the driving field amplitude. At the leading edge of the laser pulse, where the field amplitude is increasing, the collisions at a given velocity occur periodically in time but are shifted monotonically in phase with each optical cycle, the sign of the shift depending on the delay time at which freed electrons are detached from the atoms after tunneling. The relationship between the dynamics of the $|\Psi|^2$ function and the possibility to produce a frequency shift in the presence of a monotonously varying amplitude pulse is discussed.

INTRODUCTION

The effect of high-order harmonic generation during intense laser pulse interactions with gases has recently been proposed to be a direct result of the atom ionization in the presence of a strong linearly-polarized optical field (1,2). Though general spectral and temporal features of high-order harmonic generation have been explained by studying the freed electron dynamics, more macroscopic evidences are needed to adopt ultimately this idea. In this paper, we demonstrate that the ionization-induced high-order harmonic emission mechanism can be confirmed by observation of a <u>new frequency-shifting effect</u> which is present directly in the single-atom response at fixed laser frequency and monotonically varying laser intensity. The origin of this shift is to be found in the time delay which characterizes the electron detachment from the atom with respect to the instants of maximum (time varying) electric field amplitude.

In the case of the leading edge of a short laser pulse, it turns out that, at sufficiently short wavelengths, a blue-shifting of the high-order harmonic spectrum is predicted, which is distinct from the collective ionization blue-shifting observed in previous experiments (3-5).

The physical mechanism of this frequency-shifting is analyzed in the frame of the Schroedinger equation, and a basis is provided for experimental distinction of the new effect, which can prove, at the macroscopic level, our understanding of the nonlinear atomic phenomena.

CLASSICAL VIEW OF ATOM IONIZATION DUE TO A MONOTONICALLY VARYING FIELD AMPLITUDE

Let us consider the classical one-dimensional trajectory of an electron of mass m and charge $q = -e < 0$, under the action of an oscillating electric field with the slowly increasing $E_0(1+\alpha t)$ amplitude and the frequency ω_E, that is:

$$m\frac{d^2X}{dt^2} = qE_0(1+\alpha t)\cos\omega_E t \qquad (1)$$

with $\delta = |\alpha|/\omega_E \ll 1$.

As it has been shown in Refs.1 and 2, during half of a radiation cycle, each electron freed from an ion will collide back with the same ion from which it was originated, producing a wide discrete bremsstrahlung spectrum. The emission spectrum shows a sharp cut-off at a frequency corresponding to the maximum energy of collision, i.e. $3.17 U_p + E_i$, where $U_p = q^2 E_0^2/4m\omega_E^2$ is the ponderomotive potential, and E_i is the ionization potential.

The integration of Eq.(1) gives the electron position

$$\frac{X(\varphi,\varphi_0)}{qE_0/m\omega_E^2} = -\cos\varphi + \cos\varphi_0 - (\varphi-\varphi_0)\sin\varphi_0 +$$
$$+\delta\left[-\varphi(\cos\varphi+\cos\varphi_0) + 2\varphi_0\cos\varphi_0 + 2\sin\varphi - (2+\varphi_0\varphi-\varphi_0^2)\sin\varphi_0\right] \qquad (2)$$

and the electron velocity

$$\frac{V(\varphi,\varphi_0)}{qE_0/m\omega_E} = \sin\varphi - \sin\varphi_0 + \delta[\varphi\sin\varphi - \varphi_0\sin\varphi_0 + \cos\varphi - \cos\varphi_0] \qquad (3)$$

as functions of the dimensionless time, $\varphi = \omega_E t$, and initial phase, $\varphi_0 = \omega_E t_0$. Here, atomic units are used to express lengths and velocities.

When the collision event occurs, $X(\varphi = \varphi_c, \varphi_0) \equiv 0$. Moreover, the maximum energy an electron can have at collision is determined by the condition $dV/d\varphi_0|_{\varphi=\varphi_c} \equiv 0$. In Fig.1 the phase of collision and the energy at collision are plotted versus the phase of ionization, counted from the maximum electric field (which corresponds to $\varphi_0=0$), for $\delta=0, 0.01, 0.02, 0.03$.

In the case of slightly varying field amplitudes, i.e. for $\delta \to 0$, the event of collision will be delayed or anticipated according to the law

$$\left.\frac{\partial \varphi_c}{\partial \delta}\right|_{\delta=0} = \frac{\left[(\varphi_c - \varphi_0)\varphi_c - 2\right]\sin\varphi_c + 2\sin\varphi_0 - \varphi_0\cos\varphi_0 + (2\varphi_c - \varphi_0)\cos\varphi_c}{\sin\varphi_c - \sin\varphi_0 - (\varphi_c - \varphi_0)\cos\varphi_c} \quad (4)$$

For an increasing field amplitude, $\delta > 0$, Eq.(4) foresees a delay of the collision for $\varphi_0 < \overline{\varphi}_0$, and its anticipation for $\varphi_0 > \overline{\varphi}_0$, where $\overline{\varphi}_0 \approx 18°$ is the phase of collision with the maximum electron velocity. This change in the time of collision is expected to produce a frequency shift in the harmonics of the emission spectrum. The sign of this shift depends on the delay, counted from the time of the maximum field intensity, the electrons are detached from the atoms after tunneling, i.e. if $\varphi_0 < \overline{\varphi}_0$ or $\varphi_0 > \overline{\varphi}_0$, a red or blue shift will be produced.

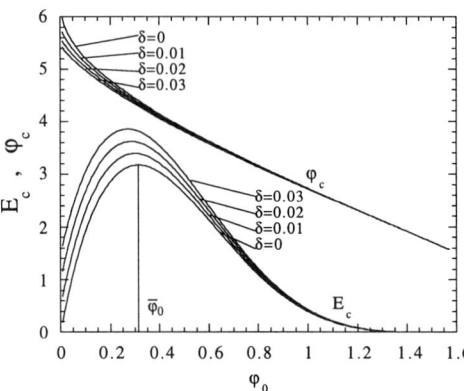

FIGURE 1. The electron energy at collision, E_c, and the phase at which the collision occurs, φ_c, are plotted versus the phase φ_0 at which the electron is detached from the atom. Four values of the rate at which the laser pulse electric field grows in time ($\delta=0$, 0.01, 0.02 0.03) have been considered. The time $\overline{\varphi}_0$ at which the back collision occurs with the maximum velocity is indicated.

For sufficiently low frequency laser radiation, the delay is expected to be negligible, the maximum probability of ionization occurring in correspondence of the maximum field amplitude, according to the Keldish formula (6), and a red shift should be observed. For rather short radiation wavelengths, most of the electrons contributing to the high-order harmonic emission are freed from the atoms at later times, that is at $\varphi_0 > \overline{\varphi}_0$. Then a blue shift is produced.

THE INERTIA OF TUNNELING IONIZATION

Let us first normalize time, space, energy, and field to atomic units, that is to $\hbar^3/me^4$, $\hbar^2/me^2$, $me^4/\hbar^2$, $m^2e^5/\hbar^4$. In order to investigate consistently the harmonic spectrum production, we have solved the one-dimensional Schroedinger equation for an electron in a static interatomic potential of the Eberly-type (7), that is $V(x) = -(1+x^2)^{-1/2}$, in the presence of an oscillating electric potential, $xE(t)$, which represents the linearly polarized laser pulse in the long-wavelength limit. The relevant equation reads:

$$i\frac{\partial \Psi}{\partial t} = -\frac{1}{2}\frac{\partial^2 \Psi}{\partial x^2} + \left[xE(t) - \frac{1}{\sqrt{1+x^2}}\right]\Psi(x,t) \qquad (5)$$

with the same field $E(t)$ as in Eq.(1). The polarization spectrum is computed by time-Fourier transforming the atom response given by

$$R(t) = \int_{-\infty}^{+\infty} \frac{\partial V}{\partial x} |\Psi(x,t)|^2 dx. \qquad (6)$$

In Fig.2 the space distribution of the probability density $|\Psi|^2$ is plotted at four successive times corresponding to the following fractions of the same oscillation period: 0.25 (*a*), 0.375 (*b*), 0.5 (*c*), 0.625 (*d*). They refer to $\omega_E = 0.08$ and $E_0 = 0.1$, and the field amplitude is stationary, i.e. $\delta = 0$. The bulk of the $|\Psi|^2$ function is centered at x=0. The three labelled peaks correspond to electrons which tunnel through the potential barrier and leave the atom with phases which make them to collide back with the same ion within half a cycle. On the basis of a quasi-stationary tunneling ionization model (8), the maximum ionization probability is expected to occur when the field reaches its maximum amplitude. Correspondingly, the maximum elongation of the electron trajectory is attained, making the electron to reach $x_{max} = 2E_0/\omega_E^2 \cong 31$. Following the dynamics of the peaks 1-3, it can be seen that they are distributed over the interval starting from $|x| \cong 25$ down to the bulk of the $|\Psi|^2$ function, and none of them reaches the maximum elongation predicted by the classical model. This behaviour can be interpreted by assuming that a delay in the ionization event takes place, due to the tunneling inertia which manifests itself more pronouncedly at higher frequencies.

As it was said previously, the tunneling inertia can be put in evidence by observing the dipole radiation spectra emitted during the interaction between a short laser pulse, with monotonically varying intensity, and a single atom. In Fig.3 the discrete bremsstrahlung spectra relevant to a radiation pulse which has a steep leading edge lasting one period, a plateau of four periods, and a trailing edge of one period are shown. The same parameters as in Fig.2 are used.

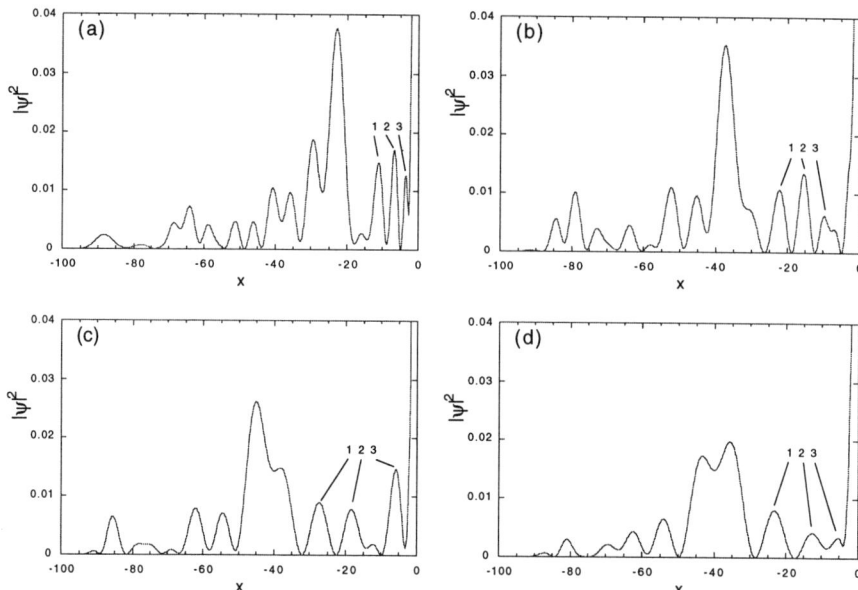

FIGURE 2. $|\Psi|^2$ is plotted versus x at 0.25 (a), 0.375 (b), 0.5 (c), and 0.625 (d) times $2\pi/\omega_E$, for $\omega_E = 0.08$ and $E_0 = 0.1$. The peaks of the $|\Psi|^2$ function, labelled with 1,2,3, reach the maximum elongation at times earlier than $0.5 \times 2\pi/\omega_E$ (c), then collide back with the bulk of $|\Psi|^2$, contributing to the discrete bremsstrahlung spectrum.

Thin lines correspond to a constant amplitude plateau, while thick lines refer to a slightly decreasing (a, δ=-0.05) and slightly increasing (b, δ=+0.05) amplitudes in the plateau region of the pulse. A red-shift in the case a and a blue-shift in the case b are observed, which, according to the discussion of the previous section, could indicate that a phase-delay in tunneling larger than $\overline{\varphi}_0$ has occurred.

In Fig.4 $|R(\omega)|^2$ is shown for $\omega_E = 0.03$, $E = 0.05$, and δ=0 (dashed line), -0.025 (thin line), -0.05 (thick line). An enhanced red-shift with increasing amplitude variation rate is observed. The larger is the field amplitude, the higher is the intensity of the radiation at a given frequency interval. The spectrum, wider than that in Fig.3, shows a frequency shift of more than one harmonic.

In order to observe the above discussed frequency shift, it should be distinguished from the well known collective ionization-induced blue-shift which takes place, as well. The latter turns out to be proportional to the thickness d of the interaction region, or to the total number of ionized particles along the path of propagation. This introduces an upper limit on d. For a Ti:Sa laser pulse ($\lambda = 0.8$ μm), with a duration of 200 fs, and a maximum intensity of 10^{15} W/cm^2, when the ionization takes place during ≈ 30 fs at the leading front of the pulse, we have $\Delta f/f \approx 1\%$ for H gas at 10 Torr and jet-thickness $d \approx 1$ mm ($\omega_E \approx 0.08$). By assuming a field amplitude increase of 5% per cycle, we estimate $\Delta f/f \approx 10\%$ for harmonic numbers around 30, well above the collective effect.

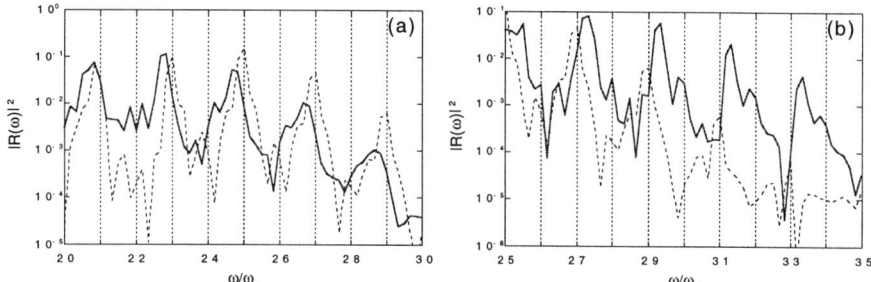

FIGURE 3. Dipole radiation spectra are shown for a finite duration laser pulse with $\omega_E = 0.08$ and $E_0 = 0.1$. Thin lines refer to the laser pulse which consists of a rump-up phase lasting one cycle, a constant-field central plateau, lasting four cycles, and a slowing-down phase, lasting one cycle. Thick lines refer to the pulse with the a slowly decreasing (a) and a slowly increasing (b) plateau, with a rate of ±5% over four cycles.

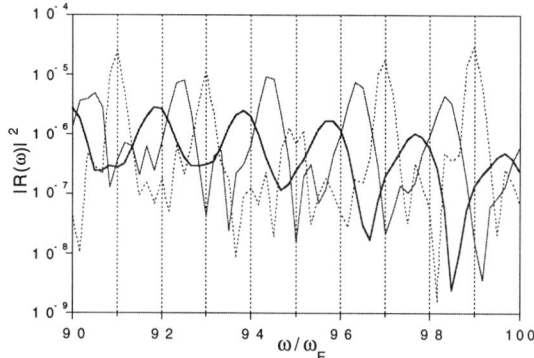

FIGURE 4. $|R(\omega)|^2$ is plotted for $\omega_E = 0.03$ and $E = 0.05$, for $\delta=0$ (dashed line), -0.025 (thin line), and -0.05 (thick line).

REFERENCES

1. Corkum, P.B., *Phys. Rev. Lett.* **71**, 1994 (1993).
2. Vanin, E.V., Kim, A.V., Sergeev, A.M., Downer, M.C., *JETP Lett.* **58**, 900 (1993).
3. Wood, W.M., Focht, G., Downer, M.C., *Opt. Lett.* **13**, 984 (1988).
4. LeBlanc, S.P., Sauerbrey, R., Rae, S.C., Burnett, K., *JOSA* **B10**, 1801 (1993).
5. Wahlstrom, C.-G., Larsson, J., Persson, A., et al., *Phys. Rev. A* **48**, 4709 (1993).
6. Keldysh, L.V., *JETP* **20**, 1307 (1965).
7. Eberly, J.H., Su, Q., Javanainen, J., *Phys. Rev. Lett.* **62**, 881 (1989).
8. Landau, L.D., Lifshitz, E.M., *Quantum Mechanics*, London: Pergamon Press, 1965.

AUTHOR INDEX

A

Adachi, M., 1189
Afanasiev, Y. V., 1274
Ahn, H., 666
Akazaki, M., 1177
Akiyama, H., 1090
Al-Khateeb, A., 565
Amano, M., 878
Amano, S., 849
Anderson, A., 200
Anderson, D., 609
Ando, K., 755
Ando, R., 1222
Andreev, A. A., 639, 713
Andreev, N. E., 603, 707
Anisimov, S. I., 302, 1197
Antonetti, A., 584
Antonov, V. M., 357
Aoki, T., 347, 1048, 1133
Aoyagi, Y., 755
Aoyama, M., 755
Asakawa, M., 814, 820, 830
Asakuma, T., 814
Audebert, P., 584
Avrorin, E. N., 194
Azechi, H., 101, 108, 131, 136, 225, 550
Azuma, K., 1153

B

Bailey, J. E., 1042
Barbee, Jr., T. W., 737
Barber, T. L., 1042
Barnett, D. M., 447
Barnouin, O. M., 1009
Bartnik, A., 772
Barty, C. P. J., 672, 784, 939
Barzilov, A. P., 933
Basov, N., 906
Batyrbekov, E., 883
Bauer, D., 565
Baumung, K., 1035, 1060
Bayanov, V. I., 639
Beck, J. B., 160
Bell, P. M., 672
Bernat, T., 200
Blenski, T., 398
Bluhm, H., 1035
Blyth, W. J., 559
Bocher, J. L., 113
Boehly, T. R., 71
Bolkhovitinov, E. A., 465
Bolton, P. R., 1234
Bonyushkin, E. K., 906
Boreham, B. W., 1234
Borisenko, N. G., 1228
Bradley, D. K., 71
Bradley, P., 186
Budil, K. S., 155
Bufetov, I. A., 533
Bufetova, G. A., 533
Bulanov, S. V., 701
Bychenkov, V. Y., 465

C

Cable, M. D., 89, 95, 527
Caird, J. A., 89
Campbell, E. M., 53
Carlson, A. L., 1042
Cauble, R., 737
Celliers, P., 737
Chandler, G. A., 1042
Chen, C., 136
Chen, J., 814, 820, 830
Chen, Y.-W., 538, 1165
Chernyak, V. M., 969
Chernyakov, V. E., 194
Chizhov, S. A., 639
Choi, C. K., 160
Chrien, R., 95
Cobble, J. A., 255, 380
Collier, J., 1029
Conder, A. D., 652
Cook, D. L., 1042
Craxton, R. S., 71
Cray, M., 53
Cros, B., 477, 684

D

Daido, H., 743, 951
DaSilva, L. B., 737
D'Avanzo, J., 1115
Delamater, N. D., 95, 149, 255, 263
Delettrez, J. A., 71
DelMedico, S. G., 1009
Dendooven, P. G., 527
Derzon, M. S., 1042
Ding, Y., 483
Djaoui, A., 559
Dos Santos, A., 584
Downer, M. C., 728
Dubau, J., 404
Dukart, R. J., 1042
Dunn, J., 652
Dyabilin, K., 428
Dyachenko, P. P., 894, 933
Dyuzhov, Y. A., 927

E

Eimerl, D., 969
Eliezer, S., 172, 283
Endou, T., 1133
Epstein, R., 71
Esashi, M., 1268
Esirkepov, T. Z., 701
Estabrook, K., 615
Evans, S. C., 380
Ezaki, M., 1262

F

Failor, B. H., 255
Fallies, F., 584
Fan, P., 749
Farina, D., 728
Farny, J., 521
Faucheux, G., 113
Fedenev, A. V., 860, 915
Fedorov, V. B., 533
Feoktistov, L. P., 207
Fernández, J. C., 255, 380
Fiedorowicz, H., 772
Fill, E., 772
Förster, E., 544

Fortov, V. E., 428, 1060
Foster, J. M., 125
Fujii, K., 849, 872, 878, 1133
Fujii, T., 1048
Fujimoto, T., 404, 790
Fujimoto, Y., 992
Fujita, H., 1004, 1189
Fujita, K., 544
Fujita, M., 814, 820, 830
Fukuda, T., 511
Fukumoto, H., 1250
Fukumura, H., 1250
Fukuyama, A., 386
Furukawa, H., 136, 392, 447
Furutani, Y., 386
Fusuki, T., 1171

G

Gauthier, J. C., 584
Geindre, J. P., 584
Gifford, K., 95
Go, K., 621, 796
Goel, B., 428, 1035, 1060
Goncharov, S. F., 499
Gorbunov, L. M., 603, 707
Gordon III, C. L., 672, 784, 939
Grabovskij, E., 428
Grillon, G., 584
Grishanin, B. A., 1256
Gulevich, A. V., 933
Guo, X. M., 1096
Gus'kov, S. Y., 219, 906

H

Haan, S. W., 155
Hamada, Y., 1222
Hammel, B., 149
Hammer, J., 590
Han, S., 749
Hanabusa, M., 1280
Hanasawa, Y., 849
Hankla, A. K., 652

Hara, T., 755
Harris, D. B., 95
Harris, S. E., 784
Hasegawa, J., 1066
Hasegawa, N., 660
Hasegawa, S., 237
Haseroth, H., 1029
Hashimoto, H., 131
Hatchett, S. P., 89
Hattori, T., 1121
Hauer, A. A., 95, 263
He, X. T., 321
Hebron, D. E., 1042
Heya, M., 108, 143
Hoffman, N. M., 160, 166
Hoffmann, D. H. H., 80, 333
Honda, C., 1177
Honda, H., 1205
Honda, M., 179, 225
Honrubia, J. J., 113, 172, 283
Höpfl, R., 172
Hoppé, P., 1035
Hora, H., 172, 283, 443, 1234
Horioka, K., 1048, 1066, 1133
Hosokai, T., 1133
Hotta, T., 1138
Hsing, W. H., 160
Hsing, W. W., 255

I

Iguchi, T., 398, 900
Ihara, S., 1177
Iida, T., 550
Ikeda, A., 1222
Il'kaev, R. I., 906
Imasaki, K., 814, 820, 830, 1048, 1072
Inagaki, M., 998
Inoue, M., 1127
Ishibashi, H., 1286
Ishibashi, M., 878
Ishii, K., 1291
Ishii, Y., 808
Ishikawa, K., 398
Ishikubo, Y., 1189
Ishimine, M., 1177
Ishizaki, R., 243
Isobe, S., 1205
Ito, A., 1314

Ito, T., 1121
Itoh, H., 477
Itoh, Y., 1054
Ivanov, V. V., 471, 957
Iwamoto, K., 489
Iwasaki, H., 1066
Iwase, O., 1133
Iwashita, Y., 1127
Iwata, Y., 1303
Izawa, Y., 836, 951, 1022, 1165, 1189
Izumi, N., 108, 550

J

Jaanimagi, P. A., 71
Jacobs, S. D., 71
Jacoby, J., 80
Jassby, D. L., 883
Jitsuno, T., 963, 975, 981, 1309
Johnson, D. J., 1042
Johzaki, T., 315
Jovanović, M. S., 363

K

Kachanov, B. V., 933
Kado, M., 459
Kakuta, T., 900
Kamada, K., 1222
Kamiya, T., 1054
Kanabe, T., 101, 136, 963, 992
Kanavin, A. P., 1274
Kanazawa, H., 1054
Kanda, Y., 1171
Kandiev, Y. Z., 194
Kanel, G. I., 1035, 1060
Kapin, V., 1127
Karashima, S., 621, 796
Karlykhanov, N. G., 194
Karow, H., 1060
Karow, H. U., 1035
Kasai, T., 550
Kasuya, K., 1048, 1054
Katagiri, M., 900
Katahira, O., 1177
Kato, H., 872, 878
Kato, S., 179, 627, 633, 1109

Kato, T., 404
Kato, Y., 101, 108, 131, 143, 544, 550, 743, 778, 951
Katschinski, U., 1090
Katsuki, S., 1090
Kauffman, R. L., 95, 125, 263
Kawachi, T., 404, 790
Kawakita, Y., 1054
Kawamura, T., 808
Kawarasaki, Y., 1244
Kawasaki, Z., 1189
Kawase, K., 690
Kawata, S., 327, 339, 1103, 1109
Kayukov, S. V., 1274
Keane, C., 149
Keck, R. L., 71
Kelly, J. H., 71
Kessler, G., 1060
Kessler, T. J., 71
Key, M. H., 559
Khoda-Bakhsh, R., 289
Kilkenny, J. D., 89, 155
Kilpio, A. V., 465
Kim, A. V., 609, 728
Kim, H., 71
Kimura, T., 836, 1022
Kinoshita, F., 1177
Kirillov, G. A., 866, 1332
Kiriyama, H., 836, 1022
Kirkwood, R. K., 125
Kirsanov, V. I., 603
Kiselev, N. G., 465
Kitada, T., 951
Kitagawa, Y., 108, 136, 690
Kitatani, H., 872, 878
Kitazawa, H., 621, 796
Kiyono, H., 849
Kmetik, V., 1004
Knauer, J. P., 71
Knyazev, A. K., 471
Kobayashi, K., 1262
Kobayashi, T., 1286
Kobayashi, Y., 796, 945
Kochemasov, G. G., 866
Kochiev, D. G., 465
Kodama, R., 101, 131, 459, 743
Koguchi, H., 1222
Koike, F., 743
Kolyada, S. G., 933
Komarov, V. A., 639

Kondo, K., 945
Korn, G., 939
Kornblum, H. N., 89, 125
Kosaka, K., 410
Koshevoi, M. O., 465
Koutsenko, A. V., 471
Koyama, K., 1303
Kozlov, A. A., 639
Kozmanov, M. Y., 194
Kozubskaya, T., 219
Krasyuk, I. K., 369
Krauser, W. J., 166, 186, 255
Kravtsov, S. B., 533
Kremens, R. L., 71
Kruer, W. L., 590
Kudo, K., 309, 315
Kukharchuk, O. F., 933
Kulikov, S. M., 866
Kumagai, H., 1262
Kumar, A., 1205
Kuminaka, H., 1211
Kumpan, S. A., 71
Kurawaki, K., 327
Kuriki, K., 1211
Kurnin, I. V., 639
Kuroda, H., 760
Kuwabara, M., 621
Kuwako, A., 1159

L

LaGattuta, K. J., 416
Landen, N., 149
Landen, O. L., 155
Lane, S. M., 89
Langbein, K. L., 1029
Láska, L., 521, 1029
Latkowski, J., 200
Laumann, C., 89
Lazhintsev, B. V., 906
Lebedev, M., 428
Lebo, I. G., 207, 219
Leeper, R. J., 1042
Lemoff, B. E., 672, 784, 939
Lerche, R. A., 89, 527
Letzring, S. A., 71

Levedahl, W. K., 125
Li, R., 749
Li, Y., 772
Li, Y. S., 321
Libby, S. B., 737
Light, V., 1060
Lindman, E. L., 95, 255, 263
Lisak, M., 609
Liu, S., 483
Liu, Y., 749
Logan, B. G., 269
London, R. A., 737
Lontano, M., 728, 1115
Lu, P., 749
Luches, A., 1197
Luk'yanchuk, B. S., 1197
Luo, C. M., 1096
Lykov, V. A., 194

M

Maeda, S., 1090
Magelssen, G. R., 95, 263
Maiguma, T., 1177
Marinak, M. M., 155
Marshall, C., 200
Marshall, F. J., 71
Martínez-Val, J. M., 113, 172, 277, 283
Mašek, K., 521, 1029
Masuhara, H., 1250
Masumura, Y., 878
Masuzaki, M., 1222
Matoba, T., 511
Matsumoto, Y., 854
Matsuoka, S., 131, 963
Matsushima, I., 854
Matsushita, T., 131
Matsu-ura, K., 1189
Matthews, D. L., 737
Matthieussent, B., 684
Matzen, M. K., 1042
Matzveiko, A. A., 471
Mavlyutov, A. A., 921
Mayle, R. W., 778
McArthur, D., 883
McCrory, R. L., 71, 1323
McCutcheon, M. J., 849
McKenty, P. W., 71
Mehlhorn, T. A., 1042

Meisel, G., 1035
Melekhov, A. V., 357
Merkuliev, Y. A., 1228
Meyer-ter-Vehn, J., 213, 302, 347
Mibuka, N., 1250
Midorikawa, K., 374
Mikhailov, Y. A., 471, 957
Miley, G. H., 883, 1009, 1334
Mima, K., 101, 108, 131, 179, 225, 459, 627, 633, 646, 830
Minami, K., 1268
Minamiguchi, T., 814
Mínguez, E., 283
Mironov, V. A., 609
Mirov, S. B., 849
Mis'kevich, A. I., 921
Miura, E., 854
Miyagawa, S., 119
Miyajima, K., 1222
Miyamoto, A., 998
Miyamoto, S., 108, 179, 627, 633, 1048, 1072
Miyanaga, N., 101, 131, 136, 143, 963, 975
Mizuguchi, S., 1297
Mizuno, K., 615
Mizutani, T., 119
Moats, A. R., 1042
Montgomery, D. S., 125
Moon, A., 814
Moore, J. B., 95
Moreno, J. C., 737
Mori, Y., 998
Morovov, A. P., 906
Morse, S. F. B., 71
Mrowka, S., 737
Mróz, W., 521, 1029
Mulser, P., 565
Murai, K., 743, 951
Murakami, K., 1133
Murakami, M., 131, 243, 297, 989, 1048
Muraoka, K., 1177
Murashkina, V. A., 194
Murphy, T. J., 89, 95, 263
Murray, J., 89
Mysyrowicz, A., 584

N

Nabekawa, Y., 945
Nagamine, Y., 1171
Nagata, Y., 374
Naito, K., 1109
Nakagawa, H., 660
Nakai, M., 101, 131, 136, 459
Nakai, S., 30, 101, 108, 131, 136, 143, 225, 459, 544, 550, 690, 814, 820, 830, 836, 951, 963, 975, 981, 992, 998, 1004, 1022, 1048, 1072, 1165, 1309
Nakaji, S., 131, 459
Nakajima, M., 1066, 1133
Nakajima, S., 998
Nakamura, T., 1133
Nakano, H., 666
Nakano, T., 550
Nakao, Y., 309, 315, 1171
Nakashima, H., 309, 315, 1171
Nakashima, N., 1309
Nakasuji, M., 108
Nakatsuka, H., 1286
Nakatsuka, M., 101, 836, 951, 963, 975, 981, 992, 1309
Nakayama, S., 849
Nakazawa, M., 398, 900
Nash, T. J., 1042
Naumova, N. M., 701
Neely, D., 559
Nelson, M. B., 89, 95
Nikishin, V., 219
Nikitin, S. A., 357
Ninomiya, S., 743, 951
Ninomiya, Y., 1314
Nishi, N., 981, 1309
Nishida, Y., 477, 597, 678, 684
Nishiguchi, A., 225, 459, 633
Nishihara, K., 101, 108, 136, 143, 231, 237, 243, 297, 447, 459, 1048
Nishikawa, T., 422, 666
Nishikawa, Y., 1297
Nishimura, H., 101, 131, 544
Nishiyama, K., 1211
Nishiyama, S., 1103, 1109
Niu, K., 1084
Noda, A., 1127
Nomaru, K., 1165
Nomoto, K., 489
Norimatsu, T., 101, 136, 459

Nozawa, O., 796

O

Obara, M., 374, 1262, 1291, 1314
Ochi, T., 1048, 1072
Oda, A., 315
Oda, Y., 1153
Oertel, J. A., 380
Offenberger, A. A., 559
Ogawa, M., 1133
Oguri, Y., 1121
Ohigashi, N., 814
Ohmi, M., 836, 1022
Ohnishi, T., 131
Okada, M., 1121
Okada, T., 505, 1078, 1138
Okamura, M., 1121
Okishev, A., 71
Okuda, I., 854
Olsen, R. W., 1042
Olson, R. E., 1042
Onozuka, M., 1153
Oouchi, K., 872, 878
Oparin, A., 347
Oparin, A. M., 302
Orimo, S., 760
Orzechowski, T. J., 125
Osetrov, V. P., 471, 957
Oshikane, Y., 131
Osipov, M. V., 465
Osterheld, A. L., 778
Owadano, Y., 854
Ozaki, A., 131
Ozaki, T., 760

P

Parys, P., 521, 1029
Pashin, E. A., 933
Pashinin, P. P., 369, 465
Pavlovskii, A. I., 906
Peng, H., 61
Pergament, M. I., 969
Peria, M., 172
Perlado, M., 283
Petra, M., 883, 1009
Pevny, S. N., 866

Phillion, D. W., 89
Piera, M., 277, 283
Poletaev, E. D., 927
Pollaine, S. M., 95, 263
Pollak, G., 149
Ponomarenko, A. G., 357
Popov, A. I., 471, 957
Popov, I., 219
Porter, J. L., 1042
Posukh, V. G., 357
Powell, H., 89
Powers, L. V., 95, 125, 263
Preston, S. G., 559
Pretzler, G., 772
Pukhov, A., 213

Q

Quintenz, J. P., 1042

R

Raimondi, F., 1115
Rajković, M. R., 363
Raksi, R., 939
Razorenov, S. V., 1035
Remington, B. A., 155
Ress, D., 89, 737
Richard, A., 136
Rivlin, L. A., 766
Rohlena, K., 521, 1029
Rosanov, V. B., 906
Rosen, P., 125
Rose-Petruck, C., 939
Rousse, A., 584
Roy, P. K., 814
Rozanov, V. B., 207, 219
Ruhl, H., 575
Ruiz, C. L., 1042
Rupasov, A. A., 465
Rusch, D., 1035

S

Saemann, A., 565
Safronova, U., 404
Sagisaka, A., 945
Saito, H., 872, 878
Saito, M., 108
Saito, N., 1303
Sakagami, H., 243, 646
Sakagami, Y., 119
Sanjou, S., 678
Sasa, K., 1121
Sasaki, A., 802
Sasaki, T., 998
Sato, M., 136
Satoh, S., 1177
Satori, S., 1211
Satou, K., 505
Sawai, K., 690
Schneider, R., 565
Seka, W., 71
Sekimura, M., 119
Semenov, A. Y., 369, 434, 499
Sergeev, A. M., 609, 728
Shaikhislamov, I. F., 357
Shamaev, O. B., 1029
Sharkov, B. Y., 1029
Sharma, L. B., 951
Shashkov, E. V., 465
Shepard, T. D., 125
Sherwood, T. R., 1029, 1145
Shigeyama, T., 489
Shikanov, A. S., 465
Shimada, K., 131
Shimada, Y., 1189
Shimizu, K., 621, 796
Shimoide, M., 297
Shimuta, Y., 231
Shioda, K., 1054
Shiraga, H., 101, 108, 131, 143
Short, R. W., 71
Shoyama, H., 1171
Shumshurov, A. V., 1029
Singh, M., 200
Skakum, V. S., 860
Skeldon, M. D., 71
Sklizkov, G. V., 471, 957
Škorić, M. M., 363
Skupsky, S., 71
Sluyter, M. M., 21
Smirnov, R. V., 969
Smirnov, V., 428
Smith, C. G., 559
Sokolov, V. I., 969

Solovyev, N. A., 639
Soures, J. M., 71
Squier, J., 939
Srinivasan, N., 836, 1022
Stark, M. A., 1042
Starodub, A. N., 471
Stenflo, L., 609
Stewart, R. E., 652
Stoltz, O., 1035
Stone, G. F., 125
Stoyanovsky, V. O., 357
Suchkov, Y. A., 465
Sueda, T., 1090
Sugawa, M., 1205, 1216
Sugaya, R., 1205, 1216
Sukharev, S. A., 866
Sumiyoshi, T., 1291
Sunahara, A., 249
Suter, L. J., 95, 125, 263
Sutyagin, A. N., 713
Suzuki, M., 872, 878
Suzuki, T., 1054
Swenson, F., 186
Szczurek, M., 772

Tanaka, K. A., 101, 131, 136, 459
Tanaka, T., 1177
Tang, D., 483
Tanimoto, M., 1303
Tarasenko, V. F., 860, 915
Tikhonchuk, V. T., 465
Tishkin, V. F., 207, 219
Tobin, M., 200
Tokita, Y., 849
Tokumura, K., 1309
Tomie, T., 854
Tomita, H., 1291, 1314
Torres, J. A., 1042
Toyoda, K., 374, 1262
Tracy, M., 71
Trebes, J. E., 737
Tsintsadze, L. N., 719
Tsubakimoto, K., 131, 963, 975, 992
Tsuji, R., 849, 878
Tsuji, T., 1177
Tsujihara, K., 1280
Tsukamoto, M., 131, 459
Tsunawaki, Y., 814

T

Tabak, M., 590
Tabaru, Y., 309
Taguchi, A., 998
Tahir, N. A., 333
Tajima, T., 447
Takabe, H., 101, 108, 225, 249, 489, 544, 743
Takagi, M., 101
Takahashi, A., 1291
Takahashi, E., 854
Takahashi, H., 398, 900
Takahashi, R., 1138
Takahashi, T., 872, 878
Takahata, S., 1222
Takakuwa, H., 621, 796
Takenaka, H., 808
Takeuchi, H., 249
Takeuchi, S., 695
Takigawa, Y., 1309
Takuma, H., 660, 802
Takuma, T., 1177
Tanaka, K., 143

U

Uchida, S., 1189
Uchiumi, M., 1177
Uchiyama, Y., 849
Ueda, K., 660, 802
Uematsu, H., 951
Uenoyama, T., 843
Ueshima, Y., 447
Uesugi, N., 666
Urai, H., 1048
Urai, M., 1072
Urano, T., 108
Uschmann, I., 544
Utkin, A. V., 1035, 1060

V

Vanin, E. V., 728
Vankov, A. B., 639
Varnum, W. S., 166
Velarde, G., 172, 283

Velarde, P. M., 277, 283
Verdon, C. P., 71
Vergunova, G., 219
Vorobiev, O. Y., 428, 1060
Vovchenko, V. I., 369

W

Wallace, R. J., 95, 125, 155
Wan, A. S., 737, 778
Wang, D., 1189
Wang, X., 749
Wark, J. S., 559
Watanabe, M., 1054
Watanabe, S., 690, 945
Watanabe, T., 1159
Watt, R. G., 380
Weber, F., 737
Weber, S. V., 155
Wenger, D. F., 1042
Whinnery, J. R., 849
White, W. E., 652
Wilde, B. H., 95, 166, 186, 255, 263, 380
Wilks, S. C., 590
Wilson, D. C., 166, 186
Wilson, K. R., 939
Wołowski, J., 521, 1029
Woryna, E., 521, 1029

X

Xu, G., 849
Xu, X., 477
Xu, Z., 749

Y

Yaakobi, B., 71
Yakovlev, V. V., 939
Yamagajo, T., 108, 550

Yamagishi, H., 900
Yamakawa, K., 939
Yamanaka, C., 3, 814, 820, 830, 836, 951, 975, 1022, 1048, 1072, 1165, 1189
Yamanaka, M., 836, 1022
Yamanaka, N., 900
Yamanaka, T., 101, 136, 143, 459, 836, 1004, 1189
Yamashita, T., 1048, 1072
Yashin, V. E., 639
Yashiro, H., 854
Yasuda, H., 1189
Yasufuku, K., 119
Yasuike, K., 1048, 1072
Yin, G. Y., 672, 784
Yoguchi, I., 1159
Yoneda, H., 660, 802
Yoshida, H., 119, 998, 1004, 1222
Yoshida, K., 1004
Yoshida, T., 1159
Yoshida, Y., 1297
Yoshikawa, M., 1222
Yoshimi, N., 1171
Young, B. K. F., 652
Young, P. E., 590, 615
Yu, M., 321
Yugami, N., 477, 678, 684

Z

Zadkov, V. N., 1256
Zakharov, Y. P., 357
Zavestovskaya, I. N., 1274
Zepf, M., 559
Zhang, J. T., 453
Zhang, L., 749
Zhang, T., 951
Zhang, Z., 749
Zheng, Z., 483
Zmitrenko, N. V., 219, 906
Zrodnikov, A. V., 933